Conversion Factors

Length

1 in. = 2.54 cm

1 ft = 0.3048 m

1 mi = 5280 ft = 1.609 km

1 m = 3.281 ft

1 km = 0.6214 mi

1 angstrom (Å) = 10^{-10} m

Mass

1 slug = 14.59 kg

1 kg = 1000 grams = 6.852×10^{-2} slug

1 atomic mass unit (u) = 1.6605×10^{-27} kg

(1 kg has a weight of 2.205 lb where the acceleration due to gravity is 32.174 ft/s^2)

Time

1 day = 24 h = 1.44×10^3 min = 8.64×10^4 s

1 yr = 365.24 days = 3.156×10^7 s

Speed

1 mi/h = 1.609 km/h = 1.467 ft/s = 0.4470 m/s

1 km/h = 0.6214 mi/h = 0.2778 m/s = 0.9113 ft/s

Force

1 lb = 4.448 N

1 N = 10^5 dynes = 0.2248 lb

Work and Energy

1 J = 0.7376 ft·lb = 10^7 ergs

1 kcal = 4186 J

1 Btu = 1055 J

1 kWh = 3.600×10^6 J

1 eV = 1.602×10^{-19} J

Power

1 hp = 550 ft·lb/s = 745.7 W

1 W = 0.7376 ft·lb/s

Pressure

1 Pa = 1 N/m^2 = 1.450×10^{-4} lb/in.^{2}

1 lb/in.^{2} = 6.895×10^3 Pa

1 atm = 1.013×10^5 Pa = 1.013 bar = 14.70 lb/in.^{2} = 760 torr

Volume

1 liter = 10^{-3} m^3 = 1000 cm^3 = 0.03531 ft^3

1 ft^3 = 0.02832 m^3 = 7.481 U.S. gallons

1 U.S. gallon = 3.785×10^{-3} m^3 = 0.1337 ft^3

Angle

1 radian = 57.30°

1° = 0.01745 radian

Standard Prefixes Used to Denote Multiples of Ten

Prefix	Symbol	Factor
Tera	T	10^{12}
Giga	G	10^{9}
Mega	M	10^{6}
Kilo	k	10^{3}
Hecto	h	10^{2}
Deka	da	10^{1}
Deci	d	10^{-1}
Centi	c	10^{-2}
Milli	m	10^{-3}
Micro	μ	10^{-6}
Nano	n	10^{-9}
Pico	p	10^{-12}
Femto	f	10^{-15}

Basic Mathematical Formulae

Area of a circle = πr^2

Circumference of a circle = $2\pi r$

Surface area of a sphere = $4\pi r^2$

Volume of a sphere = $\frac{4}{3}\pi r^3$

Pythagorean theorem: $h^2 = h_o^2 + h_a^2$

Sine of an angle: $\sin\theta = h_o/h$

Cosine of an angle: $\cos\theta = h_a/h$

Tangent of an angle: $\tan\theta = h_o/h_a$

Law of cosines: $c^2 = a^2 + b^2 - 2ab\cos\gamma$

Law of sines: $a/\sin\alpha = b/\sin\beta = c/\sin\gamma$

Quadratic formula:

If $ax^2 + bx + c = 0$, then, $x = (-b \pm \sqrt{b^2 - 4ac})/(2a)$

🖱 Web site www.wiley.com/college/cutnell

The *Physics,* 6th edition, Web site offers a range of study aids that support your study of physics. Most of these are cross-referenced in the text so that you always know the right time to go to the Web site. The following pages will give you an idea of the resources available to you:

- **Solutions to selected end-of-chapter problems**

- **Interactive Solutions**

- **Interactive LearningWare examples**

- **Self-Assessment Tests**

- **Concept Simulations**

- **MCAT Self-Quizzes**

- **Web Links to other physics tutorials**

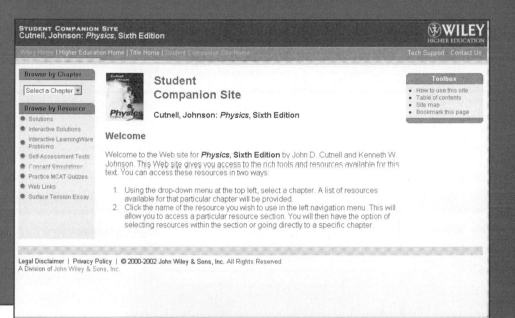

CHAPTER 2 | KINEMATICS IN ONE DIMENSION

9.　　**REASONING**　Since the woman runs for a known distance at a known constant speed, we can find the time it takes for her to reach the water from Equation 2.1. We can then use Equation 2.1 to determine the total distance traveled by the dog in this time.

SOLUTION　The time required for the woman to reach the water is

$$\text{Elapsed time} = \frac{d_{\text{woman}}}{v_{\text{woman}}} = \left(\frac{4.0 \text{ km}}{2.5 \text{ m/s}}\right)\left(\frac{1000 \text{ m}}{1.0 \text{ km}}\right) = 1600 \text{ s}$$

In 1600 s, the dog travels a total distance of

$$d_{\text{dog}} = v_{\text{dog}}t = (4.5 \text{ m/s})(1600 \text{ s}) = \boxed{7.2 \times 10^3 \text{ m}}$$

● **Solutions** These solutions to selected end-of-chapter problems, taken from the Student Solutions Manual, are identified with a **www** icon in the text. They have been written by the authors of the text and follow the problem-solving methodology presented in the text.

→

● **Interactive Solutions** These solutions are designed to serve as models for an addition-al 111 end-of-chapter problems, all labeled in the text. The solutions are structured in an interactive tutorial format and follow the problem-solving methodology presented in the text. You can find them easily on the Web site because the numbers of the Interactive Solutions match the numbers of the corre-sponding problems.

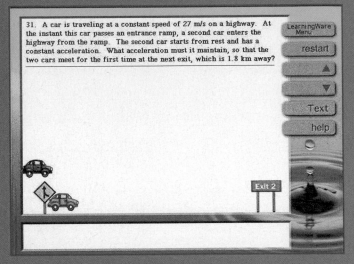

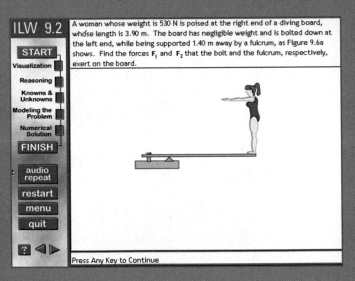

● **Interactive LearningWare** These interactive calculational examples will help you develop your problem-solving skills. References to these examples are contained in nearly every chapter of the text. Their user-friendly clarity allows you to see how conceptual understanding and quantitative analysis work together in a problem-solving environment.

● **Self-Assessment Tests** The Self-Assessment Tests are linked directly to specific sections of the book. After you study these text sections, the tests allow you to assess your understand-ing of the material immediately. Each test contains thought-provoking conceptual questions as well as calculational problems. You can take the tests as often as you wish. They provide extensive feedback in the form of helpful hints and suggestions when incorrect answers are selected.

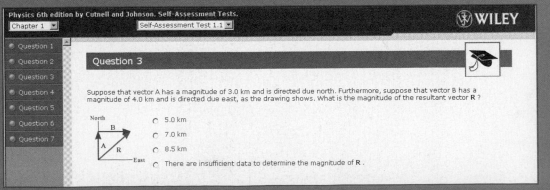

● Concept Simulations This section consists of 61 interactive simulations. Many concepts in physics—such as relative velocities, collisions, and ray tracing—can be better understood when they are simulated. With each simulation, you can experiment, since one or more parameters are under user control.

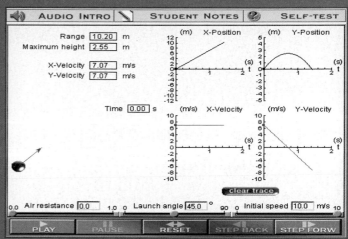

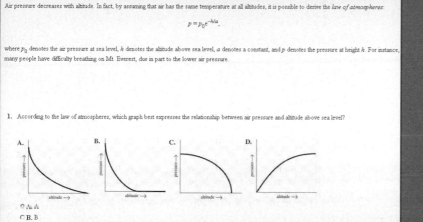

● MCAT Self-Quizzes These questions are similar to those you will find on the MCAT exam. They are presented in a multiple-choice, self-test format.

● Web Links These links, organized by chapter, provide a rich resource of on-line study aids.

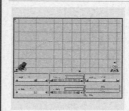

Volume One

Physics

Sixth Edition

WILEY

John Wiley & Sons, Inc.

John D. Cutnell
Kenneth W. Johnson

Southern Illinois University at Carbondale

Volume One

Physics

Sixth Edition

ACQUISITION EDITOR	*Stuart Johnson*
DIRECTOR OF DEVELOPMENT	*Barbara Heaney*
MARKETING MANAGER	*Robert Smith*
PRODUCTION EDITOR	*Barbara Russiello*
COVER AND TEXT DESIGNER	*Madelyn Lesure*
SENIOR ILLUSTRATION EDITOR	*Sigmund Malinowski*
ELECTRONIC ILLUSTRATIONS	*Precision Graphics*
PHOTO EDITOR	*Hilary Newman*
COVER PHOTO	© *John Kelly/The Image Bank/Getty Images*

This book was set in 10/12 Times Roman by Progressive Information
Technologies and printed and bound by Von Hoffmann Press, Inc. The cover was
printed by Von Hoffmann Press.

This book is printed on acid free paper. ∞

Main text (hardcover)	ISBN 0471-15183-1
Volume 1 (paperback)	ISBN 0-471-20940-6
Volume 2 (paperback)	ISBN 0-471-20939-2
Wiley International Edition	ISBN 0471-44895-8

Printed in the United States of America

10 9 8 7 6 5 4 3 2 1

To my wife Joan Cutnell,
a patient friend and my support
throughout this project.

To Anne Johnson,
my wonderful wife, a caring person,
and a great friend.

Brief Contents

Contents

Preface

We have written this text for students and teachers who are partners in a one-year algebra-based physics course. The single most important pedagogical issue in this course is the synergistic relationship between conceptual understanding and problem solving; we have substantially strengthened this relationship in the sixth edition. Our efforts were helped immeasurably by the many insights and suggestions provided by users of the fifth edition, as well as by the work of physics-education researchers. In taking a fresh look at the text, we have added new features and material, refined areas in need of improvement, and worked toward an even greater clarity in the design of the book. In particular, the high level of student activity on our Web site indicates that students are quite interested in on-line help, so we have greatly increased our on-line learning aids and facilitated their use with carefully placed text references to them.

Goals

Conceptual Understanding   Helping students develop a conceptual understanding of physics principles is a primary goal of this text. This is a challenging task, because physics is often regarded as a collection of equations that can be used blindly to solve problems. However, a good problem-solving technique does not begin with equations. It starts with a firm grasp of the concepts and how they fit together to provide a coherent description of the physical world. The features in the text that work toward this goal are *Check Your Understanding* questions (a new feature), *Self-Assessment Tests* (a new on-line feature), *Concept Simulations* (a new online feature), *Concepts & Calculations* sections, *Conceptual Examples,* and *Concepts-at-a-Glance* charts.

Reasoning   The ability to reason in an organized manner is essential to solving problems, and helping students to improve their reasoning ability is also one of our primary goals. To this end, we have included *Interactive LearningWare* (an on-line tutorial that steps students through 60 different examples), *Interactive Solutions* (111 on-line solutions that explore the reasoning pertinent to certain homework problems), *explicit reasoning steps* in all examples, and *Reasoning Strategies* that summarize the logical steps involved in solving certain classes of problems. A strong reasoning ability, combined with firm conceptual understanding, is the legacy that we wish to leave to our students.

Relevance   Finally, we want to show students that physics principles come into play over and over again in their lives. It is always easier to learn something new if it is directly related to day-to-day living. Direct applications of physics principles are identified in the margin by the label *The Physics of....* Many of these applications are biomedical in nature and deal with human physiology. Others, such as CDs and DVDs and digital photography, deal with modern technology. Still others deal with things that we take for granted in our lives, such as household plumbing. We have also incorporated real-world situations into many of the worked-out examples and the homework material at the end of each chapter.

Organization and Coverage

The text consists of 32 chapters and is organized in a fairly standard fashion according to the following sequence: *Mechanics, Thermal Physics, Wave Motion, Electricity and Magnetism, Light and Optics,* and *Modern Physics.* The text is available in two formats: in a complete volume, consisting of all 32 chapters, and in two volumes (Chapters 1 – 17 and Chapters 18 – 32). **Chapter sections marked with an asterisk can be omitted with little impact on the overall development of the material.**

Based on feedback from users and reviewers, we have added new material, made judicial deletions, and revised other material. In particular, a number of the examples have been modified to improve clarity and present topics of current interest. The new material includes completely rewritten chapter-ending summaries for each chapter, a discussion of the relation between relativistic total energy and momentum (Section 28.6), a discussion of light emitting diodes (Section 23.5), and a table of the isotopes (Appendix F). We have also updated our applications of physics principles and added new ones. Because many students are interested in health-related careers, a number of the new applications are biomedical in nature, such as wireless capsule endoscopy, transcranial magnetic stimulation, a fetal oxygen monitor, and exercise thallium heart scans. We have also presented a variety of new applications relating to modern technology, including, for example, ion propulsion drives for space probes, special relativity and the global positioning system, 3-D films, and light emitting diodes.

For instructors who wish to cover surface tension, we have included a module on our Web site (**www.wiley.com/college/cutnell**). This module discusses the nature of surface tension, capillary action, and the pressure inside a soap bubble and a liquid drop. Homework problems are also included for assignment.

Multimedia Version of the Sixth Edition

The Multimedia Book This hypertext version of the text is linked to the Study Guide, the Student Solutions Manual, the Self-Assessment Tests, the Interactive LearningWare, the Interactive Solutions, the Concept Simulations, and other multimedia. The Multimedia Book can be used by students in many ways, but it is preeminently a *very powerful study aid*. The algebra-based introductory physics course is a problem-solving course, and the Multimedia Book provides more problem-solving help than any other source available.

NEW!

eGrade Plus By combining two powerful Web-based programs—eGrade and the Multimedia Book—we can offer physics teachers and physics students a unique teaching and learning environment.

Helping Teachers Teach: eGrade is a homework management system that allows the teacher to assign, collect, and grade homework automatically. eGrade keeps the students "on task" without the teacher having to invest hours every week grading homework.

Helping Students Learn: All problems assigned in eGrade have a link to an on-line version of the complete text and the associated student supplements. The links give the students quick access to focused problem-solving help. Context-sensitive help—just a click away.

Features of the Sixth Edition

NEW!

Check Your Understanding This feature consists of 130 questions found at key locations throughout the text. Each is a carefully selected question that is designed to test a student's understanding of the concepts. The questions are in either a multiple-choice or free-response format, and the answers are provided at the end of the book.

> ✔ **Check Your Understanding 1**
>
> A suitcase is hanging straight down from your hand as you ride an escalator. Your hand exerts a force on the suitcase, and this force does work. Which one of the following statements is correct? (a) The work is negative when you ride up the escalator and positive when you ride down the escalator. (b) The work is positive when you ride up the escalator and negative when you ride down the escalator. (c) The work is positive irrespective of whether you ride up or down the escalator. (d) The work is negative irrespective of whether you ride up or down the escalator. *(The answer is given at the end of the book.)*
>
> **Background:** This question deals with work and its relationship to force and displacement. Pay close attention to the directions of the force and displacement vectors.
>
> **For similar questions (including calculational counterparts), consult Self-Assessment Test 6.1. This test is described at the end of Section 6.2.**

NEW!

Self-Assessment Tests The self-assessment tests are on-line tests (**www.wiley.com/college/cutnell**) that are linked directly to specific sections of the book. There are about two self-assessment tests per chapter, each containing 8–10 questions and problems. The tests are designed so that students, after studying the designated sections, can immediately assess their understanding of the material. Each contains thought-provoking conceptual questions, like the Check-Your-Understanding questions, as well as calculational problems. The Self-Assessment tests are user-friendly, for they can be taken as often as the student wishes, and they provide extensive feedback in the form of helpful hints and suggestions when incorrect answers are selected.

> **Self-Assessment Test 2.3**
>
> Test your understanding of the material in Sections 2.6 and 2.7:
>
> • Freely Falling Bodies   • Graphical Analysis of Velocity and Acceleration
>
> Go to **www.wiley.com/college/cutnell**

Concept Simulation 4.1

Three horizontally directed forces are applied to an object, and the resulting motion of the object is displayed, along with synchronized graphs of position, velocity, and acceleration versus time. The mass of the object and the magnitudes and directions of the forces are under your control, so that you can explore how these quantities affect the motion.

Related Homework: Problems 2, 64

Go to
www.wiley.com/college/cutnell

NEW! *Concept Simulations* Often, many concepts in physics—such as relative velocities, collisions, and ray tracing—can be better understood when they are simulated. At appropriate locations throughout the text, we have placed references to 61 on-line simulations (**www.wiley.com/college/cutnell**). With each simulation, the student can experiment, since one or more parameters are under user control. Many of the simulations provide references to related homework material, which, in turn, is cross-referenced to the simulation.

NEW! *Concept Summaries* All chapter-ending summaries have been rewritten so that they present an abridged, but complete, version of the material on a section-by-section basis, including important equations. To help students review the material, the discussion of each section is flanked by two lists. On one side, a list of topics provides a quick-reference option for locating a given topic. On the other side, a color-coded list of the text examples and the on-line learning aids (Concept Simulations, Interactive LearningWare, and Interactive Solutions) shows students the related learning aids. The summary also identifies the on-line Self-Assessment Tests that are keyed to the sections, so that students can evaluate their understanding of the chapter topics.

Concept Summary

This summary presents an abridged version of the chapter, including the important equations and all available learning aids. For convenient reference, the learning aids (including the text's examples) are placed next to or immediately after the relevant equation or discussion. The following Learning Aids may be found on-line at **www.wiley.com/college/cutnell**:

Topic	Discussion	Learning Aids
	5.1 Uniform Circular Motion	
	Uniform circular motion is the motion of an object traveling at a constant (uniform) speed on a circular path.	
Period of the motion	The period T is the time required for the object to travel once around the circle. The speed v of the object is related to the period and the radius r of the circle by	**Example 1**
	$$v = \frac{2\pi r}{T} \qquad (5.1)$$	
	5.2 Centripetal Acceleration	
	An object in uniform circular motion experiences an acceleration, known as centripetal acceleration. The magnitude a_c of the centripetal acceleration is	**Examples 2, 3, 4, 15**
		Interactive LearningWare 5.1
Centripetal acceleration (magnitude)	$$a_c = \frac{v^2}{r} \qquad (5.2)$$	**Interactive Solution 5.3**
Centripetal acceleration (direction)	where v is the speed of the object and r is the radius of the circle. The direction of the centripetal acceleration vector always points toward the center of the circle and continually changes as the object moves.	
	5.3 Centripetal Force	
	To produce a centripetal acceleration, a net force pointing toward the center of the circle is required. This net force is called the centripetal force, and its magnitude F_c is	**Examples 5, 6, 7, 16**
		Interactive LearningWare 5.2
Centripetal force (magnitude)	$$F_c = \frac{mv^2}{r} \qquad (5.3)$$	**Concept Simulations 5.1, 5.2**
	where m and v are the mass and speed of the object, and r is the radius of the	**Interactive Solution 5.19**
Centripetal force (direction)	circle. The direction of the centripetal force vector, like that of the centripetal acceleration vector, always points toward the center of the circle.	

Use Self-Assessment Test 5.1 to evaluate your understanding of Sections 5.1–5.3.

NEW! *Interactive Solutions* Often, students experience difficulty in getting started with the more difficult homework problems. The Interactive Solutions feature consists of 111 on-line problems (**www.wiley.com/college/cutnell**), each of which is allied with a particular homework problem in the text. Each Interactive Solution is worked out by the student in an interactive manner, and the solution is designed to serve as a model for the associated homework problem. Those homework problems that are associated with Interactive Solutions are identified in the text.

Concepts & Calculations

Conceptual understanding is an essential part of a good problem-solving technique. To emphasize the role of conceptual understanding, every chapter includes a Concepts & Calculations section. These sections are organized around a special type of example, each of which begins with several conceptual questions that are answered before the quantitative problem is worked out. The purpose of the questions is to put into sharp focus the concepts with which the problem deals. We hope that these examples also provide mini-reviews of material studied earlier in the chapter and in previous chapters.

Concepts & Calculations Example 20
Velocity, Acceleration, and Newton's Second Law of Motion ▼

Figure 4.37 shows two forces, $F_1 = +3000$ N and $F_2 = +5000$ N acting on a spacecraft, where the plus signs indicate that the forces are directed along the $+x$ axis. A third force F_3 also acts on the spacecraft but is not shown in the drawing. The craft is moving with a constant velocity of $+850$ m/s. Find the magnitude and direction of F_3.

Concept Questions and Answers Suppose the spacecraft were stationary. What would be the direction of F_3?

Answer If the spacecraft is stationary, its acceleration is zero. According to Newton's second law, the acceleration of an object is proportional to the net force acting on it. Thus, the net force must also be zero. But the net force is the vector sum of the three forces in this case. Therefore, the force F_3 must have a direction such that it balances to zero the forces F_1 and F_2. Since F_1 and F_2 point along the $+x$ axis in Figure 4.37, F_3 must then point along the $-x$ axis.

When the spacecraft is moving at a constant velocity of $+850$ m/s, what is the direction of F_3?

Answer Since the velocity is constant, the acceleration is still zero. As a result, everything we said in the stationary case applies again here. The net force is zero, and the force F_3 must point along the $-x$ axis in Figure 4.37.

Solution Since the velocity is constant, the acceleration is zero. The net force must also be zero, so that

$$\Sigma F_x = F_1 + F_2 + F_3 = 0$$

Solving for F_3 yields

$$F_3 = -(F_1 + F_2) = -(3000 \text{ N} + 5000 \text{ N}) = \boxed{-8000 \text{ N}}$$

Conceptual Examples A good problem-solving technique begins with a foundation of conceptual understanding. Therefore, every chapter includes examples that are entirely conceptual in nature. These examples are worked out in a rigorous but qualitative fashion, with no (or very few) equations. We have added art to many of these as a way to help students visualize the concepts. Our intent is to provide students with explicit models of how to "think through" a problem before attempting to solve it numerically. The *Conceptual Examples* deal with a wide range of topics and a large number of issues that often confuse students. Wherever possible, we have focused on real-world situations and have structured the examples so that they lead naturally to homework material found at the ends of the chapters. The related homework material is explicitly identified, and that material contains a cross reference that encourages students to review the pertinent *Conceptual Example*.

Conceptual Example 7 **Mass Versus Weight**
▼

A vehicle is being designed for use in exploring the moon's surface and is being tested on earth, where it weighs roughly six times more than it will on the moon. In one test, the acceleration of the vehicle along the ground is measured. To achieve the same acceleration on the moon, will the net force acting on the vehicle be greater than, less than, or the same as that required on earth?

Reasoning and Solution The net force ΣF required to accelerate the vehicle is specified by Newton's second law as $\Sigma F = ma$, where m is the vehicle's mass and a is the acceleration along the ground. For a given acceleration, the net force depends only on the mass. But the mass is an intrinsic property of the vehicle and is the same on the moon as it is on the earth. Therefore, the same net force would be required for a given acceleration on the moon as on the earth. Do not be misled by the fact that the vehicle weighs more on earth. The greater weight occurs only because the earth's mass and radius are different than the moon's. In any event, *in Newton's second law, the net force is proportional to the vehicle's mass, not its weight.*

Related Homework: *Problems 21, 22*
▲

Need more practice?

Interactive LearningWare 6.1
A cyclist coasts down a hill, losing 20.0 m in height, and then coasts up an identical hill to the same initial vertical height. The bicycle and rider have a mass of 73.5 kg. Two forces act on the bicycle/rider system: the weight of the bicycle and rider and the normal force exerted on the tires by the ground. Determine the work done by each of these forces during (a) the downward part of the motion and (b) the upward part of the motion.

Related Homework: *Problem 6*

Go to
www.wiley.com/college/cutnell
for an interactive solution.

Interactive LearningWare This feature consists of interactive calculational examples presented on our Web site (**www.wiley.com/college/cutnell**). References to the Learning-Ware are contained in nearly every chapter. Each example is designed in an interactive format that consists of five steps. We believe that these steps are essential to the development of problem-solving skills and that students will benefit from seeing them explicitly identified in each example.

1. The *Visualization* step uses animations and drawings to help students understand the general nature of the situation being analyzed.
2. The *Reasoning* step establishes the conceptual basis for the solution with the aid of multiple-choice questions and answers.
3. The *Knowns-and-Unknowns* step deals with the important process of identifying all the variables and the symbols that represent them. This information is not simply given to the students but, rather, is interactively developed in the form of tables that evolve as the example proceeds.
4. In the *Modeling* step, the student, using the results of the previous steps, models the problem with equations developed in the text.
5. Finally, in the *Solution* step, the solution to the problem is obtained.

In summary, the hallmark of these Interactive LearningWare examples is a user-friendly clarity that allows the student to see how conceptual understanding and quantitative analysis work together in a problem-solving environment. Related homework material is identified and contains a cross-reference to the LearningWare.

Explicit Reasoning Steps Since reasoning is the cornerstone of problem solving, we believe that students will benefit from seeing the reasoning stated explicitly. Therefore, the format in which examples are worked out includes an explicit reasoning step. In this step we explain what motivates our procedure for solving the problem before any algebraic or numerical work is done. In the *Concepts & Calculations* examples, the reasoning is presented in a question-and-answer format.

Reasoning Strategy

Applying the Equations of Kinematics

1. Make a drawing to represent the situation being studied. A drawing helps us to see what's happening.
2. Decide which directions are to be called positive (+) and negative (−) relative to a conveniently chosen coordinate origin. Do not change your decision during the course of a calculation.
3. In an organized way, write down the values (with appropriate plus and minus signs) that are given for any of the five kinematic variables (x, a, v, v_0, and t). Be on the alert for "implied data," such as the phrase "starts from rest," which means that the value of the initial velocity is $v_0 = 0$ m/s. The data summary boxes used in the examples in the text are a good way of keeping track of this information. In addition, identify the variables that you are being asked to determine.
4. Before attempting to solve a problem, verify that the given information contains values for at least three of the five kinematic variables. Once the three known variables are identified along with the desired unknown variable, the appropriate relation from Table 2.1 can be selected. Remember that the motion of two objects may be interrelated, so they may share a common variable. The fact that the motions are interrelated is an important piece of information. In such cases, data for only two variables need be specified for each object.
5. When the motion of an object is divided into segments, as in Example 9, remember that the final velocity of one segment is the initial velocity for the next segment.
6. Keep in mind that there may be two possible answers to a kinematics problem as, for instance, in Example 8. Try to visualize the different physical situations to which the answers correspond.

Reasoning Strategies A number of the examples in the text deal with well-defined strategies for solving certain types of problems. In such cases, we have included summaries of the steps involved. These summaries, which are titled *Reasoning Strategies,* encourage frequent review of the techniques used and help students focus on the related concepts.

Concepts at a Glance To provide a coherent picture of how new concepts are built upon previous ones and to reinforce fundamental unifying ideas, the text contains 66 Concepts-at-a-Glance flowcharts. These charts show in a visual way the conceptual development of physics principles. Within a chart, new concepts are placed in gold panels, and previously introduced and related concepts are placed in light blue panels. Each chart is discussed in a separate and highlighted paragraph that helps students connect the new concept with the real world.

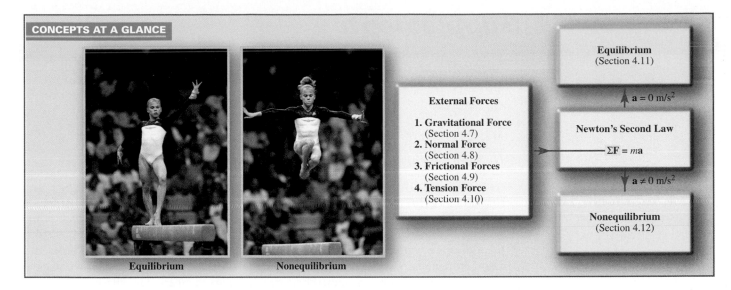

CONCEPTS AT A GLANCE

Equilibrium   Nonequilibrium

External Forces

1. **Gravitational Force** (Section 4.7)
2. **Normal Force** (Section 4.8)
3. **Frictional Forces** (Section 4.9)
4. **Tension Force** (Section 4.10)

Equilibrium (Section 4.11)

$a = 0 \ m/s^2$

Newton's Second Law

$\Sigma F = m a$

$a \neq 0 \ m/s^2$

Nonequilibrium (Section 4.12)

The Physics of . . . This edition contains 239 real-world applications that reflect our commitment to show students how prevalent physics is in their lives. Each application is identified in the margin with the label *The Physics of*, and those that deal with biomedical material are further marked with an icon in the shape of a caduceus. A complete list of the applications can be found after the Acknowledgments section.

> *The physics of*
> **catapulting a jet from an aircraft carrier.**

Problem Solving Insights To reinforce the problem-solving techniques illustrated in the worked-out examples, we have included short statements in the margins, identified by the label *Problem Solving Insight.* These statements help students to develop good problem-solving skills by providing the kind of advice that a teacher gives when explaining a calculation in detail.

> *Problem solving insight*
> **"Implied data" are important. For instance, in Example 6 the phrase "starting from rest" means that the initial velocity is zero ($v_0 = 0$ m/s).**

Homework Problems and Conceptual Questions The sixth edition contains an even greater amount of homework material than does the fifth edition; about 17% of the problems are new or modified. In providing this large number of problems and questions, we have used a wide variety of real-world situations with realistic data. The problems are ranked according to difficulty, with the most difficult problems marked with a double asterisk (**) and those of intermediate difficulty marked with a single asterisk (*). The easiest problems are unmarked. Those whose solutions appear in the *Student Solutions Manual* are identified with the label **ssm**. Those whose solutions are available on the Web (**www.wiley.com/college/cutnell**) are marked with the label **www**. Problems and questions that are biomedical in nature are identified with an icon in the shape of a caduceus. Since teachers sometimes want to assign homework without identifying a particular section, we have provided a group of problems without references to sections under the heading *Additional Problems. Concepts & Calculations—Group Learning Problems* were introduced in the fifth edition and have been very well received. Therefore, we have expanded this grouping to include about 30% more problems. These problems are modeled on the examples in the *Concepts & Calculations* section found at the end of each chapter and consist of a series of concept questions followed by a related quantitative problem. The questions are designed to help students identify the concepts that are used in solving the problem and provide a focus for group discussions.

In spite of our best efforts to produce an error-free book, errors no doubt remain. They are solely our responsibility, and we would appreciate hearing of any that you find. We hope that this text makes learning and teaching physics easier and more enjoyable and look forward to hearing about your experiences with it. Please feel free to write us care of Physics Editor, Higher Education Division, John Wiley & Sons, Inc., 111 River Street, Hoboken, NJ 07030 or contact us at **www.wiley.com/college/cutnell.**

Supplements

An extensive package of supplements to accompany *Physics,* sixth edition, is available to assist both the teacher and the student.

Instructor's Supplements
Helping Teachers Teach

Instructor's Resource Guide by David T. Marx, Southern Illinois University of Carbondale. This guide contains an extensive listing of Web-based physics education resources. It also includes teaching ideas, lecture notes, demonstration suggestions, alternative syllabi for courses of different lengths and emphasis, as well as conversion notes that allow instructors to use their class notes from other texts. A Problem Locator Guide provides an easy way to correlate fifth-edition problem numbers with the corresponding sixth-edition numbers.

Instructor's Solutions Manual by John D. Cutnell and Kenneth W. Johnson. This manual provides worked-out solutions for all end-of-chapter Conceptual Questions and Problems. Volume 1 contains the solutions for Chapter 1-17. Volume 2 contains the solutions for Chapters 18–32.

Test Bank by David T. Marx, Southern Illinois University at Carbondale. This manual includes more than 2200 multiple-choice questions. These items are also available in the *Computerized Test Bank* (see below).

Instructor's Resource CD-ROM This CD contains:

- The entire *Instructor's Solutions Manual* in both Microsoft Word © (Windows and Macintosh) and PDF files.

- A *Computerized Test Bank,* for use with both Windows and Macintosh PC's with full editing features to help you customize tests.

- Multiple-choice format for 1200 of the end-of-chapter problems in the text. This material allows instructors to assign homework in a machine-gradable format.

- All text illustrations, suitable for classroom projection, printing, and web posting.

Wiley Physics Simulations CD-ROM This CD contains 50 interactive simulations (Java applets) that can be used for classroom demonstrations.

Transparencies More than 200 four-color illustrations from the text are provided in a form suitable for projection in the classroom.

On-line Homework and Quizzing *Physics,* sixth edition, supports WebAssign, CAPA, and eGrade, which are programs that give instructors the ability to deliver and grade homework and quizzes over the Internet.

WebCT and Blackboard Many of the instructor and student supplements have been coded for easy integration into WebCT and Blackboard, which are course management programs that allow instructors to set up a complete on-line course with chat rooms, bulletin boards, quizzing, built-in student tracking tools, etc.

Student's Supplements

Helping Students Learn

Student Web Site This Web site (**www.wiley.com/college/cutnell**) was developed specifically for *Physics,* sixth edition, and is designed to assist students further in the study of physics. At this site, students can access the following resources; the first few pages of the text show some examples of these resources.

- Self-Assessment Tests

- Solutions to selected end-of-chapter problems that are identified with a **www** icon in the text.

- Interactive Solutions

- Concept Simulations

- Interactive LearningWare examples

- Review quizzes for the MCAT exam

- Links to other Web sites that offer physics tutorial help

Student Study Guide by John D. Cutnell and Kenneth W. Johnson. This student study guide consists of traditional print materials; with the Student Web site, it provides a rich, interactive environment for review and study.

Student Solutions Manual by John D. Cutnell and Kenneth W. Johnson. This manual provides students with complete worked-out solutions for approximately 600 of the odd-numbered end-of-chapter problems. These problems are indicated in the text with an **ssm** icon.

Acknowledgments

We have had the help of many individuals in putting together this text. Working with them and witnessing their talents, dedication, and professionalism has been an exhilarating experience. It is a genuine pleasure to acknowledge their contributions.

Special thanks go to our editor, Stuart Johnson, for his expert and on-target advice at all stages of the project. We are most fortunate to have his help and encouragement. His keen understanding of college publishing and his vision for its future are assets to be treasured.

Having Barbara Heaney, Director of Product and Market Development, as our developmental editor is like winning the lottery. Her impact on the project has been enormous, including an infusion of numerous ideas, an unwavering commitment to excellence, and a careful attention to detail.

We are especially indebted to Barbara Russiello, Production Editor, for coordinating the all-electronic production of the book. She was always there for us with answers to our questions, help, and encouragement. Her tireless efforts made it possible to produce a book that meets the highest production standards.

In modifying and adding to the illustrations for this edition, we worked with Sigmund Malinowski of the illustrations department. He was always ready and willing to help with advice and suggestions. We are grateful for his efforts and enjoy working with him immensely.

The design of the text and its cover is the work of Maddy Lesure, Design Director. The clear and pleasing presentation of the text's many features and elements is due to her immense talent in coordinating color, design, and fonts. It has been a pleasure, Maddy.

We are grateful to Hilary Newman, Photo Department Manager, for the photos in the text. She was able to find great photos that express just the right physics. We think that you will like the photos as much as we do.

Our admiration and appreciation are extended to Martin Batey and Tom Kulesa, for their work on the many media components; they have made this complex process much easier for us.

Outstanding proofreaders are essential in producing an error-free book. Our thanks go to Gloria Hamilton for an extraordinary job in catching errors of all types, from awkwardly worded sentences to wrong fonts.

Our sincere gratitude goes to Geraldine Osnato for coordinating the extensive supplements package, to Alec Borenstein, Editorial Assistant, and to Helen Walden for copyediting the manuscript.

The sales representatives of John Wiley & Sons, Inc. are a very special group. We deeply appreciate their efforts on our behalf as they are in constant contact with physics departments throughout the country.

Many physicists have generously shared their ideas with us about good pedagogy and helped us by pointing out errors in our presentation. For all of their suggestions we are grateful. They have helped us to write more clearly and accurately and have influenced markedly the evolution of this text. To the reviewers of this and previous editions, we especially owe a large debt of gratitude. Specifically, we thank:

Joseph Alward,
University of the Pacific

Zaven Altounian,
McGill University

Chi Kwan Au,
University of South Carolina

Donald Ballegeer,
University of Wisconsin

David Bannon,
Oregon State University

William A. Barker, *formerly at*
Santa Clara University

Paul D. Beale,
University of Colorado at Boulder

Edward E. Beasley,
Gallaudet University

Roger Bland,
San Francisco State University

Edward R. Borchardt,
Mankato State University

Treasure Brasher,
West Texas State University

Robert Brehme,
Wake Forest University

Michael Bretz,
University of Michigan at Ann Arbor

Carl Bromberg,
Michigan State University

Michael E. Browne,
University of Idaho

Ronald W. Canterna,
University of Wyoming

Neal Cason,
University of Wyoming

Marvin Chester,
University of California at Los Angeles

William S. Chow, *emeritus*
University of Cincinnati

Lowell Christensen, *formerly at*
American River College

Albert C. Claus,
Loyola University of Chicago

Thomas Berry Cobb,
Bowling Green State University

Lawrence Coleman,
University of California at Davis

Lattie F. Collins,
East Tennessee State University

Mark Comella, *formerly at*
Duquesne University

Biman Das,
SUNY-Potsdam

Doyle Davis,
New Hampshire Community Technical College

Steven Davis,
University of Arkansas at Little Rock

James E. Dixon,
Iowa State University

Duane Doty,
California State University at Northridge

Miles J. Dressler, *emeritus*
Washington State University

Dewey Dykstra,
Boise State University

Robert J. Endorf,
University of Cincinnati

Lewis Ford,
Texas A&M University

Greg Francis,
Montana State University

Roger Freedman,
University of California of Santa Barbara

Roger J. Friauf,
University of Kansas

C. Sherman Frye, Jr.,
Northern Virginia Community College

John Gagliardi,
Rutgers University

Simon George,
California State University at Long Beach

James B. Gerhart,
University of Washington

John Gieniec, *formerly at*
Central Missouri State University

Barry Gilbert,
Rhode Island College

Roy Goodrich,
Louisiana State University

D. Wayne Green,
Knox College

William Gregg,
Louisiana State University

David Griffiths,
Oregon State University

Grant Hart,
Brigham Young University

Thomas Herrmann,
East Oregon University

Lawrence A. Hitchingham, *formerly at*
Jackson Community College

Paul R. Holody,
Henry Ford Community College

Darrell O. Huwe, *formerly at*
Ohio University

Mario Iona, *emeritus*
University of Denver

David A. Jerde,
St. Cloud State University

Larry Josbeno,
Corning Community College

R. Lee Kernell, *emeritus*
Old Dominion University

Gary Kessler, *formerly at*
Illinois Wesleyan University

Randy Kobes,
University of Winnipeg

K. Kothari,
Tuskegee University

Robert A. Kromhout,
Florida State University

Theodore Kruse,
Rutgers University

Pradeep Kumar,
University of Florida at Gainesville

Rubin H. Landau,
Oregon State University

Christopher P. Landee,
Clark University

Alfredo Louro,
University of Calgary

Daines Lund, *formerly at*
Utah State University

R. Wayne Major, *formerly at*
University of Richmond

A. John Mallinckrodt,
California State Polytechnic University,
Pomona

Thomas P. Marvin,
Southern Oregon State College

John McClain, *formerly at*
Southern Illinois University at Carbondale

John McCullen,
University of Arizona

Laurence McIntyre,
University of Arizona

B. Wieb Van Der Meer,
Western Kentucky University

Donald D. Miller,
Central Missouri State University

James Miller, *formerly at*
East Tennessee State University

Paul Morris,
Abilene Christian University

Robert Morris,
Coppin State College

Richard A. Morrow,
University of Maine

Kenneth Mucker,
Bowling Green State University

Rod Nave,
Georgia State University

David Newton,
DeAnza College

R. Chris Olsen,
Naval Postgraduate School, Monterey

Robert F. Petry, *formerly at*
University of Oklahoma

Peter John Polito,
Springfield College

Jon Pumplin,
Michigan State University

Talat Rahman,
Kansas State University

Michael Ram,
State University of New York at Buffalo

Vallabhaneni Rao, *formerly at*
Memorial University of Newfoundland

Jacobo Rapaport,
Ohio University

Wayne W. Repko,
Michigan State University

Barry C. Robertson,
Queens University (Ontario)

Harold Romero, *formerly at*
University of Southern Mississippi

Larry Rowan,
University of North Carolina at Chapel Hill

Roy S. Rubins,
University of Texas at Arlington

O.M.P. Rustgi, *formerly at*
State University of New York at Buffalo

Al Sanders,
University of Tennessee at Knoxville

Charles Scherr, *formerly at*
University of Texas at Austin

Wesley Shanholtzer,
Marshall University

Marc Sher,
College of William & Mary

James Simmons, *formerly at*
Concordia College

John J. Sinai,
University of Louisville

Virgil Stubblefield,
John A. Logan College

Ronald G. Tabak,
Youngstown State University

Patrick Tam,
Humboldt State University

Robert Tyson,
University of North Carolina at Charlotte

Timothy Usher,
California State University at San Bernardino

Rolf Vatne,
Portland Community College

Howard G. Voss,
Arizona State University

James M. Wallace, *formerly at*
Jackson Community College

Henry White,
University of Missouri

Jerry H. Willson, *formerly at*
Metropolitan State College

Linn Van Woerkom,
Ohio State University

The physics of

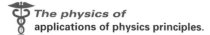

The physics of
applications of physics principles.

To show students that physics has a widespread impact on their lives, we have included a large number of applications of physics principles. Many of these applications are not found in other texts. The most important ones are listed below along with the page number locating the corresponding discussion. They are identified in the margin of the page on which they occur with the label "*The physics of*" Biomedical applications are marked with an icon in the shape of a caduceus ⚕. The discussions are integrated into the text, so that they occur as a natural part of the physics being presented. It should be noted that the list is not a complete list of all the applications of physics principles to be found in the text. There are many additional applications that are discussed only briefly or that occur in the homework questions and problems.

Chapter 1

Introduction and Mathematical Concepts

In the movie *Spider-Man* many of the background scenes were created using animation techniques that rely on computers and mathematical concepts such as trigonometry and vectors. These mathematical tools will also be useful throughout this book in dealing with the laws of physics. (© Photofest)

1.1 The Nature of Physics

The science of physics has developed out of the efforts of men and women to explain our physical environment. These efforts have been so successful that the laws of physics now encompass a remarkable variety of phenomena, including planetary orbits, radio and TV waves, magnetism, and lasers, to name just a few.

The exciting feature of physics is its capacity for predicting how nature will behave in one situation on the basis of experimental data obtained in another situation. Such predictions place physics at the heart of modern technology and, therefore, can have a tremendous impact on our lives. Rocketry and the development of space travel have their roots firmly planted in the physical laws of Galileo Galilei (1564–1642) and Isaac Newton (1642–1727). The transportation industry relies heavily on physics in the development of engines and the design of aerodynamic vehicles. Entire electronics and computer industries owe their existence to the invention of the transistor, which grew directly out of the laws of physics that describe the electrical behavior of solids. The telecommunications industry depends extensively on electromagnetic waves, whose existence was predicted by James Clerk Maxwell (1831–1879) in his theory of electricity and magnetism. The medical profession uses X-ray, ultrasonic, and magnetic resonance methods for obtaining images of the interior of the human body, and physics lies at the core of all these. Perhaps the most widespread impact in modern technology is that due to the laser. Fields ranging from space exploration to medicine benefit from this incredible device, which is a direct application of the principles of atomic physics.

Because physics is so fundamental, it is a required course for students in a wide range of major areas. We welcome you to the study of this fascinating topic. You will learn how to see the world through the "eyes" of physics and to reason as a physicist does. In the process, you will learn how to apply physics principles to a wide range of problems. We hope that you will come to recognize that physics has important things to say about your environment.

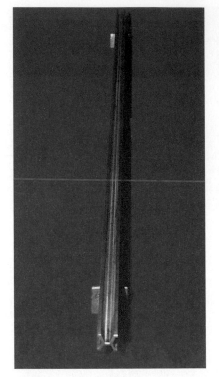

Figure 1.1 The standard platinum–iridium meter bar. (Courtesy Bureau International des Poids et Mesures, France)

1.2 Units

Physics experiments involve the measurement of a variety of quantities, and a great deal of effort goes into making these measurements as accurate and reproducible as possible. The first step toward ensuring accuracy and reproducibility is defining the units in which the measurements are made.

In this text, we emphasize the system of units known as **SI units,** which stands for the French phrase "Le **S**ystème **I**nternational d'Unités." By international agreement, this system employs the ***meter*** (m) as the unit of length, the ***kilogram*** (kg) as the unit of mass, and the ***second*** (s) as the unit of time. Two other systems of units are also in use, however. The CGS system utilizes the centimeter (cm), the gram (g), and the second for length, mass, and time, respectively, and the BE or British Engineering system (the gravitational version) uses the foot (ft), the slug (sl), and the second. Table 1.1 summarizes the units used for length, mass, and time in the three systems.

Originally, the meter was defined in terms of the distance measured along the earth's surface between the north pole and the equator. Eventually, a more accurate measurement standard was needed, and by international agreement the meter became the distance between two marks on a bar of platinum–iridium alloy (see Figure 1.1) kept at a tempera-

Figure 1.2 The standard platinum–iridium kilogram is kept at the International Bureau of Weights and Measures in Sèvres, France. (Courtesy Bureau International des Poids et Mesures, France)

Table 1.1 *Units of Measurement*

	System		
	SI	CGS	BE
Length	meter (m)	centimeter (cm)	foot (ft)
Mass	kilogram (kg)	gram (g)	slug (sl)
Time	second (s)	second (s)	second (s)

ture of 0 °C. Today, to meet further demands for increased accuracy, the meter is defined as the distance that light travels in a vacuum in a time of 1/299 792 458 second. This definition arises because the speed of light is a universal constant that is defined to be 299 792 458 m/s.

The definition of a kilogram as a unit of mass has also undergone changes over the years. As Chapter 4 discusses, the mass of an object indicates the tendency of the object to continue in motion with a constant velocity. Originally, the kilogram was expressed in terms of a specific amount of water. Today, one kilogram is defined to be the mass of a standard cylinder of platinum–iridium alloy, like the one in Figure 1.2.

As with the units for length and mass, the present definition of the second as a unit of time is different from the original definition. Originally, the second was defined according to the average time for the earth to rotate once about its axis, one day being set equal to 86 400 seconds. The earth's rotational motion was chosen because it is naturally repetitive, occurring over and over again. Today, we still use a naturally occurring repetitive phenomenon to define the second, but of a very different kind. We use the electromagnetic waves emitted by cesium-133 atoms in an atomic clock like that in Figure 1.3. One second is defined as the time needed for 9 192 631 770 wave cycles to occur.*

The units for length, mass, and time, along with a few other units that will arise later, are regarded as **base** SI units. The word "base" refers to the fact that these units are used along with various laws to define additional units for other important physical quantities, such as force and energy. The units for such other physical quantities are referred to as **derived** units, since they are combinations of the base units. Derived units will be introduced from time to time, as they arise naturally along with the related physical laws.

The value of a quantity in terms of base or derived units is sometimes a very large or very small number. In such cases, it is convenient to introduce larger or smaller units that are related to the normal units by multiples of ten. Table 1.2 summarizes the prefixes that are used to denote multiples of ten. For example, 1000 or 10^3 meters are referred to as 1 kilometer (km), and 0.001 or 10^{-3} meter is called 1 millimeter (mm). Similarly, 1000 grams and 0.001 gram are referred to as 1 kilogram (kg) and 1 milligram (mg), respectively. Appendix A contains a discussion of scientific notation and powers of ten, such as 10^3 and 10^{-3}.

Figure 1.3 This atomic clock, the NIST-F1, is considered one of the world's most accurate clocks. It keeps time with an uncertainty of about one second in twenty million years.
(© Geoffrey Wheeler)

1.3 The Role of Units in Problem Solving
THE CONVERSION OF UNITS

Since any quantity, such as length, can be measured in several different units, it is important to know how to convert from one unit to another. For instance, the foot can be used to express the distance between the two marks on the standard platinum–iridium meter bar. There are 3.281 feet in one meter, and this number can be used to convert from meters to feet, as the following example demonstrates.

Example 1 The World's Highest Waterfall

The highest waterfall in the world is Angel Falls in Venezuela, with a total drop of 979.0 m (see Figure 1.4). Express this drop in feet.

Reasoning When converting between units, we write down the units explicitly in the calculations and treat them like any algebraic quantity. In particular, we will take advantage of the following algebraic fact: Multiplying or dividing an equation by a factor of 1 does not alter an equation.

Solution Since 3.281 feet = 1 meter, it follows that (3.281 feet)/(1 meter) = 1. Using this factor of 1 to multiply the equation "Length = 979.0 meters," we find that

$$\text{Length} = (979.0 \text{ m})(1) = (979.0 \text{ meters})\left(\frac{3.281 \text{ feet}}{1 \text{ meter}}\right) = \boxed{3212 \text{ feet}}$$

Table 1.2 *Standard Prefixes Used to Denote Multiples of Ten*

Prefix	Symbol	Factor[a]
Tera	T	10^{12}
Giga[b]	G	10^{9}
Mega	M	10^{6}
Kilo	k	10^{3}
Hecto	h	10^{2}
Deka	da	10^{1}
Deci	d	10^{-1}
Centi	c	10^{-2}
Milli	m	10^{-3}
Micro	μ	10^{-6}
Nano	n	10^{-9}
Pico	p	10^{-12}
Femto	f	10^{-15}

[a] Appendix A contains a discussion of powers of ten and scientific notation.

[b] Pronounced jig′a.

* See Chapter 16 for a discussion of waves in general and Chapter 24 for a discussion of electromagnetic waves in particular.

Figure 1.4 Angel Falls in Venezuela is the highest waterfall in the world. (© Kevin Schafer/The Image Bank/Getty Images)

The colored lines emphasize that the units of meters behave like any algebraic quantity and cancel when the multiplication is performed, leaving only the desired unit of feet to describe the answer. In this regard, note that 3.281 feet = 1 meter also implies that (1 meter)/(3.281 feet) = 1. However, we chose not to multiply by a factor of 1 in this form, because the units of meters would not have canceled.

A calculator gives the answer as 3212.099 feet. Standard procedures for significant figures, however, indicate that the answer should be rounded off to four significant figures, since the value of 979.0 meters is accurate to only four significant figures. In this regard, the "1 meter" in the denominator does not limit the significant figures of the answer, because this number is precisely one meter by definition of the conversion factor. Appendix B contains a review of significant figures.

In any conversion, if the units do not combine algebraically to give the desired result, the conversion has not been carried out properly. With this in mind, the next example stresses the importance of writing down the units and illustrates a typical situation in which several conversions are required.

Example 2 Interstate Speed Limit

Express the speed limit of 65 miles/hour in terms of meters/second.

Reasoning As in Example 1, it is important to write down the units explicitly in the calculations and treat them like any algebraic quantity. Here, two well-known relationships come into play—namely, 5280 feet = 1 mile and 3600 seconds = 1 hour. As a result, (5280 feet)/(1 mile) = 1 and (3600 seconds)/(1 hour) = 1. In our solution we will use the fact that multiplying and dividing by these factors of unity does not alter an equation.

Solution Multiplying and dividing by factors of unity, we find the speed limit in feet per second as shown below:

$$\text{Speed} = \left(65\ \frac{\text{miles}}{\text{hour}}\right)(1)(1) = \left(65\ \frac{\cancel{\text{miles}}}{\cancel{\text{hour}}}\right)\left(\frac{5280\ \text{feet}}{1\ \cancel{\text{mile}}}\right)\left(\frac{1\ \cancel{\text{hour}}}{3600\ \text{s}}\right) = 95\ \frac{\text{feet}}{\text{second}}$$

To convert feet into meters, we use the fact that (1 meter)/(3.281 feet) = 1:

$$\text{Speed} = \left(95\ \frac{\text{feet}}{\text{second}}\right)(1) = \left(95\ \frac{\cancel{\text{feet}}}{\text{second}}\right)\left(\frac{1\ \text{meter}}{3.281\ \cancel{\text{feet}}}\right) = \boxed{29\ \frac{\text{meters}}{\text{second}}}$$

Problem solving insight

In addition to their role in guiding the use of conversion factors, units serve a useful purpose in solving problems. They can provide an internal check to eliminate errors, if they are carried along during each step of a calculation and treated like any algebraic factor. In particular, remember that *only quantities with the same units can be added or subtracted.* Thus, at one point in a calculation, if you find yourself adding 12 miles to 32 kilometers, stop and reconsider. Either miles must be converted into kilometers or kilometers must be converted into miles before the addition can be carried out.

A collection of useful conversion factors is given on the page facing the inside of the front cover. The reasoning strategy that we have followed in Examples 1 and 2 for converting between units is outlined as follows:

Reasoning Strategy

Converting Between Units

1. In all calculations, write down the units explicitly.

2. Treat all units as algebraic quantities. In particular, when identical units are divided, they are eliminated algebraically.

3. Use the conversion factors located on the page facing the inside of the front cover. Be guided by the fact that multiplying or dividing an equation by a factor of 1 does not alter the equation. For instance, the conversion factor of 3.281 feet = 1 meter might be applied in the form (3.281 feet)/(1 meter) = 1.

This factor of 1 would be used to multiply an equation such as "Length = 5.00 meters" in order to convert meters to feet.

4. Check to see that your calculations are correct by verifying that the units combine algebraically to give the desired unit for the answer. Only quantities with the same units can be added or subtracted.

DIMENSIONAL ANALYSIS

We have seen that many quantities are denoted by specifying both a number and a unit. For example, the distance to the nearest telephone may be 8 meters, or the speed of a car might be 25 meters/second. Each quantity, according to its physical nature, requires a certain *type* of unit. Distance must be measured in a length unit such as meters, feet, or miles, and a time unit will not do. Likewise, the speed of an object must be specified as a length unit divided by a time unit. In physics, the term **dimension** is used to refer to the physical nature of a quantity and the type of unit used to specify it. Distance has the dimension of length, which is symbolized as [L], while speed has the dimensions of length [L] divided by time [T], or [L/T]. Many physical quantities can be expressed in terms of a combination of fundamental dimensions such as length [L], time [T], and mass [M]. Later on, we will encounter certain other quantities, such as temperature, which are also fundamental. A fundamental quantity like temperature cannot be expressed as a combination of the dimensions of length, time, mass, or any other fundamental dimension.

Dimensional analysis is used to check mathematical relations for the consistency of their dimensions. As an illustration, consider a car that starts from rest and accelerates to a speed v in a time t. Suppose we wish to calculate the distance x traveled by the car but are not sure whether the correct relation is $x = \frac{1}{2}vt^2$ or $x = \frac{1}{2}vt$. We can decide by checking the quantities on both sides of the equals sign to see whether they have the same dimensions. If the dimensions are not the same, the relation is incorrect. For $x = \frac{1}{2}vt^2$, we use the dimensions for distance [L], time [T], and speed [L/T] in the following way:

$$x = \tfrac{1}{2}vt^2$$

Dimensions
$$[L] \overset{?}{=} \left[\frac{L}{\cancel{T}}\right][T]^{\cancel{2}} = [L][T]$$

Dimensions cancel just like algebraic quantities, and pure numerical factors like $\frac{1}{2}$ have no dimensions, so they can be ignored. The dimension on the left of the equals sign does not match those on the right, so the relation $x = \frac{1}{2}vt^2$ cannot be correct. On the other hand, applying dimensional analysis to $x = \frac{1}{2}vt$, we find that

$$x = \tfrac{1}{2}vt$$

Dimensions
$$[L] \overset{?}{=} \left[\frac{L}{\cancel{T}}\right][\cancel{T}] = [L]$$

Problem solving insight
You can check for errors that may have arisen during algebraic manipulations by doing a dimensional analysis on the final expression.

The dimension on the left of the equals sign matches that on the right, so this relation is dimensionally correct. If we know that one of our two choices is the right one, then $x = \frac{1}{2}vt$ is it. In the absence of such knowledge, however, dimensional analysis cannot identify the correct relation. It can only identify which choices *may be* correct, since it does not account for numerical factors like $\frac{1}{2}$ or for the manner in which an equation was derived from physics principles.

1.4 *Trigonometry*

Scientists use mathematics to help them describe how the physical universe works, and trigonometry is an important branch of mathematics. Three trigonometric functions are utilized throughout this text. They are the sine, the cosine, and the tangent of the angle θ

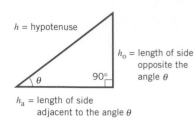

h = hypotenuse

h_o = length of side opposite the angle θ

90°

θ

h_a = length of side adjacent to the angle θ

Figure 1.5 A right triangle.

(Greek theta), abbreviated as sin θ, cos θ, and tan θ, respectively. These functions are defined below in terms of the symbols given along with the right triangle in Figure 1.5.

■ **DEFINITION OF SIN θ, COS θ, AND TAN θ**

$$\sin \theta = \frac{h_o}{h} \tag{1.1}$$

$$\cos \theta = \frac{h_a}{h} \tag{1.2}$$

$$\tan \theta = \frac{h_o}{h_a} \tag{1.3}$$

h = length of the **hypotenuse** of a right triangle
h_o = length of the side **opposite** the angle θ
h_a = length of the side **adjacent** to the angle θ

The sine, cosine, and tangent of an angle are numbers without units, because each is the ratio of the lengths of two sides of a right triangle. Example 3 illustrates a typical application of Equation 1.3.

Example 3   Using Trigonometric Functions

On a sunny day, a tall building casts a shadow that is 67.2 m long. The angle between the sun's rays and the ground is $\theta = 50.0°$, as Figure 1.6 shows. Determine the height of the building.

Reasoning We want to find the height of the building. Therefore, we begin with the colored right triangle in Figure 1.6 and identify the height as the length h_o of the side opposite the angle θ. The length of the shadow is the length h_a of the side that is adjacent to the angle θ. The ratio of the length of the opposite side to the length of the adjacent side is the tangent of the angle θ, which can be used to find the height of the building.

Solution We use the tangent function in the following way, with $\theta = 50.0°$ and $h_a = 67.2$ m:

$$\tan \theta = \frac{h_o}{h_a} \tag{1.3}$$

$$h_o = h_a \tan \theta = (67.2 \text{ m})(\tan 50.0°) = (67.2 \text{ m})(1.19) = \boxed{80.0 \text{ m}}$$

The value of tan 50.0° is found by using a calculator.

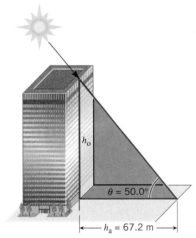

h_o

$\theta = 50.0°$

$h_a = 67.2$ m

Figure 1.6 From a value for the angle θ and the length h_a of the shadow, the height h_o of the building can be found using trigonometry.

Problem solving insight

The sine, cosine, or tangent may be used in calculations such as that in Example 3, depending on which side of the triangle has a known value and which side is asked for. However, *the choice of which side of the triangle to label h_o (opposite) and which to label h_a (adjacent) can be made only after the angle θ is identified.*

Often the values for two sides of the right triangle in Figure 1.5 are available, and the value of the angle θ is unknown. The concept of *inverse trigonometric functions* plays an important role in such situations. Equations 1.4–1.6 give the inverse sine, inverse cosine, and inverse tangent in terms of the symbols used in the drawing. For instance, Equation 1.4 is read as "θ equals the angle whose sine is h_o/h."

$$\theta = \sin^{-1}\left(\frac{h_o}{h}\right) \tag{1.4}$$

$$\theta = \cos^{-1}\left(\frac{h_a}{h}\right) \tag{1.5}$$

$$\theta = \tan^{-1}\left(\frac{h_o}{h_a}\right) \tag{1.6}$$

The use of "-1" as an exponent in Equations 1.4–1.6 *does not mean* "take the reciprocal." For instance, $\tan^{-1}(h_o/h_a)$ does not equal $1/\tan(h_o/h_a)$. Another way to express the inverse trigonometric functions is to use arc sin, arc cos, and arc tan instead of $\sin^{-1}$, $\cos^{-1}$, and $\tan^{-1}$. Example 4 illustrates the use of an inverse trigonometric function.

Example 4 Using Inverse Trigonometric Functions

A lakefront drops off gradually at an angle θ, as Figure 1.7 indicates. For safety reasons, it is necessary to know how deep the lake is at various distances from the shore. To provide some information about the depth, a lifeguard rows straight out from the shore a distance of 14.0 m and drops a weighted fishing line. By measuring the length of the line, the lifeguard determines the depth to be 2.25 m. (a) What is the value of θ? (b) What would be the depth d of the lake at a distance of 22.0 m from the shore?

Reasoning Near the shore, the lengths of the opposite and adjacent sides of the right triangle in Figure 1.7 are $h_o = 2.25$ m and $h_a = 14.0$ m, relative to the angle θ. Having made this identification, we can use the inverse tangent to find the angle in part (a). For part (b) the opposite and adjacent sides farther from the shore become $h_o = d$ and $h_a = 22.0$ m. With the value for θ obtained in part (a), the tangent function can be used to find the unknown depth. Considering the way in which the lake bottom drops off in Figure 1.7, we expect the unknown depth to be greater than 2.25 m.

Solution

(a) Using the inverse tangent given in Equation 1.6, we find that

$$\theta = \tan^{-1}\left(\frac{h_o}{h_a}\right) = \tan^{-1}\left(\frac{2.25 \text{ m}}{14.0 \text{ m}}\right) = \boxed{9.13°}$$

(b) With $\theta = 9.13°$, the tangent function given in Equation 1.3 can be used to find the unknown depth farther from the shore, where $h_o = d$ and $h_a = 22.0$ m. Since $\tan \theta = h_o/h_a$, it follows that

$$h_o = h_a \tan \theta$$
$$d = (22.0 \text{ m})(\tan 9.13°) = \boxed{3.54 \text{ m}}$$

which is greater than 2.25 m, as expected.

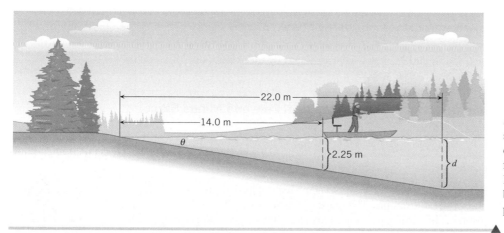

Figure 1.7 If the distance from the shore and the depth of the water at any one point are known, the angle θ can be found with the aid of trigonometry. Knowing the value of θ is useful, because then the depth d at another point can be determined.

The right triangle in Figure 1.5 provides the basis for defining the various trigonometric functions according to Equations 1.1–1.3. These functions always involve an angle and two sides of the triangle. There is also a relationship among the lengths of the three sides of a right triangle. This relationship is known as the ***Pythagorean theorem*** and is used often in this text.

■ **PYTHAGOREAN THEOREM**

The square of the length of the hypotenuse of a right triangle is equal to the sum of the squares of the lengths of the other two sides:

$$h^2 = h_o{}^2 + h_a{}^2 \qquad (1.7)$$

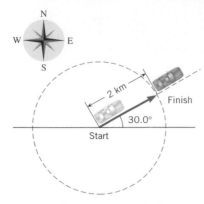

Figure 1.8 A vector quantity has a magnitude and a direction. The arrow in this drawing represents a displacement vector.

The velocity of this in-line skater is another example of a vector quantity, because the velocity has a magnitude (the speed of the skater) and a direction. (© Scott Markewitz/Taxi/Getty Images)

1.5 Scalars and Vectors

The volume of water in a swimming pool might be 50 cubic meters, or the winning time of a race could be 11.3 seconds. In cases like these, only the size of the numbers matters. In other words, *how much* volume or time is there? The "50" specifies the amount of water in units of cubic meters, while the "11.3" specifies the amount of time in seconds. Volume and time are examples of scalar quantities. A **scalar quantity** is one that can be described with a single number (including any units) giving its size or magnitude. Some other common scalars are temperature (e.g., 20 °C) and mass (e.g., 85 kg).

While many quantities in physics are scalars, there are also many that are not, and for these quantities the magnitude tells only part of the story. Consider Figure 1.8, which depicts a car that has moved 2 km along a straight line from start to finish. When describing the motion, it is incomplete to say that "the car moved a distance of 2 km." This statement would indicate only that the car ends up somewhere on a circle whose center is at the starting point and whose radius is 2 km. A complete description must include the direction along with the distance, as in the statement "the car moved a distance of 2 km in a direction 30° north of east." A quantity that deals inherently with both magnitude and direction is called a **vector quantity.** Because direction is an important characteristic of vectors, arrows are used to represent them; *the direction of the arrow gives the direction of the vector.* The colored arrow in Figure 1.8, for example, is called the displacement vector, because it shows how the car is displaced from its starting point. Chapter 2 discusses this particular vector.

The length of the arrow in Figure 1.8 represents the magnitude of the displacement vector. If the car had moved 4 km instead of 2 km from the starting point, the arrow would have been drawn twice as long. *By convention, the length of a vector arrow is proportional to the magnitude of the vector.*

In physics there are many important kinds of vectors, and the practice of using the length of an arrow to represent the magnitude of a vector applies to each of them. All forces, for instance, are vectors. In common usage a force is a push or a pull, and the direction in which a force acts is just as important as the strength or magnitude of the force. The magnitude of a force is measured in SI units called newtons (N). An arrow representing a force of 20 newtons is drawn twice as long as one representing a force of 10 newtons.

The fundamental distinction between scalars and vectors is the characteristic of direction. Vectors have it, and scalars do not. Conceptual Example 5 helps to clarify this distinction and explains what is meant by the "direction" of a vector.

Conceptual Example 5
Vectors, Scalars, and the Role of Plus and Minus Signs

There are places where the temperature is +20 °C at one time of the year and −20 °C at another time. Do the plus and minus signs that signify positive and negative temperatures imply that temperature is a vector quantity?

Reasoning and Solution A vector has a physical direction associated with it, due east or due west, for example. The question, then, is whether such a direction is associated with temperature. In particular, do the plus and minus signs that go along with temperature imply this kind of direction? On a thermometer, the algebraic signs simply mean that the temperature is a number less than or greater than zero on the scale and have nothing to do with east, west, or any other physical direction. Temperature, then, is not a vector. It is a scalar, and scalars can sometimes be negative. *The fact that a quantity is positive or negative does not necessarily mean that the quantity is a scalar or a vector.*

Often, for the sake of convenience, quantities such as volume, time, displacement, and force are represented by symbols. This text follows the usual practice of writing vectors in boldface symbols* **(this is boldface)** and writing scalars in italic symbols (*this is italic*). Thus, a displacement vector is written as "**A** = 750 m, due east," where the **A** is a

---

* A vector quantity can also be represented without boldface symbols, by including an arrow above the symbol—e.g., $\vec{A}$.

boldface symbol. By itself, however, separated from the direction, the magnitude of this vector is a scalar quantity. Therefore, the magnitude is written as "$A = 750$ m," where the A is an italic symbol.

✔ **Check Your Understanding 1**

Which of the following statements, if any, involves a vector? (a) I walked 2 miles along the beach. (b) I walked 2 miles due north along the beach. (c) I jumped off a cliff and hit the water traveling at 17 miles per hour. (d) I jumped off a cliff and hit the water traveling straight down at 17 miles per hour. (e) My bank account shows a negative balance of −25 dollars. *(The answers are given at the end of the book.)*

Background: These questions deal with the concepts of vectors and scalars, and the difference between them.

For similar questions (including calculational counterparts), consult Self-Assessment Test 1.1. This test is described at the end of Section 1.6.

1.6 *Vector Addition and Subtraction*

ADDITION

Often it is necessary to add one vector to another, and the process of addition must take into account both the magnitude and the direction of the vectors. The simplest situation occurs when the vectors point along the same direction—that is, when they are colinear, as in Figure 1.9. Here, a car first moves along a straight line, with a displacement vector **A** of 275 m, due east. Then, the car moves again in the same direction, with a displacement vector **B** of 125 m, due east. These two vectors add to give the total displacement vector **R**, which would apply if the car had moved from start to finish in one step. The symbol **R** is used because the total vector is often called the *resultant vector*. With the tail of the second arrow located at the head of the first arrow, the two lengths simply add to give the length of the total displacement. This kind of vector addition is identical to the familiar addition of two scalar numbers $(2 + 3 = 5)$ *and can be carried out here only because the vectors point along the same direction.* In such cases we add the individual magnitudes to get the magnitude of the total, knowing in advance what the direction must be. Formally, the addition is written as follows:

$$\mathbf{R} = \mathbf{A} + \mathbf{B}$$
$$\mathbf{R} = 275 \text{ m, due east} + 125 \text{ m, due east} = 400 \text{ m, due east}$$

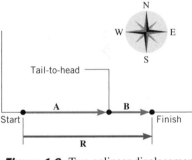

Figure 1.9 Two colinear displacement vectors **A** and **B** add to give the resultant displacement vector **R**.

Perpendicular vectors are frequently encountered, and Figure 1.10 indicates how they can be added. This figure applies to a car that first travels with a displacement vector **A** of 275 m, due east, and then with a displacement vector **B** of 125 m, due north. The two vectors add to give a resultant displacement vector **R**. Once again, the vectors to be added are arranged in a tail-to-head fashion, and the resultant vector points from the tail of the first to the head of the last vector added. The resultant displacement is given by the vector equation

$$\mathbf{R} = \mathbf{A} + \mathbf{B}$$

The addition in this equation cannot be carried out by writing $R = 275$ m $+ 125$ m, because the vectors have different directions. Instead, we take advantage of the fact that the triangle in Figure 1.10 is a right triangle and use the Pythagorean theorem (Equation 1.7). According to this theorem, the magnitude of **R** is

$$R = \sqrt{(275 \text{ m})^2 + (125 \text{ m})^2} = 302 \text{ m}$$

The angle θ in Figure 1.10 gives the direction of the resultant vector. Since the lengths of all three sides of the right triangle are now known, either $\sin \theta$, $\cos \theta$, or $\tan \theta$ can be used to determine θ. Noting that $\tan \theta = B/A$ and using the inverse trigonometric function, we find that:

$$\theta = \tan^{-1}\left(\frac{B}{A}\right) = \tan^{-1}\left(\frac{125 \text{ m}}{275 \text{ m}}\right) = 24.4°$$

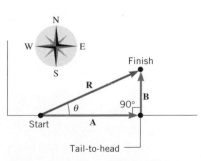

Figure 1.10 The addition of two perpendicular displacement vectors **A** and **B** gives the resultant vector **R**.

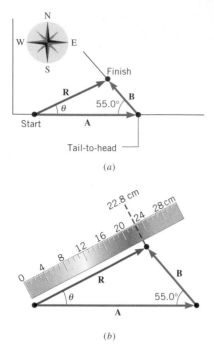

Figure 1.11 (a) The two displacement vectors **A** and **B** are neither colinear nor perpendicular but even so they add to give the resultant vector **R**. (b) In one method for adding them together, a graphical technique is used.

Thus, the resultant displacement of the car has a magnitude of 302 m and points north of east at an angle of 24.4°. This displacement would bring the car from the start to the finish in Figure 1.10 in a single straight-line step.

When two vectors to be added are not perpendicular, the tail-to-head arrangement does not lead to a right triangle, and the Pythagorean theorem cannot be used. Figure 1.11a illustrates such a case for a car that moves with a displacement **A** of 275 m, due east, and then with a displacement **B** of 125 m in a direction 55.0° north of west. As usual, the resultant displacement vector **R** is directed from the tail of the first to the head of the last vector added. The vector addition is still given according to

$$R = A + B$$

However, the magnitude of **R** is not $R = \sqrt{A^2 + B^2}$, because the vectors **A** and **B** are not perpendicular and the Pythagorean theorem does not apply. Some other means must be used to find the magnitude and direction of the resultant vector.

One approach uses a graphical technique. In this method, a diagram is constructed in which the arrows are drawn tail to head. The lengths of the vector arrows are drawn to scale, and the angles are drawn accurately (with a protractor, perhaps). Then, the length of the arrow representing the resultant vector is measured with a ruler. This length is converted to the magnitude of the resultant vector by using the scale factor with which the drawing is constructed. In Figure 1.11b, for example, a scale of one centimeter of arrow length for each 10.0 m of displacement is used, and it can be seen that the length of the arrow representing **R** is 22.8 cm. Since each centimeter corresponds to 10.0 m of displacement, the magnitude of **R** is 228 m. The angle θ, which gives the direction of **R**, can be measured with a protractor to be θ = 26.7° north of east.

✔ **Check Your Understanding 2**

Two vectors, **A** and **B**, are added by means of vector addition to give a resultant vector **R**: **R** = **A** + **B**. The magnitudes of **A** and **B** are 3 and 8 m, but they can have any orientation. What is (a) the maximum possible value and (b) the minimum possible value for the magnitude of **R**? *(The answers are given at the end of the book.)*

Background: These questions deal with adding two vectors by means of the tail-to-head method. They illustrate how the two vectors must be oriented relative to each other to produce a resultant vector that has the greatest possible magnitude and one that has the least possible magnitude.

For similar questions (including conceptual counterparts), consult Self-Assessment Test 1.1. The test is described at the end of this section.

SUBTRACTION

The subtraction of one vector from another is carried out in a way that depends on the following fact. *When a vector is multiplied by −1, the magnitude of the vector remains the same, but the direction of the vector is reversed.* Conceptual Example 6 illustrates the meaning of this statement.

Conceptual Example 6 **Multiplying a Vector by −1**

Consider the two vectors described as follows:

1. A woman climbs 1.2 m up a ladder, so that her displacement vector **D** is 1.2 m, upward along the ladder, as in Figure 1.12a.

2. A man is pushing with 450 N of force on his stalled car, trying to move it eastward. The force vector **F** that he applies to the car is 450 N, due east, as in Figure 1.13a.

What are the physical meanings of the vectors −**D** and −**F**?

Reasoning and Solution A displacement vector of −**D** is (−1)**D** and has the same magnitude as the vector **D**, but is opposite in direction. Thus, −**D** would represent the displacement of a woman climbing 1.2 m down the ladder, as in Figure 1.12b. Similarly, a force vector of −**F** has the same magnitude as the vector **F** but has the opposite direction. As a result, −**F**

(a)   (b)

Figure 1.12 (a) The displacement vector for a woman climbing 1.2 m up a ladder is **D**. (b) The displacement vector for a woman climbing 1.2 m down a ladder is −**D**.

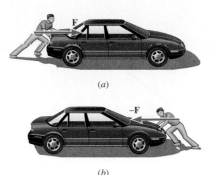

(a)

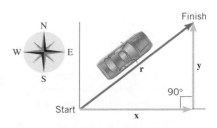

(b)

Figure 1.13 (a) The force vector for a man pushing on a car with 450 N of force in a direction due east is **F**. (b) The force vector for a man pushing on a car with 450 N of force in a direction due west is −**F**.

would represent a force of 450 N applied to the car in a direction of due west, as in Figure 1.13b.

Related Homework: *Conceptual Question 15, Problem 62*

In practice, vector subtraction is carried out exactly like vector addition, except that one of the vectors added is multiplied by a scalar factor of −1. To see why, look at the two vectors **A** and **B** in Figure 1.14a. These vectors add together to give a third vector **C**, according to **C** = **A** + **B**. Therefore, we can calculate vector **A** as **A** = **C** − **B**, which is an example of vector subtraction. However, we can also write this result as **A** = **C** + (−**B**) and treat it as vector addition. Figure 1.14b shows how to calculate vector **A** by adding the vectors **C** and −**B**. Notice that vectors **C** and −**B** are arranged tail to head and that any suitable method of vector addition can be employed to determine **A**.

Self-Assessment Test 1.1

Test your understanding of the concepts discussed in Sections 1.5 and 1.6:

• Adding and Subtracting Vectors by the Tail-to-Head Method   • Using the Pythagorean Theorem and Trigonometry to Find the Magnitude and Direction of the Resultant Vector

Go to **www.wiley.com/college/cutnell**

1.7 *The Components of a Vector*
VECTOR COMPONENTS

Suppose a car moves along a straight line from start to finish in Figure 1.15, the corresponding displacement vector being **r**. The magnitude and direction of the vector **r** give the distance and direction traveled along the straight line. However, the car could also arrive at the finish point by first moving due east, turning through 90°, and then moving due north. This alternative path is shown in the drawing and is associated with the two displacement vectors **x** and **y**. The vectors **x** and **y** are called the *x* vector component and the *y* vector component of **r**.

Vector components are very important in physics and have two basic features that are apparent in Figure 1.15. One is that the components add together to equal the original vector:

$$\mathbf{r} = \mathbf{x} + \mathbf{y}$$

The components **x** and **y**, when added vectorially, convey exactly the same meaning as does the original vector **r**: they indicate how the finish point is displaced relative to the starting point. In general, *the components of any vector can be used in place of the vector itself in any calculation where it is convenient to do so.* The other feature of vector components that is apparent in Figure 1.15 is that **x** and **y** are not just any two vectors that add together to give the original vector **r**: they are perpendicular vectors. This perpendicularity is a valuable characteristic, as we will soon see.

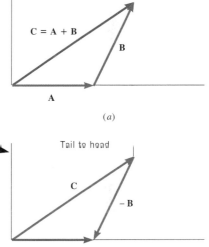

(a)

(b)

Figure 1.14 (a) Vector addition according to **C** = **A** + **B**. (b) Vector subtraction according to **A** = **C** − **B** = **C** + (−**B**).

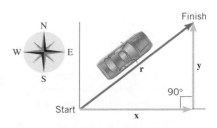

Figure 1.15 The displacement vector **r** and its vector components **x** and **y**.

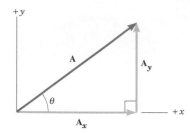

Figure 1.16 An arbitrary vector **A** and its vector components **A**$_x$ and **A**$_y$.

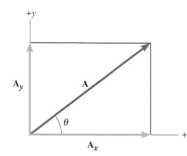

Figure 1.17 This alternative way of drawing the vector **A** and its vector components is completely equivalent to that shown in Figure 1.16.

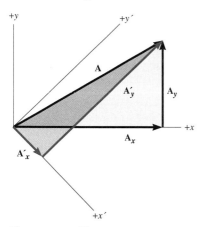

Figure 1.18 The vector components of the vector depend on the orientation of the axes used as a reference.

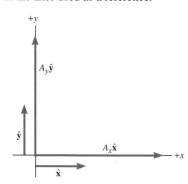

Figure 1.19 The dimensionless unit vectors $\hat{\mathbf{x}}$ and $\hat{\mathbf{y}}$ have magnitudes equal to 1, and they point in the $+x$ and $+y$ directions, respectively. Expressed in terms of unit vectors, the vector components of the vector **A** are $A_x\hat{\mathbf{x}}$ and $A_y\hat{\mathbf{y}}$.

Any type of vector may be expressed in terms of its components, in a way similar to that illustrated for the displacement vector in Figure 1.15. Figure 1.16 shows an arbitrary vector **A** and its vector components **A**$_x$ and **A**$_y$. The components are drawn parallel to convenient x and y axes and are perpendicular. They add vectorially to equal the original vector **A**:

$$\mathbf{A} = \mathbf{A}_x + \mathbf{A}_y$$

There are times when a drawing such as Figure 1.16 is not the most convenient way to represent vector components, and Figure 1.17 presents an alternative method. The disadvantage of this alternative is that the tail-to-head arrangement of **A**$_x$ and **A**$_y$ is missing, an arrangement that is a nice reminder that **A**$_x$ and **A**$_y$ add together to equal **A**.

The definition that follows summarizes the meaning of vector components:

> ■ **DEFINITION OF VECTOR COMPONENTS**
>
> In two dimensions, the vector components of a vector **A** are two perpendicular vectors **A**$_x$ and **A**$_y$ that are parallel to the x and y axes, respectively, and add together vectorially so that $\mathbf{A} = \mathbf{A}_x + \mathbf{A}_y$.

The values calculated for vector components depend on the orientation of the vector relative to the axes used as a reference. Figure 1.18 illustrates this fact for a vector **A** by showing two sets of axes, one set being rotated clockwise relative to the other. With respect to the black axes, vector **A** has perpendicular vector components **A**$_x$ and **A**$_y$; with respect to the colored rotated axes, vector **A** has different vector components **A**$_x'$ and **A**$_y'$. The choice of which set of components to use is purely a matter of convenience.

SCALAR COMPONENTS

It is often easier to work with the ***scalar components,*** A_x and A_y (note the italic symbols), rather than the vector components **A**$_x$ and **A**$_y$. Scalar components are positive or negative numbers (with units) that are defined as follows. The component A_x has a magnitude that is equal to that of **A**$_x$ and is given a positive sign if **A**$_x$ points along the $+x$ axis and a negative sign if **A**$_x$ points along the $-x$ axis. The component A_y is defined in a similar manner. The following table shows an example of vector and scalar components:

Vector Components	Scalar Components	Unit Vectors
A$_x$ = 8 meters, directed along the $+x$ axis	$A_x = +8$ meters	**A**$_x$ = (+8 meters) $\hat{\mathbf{x}}$
A$_y$ = 10 meters, directed along the $-y$ axis	$A_y = -10$ meters	**A**$_y$ = (−10 meters) $\hat{\mathbf{y}}$

In this text, when we use the term "component," we will be referring to a scalar component, unless otherwise indicated.

Another method of expressing vector components is to use unit vectors. A ***unit vector*** is a vector that has a magnitude of 1, but no dimensions. We will use a caret (^) to distinguish it from other vectors. Thus,

$\hat{\mathbf{x}}$ is a dimensionless unit vector of length 1 that points in the positive x direction, and

$\hat{\mathbf{y}}$ is a dimensionless unit vector of length 1 that points in the positive y direction.

These unit vectors are illustrated in Figure 1.19. With the aid of unit vectors, the vector components of an arbitrary vector **A** can be written as $\mathbf{A}_x = A_x\,\hat{\mathbf{x}}$ and $\mathbf{A}_y = A_y\,\hat{\mathbf{y}}$, where A_x and A_y are its scalar components (see the drawing and third column of the table above). The vector **A** is then written as $\mathbf{A} = A_x\,\hat{\mathbf{x}} + A_y\,\hat{\mathbf{y}}$.

✔ **Check Your Understanding 3**

Two vectors, **A** and **B**, are shown in the drawing that follows. (a) What are the signs (+ or −) of the scalar components A_x and A_y of vector **A**? (b) What are the signs of the scalar components B_x and B_y of vector **B**? (c) What are the signs of the scalar components R_x and R_y of the resultant vector **R**, where $\mathbf{R} = \mathbf{A} + \mathbf{B}$? (*The answers are given at the end of the book.*)

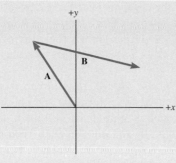

Background: Two concepts play a role in this question: the scalar components of a vector, and how two vectors are added by means of the tail-to-head method to produce a resultant vector.

For similar questions (including calculational counterparts), consult Self-Assessment Test 1.2. This test is described at the end of Section 1.8.

RESOLVING A VECTOR INTO ITS COMPONENTS

If the magnitude and direction of a vector are known, it is possible to find the components of the vector. The process of finding the components is called "resolving the vector into its components." As Example 7 illustrates, this process can be carried out with the aid of trigonometry, because the two perpendicular vector components and the original vector form a right triangle.

Example 7 Finding the Components of a Vector

A displacement vector **r** has a magnitude of $r = 175$ m and points at an angle of 50.0° relative to the x axis in Figure 1.20. Find the x and y components of this vector.

Reasoning We will base our solution on the fact that the triangle formed in Figure 1.20 by the vector **r** and its components **x** and **y** is a right triangle. This fact enables us to use the trigonometric sine and cosine functions, as defined in Equations 1.1 and 1.2.

Solution 1 The y component can be obtained using the 50.0° angle and Equation 1.1, $\sin \theta = y/r$:

$$y = r \sin \theta = (175 \text{ m})(\sin 50.0°) = \boxed{134 \text{ m}}$$

In a similar fashion, the x component can be obtained using the 50.0° angle and Equation 1.2, $\cos \theta = x/r$:

$$x = r \cos \theta = (175 \text{ m})(\cos 50.0°) = \boxed{112 \text{ m}}$$

Solution 2 The angle α in Figure 1.20 can also be used to find the components. Since $\alpha + 50.0° = 90.0°$, it follows that $\alpha = 40.0°$. The solution using α yields the same answers as in Solution 1:

$$\cos \alpha = \frac{y}{r}$$

$$y = r \cos \alpha = (175 \text{ m})(\cos 40.0°) = \boxed{134 \text{ m}}$$

$$\sin \alpha = \frac{x}{r}$$

$$x = r \sin \alpha = (175 \text{ m})(\sin 40.0°) = \boxed{112 \text{ m}}$$

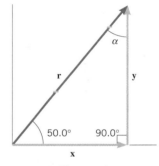

Figure 1.20 The x and y components of the displacement vector **r** can be found using trigonometry.

Problem solving insight
Either acute angle of a right triangle can be used to determine the components of a vector. The choice of angle is a matter of convenience.

Since the vector components and the original vector form a right triangle, the Pythagorean theorem can be applied to check the validity of calculations such as those in Example 7. Thus, with the components obtained in Example 7, the theorem can be used to verify that the magnitude of the original vector is indeed 175 m, as given initially:

$$r = \sqrt{(112 \text{ m})^2 + (134 \text{ m})^2} = 175 \text{ m}$$

Problem solving insight
You can check to see whether the components of a vector are correct by substituting them into the Pythagorean theorem and verifying that the result is the magnitude of the original vector.

It is possible for one of the components of a vector to be zero. This does not mean that the vector itself is zero, however. ***For a vector to be zero, every vector component must individually be zero.*** Thus, in two dimensions, saying that $\mathbf{A} = 0$ is equivalent to saying that $\mathbf{A}_x = 0$ and $\mathbf{A}_y = 0$. Or, stated in terms of scalar components, if $\mathbf{A} = 0$, then $A_x = 0$ and $A_y = 0$.

Two vectors are equal if, and only if, they have the same magnitude and direction. Thus, if one displacement vector points east and another points north, they are *not* equal, even if each has the same magnitude of 480 m. In terms of vector components, two vectors, $\mathbf{A}$ and $\mathbf{B}$, are equal if, and only if, each vector component of one is equal to the corresponding vector component of the other. In two dimensions, if $\mathbf{A} = \mathbf{B}$, then $\mathbf{A}_x = \mathbf{B}_x$ and $\mathbf{A}_y = \mathbf{B}_y$. Alternatively, using scalar components, we write that $A_x = B_x$ and $A_y = B_y$.

1.8 Addition of Vectors by Means of Components

The components of a vector provide the most convenient and accurate way of adding (or subtracting) any number of vectors. For example, suppose that vector $\mathbf{A}$ is added to vector $\mathbf{B}$. The resultant vector is $\mathbf{C}$, where $\mathbf{C} = \mathbf{A} + \mathbf{B}$. Figure 1.21*a* illustrates this vector addition, along with the x and y vector components of $\mathbf{A}$ and $\mathbf{B}$. In part *b* of the drawing, the vectors $\mathbf{A}$ and $\mathbf{B}$ have been removed, because we can use the vector components of these vectors in place of them. The vector component $\mathbf{B}_x$ has been shifted downward and arranged tail to head with the vector component $\mathbf{A}_x$. Similarly, the vector component $\mathbf{A}_y$ has been shifted to the right and arranged tail to head with the vector component $\mathbf{B}_y$. The x components are colinear and add together to give the x component of the resultant vector $\mathbf{C}$. In like fashion, the y components are colinear and add together to give the y component of $\mathbf{C}$. In terms of scalar components, we can write

$$C_x = A_x + B_x \quad \text{and} \quad C_y = A_y + B_y$$

The vector components $\mathbf{C}_x$ and $\mathbf{C}_y$ of the resultant vector form the sides of the right triangle shown in Figure 1.21*c*. Thus, we can find the magnitude of $\mathbf{C}$ by using the Pythagorean theorem:

$$C = \sqrt{C_x^2 + C_y^2}$$

The angle θ that $\mathbf{C}$ makes with the x axis is given by $\theta = \tan^{-1}(C_y/C_x)$. Example 8 illustrates how to add several vectors using the component method.

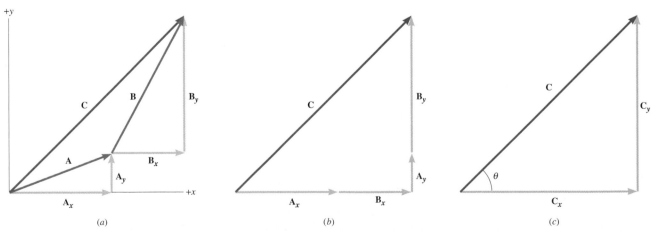

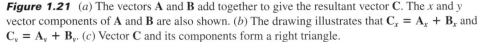

Figure 1.21 (*a*) The vectors $\mathbf{A}$ and $\mathbf{B}$ add together to give the resultant vector $\mathbf{C}$. The x and y vector components of $\mathbf{A}$ and $\mathbf{B}$ are also shown. (*b*) The drawing illustrates that $C_x = A_x + B_x$ and $C_y = A_y + B_y$. (*c*) Vector $\mathbf{C}$ and its components form a right triangle.

Example 8 The Component Method of Vector Addition

A jogger runs 145 m in a direction 20.0° east of north (displacement vector **A**) and then 105 m in a direction 35.0° south of east (displacement vector **B**). Determine the magnitude and direction of the resultant vector **C** for these two displacements.

Reasoning Figure 1.22*a* shows the vectors **A** and **B**, assuming that the *y* axis corresponds to the direction due north. Since the vectors are not given in component form, we will begin by using the given magnitudes and directions to find the components. Then, the components of **A** and **B** can be used to find the components of the resultant **C**. Finally, with the aid of the Pythagorean theorem and trigonometry, the components of **C** can be used to find its magnitude and direction.

Solution The first two rows of the following table give the *x* and *y* components of the vectors **A** and **B**. Note that the component B_y is negative, because B_y points downward, in the negative *y* direction in the drawing.

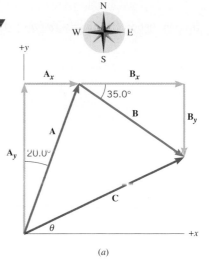

Vector	*x* Component	*y* Component
A	$A_x = (145 \text{ m}) \sin 20.0° = 49.6 \text{ m}$	$A_y = (145 \text{ m}) \cos 20.0° = 136 \text{ m}$
B	$B_x = (105 \text{ m}) \cos 35.0° = 86.0 \text{ m}$	$B_y = -(105 \text{ m}) \sin 35.0° = -60.2 \text{ m}$
C	$C_x = A_x + B_x = 135.6 \text{ m}$	$C_y = A_y + B_y = 76 \text{ m}$

The third row in the table gives the *x* and *y* components of the resultant vector **C**: $C_x = A_x + B_x$ and $C_y = A_y + B_y$. Part *b* of the drawing shows **C** and its vector components. The magnitude of **C** is given by the Pythagorean theorem as

$$C = \sqrt{C_x^2 + C_y^2} = \sqrt{(135.6 \text{ m})^2 + (76 \text{ m})^2} = \boxed{155 \text{ m}}$$

The angle θ that **C** makes with the *x* axis is

$$\theta = \tan^{-1}\left(\frac{C_y}{C_x}\right) = \tan^{-1}\left(\frac{76 \text{ m}}{135.6 \text{ m}}\right) = \boxed{29°}$$

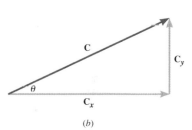

Figure 1.22 (*a*) The vectors **A** and **B** add together to give the resultant vector **C**. The vector components of **A** and **B** are also shown. (*b*) The resultant vector **C** can be obtained once its components have been found.

✔ **Check Your Understanding 4**

Two vectors, **A** and **B**, have vector components that are shown (to the same scale) in the first row of drawings. Which vector **R** in the second row of drawings is the vector sum of **A** and **B**? *(The answer is given at the end of the book.)*

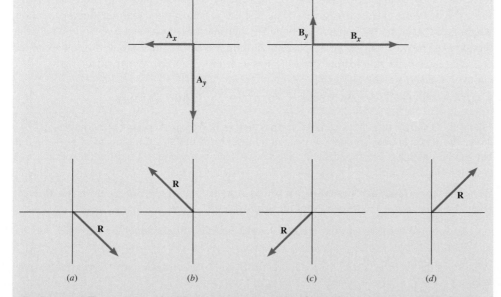

Background: The concept of adding vectors by means of vector components is featured in this question. Note that the *x* components of **A** and **B** point in opposite directions, as do the *y* components.

For similar questions (including calculational counterparts), consult Self-Assessment Test 1.2. The test is described at the end of this section.

In later chapters we will often use the component method for vector addition. For future reference, the main features of the reasoning strategy used in this technique are summarized below.

Reasoning Strategy

The Component Method of Vector Addition

1. For each vector to be added, determine the x and y components relative to a conveniently chosen x, y coordinate system. Be sure to take into account the directions of the components by using plus and minus signs to denote whether the components point along the positive or negative axes.

2. Find the algebraic sum of the x components, which is the x component of the resultant vector. Similarly, find the algebraic sum of the y components, which is the y component of the resultant vector.

3. Use the x and y components of the resultant vector and the Pythagorean theorem to determine the magnitude of the resultant vector.

4. Use either the inverse sine, inverse cosine, or inverse tangent function to find the angle that specifies the direction of the resultant vector.

1.9 Concepts & Calculations

This chapter has presented an introduction to the mathematics of trigonometry and vectors, which will be used throughout this text. Therefore, in this last section we consider several examples in order to review some of the important features of this mathematics. The three-part format of these examples stresses the role of conceptual understanding in problem solving. First, the problem statement is given. Then, there is a concept question-and-answer section, which is followed by the solution section. The purpose of the concept question-and-answer section is to provide help in understanding the solution and to illustrate how a review of the concepts can help in anticipating some of the characteristics of the numerical answers.

Concepts & Calculations Example 9   Equal Vectors

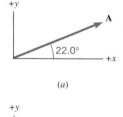

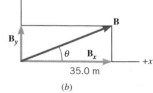

Figure 1.23 The two displacement vectors **A** and **B** are equal. Example 9 discusses what this equality means.

Figure 1.23 shows two displacement vectors **A** and **B**. Vector **A** points at an angle of 22.0° above the x axis but has an unknown magnitude. Vector **B** has an x component of $B_x = 35.0$ m but has an unknown y component B_y. These two vectors are equal. Find the magnitude of **A** and the value of B_y.

Concept Questions and Answers   What does the fact that vector **A** equals vector **B** imply about the magnitudes and directions of the vectors?

> *Answer*   When two vectors are equal, each has the same magnitude and each has the same direction.

What does the fact that vector **A** equals vector **B** imply about the x and y components of the vectors?

> *Answer*   When two vectors are equal, the x component of vector **A** equals the x component of vector **B** ($A_x = B_x$) and the y component of vector **A** equals the y component of vector **B** ($A_y = B_y$).

Solution   We focus on the fact that the x components of the vectors are the same and the y components of the vectors are the same. This allows us to write that

$$A \cos 22.0° \quad = \quad 35.0 \text{ m} \tag{1.8}$$

$$\underbrace{}_{\substack{\text{Component } A_x \\ \text{of vector } \mathbf{A}}} \quad \underbrace{\phantom{35.0 \text{ m}}}_{\substack{\text{Component } B_x \\ \text{of vector } \mathbf{B}}}$$

$$A \sin 22.0° \quad = \quad B_y \tag{1.9}$$

$$\underbrace{}_{\substack{\text{Component } A_y \\ \text{of vector } \mathbf{A}}} \quad \underbrace{}_{\substack{\text{Component } B_y \\ \text{of vector } \mathbf{B}}}$$

Dividing Equation 1.9 by Equation 1.8 shows that

$$\frac{A \sin 22.0°}{A \cos 22.0°} = \frac{B_y}{35.0 \text{ m}}$$

$$B_y = (35.0 \text{ m}) \frac{\sin 22.0°}{\cos 22.0°} = (35.0 \text{ m}) \tan 22.0° = \boxed{14.1 \text{ m}}$$

Solving Equation 1.8 directly for A gives

$$A = \frac{35.0 \text{ m}}{\cos 22.0°} = \boxed{37.7 \text{ m}}$$

Concepts & Calculations Example 10 Adding Vectors

Figure 1.24*a* shows two displacement vectors **A** and **B**, which add together to give a resultant displacement **C**. Find the magnitude and direction of **C**.

Concept Questions and Answers Does the Pythagorean theorem apply directly to this problem, so that the magnitude of vector **C** is given by $C = \sqrt{A^2 + B^2}$?

Answer The magnitude of the vector **C** is not given by the Pythagorean theorem in the form $C = \sqrt{A^2 + B^2}$, because the vectors **A** and **B** are not perpendicular (see Figure 1.24*a*). It is only when the two vectors are perpendicular that this equation for the magnitude applies.

Does the component method for vector addition apply to this problem, even though the vectors are not perpendicular?

Answer Yes. The component method applies whether or not the vectors are perpendicular, and it applies to any number of vectors. We will use it in our solution.

Solution The components of the vectors **A** and **B** can be determined from the data in Figure 1.24*a* and are listed in the first two rows of the following table. Note that the vector **A** points along the *x* axis and, therefore, has no *y* component.

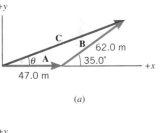

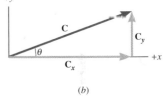

Figure 1.24 (*a*) Two displacement vectors **A** and **B** add together to give a resultant vector **C**. (*b*) The components of **C** are $\mathbf{C}_x$ and $\mathbf{C}_y$.

Vector	*x* Component	*y* Component
A	$A_x = 47.0 \text{ m}$	$A_y = 0 \text{ m}$
B	$B_x = (62.0 \text{ m}) \cos 35.0° = 50.8 \text{ m}$	$B_y = (62.0 \text{ m}) \sin 35.0° = 35.6 \text{ m}$
C	$C_x = A_x + B_x = 97.8 \text{ m}$	$C_y = A_y + B_y = 35.6 \text{ m}$

Figure 1.24*b* shows the components of the resultant vector, and the third row in the table gives their values. The components C_x and C_y are perpendicular, so that we can use the Pythagorean theorem to find the magnitude of **C**:

$$C = \sqrt{C_x^2 + C_y^2} = \sqrt{(97.8 \text{ m})^2 + (35.6 \text{ m})^2} = \boxed{104 \text{ m}}$$

The value of the directional angle θ is

$$\theta = \tan^{-1}\left(\frac{C_y}{C_x}\right) = \tan^{-1}\left(\frac{35.6 \text{ m}}{97.8 \text{ m}}\right) = \boxed{20.0°}$$

At the end of the problem set for this chapter, you will find homework problems that contain both conceptual and quantitative parts. These problems are grouped under the heading *Concepts & Calculations, Group Learning Problems*. They are designed for use by students working alone or in small learning groups. The conceptual part of each problem provides a convenient focus for group discussions.

Concept Summary

This summary presents an abridged version of the chapter, including the important equations and all available learning aids. For convenient reference, the learning aids (including the text's examples) are placed next to or immediately after the relevant equation or discussion. The following learning aids may be found on-line at **www.wiley.com/college/cutnell**:

Interactive LearningWare examples are solved according to a five-step interactive format that is designed to help you develop problem-solving skills.	**Concept Simulations** are animated versions of text figures or animations that illustrate important concepts. You can control parameters that affect the display, and we encourage you to experiment.
Interactive Solutions offer specific models for certain types of problems in the chapter homework. The calculations are carried out interactively.	**Self-Assessment Tests** include both qualitative and quantitative questions. Extensive feedback is provided for both incorrect and correct answers, to help you evaluate your understanding of the material.

Topic	*Discussion*	*Learning Aids*

1.2 Units

The SI system of units includes the meter (m), the kilogram (kg), and the second (s) as the base units for length, mass, and time, respectively.

Meter One meter is the distance that light travels in a vacuum in a time of 1/299 792 458 second.

Kilogram One kilogram is the mass of a standard cylinder of platinum–iridium alloy kept at the International Bureau of Weights and Measures.

Second One second is the time for a certain type of electromagnetic wave emitted by cesium-133 atoms to undergo 9 192 631 770 wave cycles.

1.3 The Role of Units in Problem Solving

Conversion of units To convert a number from one unit to another, multiply the number by the ratio of the two units. For instance, to convert 979 meters to feet, multiply 979 meters by the factor (3.281 foot/1 meter).   **Examples 1, 2**

Dimension The dimension of a quantity represents its physical nature and the type of unit used to specify it. Three such dimensions are length [L], mass [M], time [T].

Dimensional analysis A method for checking mathematical relations for the consistency of their dimensions.

1.4 Trigonometry

Sine, cosine, and tangent of an angle θ The sine, cosine, and tangent functions of an angle θ are defined in terms of a right triangle that contains θ: **Example 3**
Interactive Solution 1.17

$$\sin \theta = \frac{h_o}{h} \ (1.1) \qquad \cos \theta = \frac{h_a}{h} \ (1.2) \qquad \tan \theta = \frac{h_o}{h_a} \ (1.3)$$

where h_o and h_a are, respectively, the lengths of the sides opposite and adjacent to the angle θ, and h is the length of the hypotenuse.

Inverse trigonometric functions The inverse sine, inverse cosine, and inverse tangent functions are   **Example 4**

$$\theta = \sin^{-1}\left(\frac{h_o}{h}\right) \ (1.4) \qquad \theta = \cos^{-1}\left(\frac{h_a}{h}\right) \ (1.5)$$

$$\theta = \tan^{-1}\left(\frac{h_o}{h_a}\right) \ (1.6)$$

Pythagorean theorem The Pythagorean theorem states that the square of the length of the hypotenuse of a right triangle is equal to the sum of the squares of the lengths of the other two sides:

$$h^2 = h_o{}^2 + h_a{}^2 \qquad\qquad (1.7)$$

1.5 Scalars and Vectors

Scalars and vectors A scalar quantity is described completely by its size, which is also called its magnitude. A vector quantity has both a magnitude and a direction. Vectors are often represented by arrows, the length of the arrow being proportional to the magnitude of the vector and the direction of the arrow indicating the direction of the vector.   **Example 5**

Topic	Discussion	Learning Aids

1.6 Vector Addition and Subtraction

Graphical method of vector addition and subtraction

One procedure for adding vectors utilizes a graphical technique, in which the vectors to be added are arranged in a tail-to-head fashion. The resultant vector is drawn from the tail of the first vector to the head of the last vector.

The subtraction of a vector is treated as the addition of a vector that has been multiplied by a scalar factor of -1. Multiplying a vector by -1 reverses the direction of the vector.

Example 6
Interactive Solution 1.29

Use Self-Assessment Test 1.1 to evaluate your understanding of Sections 1.5 and 1.6.

1.7 The Components of a Vector

Vector components

In two dimensions, the vector components of a vector **A** are two perpendicular vectors $\mathbf{A}_x$ and $\mathbf{A}_y$ that are parallel to the x and y axes, respectively, and that add together vectorially so that $\mathbf{A} = \mathbf{A}_x + \mathbf{A}_y$.

Scalar components

The scalar component A_x has a magnitude that is equal to that of $\mathbf{A}_x$ and is given a positive sign if $\mathbf{A}_x$ points along the $+x$ axis and a negative sign if $\mathbf{A}_x$ points along the $-x$ axis. The scalar component A_y is defined in a similar manner.

Interactive Solution 1.37
Example 7

Condition for a vector to be zero

A vector is zero if, and only if, each of its vector components is zero.

Condition for two vectors to be equal

Two vectors are equal if, and only if, they have the same magnitude and direction. Alternatively, two vectors are equal in two dimensions if the x vector components of each are equal and the y vector components of each are equal.

1.8 Addition of Vectors by Means of Components

If two vectors **A** and **B** are added to give a resultant vector **C** such that $\mathbf{C} = \mathbf{A} + \mathbf{B}$, then

Example 8
Concept Simulation 1.1

$$C_x = A_x + B_x \quad \text{and} \quad C_y = A_y + B_y$$

Interactive Solution 1.49

where C_x, A_x, and B_x are the scalar components of the vectors along the x direction, and C_y, A_y, and B_y are the scalar components of the vectors along the y direction.

Use Self-Assessment Test 1.2 to evaluate your understanding of Sections 1.7 and 1.8.

Conceptual Questions

1. The following table lists four variables along with their units:

Variable	Units
x	meters (m)
v	meters per second (m/s)
t	seconds (s)
a	meters per second squared (m/s^2)

These variables appear in the following equations, along with a few numbers that have no units. In which of the equations are the units on the left side of the equals sign consistent with the units on the right side?

(a) $x = vt$

(b) $x = vt + \frac{1}{2}at^2$

(c) $v = at$

(d) $v = at + \frac{1}{2}at^3$

(e) $v^3 = 2ax^2$

(f) $t = \sqrt{\dfrac{2x}{a}}$

2. The variables x and v have the units shown in the table that accompanies question 1. Is it possible for x and v to be related to an angle θ according to $\tan\theta = x/v$? Account for your answer.

3. You can always add two numbers that have the same units. However, you cannot always add two numbers that have the same dimensions. Explain why not, and include an example in your explanation.

4. (a) Is it possible for two quantities to have the same dimensions but different units? (b) Is it possible for two quantities to have the same units but different dimensions? In each case, support your answer with an example and an explanation.

5. In the equation $y = c^n at^2$ you wish to determine the integer value (1, 2, etc.) of the exponent n. The dimensions of y, a, and t are known. It is also known that c has no dimensions. Can dimensional analysis be used to determine n? Account for your answer.

6. Using your calculator, verify that $\sin\theta$ divided by $\cos\theta$ is equal to $\tan\theta$, for an angle θ. Try 30°, for example. Prove that this result is true in general by using the definitions for $\sin\theta$, $\cos\theta$, and $\tan\theta$ given in Equations 1.1–1.3.

7. Sin θ and cos θ are called sinusoidal functions of the angle θ. The way in which these functions change as θ changes leads to a characteristic pattern when they are graphed. This pattern arises many times in physics. (a) To familiarize yourself with the sinusoidal pattern, use a calculator and construct a graph, with $\sin\theta$ plotted on the vertical axis and θ on the horizontal axis. Use $15°$ increments for θ between $0°$ and $720°$. (b) Repeat for $\cos \theta$.

8. Which of the following quantities (if any) can be considered a vector: (a) the number of people attending a football game, (b) the number of days in a month, and (c) the number of pages in a book? Explain your reasoning.

9. Which of the following displacement vectors (if any) are equal? Explain your reasoning.

Vector	Magnitude	Direction
A	100 m	30° north of east
B	100 m	30° south of west
C	50 m	30° south of west
D	100 m	60° east of north

10. Are two vectors with the same magnitude necessarily equal? Give your reasoning.

11. A cube has six faces and twelve edges. You start at one corner, are allowed to move only along the edges, and may not retrace your path along any edge. Consistent with these rules, there are a number of ways to arrive back at your starting point. For instance, you could move around the four edges that make up one of the square faces. The four corresponding displacement vectors would add to zero. How many ways are there to arrive back at your starting point that involve *eight* displacement vectors that add to zero? Describe each possibility, using drawings for clarity.

12. (a) Is it possible for one component of a vector to be zero, while the vector itself is not zero? (b) Is it possible for a vector to be zero, while one component of the vector is not zero? Explain.

13. Can two nonzero perpendicular vectors be added together so their sum is zero? Explain.

14. Can three or more vectors with unequal magnitudes be added together so their sum is zero? If so, show by means of a tail-to-head arrangement of the vectors how this could occur.

15. In preparation for this question, review Conceptual Example 6. Vectors **A** and **B** satisfy the vector equation $\mathbf{A} + \mathbf{B} = \mathbf{0}$. (a) How does the magnitude of **B** compare with the magnitude of **A**? (b) How does the direction of **B** compare with the direction of **A**? Give your reasoning.

16. Vectors **A**, **B**, and **C** satisfy the vector equation $\mathbf{A} + \mathbf{B} = \mathbf{C}$, and their magnitudes are related by the scalar equation $A^2 + B^2 = C^2$. How is vector **A** oriented with respect to vector **B**? Account for your answer.

17. Vectors **A**, **B**, and **C** satisfy the vector equation $\mathbf{A} + \mathbf{B} = \mathbf{C}$, and their magnitudes are related by the scalar equation $A + B = C$. How is vector **A** oriented with respect to vector **B**? Explain your reasoning.

18. The magnitude of a vector has doubled, its direction remaining the same. Can you conclude that the magnitude of each component of the vector has doubled? Explain your answer.

19. The tail of a vector is fixed to the origin of an x, y axis system. Originally the vector points along the $+x$ axis. As time passes, the vector rotates counterclockwise. Describe how the sizes of the x and y components of the vector compare to the size of the original vector for rotational angles of (a) $90°$, (b) $180°$, (c) $270°$, and (d) $360°$.

20. A vector has a component of zero along the x axis of a certain axis system. Does this vector necessarily have a component of zero along the x axis of another (rotated) axis system? Use a drawing to justify your answer.

Problems

Problems that are not marked with a star are considered the easiest to solve. Problems that are marked with a single star (*) are more difficult, while those marked with a double star (**) are the most difficult.

ssm   Solution is in the Student Solutions Manual.   www   Solution is available on the World Wide Web at www.wiley.com/college/cutnell
⚕   This icon represents a biomedical application.

Section 1.2 Units, Section 1.3 The Role of Units in Problem Solving

1. ssm The mass of the parasitic wasp *Caraphractus cintus* can be as small as 5×10^{-6} kg. What is this mass in (a) grams (g), (b) milligrams (mg), and (c) micrograms (μg)?

2. Vesna Vulovic survived the longest fall on record without a parachute when her plane exploded and she fell 6 miles, 551 yards. What is this distance in meters?

3. How many seconds are there in (a) one hour and thirty-five minutes and (b) one day?

4. Bicyclists in the Tour de France reach speeds of 34.0 miles per hour (mi/h) on flat sections of the road. What is this speed in (a) kilometers per hour (km/h) and (b) meters per second (m/s)?

5. ssm The largest diamond ever found had a size of 3106 carats. One carat is equivalent to a mass of 0.200 g. Use the fact that 1 kg (1000 g) has a weight of 2.205 lb under certain conditions, and determine the weight of this diamond in pounds.

6. A bottle of wine known as a magnum contains a volume of 1.5 liters. A bottle known as a jeroboam contains 0.792 U.S. gallons. How many magnums are there in one jeroboam?

7. The following are dimensions of various physical parameters that will be discussed later on in the text. Here [L], [T], and [M] denote, respectively, dimensions of length, time, and mass.

	Dimension		Dimension
Distance (x)	[L]	Acceleration (a)	$[L]/[T]^2$
Time (t)	[T]	Force (F)	$[M][L]/[T]^2$
Mass (m)	[M]	Energy (E)	$[M][L]^2/[T]^2$
Speed (v)	[L]/[T]		

Which of the following equations are dimensionally correct?

(a) $F = ma$ (d) $E = max$
(b) $x = \frac{1}{2}at^3$ (e) $v = \sqrt{Fx/m}$
(c) $E = \frac{1}{2}mv$

8. The variables x, v, and a have the dimensions of [L], [L]/[T], and $[L]/[T]^2$, respectively. These variables are related by an equation that has the form $v^n = 2ax$, where n is an integer constant (1, 2, 3, etc.) without dimensions. What must be the value of n, so that both sides of the equation have the same dimensions? Explain your reasoning.

* **9. ssm** The depth of the ocean is sometimes measured in fathoms (1 fathom = 6 feet). Distance on the surface of the ocean is sometimes measured in nautical miles (1 nautical mile = 6076 feet). The water beneath a surface rectangle 1.20 nautical miles by 2.60 nautical miles has a depth of 16.0 fathoms. Find the volume of water (in cubic meters) beneath this rectangle.

* **10.** A spring is hanging down from the ceiling, and an object of mass *m* is attached to the free end. The object is pulled down, thereby stretching the spring, and then released. The object oscillates up and down, and the time *T* required for one complete up-and-down oscillation is given by the equation $T = 2\pi\sqrt{m/k}$, where *k* is known as the spring constant. What must be the dimension of *k* for this equation to be dimensionally correct?

Section 1.4 Trigonometry

11. You are driving into St. Louis, Missouri, and in the distance you see the famous Gateway-to-the-West arch. This monument rises to a height of 192 m. You estimate your line of sight with the top of the arch to be 2.0° above the horizontal. Approximately how far (in kilometers) are you from the base of the arch?

12. An observer, whose eyes are 1.83 m above the ground, is standing 32.0 m away from a tree. The ground is level, and the tree is growing perpendicular to it. The observer's line of sight with the treetop makes an angle of 20.0° above the horizontal. How tall is the tree?

13. ssm www A highway is to be built between two towns, one of which lies 35.0 km south and 72.0 km west of the other. What is the shortest length of highway that can be built between the two towns, and at what angle would this highway be directed with respect to due west?

14. The two hot-air balloons in the drawing are 48.2 and 61.0 m above the ground. A person in the left balloon observes that the right balloon is 13.3° above the horizontal. What is the horizontal distance *x* between the two balloons?

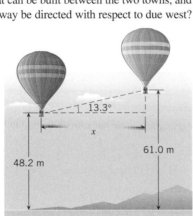

Problem 14

15. The silhouette of a Christmas tree is an isosceles triangle. The angle at the top of the triangle is 30.0°, and the base measures 2.00 m across. How tall is the tree?

* **16.** The drawing shows sodium and chlorine ions positioned at the corners of a cube that is part of the crystal structure of sodium chloride (common table salt). The edge of the cube is 0.281 nm (1 nm = 1 nanometer = 10^{-9} m) in length. Find the distance (in nanometers) between the sodium ion located at one corner of the cube and the chlorine ion located on the diagonal at the opposite corner.

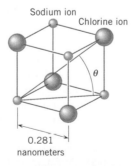

* **17. Interactive Solution 1.17** at **www.wiley.com/college/cutnell** presents a method for modeling this problem. What is the value of the angle θ in the drawing that accompanies problem 16?

* **18.** A person is standing at the edge of the water and looking out at the ocean (see the drawing). The height of the person's eyes above the water is *h* = 1.6 m, and the radius of the earth is $R = 6.38 \times 10^6$ m. (a)

How far is it to the horizon? In other words, what is the distance *d* from the person's eyes to the horizon? (*Note: At the horizon the angle between the line of sight and the radius of the earth is 90°.*) (b) Express this distance in miles.

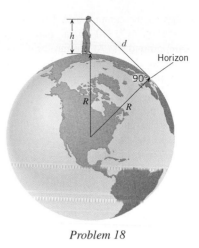

Problem 18

* **19. ssm** What is the value of each of the angles of a triangle whose sides are 95, 150, and 190 cm in length? (*Hint: Consider using the law of cosines given in Appendix E.*)

** **20.** A regular tetrahedron is a three-dimensional object that has four faces, each of which is an equilateral triangle. Each of the edges of such an object has a length *L*. The height *H* of a regular tetrahedron is the perpendicular distance from one corner to the center of the opposite triangular face. Show that the ratio between *H* and *L* is $H/L = \sqrt{2/3}$.

Section 1.6 Vector Addition and Subtraction

21. A force vector $\mathbf{F}_1$ points due east and has a magnitude of 200 newtons. A second force $\mathbf{F}_2$ is added to $\mathbf{F}_1$. The resultant of the two vectors has a magnitude of 400 newtons and points along the east/west line. Find the magnitude and direction of $\mathbf{F}_2$. Note that there are two answers.

22. (a) Two workers are trying to move a heavy crate. One pushes on the crate with a force **A**, which has a magnitude of 445 newtons and is directed due west. The other pushes with a force **B**, which has a magnitude of 325 newtons and is directed due north. What are the magnitude and direction of the resultant force **A** + **B** applied to the crate? (b) Suppose that the second worker applies a force −**B** instead of **B**. What then are the magnitude and direction of the resultant force **A** − **B** applied to the crate? In both cases express the direction relative to due west.

23. ssm Displacement vector **A** points due east and has a magnitude of 2.00 km. Displacement vector **B** points due north and has a magnitude of 3.75 km. Displacement vector **C** points due west and has a magnitude of 2.50 km. Displacement vector **D** points due south and has a magnitude of 3.00 km. Find the magnitude and direction (relative to due west) of the resultant vector **A** + **B** + **C** + **D**.

24. The drawing shows a triple jump on a checkerboard, starting at the center of square A and ending on the center of square B. Each side of a square measures 4.0 cm. What is the magnitude of the displacement of the colored checker during the triple jump?

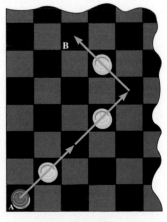

Problem 24

25. ssm www One displacement vector **A** has a magnitude of 2.43 km and points due north. A second displacement vector **B** has a magnitude of 7.74 km and also points due north. (a) Find the magnitude and direction of **A** − **B**. (b) Find the magnitude and direction of **B** − **A**.

26. Two bicyclists, starting at the same place, are riding toward the same campground by two different routes. One cyclist rides 1080 m due east and then turns due north and travels another 1430 m before reaching the campground. The second cyclist starts out by heading due north for 1950 m and then turns and heads directly toward the campground. (a) At the turning point, how far is the second cyclist from the campground? (b) What direction (measured relative to due east) must the second cyclist head during the last part of the trip?

* **27. ssm www** A car is being pulled out of the mud by two forces that are applied by the two ropes shown in the drawing. The

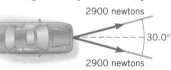

2900 newtons

30.0°

2900 newtons

dashed line in the drawing bisects the 30.0° angle. The magnitude of the force applied by each rope is 2900 newtons. Arrange the force vectors tail to head and use the graphical technique to answer the following questions. (a) How much force would a single rope need to apply to accomplish the same effect as the two forces added together? (b) How would the single rope be directed relative to the dashed line?

* **28.** In wandering, a grizzly bear makes a displacement of 1563 m due west, followed by a displacement of 3348 m in a direction 32.0° north of west. What are (a) the magnitude and (b) the direction of the displacement needed for the bear to *return to its starting point?* Specify the direction relative to due east.

* **29.** Before starting this problem, review **Interactive Solution 1.29** at **www.wiley.com/college/cutnell.** Vector **A** has a magnitude of 12.3 units and points due west. Vector **B** points due north. (a) What is the magnitude of **B** if **A** + **B** has a magnitude of 15.0 units? (b) What is the direction of **A** + **B** relative to due west? (c) What is the magnitude of **B** if **A** − **B** has a magnitude of 15.0 units? (d) What is the direction of **A** − **B** relative to due west?

* **30.** At a picnic, there is a contest in which hoses are used to shoot water at a beach ball from three directions. As a result, three forces act on the ball, **F₁**, **F₂**, and **F₃** (see the drawing). The magnitudes of **F₁** and **F₂** are **F₁** = 50.0 newtons and F_2 = 90.0 newtons. Using a scale drawing and the graphical technique, determine (a) the magnitude of **F₃** and (b) the angle θ such that the resultant force acting on the ball is zero.

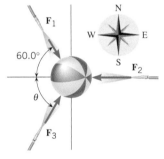

Section 1.7 The Components of a Vector

31. ssm The speed of an object and the direction in which it moves constitute a vector quantity known as the velocity. An ostrich is running at a speed of 17.0 m/s in a direction of 68.0° north of west. What is the magnitude of the ostrich's velocity component that is directed (a) due north and (b) due west?

32. Your friend has slipped and fallen. To help her up, you pull with a force **F**, as the drawing shows. The vertical component of this force is 130 newtons, and the horizontal component is 150 newtons. Find (a) the magnitude of **F** and (b) the angle θ.

33. An ocean liner leaves New York City and travels 18.0° north of east for 155 km. How far east and how far north has it gone? In other words, what are the magnitudes of the components of the ship's displacement vector in the directions (a) due east and (b) due north?

34. Soccer player #1 is 8.6 m from the goal, as the drawing shows. If she kicks the ball directly into the net, the ball has a displacement labeled **A**. If, on the other hand, she first kicks it to player #2, who then kicks it into the net, the ball undergoes two successive displacements, **Aᵧ** and **Aₓ**. What are the magnitude and direction of **Aₓ** and **Aᵧ**?

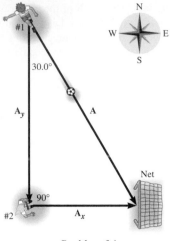

Problem 34

35. ssm Two ropes are attached to a heavy box to pull it along the floor. One rope applies a force of 475 newtons in a direction due west; the other applies a force of 315 newtons in a direction due south. As we will see later in the text, force is a vector quantity. (a) How much force should be applied by a single rope, and (b) in what direction (relative to due west), if it is to accomplish the same effect as the two forces added together?

36. On takeoff, an airplane climbs with a speed of 180 m/s at an angle of 34° above the horizontal. The speed and direction of the airplane constitute a vector quantity known as the velocity. The sun is shining directly overhead. How fast is the shadow of the plane moving along the ground? (That is, what is the magnitude of the horizontal component of the plane's velocity?)

* **37.** To review the solution to a similar problem, consult **Interactive Solution 1.37** at **www.wiley.com/college/cutnell.** The magnitude of the force vector **F** is 82.3 newtons. The x component of this vector is directed along the $+x$ axis and has a magnitude of 74.6 newtons. The y component points along the $+y$ axis. (a) Find the direction of **F** relative to the $+x$ axis. (b) Find the component of **F** along the $+y$ axis.

* **38.** A force vector points at an angle of 52° above the $+x$ axis. It has a y component of $+290$ newtons. Find (a) the magnitude and (b) the x component of the force vector.

** **39. ssm www** The drawing shows a force vector that has a magnitude of 475 newtons. Find the (a) x, (b) y, and (c) z components of the vector.

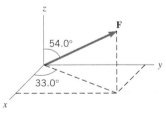

Section 1.8 Addition of Vectors by Means of Components

40. As an aid in working this problem, consult **Concept Simulation 1.1** at **www.wiley.com/college/cutnell.** Two forces are applied to a tree stump to pull it out of the ground. Force **Fₐ** has a magnitude of 2240 newtons and points 34.0° south of east, while force **F_B** has a magnitude of 3160 newtons and points due south. Using the component method, find the magnitude and direction of the resultant force **Fₐ** + **F_B** that is applied to the stump. Specify the direction with respect to due east.

41. ssm A golfer, putting on a green, requires three strokes to "hole the ball." During the first putt, the ball rolls 5.0 m due east. For the second putt, the ball travels 2.1 m at an angle of 20.0° north of east. The third putt is 0.50 m due north. What displacement (magnitude and direction relative to due east) would have been needed to "hole the ball" on the very first putt?

42. You are on a treasure hunt and your map says "Walk due west for 52 paces, then walk 30.0° north of west for 42 paces, and finally walk due north for 25 paces." What is the magnitude of the component of your displacement in the direction (a) due north and (b) due west?

43. Find the resultant of the three displacement vectors in the drawing by means of the component method. The magnitudes of the vectors are $A = 5.00$ m, $B = 5.00$ m, and $C = 4.00$ m.

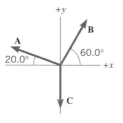

Problem 43

44. A baby elephant is stuck in a mud hole. To help pull it out, game keepers use a rope to apply force $\mathbf{F_A}$, as part a of the drawing shows. By itself, however, force $\mathbf{F_A}$ is insufficient. Therefore, two additional forces $\mathbf{F_B}$ and $\mathbf{F_C}$ are applied, as in part b of the drawing. Each of these additional forces has the same magnitude F. The magnitude of the resultant force acting on the elephant in part b of the drawing is twice that in part a. Find the ratio F/F_A.

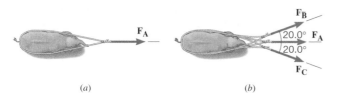

(a)   (b)

45. As preparation for this problem, consult **Concept Simulation 1.1** at **www.wiley.com/college/cutnell**. On a safari, a team of naturalists sets out toward a research station located 4.8 km away in a direction 42° north of east. After traveling in a straight line for 2.4 km, they stop and discover that they have been traveling 22° north of east, because their guide misread his compass. What are (a) the magnitude and (b) the direction (relative to due east) of the displacement vector now required to bring the team to the research station?

* **46.** Three forces are applied to an object, as indicated in the drawing. Force $\mathbf{F_1}$ has a magnitude of 21.0 newtons (21.0 N) and is directed 30.0° to the left of the $+y$ axis. Force $\mathbf{F_2}$ has a magnitude of 15.0 N and points along the $+x$ axis. What must be the magnitude and direction (specified by the angle θ in the drawing) of the third force $\mathbf{F_3}$ such that the vector sum of the three forces is 0 N?

* **47. ssm** Vector $\mathbf{A}$ has a magnitude of 6.00 units and points due east. Vector $\mathbf{B}$ points due north. (a) What is the magnitude of $\mathbf{B}$, if the vector $\mathbf{A} + \mathbf{B}$ points 60.0° north of east? (b) Find the magnitude of $\mathbf{A} + \mathbf{B}$.

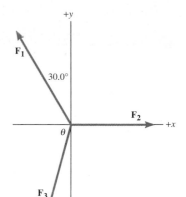

Problem 46

* **48.** The route followed by a hiker consists of three displacement vectors $\mathbf{A}$, $\mathbf{B}$, and $\mathbf{C}$. Vector $\mathbf{A}$ is along a measured trail and is 1550 m in a direction 25.0° north of east. Vector $\mathbf{B}$ is not along a measured trail, but the hiker uses a compass and knows that the direction is 41.0° east of south. Similarly, the direction of vector $\mathbf{C}$ is 35.0° north of west. The hiker ends up back where she started, so the resultant displacement is zero, or $\mathbf{A} + \mathbf{B} + \mathbf{C} = 0$. Find the magnitudes of (a) vector $\mathbf{B}$ and (b) vector $\mathbf{C}$.

* **49. Interactive Solution 1.49** at **www.wiley.com/college/cutnell** presents the solution to a problem that is similar to this one. Vector $\mathbf{A}$ has a magnitude of 145 units and points 35.0° north of west. Vector $\mathbf{B}$ points 65.0° east of north. Vector $\mathbf{C}$ points 15.0° west of south. These three vectors add to give a resultant vector that is zero. Using components, find the magnitudes of (a) vector $\mathbf{B}$ and (b) vector $\mathbf{C}$.

50. A grasshopper makes four jumps. The displacement vectors are (1) 27.0 cm, due west; (2) 23.0 cm, 35.0° south of west; (3) 28.0 cm, 55.0° south of east; and (4) 35.0 cm, 63.0° north of east. Find the magnitude and direction of the resultant displacement. Express the direction with respect to due west.

Additional Problems

51. A chimpanzee sitting against his favorite tree gets up and walks 51 m due east and 39 m due south to reach a termite mound, where he eats lunch. (a) What is the shortest distance between the tree and the termite mound? (b) What angle does the shortest distance make with respect to due east?

52. The gondola ski lift at Keystone, Colorado, is 2830 m long. On average, the ski lift rises 14.6° above the horizontal. How high is the top of the ski lift relative to the base?

53. ssm Vector $\mathbf{A}$ points along the $+y$ axis and has a magnitude of 100.0 units. Vector $\mathbf{B}$ points at an angle of 60.0° above the $+x$ axis and has a magnitude of 200.0 units. Vector $\mathbf{C}$ points along the $+x$ axis and has a magnitude of 150.0 units. Which vector has (a) the largest x component and (b) the largest y component?

54. Consider the equation $v = \frac{1}{3}zxt^2$. The dimensions of the variables x, v, and t are [L], [L]/[T], and [T], respectively. What must be the dimensions of the variable z, such that both sides of the equation have the same dimensions? Show how you determined your answer.

55. As an aid in visualizing the concepts in this problem, consult **Concept Simulation 1.1** at **www.wiley.com/college/cutnell**. A football player runs the pattern given in the drawing by the three displacement vectors $\mathbf{A}$, $\mathbf{B}$, and $\mathbf{C}$. The magnitudes of these vectors are $A = 5.00$ m, $B = 15.0$ m, and $C = 18.0$ m. Using the component method, find the magnitude and direction θ of the resultant vector $\mathbf{A} + \mathbf{B} + \mathbf{C}$.

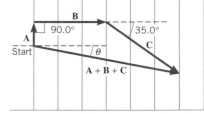

56. A circus performer begins his act by walking out along a nearly horizontal high wire. He slips and falls to the safety net, 25.0 ft below. The magnitude of his displacement from the beginning of the walk to the net is 26.7 ft. (a) How far out along the high wire did he walk? (b) Find the angle that his displacement vector makes below the horizontal.

57. ssm The x vector component of a displacement vector $\mathbf{r}$ has a magnitude of 125 m and points along the negative x axis. The y vector component has a magnitude of 184 m and points along the negative y axis. Find the magnitude and direction of $\mathbf{r}$. Specify the direction with respect to the negative x axis.

* **58.** The vector **A** in the drawing has a magnitude of 750 units. Determine the magnitude and direction of the x and y components of the vector **A**, relative to (a) the black axes and (b) the colored axes.

* **59. ssm** A sailboat race course consists of four legs, defined by the displacement vectors **A**, **B**, **C**, and **D**, as the drawing indicates. The magnitudes of the first three vectors are $A = 3.20$ km, $B = 5.10$ km, and $C = 4.80$ km. The finish line of the course coincides with the starting line. Using the data in the drawing, find the distance of the fourth leg and the angle θ.

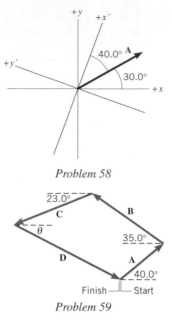

Problem 58

Problem 59

* **60.** A jogger travels a route that has two parts. The first is a displacement **A** of 2.50 km due south, and the second involves a displacement **B** that points due east. (a) The resultant displacement **A** + **B** has a magnitude of 3.75 km. What is the magnitude of **B**,

and what is the direction of **A** + **B** relative to due south? (b) Suppose that **A** − **B** had a magnitude of 3.75 km. What then would be the magnitude of **B**, and what is the direction of **A** − **B** relative to due south?

* **61.** Three deer, A, B, and C, are grazing in a field. Deer B is located 62 m from deer A at an angle of 51° north of west. Deer C is located 77° north of east relative to deer A. The distance between deer B and C is 95 m. What is the distance between deer A and C? *(Hint: Consider the law of cosines given in Appendix E.)*

* **62.** Before starting this problem, review Conceptual Example 6. The force vector **F**$_A$ has a magnitude of 90.0 newtons and points due east. The force vector **F**$_B$ has a magnitude of 135 newtons and points 75° north of east. Use the graphical method and find the magnitude and direction of (a) **F**$_A$ − **F**$_B$ (give the direction with respect to due east) and (b) **F**$_B$ − **F**$_A$ (give the direction with respect to due west).

** **63.** What are the x and y components of the vector that must be added to the following three vectors, so that the sum of the four vectors is zero? Due east is the $+x$ direction, and due north is the $+y$ direction.

$$A = 113 \text{ units, } 60.0° \text{ south of west}$$
$$B = 222 \text{ units, } 35.0° \text{ south of east}$$
$$C = 177 \text{ units, } 23.0° \text{ north of east}$$

Concepts & Calculations   Group Learning Problems

Note: Each of these problems consists of Concept Questions followed by a related quantitative Problem. They are designed for use by students working alone or in small learning groups. The Concept Questions involve little or no mathematics and are intended to stimulate group discussions. They focus on the concepts with which the problems deal. Recognizing the concepts is the essential initial step in any problem-solving technique.

64. Concept Questions (a) Considering the fact that 3.28 ft = 1 m, which is the larger unit for measuring area, 1 ft^2 or 1 m^2? (b) Consider a 1330-ft^2 apartment. With your answer to part (a) in mind and without doing any calculations, decide whether this apartment has an area that is greater than or less than 1330 m^2.

Problem In a 1330-ft^2 apartment, how many square meters of area are there? Be sure that your answer is consistent with your answers to the Concept Questions.

65. Concept Question The corners of a square lie on a circle of diameter D. Each side of the square has a length L. Is L smaller or larger than D? Explain your reasoning using the Pythagorean theorem.

Problem The diameter D of the circle is 0.35 m. Each side of the square has a length L. Find L. Be sure that your answer is consistent with your answer to the Concept Question.

66. Concept Question Can the x or y component of a vector ever have a greater magnitude than the vector itself has? Give your reasoning.

Problem A force vector has a magnitude of 575 newtons and points at an angle of 36.0° below the positive x axis. What are (a) the x scalar component and (b) the y scalar component of the vector? Verify that your answers are consistent with your answer to the Concept Question.

67. Concept Questions The components of vector **A** are A_x and A_y, and the angle that it makes with respect to the positive x axis is θ. (a) Does increasing the component A_x (while holding A_y constant) in-

crease or decrease the angle θ? (b) Does increasing the component A_y (while holding A_x constant) increase or decrease the angle θ? Account for your answers.

Problem (a) The components of displacement vector **A** are $A_x = 12$ m and $A_y = 12$ m. Find θ. (b) The components of displacement vector **A** are $A_x = 17$ m and $A_y = 12$ m. Find θ. (c) The components of displacement vector **A** are $A_x = 12$ m and $A_y = 17$ m. Find θ. Be sure that your answers are consistent with your answers to the Concept Questions.

68. Concept Questions Vector **A** points due west, while vector **B** points due south. (a) Does the direction **A** + **B** point north or south of due west? (b) Does the direction of **A** − **B** point north or south of due west? Give your reasoning in each case.

Problem Vector **A** has a magnitude of 63 units and points due west, while vector **B** has the same magnitude and points due south. Find the magnitude and direction of (a) **A** + **B** and (b) **A** − **B**. Specify the directions relative to due west. Verify that your answers agree with your answers to the Concept Questions.

69. Concept Questions A pilot flies her route in two straight-line segments. The displacement vector **A** for the first segment has a magnitude of 244 km and a direction 30.0° north of east. The displacement vector **B** for the second segment has a magnitude of 175 km and a direction due west. The resultant displacement vector is **R** = **A** + **B** and makes an angle θ with the direction due east. Make a drawing to scale showing the vectors **A** and **B** placed tail to head and the resultant vector **R**. Without doing any calculations decide whether (a) the magnitude of **R** is greater or smaller than the magnitude of **A**, (b) the magnitude of **R** is greater or smaller than the magnitude of **B**, and (c) the angle θ is greater than, smaller than, or equal to 30.0°.

Problem Using the component method, find the magnitude of **R** and the directional angle θ. Check to see that your answers are consistent with your answers to the Concept Questions.

Chapter 2

Kinematics in One Dimension

As this bald eagle comes in for a landing, it is slowing down while moving forward. To describe such motion, this chapter presents the concepts of displacement, velocity, and acceleration. (© Frank Oberle/ Stone/Getty Images)

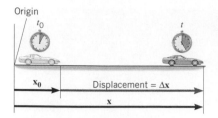

Figure 2.1 The displacement $\Delta\mathbf{x}$ is a vector that points from the initial position $\mathbf{x_0}$ to the final position $\mathbf{x}$.

2.1 Displacement

There are two aspects to any motion. In a purely descriptive sense, there is the movement itself. Is it rapid or slow, for instance? Then, there is the issue of what causes the motion or what changes it, which requires that forces be considered. **Kinematics** deals with the concepts that are needed to describe motion, without any reference to forces. The present chapter discusses these concepts as they apply to motion in one dimension, and the next chapter treats two-dimensional motion. **Dynamics** deals with the effect that forces have on motion, a topic that is considered in Chapter 4. Together, kinematics and dynamics form the branch of physics known as **mechanics.** We turn now to the first of the kinematics concepts to be discussed, which is displacement.

To describe the motion of an object, we must be able to specify the location of the object at all times, and Figure 2.1 shows how to do this for one-dimensional motion. In this drawing, the initial position of a car is indicated by the vector labeled $\mathbf{x_0}$. The length of $\mathbf{x_0}$ is the distance of the car from an arbitrarily chosen origin. At a later time the car has moved to a new position, which is indicated by the vector $\mathbf{x}$. The **displacement** of the car $\Delta\mathbf{x}$ (read as "delta x" or "the change in x") is a vector drawn from the initial position to the final position. Displacement is a vector quantity in the sense discussed in Section 1.5, for it conveys both a magnitude (the distance between the initial and final positions) and a direction. The displacement can be related to $\mathbf{x_0}$ and $\mathbf{x}$ by noting from the drawing that

$$\mathbf{x_0} + \Delta\mathbf{x} = \mathbf{x} \quad \text{or} \quad \Delta\mathbf{x} = \mathbf{x} - \mathbf{x_0}$$

Thus, the displacement $\Delta\mathbf{x}$ is the difference between $\mathbf{x}$ and $\mathbf{x_0}$, and the Greek letter delta (Δ) is used to signify this difference. It is important to note that the change in any variable is always the final value minus the initial value.

■ **DEFINITION OF DISPLACEMENT**

The displacement is a vector that points from an object's initial position to its final position and has a magnitude that equals the shortest distance between the two positions.

SI Unit of Displacement: meter (m)

The SI unit for displacement is the meter (m), but there are other units as well, such as the centimeter and the inch. When converting between centimeters (cm) and inches (in.), remember that 2.54 cm = 1 in.

Often, we will deal with motion along a straight line. In such a case, a displacement in one direction along the line is assigned a positive value, and a displacement in the opposite direction is assigned a negative value. For instance, assume that a car is moving along an east/west direction and that a positive (+) sign is used to denote a direction due east. Then, $\Delta\mathbf{x} = +500$ m represents a displacement that points to the east and has a magnitude of 500 meters. Conversely, $\Delta\mathbf{x} = -500$ m is a displacement that has the same magnitude but points in the opposite direction, due west.

2.2 Speed and Velocity

AVERAGE SPEED

One of the most obvious features of an object in motion is how fast it is moving. If a car travels 200 meters in 10 seconds, we say its average speed is 20 meters per second, the **average speed** being the distance traveled divided by the time required to cover the distance:

$$\text{Average speed} = \frac{\text{Distance}}{\text{Elapsed time}} \tag{2.1}$$

Equation 2.1 indicates that the unit for average speed is the unit for distance divided by the unit for time, or meters per second (m/s) in SI units. Example 1 illustrates how the idea of average speed is used.

Example 1 Distance Run by a Jogger

How far does a jogger run in 1.5 hours (5400 s) if his average speed is 2.22 m/s?

Reasoning The average speed of the jogger is the average distance per second that he travels. Thus, the distance covered by the jogger is equal to the average distance per second (his average speed) multiplied by the number of seconds (the elapsed time) that he runs.

Solution To find the distance run, we rewrite Equation 2.1 as

$$\text{Distance} = (\text{Average speed})(\text{Elapsed time}) = (2.22 \text{ m/s})(5400 \text{ s}) = \boxed{12\ 000 \text{ m}}$$

Speed is a useful idea, because it indicates how fast an object is moving. However, speed does not reveal anything about the direction of the motion. To describe both how fast an object moves and the direction of its motion, we need the vector concept of velocity.

AVERAGE VELOCITY

▶ CONCEPTS AT A GLANCE   To define the velocity of an object, we will use two concepts that we have already encountered, displacement and time. The building of new concepts from more basic ones is a common theme in physics. In fact, the great strength of physics as a science is that it builds a coherent understanding of nature through the development of interrelated concepts. This development is so important that it is emphasized throughout this text by Concepts-at-a-Glance charts, such as that in Figure 2.2. These charts illustrate diagrammatically how useful ideas (e.g., velocity) emerge from more basic ones (e.g., displacement and time). ◀

Suppose that the initial position of the car in Figure 2.1 is x_0 when the time is t_0. A little later the car arrives at the final position x at the time t. The difference between these times is the time required for the car to travel between the two positions. We denote this difference by the shorthand notation Δt (read as "delta t"), where Δt represents the final time t minus the initial time t_0.

$$\Delta t = \underbrace{t - t_0}_{\text{Elapsed time}}$$

Note that Δt is defined in a manner analogous to Δx, which is the final position minus the initial position ($\Delta x = x - x_0$). Dividing the displacement Δx of the car by the elapsed time Δt gives the *average velocity* of the car. It is customary to denote the average value

Figure 2.2 CONCEPTS AT A GLANCE The concepts of displacement and time are brought together to formulate the concept of velocity. The greater the velocity of these skiers, the greater is their displacement per unit time. (© Mark Junak/Stone/Getty Images)

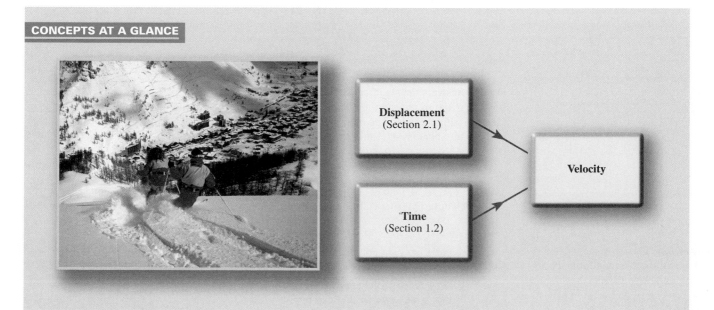

CONCEPTS AT A GLANCE

Displacement (Section 2.1)

Time (Section 1.2)

Velocity

Figure 2.3 In this time-lapse photo of traffic on the Los Angeles Freeway in California, the velocity of a car in the left lane (white headlights) is opposite to that of an adjacent car in the right lane (red taillights). (© Peter Essick/Aurora & Quanta Productions)

of a quantity by placing a horizontal bar above the symbol representing the quantity. The average velocity, then, is written as $\bar{v}$, as specified in Equation 2.2:

■ **DEFINITION OF AVERAGE VELOCITY**

$$\text{Average velocity} = \frac{\text{Displacement}}{\text{Elapsed time}}$$

$$\bar{v} = \frac{x - x_0}{t - t_0} = \frac{\Delta x}{\Delta t} \qquad (2.2)$$

SI Unit of Average Velocity: meter per second (m/s)

Equation 2.2 indicates that the unit for average velocity is the unit for length divided by the unit for time, or meters per second (m/s) in SI units. Velocity can also be expressed in other units, such as kilometers per hour (km/h) or miles per hour (mi/h).

Average velocity is a vector that points in the same direction as the displacement in Equation 2.2. Figure 2.3 illustrates that the velocity of a car confined to move along a line can point either in one direction or in the opposite direction. As with displacement, we will use plus and minus signs to indicate the two possible directions. If the displacement points in the positive direction, the average velocity is positive. Conversely, if the displacement points in the negative direction, the average velocity is negative. Example 2 illustrates these features of average velocity.

Example 2 The World's Fastest Jet-Engine Car

Andy Green in the car *ThrustSSC* set a world record of 341.1 m/s (763 mi/h) in 1997. The car was powered by two jet engines, and it was the first one officially to exceed the speed of sound. To establish such a record, the driver makes two runs through the course, one in each direction, to nullify wind effects. Figure 2.4*a* shows that the car first travels from left to right and covers a distance of 1609 m (1 mile) in a time of 4.740 s. Figure 2.4*b* shows that in the reverse direction, the car covers the same distance in 4.695 s. From these data, determine the average velocity for each run.

Reasoning Average velocity is defined as the displacement divided by the elapsed time. In using this definition we recognize that the displacement is not the same as the distance traveled. Displacement takes the direction of the motion into account, and distance does not. During both runs, the car covers the same distance of 1609 m. However, for the first run the displacement is $\Delta x = +1609$ m, while for the second it is $\Delta x = -1609$ m. The plus and minus signs are essential, because the first run is to the right, which is the positive direction, and the second run is in the opposite or negative direction.

Solution According to Equation 2.2, the average velocities are

Run 1
$$\bar{v} = \frac{\Delta x}{\Delta t} = \frac{+1609 \text{ m}}{4.740 \text{ s}} = \boxed{+339.5 \text{ m/s}}$$

Run 2
$$\bar{v} = \frac{\Delta x}{\Delta t} = \frac{-1609 \text{ m}}{4.695 \text{ s}} = \boxed{-342.7 \text{ m/s}}$$

In these answers the algebraic signs convey the directions of the velocity vectors. In particular, for Run 2 the minus sign indicates that the average velocity, like the displacement, points to the left in Figure 2.4*b*. The magnitudes of the velocities are 339.5 and 342.7 m/s. The average of these numbers is 341.1 m/s and is recorded in the record book.

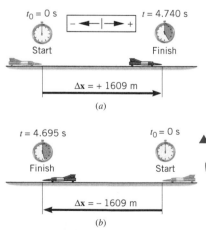

Figure 2.4 The arrows in the box at the top of the drawing indicate the positive and negative directions for the displacements of the car, as explained in Example 2.

✔ **Check Your Understanding 1**

A straight track is 1600 m in length. A runner begins at the starting line, runs due east for the full length of the track, turns around, and runs halfway back. The time for this run is five minutes. What is the runner's average velocity, and what is his average speed? *(The answers are given at the end of the book.)*

Background: Displacement, distance, velocity, and speed play roles in this problem. Displacement and distance are not the same thing; neither are velocity and speed.

For similar questions (including conceptual counterparts), consult Self-Assessment Test 2.1. This test is described at the end of Section 2.3.

INSTANTANEOUS VELOCITY

Suppose the magnitude of your average velocity for a long trip was 20 m/s. This value, being an average, does not convey any information about how fast you were moving at any instant during the trip. Surely there were times when your car traveled faster than 20 m/s and times when it traveled more slowly. The ***instantaneous velocity*** **v** of the car indicates how fast the car moves and the direction of the motion at each instant of time. The magnitude of the instantaneous velocity is called the ***instantaneous speed,*** and it is the number (with units) indicated by the speedometer.

The instantaneous velocity at any point during a trip can be obtained by measuring the time interval Δt for the car to travel a *very small* displacement $\Delta \mathbf{x}$. We can then compute the average velocity over this interval. If the time Δt is small enough, the instantaneous velocity does not change much during the measurement. Then, the instantaneous velocity **v** at the point of interest is approximately equal to ($\approx$) the average velocity $\overline{\mathbf{v}}$ computed over the interval, or $\mathbf{v} \approx \overline{\mathbf{v}} = \Delta \mathbf{x}/\Delta t$ (for sufficiently small Δt). In fact, in the limit that Δt becomes infinitesimally small, the instantaneous velocity and the average velocity become equal, so that

$$\mathbf{v} = \lim_{\Delta t \to 0} \frac{\Delta \mathbf{x}}{\Delta t} \tag{2.3}$$

The notation $\lim_{\Delta t \to 0} (\Delta \mathbf{x}/\Delta t)$ means that the ratio $\Delta \mathbf{x}/\Delta t$ is defined by a limiting process in which smaller and smaller values of Δt are used, so small that they approach zero. As smaller values of Δt are used, $\Delta \mathbf{x}$ also becomes smaller. However, the ratio $\Delta \mathbf{x}/\Delta t$ does *not* become zero but, rather, approaches the value of the instantaneous velocity. For brevity, we will use the word *velocity* to mean "instantaneous velocity" and *speed* to mean "instantaneous speed."

In a wide range of motions, the velocity changes from moment to moment. To describe the manner in which it changes, the concept of acceleration is needed.

2.3 Acceleration

The velocity of a moving object may change in a number of ways. For example, it may increase, as it does when the driver of a car steps on the gas pedal to pass the car ahead. Or it may decrease, as it does when the driver applies the brakes to stop at a red light. In either case, the change in velocity may occur over a short or a long time interval.

▶ CONCEPTS AT A GLANCE To describe how the velocity of an object changes during a given time interval, we now introduce the new idea of acceleration. As the Concepts-at-a-Glance chart in Figure 2.5 illustrates, this idea depends on two concepts that we have previously encountered, velocity and time. Specifically, the notion of acceleration emerges when the *change* in the velocity is combined with the time during which the change occurs. ◀

Figure 2.5 CONCEPTS AT A GLANCE To formulate the concept of acceleration, the change in velocity is combined with the time required for the change to occur. As speed skater Jennifer Rodriguez competes in this race, she is accelerating when her velocity is changing. (© Ed Purcell/ Aurora & Quanta Productions)

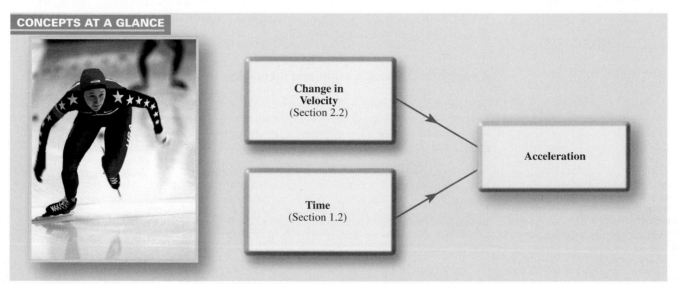

CONCEPTS AT A GLANCE

Change in
Velocity
(Section 2.2)

Time
(Section 1.2)

Acceleration

Figure 2.6 During takeoff, the plane accelerates from an initial velocity $\mathbf{v_0}$ to a final velocity $\mathbf{v}$ during the time interval $\Delta t = t - t_0$.

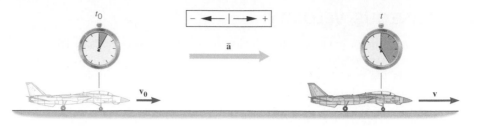

The meaning of *average acceleration* can be illustrated by considering a plane during takeoff. Figure 2.6 focuses attention on how the plane's velocity changes along the runway. During an elapsed time interval $\Delta t = t - t_0$, the velocity changes from an initial value of $\mathbf{v_0}$ to a final value of $\mathbf{v}$. The change $\Delta\mathbf{v}$ in the plane's velocity is its final velocity minus its initial velocity, so that $\Delta\mathbf{v} = \mathbf{v} - \mathbf{v_0}$. The average acceleration $\bar{\mathbf{a}}$ is defined in the following manner, to provide a measure of how much the velocity changes per unit of elapsed time.

■ **DEFINITION OF AVERAGE ACCELERATION**

$$\text{Average acceleration} = \frac{\text{Change in velocity}}{\text{Elapsed time}}$$

$$\bar{\mathbf{a}} = \frac{\mathbf{v} - \mathbf{v_0}}{t - t_0} = \frac{\Delta\mathbf{v}}{\Delta t} \qquad (2.4)$$

SI Unit of Average Acceleration: meter per second squared (m/s^2)

These thoroughbreds accelerate out of the starting gate. (© Robert Maass/Corbis Images)

The average acceleration $\bar{\mathbf{a}}$ is a vector that points in the same direction as $\Delta\mathbf{v}$, the change in the velocity. Following the usual custom, plus and minus signs indicate the two possible directions for the acceleration vector when the motion is along a straight line.

We are often interested in an object's acceleration at a particular instant of time. The *instantaneous acceleration* $\mathbf{a}$ can be defined by analogy with the procedure used in Section 2.2 for instantaneous velocity:

$$\mathbf{a} = \lim_{\Delta t \to 0} \frac{\Delta\mathbf{v}}{\Delta t} \qquad (2.5)$$

Equation 2.5 indicates that the instantaneous acceleration is a limiting case of the average acceleration. When the time interval Δt for measuring the acceleration becomes extremely small (approaching zero in the limit), the average acceleration and the instantaneous acceleration become equal. Moreover, in many situations the acceleration is constant, so the acceleration has the same value at any instant of time. In the future, we will use the word *acceleration* to mean "instantaneous acceleration." Example 3 deals with the acceleration of a plane during takeoff.

Example 3   Acceleration and Increasing Velocity

Suppose the plane in Figure 2.6 starts from rest ($\mathbf{v_0} = 0$ m/s) when $t_0 = 0$ s. The plane accelerates down the runway and at $t = 29$ s attains a velocity of $\mathbf{v} = +260$ km/h, where the plus sign indicates that the velocity points to the right. Determine the average acceleration of the plane.

Reasoning The average acceleration of the plane is defined as the change in its velocity divided by the elapsed time. The change in the plane's velocity is its final velocity $\mathbf{v}$ minus its initial velocity $\mathbf{v_0}$, or $\mathbf{v} - \mathbf{v_0}$. The elapsed time is the final time t minus the initial time t_0, or $t - t_0$.

Problem solving insight
The change in any variable is the final value minus the initial value: for example, the change in velocity is $\Delta\mathbf{v} = \mathbf{v} - \mathbf{v_0}$, and the change in time is $\Delta t = t - t_0$.

Solution The average acceleration is expressed by Equation 2.4 as

$$\bar{\mathbf{a}} = \frac{\mathbf{v} - \mathbf{v_0}}{t - t_0} = \frac{260 \text{ km/h} - 0 \text{ km/h}}{29 \text{ s} - 0 \text{ s}} = \boxed{+9.0 \, \frac{\text{km/h}}{\text{s}}}$$

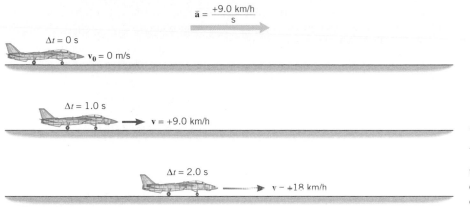

Figure 2.7 An acceleration of $+9.0 \dfrac{\text{km/h}}{\text{s}}$ means that the velocity of the plane changes by $+9.0$ km/h during each second of the motion. The "+" direction for **a** and **v** is to the right.

The average acceleration calculated in Example 3 is read as "nine kilometers per hour per second." Assuming the acceleration of the plane is constant, a value of $+9.0 \dfrac{\text{km/h}}{\text{s}}$ means the velocity changes by $+9.0$ km/h during each second of the motion. During the first second, the velocity increases from 0 to 9.0 km/h; during the next second, the velocity increases by another 9.0 km/h to 18 km/h, and so on. Figure 2.7 illustrates how the velocity changes during the first two seconds. By the end of the 29th second, the velocity is 260 km/h.

It is customary to express the units for acceleration solely in terms of SI units. One way to obtain SI units for the acceleration in Example 3 is to convert the velocity units from km/h to m/s:

$$\left(260 \, \frac{\text{km}}{\text{h}} \right) \left(\frac{1000 \text{ m}}{1 \text{ km}} \right) \left(\frac{1 \text{ h}}{3600 \text{ s}} \right) = 72 \, \frac{\text{m}}{\text{s}}$$

The average acceleration then becomes

$$\bar{a} = \frac{72 \text{ m/s} - 0 \text{ m/s}}{29 \text{ s} - 0 \text{ s}} = +2.5 \text{ m/s}^2$$

where we have used $2.5 \, \dfrac{\text{m/s}}{\text{s}} = 2.5 \, \dfrac{\text{m}}{\text{s} \cdot \text{s}} = 2.5 \, \dfrac{\text{m}}{\text{s}^2}$. An acceleration of $2.5 \, \dfrac{\text{m}}{\text{s}^2}$ is read as "2.5 meters per second per second" (or "2.5 meters per second squared") and means that the velocity changes by 2.5 m/s during each second of the motion.

Example 4 deals with a case where the motion becomes slower as time passes.

Example 4   Acceleration and Decreasing Velocity

A drag racer crosses the finish line, and the driver deploys a parachute and applies the brakes to slow down, as Figure 2.8 illustrates. The driver begins slowing down when $t_0 = 9.0$ s and

Figure 2.8 (*a*) To slow down, a drag racer deploys a parachute and applies the brakes. (*b*) The velocity of the car is decreasing, giving rise to an average acceleration $\bar{a}$ that points opposite to the velocity. (© Geoff Stunkard)

(*a*)

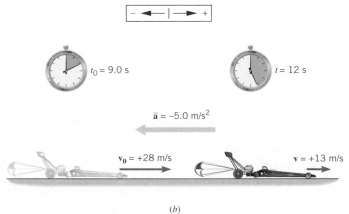

(*b*)

the car's velocity is $\mathbf{v_0} = +28$ m/s. When $t = 12.0$ s, the velocity has been reduced to $\mathbf{v} = +13$ m/s. What is the average acceleration of the dragster?

Reasoning The average acceleration of an object is always specified as its change in velocity, $\mathbf{v} - \mathbf{v_0}$, divided by the elapsed time, $t - t_0$. This is true whether the final velocity is less than the initial velocity or greater than the initial velocity.

Solution The average acceleration is, according to Equation 2.4,

$$\overline{\mathbf{a}} = \frac{\mathbf{v} - \mathbf{v_0}}{t - t_0} = \frac{13 \text{ m/s} - 28 \text{ m/s}}{12.0 \text{ s} - 9.0 \text{ s}} = \boxed{-5.0 \text{ m/s}^2}$$

Figure 2.9 shows how the velocity of the dragster changes during the braking, assuming that the acceleration is constant throughout the motion. The acceleration calculated in Example 4 is negative, indicating that the acceleration points to the left in the drawing. As a result, the acceleration and the velocity point in *opposite* directions. ***Whenever the acceleration and velocity vectors have opposite directions, the object slows down and is said to be "decelerating."*** In contrast, the acceleration and velocity vectors in Figure 2.7 point in the *same* direction, and the object speeds up.

✓ **Check Your Understanding 2**

Two cars are moving on straight sections of a highway. The acceleration of the first car is greater than the acceleration of the second car, and both accelerations have the same direction. Which one of the following is true? (a) The velocity of the first car is always greater than the velocity of the second car. (b) The velocity of the second car is always greater than the velocity of the first car. (c) In the same time interval, the velocity of the first car changes by a greater amount than the velocity of the second car does. (d) In the same time interval, the velocity of the second car changes by a greater amount than the velocity of the first car does. *(The answer is given at the end of the book.)*

Background: Velocity and acceleration are the focus of this question. They are not the same concept.

For similar questions (including calculational counterparts), consult Self-Assessment Test 2.1. The test is described at the end of this section.

Self-Assessment Test 2.1

Test your understanding of the key concepts in Sections 2.1–2.3:

• Displacement   • Velocity   • Acceleration

Go to **www.wiley.com/college/cutnell**

$\overline{\mathbf{a}} = -5.0$ m/s^2

$\Delta t = 0$ s $\mathbf{v_0} = +28$ m/s

$\Delta t = 1.0$ s $\mathbf{v} = +23$ m/s

$\Delta t = 2.0$ s $\mathbf{v} = +18$ m/s

Figure 2.9 Here, an acceleration of -5.0 m/s^2 means the velocity decreases by 5.0 m/s during each second of elapsed time.

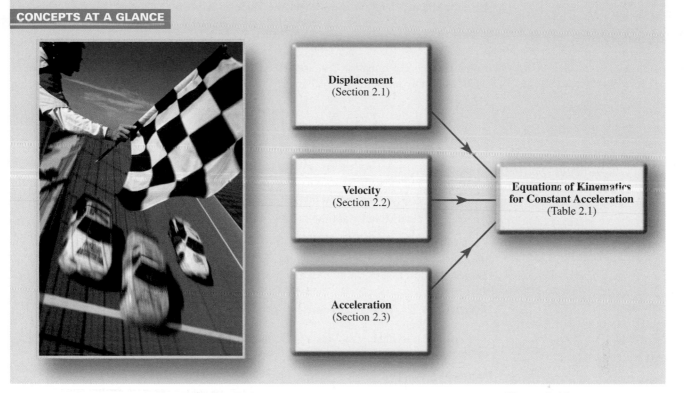

CONCEPTS AT A GLANCE

Figure 2.10 CONCEPTS AT A GLANCE The equations of kinematics for constant acceleration are obtained by combining the concepts of displacement, velocity, and acceleration. Whenever the drivers maintained a constant acceleration during the race, the equations of kinematics could have been used to calculate the displacement and velocity of these cars as a function of time. (© Mark Scott/Taxi/Getty Images)

2.4 *Equations of Kinematics for Constant Acceleration*

▶ CONCEPTS AT A GLANCE It is now possible to describe the motion of an object traveling with a constant acceleration along a straight line. To do so, we will use a set of equations known as the equations of kinematics for constant acceleration. These equations entail no new concepts, because they will be obtained by combining the familiar ideas of displacement, velocity, and acceleration, as the Concepts-at-a-Glance chart in Figure 2.10 shows. However, they will provide a very convenient way to determine certain aspects of the motion, such as the final position and velocity of a moving object. ◄

In discussing the equations of kinematics, it will be convenient to assume that the object is located at the origin $x_0 = 0$ m when $t_0 = 0$ s. With this assumption, the displacement $\Delta x = x - x_0$ becomes $\Delta x = x$. Furthermore, it is customary to dispense with the use of boldface symbols for the displacement, velocity, and acceleration vectors in the equations that follow. We will, however, continue to convey the directions of these vectors with plus or minus signs.

Consider an object that has an initial velocity of v_0 at time $t_0 = 0$ s and moves for a time t with a constant acceleration a. For a complete description of the motion, it is also necessary to know the final velocity and displacement at time t. The final velocity v can be obtained directly from Equation 2.4:

$$\bar{a} = a = \frac{v - v_0}{t} \quad \text{or} \quad v = v_0 + at \qquad \text{(constant acceleration)} \qquad (2.4)$$

The displacement x at time t can be obtained from Equation 2.2, if a value for the average velocity $\bar{v}$ can be obtained. Considering the assumption that $x_0 = 0$ m at $t_0 = 0$ s, we have

$$\bar{v} = \frac{x - x_0}{t - t_0} = \frac{x}{t} \quad \text{or} \quad x = \bar{v}t \qquad (2.2)$$

Because the acceleration is constant, the velocity increases at a constant rate. Thus, the average velocity $\bar{v}$ is midway between the initial and final velocities:

$$\bar{v} = \tfrac{1}{2}(v_0 + v) \qquad \text{(constant acceleration)} \qquad (2.6)$$

Equation 2.6, like Equation 2.4, applies only if the acceleration is constant and cannot be used when the acceleration is changing. The displacement at time t can now be determined as

$$x = \bar{v}t = \tfrac{1}{2}(v_0 + v)t \qquad \text{(constant acceleration)} \qquad (2.7)$$

Notice in Equations 2.4 ($v = v_0 + at$) and 2.7 [$x = \tfrac{1}{2}(v_0 + v)t$] that there are five kinematic variables:

1. x = displacement
2. $a = \bar{a}$ = acceleration (constant)
3. v = final velocity at time t
4. v_0 = initial velocity at time $t_0 = 0$ s
5. t = time elapsed since $t_0 = 0$ s

Each of the two equations contains four of these variables, so if three of them are known, the fourth variable can always be found. Example 5 illustrates how Equations 2.4 and 2.7 are used to describe the motion of an object.

Example 5 The Displacement of a Speedboat

The speedboat in Figure 2.11 has a constant acceleration of $+2.0$ m/s². If the initial velocity of the boat is $+6.0$ m/s, find its displacement after 8.0 seconds.

Reasoning Numerical values for the three known variables are listed in the data table below. We wish to determine the displacement x of the speedboat, so it is an unknown variable. Therefore, we have placed a question mark in the displacement column of the data table.

Speedboat Data				
x	a	v	v_0	t
?	$+2.0$ m/s²		$+6.0$ m/s	8.0 s

We can use $x = \tfrac{1}{2}(v_0 + v)t$ to find the displacement of the boat if a value for the final velocity v can be found. To find the final velocity, it is necessary to use the value given for the acceleration, because it tells us how the velocity changes, according to $v = v_0 + at$.

Solution The final velocity is

$$v = v_0 + at = 6.0 \text{ m/s} + (2.0 \text{ m/s}^2)(8.0 \text{ s}) = +22 \text{ m/s} \qquad (2.4)$$

The displacement of the boat can now be obtained:

$$x = \tfrac{1}{2}(v_0 + v)t = \tfrac{1}{2}(6.0 \text{ m/s} + 22 \text{ m/s})(8.0 \text{ s}) = \boxed{+110 \text{ m}} \qquad (2.7)$$

A calculator would give the answer as 112 m, but this number must be rounded to 110 m, since the data are accurate to only two significant figures.

Figure 2.11 (a) An accelerating speedboat. (b) The boat's displacement x can be determined if the boat's acceleration, initial velocity, and time of travel are known. (© Onne van der Wal/Corbis Images)

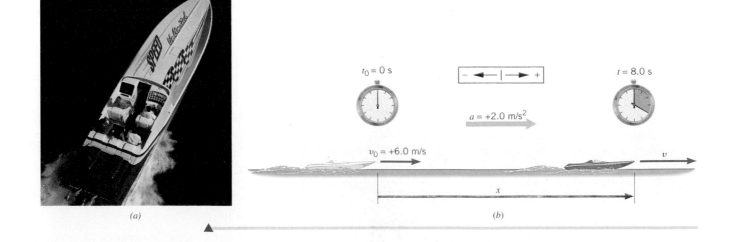

The solution to Example 5 involved two steps: finding the final velocity v and then calculating the displacement x. It would be helpful if we could find an equation that allows us to determine the displacement in a single step. Using Example 5 as a guide, we can obtain such an equation by substituting the final velocity v from Equation 2.4 ($v = v_0 + at$) into Equation 2.7 [$x = \frac{1}{2}(v_0 + v)t$]:

$$x = \tfrac{1}{2}(v_0 + v)t = \tfrac{1}{2}(v_0 + \boxed{v_0 + at})t = \tfrac{1}{2}(2v_0 t + at^2)$$

$$x = v_0 t + \tfrac{1}{2}at^2 \qquad \text{(constant acceleration)} \qquad (2.8)$$

You can verify that Equation 2.8 gives the displacement of the speedboat directly without the intermediate step of determining the final velocity. The first term ($v_0 t$) on the right side of this equation represents the displacement that would result if the acceleration were zero and the velocity remained constant at its initial value of v_0. The second term ($\frac{1}{2}at^2$) gives the additional displacement that arises because the velocity changes (a is not zero) to values that are different from its initial value. We now turn to another example of accelerated motion.

Example 6 Catapulting a Jet

A jet is taking off from the deck of an aircraft carrier, as Figure 2.12 shows. Starting from rest, the jet is catapulted with a constant acceleration of $+31$ m/s^2 along a straight line and reaches a velocity of $+62$ m/s. Find the displacement of the jet.

Reasoning The data are as follows:

Jet Data				
x	a	v	v_0	t
?	$+31$ m/s^2	$+62$ m/s	0 m/s	

The initial velocity v_0 is zero, since the jet starts from rest. The displacement x of the aircraft can be obtained from $x = \frac{1}{2}(v_0 + v)t$, if we can determine the time t during which the plane is being accelerated. But t is controlled by the value of the acceleration. With larger accelerations, the jet reaches its final velocity in shorter times, as can be seen by solving Equation 2.4 ($v = v_0 + at$) for t.

Solution Solving Equation 2.4 for t, we find

$$t = \frac{v - v_0}{a} = \frac{62 \text{ m/s} - 0 \text{ m/s}}{31 \text{ m/s}^2} = 2.0 \text{ s}$$

Since the time is now known, the displacement can be found by using Equation 2.7:

$$x = \tfrac{1}{2}(v_0 + v)t = \tfrac{1}{2}(0 \text{ m/s} + 62 \text{ m/s})(2.0 \text{ s}) = \boxed{+62 \text{ m}} \qquad (2.7)$$

Concept Simulation 2.1

This simulation compares motion at a constant velocity to motion at a constant acceleration. You can see the effect of setting the acceleration to different values.

Related Homework: Problems 24, 26

Go to
www.wiley.com/college/cutnell

The physics of **catapulting a jet from an aircraft carrier.**

Problem solving insight
"Implied data" are important. For instance, in Example 6 the phrase "starting from rest" means that the initial velocity is zero ($v_0 = \mathbf{0}$ m/s).

Figure 2.12 (*a*) A plane is being launched from an aircraft carrier. (*b*) During the launch, a catapult accelerates the jet down the flight deck. (© George Hall/Corbis Images)

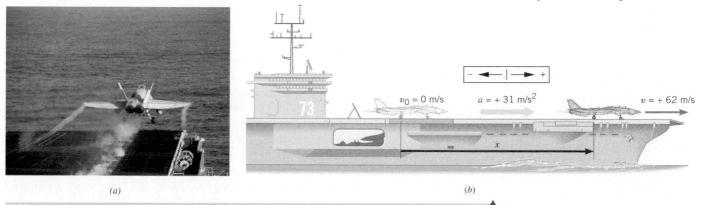

$v_0 = 0$ m/s $a = +31$ m/s^2 $v = +62$ m/s

x

(a)   *(b)*

Table 2.1 *Equations of Kinematics for Constant Acceleration*

Equation Number	Equation	Variables				
		x	a	v	v_0	t
(2.4)	$v = v_0 + at$	—	✓	✓	✓	✓
(2.7)	$x = \frac{1}{2}(v_0 + v)t$	✓	—	✓	✓	✓
(2.8)	$x = v_0 t + \frac{1}{2}at^2$	✓	✓	—	✓	✓
(2.9)	$v^2 = v_0^2 + 2ax$	✓	✓	✓	✓	—

When a, v, and v_0 are known, but the time t is not known, as in Example 6, it is possible to calculate the displacement x in a single step. Solving Equation 2.4 for the time $[t = (v - v_0)/a]$ and then substituting into Equation 2.7 $[x = \frac{1}{2}(v_0 + v)t]$ reveals that

$$x = \tfrac{1}{2}(v_0 + v)t = \tfrac{1}{2}(v_0 + v)\boxed{\frac{v - v_0}{a}} = \frac{v^2 - v_0^2}{2a}$$

Solving for v^2 shows that

$$v^2 = v_0^2 + 2ax \qquad \text{(constant acceleration)} \tag{2.9}$$

It is a straightforward exercise to verify that Equation 2.9 can be used to find the displacement of the jet in Example 6 without having to solve first for the time.

Table 2.1 presents a summary of the equations that we have been considering. These equations are called the ***equations of kinematics.*** Each equation contains four variables, as indicated by the check marks (✓) in the table. The next section shows how to apply the equations of kinematics.

✔ **Check Your Understanding 3**

A motorcycle starts from rest and has a constant acceleration. In a certain time interval, its displacement triples. In the same time interval, by what factor does its velocity increase? *(The answer is given at the end of the book.)*

Background: When the acceleration is constant, the equations of kinematics given in Table 2.1 apply.

For similar questions (including calculational counterparts), consult Self-Assessment Test 2.2. This test is described at the end of Section 2.5.

2.5 *Applications of the Equations of Kinematics*

The equations of kinematics can be used for any moving object, as long as the acceleration of the object is constant. However, to avoid errors when using these equations, it helps to follow a few sensible guidelines and to be alert for a few situations that can arise during your calculations.

Decide at the start which directions are to be called positive (+) and negative (−) relative to a conveniently chosen coordinate origin. This decision is arbitrary, but important because displacement, velocity, and acceleration are vectors, and their directions must always be taken into account. In the examples that follow, the positive and negative directions will be shown in the drawings that accompany the problems. It does not matter which direction is chosen to be positive. However, once the choice is made, it should not be changed during the course of the calculation.

As you reason through a problem before attempting to solve it, be sure to interpret the terms "decelerating" or "deceleration" correctly, should they occur in the problem statement. These terms are the source of frequent confusion, and Conceptual Example 7 offers help in understanding them.

Need more practice?

Interactive LearningWare 2.1
Here we revisit Example 5, in which the positive direction was assumed to point toward the right (see Figure 2.11). Now, however, we assume that the positive direction points to the left and will show how to deal with such a choice and how to interpret the results.

Go to
www.wiley.com/college/cutnell
for an interactive solution.

Conceptual Example 7 Deceleration Versus Negative Acceleration

A car is traveling along a straight road and is decelerating. Does the car's acceleration *a* necessarily have a negative value?

Reasoning and Solution We begin with the meaning of the term "decelerating," which has nothing to do with whether the acceleration *a* is positive or negative. The term means only that the acceleration vector points opposite to the velocity vector and indicates that the moving object is slowing down. When a moving object slows down, its instantaneous speed (the magnitude of the instantaneous velocity) decreases. One possibility is that the velocity vector of the car points to the right, in the positive direction, as Figure 2.13*a* shows. The term "decelerating" implies that the acceleration vector points opposite, or to the left, which is the negative direction. Here, the value of the acceleration *a* would indeed be negative. However, there is another possibility. The car could be traveling to the left, as in Figure 2.13*b*. Now, since the velocity vector points to the left, the acceleration vector would point opposite or to the right, according to the meaning of the term "decelerating." But right is the positive direction, so the acceleration *a* would have a positive value in Figure 2.13*b*. We see, then, that *a decelerating object does not necessarily have a negative acceleration.*

Related Homework: *Problems 20, 38*

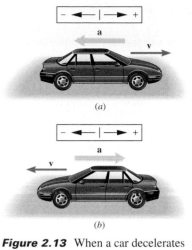

Figure 2.13 When a car decelerates along a straight road, the acceleration vector points opposite to the velocity vector, as Conceptual Example 7 discusses.

Sometimes there are two possible answers to a kinematics problem, each answer corresponding to a different situation. Example 8 discusses one such case.

Example 8 An Accelerating Spacecraft

The spacecraft shown in Figure 2.14*a* is traveling with a velocity of $+3250$ m/s. Suddenly the retrorockets are fired, and the spacecraft begins to slow down with an acceleration whose mag-

The physics of **the acceleration caused by a retrorocket.**

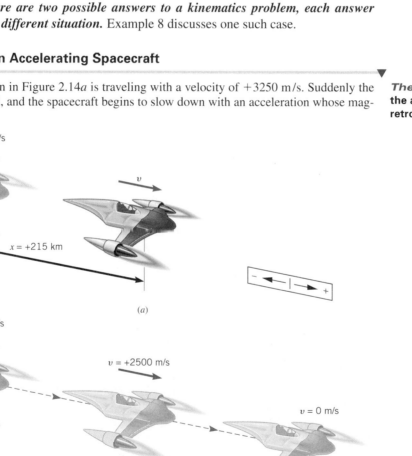

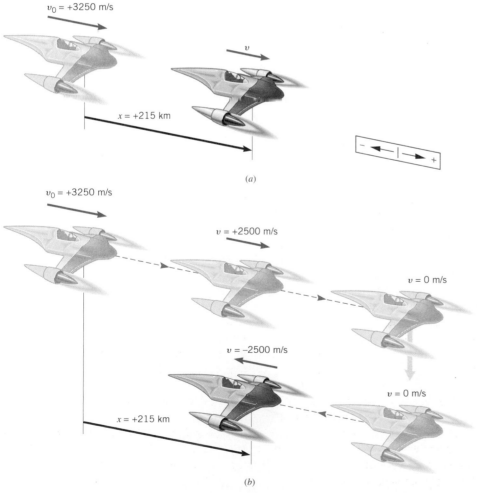

Figure 2.14 (*a*) Because of an acceleration of -10.0 m/s^2, the spacecraft changes its velocity from v_0 to v. (*b*) Continued firing of the retrorockets changes the direction of the craft's motion.

nitude is 10.0 m/s². What is the velocity of the spacecraft when the displacement of the craft is +215 km, relative to the point where the retrorockets began firing?

Reasoning Since the spacecraft is slowing down, the acceleration must be opposite to the velocity. The velocity points to the right in the drawing, so the acceleration points to the left, in the negative direction; thus, $a = -10.0$ m/s². The three known variables are listed as follows:

Spacecraft Data				
x	a	v	v_0	t
+215 000 m	-10.0 m/s²	?	+3250 m/s	

The final velocity v of the spacecraft can be calculated using Equation 2.9, since it contains the four pertinent variables.

Solution From Equation 2.9 ($v^2 = v_0^2 + 2ax$), we find that

$$v = \pm\sqrt{v_0^2 + 2ax} = \pm\sqrt{(3250 \text{ m/s})^2 + 2(-10.0 \text{ m/s}^2)(215\ 000 \text{ m})}$$
$$= \boxed{+2500 \text{ m/s}} \quad \text{and} \quad \boxed{-2500 \text{ m/s}}$$

Both of these answers correspond to the *same* displacement ($x = +215$ km), but each arises in a different part of the motion. The answer $v = +2500$ m/s corresponds to the situation in Figure 2.14a, where the spacecraft has slowed to a speed of 2500 m/s, but is still traveling to the right. The answer $v = -2500$ m/s arises because the retrorockets eventually bring the spacecraft to a momentary halt and cause it to reverse its direction. Then it moves to the left, and its speed increases due to the continually firing rockets. After a time, the velocity of the craft becomes $v = -2500$ m/s, giving rise to the situation in Figure 2.14b. In both parts of the drawing the spacecraft has the same displacement, but a greater travel time is required in part b compared to part a.

The motion of two objects may be interrelated, so they share a common variable. The fact that the motions are interrelated is an important piece of information. In such cases, data for only two variables need be specified for each object. See Interactive LearningWare 2.2 for an example that illustrates this point.

Often the motion of an object is divided into segments, each with a different acceleration. When solving such problems, it is important to realize that the final velocity for one segment is the initial velocity for the next segment, as Example 9 illustrates.

Example 9   A Motorcycle Ride

A motorcycle, starting from rest, has an acceleration of +2.6 m/s². After the motorcycle has traveled a distance of 120 m, it slows down with an acceleration of -1.5 m/s² until its velocity is +12 m/s (see Figure 2.15). What is the total displacement of the motorcycle?

Reasoning The total displacement is the sum of the displacements for the first ("speeding up") and second ("slowing down") segments. The displacement for the first segment is +120 m. The displacement for the second segment can be found if the initial velocity for this segment can be determined, since values for two other variables are already known ($a = -1.5$ m/s² and $v = +12$ m/s). The initial velocity for the second segment can be determined, since it is the final velocity of the first segment.

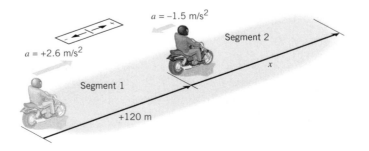

Figure 2.15 This motorcycle ride consists of two segments, each with a different acceleration.

Solution Recognizing that the motorcycle starts from rest ($v_0 = 0$ m/s), we can determine the final velocity v of the first segment from the given data:

Segment 1 Data				
x	a	v	v_0	t
$+120$ m	$+2.6$ m/s^2	?	0 m/s	

From Equation 2.9 ($v^2 = v_0^2 + 2ax$), it follows that

$$v = \pm\sqrt{v_0^2 + 2ax} = \pm\sqrt{(0 \text{ m/s})^2 + 2(2.6 \text{ m/s}^2)(120 \text{ m})} = +25 \text{ m/s}$$

Now we can use $+25$ m/s as the initial velocity for the second segment, along with the remaining data listed below:

Segment 2 Data				
x	a	v	v_0	t
?	-1.5 m/s^2	$+12$ m/s	$+25$ m/s	

The displacement for segment 2 can be obtained by solving $v^2 = v_0^2 + 2ax$ for x.

$$x = \frac{v^2 - v_0^2}{2a} = \frac{(12 \text{ m/s})^2 - (25 \text{ m/s})^2}{2(-1.5 \text{ m/s}^2)} = +160 \text{ m}$$

The total displacement of the motorcycle is 120 m + 160 m = $\boxed{280 \text{ m}}$.

As this Harris hawk comes in for a landing on its trainer's wrist, its velocity vector points in the same direction as its motion. However, since the hawk is slowing down, its acceleration vector points opposite to the velocity vector. (© Carl D. Walsh/ Aurora & Quanta Productions)

Now that we have seen how the equations of kinematics are applied to various situations, it's a good idea to summarize the reasoning strategy that has been used. This strategy, which is outlined below, will also be used when we consider freely falling bodies in Section 2.6 and two-dimensional motion in Chapter 3.

Reasoning Strategy

Applying the Equations of Kinematics

1. Make a drawing to represent the situation being studied. A drawing helps us to see what's happening.

2. Decide which directions are to be called positive ($+$) and negative ($-$) relative to a conveniently chosen coordinate origin. Do not change your decision during the course of a calculation.

3. In an organized way, write down the values (with appropriate plus and minus signs) that are given for any of the five kinematic variables (x, a, v, v_0, and t). Be on the alert for "implied data," such as the phrase "starts from rest," which means that the value of the initial velocity is $v_0 = 0$ m/s. The data summary boxes used in the examples in the text are a good way of keeping track of this information. In addition, identify the variables that you are being asked to determine.

4. Before attempting to solve a problem, verify that the given information contains values for at least three of the five kinematic variables. Once the three known variables are identified along with the desired unknown variable, the appropriate relation from Table 2.1 can be selected. Remember that the motion of two objects may be interrelated, so they may share a common variable. The fact that the motions are interrelated is an important piece of information. In such cases, data for only two variables need be specified for each object.

5. When the motion of an object is divided into segments, as in Example 9, remember that the final velocity of one segment is the initial velocity for the next segment.

6. Keep in mind that there may be two possible answers to a kinematics problem as, for instance, in Example 8. Try to visualize the different physical situations to which the answers correspond.

Air-filled tube
(a)

Evacuated tube
(b)

Figure 2.16 (*a*) In the presence of air resistance, the acceleration of the rock is greater than that of the paper. (*b*) In the absence of air resistance, both the rock and the paper have the same acceleration.

Self-Assessment Test 2.2

Test your understanding of the equations of kinematics for constant acceleration, which are discussed in Sections 2.4 and 2.5.

Go to **www.wiley.com/college/cutnell**

2.6 Freely Falling Bodies

Everyone has observed the effect of gravity as it causes objects to fall downward. In the absence of air resistance, it is found that all bodies at the same location above the earth fall vertically with the same acceleration. Furthermore, if the distance of the fall is small compared to the radius of the earth, the acceleration remains essentially constant throughout the descent. This idealized motion, in which air resistance is neglected and the acceleration is nearly constant, is known as ***free-fall.*** Since the acceleration is constant in free-fall, the equations of kinematics can be used.

The acceleration of a freely falling body is called the ***acceleration due to gravity,*** and its magnitude (without any algebraic sign) is denoted by the symbol g. The acceleration due to gravity is directed downward, toward the center of the earth. Near the earth's surface, g is approximately

$$g = 9.80 \text{ m/s}^2 \quad \text{or} \quad 32.2 \text{ ft/s}^2$$

Unless circumstances warrant otherwise, we will use either of these values for g in subsequent calculations. In reality, however, g decreases with increasing altitude and varies slightly with latitude.

Figure 2.16*a* shows the well-known phenomenon of a rock falling faster than a sheet of paper. The effect of air resistance is responsible for the slower fall of the paper, for when air is removed from the tube, as in Figure 2.16*b*, the rock and the paper have exactly the same acceleration due to gravity. In the absence of air, the rock and the paper both exhibit free-fall motion. Free-fall is closely approximated for objects falling near the surface of the moon, where there is no air to retard the motion. A nice demonstration of lunar free-fall was performed by astronaut David Scott, who dropped a hammer and a feather simultaneously from the same height. Both experienced the same acceleration due to lunar gravity and consequently hit the ground at the same time. The acceleration due to gravity near the surface of the moon is approximately one-sixth as large as that on the earth.

When the equations of kinematics are applied to free-fall motion, it is natural to use the symbol y for the displacement, since the motion occurs in the vertical or y direction. Thus, when using the equations in Table 2.1 for free-fall motion, we will simply replace x with y. There is no significance to this change. The equations have the same algebraic form for either the horizontal or vertical direction, provided that the acceleration remains constant during the motion. We now turn our attention to several examples that illustrate how the equations of kinematics are applied to freely falling bodies.

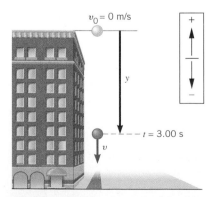

$v_0 = 0$ m/s

y

$t = 3.00$ s

v

Figure 2.17 The stone, starting with zero velocity at the top of the building, is accelerated downward by gravity.

Example 10 A Falling Stone

A stone is dropped from rest from the top of a tall building, as Figure 2.17 indicates. After 3.00 s of free-fall, what is the displacement y of the stone?

Reasoning The upward direction is chosen as the positive direction. The three known variables are shown in the box below. The initial velocity v_0 of the stone is zero, because the stone is dropped from rest. The acceleration due to gravity is negative, since it points downward in the negative direction.

Stone Data				
y	a	v	v_0	t
?	-9.80 m/s^2		0 m/s	3.00 s

Equation 2.8 contains the appropriate variables and offers a direct solution to the problem. Since the stone moves downward, and upward is the positive direction, we expect the displacement *y* to have a negative value.

Solution Using Equation 2.8, we find that

$$y = v_0 t + \tfrac{1}{2}at^2 = (0 \text{ m/s})(3.00 \text{ s}) + \tfrac{1}{2}(-9.80 \text{ m/s}^2)(3.00 \text{ s})^2 = \boxed{-44.1 \text{ m}}$$

The answer for *y* is negative, as expected.

▲

Example 11 The Velocity of a Falling Stone

After 3.00 s of free fall, what is the velocity *v* of the stone in Figure 2.17?

Reasoning Because of the acceleration due to gravity, the magnitude of the stone's downward velocity increases by 9.80 m/s during each second of free-fall. The data for the stone are the same as in Example 10, and Equation 2.4 offers a direct solution for the final velocity. Since the stone is moving downward in the negative direction, the value determined for *v* should be negative.

Solution Using Equation 2.4, we obtain

$$v = v_0 + at = 0 \text{ m/s} + (-9.80 \text{ m/s}^2)(3.00 \text{ s}) = \boxed{-29.4 \text{ m/s}}$$

The velocity is negative, as expected.

▲

The acceleration due to gravity is always a downward-pointing vector. It describes how the speed increases for an object that is falling freely downward. This same acceleration also describes how the speed decreases for an object moving upward under the influence of gravity alone, in which case the object eventually comes to a momentary halt and then falls back to earth. Examples 12 and 13 show how the equations of kinematics are applied to an object that is moving upward under the influence of gravity.

Example 12 How High Does It Go?

A football game customarily begins with a coin toss to determine who kicks off. The referee tosses the coin up with an initial speed of 5.00 m/s. In the absence of air resistance, how high does the coin go above its point of release?

Reasoning The coin is given an upward initial velocity, as in Figure 2.18. But the acceleration due to gravity points downward. Since the velocity and acceleration point in opposite directions, the coin slows down as it moves upward. Eventually, the velocity of the coin becomes *v* = 0 m/s at the highest point. Assuming that the upward direction is positive, the data can be summarized as shown below:

Coin Data

y	*a*	*v*	v_0	*t*
?	-9.80 m/s^2	0 m/s	+5.00 m/s	

With these data, we can use Equation 2.9 ($v^2 = v_0^2 + 2ay$) to find the maximum height y.

Solution Rearranging Equation 2.9, we find that the maximum height of the coin above its release point is

$$y = \frac{v^2 - v_0^2}{2a} = \frac{(0 \text{ m/s})^2 - (5.00 \text{ m/s})^2}{2(-9.80 \text{ m/s}^2)} = \boxed{1.28 \text{ m}}$$

Example 13 How Long Is It in the Air?

In Figure 2.18, what is the total time the coin is in the air before returning to its release point?

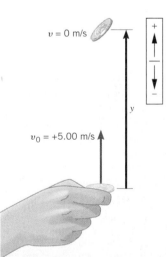

▼

Problem solving insight
"Implied data" are important. In Example 12, for instance, the phrase "how high does the coin go" refers to the maximum height, which occurs when the final velocity *v* in the vertical direction is *v* = 0 m/s.

v = 0 m/s

v_0 = +5.00 m/s

▲ **Figure 2.18** At the start of a football game, a referee tosses a coin upward with an initial velocity of v_0 = +5.00 m/s. The velocity of the coin is momentarily zero when the coin reaches its maximum height.

Reasoning During the time the coin travels upward, gravity causes its speed to decrease to zero. On the way down, however, gravity causes the coin to regain the lost speed. Thus, the time for the coin to go up is equal to the time for it to come down. In other words, the total travel time is twice the time for the upward motion. The data for the coin during the upward trip are the same as in Example 12. With these data, we can use Equation 2.4 ($v = v_0 + at$) to find the upward travel time.

Solution Rearranging Equation 2.4, we find that

$$t = \frac{v - v_0}{a} = \frac{0 \text{ m/s} - 5.00 \text{ m/s}}{-9.80 \text{ m/s}^2} = 0.510 \text{ s}$$

The total up-and-down time is twice this value, or $\boxed{1.02 \text{ s}}$.

It is possible to determine the total time by another method. When the coin is tossed upward and returns to its release point, the displacement for the *entire trip* is $y = 0$ m. With this value for the displacement, Equation 2.8 ($y = v_0 t + \frac{1}{2}at^2$) can be used to find the time for the entire trip directly.

Examples 12 and 13 illustrate that the expression "freely falling" does not necessarily mean an object is falling down. A freely falling object is any object moving either upward or downward under the influence of gravity alone. In either case, the object always experiences the same *downward acceleration* due to gravity, a fact that is the focus of the next example.

Conceptual Example 14   **Acceleration Versus Velocity**

There are three parts to the motion of the coin in Figure 2.18. On the way up, the coin has a velocity vector that is directed upward and has a decreasing magnitude. At the top of its path, the coin momentarily has a zero velocity. On the way down, the coin has a downward-pointing velocity vector with an increasing magnitude. In the absence of air resistance, does the acceleration of the coin, like the velocity, change from one part of the motion to another?

Reasoning and Solution Since air resistance is absent, the coin is in free-fall motion. Therefore, the acceleration vector is that due to gravity and has the same magnitude and the same direction at all times. It has a magnitude of 9.80 m/s^2 and points downward during both the upward and downward portions of the motion. Furthermore, just because the coin's instantaneous velocity is zero at the top of the motional path, don't think that the acceleration vector is also zero there. Acceleration is the rate at which velocity changes, and the velocity at the top is changing, even though at one instant it is zero. In fact, the acceleration at the top has the same magnitude of 9.80 m/s^2 and the same downward direction as during the rest of the motion. Thus, *the coin's velocity vector changes from moment to moment, but its acceleration vector does not change.*

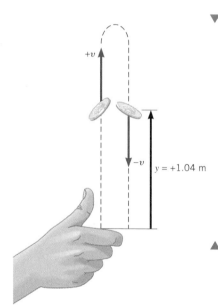

Figure 2.19 For a given displacement along the motional path, the upward speed of the coin is equal to its downward speed, but the two velocities point in opposite directions.

Concept Simulation 2.3

In Figure 2.19 changing the initial velocity affects how high the coin goes and its total flight time. Here you can change the initial velocity and see how the motion of the coin changes.

Related Homework: Conceptual Question 13, Problems 39, 42, 50, 86

Go to
www.wiley.com/college/cutnell

The motion of an object that is thrown upward and eventually returns to earth contains a symmetry that is useful to keep in mind from the point of view of problem solving. The calculations just completed indicate that a time symmetry exists in free-fall motion, in the sense that the time required for the object to reach maximum height equals the time for it to return to its starting point.

A type of symmetry involving the speed also exists. Figure 2.19 shows the coin considered in Examples 12 and 13. At any displacement y above the point of release, the coin's speed during the upward trip equals the speed at the same point during the downward trip. For instance, when $y = +1.04$ m, Equation 2.9 gives two possible values for the final velocity v, assuming that the initial velocity is $v_0 = +5.00$ m/s:

$$v^2 = v_0^2 + 2ay = (5.00 \text{ m/s})^2 + 2(-9.80 \text{ m/s}^2)(1.04 \text{ m}) = 4.62 \text{ m}^2/\text{s}^2$$
$$v = \pm 2.15 \text{ m/s}$$

The value $v = +2.15$ m/s is the velocity of the coin on the upward trip, and $v = -2.15$ m/s is the velocity on the downward trip. The speed in both cases is identical and equals 2.15 m/s. Likewise, the speed just as the coin returns to its point of release is 5.00 m/s, which equals the initial speed. This symmetry involving the speed arises because the coin loses 9.80 m/s in speed each second on the way up and gains back the same amount each second on the way down. In Conceptual Example 15, we use just this kind of symmetry to guide our reasoning as we analyze the motion of a pellet shot from a gun.

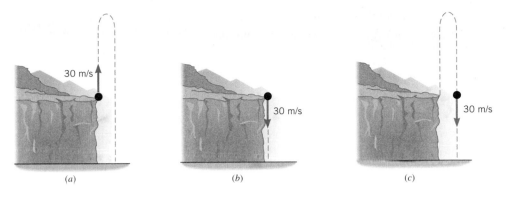

Figure 2.20 (*a*) From the edge of a cliff, a pellet is fired straight upward from a gun. The pellet's initial speed is 30 m/s. (*b*) The pellet is fired straight downward with an initial speed of 30 m/s. (*c*) In Conceptual Example 15 this drawing plays the central role in reasoning that is based on symmetry.

Conceptual Example 15 **Taking Advantage of Symmetry**

Figure 2.20*a* shows a pellet, having been fired from a gun, moving straight upward from the edge of a cliff. The initial speed of the pellet is 30 m/s. It goes up and then falls back down, eventually hitting the ground beneath the cliff. In Figure 2.20*b* the pellet has been fired straight downward at the same initial speed. In the absence of air resistance, does the pellet in part *b* strike the ground beneath the cliff with a smaller, a greater, or the same speed as the pellet in part *a*?

Reasoning and Solution Because air resistance is absent, the motion is that of free-fall, and the symmetry inherent in free-fall motion offers an immediate answer to the question. Figure 2.20*c* shows why. This part of the drawing shows the pellet after it has been fired upward and then fallen back down to its starting point. Symmetry indicates that the speed in part *c* is the same as in part *a*—namely, 30 m/s. Thus, part *c* is just like part *b*, where the pellet is actually fired downward with a speed of 30 m/s. Consequently, whether the pellet is fired as in part *a* or part *b*, it starts to move downward from the cliff edge at a speed of 30 m/s. In either case, there is the same acceleration due to gravity and the same displacement from the cliff edge to the ground below. Under these conditions, ***the pellet reaches the ground with the same speed no matter in which vertical direction it is fired initially.***

Related Homework: *Problems 43, 46*

✔ **Check Your Understanding 4**

A ball is thrown straight upward with a velocity v_0 and in a time t reaches the top of its flight path, which is a displacement **y** above the launch point. With a launch velocity of $2v_0$, what would be the time required to reach the top of its flight path and what would be the displacement of the top point above the launch point? *(The answer is given at the end of the book.)*

    (a) 4*t* and 2**y**   (b) 2*t* and 4**y**   (c) 2*t* and 2**y**   (d) 4*t* and 4**y**   (e) *t* and 2**y**

Background: The acceleration due to gravity and the equations of kinematics given in Table 2.1 are the focus of this question.

For similar questions (including calculational counterparts), consult Self-Assessment Test 2.3. This test is described at the end of Section 2.7.

2.7 *Graphical Analysis of Velocity and Acceleration*

Graphical techniques are helpful in understanding the concepts of velocity and acceleration. Suppose a bicyclist is riding with a constant velocity of $v = +4$ m/s. The position x of the bicycle can be plotted along the vertical axis of a graph, while the time t is plotted along the horizontal axis. Since the position of the bike increases by 4 m every second, the graph of x versus t is a straight line. Furthermore, if the bike is assumed to be at $x = 0$ m when $t = 0$ s, the straight line passes through the origin, as Figure 2.21 shows. Each point on this line gives the position of the bike at a particular time. For instance, at $t = 1$ s the position is 4 m, while at $t = 3$ s the position is 12 m.

    In constructing the graph in Figure 2.21, we used the fact that the velocity was +4 m/s. Suppose, however, that we were given this graph, but did not have prior knowledge of the velocity. The velocity could be determined by considering what happens to the bike between the times of 1 and 3 s, for instance. The change in time is $\Delta t = 2$ s. During this

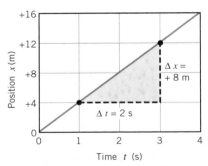

Figure 2.21 A graph of position vs. time for an object moving with a constant velocity of $v = \Delta x/\Delta t = +4$ m/s.

time interval, the position of the bike changes from $+4$ to $+12$ m, and the change in position is $\Delta x = +8$ m. The ratio $\Delta x / \Delta t$ is called the *slope* of the straight line.

$$\text{Slope} = \frac{\Delta x}{\Delta t} = \frac{+8 \text{ m}}{2 \text{ s}} = +4 \text{ m/s}$$

Notice that the slope is equal to the velocity of the bike. This result is no accident, because $\Delta x / \Delta t$ is the definition of average velocity (see Equation 2.2). Thus, for an object moving with a constant velocity, the slope of the straight line in a position–time graph gives the velocity. Since the position–time graph is a straight line, any time interval Δt can be chosen to calculate the velocity. Choosing a different Δt will yield a different Δx, but the velocity $\Delta x / \Delta t$ will not change. In the real world, objects rarely move with a constant velocity at all times, as the next example illustrates.

Example 16   A Bicycle Trip

A bicyclist maintains a constant velocity on the outgoing leg of a trip, zero velocity while stopped, and another constant velocity on the way back. Figure 2.22 shows the corresponding position–time graph. Using the time and position intervals indicated in the drawing, obtain the velocities for each segment of the trip.

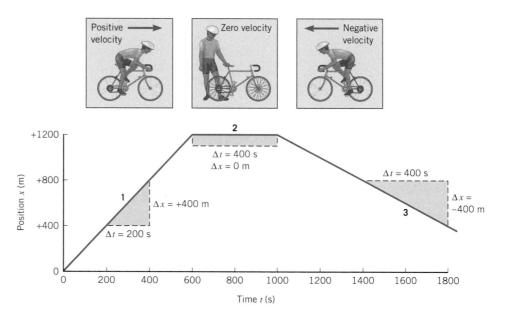

Figure 2.22 This position-vs.-time graph consists of three straight-line segments, each corresponding to a different constant velocity.

Reasoning The average velocity $\overline{v}$ is equal to the displacement Δx divided by the elapsed time Δt, $\overline{v} = \Delta x / \Delta t$. The displacement is the final position minus the initial position, which is a positive number for segment 1 and a negative number for segment 3. Note for segment 2 that $\Delta x = 0$ m, since the bicycle is at rest. The drawing shows values for Δx and Δt for each of the three segments.

Solution The average velocities for the three segments are

Segment 1 $\overline{v} = \dfrac{\Delta x}{\Delta t} = \dfrac{800 \text{ m} - 400 \text{ m}}{400 \text{ s} - 200 \text{ s}} = \dfrac{+400 \text{ m}}{200 \text{ s}} = \boxed{+2 \text{ m/s}}$

Segment 2 $\overline{v} = \dfrac{\Delta x}{\Delta t} = \dfrac{1200 \text{ m} - 1200 \text{ m}}{1000 \text{ s} - 600 \text{ s}} = \dfrac{0 \text{ m}}{400 \text{ s}} = \boxed{0 \text{ m/s}}$

Segment 3 $\overline{v} = \dfrac{\Delta x}{\Delta t} = \dfrac{400 \text{ m} - 800 \text{ m}}{1800 \text{ s} - 1400 \text{ s}} = \dfrac{-400 \text{ m}}{400 \text{ s}} = \boxed{-1 \text{ m/s}}$

In the second segment of the journey the velocity is zero, reflecting the fact that the bike is stationary. Since the position of the bike does not change, segment 2 is a horizontal line that has a zero slope. In the third part of the motion the velocity is negative, because the position of the bike decreases from $x = +800$ m to $x = +400$ m during the 400-s interval shown in the graph. As a result, segment 3 has a negative slope, and the velocity is negative.

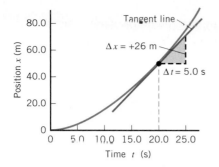

Figure 2.23 When the velocity is changing, the position-vs.-time graph is a curved line. The slope $\Delta x/\Delta t$ of the tangent line drawn to the curve at a given time is the instantaneous velocity at that time.

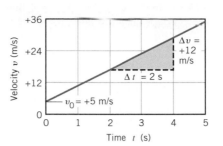

Figure 2.24 A velocity-vs.-time graph that applies to an object with an acceleration of $\Delta v/\Delta t = +6$ m/s². The initial velocity is $v_0 = +5$ m/s when $t = 0$ s.

If the object is accelerating, its velocity is changing. When the velocity is changing, the x-versus-t graph is not a straight line, but is a curve, perhaps like that in Figure 2.23. This curve was drawn using Equation 2.8 ($x = v_0 t + \frac{1}{2}at^2$), assuming an acceleration of $a = 0.26$ m/s² and an initial velocity of $v_0 = 0$ m/s. The velocity at any instant of time can be determined by measuring the slope of the curve at that instant. The slope at any point along the curve is defined to be the slope of the tangent line drawn to the curve at that point. For instance, in Figure 2.23 a tangent line is drawn at $t = 20.0$ s. To determine the slope of the tangent line, a triangle is constructed using an arbitrarily chosen time interval of $\Delta t = 5.0$ s. The change in x associated with this time interval can be read from the tangent line as $\Delta x = +26$ m. Therefore,

$$\text{Slope of tangent line} = \frac{\Delta x}{\Delta t} = \frac{+26 \text{ m}}{5.0 \text{ s}} = +5.2 \text{ m/s}$$

The slope of the tangent line is the instantaneous velocity, which in this case is $v = +5.2$ m/s. This graphical result can be verified by using Equation 2.4 with $v_0 = 0$ m/s. $v = at = (+0.26 \text{ m/s}^2)(20.0 \text{ s}) = +5.2$ m/s.

Insight into the meaning of acceleration can also be gained with the aid of a graphical representation. Consider an object moving with a constant acceleration of $a = +6$ m/s². If the object has an initial velocity of $v_0 = +5$ m/s, its velocity at any time is represented by Equation 2.4 as

$$v = v_0 + at = 5 \text{ m/s} + (6 \text{ m/s}^2)t$$

This relation is plotted as the velocity-versus-time graph in Figure 2.24. The graph of v versus t is a straight line that intercepts the vertical axis at $v_0 = 5$ m/s. The slope of this straight line can be calculated from the data shown in the drawing:

$$\text{Slope} = \frac{\Delta v}{\Delta t} = \frac{+12 \text{ m/s}}{2 \text{ s}} = +6 \text{ m/s}^2$$

The ratio $\Delta v/\Delta t$ is, by definition, equal to the average acceleration (Equation 2.4), so the slope of the straight line in a velocity–time graph is the average acceleration.

Concept Simulation 2.4

This simulation constructs the position-versus-time and velocity-versus-time graphs for a moving cart, according to your choice of values for the initial position, initial velocity, and acceleration. The graphs are created in synch with the cart's motion.

Related Homework: *Problem 64*

Go to **www.wiley.com/college/cutnell**

Self-Assessment Test 2.3

Test your understanding of the material in Sections 2.6 and 2.7:

• Freely Falling Bodies   • Graphical Analysis of Velocity and Acceleration

Go to **www.wiley.com/college/cutnell**

2.8 *Concepts & Calculations*

In this chapter we have studied the displacement, velocity, and acceleration vectors. We conclude now by presenting several examples that review some of the important features of these concepts. The three-part format of these examples stresses the role of conceptual understanding in problem solving. First, the problem statement is given. Then, there is a

concept question-and-answer section, followed by the solution section. The purpose of the concept question-and-answer section is to provide help in understanding the solution and to illustrate how a review of the concepts can help in anticipating some of the characteristics of the numerical answers.

Concepts & Calculations Example 17   Skydiving

A skydiver is falling straight down, along the negative *y* direction. (a) During the initial part of the fall, her speed increases from 16 to 28 m/s in 1.5 s, as in Figure 2.25*a*. (b) Later, her parachute opens, and her speed decreases from 48 to 26 m/s in 11 s, as in part *b* of the drawing. In both instances, determine the magnitude and direction of her average acceleration.

Concept Questions and Answers Is her average acceleration positive or negative when her speed is increasing?

Answer   Since her speed is increasing, the acceleration vector must point in the same direction as the velocity vector, which points in the negative *y* direction. Thus, the acceleration is negative.

Is her average acceleration positive or negative when her speed is decreasing?

Answer   Since her speed is decreasing, the acceleration vector must point opposite to the velocity vector. Since the velocity vector points in the negative *y* direction, the acceleration must point in the positive *y* direction. Thus, the acceleration is positive.

Solution

(a) Since the skydiver is moving in the negative *y* direction, her initial velocity is $v_0 = -16$ m/s and her final velocity is $v = -28$ m/s. Her average acceleration $\bar{a}$ is the change in the velocity divided by the elapsed time:

$$\bar{a} = \frac{v - v_0}{t} = \frac{-28 \text{ m/s} - (-16 \text{ m/s})}{1.5 \text{ s}} = \boxed{-8.0 \text{ m/s}^2} \tag{2.4}$$

As expected, her average acceleration is negative. Note that her acceleration is not that due to gravity (-9.8 m/s^2) because of wind resistance.

(b) Now the skydiver is slowing down, but still falling along the negative *y* direction. Her initial and final velocities are $v_0 = -48$ m/s and $v = -26$ m/s, respectively. The average acceleration for this phase of the motion is

$$\bar{a} = \frac{v - v_0}{t} = \frac{-26 \text{ m/s} - (-48 \text{ m/s})}{11 \text{ s}} = \boxed{+2.0 \text{ m/s}^2} \tag{2.4}$$

Now, as anticipated, her average acceleration is positive.

Concepts & Calculations Example 18   A Top-Fuel Dragster

A top-fuel dragster starts from rest and has a constant acceleration of 40.0 m/s^2. What are the (a) final velocities and (b) displacements of the dragster at the end of 2.0 s and at the end of twice this time, or 4.0 s?

Concept Questions and Answers At a time *t* the dragster has a certain velocity. When the time doubles to 2*t*, does the velocity also double?

Answer   Because the dragster has an acceleration of 40.0 m/s^2, its velocity changes by 40.0 m/s during each second of travel. Therefore, since the dragster starts from rest, the velocity is 40.0 m/s at the end of the 1st second, 2 × 40.0 m/s at the end of the 2nd second, 3 × 40.0 m/s at the end of the 3rd second, and so on. Thus, when the time doubles, the velocity also doubles.

When the time doubles to 2*t*, does the displacement of the dragster also double?

Answer   The displacement of the dragster is equal to its average velocity multiplied by the elapsed time. The average velocity $\bar{v}$ is just one-half the sum of the initial and final velocities, or $\bar{v} = \frac{1}{2}(v_0 + v)$. Since the initial velocity is zero, $v_0 = 0$ m/s and the average velocity is just one-half the final velocity, or $\bar{v} = \frac{1}{2}v$. However, as we have seen, the final velocity is proportional to the elapsed time, since when the time doubles, the final velocity also doubles. Therefore, the displacement, being the product of the average ve-

Figure 2.25 (*a*) A skydiver falls initially with her parachute unopened. (*b*) Later on, she opens her parachute. Her acceleration is different in the two parts of the motion. The initial and final velocities are v_0 and v, respectively.

locity and the time, is proportional to the time squared, or t^2. Consequently, as the time doubles, the displacement does not double, but increases by a factor of four.

Solution

(**a**) According to Equation 2.4, the final velocity v, the initial velocity v_0, the acceleration a, and the elapsed time t are related by $v = v_0 + at$. The final velocities at the two times are

[**$t = 2.0\ s$**] $v = v_0 + at = 0\ \text{m/s} + (40.0\ \text{m/s}^2)(2.0\ \text{s}) = \boxed{80\ \text{m/s}}$

[**$t = 4.0\ s$**] $v = v_0 + at = 0\ \text{m/s} + (40.0\ \text{m/s}^2)(4.0\ \text{s}) = \boxed{160\ \text{m/s}}$

We see that the velocity doubles when the time doubles, as expected.

(**b**) The displacement x is equal to the average velocity multiplied by the time, so

$$x = \underbrace{\tfrac{1}{2}(v_0 + v)t}_{\text{Average velocity}} = \tfrac{1}{2}vt$$

where we have used the fact that $v_0 = 0$ m/s. According to Equation 2.4, the final velocity is related to the acceleration by $v = v_0 + at$, or $v = at$, since $v_0 = 0$ m/s. Therefore, the displacement can be written as $x = \tfrac{1}{2}vt = \tfrac{1}{2}(at)t = \tfrac{1}{2}at^2$. The displacements at the two times are then

[**$t = 2.0\ s$**] $x = \tfrac{1}{2}at^2 = \tfrac{1}{2}(40.0\ \text{m/s}^2)(2.0\ \text{s})^2 = \boxed{80\ \text{m}}$

[**$t = 4.0\ s$**] $x = \tfrac{1}{2}at^2 = \tfrac{1}{2}(40.0\ \text{m/s}^2)(4.0\ \text{s})^2 = \boxed{320\ \text{m}}$

As predicted, the displacement at $t = 4.0$ s is four times the displacement at $t = 2.0$ s.

▲

At the end of the problem set for this chapter, you will find homework problems that contain both conceptual and quantitative parts. These problems are grouped under the heading *Concepts & Calculations, Group Learning Problems*. They are designed for use by students working alone or in small learning groups. The conceptual part of each problem provides a convenient focus for group discussions.

Concept Summary

This summary presents an abridged version of the chapter, including the important equations and all available learning aids. For convenient reference, the learning aids (including the text's examples) are placed next to or immediately after the relevant equation or discussion. The following learning aids may be found on-line at **www.wiley.com/college/cutnell**:

Interactive LearningWare examples are solved according to a five-step interactive format that is designed to help you develop problem-solving skills.	**Concept Simulations** are animated versions of text figures or animations that illustrate important concepts. You can control parameters that affect the display, and we encourage you to experiment.
Interactive Solutions offer specific models for certain types of problems in the chapter homework. The calculations are carried out interactively.	**Self-Assessment Tests** include both qualitative and quantitative questions. Extensive feedback is provided for both incorrect and correct answers, to help you evaluate your understanding of the material.

Topic	**Discussion**	**Learning Aids**
	2.1 Displacement	
Displacement	Displacement is a vector that points from an object's initial position to its final position. The magnitude of the displacement is the shortest distance between the two positions.	
	2.2 Speed and Velocity	
	The average speed of an object is the distance traveled by the object divided by the time required to cover the distance:	
Average speed	$$\text{Average speed} = \frac{\text{Distance}}{\text{Elapsed time}}$$	(2.1) **Example 1**
	The average velocity $\bar{\mathbf{v}}$ of an object is the object's displacement $\Delta\mathbf{x}$ divided by the elapsed time Δt:	
Average velocity	$$\bar{\mathbf{v}} = \frac{\Delta\mathbf{x}}{\Delta t}$$	(2.2) **Example 2**

Topic	Discussion	Learning Aids
	Average velocity is a vector that has the same direction as the displacement. When the elapsed time becomes infinitesimally small, the average velocity becomes equal to the instantaneous velocity **v**, the velocity at an instant of time:	
Instantaneous velocity	$$\mathbf{v} = \lim_{\Delta t \to 0} \frac{\Delta \mathbf{x}}{\Delta t} \qquad (2.3)$$	

2.3 Acceleration

Topic	Discussion	Learning Aids
	The average acceleration $\overline{\mathbf{a}}$ is a vector. It equals the change $\Delta \mathbf{v}$ in the velocity divided by the elapsed time Δt, the change in the velocity being the final minus the initial velocity:	
Average acceleration	$$\overline{\mathbf{a}} = \frac{\Delta \mathbf{v}}{\Delta t} \qquad (2.4)$$	**Examples 3, 4, 17**
	When Δt becomes infinitesimally small, the average acceleration becomes equal to the instantaneous acceleration **a**:	**Interactive Solution 2.17**
Instantaneous acceleration	$$\mathbf{a} = \lim_{\Delta t \to 0} \frac{\Delta \mathbf{v}}{\Delta t} \qquad (2.5)$$	
	Acceleration is the rate at which the velocity is changing.	

Use Self-Assessment Test 2.1 to evaluate your understanding of Sections 2.1–2.3.

2.4 Equations of Kinematics for Constant Acceleration

2.5 Applications of the Equations of Kinematics

Topic	Discussion	Learning Aids
	The equations of kinematics apply when an object moves with a constant acceleration along a straight line. These equations relate the displacement $x - x_0$, the acceleration a, the final velocity v, the initial velocity v_0, and the elapsed time $t - t_0$. Assuming that $x_0 = 0$ m at $t_0 = 0$ s, the equations of kinematics are	**Examples 5–9, 18**
		Concept Simulations 2.1, 2.2
Equations of kinematics	$$v = v_0 + at \qquad (2.4)$$	
	$$x = \tfrac{1}{2}(v_0 + v)t \qquad (2.7)$$	**Interactive LearningWare 2.1, 2.2**
	$$x = v_0 t + \tfrac{1}{2}at^2 \qquad (2.8)$$	
	$$v^2 = v_0^2 + 2ax \qquad (2.9)$$	**Interactive Solutions 2.29, 2.31**

Use Self-Assessment Test 2.2 to evaluate your understanding of Sections 2.4 and 2.5.

2.6 Freely Falling Bodies

Topic	Discussion	Learning Aids
	In free-fall motion, an object experiences negligible air resistance and a constant acceleration due to gravity. All objects at the same location above the earth have the same acceleration due to gravity. The acceleration due to gravity is directed toward the center of the earth and has a magnitude of approximately 9.80 m/s^2 near the earth's surface.	**Examples 10–15**
Acceleration due to gravity		**Concept Stimulation 2.3**
		Interactive Solutions 2.47, 2.49

2.7 Graphical Analysis of Velocity and Acceleration

Topic	Discussion	Learning Aids
	The slope of a plot of position versus time for a moving object gives the object's velocity. The slope of a plot of velocity versus time gives the object's acceleration.	**Example 16**
		Concept Simulation 2.4

Use Self-Assessment Test 2.3 to evaluate your understanding of Sections 2.6 and 2.7.

Conceptual Questions

1. A honeybee leaves the hive and travels 2 km before returning. Is the displacement for the trip the same as the distance traveled? If not, why not?

2. Two buses depart from Chicago, one going to New York and one to San Francisco. Each bus travels at a speed of 30 m/s. Do they have equal velocities? Explain.

3. Is the average speed of a vehicle a vector or a scalar quantity? Provide a reason for your answer.

4. Often, traffic lights are timed so that if you travel at a certain constant speed, you can avoid all red lights. Discuss how the timing of the lights is determined, considering that the distance between them varies from one light to the next.

5. One of the following statements is incorrect. (a) The car traveled around the track at a constant velocity. (b) The car traveled around the track at a constant speed. Which statement is incorrect and why?

6. Give an example from your own experience in which the velocity of an object is zero for just an instant of time, but its acceleration is not zero.

7. At a given instant of time, a car and a truck are traveling side by side in adjacent lanes of a highway. The car has a greater velocity than the truck. Does the car necessarily have a greater acceleration? Explain.

8. The average velocity for a trip has a positive value. Is it possible for the instantaneous velocity at any point during the trip to have a negative value? Justify your answer.

9. A runner runs half the remaining distance to the finish line every ten seconds. She runs in a straight line and does not ever reverse her direction. Does her acceleration have a constant magnitude? Give a reason for your answer.

10. An object moving with a constant acceleration can certainly slow down. But can an object ever come to a permanent halt if its acceleration truly remains constant? Explain.

11. Review **Concept Simulation 2.2** at **www.wiley.com/college/ cutnell** before answering this question. An experimental vehicle slows down and comes to a halt with an acceleration whose magnitude is 9.80 m/s². After reversing direction in a negligible amount of time, the vehicle speeds up with an acceleration of 9.80 m/s². Other

than being horizontal, how is this motion different, if at all, from the motion of a ball that is thrown straight upward, comes to a halt, and falls back to earth?

12. A ball is dropped from rest from the top of a building and strikes the ground with a speed v_f. From ground level, a second ball is thrown straight upward at the same instant that the first ball is dropped. The initial speed of the second ball is $v_0 = v_f$, the same speed with which the first ball will eventually strike the ground. Ignoring air resistance, decide whether the balls cross paths at half the height of the building, above the halfway point, or below the halfway point. Give your reasoning.

13. Review **Concept Simulation 2.3** at **www.wiley.com/college/ cutnell** before answering this question. Two objects are thrown vertically upward, first one, and then, a bit later, the other. Is it possible that both reach the same maximum height at the same instant? Account for your answer.

14. The muzzle velocity of a gun is the velocity of the bullet when it leaves the barrel. The muzzle velocity of one rifle with a short barrel is greater than the muzzle velocity of another rifle that has a longer barrel. In which rifle is the acceleration of the bullet larger? Explain your reasoning.

Problems

ssm Solution is in the Student Solutions Manual.   **www** Solution is available on the World Wide Web at www.wiley.com/college/cutnell
☤ **This icon represents a biomedical application.**

Section 2.1 Displacement, Section 2.2 Speed and Velocity

1. ssm A plane is sitting on a runway, awaiting takeoff. On an adjacent parallel runway, another plane lands and passes the stationary plane at a speed of 45 m/s. The arriving plane has a length of 36 m. By looking out of a window (very narrow), a passenger on the stationary plane can see the moving plane. For how long a time is the moving plane visible?

2. One afternoon, a couple walks three-fourths of the way around a circular lake, the radius of which is 1.50 km. They start at the west side of the lake and head due south to begin with. (a) What is the distance they travel? (b) What are the magnitude and direction (relative to due east) of the couple's displacement?

3. ssm A whale swims due east for a distance of 6.9 km, turns around and goes due west for 1.8 km, and finally turns around again and heads 3.7 km due east. (a) What is the total distance traveled by the whale? (b) What are the magnitude and direction of the displacement of the whale?

4. The Space Shuttle travels at a speed of about 7.6×10^3 m/s. The blink of an astronaut's eye lasts about 110 ms. How many football fields (length = 91.4 m) does the Shuttle cover in the blink of an eye?

5. As the earth rotates through one revolution, a person standing on the equator traces out a circular path whose radius is equal to the radius of the earth (6.38×10^6 m). What is the average speed of this person in (a) meters per second and (b) miles per hour?

6. In 1954 the English runner Roger Bannister broke the four-minute barrier for the mile with a time of 3:59.4 s (3 min and 59.4 s). In 1999 the Moroccan runner Hicham el-Guerrouj set a record of 3:43.13 s for the mile. If these two runners had run in the same race, each running the entire race at the average speed that earned him a place in the record books, el-Guerrouj would have won. By how many meters?

7. A tourist being chased by an angry bear is running in a straight

line toward his car at a speed of 4.0 m/s. The car is a distance d away. The bear is 26 m behind the tourist and running at 6.0 m/s. The tourist reaches the car safely. What is the maximum possible value for d?

* 8. In reaching her destination, a backpacker walks with an average velocity of 1.34 m/s, due west. This average velocity results because she hikes for 6.44 km with an average velocity of 2.68 m/s, due west, turns around, and hikes with an average velocity of 0.447 m/s, due east. How far east did she walk?

* 9. ssm www A woman and her dog are out for a morning run to the river, which is located 4.0 km away. The woman runs at 2.5 m/s in a straight line. The dog is unleashed and runs back and forth at 4.5 m/s between his owner and the river, until she reaches the river. What is the total distance run by the dog?

* 10. A car makes a trip due north for three-fourths of the time and due south one-fourth of the time. The average northward velocity has a magnitude of 27 m/s, and the average southward velocity has a magnitude of 17 m/s. What is the average velocity, magnitude and direction, for the entire trip?

** 11. You are on a train that is traveling at 3.0 m/s along a level straight track. Very near and parallel to the track is a wall that slopes upward at a 12° angle with the horizontal. As you face the window (0.90 m high, 2.0 m wide) in your compartment, the train is moving to the left, as the drawing indicates. The top edge of the wall first appears at window corner A and eventually disappears at window corner B. How much time passes between appearance and disappearance of the upper edge of the wall?

Section 2.3 Acceleration

12. For a standard production car, the highest road-tested acceleration ever reported occurred in 1993, when a Ford RS200 Evolution went from zero to 26.8 m/s (60 mi/h) in 3.275 s. Find the magnitude of the car's acceleration.

13. ssm A motorcycle has a constant acceleration of 2.5 m/s^2. Both the velocity and acceleration of the motorcycle point in the same direction. How much time is required for the motorcycle to change its speed from (a) 21 to 31 m/s, and (b) 51 to 61 m/s?

14. NASA has developed *Deep-Space 1* (DS-1), a spacecraft that is scheduled to rendezvous with the asteroid named 1992 KD (which orbits the sun millions of miles from the earth). The propulsion system of DS-1 works by ejecting high-speed argon ions out the rear of the engine. The engine slowly increases the velocity of DS-1 by about +9.0 m/s per day. (a) How much time (in days) will it take to increase the velocity of DS-1 by +2700 m/s? (b) What is the acceleration of DS-1 (in m/s^2)?

15. ssm A runner accelerates to a velocity of 5.36 m/s due west in 3.00 s. His average acceleration is 0.640 m/s^2, also directed due west. What was his velocity when he began accelerating?

16. The land speed record of 13.9 m/s (31 mi/h) for birds is held by the Australian emu. An emu running due south in a straight line at this speed slows down to a speed of 11.0 m/s in 3.0 s. (a) What is the direction of the bird's acceleration? (b) Assuming that the acceleration remains the same, what is the bird's velocity after an additional 4.0 s has elapsed?

* **17.** Consult **Interactive Solution 2.17** at **www.wiley.com/college/cutnell** before beginning this problem. A car is traveling along a straight road at a velocity of +36.0 m/s when its engine cuts out. For the next twelve seconds the car slows down, and its average acceleration is $\bar{a}_1$. For the next six seconds the car slows down further, and its average acceleration is $\bar{a}_2$. The velocity of the car at the end of the eighteen-second period is +28.0 m/s. The ratio of the average acceleration values is $\bar{a}_1/\bar{a}_2 = 1.50$. Find the velocity of the car at the end of the initial twelve-second interval.

** **18.** Two motorcycles are traveling due east with different velocities. However, four seconds later, they have the same velocity. During this four-second interval, motorcycle A has an average acceleration of 2.0 m/s^2 due east, while motorcycle B has an average acceleration of 4.0 m/s^2 due east. By how much did the speeds *differ* at the beginning of the four-second interval, and which motorcycle was moving faster?

Section 2.4 Equations of Kinematics for Constant Acceleration, Section 2.5 Applications of the Equations of Kinematics

19. In getting ready to slam-dunk the ball, a basketball player starts from rest and sprints to a speed of 6.0 m/s in 1.5 s. Assuming that the player accelerates uniformly, determine the distance he runs.

20. Review Conceptual Example 7 as background for this problem. A car is traveling to the left, which is the negative direction. The direction of travel remains the same throughout this problem. The car's initial speed is 27.0 m/s, and during a 5.0-s interval, it changes to a final speed of (a) 29.0 m/s and (b) 23.0 m/s. In each case, find the acceleration (magnitude and algebraic sign) and state whether or not the car is decelerating.

21. ssm A VW Beetle goes from 0 to 60.0 mi/h with an acceleration of +2.35 m/s^2. (a) How much time does it take for the Beetle to reach this speed? (b) A top-fuel dragster can go from 0 to 60.0 mi/h in 0.600 s. Find the acceleration (in m/s^2) of the dragster.

22. (a) What is the magnitude of the average acceleration of a skier who, starting from rest, reaches a speed of 8.0 m/s when going down a slope for 5.0 s? (b) How far does the skier travel in this time?

23. The left ventricle of the heart accelerates blood from rest to a velocity of +26 cm/s. (a) If the displacement of the blood during the acceleration is +2.0 cm, determine its acceleration (in cm/s^2). (b) How much time does blood take to reach its final velocity?

24. Consult **Concept Simulation 2.1** at **www.wiley.com/college/cutnell** for help in preparing for this problem. A cheetah is hunting. Its prey runs for 3.0 s at a constant velocity of +9.0 m/s. Starting from rest, what constant acceleration must the cheetah maintain in order to run the same distance as its prey runs in the same time?

25. ssm A jetliner, traveling northward, is landing with a speed of 69 m/s. Once the jet touches down, it has 750 m of runway in which to reduce its speed to 6.1 m/s. Compute the average acceleration (magnitude and direction) of the plane during landing.

26. Consult **Concept Simulation 2.1** at **www.wiley.com/college/cutnell** before starting this problem. The Kentucky Derby is held at the Churchill Downs track in Louisville, Kentucky. The track is one and one-quarter miles in length. One of the most famous horses to win this event was Secretariat. In 1973 he set a Derby record that has never been broken. His average acceleration during the last four quarter-miles of the race was +0.0105 m/s^2. His velocity at the start of the final mile ($x = +1609$ m) was about +16.58 m/s. The acceleration, although small, was very important to his victory. To assess its effect, determine the difference between the time he would have taken to run the final mile at a constant velocity of +16.58 m/s and the time he actually took. Although the track is oval in shape, assume it is straight for the purpose of this problem.

27. ssm www A speed ramp at an airport is basically a large conveyor belt on which you can stand and be moved along. The belt of one ramp moves at a constant speed such that a person who stands still on it leaves the ramp 64 s after getting on. Clifford is in a real hurry, however, and skips the speed ramp. Starting from rest with an acceleration of 0.37 m/s^2, he covers the same distance as the ramp does, but in one-fourth the time. What is the speed at which the belt of the ramp is moving?

* **28.** A drag racer, starting from rest, speeds up for 402 m with an acceleration of +17.0 m/s^2. A parachute then opens, slowing the car down with an acceleration of −6.10 m/s^2. How fast is the racer moving 3.50 × 10^2 m after the parachute opens?

* **29.** Review **Interactive Solution 2.29** at **www.wiley.com/college/cutnell** in preparation for this problem. Suppose a car is traveling at 20.0 m/s, and the driver sees a traffic light turn red. After 0.530 s has elapsed (the reaction time), the driver applies the brakes, and the car decelerates at 7.00 m/s^2. What is the stopping distance of the car, as measured from the point where the driver first notices the red light?

* **30.** A speedboat starts from rest and accelerates at +2.01 m/s^2 for 7.00 s. At the end of this time, the boat continues for an additional 6.00 s with an acceleration of +0.518 m/s^2. Following this, the boat accelerates at −1.49 m/s^2 for 8.00 s. (a) What is the velocity of the boat at $t = 21.0$ s? (b) Find the total displacement of the boat.

* **31. Interactive Solution 2.31** at **www.wiley.com/college/cutnell** offers help in modeling this problem. A car is traveling at a constant speed of 33 m/s on a highway. At the instant this car passes an entrance ramp, a second car enters the highway from the ramp. The second car starts from rest and has a constant acceleration. What acceleration must it maintain, so that the two cars meet for the first time at the next exit, which is 2.5 km away?

* **32.** A cab driver picks up a customer and delivers her 2.00 km away, on a straight route. The driver accelerates to the speed limit and, on reaching it, begins to decelerate at once. The magnitude of the deceleration is three times the magnitude of the acceleration. Find the lengths of the acceleration and deceleration phases.

* 33. Along a straight road through town, there are three speed-limit signs They occur in the following order: 55, 35, and 25 mi/h, with the 35-mi/h sign being midway between the other two. Obeying these speed limits, the smallest possible time t_A that a driver can spend on this part of the road is to travel between the first and second signs at 55 mi/h and between the second and third signs at 35 mi/h. More realistically, a driver could slow down from 55 to 35 mi/h with a constant deceleration and then do a similar thing from 35 to 25 mi/h. This alternative requires a time t_B. Find the ratio t_B/t_A.

** 34. A Boeing 747 "Jumbo Jet" has a length of 59.7 m. The runway on which the plane lands intersects another runway. The width of the intersection is 25.0 m. The plane decelerates through the intersection at a rate of 5.70 m/s² and clears it with a final speed of 45.0 m/s. How much time is needed for the plane to clear the intersection?

** 35. ssm A train has a length of 92 m and starts from rest with a constant acceleration at time $t = 0$ s. At this instant, a car just reaches the end of the train. The car is moving with a constant velocity. At a time $t = 14$ s, the car just reaches the front of the train. Ultimately, however, the train pulls ahead of the car, and at time $t = 28$ s, the car is again at the rear of the train. Find the magnitudes of (a) the car's velocity and (b) the train's acceleration.

** 36. In the one-hundred-meter dash a sprinter accelerates from rest to a top speed with an acceleration whose magnitude is 2.68 m/s². After achieving top speed, he runs the remainder of the race without speeding up or slowing down. If the total race is run in 12.0 s, how far does he run during the acceleration phase?

Section 2.6 Freely Falling Bodies

37. ssm A penny is dropped from rest from the top of the Sears Tower in Chicago. Considering that the height of the building is 427 m and ignoring air resistance, find the speed with which the penny strikes the ground.

38. In preparation for this problem, review Conceptual Example 7. From the top of a cliff, a person uses a slingshot to fire a pebble straight downward, which is the negative direction. The initial speed of the pebble is 9.0 m/s. (a) What is the acceleration (magnitude and direction) of the pebble during the downward motion? Is the pebble decelerating? Explain. (b) After 0.50 s, how far beneath the cliff top is the pebble?

39. Concept Simulation 2.3 at **www.wiley.com/college/cutnell** offers a useful review of the concepts central to this problem. An astronaut on a distant planet wants to determine its acceleration due to gravity. The astronaut throws a rock straight up with a velocity of +15 m/s and measures a time of 20.0 s before the rock returns to his hand. What is the acceleration (magnitude and direction) due to gravity on this planet?

40. The drawing shows a device that you can make with a piece of cardboard, which can be used to measure a person's reaction time. Hold the card at the top and suddenly drop it. Ask a friend to try to catch the card between his or her thumb and index finger. Initially, your friend's fingers must be level with the asterisks at the bottom. By noting where your friend catches the card, you can determine his or her reaction time in milliseconds (ms). Calculate the distances d_1, d_2, and d_3.

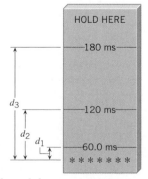

41. ssm From her bedroom window a girl drops a water-filled balloon to the ground, 6.0 m below. If the balloon is released from rest, how long is it in the air?

42. Review **Concept Simulation 2.3** at **www.wiley.com/college/cutnell** before attempting this problem. At the beginning of a basketball game, a referee tosses the ball straight up with a speed of 4.6 m/s. A player cannot touch the ball until after it reaches its maximum height and begins to fall down. What is the minimum time that a player must wait before touching the ball?

43. Review Conceptual Example 15 before attempting this problem. Two identical pellet guns are fired simultaneously from the edge of a cliff. These guns impart an initial speed of 30.0 m/s to each pellet. Gun A is fired straight upward, with the pellet going up and then falling back down, eventually hitting the ground beneath the cliff. Gun B is fired straight downward. In the absence of air resistance, how long after pellet B hits the ground does pellet A hit the ground?

44. A diver springs upward with an initial speed of 1.8 m/s from a 3.0-m board. (a) Find the velocity with which he strikes the water. [Hint: When the diver reaches the water, his displacement is $y = -3.0$ m (measured from the board), assuming that the downward direction is chosen as the negative direction.] (b) What is the highest point he reaches above the water?

45. ssm A wrecking ball is hanging at rest from a crane when suddenly the cable breaks. The time it takes for the ball to fall halfway to the ground is 1.2 s. Find the time it takes for the ball to fall from rest all the way to the ground.

46. Before working this problem, review Conceptual Example 15. A pellet gun is fired straight downward from the edge of a cliff that is 15 m above the ground. The pellet strikes the ground with a speed of 27 m/s. How far above the cliff edge would the pellet have gone had the gun been fired straight upward?

47. Consult **Interactive Solution 2.47** at **www.wiley.com/college/cutnell** before beginning this problem. A ball is thrown straight upward and rises to a maximum height of 12.0 m above its launch point. At what height above its launch point has the speed of the ball decreased to one-half of its initial value?

* 48. Two arrows are shot vertically upward. The second arrow is shot after the first one, but while the first is still on its way up. The initial speeds are such that both arrows reach their maximum heights at the same instant, although these heights are different. Suppose that the initial speed of the first arrow is 25.0 m/s and that the second arrow is fired 1.20 s after the first. Determine the initial speed of the second arrow.

* 49. Review **Interactive Solution 2.49** at **www.wiley.com/college/cutnell** before beginning this problem. A woman on a bridge 75.0 m high sees a raft floating at a constant speed on the river below. She drops a stone from rest in an attempt to hit the raft. The stone is released when the raft has 7.00 m more to travel before passing under the bridge. The stone hits the water 4.00 m in front of the raft. Find the speed of the raft.

* 50. Consult **Concept Simulation 2.3** at **www.wiley.com/college/cutnell** to review the concepts on which this problem is based. Two students, Anne and Joan, are bouncing straight up and down on a trampoline. Anne bounces twice as high as Joan does. Assuming both are in free-fall, find the ratio of the time Anne spends between bounces to the time Joan spends.

* 51. ssm A log is floating on swiftly moving water. A stone is dropped from rest from a 75-m-high bridge and lands on the log as it passes under the bridge. If the log moves with a constant speed of 5.0 m/s, what is the horizontal distance between the log and the bridge when the stone is released?

* **52.** (a) Just for fun, a person jumps from rest from the top of a tall cliff overlooking a lake. In falling through a distance H, she acquires a certain speed v. Assuming free-fall conditions, how much farther must she fall in order to acquire a speed of $2v$? Express your answer in terms of H. (b) Would the answer to part (a) be different if this event were to occur on another planet where the acceleration due to gravity had a value other than 9.80 m/s^2? Explain.

* **53. ssm www** A spelunker (cave explorer) drops a stone from rest into a hole. The speed of sound is 343 m/s in air, and the sound of the stone striking the bottom is heard 1.50 s after the stone is dropped. How deep is the hole?

* **54.** A ball is thrown upward from the top of a 25.0-m-tall building. The ball's initial speed is 12.0 m/s. At the same instant, a person is running on the ground at a distance of 31.0 m from the building. What must be the average speed of the person if he is to catch the ball at the bottom of the building?

** **55.** A ball is dropped from rest from the top of a cliff that is 24 m high. From ground level, a second ball is thrown straight upward at the same instant that the first ball is dropped. The initial speed of the second ball is exactly the same as that with which the first ball eventually hits the ground. In the absence of air resistance, the motions of the balls are just the reverse of each other. Determine how far below the top of the cliff the balls cross paths.

** **56.** Review **Interactive LearningWare 2.2** at **www.wiley.com/college/cutnell** as an aid in solving this problem. A hot air balloon is ascending straight up at a constant speed of 7.0 m/s. When the balloon is 12.0 m above the ground, a gun fires a pellet straight up from ground level with an initial speed of 30.0 m/s. Along the paths of the balloon and the pellet, there are two places where each of them has the same altitude at the same time. How far above ground level are these places?

Section 2.7 Graphical Analysis of Velocity and Acceleration

57. ssm For the first 10.0 km of a marathon, a runner averages a velocity that has a magnitude of 15.0 km/h. For the next 15.0 km, he averages 10.0 km/h, and for the last 15.0 km, he averages 5.0 km/h. Construct, to scale, the position–time graph for the runner.

58. A bus makes a trip according to the position–time graph shown in the drawing. What is the average velocity (magnitude and direction) of the bus during each of the segments labeled A, B, and C? Express your answers in km/h.

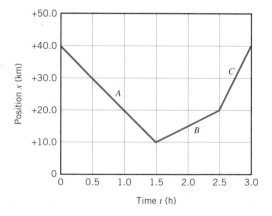

59. Concept Simulation 2.5 at **www.wiley.com/college/cutnell** provides a review of the concepts that play a role in this problem. A snowmobile moves according to the velocity–time graph shown in the drawing (see top of right column). What is the snowmobile's average acceleration during each of the segments A, B, and C?

60. A person who walks for exercise produces the position–time graph given with this problem. (a) Without doing any calculations, decide which segments of the graph (A, B, C, or D) indicate positive,

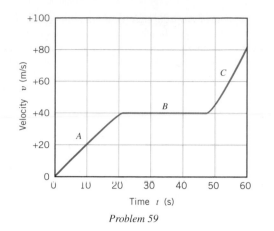

Problem 59

negative, and zero average velocities. (b) Calculate the average velocity for each segment to verify your answers to part (a).

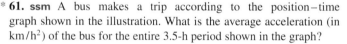

* **61. ssm** A bus makes a trip according to the position–time graph shown in the illustration. What is the average acceleration (in km/h^2) of the bus for the entire 3.5-h period shown in the graph?

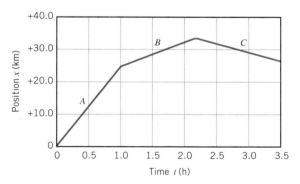

* **62.** A runner is at the position $x = 0$ m when time $t = 0$ s. One hundred meters away is the finish line. Every ten seconds, this runner runs half the remaining distance to the finish line. During each ten-second segment, the runner has a constant velocity. For the first forty seconds of the motion, construct (a) the position–time graph and (b) the velocity–time graph.

** **63.** Two runners start one hundred meters apart and run toward each other. Each runs ten meters during the first second. During each second thereafter, each runner runs ninety percent of the distance he ran in the previous second. Thus, the velocity of each person changes from second to second. However, during any one second, the velocity remains constant. Make a position–time graph for one of the runners. From this graph, determine (a) how much time passes before the runners collide and (b) the speed with which each is running at the moment of collision.

Additional Problems

64. Refer to **Concept Simulation 2.4** at **www.wiley.com/college/cutnell** for help in visualizing this problem graphically. A cart is driven by a large propeller or fan, which can accelerate or decelerate the cart. The cart starts out at the position $x = 0$ m, with an initial velocity of $+5.0$ m/s and a constant acceleration due to the fan. The direction to the right is positive. The cart reaches a maximum position of $x = +12.5$ m, where it begins to travel in the negative direction. Find the acceleration of the cart.

65. ssm The greatest height reported for a jump into an airbag is 99.4 m by stuntman Dan Koko. In 1948 he jumped from rest from the top of the Vegas World Hotel and Casino. He struck the airbag at a speed of 39 m/s (88 mi/h). To assess the effects of air resistance, determine how fast he would have been traveling on impact had air resistance been absent.

66. The three-toed sloth is the slowest moving land mammal. On the ground, the sloth moves at an average speed of 0.037 m/s, considerably slower than the giant tortoise, which walks at 0.076 m/s. After 12 minutes of walking, how much further would the tortoise have gone relative to the sloth?

67. ssm www A jogger accelerates from rest to 3.0 m/s in 2.0 s. A car accelerates from 38.0 to 41.0 m/s also in 2.0 s. (a) Find the acceleration (magnitude only) of the jogger. (b) Determine the acceleration (magnitude only) of the car. (c) Does the car travel farther than the jogger during the 2.0 s? If so, how much farther?

68. A hot-air balloon is rising upward with a constant speed of 2.50 m/s. When the balloon is 3.00 m above the ground, the balloonist accidentally drops a compass over the side of the balloon. How much time elapses before the compass hits the ground?

69. Due to continental drift, the North American and European continents are drifting apart at an average speed of about 3 cm per year. At this speed, how long (in years) will it take for them to drift apart by another 1500 m (a little less than a mile)?

70. An automobile starts from rest and accelerates to a final velocity in two stages along a straight road. Each stage occupies the same amount of time. In stage 1, the magnitude of the car's acceleration is 3.0 m/s^2. The magnitude of the car's velocity at the end of stage 2 is 2.5 times greater than it is at the end of stage 1. Find the magnitude of the acceleration in stage 2.

71. Concept Simulation 2.2 at **www.wiley.com/college/cutnell** offers a useful review of the concepts that lie at the heart of this problem. Two rockets are flying in the same direction and are side by side at the instant their retrorockets fire. Rocket A has an initial velocity of $+5800$ m/s, while rocket B has an initial velocity of $+8600$ m/s. After a time t both rockets are again side by side, the displacement of each being zero. The acceleration of rocket A is -15 m/s^2. What is the acceleration of rocket B?

* **72.** Review **Interactive LearningWare 2.2** at **www.wiley.com/college/cutnell** in preparation for this problem. A race driver has made a pit stop to refuel. After refueling, he leaves the pit area with an acceleration whose magnitude is 6.0 m/s^2; after 4.0 s he enters the main speedway. At the same instant, another car on the speedway and traveling at a constant speed of 70.0 m/s overtakes and passes the entering car. If the entering car maintains its acceleration, how much time is required for it to catch the other car?

* **73. ssm** A bicyclist makes a trip that consists of three parts, each in the same direction (due north) along a straight road. During the first part, she rides for 22 minutes at an average speed of 7.2 m/s. During the second part, she rides for 36 minutes at an average speed of 5.1 m/s. Finally, during the third part, she rides for 8.0 minutes at an average speed of 13 m/s. (a) How far has the bicyclist traveled during the entire trip? (b) What is her average velocity for the trip?

* **74.** A sky diver, with parachute unopened, falls 625 m in 15.0 s. Then she opens her parachute and falls another 356 m in 142 s. What is her average velocity (both magnitude and direction) for the entire fall?

* **75. ssm** A cement block accidentally falls from rest from the ledge of a 53.0-m-high building. When the block is 14.0 m above the ground, a man, 2.00 m tall, looks up and notices that the block is directly above him. How much time, at most, does the man have to get out of the way?

* **76.** Review **Interactive LearningWare 2.2** at **www.wiley.com/college/cutnell** in preparation for this problem. A police car is traveling at a velocity of 18.0 m/s due north, when a car zooms by at a constant velocity of 42.0 m/s due north. After a reaction time of 0.800 s the policeman begins to pursue the speeder with an acceleration of 5.00 m/s^2. Including the reaction time, how long does it take for the police car to catch up with the speeder?

* **77.** Two soccer players start from rest, 48 m apart. They run directly toward each other, both players accelerating. The first player has an acceleration whose magnitude is 0.50 m/s^2. The second player's acceleration has a magnitude of 0.30 m/s^2. (a) How much time passes before they collide? (b) At the instant they collide, how far has the first player run?

** **78.** A roof tile falls from rest from the top of a building. An observer inside the building notices that it takes 0.20 s for the tile to pass her window, whose height is 1.6 m. How far above the top of this window is the roof?

** **79. ssm** A locomotive is accelerating at 1.6 m/s^2. It passes through a 20.0-m-wide crossing in a time of 2.4 s. After the locomotive leaves the crossing, how much time is required until its speed reaches 32 m/s?

** **80.** A football player, starting from rest at the line of scrimmage, accelerates along a straight line for a time of 3.0 s. Then, during a negligible amount of time, he changes the magnitude of his acceleration to a value of 1.1 m/s^2. With this acceleration, he continues in the same direction for another 2.0 s, until he reaches a speed of 6.4 m/s. What is the value of his acceleration (assumed to be constant) during the initial 3.0-s period?

Concepts & Calculations   Group Learning Problems

Note: Each of these problems consists of Concept Questions followed by a related quantitative Problem. They are designed for use by students working alone or in small learning groups. The Concept Questions involve little or no mathematics and are intended to stimulate group discussions. They focus on the concepts with which the problems deal. Recognizing the concepts is the essential initial step in any problem-solving technique.

81. Concept Question Listed below are three pairs of initial and final positions that lie along the x axis. Using the concept of displacement discussed in Section 2.1, decide which pairs give a positive displacement and which a negative displacement. Explain.

	Initial position x_0	Final position x
(a)	+2 m	+6 m
(b)	+6 m	+2 m
(c)	−3 m	+7 m

Problem For each of the three pairs of positions listed above, determine the magnitude and direction (positive or negative) of the displacement. Verify that your answers are consistent with those determined in the Concept Question.

82. Concept Question Suppose that each pair of positions in Problem 81 represents the initial and final positions of a moving car. In each case, is the direction of the car's average velocity the same as its initial position x_0, its final position x, or the difference $x - x_0$ between its final and initial positions? Decide which pairs give a positive average velocity and which give a negative average velocity. Provide reasons for your answers.

Problem The elapsed time for each pair of positions is 0.5 s. Review the concept of average velocity in Section 2.2 and then determine the average velocity (magnitude and direction) for each of the three pairs. Check to see that the directions of the velocities agree with the directions found in the Concept Question.

83. Concept Questions (a) In general, does the average acceleration of an object have the same direction as its initial velocity v_0, its final velocity v, or the difference $v - v_0$ between its final and initial velocities? Provide a reason for your answer. (b) The following table lists four pairs of initial and final velocities for a boat traveling along the x axis. Use the concept of acceleration presented in Section 2.3 to determine the direction (positive or negative) of the average acceleration.

	Initial velocity v_0	Final velocity v
(a)	+2.0 m/s	+5.0 m/s
(b)	+5.0 m/s	+2.0 m/s
(c)	−6.0 m/s	−3.0 m/s
(d)	+4.0 m/s	−4.0 m/s

Problem The elapsed time for each of the four pairs of velocities is 2.0 s. Find the average acceleration (magnitude and direction) for each of the four pairs. Be sure that your directions agree with those found in the Concept Questions.

84. Concept Questions The initial velocity v_0 and acceleration a of four moving objects at a given instant in time are given in the table:

	Initial velocity v_0	Acceleration a
(a)	+12 m/s	+3.0 m/s^2
(b)	+12 m/s	−3.0 m/s^2
(c)	−12 m/s	+3.0 m/s^2
(d)	−12 m/s	−3.0 m/s^2

Draw vectors for v_0 and a and, in each case, state whether the *speed* of the object is increasing or decreasing in time. Account for your answers.

Problem For each of the four pairs in the table above, determine the final *speed* of the object if the elapsed time is 2.0 s. Compare your final speeds with the initial speeds and make sure that your answers are consistent with your answers to the Concept Questions.

85. Concept Questions (a) The acceleration of a NASCAR race car is zero. Does this necessarily mean that the velocity of the car is also zero? (b) If the speed of the car is constant as it goes around the track, is its average acceleration necessarily zero? Justify your answers.

Problem (a) Suppose that the race car is moving to the right with a constant velocity of +82 m/s. What is the acceleration of the car? (b) Twelve seconds later, the car is halfway around the track and traveling in the opposite direction with the same speed. What is the average acceleration of the car? Verify that your answers agree with your answers to the Concept Questions.

86. Concept Questions Concept Simulation 2.3 at **www.wiley.com/college/cutnell** provides some background for this problem. A ball is thrown vertically upward, which is the positive direction. A little later it returns to its point of release. (a) Does the acceleration of the ball reverse direction when the ball starts its downward trip? (b) What is the displacement of the ball when it returns to its point of release? Explain your answers.

Problem If the ball is in the air for a total time of 8.0 s, what is its initial velocity?

* **87. Concept Questions** Two runners cover the same distance in a straight line. Runner A covers the distance at a constant velocity. Runner B starts from rest, maintains a constant acceleration, and covers the distance in the same time as runner A. (a) Is runner B's final velocity smaller than, equal to, or greater than runner A's constant velocity? Explain. (b) Is runner B's average velocity smaller than, equal to, or greater than runner A's constant velocity? Why? (c) Given a value for the time to cover the distance and the fact that runner B starts from rest, what must be determined before runner B's acceleration can be determined? Provide a reason for your answer.

Problem Both runners cover a distance of 460 m in 210 s. Determine (a) the constant velocity of runner A, (b) the final velocity of runner B, and (c) the acceleration of runner B. Verify that your answers are consistent with your answers to the Concept Questions.

* **88. Concept Questions** Two stones are thrown simultaneously, one straight upward from the base of a cliff and the other straight downward from the top of the cliff. The stones are thrown with the same speed. (a) Does the stone thrown upward gain or lose speed as it moves upward? Why? (b) Does the stone thrown downward gain or lose speed as time passes? Explain. (c) The speed at which the stones are thrown is such that they cross paths. Where do they cross paths, above or below the point that corresponds to half the height of the cliff? Justify your answer.

Problem The height of the cliff is 6.00 m, and the speed with which the stones are thrown is 9.00 m/s. Find the location of the crossing point. Check to see that your answer is consistent with your answers to the Concept Questions.

Chapter 3

Kinematics in Two Dimensions

Once this motorcycle is airborne, it follows a familiar arc-shaped path that depends on the launch velocity and the acceleration due to gravity, assuming that the effects of air resistance can be ignored.
(© Jamie Budge/Corbis Images)

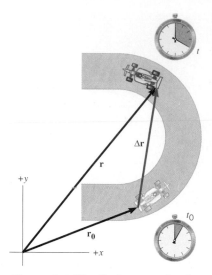

Figure 3.1 The displacement $\Delta \mathbf{r}$ of the car is a vector that points from the initial position of the car at time t_0 to the final position at time t. The magnitude of $\Delta \mathbf{r}$ is the shortest distance between the two positions.

3.1 *Displacement, Velocity, and Acceleration*

In Chapter 2 the concepts of displacement, velocity, and acceleration are used to describe an object moving in one dimension. There are also situations in which the motion is along a curved path that lies in a plane. Such two-dimensional motion can be described using the same concepts. In Grand Prix racing, for example, the course follows a curved road, and Figure 3.1 shows a race car at two different positions along it. These positions are identified by the vectors $\mathbf{r}$ and $\mathbf{r_0}$, which are drawn from an arbitrary coordinate origin. The ***displacement*** $\Delta \mathbf{r}$ of the car is the vector drawn from the initial position $\mathbf{r_0}$ at time t_0 to the final position $\mathbf{r}$ at time t. The magnitude of $\Delta \mathbf{r}$ is the shortest distance between the two positions. In the drawing, the vectors $\mathbf{r_0}$ and $\Delta \mathbf{r}$ are drawn tail to head, so it is evident that $\mathbf{r}$ is the vector sum of $\mathbf{r_0}$ and $\Delta \mathbf{r}$. (See Sections 1.5 and 1.6 for a review of vectors and vector addition.) This means that $\mathbf{r} = \mathbf{r_0} + \Delta \mathbf{r}$, or

$$\text{Displacement} = \Delta \mathbf{r} = \mathbf{r} - \mathbf{r_0}$$

The displacement here is defined as it is in Chapter 2. Now, however, the displacement vector can lie anywhere in a plane, rather than just along a straight line.

The average velocity $\overline{\mathbf{v}}$ of the car between two positions is defined in a manner similar to that in Equation 2.2, as the displacement $\Delta \mathbf{r} = \mathbf{r} - \mathbf{r_0}$ divided by the elapsed time $\Delta t = t - t_0$:

$$\overline{\mathbf{v}} = \frac{\mathbf{r} - \mathbf{r_0}}{t - t_0} = \frac{\Delta \mathbf{r}}{\Delta t} \tag{3.1}$$

Since both sides of Equation 3.1 must agree in direction, the average velocity vector has the same direction as the displacement. The velocity of the car at an instant of time is its ***instantaneous velocity*** $\mathbf{v}$. The average velocity becomes equal to the instantaneous velocity $\mathbf{v}$ in the limit that Δt becomes infinitesimally small:

$$\mathbf{v} = \lim_{\Delta t \to 0} \frac{\Delta \mathbf{r}}{\Delta t}$$

Figure 3.2 illustrates that the instantaneous velocity $\mathbf{v}$ is tangent to the path of the car. The drawing also shows the vector components $\mathbf{v}_x$ and $\mathbf{v}_y$ of the velocity, which are parallel to the x and y axes, respectively.

✔ Check Your Understanding 1

Suppose you are driving due east, traveling a distance of 1500 meters in 2 minutes. You then turn due north and travel the same distance in the same time. What can be said about the average speeds and the average velocities for the two segments of the trip? (a) The average speeds are the same, and the average velocities are the same. (b) The average speeds are the same, but the average velocities are different. (c) The average speeds are different, but the average velocities are the same. *(The answer is at the end of the book.)*

Background: This question deals with the concepts of velocity (a vector) and speed (a scalar). Think about the additional information conveyed by a vector.

For similar questions (including calculational counterparts), consult Self-Assessment Test 3.1. This test is described at the end of Section 3.3.

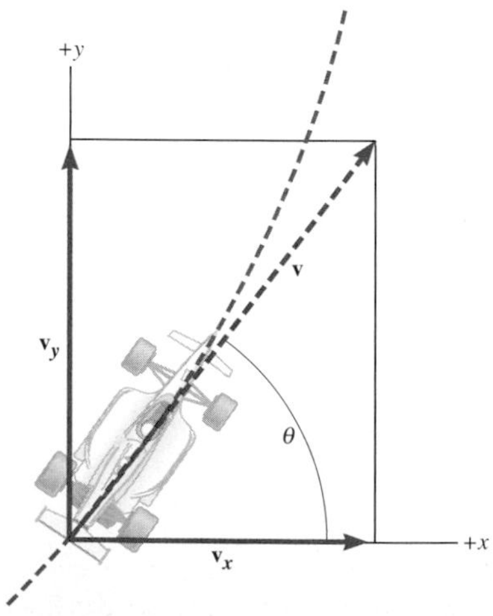

Figure 3.2 The instantaneous velocity $\mathbf{v}$ and its two vector components $\mathbf{v}_x$ and $\mathbf{v}_y$.

The ***average acceleration*** $\overline{\mathbf{a}}$ is defined just as it is for one-dimensional motion—namely, as the change in velocity, $\Delta \mathbf{v} = \mathbf{v} - \mathbf{v_0}$, divided by the elapsed time Δt:

$$\overline{\mathbf{a}} = \frac{\mathbf{v} - \mathbf{v_0}}{t - t_0} = \frac{\Delta \mathbf{v}}{\Delta t} \tag{3.2}$$

The average acceleration vector has the same direction as the change in velocity. In the limit that the elapsed time becomes infinitesimally small, the average acceleration becomes equal to the ***instantaneous acceleration*** $\mathbf{a}$:

$$\mathbf{a} = \lim_{\Delta t \to 0} \frac{\Delta \mathbf{v}}{\Delta t}$$

The acceleration has a vector component $\mathbf{a}_x$ along the x direction and a vector component $\mathbf{a}_y$ along the y direction.

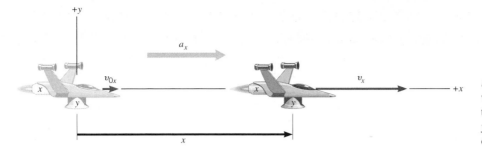

Figure 3.3 The spacecraft is moving with a constant acceleration a_x parallel to the x axis. There is no motion in the y direction, and the y engine is turned off.

3.2 Equations of Kinematics in Two Dimensions

To understand how displacement, velocity, and acceleration are applied to two-dimensional motion, consider a spacecraft equipped with two engines that are mounted perpendicular to each other. These engines produce the only forces that the craft experiences, and the spacecraft is assumed to be at the coordinate origin when $t_0 = 0$ s, so that $\mathbf{r_0} = 0$ m. At a later time t, the spacecraft's displacement is $\Delta \mathbf{r} = \mathbf{r} - \mathbf{r_0} = \mathbf{r}$. Relative to the x and y axes, the displacement $\mathbf{r}$ has vector components of $\mathbf{x}$ and $\mathbf{y}$, respectively.

In Figure 3.3 only the engine oriented along the x direction is firing, and the vehicle accelerates along this direction. It is assumed that the velocity in the y direction is zero, and it remains zero, since the y engine is turned off. The motion of the spacecraft along the x direction is described by the five kinematic variables x, a_x, v_x, v_{0x}, and t. Here the symbol "x" reminds us that we are dealing with the x components of the displacement, velocity, and acceleration vectors. (See Sections 1.7 and 1.8 for a review of vector components.) The variables x, a_x, v_x, and v_{0x} are scalar components (or "components," for short). As Section 1.7 discusses, these components are positive or negative numbers (with units), depending on whether the associated vector components point along the $+x$ or the $-x$ axis. If the spacecraft has a constant acceleration along the x direction, the motion is exactly like that described in Chapter 2, and the equations of kinematics can be used. For convenience, these equations are written in the left column of Table 3.1.

Figure 3.4 is analogous to Figure 3.3, except that now only the y engine is firing, and the spacecraft accelerates along the y direction. Such a motion can be described in terms of the kinematic variables y, a_y, v_y, v_{0y}, and t. And if the acceleration along the y direction is constant, these variables are related by the equations of kinematics, as written in the right column of Table 3.1. Like their counterparts in the x direction, the scalar components, y, a_y, v_y, and v_{0y}, may be positive ($+$) or negative ($-$) numbers (with units).

Table 3.1 *Equations of Kinematics for Constant Acceleration in Two-Dimensional Motion*

x Component		Variable	y Component	
x		Displacement	y	
a_x		Acceleration	a_y	
v_x		Final velocity	v_y	
v_{0x}		Initial velocity	v_{0y}	
t		Elapsed time	t	
$v_x = v_{0x} + a_x t$	(3.3a)		$v_y = v_{0y} + a_y t$	(3.3b)
$x = \frac{1}{2}(v_{0x} + v_x)t$	(3.4a)		$y = \frac{1}{2}(v_{0y} + v_y)t$	(3.4b)
$x = v_{0x}t + \frac{1}{2}a_x t^2$	(3.5a)		$y = v_{0y}t + \frac{1}{2}a_y t^2$	(3.5b)
$v_x^2 = v_{0x}^2 + 2a_x x$	(3.6a)		$v_y^2 = v_{0y}^2 + 2a_y y$	(3.6b)

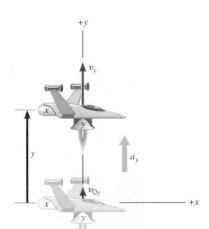

Figure 3.4 The spacecraft is moving with a constant acceleration a_y parallel to the y axis. There is no motion in the x direction, and the x engine is turned off.

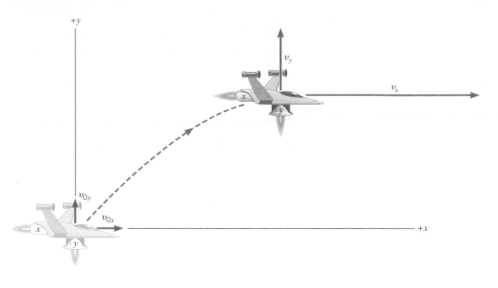

Figure 3.5 The two-dimensional motion of the spacecraft can be viewed as the combination of the separate *x* and *y* motions.

Figure 3.6 CONCEPTS AT A GLANCE In two dimensions, motion along the *x* direction and motion along the *y* direction are independent of each other. As a result, each can be analyzed separately according to the procedures for one-dimensional kinematics discussed in Chapter 2. On the space shuttle *Challenger,* motion in perpendicular directions is controlled by thrusters. The photographs show the *Challenger* in orbit with different thrusters activated. (Courtesy NASA)

If both engines of the spacecraft are firing *at the same time,* the resulting motion takes place in part along the *x* axis and in part along the *y* axis, as Figure 3.5 illustrates. The thrust of each engine gives the vehicle a corresponding acceleration component. The *x* engine accelerates the ship in the *x* direction and causes a change in the *x* component of the velocity. Likewise, the *y* engine causes a change in the *y* component of the velocity. *It is important to realize that the x part of the motion occurs exactly as it would if the y part did not occur at all. Similarly, the y part of the motion occurs exactly as it would if the x part of the motion did not exist.* In other words, the *x* and *y* motions are independent of each other.

▶ CONCEPTS AT A GLANCE   The independence of the *x* and *y* motions lies at the heart of two-dimensional kinematics. It allows us to treat two-dimensional motion as two distinct one-dimensional motions, one for the *x* direction and one for the *y* direction. As the Concepts-at-a-Glance chart in Figure 3.6 illustrates, everything that we have learned in Chapter 2 about kinematics in one dimension will now be applied separately to each of the two directions. In so doing, we will be able to describe the *x* and *y* variables separately and then bring these descriptions together to understand the two-dimensional picture. Example 1 takes this approach in dealing with a moving spacecraft. ◀

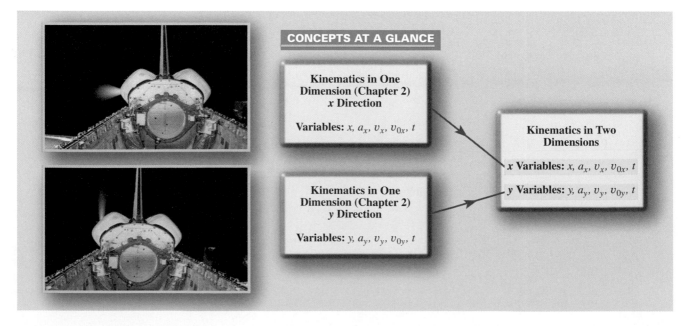

CONCEPTS AT A GLANCE

Kinematics in One Dimension (Chapter 2)
x Direction

Variables: x, a_x, v_x, v_{0x}, t

Kinematics in One Dimension (Chapter 2)
y Direction

Variables: y, a_y, v_y, v_{0y}, t

Kinematics in Two Dimensions

x **Variables:** x, a_x, v_x, v_{0x}, t

y **Variables:** y, a_y, v_y, v_{0y}, t

Example 1 A Moving Spacecraft

In the x direction, the spacecraft in Figure 3.5 has an initial velocity component of $v_{0x} = +22$ m/s and an acceleration component of $a_x = +24$ m/s^2. In the y direction, the analogous quantities are $v_{0y} = +14$ m/s and $a_y = +12$ m/s^2. The directions to the right and upward have been chosen as the positive directions. Find (a) x and v_x, (b) y and v_y, and (c) the final velocity (magnitude and direction) of the spacecraft at time $t = 7.0$ s.

Reasoning The motion in the x direction and the motion in the y direction can be treated separately, each as a one-dimensional motion. We will follow this approach in parts (a) and (b) to obtain the location and velocity components of the spacecraft. Then, in part (c) the velocity components will be combined to give the final velocity.

Solution (a) The data for the motion in the x direction are listed below.

x-Direction Data				
x	a_x	v_x	v_{0x}	t
?	$+24$ m/s^2	?	$+22$ m/s	7.0 s

The x component of the craft's displacement can be found by using Equation 3.5a.

$$x = v_{0x}t + \tfrac{1}{2}a_x t^2 = (22 \text{ m/s})(7.0 \text{ s}) + \tfrac{1}{2}(24 \text{ m/s}^2)(7.0 \text{ s})^2 = \boxed{+740 \text{ m}}$$

The velocity component v_x can be calculated with the aid of Equation 3.3a:

$$v_x = v_{0x} + a_x t = (22 \text{ m/s}) + (24 \text{ m/s}^2)(7.0 \text{ s}) = \boxed{+190 \text{ m/s}}$$

(b) The data for the motion in the y direction are listed below.

y-Direction Data				
y	a_y	v_y	v_{0y}	t
?	$+12$ m/s^2	?	$+14$ m/s	7.0 s

Proceeding in the same manner as in part (a), we find that

$$\boxed{y = +390 \text{ m}} \quad \text{and} \quad \boxed{v_y = +98 \text{ m/s}}$$

(c) Figure 3.7 shows the velocity of the vehicle and its components v_x and v_y. The magnitude v of the velocity can be found by using the Pythagorean theorem:

$$v = \sqrt{v_x{}^2 + v_y{}^2} = \sqrt{(190 \text{ m/s})^2 + (98 \text{ m/s})^2} = \boxed{210 \text{ m/s}}$$

The direction of the velocity vector is given by the angle θ in the drawing:

$$\tan \theta = \frac{v_y}{v_x} \quad \text{or} \quad \theta = \tan^{-1}\left(\frac{v_y}{v_x}\right) = \tan^{-1}\left(\frac{98 \text{ m/s}}{190 \text{ m/s}}\right) = \boxed{27^\circ}$$

After 7.0 s, the spacecraft has a velocity of 210 m/s in a direction of 27° above the positive x axis. The craft is 740 m to the right and 390 m above the origin, as in Figure 3.5.

Problem solving insight
When the motion is two-dimensional, the time variable t has the same value for both the x and y directions.

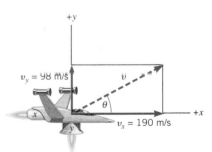

Figure 3.7 The magnitude of the velocity vector gives the speed of the spacecraft, and the angle θ gives the direction of travel relative to the positive x direction.

✔ Check Your Understanding 2

A power boat, starting from rest, maintains a constant acceleration. After a certain time t, its displacement and velocity are **r** and **v**. At a time $2t$, what would be its displacement and velocity, assuming the acceleration remains the same? *(The answer is at the end of the book.)*
    (a) 2**r** and 2**v**   (b) 2**r** and 4**v**   (c) 4**r** and 2**v**   (d) 4**r** and 4**v**

Background: When an object accelerates, its displacement and velocity depend on time. If the acceleration is constant, the equations of kinematics in Table 3.1 apply.

For similar questions (including calculational counterparts), consult Self-Assessment Test 3.1. This test is described at the end of Section 3.3.

The following Reasoning Strategy gives an overview of how the equations of kinematics are applied to describe motion in two dimensions, such as that in Example 1.

Reasoning Strategy

Applying the Equations of Kinematics in Two Dimensions

1. Make a drawing to represent the situation being studied.

2. Decide which directions are to be called positive (+) and negative (−) relative to a conveniently chosen coordinate origin. Do not change your decision during the course of a calculation.

3. Remember that the time variable t has the same value for the part of the motion along the x axis and the part along the y axis.

4. In an organized way, write down the values (with appropriate + and − signs) that are given for any of the five kinematic variables associated with the x direction and the y direction. Be on the alert for "implied data," such as the phrase "starts from rest," which means that the values of the initial velocity components are zero: $v_{0x} = 0$ m/s and $v_{0y} = 0$ m/s. The data summary boxes used in the examples are a good way of keeping track of this information. In addition, identify the variables that you are being asked to determine.

5. Before attempting to solve a problem, verify that the given information contains values for at least three of the kinematic variables. Do this for the x and the y direction of the motion. Once the three known variables are identified along with the desired unknown variable, the appropriate relations from Table 3.1 can be selected.

6. When the motion is divided into "segments," remember that the final velocity for one segment is the initial velocity for the next segment.

7. Keep in mind that a kinematics problem may have two possible answers. Try to visualize the different physical situations to which the answers correspond.

3.3 Projectile Motion

The biggest thrill in baseball is a home run. The motion of the ball on its curving path into the stands is a common type of two-dimensional motion called "projectile motion." A good description of such motion can often be obtained with the assumption that air resistance is absent.

▶ **CONCEPTS AT A GLANCE** Following the approach outlined in Figure 3.6, we consider the horizontal and vertical parts of the motion separately. In the horizontal or x direction, the moving object (the projectile) does not slow down in the absence of air resistance. Thus, the x component of the velocity remains constant at its initial value or $v_x = v_{0x}$, and the x component of the acceleration is $a_x = 0$ m/s^2. In the vertical or y direction, however, the projectile experiences the effect of gravity. As a result, the y component of the velocity v_y is not constant, but changes. The y component of the acceleration a_y is the downward acceleration due to gravity. If the path or trajectory of the projectile is near the earth's surface, a_y has a magnitude of 9.80 m/s^2. In this text, then, the phrase "projectile motion" means that $a_x = 0$ m/s^2 and a_y equals the acceleration due to gravity, as the Concept-at-a-Glance chart in Figure 3.8 summarizes. Example 2 and other examples in this section illustrate how the equations of kinematics are applied to projectile motion. ◀

Example 2   A Falling Care Package

Figure 3.9 shows an airplane moving horizontally with a constant velocity of $+115$ m/s at an altitude of 1050 m. The directions to the right and upward have been chosen as the positive directions. The plane releases a "care package" that falls to the ground along a curved trajectory. Ignoring air resistance, determine the time required for the package to hit the ground.

Reasoning The time required for the package to hit the ground is the time it takes for the package to fall through a vertical distance of 1050 m. In falling, it moves to the right, as well as downward, but these two parts of the motion occur independently. Therefore, we can focus solely on the vertical part. We note that the package is moving initially in the horizontal or x direction, not in the y direction, so that $v_{0y} = 0$ m/s. Furthermore, when the package hits the

CONCEPTS AT A GLANCE

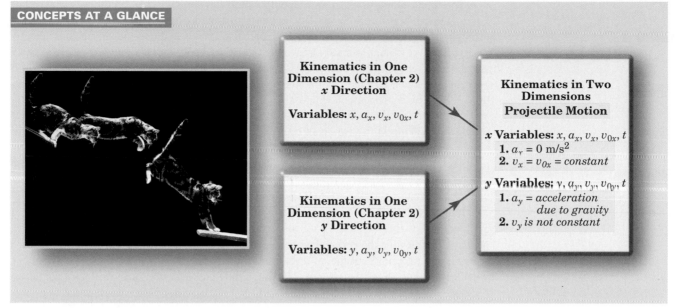

Kinematics in One Dimension (Chapter 2) x Direction

Variables: x, a_x, v_x, v_{0x}, t

Kinematics in Two Dimensions Projectile Motion

x Variables: x, a_x, v_x, v_{0x}, t
1. $a_x = 0 \text{ m/s}^2$
2. $v_x = v_{0x} = constant$

y Variables: y, a_y, v_y, v_{0y}, t
1. $a_y = acceleration$
   $due \ to \ gravity$
2. $v_y \ is \ not \ constant$

Kinematics in One Dimension (Chapter 2) y Direction

Variables: y, a_y, v_y, v_{0y}, t

ground, the y component of its displacement is $y = -1050$ m, as the drawing shows. The acceleration is that due to gravity, so $a_y = -9.80 \text{ m/s}^2$. These data are summarized as follows:

y-Direction Data				
y	a_y	v_y	v_{0y}	t
-1050 m	-9.80 m/s^2		0 m/s	?

With these data, Equation 3.5b ($y = v_{0y}t + \frac{1}{2}a_y t^2$) can be used to find the fall time.

Solution Since $v_{0y} = 0$ m/s, it follows from Equation 3.5b that $y = \frac{1}{2}a_y t^2$ and

$$t = \sqrt{\frac{2y}{a_y}} = \sqrt{\frac{2(-1050 \text{ m})}{-9.80 \text{ m/s}^2}} = \boxed{14.6 \text{ s}}$$

Figure 3.8 CONCEPTS AT A GLANCE In projectile motion, the horizontal or x component of the acceleration is zero, and the vertical or y component of the acceleration is the acceleration due to gravity. In this time-lapse photograph, the cat exhibits projectile motion while in the air, assuming that the effects of air resistance can be ignored. (© Stephen Dalton/Photo Researchers)

Problem solving insight
The variables y, a_y, v_y, and v_{0y} are scalar components. Therefore, an algebraic sign (+ or −) must be included with each one to denote direction.

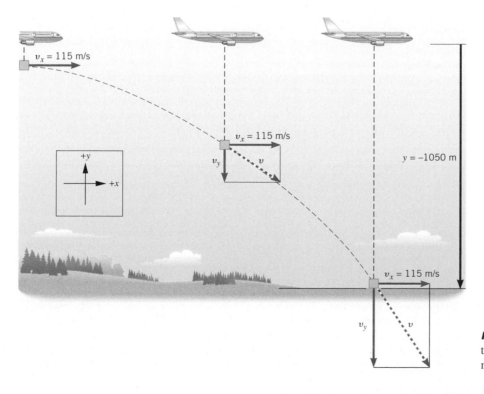

Figure 3.9 The package falling from the plane is an example of projectile motion, as Examples 2 and 3 discuss.

The freely falling package in Example 2 picks up vertical speed on the way down. The horizontal component of the velocity, however, retains its initial value of $v_{0x} = +115$ m/s throughout the entire descent. Since the plane also travels at a constant horizontal velocity of $+115$ m/s, it remains directly above the falling package. The pilot always sees the package directly beneath the plane, as the dashed vertical lines in Figure 3.9 show. This result is a direct consequence of the fact that the package has no acceleration in the horizontal direction. In reality, air resistance would slow down the package, and it would not remain directly beneath the plane during the descent. Figure 3.10 further clarifies this point by illustrating what happens to two packages that are released simultaneously from the same height. Package B is given an initial velocity component of $v_{0x} = +115$ m/s in the horizontal direction, as in Example 2, and the package follows the path shown in the figure. Package A, on the other hand, is dropped from a stationary balloon and falls straight down toward the ground, since $v_{0x} = 0$ m/s. Both packages hit the ground at the same time.

Not only do the packages in Figure 3.10 reach the ground at the same time, but the y components of their velocities are also equal at all points on the way down. However, package B does hit the ground with a greater speed than does package A. Remember, speed is the magnitude of the velocity vector, and the velocity of B has an x component, whereas the velocity of A does not. The magnitude and direction of the velocity vector for package B at the instant just before the package hits the ground is computed in Example 3.

Example 3   The Velocity of the Care Package

For the situation shown in Figure 3.10, find the speed of package B and the direction of the velocity vector just before package B hits the ground.

Reasoning Since the speed v of the package is given by $v = \sqrt{v_x^2 + v_y^2}$, it is necessary to know values for v_x and v_y at the instant before impact. The component v_x is constant and has a value of 115 m/s. The component v_y can be determined by using Equation 3.3b and the data from Example 2 ($a_y = -9.80$ m/s^2, $v_{0y} = 0$ m/s, $t = 14.6$ s). Thus, we expect that the final speed v of package B will be greater than 115 m/s.

Solution From Equation 3.3b, it follows that

$$v_y = v_{0y} + a_y t = 0 \text{ m/s} + (-9.80 \text{ m/s}^2)(14.6 \text{ s}) = -143 \text{ m/s}$$

The speed of package B at the instant before impact is

$$v = \sqrt{(115 \text{ m/s})^2 + (-143 \text{ m/s})^2} = \boxed{184 \text{ m/s}}$$

Problem solving insight
The speed of a projectile at any location along its path is the magnitude v of its velocity at that location: $v = \sqrt{v_x^2 + v_y^2}$. Both the horizontal and vertical velocity components contribute to the speed.

which is greater than 115 m/s, as expected. The velocity vector makes an angle θ with the horizontal, as Figure 3.10 indicates:

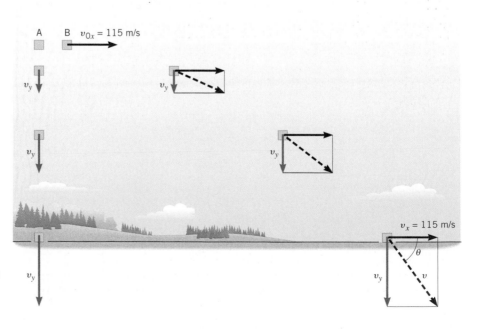

Figure 3.10 Package A and package B are released simultaneously at the same height and strike the ground at the same time because their y variables (y, a_y, and v_{0y}) are the same.

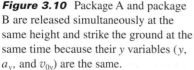

$$\cos \theta = \frac{v_x}{v} \quad \text{or} \quad \theta = \cos^{-1}\left(\frac{v_x}{v}\right) = \cos^{-1}\left(\frac{115 \text{ m/s}}{184 \text{ m/s}}\right) = \boxed{51.3°}$$

An important feature of projectile motion is that there is no acceleration in the horizontal or x direction. Conceptual Example 4 discusses an interesting implication of this feature.

Conceptual Example 4 I Shot a Bullet into the Air. . . .

Suppose you are driving in a convertible with the top down. The car is moving to the right at a constant velocity. As Figure 3.11 illustrates, you point a rifle straight upward and fire it. In the absence of air resistance, where would the bullet land—behind you, ahead of you, or in the barrel of the rifle?

Reasoning and Solution If air resistance were present, it would slow down the bullet and cause it to land behind you, toward the rear of the car. However, air resistance is absent, so we must consider the bullet's motion more carefully. Before the rifle is fired, the bullet, rifle, and car are moving together, so the bullet and rifle have the same horizontal velocity as the car. When the rifle is fired, the bullet is given an additional velocity component in the vertical direction; the bullet retains the velocity of the car as its initial horizontal velocity component, since the rifle is pointed straight up. Because there is no air resistance to slow it down, the bullet experiences no horizontal acceleration. Thus, the bullet's horizontal velocity component does not change. It retains its initial value, and remains matched to that of the rifle and the car. As a result, *the bullet remains directly above the rifle at all times and would fall directly back into the barrel of the rifle,* as the drawing indicates. This situation is analogous to that in Figure 3.9, where the care package, as it falls, remains directly below the plane.

Related Homework: *Conceptual Question 12, Problem 34*

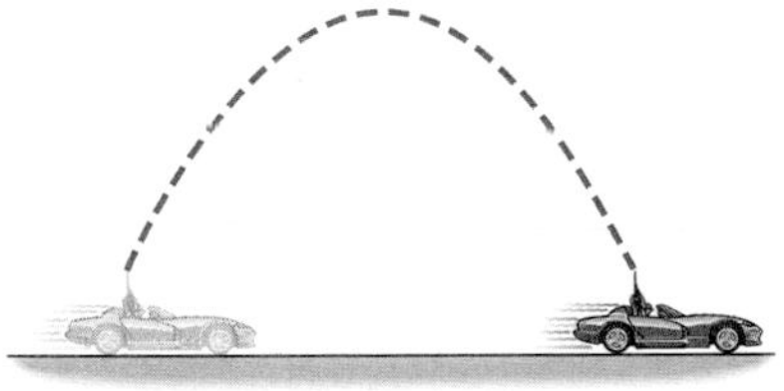

Figure 3.11 The car is moving with a constant velocity to the right, and the rifle is pointed straight up. In the absence of air resistance, a bullet fired from the rifle has no acceleration in the horizontal direction. As a result, the bullet would land back in the barrel of the rifle.

Concept Simulation 3.1

In this simulation, which is discussed in Conceptual Example 4 and illustrated in Figure 3.11, you can alter the trajectory of the bullet by changing the speed of the car and the initial speed of the bullet. In addition, the simulation shows the x and y components of the bullet's velocity as the bullet moves through the air.

Related Homework: *Conceptual Question 7, Problem 34*

Go to **www.wiley.com/college/cutnell**

Often projectiles, like footballs and baseballs, are sent into the air at an angle with respect to the ground. From a knowledge of the projectile's initial velocity, a wealth of information can be obtained about the motion. For instance, Example 5 demonstrates how to calculate the maximum height reached by the projectile.

Example 5 The Height of a Kickoff

A placekicker kicks a football at an angle of $\theta = 40.0°$ above the horizontal axis, as Figure 3.12 shows. The initial speed of the ball is $v_0 = 22$ m/s. Ignore air resistance and find the maximum height H that the ball attains.

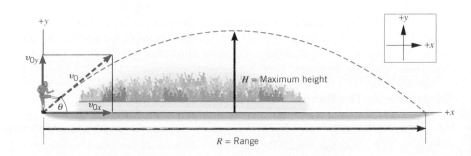

Figure 3.12 A football is kicked with an initial speed of v_0 at an angle of θ above the ground. The ball attains a maximum height H and a range R.

Reasoning The maximum height is a characteristic of the vertical part of the motion, which can be treated separately from the horizontal part. In preparation for making use of this fact, we calculate the vertical component of the initial velocity:

$$v_{0y} = v_0 \sin \theta = +(22 \text{ m/s}) \sin 40.0° = +14 \text{ m/s}$$

The vertical component of the velocity, v_y, decreases as the ball moves upward. Eventually, $v_y = 0$ m/s at the maximum height H. The data below can be used in Equation 3.6b ($v_y^2 = v_{0y}^2 + 2a_y y$) to find the maximum height:

y-Direction Data				
y	a_y	v_y	v_{0y}	t
$H = ?$	-9.80 m/s^2	0 m/s	+14 m/s	?

Solution From Equation 3.6b, we find that

$$y = H = \frac{v_y^2 - v_{0y}^2}{2a_y} = \frac{(0 \text{ m/s})^2 - (14 \text{ m/s})^2}{2(-9.80 \text{ m/s}^2)} = \boxed{+10 \text{ m}}$$

The height H depends only on the y variables; the same height would have been reached had the ball been thrown *straight up* with an initial velocity of $v_{0y} = +14$ m/s.

It is also possible to find the total time or "hang time" during which the football in Figure 3.12 is in the air. Example 6 shows how to determine this time.

Example 6 The Time of Flight of a Kickoff

For the motion illustrated in Figure 3.12, ignore air resistance and use the data from Example 5 to determine the time of flight between kickoff and landing.

Reasoning Given the initial velocity, it is the acceleration due to gravity that determines how long the ball stays in the air. Thus, to find the time of flight we deal with the vertical part of the motion. Since the ball starts at and returns to ground level, the displacement in the y direction is zero. The initial velocity component in the y direction is the same as that in Example 5; that is, $v_{0y} = +14$ m/s. Therefore, we have

y-Direction Data				
y	a_y	v_y	v_{0y}	t
0 m	-9.80 m/s^2		+14 m/s	?

The time of flight can be determined from Equation 3.5b ($y = v_{0y}t + \frac{1}{2}a_y t^2$).

Solution Using Equation 3.5b, we find

$$0 \text{ m} = (14 \text{ m/s})t + \tfrac{1}{2}(-9.80 \text{ m/s}^2)t^2 = [(14 \text{ m/s}) + \tfrac{1}{2}(-9.80 \text{ m/s}^2)t]t$$

There are two solutions to this equation. One is given by

$$(14 \text{ m/s}) + \tfrac{1}{2}(-9.80 \text{ m/s}^2)t = 0 \quad \text{or} \quad t = 2.9 \text{ s}$$

The other is given by $t = 0$ s. The solution we seek is $\boxed{t = 2.9 \text{ s}}$, because $t = 0$ s corresponds to the initial kickoff.

Another important feature of projectile motion is called the "range." The range, as Figure 3.12 shows, is the horizontal distance traveled between launching and landing, assuming the projectile returns to the *same vertical level* at which it was fired. Example 7 shows how to obtain the range.

Example 7 The Range of a Kickoff

For the motion shown in Figure 3.12 and discussed in Examples 5 and 6, ignore air resistance and calculate the range R of the projectile.

Reasoning The range is a characteristic of the horizontal part of the motion. Thus, our starting point is to determine the horizontal component of the initial velocity:

$$v_{0x} = v_0 \cos\theta = +(22 \text{ m/s}) \cos 40.0° = +17 \text{ m/s}$$

Recall from Example 6 that the time of flight is $t = 2.9$ s. Since there is no acceleration in the x direction, v_x remains constant, and the range is simply the product of $v_x = v_{0x}$ and the time.

Solution The range is

$$x = R = v_{0x}t = +(17 \text{ m/s})(2.9 \text{ s}) = \boxed{+49 \text{ m}}$$

The range in the previous example depends on the angle θ at which the projectile is fired above the horizontal. When air resistance is absent, the maximum range results when $\theta = 45°$.

Concept Simulation 3.2

This simulation will let you explore projectile motion, along with the concepts of maximum height and range. You can control the initial speed and angle of a ball and then see how its velocity components change with time as it moves along the curved path. The simulation also displays graphs of position and velocity as functions of time.

Related Homework: Problems 18, 73

Go to
www.wiley.com/college/cutnell

✔ **Check Your Understanding 3**

A projectile is fired into the air, and it follows the parabolic path shown in the drawing. There is no air resistance. At any instant, the projectile has a velocity **v** and an acceleration **a**. Which one or more of the drawings could *not* represent the directions for **v** and **a** at any point on the trajectory? *(The answer is given at the end of the book.)*

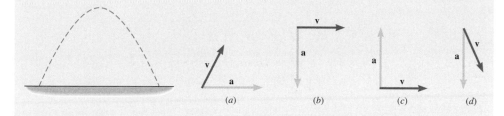

(a)   *(b)*   *(c)*   *(d)*

Background: The fundamental nature of projectile motion lies at the heart of this question, as does the nature of gravity.

For similar questions (including calculational counterparts), consult Self-Assessment Test 3.1. The test is described at the end of this section.

The examples considered thus far have used information about the initial location and velocity of a projectile to determine the final location and velocity. Example 8 deals with the opposite situation and illustrates how the final parameters can be used with the equations of kinematics to determine the initial parameters.

Example 8 A Home Run

A baseball player hits a home run, and the ball lands in the left-field seats, 7.5 m above the point at which it was hit. It lands with a velocity of 36 m/s at an angle of 28° below the horizontal (see Figure 3.13). Ignoring air resistance, find the initial velocity with which the ball leaves the bat.

Reasoning To find the initial velocity, we must determine its magnitude (the initial speed v_0) and its direction (the angle θ in the drawing). These quantities are related to the horizontal and vertical components of the initial velocity (v_{0x} and v_{0y}) by the relations

$$v_0 = \sqrt{v_{0x}^2 + v_{0y}^2} \quad \text{and} \quad \theta = \tan^{-1}\left(\frac{v_{0y}}{v_{0x}}\right)$$

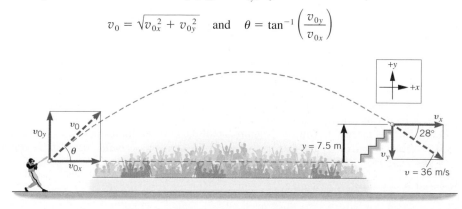

Figure 3.13 The velocity and location of the baseball upon landing can be used to determine its initial velocity, as Example 8 illustrates.

Need more practice?

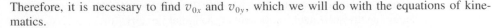

Interactive LearningWare 3.2
In 1971 astronaut Alan Shepard walked on the moon's surface. In a moment of whimsy, he hit a golf ball, which was launched upward and followed the familiar trajectory. However, the trajectory differed from what it would have been on earth, because the acceleration due to gravity on the moon is about six times smaller than that on earth. Consider the same ball on earth, launched at the same angle, with the same speed. Find the ratio of (a) the maximum height of the moon-ball to the maximum height of the earth-ball and (b) the range of the moon-ball to the range of the earth-ball.

Related Homework: *Problems 19, 23*

Go to
www.wiley.com/college/cutnell
for an interactive solution.

Therefore, it is necessary to find v_{0x} and v_{0y}, which we will do with the equations of kinematics.

Solution Since air resistance is being ignored, the horizontal component of the velocity v_x remains constant throughout the motion. Thus,

$$v_{0x} = v_x = +(36 \text{ m/s}) \cos 28° = +32 \text{ m/s}$$

The value for v_{0y} can be obtained from Equation 3.6b and the data displayed below (see Figure 3.13 for the positive and negative directions):

y-Direction Data				
y	a_y	v_y	v_{0y}	t
+7.5 m	−9.80 m/s²	(−36 sin 28°) m/s	?	

$$v_y^2 = v_{0y}^2 + 2a_y y \quad \text{or} \quad v_{0y} = +\sqrt{v_y^2 - 2a_y y}$$

$$v_{0y} = +\sqrt{[(-36 \sin 28°) \text{ m/s}]^2 - 2(-9.80 \text{ m/s}^2)(7.5 \text{ m})} = +21 \text{ m/s}$$

In determining v_{0y} we choose the plus sign for the square root, because the vertical component of the initial velocity points upward in Figure 3.13, which is the positive direction. The initial speed v_0 and angle θ of the baseball are

$$v_0 = \sqrt{v_{0x}^2 + v_{0y}^2} = \sqrt{(32 \text{ m/s})^2 + (21 \text{ m/s})^2} = \boxed{38 \text{ m/s}}$$

$$\theta = \tan^{-1}\left(\frac{v_{0y}}{v_{0x}}\right) = \tan^{-1}\left(\frac{21 \text{ m/s}}{32 \text{ m/s}}\right) = \boxed{33°}$$

In projectile motion, the magnitude of the acceleration due to gravity affects the trajectory in a significant way. For example, a baseball or a golf ball would travel much farther and higher on the moon than on the earth, when launched with the same initial velocity. The reason is that the moon's gravity is only about one-sixth as strong as the earth's.

Section 2.6 points out that certain types of symmetry with respect to time and speed are present for freely falling bodies. These symmetries are also found in projectile motion, since projectiles are falling freely in the vertical direction. In particular, the time required for a projectile to reach its maximum height H is equal to the time spent returning to the ground. In addition, Figure 3.14 shows that the speed v of the object at any height above the ground on the upward part of the trajectory is equal to the speed v at the same height on the downward part. Although the two speeds are the same, the velocities are different, because they point in different directions. Conceptual Example 9 shows how to use this type of symmetry in your reasoning.

Figure 3.14 The speed v of a projectile at a given height above the ground is the same on the upward and downward parts of the trajectory. The velocities are different, however, since they point in different directions.

Conceptual Example 9   Two Ways to Throw a Stone

From the top of a cliff overlooking a lake, a person throws two stones. The stones have identical initial speeds v_0, but stone 1 is thrown downward at an angle θ below the horizontal, while stone 2 is thrown upward at the same angle above the horizontal, as Figure 3.15 shows. Neglect air resistance and decide which stone, if either, strikes the water with the greater velocity.

Reasoning and Solution We might guess that stone 1, being hurled downward, would strike the water with the greater velocity. To show that this is not true, let's follow the path of stone 2 as it rises to its maximum height and falls back to earth. Notice point P in the drawing, where stone 2 returns to its initial height; here the speed of stone 2 is v_0, but its velocity is directed at an angle θ below the horizontal. This is exactly the type of projectile symmetry illustrated in Figure 3.14. At this point, then, stone 2 has a velocity that is identical to the velocity with which stone 1 is thrown downward from the top of the cliff. From this point on, the velocity of stone 2 changes in exactly the same way as that for stone 1, so ***both stones strike the water with the same velocity.***

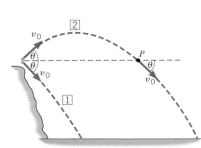

Figure 3.15 Two stones are thrown off the cliff with identical initial speeds v_0, but at equal angles θ that are below and above the horizontal. Conceptual Example 9 compares the velocities with which the stones hit the water below.

Related Homework: *Problems 37, 65*

In all the examples in this section, the projectiles follow a curved trajectory. In general, if the only acceleration is that due to gravity, the shape of the path can be shown to be a *parabola*.

3.4 Relative Velocity

To someone hitchhiking along a highway, two cars speeding by in adjacent lanes seem like a blur. But if the cars have the same velocity, each driver sees the other remaining in place, one lane away. The hitchhiker observes a velocity of perhaps 30 m/s, but each driver observes the other's velocity to be zero. Clearly, the velocity of an object is relative to the observer who is making the measurement.

Figure 3.16 illustrates the concept of relative velocity by showing a passenger walking toward the front of a moving train. The people sitting on the train see the passenger walking with a velocity of +2.0 m/s, where the plus sign denotes a direction to the right. Suppose the train is moving with a velocity of +9.0 m/s relative to an observer standing on the ground. Then the ground-based observer would see the passenger moving with a velocity of +11 m/s, due in part to the walking motion and in part to the train's motion. As an aid in describing relative velocity, let us define the following symbols:

v_{PT} = velocity of the $\boxed{\text{Passenger}}$ relative to the $\boxed{\text{Train}}$ = +2.0 m/s

v_{TG} = velocity of the $\boxed{\text{Train}}$ relative to the $\boxed{\text{Ground}}$ = +9.0 m/s

v_{PG} = velocity of the $\boxed{\text{Passenger}}$ relative to the $\boxed{\text{Ground}}$ = +11 m/s

In terms of these symbols, the situation in Figure 3.16 can be summarized as follows:

$$v_{PG} = v_{PT} + v_{TG} \qquad (3.7)$$

or

$$v_{PG} = (2.0 \text{ m/s}) + (9.0 \text{ m/s}) = +11 \text{ m/s}$$

According to Equation 3.7, v_{PG} is the vector sum of v_{PT} and v_{TG}, and this sum is shown in the drawing. Had the passenger been walking toward the rear of the train, rather than the front, the velocity relative to the ground-based observer would have been $v_{PG} = (-2.0 \text{ m/s}) + (9.0 \text{ m/s}) = +7.0 \text{ m/s}$.

Each velocity symbol in Equation 3.7 contains a two-letter subscript. The first letter in the subscript refers to the body that is moving, while the second letter indicates the object relative to which the velocity is measured. For example, v_{TG} and v_{PG} are the velocities of the **T**rain and **P**assenger measured relative to the **G**round. Similarly, v_{PT} is the velocity of the **P**assenger measured by an observer sitting on the **T**rain.

The ordering of the subscript symbols in Equation 3.7 follows a definite pattern. The first subscript (P) on the left side of the equation is also the first subscript on the right

Midair refueling offers an interesting example of relative velocity. To refuel, the lower plane matches its velocity to that of the tanker (the larger aircraft) and couples to the tanker's delivery tube. During refueling, the relative velocity of the two planes is zero. (© George Hall/Corbis Images)

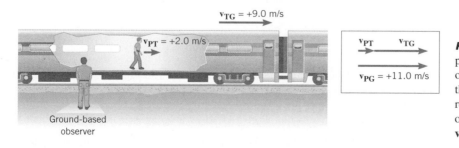

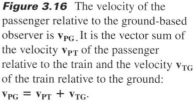

Figure 3.16 The velocity of the passenger relative to the ground-based observer is v_{PG}. It is the vector sum of the velocity v_{PT} of the passenger relative to the train and the velocity v_{TG} of the train relative to the ground: $v_{PG} = v_{PT} + v_{TG}$.

side of the equation. Likewise, the last subscript (G) on the left side is also the last subscript on the right side. The third subscript (T) appears only on the right side of the equation as the two "inner" subscripts. The colored boxes below emphasize the pattern of the symbols in the subscripts:

$$v_{\boxed{PG}} = v_{\boxed{P}T} + v_{T\boxed{G}}$$

In other situations, the subscripts will not necessarily be P, G, and T, but will be compatible with the names of the objects involved in the motion.

✔ Check Your Understanding 4

Three cars, A, B, and C, are moving along a straight section of a highway. The velocity of A relative to B is v_{AB}, the velocity of A relative to C is v_{AC}, and the velocity of C relative to B is v_{CB}. Fill in the missing velocities in the table. *(The answers are given at the end of the book.)*

	v_{AB}	v_{AC}	v_{CB}
1.	?	+40 m/s	+30 m/s
2.	?	+50 m/s	−20 m/s
3.	+60 m/s	+20 m/s	?
4.	−50 m/s	?	+10 m/s

Background: The relative velocities of three (or more) objects are related by means of vector addition. Consider how the subscripts are ordered in this addition.

For similar questions (including conceptual counterparts), consult Self-Assessment Test 3.2. The test is described at the end of this section.

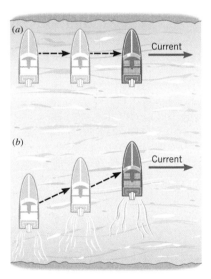

Figure 3.17 (*a*) A boat with its engine turned off is carried along by the current. (*b*) With the engine turned on, the boat moves across the river in a diagonal fashion.

Equation 3.7 has been presented in connection with one-dimensional motion, but the result is also valid for two-dimensional motion. Figure 3.17 depicts a common situation that deals with relative velocity in two dimensions. Part *a* of the drawing shows a boat being carried downstream by a river; the engine of the boat is turned off. In part *b*, the engine has been turned on, and now the boat moves across the river in a diagonal fashion because of the combined motion produced by the current and the engine. The list below gives the velocities for this type of motion and the objects relative to which they are measured:

$$v_{\boxed{BW}} = \text{velocity of the } \boxed{\text{Boat}} \text{ relative to the } \boxed{\text{Water}}$$

$$v_{\boxed{WS}} = \text{velocity of the } \boxed{\text{Water}} \text{ relative to the } \boxed{\text{Shore}}$$

$$v_{\boxed{BS}} = \text{velocity of the } \boxed{\text{Boat}} \text{ relative to the } \boxed{\text{Shore}}$$

The velocity v_{BW} of the boat relative to the water is the velocity measured by an observer who, for instance, is floating on an inner tube and drifting downstream with the current. When the engine is turned off, the boat also drifts downstream with the current, and v_{BW} is zero. When the engine is turned on, however, the boat can move relative to the water, and v_{BW} is no longer zero. The velocity v_{WS} of the water relative to the shore is the velocity of the current measured by an observer on the shore. The velocity v_{BS} of the boat relative to the shore is due to the combined motion of the boat relative to the water and the motion of the water relative to the shore. In symbols,

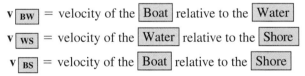

The ordering of the subscripts in this equation is identical to that in Equation 3.7, although the letters have been changed to reflect a different physical situation. Example 10 illustrates the concept of relative velocity in two dimensions.

Example 10   Crossing a River

The engine of a boat drives it across a river that is 1800 m wide. The velocity v_{BW} of the boat relative to the water is 4.0 m/s, directed perpendicular to the current, as in Figure 3.18. The velocity v_{WS} of the water relative to the shore is 2.0 m/s. (a) What is the velocity v_{BS} of the boat relative to the shore? (b) How long does it take for the boat to cross the river?

Reasoning

(a) The velocity of the boat relative to the shore is $\mathbf{v}_{BS}$. It is the vector sum of the velocity $\mathbf{v}_{BW}$ of the boat relative to the water and the velocity $\mathbf{v}_{WS}$ of the water relative to the shore: $\mathbf{v}_{BS} = \mathbf{v}_{BW} + \mathbf{v}_{WS}$. Since $\mathbf{v}_{BW}$ and $\mathbf{v}_{WS}$ are both known, we can use this relation among the relative velocities, with the aid of trigonometry, to find the magnitude and directional angle of $\mathbf{v}_{BS}$.

(b) The component of $\mathbf{v}_{BS}$ that is parallel to the width of the river (see Figure 3.18) determines how fast the boat moves across the river; this parallel component is $v_{BS} \sin \theta = v_{BW} = 4.0$ m/s. The time for the boat to cross the river is equal to the width of the river divided by the magnitude of this velocity component.

Solution

(a) Since the vectors $\mathbf{v}_{BW}$ and $\mathbf{v}_{WS}$ are perpendicular (see Figure 3.18), the magnitude of $\mathbf{v}_{BS}$ can be determined by using the Pythagorean theorem:

$$v_{BS} = \sqrt{v_{BW}^2 + v_{WS}^2} = \sqrt{(4.0 \text{ m/s})^2 + (2.0 \text{ m/s})^2} = \boxed{4.5 \text{ m/s}}$$

Thus, the boat moves at a speed of 4.5 m/s with respect to an observer on shore. The direction of the boat relative to the shore is given by the angle θ in the drawing:

$$\tan \theta = \frac{v_{RW}}{v_{WS}} \quad \text{or} \quad \theta = \tan^{-1}\left(\frac{v_{BW}}{v_{WS}}\right) - \tan^{-1}\left(\frac{4.0 \text{ m/s}}{2.0 \text{ m/s}}\right) = \boxed{63°}$$

(b) The time t for the boat to cross the river is

$$t = \frac{\text{Width}}{v_{BS} \sin \theta} = \frac{1800 \text{ m}}{4.0 \text{ m/s}} = \boxed{450 \text{ s}}$$

▲

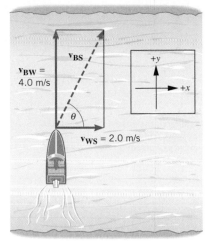

Figure 3.18 The velocity of the boat relative to the shore is $\mathbf{v}_{BS}$. It is the vector sum of the velocity $\mathbf{v}_{BW}$ of the boat relative to the water and the velocity $\mathbf{v}_{WS}$ of the water relative to the shore: $\mathbf{v}_{BS} = \mathbf{v}_{BW} + \mathbf{v}_{WS}$.

Occasionally, situations arise when two vehicles are in relative motion, and it is useful to know the relative velocity of one with respect to the other. Example 11 considers this type of relative motion.

Example 11 Approaching an Intersection

Figure 3.19a shows two cars approaching an intersection along perpendicular roads. The cars have the following velocities:

$\mathbf{v}_{AG}$ = velocity of car A relative to the Ground = 25.0 m/s, eastward

$\mathbf{v}_{BG}$ = velocity of car B relative to the Ground = 15.8 m/s, northward

Find the magnitude and direction of $\mathbf{v}_{AB}$, where

$\mathbf{v}_{AB}$ = velocity of car A as measured by a passenger in car B

Concept Simulation 3.3

This simulation illustrates the concept of relative velocity by considering a boat traveling across a flowing river. You can change the speed and direction of the boat relative to the water, as well as the velocity of the water. The simulation then shows the velocity of the boat as viewed by a person standing on the shore.

Go to
www.wiley.com/college/cutnell

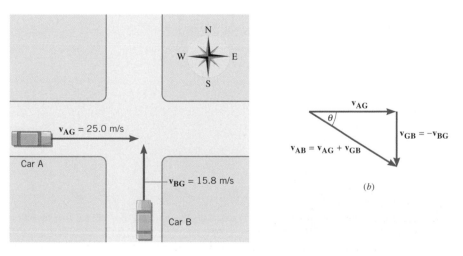

(a)

$\mathbf{v}_{AG} = 25.0$ m/s

Car A

$\mathbf{v}_{BG} = 15.8$ m/s

Car B

$\mathbf{v}_{AG}$

θ

$\mathbf{v}_{GB} = -\mathbf{v}_{BG}$

$\mathbf{v}_{AB} = \mathbf{v}_{AG} + \mathbf{v}_{GB}$

(b)

Figure 3.19 Two cars are approaching an intersection along perpendicular roads.

Reasoning To find $\mathbf{v}_{AB}$, we use an equation whose subscripts follow the order outlined earlier. Thus,

$$\mathbf{v}_{\boxed{AB}} = \mathbf{v}_{\boxed{A}G} + \mathbf{v}_{G\boxed{B}}$$

In this equation, the term $\mathbf{v}_{GB}$ is the velocity of the ground relative to a passenger in car B, rather than $\mathbf{v}_{BG}$, which is given as 15.8 m/s, northward. In other words, the subscripts are reversed. However, $\mathbf{v}_{GB}$ is related to $\mathbf{v}_{BG}$ according to

$$\mathbf{v}_{GB} = -\mathbf{v}_{BG}$$

This relationship reflects the fact that a passenger in car B, moving northward relative to the ground, looks out the car window and sees the ground moving southward, in the opposite direction. Therefore, the equation $\mathbf{v}_{AB} = \mathbf{v}_{AG} + \mathbf{v}_{GB}$ may be used to find $\mathbf{v}_{AB}$, provided we recognize $\mathbf{v}_{GB}$ as a vector that points opposite to the given velocity $\mathbf{v}_{BG}$. With this in mind, Figure 3.19b illustrates how $\mathbf{v}_{AG}$ and $\mathbf{v}_{GB}$ are added vectorially to give $\mathbf{v}_{AB}$.

Solution From the vector triangle in Figure 3.19b, the magnitude and direction of $\mathbf{v}_{AB}$ can be calculated as

$$v_{AB} = \sqrt{v_{AG}^2 + v_{GB}^2} = \sqrt{(25.0 \text{ m/s})^2 + (-15.8 \text{ m/s})^2} = \boxed{29.6 \text{ m/s}}$$

and

$$\cos\theta = \frac{v_{AG}}{v_{AB}} \quad \text{or} \quad \theta = \cos^{-1}\left(\frac{v_{AG}}{v_{AB}}\right) = \cos^{-1}\left(\frac{25.0 \text{ m/s}}{29.6 \text{ m/s}}\right) = \boxed{32.4°}$$

Problem solving insight
In general, the velocity of object R relative to object S is always the negative of the velocity of object S relative to R:
$\mathbf{v}_{RS} = -\mathbf{v}_{SR}$.

The physics of raindrops falling on car windows.

While driving a car, have you ever noticed that the rear window sometimes remains dry, even though rain is falling? This phenomenon is a consequence of relative velocity, as Figure 3.20 helps to explain. Part a shows a car traveling horizontally with a velocity of $\mathbf{v}_{CG}$ and a raindrop falling vertically with a velocity of $\mathbf{v}_{RG}$. Both velocities are measured relative to the ground. To determine whether the raindrop hits the window, however, we need to consider the velocity of the raindrop relative to the car, not the ground. This velocity is $\mathbf{v}_{RC}$, and we know that

$$\mathbf{v}_{RC} = \mathbf{v}_{RG} + \mathbf{v}_{GC} = \mathbf{v}_{RG} - \mathbf{v}_{CG}$$

Here, we have used the fact that $\mathbf{v}_{GC} = -\mathbf{v}_{CG}$. Part b of the drawing shows the tail-to-head arrangement corresponding to this vector addition and indicates that the direction of $\mathbf{v}_{RC}$ is given by the angle θ_R. In comparison, the rear window is inclined at an angle θ_W with respect to the vertical (see the blowup in part a). When θ_R is greater than θ_W, the raindrop will miss the window. However, θ_R is determined by the speed v_{RG} of the raindrop and the speed v_{CG} of the car, according to $\theta_R = \tan^{-1}(v_{CG}/v_{RG})$. At higher car speeds, the angle θ_R becomes too large for the drop to hit the window. At a high enough speed, then, the car simply drives out from under each falling drop!

Self-Assessment Test 3.2

Test your understanding of the key ideas in Section 3.4:

• Relative Velocity   • Vector Addition of Relative Velocities

Go to **www.wiley.com/college/cutnell**

Figure 3.20 (a) With respect to the ground, a car is traveling at a velocity of $\mathbf{v}_{CG}$ and a raindrop is falling at a velocity of $\mathbf{v}_{RG}$. The rear window of the car is inclined at an angle θ_W with respect to the vertical. (b) This tail-to-head arrangement of vectors corresponds to the equation $\mathbf{v}_{RC} = \mathbf{v}_{RG} - \mathbf{v}_{CG}$.

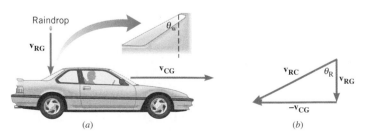

3.5 *Concepts & Calculations*

A primary focus of this chapter has been projectile motion. This section presents two additional examples that serve as a review of the basic features of this type of motion. Example 12 deals with the fact that projectile motion consists of a horizontal and a vertical part, which occur independently of one another. Example 13 stresses the fact that the time variable has the same value for both the horizontal and vertical parts of the motion.

Concepts & Calculations Example 12 Projectile Motion

In a circus act, Burpy the clown is fired from a cannon at an initial velocity $\mathbf{v_0}$ directed at an angle θ above the horizontal, as Figure 3.21 shows. Simultaneously, two other clowns are also launched. Bingo is fired straight upward at a speed of 10.0 m/s and reaches the same maximum height at the same instant as Burpy. Bongo, however, is launched horizontally on roller skates at a speed of 4.6 m/s. He rolls along the ground while Burpy flies through the air. When Burpy returns to the ground, he lands side by side with his roller-skating friend, who is gliding by just at the instant of landing. Ignore air resistance, and assume that the roller skates are unimpeded by friction. Find the speed v_0 and the angle θ for Burpy.

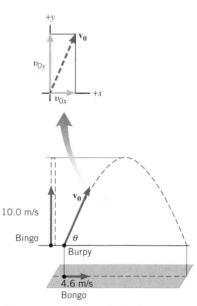

Figure 3.21 Example 12 discusses the projectile motion shown here, in which three circus clowns, Burpy, Bongo, and Bingo, are launched simultaneously.

Concept Questions and Answers Is Bongo's left-to-right motion the same as or different from the horizontal part of Burpy's motion along his trajectory?

Answer The horizontal and vertical parts of projectile motion occur independently of one another. Therefore, since Burpy lands side by side with Bongo, Bongo's left-to-right motion is identical to the horizontal part of Burpy's motion along his trajectory.

Is Burpy's motion in the horizontal direction determined by the initial velocity $\mathbf{v_0}$, just its horizontal component v_{0x}, or just its vertical component v_{0y}?

Answer Just the horizontal component v_{0x} determines Burpy's motion in the horizontal direction.

Is Bingo's up-and-down motion the same as or different from the vertical part of Burpy's motion along his trajectory?

Answer The horizontal and vertical parts of projectile motion occur independently of one another. Therefore, since they reach the same maximum height at the same instant, Bingo's up-and-down motion is identical to the vertical part of Burpy's motion along his trajectory.

Is Burpy's initial motion in the vertical direction determined by the initial velocity $\mathbf{v_0}$, just its horizontal component v_{0x}, or just its vertical component v_{0y}?

Answer Just the vertical component v_{0y} determines Burpy's motion in the vertical direction.

Solution Based on the Concept Questions and Answers, we can identify the x and y components of Burpy's initial velocity as follows:

$$v_{0x} = 4.6 \text{ m/s} \quad \text{and} \quad v_{0y} = 10.0 \text{ m/s}$$

The initial speed v_0 is the magnitude of the initial velocity, and it, along with the directional angle θ, can be determined from the components:

$$v_0 = \sqrt{v_{0x}^2 + v_{0y}^2} = \sqrt{(4.6 \text{ m/s})^2 + (10.0 \text{ m/s})^2} = \boxed{11 \text{ m/s}}$$

$$\theta = \tan^{-1}\left(\frac{v_{0y}}{v_{0x}}\right) = \tan^{-1}\left(\frac{10.0 \text{ m/s}}{4.6 \text{ m/s}}\right) = \boxed{65°}$$

Concepts & Calculations Example 13 Time and Projectile Motion

A projectile is launched from and returns to ground level, as Figure 3.22 shows. Air resistance is absent. The horizontal range of the projectile is $R = 175$ m, and the horizontal component of the launch velocity is $v_{0x} = 25$ m/s. Find the vertical component v_{0y} of the launch velocity.

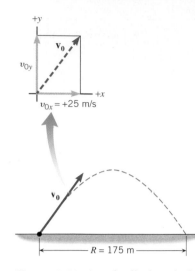

Figure 3.22 A projectile, launched with a velocity whose horizontal component is $v_{0x} = +25$ m/s, has a range of $R = 175$ m. From these data the vertical component v_{0y} of the initial velocity can be determined.

Concept Questions and Answers What is the final value of the horizontal component v_x of the projectile's velocity?

Answer   The final value v_x of the horizontal component of the projectile's velocity is the same as the initial value in the absence of air resistance. In other words, the horizontal motion occurs at a constant velocity of 25 m/s.

Can the time be determined for the horizontal part of the motion?

Answer   Yes. In constant-velocity motion, the time is just the horizontal distance (the range) divided by the magnitude of the horizontal component of the projectile's velocity.

Is the time for the horizontal part of the motion the same as the time for the vertical part of the motion?

Answer   Yes. The value for the time calculated for the horizontal part of the motion can be used to analyze the vertical part of the motion.

For the vertical part of the motion, what is the displacement of the projectile?

Answer   Since the projectile is launched from and returns to ground level, the vertical displacement is zero.

Solution From the constant-velocity horizontal motion, we find that the time is

$$t = \frac{R}{v_{0x}} = \frac{175 \text{ m}}{25 \text{ m/s}} = 7.0 \text{ s}$$

For the vertical part of the motion, we know that the displacement is zero and that the acceleration due to gravity is -9.80 m/s^2, assuming that upward is the positive direction. Therefore, we can use Equation 3.5b to find the initial y component of the velocity:

$$y = v_{0y}t + \tfrac{1}{2}a_y t^2 \quad \text{or} \quad 0 \text{ m} = v_{0y}t + \tfrac{1}{2}a_y t^2$$
$$v_{0y} = -\tfrac{1}{2}a_y t = -\tfrac{1}{2}(-9.80 \text{ m/s}^2)(7.0 \text{ s}) = \boxed{34 \text{ m/s}}$$

▲

At the end of the problem set for this chapter, you will find homework problems that contain both conceptual and quantitative parts. These problems are grouped under the heading *Concepts & Calculations, Group Learning Problems*. They are designed for use by students working alone or in small learning groups. The conceptual part of each problem provides a convenient focus for group discussions.

Concept Summary

This summary presents an abridged version of the chapter, including the important equations and all available learning aids. For convenient reference, the learning aids (including the text's examples) are placed next to or immediately after the relevant equation or discussion. The following learning aids may be found on-line at **www.wiley.com/college/cutnell:**

Interactive LearningWare examples are solved according to a five-step interactive format that is designed to help you develop problem-solving skills.	**Concept Simulations** are animated versions of text figures or animations that illustrate important concepts. You can control parameters that affect the display, and we encourage you to experiment.
Interactive Solutions offer specific models for certain types of problems in the chapter homework. The calculations are carried out interactively.	**Self-Assessment Tests** include both qualitative and quantitative questions. Extensive feedback is provided for both incorrect and correct answers, to help you evaluate your understanding of the material.

Topic	Discussion	Learning Aids

3.1 Displacement, Velocity, and Acceleration

Displacement vector   The position of an object is located with a vector **r** drawn from the coordinate origin to the object. The displacement $\Delta\mathbf{r}$ of the object is defined as $\Delta\mathbf{r} = \mathbf{r} - \mathbf{r_0}$, where **r** and $\mathbf{r_0}$ specify its final and initial positions, respectively.

Average velocity   The average velocity $\bar{\mathbf{v}}$ of an object moving between two positions is defined as   **Interactive Solution 3.11** its displacement $\Delta\mathbf{r} = \mathbf{r} - \mathbf{r_0}$ divided by the elapsed time $\Delta t = t - t_0$:

$$\bar{\mathbf{v}} = \frac{\mathbf{r} - \mathbf{r_0}}{t - t_0} = \frac{\Delta\mathbf{r}}{\Delta t} \tag{3.1}$$

Topic	Discussion	Learning Aids
Instantaneous velocity	The instantaneous velocity $\mathbf{v}$ is the velocity at an instant of time. The average velocity becomes equal to the instantaneous velocity in the limit that the elapsed time Δt becomes infinitesimally small ($\Delta t \rightarrow 0$ s): $$\mathbf{v} = \lim_{\Delta t \to 0} \frac{\Delta \mathbf{r}}{\Delta t}$$	
Average acceleration	The average acceleration $\bar{\mathbf{a}}$ of an object is the change in its velocity $\Delta \mathbf{v} = \mathbf{v} - \mathbf{v}_0$ divided by the elapsed time $\Delta t = t - t_0$: $$\bar{\mathbf{a}} = \frac{\mathbf{v} - \mathbf{v}_0}{t - t_0} = \frac{\Delta \mathbf{v}}{\Delta t} \qquad (3.2)$$	Interactive LearningWare 3.1
Instantaneous acceleration	The instantaneous acceleration $\mathbf{a}$ is the acceleration at an instant of time. The average acceleration becomes equal to the instantaneous acceleration in the limit that the elapsed time Δt becomes infinitesimally small: $$\mathbf{a} = \lim_{\Delta t \to 0} \frac{\Delta \mathbf{v}}{\Delta t}$$	

3.2 Equations of Kinematics in Two Dimensions

	Motion in two dimensions can be described in terms of the time t and the x and y components of four vectors: the displacement, the acceleration, and the initial and final velocities.	Example 1
Independence of the x and y parts of the motion	The x part of the motion occurs exactly as it would if the y part did not occur at all. Similarly, the y part of the motion occurs exactly as it would if the x part of the motion did not exist. The motion can be analyzed by treating the x and y components of the four vectors separately and realizing that the time t is the same for each component.	
Equations of kinematics for constant acceleration	When the acceleration is constant, the x components of the displacement, the acceleration, and the initial and final velocities are related by the equations of kinematics, and so are the y components:	

x Component		y Component	
$v_x = v_{0x} + a_x t$	(3.3a)	$v_y = v_{0y} + a_y t$	(3.3b)
$x = \frac{1}{2}(v_{0x} + v_x)t$	(3.4a)	$y = \frac{1}{2}(v_{0y} + v_y)t$	(3.4b)
$x = v_{0x}t + \frac{1}{2}a_x t^2$	(3.5a)	$y = v_{0y}t + \frac{1}{2}a_y t^2$	(3.5b)
$v_x^2 = v_{0x}^2 + 2a_x x$	(3.6a)	$v_y^2 = v_{0y}^2 + 2a_y y$	(3.6b)

The directions of these components are conveyed by assigning a plus ($+$) or minus ($-$) sign to each one.

3.3 Projectile Motion

Topic	Discussion	Learning Aids
Acceleration in projectile motion	Projectile motion is an idealized kind of motion that occurs when a moving object (the projectile) experiences only the acceleration due to gravity, which acts vertically downward. If the trajectory of the projectile is near the earth's surface, a_y has a magnitude of 9.80 m/s². The acceleration has no horizontal component ($a_x = 0$ m/s²), the effects of air resistance being negligible.	Examples 2–8, 12, 13 Concept Simulations 3.1, 3.2 Interactive LearningWare 3.2 Interactive Solutions 3.35, 3.39, 3.67
Symmetries in projectile motion	There are several symmetries in projectile motion: (1) The time to reach maximum height from any point is equal to the time spent returning from the maximum height to that point. (2) The speed of a projectile depends only on its height above its launch point, and not on whether it is moving upward or downward.	Example 9

Use Self-Assessment Test 3.1 to evaluate your understanding of Sections 3.1–3.3.

3.4 Relative Velocity

Topic	Discussion	Learning Aids
Adding relative velocities	The velocity of object A relative to object B is written as $\mathbf{v}_{AB}$, and the velocity of object B relative to object C is $\mathbf{v}_{BC}$. The velocity of A relative to C is (note the ordering of the subscripts) $$\mathbf{v}_{AC} = \mathbf{v}_{AB} + \mathbf{v}_{BC}$$	Example 10 Concept Simulation 3.3 Interactive Solution 3.55

Topic	Discussion	Learning Aids
	While the velocity of object A relative to object B is $\mathbf{v}_{AB}$, the velocity of B relative to A is $\mathbf{v}_{BA} = -\mathbf{v}_{AB}$.	Example 11

Use Self-Assessment Test 3.2 to evaluate your understanding of Section 3.4.

Conceptual Questions

1. Suppose an object could move in three dimensions. What additions to the equations of kinematics in Table 3.1 would be necessary to describe three-dimensional motion?

2. An object is thrown upward at an angle θ above the ground, eventually returning to earth. (a) Is there any place along the trajectory where the velocity and acceleration are perpendicular? If so, where? (b) Is there any place where the velocity and acceleration are parallel? If so, where? In each case, explain.

3. Is the acceleration of a projectile equal to zero when it reaches the top of its trajectory? If not, why not?

4. In baseball, the pitcher's mound is raised to compensate for the fact that the ball falls downward as it travels from the pitcher toward the batter. If baseball were played on the moon, would the pitcher's mound have to be higher than, lower than, or the same height as it is on earth? Give your reasoning.

5. A tennis ball is hit upward into the air and moves along an arc. Neglecting air resistance, where along the arc is the speed of the ball (a) a minimum and (b) a maximum? Justify your answers.

6. Suppose there is a wind blowing parallel to the ground and toward the kicker in Figure 3.12. Then the acceleration component in the horizontal direction would not be zero. How would you expect the time of flight of the football to be affected, if at all? Explain.

7. **Concept Simulation 3.1** at **www.wiley.com/college/cutnell** provides a review of the concepts that are important in this question. A wrench is accidentally dropped from the top of the mast on a sailboat. Will the wrench hit at the same place on the deck whether the sailboat is at rest or moving with a constant velocity? Justify your answer.

8. A rifle, at a height H above the ground, fires a bullet parallel to the ground. At the same instant and at the same height, a second bullet is dropped from rest. In the absence of air resistance, which bullet strikes the ground first? Explain.

9. Two projectiles are launched from ground level at the same angle above the horizontal, and both return to ground level. Projectile A has a launch speed that is twice that of projectile B. Assuming that air resistance is absent, sketch the trajectories of both projectiles. If your drawings are to be accurate, what should be the ratio of the maximum heights in your drawings and what should be the ratio of the ranges? Justify your answers.

10. A stone is thrown horizontally from the top of a cliff and eventually hits the ground below. A second stone is dropped from rest from the same cliff, falls through the same height, and also hits the ground below. Ignore air resistance. Discuss whether each of the following quantities is different or the same in the two cases; if there is a difference, describe the difference: (a) displacement, (b) speed just before impact with the ground, and (c) time of flight.

11. A leopard springs upward at a 45° angle and then falls back to the ground. Does the leopard, at any point on its trajectory, ever have a speed that is one-half its initial value? Give your reasoning.

12. As background for this question, review Conceptual Example 4. A football quarterback throws a pass on the run and then keeps running without changing his velocity. Can he throw the pass and then catch it himself? Give your reasoning.

13. On a riverboat cruise, a plastic bottle is accidentally dropped overboard. A passenger on the boat estimates that the boat pulls ahead of the bottle by 5 meters each second. Is it possible to conclude that the boat is moving at 5 m/s with respect to the shore? Account for your answer.

14. A plane takes off at St. Louis, flies straight to Denver, and then returns the same way. The plane flies at the same speed with respect to the ground during the entire flight, and there are no head winds or tail winds. Since the earth revolves around its axis once a day, you might expect that the times for the outbound trip and the return trip differ, depending on whether the plane flies against the earth's rotation or with it. However, under the conditions given, the two flight times are identical. Explain why.

15. A child is playing on the floor of a recreational vehicle (RV) as it moves along the highway at a constant velocity. He has a toy cannon, which shoots a marble at a fixed angle and speed with respect to the floor. The cannon can be aimed toward the front or the rear of the RV. Is the range toward the front the same as, less than, or greater than the range toward the rear? Answer this question (a) from the child's point of view and (b) from the point of view of an observer standing still on the ground. Justify your answers.

16. Three swimmers can swim equally fast relative to the water. They have a race to see who can swim across a river in the least time. Swimmer A swims perpendicular to the current and lands on the far shore downstream, because the current has swept him in that direction. Swimmer B swims upstream at an angle to the current and lands on the far shore directly opposite the starting point. Swimmer C swims downstream at an angle to the current in an attempt to take advantage of the current. Who crosses the river in the least time? Account for your answer.

Problems

ssm   Solution is in the Student Solutions Manual.   www   Solution is available on the World Wide Web at www.wiley.com/college/cutnell
☤   This icon represents a biomedical application.

Section 3.1 Displacement, Velocity, and Acceleration

1. ssm In diving to a depth of 750 m, an elephant seal also moves 460 m due east of his starting point. What is the magnitude of the seal's displacement?

2. A baseball player hits a triple and ends up on third base. A baseball "diamond" is a square, each side of length 27.4 m, with home plate and the three bases on the four corners. What is the magnitude of his displacement?

3. A mountain-climbing expedition establishes two intermediate camps, labeled A and B in the drawing, above the base camp. What is the magnitude Δr of the displacement between camp A and camp B?

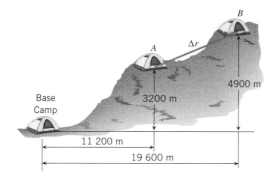

4. The altitude of a hang glider is increasing at a rate of 6.80 m/s. At the same time, the shadow of the glider moves along the ground at a speed of 15.5 m/s when the sun is directly overhead. Find the magnitude of the glider's velocity.

5. ssm A jetliner is moving at a speed of 245 m/s. The vertical component of the plane's velocity is 40.6 m/s. Determine the magnitude of the horizontal component of the plane's velocity.

6. In a football game a kicker attempts a field goal. The ball remains in contact with the kicker's foot for 0.050 s, during which time it experiences an acceleration of 340 m/s². The ball is launched at an angle of 51° above the ground. Determine the horizontal and vertical components of the launch velocity.

7. A dolphin leaps out of the water at an angle of 35° above the horizontal. The horizontal component of the dolphin's velocity is 7.7 m/s. Find the magnitude of the vertical component of the velocity.

8. A skateboarder, starting from rest, rolls down a 12.0-m ramp. When she arrives at the bottom of the ramp her speed is 7.70 m/s. (a) Determine the magnitude of her acceleration, assumed to be constant. (b) If the ramp is inclined at 25.0° with respect to the ground, what is the component of her acceleration that is parallel to the ground?

9. ssm www In a mall, a shopper rides up an escalator between floors. At the top of the escalator, the shopper turns right and walks 9.00 m to a store. The magnitude of the shopper's displacement from the bottom of the escalator is 16.0 m. The vertical distance between the floors is 6.00 m. At what angle is the escalator inclined above the horizontal?

* 10. Interactive LearningWare 3.1 at www.wiley.com/college/cutnell reviews the approach taken in problems such as this one. A bird watcher meanders through the woods, walking 0.50 km due east,

0.75 km due south, and 2.15 km in a direction 35.0° north of west. The time required for this trip is 2.50 h. Determine the magnitude and direction (relative to due west) of the bird watcher's (a) displacement and (b) average velocity. Use kilometers and hours for distance and time, respectively.

* 11. Interactive Solution 3.11 at www.wiley.com/college/cutnell presents a model for solving this problem. The earth moves around the sun in a nearly circular orbit of radius 1.50×10^{11} m. During the three summer months (an elapsed time of 7.89×10^6 s), the earth moves one-fourth of the distance around the sun. (a) What is the average speed of the earth? (b) What is the magnitude of the average velocity of the earth during this period?

Section 3.2 Equations of Kinematics in Two Dimensions, Section 3.3 Projectile Motion

12. A spacecraft is traveling with a velocity of $v_{0x} = 5480$ m/s along the $+x$ direction. Two engines are turned on for a time of 842 s. One engine gives the spacecraft an acceleration in the $+x$ direction of $a_x = 1.20$ m/s², while the other gives it an acceleration in the $+y$ direction of $a_y = 8.40$ m/s². At the end of the firing, find (a) v_x and (b) v_y.

13. A tennis ball is struck such that it leaves the racket horizontally with a speed of 28.0 m/s. The ball hits the court at a horizontal distance of 19.6 m from the racket. What is the height of the tennis ball when it leaves the racket?

14. A volleyball is spiked so that it has an initial velocity of 15 m/s directed downward at an angle of 55° below the horizontal. What is the horizontal component of the ball's velocity when the opposing player fields the ball?

15. ssm A golf ball rolls off a horizontal cliff with an initial speed of 11.4 m/s. The ball falls a vertical distance of 15.5 m into a lake below. (a) How much time does the ball spend in the air? (b) What is the speed v of the ball just before it strikes the water?

16. A rock climber throws a small first aid kit to another climber who is higher up the mountain. The initial velocity of the kit is 11 m/s at an angle of 65° above the horizontal. At the instant when the kit is caught, it is traveling horizontally, so its vertical speed is zero. What is the vertical height between the two climbers?

17. ssm www A diver runs horizontally with a speed of 1.20 m/s off a platform that is 10.0 m above the water. What is his speed just before striking the water?

18. Concept Simulation 3.2 at www.wiley.com/college/cutnell reviews the concepts that are important in this problem. A golfer imparts a speed of 30.3 m/s to a ball, and it travels the maximum possible distance before landing on the green. The tee and the green are at the same elevation. (a) How much time does the ball spend in the air? (b) What is the longest "hole in one" that the golfer can make, if the ball does not roll when it hits the green?

19. Review **Interactive LearningWare 3.2** at www.wiley.com/college/cutnell in preparation for this problem. The acceleration due to gravity on the moon has a magnitude of 1.62 m/s². Examples 5–7 deal with a placekicker kicking a football. Assume that the ball is kicked on the moon instead of on the earth. Find (a) the maximum height H and (b) the range that the ball would attain on the moon.

20. Michael Jordan, formerly of the Chicago Bulls basketball team, has some fanatic fans. They claim that he is able to jump and remain in the air for two full seconds from launch to landing. Evaluate this claim by calculating the maximum height that such a jump would attain. For comparison, Jordan's maximum jump height has been estimated at about one meter.

21. ssm A fire hose ejects a stream of water at an angle of 35.0° above the horizontal. The water leaves the nozzle with a speed of 25.0 m/s. Assuming that the water behaves like a projectile, how far from a building should the fire hose be located to hit the highest possible fire?

22. A car drives straight off the edge of a cliff that is 54 m high. The police at the scene of the accident note that the point of impact is 130 m from the base of the cliff. How fast was the car traveling when it went over the cliff?

23. Interactive LearningWare 3.2 at **www.wiley.com/college/cutnell** provides a review of the concepts that are important in this problem. On a distant planet, golf is just as popular as it is on earth. A golfer tees off and drives the ball 3.5 times as far as he would have on earth, given the same initial velocities on both planets. The ball is launched at a speed of 45 m/s at an angle of 29° above the horizontal. When the ball lands, it is at the same level as the tee. On the distant planet, what is (a) the maximum height and (b) the range of the ball?

24. The 1994 Winter Olympics included the aerials competition in skiing. In this event skiers speed down a ramp that slopes sharply upward at the end. The sharp upward slope launches them into the air, where they perform acrobatic maneuvers. In the women's competition, the end of a typical launch ramp is directed 63° above the horizontal. With this launch angle, a skier attains a height of 13 m above the end of the ramp. What is the skier's launch speed?

25. ssm An eagle is flying horizontally at 6.0 m/s with a fish in its claws. It accidentally drops the fish. (a) How much time passes before the fish's speed doubles? (b) How much additional time would be required for the fish's speed to double again?

26. A major-league pitcher can throw a baseball in excess of 41.0 m/s. If a ball is thrown horizontally at this speed, how much will it drop by the time it reaches a catcher who is 17.0 m away from the point of release?

27. A motorcycle daredevil is attempting to jump across as many buses as possible (see the drawing). The takeoff ramp makes an angle of 18.0° above the horizontal, and the landing ramp is identical to the takeoff ramp. The buses are parked side by side, and each bus is 2.74 m wide. The cyclist leaves the ramp with a speed of 33.5 m/s. What is the maximum number of buses over which the cyclist can jump?

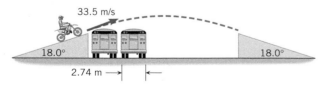

28. Suppose the water at the top of Niagara Falls has a horizontal speed of 2.7 m/s just before it cascades over the edge of the falls. At what vertical distance below the edge does the velocity vector of the water point downward at a 75° angle below the horizontal?

29. ssm A hot-air balloon is rising straight up with a speed of 3.0 m/s. A ballast bag is released from rest relative to the balloon when it is 9.5 m above the ground. How much time elapses before the ballast bag hits the ground?

30. A rocket is fired at a speed of 75.0 m/s from ground level, at an angle of 60.0° above the horizontal. The rocket is fired toward an 11.0-m-high wall, which is located 27.0 m away. By how much does the rocket clear the top of the wall?

31. A horizontal rifle is fired at a bull's-eye. The muzzle speed of the bullet is 670 m/s. The barrel is pointed directly at the center of the bull's-eye, but the bullet strikes the target 0.025 m below the center. What is the horizontal distance between the end of the rifle and the bull's-eye?

* **32.** An airplane with a speed of 97.5 m/s is climbing upward at an angle of 50.0° with respect to the horizontal. When the plane's altitude is 732 m, the pilot releases a package. (a) Calculate the distance along the ground, measured from a point directly beneath the point of release, to where the package hits the earth. (b) Relative to the ground, determine the angle of the velocity vector of the package just before impact.

* **33. ssm** An Olympic long jumper leaves the ground at an angle of 23° and travels through the air for a horizontal distance of 8.7 m before landing. What is the takeoff speed of the jumper?

* **34.** Review Conceptual Example 4 and **Concept Simulation 3.1** at **www.wiley.com/college/cutnell** before beginning this problem. You are traveling in a convertible with the top down. The car is moving at a constant velocity of 25 m/s, due east along flat ground. You throw a tomato straight upward at a speed of 11 m/s. How far has the car moved when you get a chance to catch the tomato?

* **35.** As an aid in working this problem, consult **Interactive Solution 3.35** at **www.wiley.com/college/cutnell**. A soccer player kicks the ball toward a goal that is 16.8 m in front of him. The ball leaves his foot at a speed of 16.0 m/s and an angle of 28.0° above the ground. Find the speed of the ball when the goalie catches it in front of the net.

* **36.** In the javelin throw at a track-and-field event, the javelin is launched at a speed of 29 m/s at an angle of 36° above the horizontal. As the javelin travels upward, its velocity points above the horizontal at an angle that decreases as time passes. How much time is required for the angle to be reduced from 36° at launch to 18°?

* **37. ssm www** As preparation for this problem, review Conceptual Example 9. The two stones described there have identical initial speeds of $v_0 = 13.0$ m/s and are thrown at an angle $\theta = 30.0°$, one below the horizontal and one above the horizontal. What is the distance *between* the points where the stones strike the ground?

* **38.** Stones are thrown horizontally with the same velocity from the tops of two different buildings. One stone lands twice as far from the base of the building from which it was thrown as does the other stone. Find the ratio of the height of the taller building to the height of the shorter building.

* **39.** Before beginning this problem, review **Interactive Solution 3.39** at **www.wiley.com/college/cutnell**. After leaving the end of a ski ramp, a ski jumper lands downhill at a point that is displaced 51.0 m horizontally from the end of the ramp. His velocity, just before landing, is 23.0 m/s and points in a direction 43.0° below the horizontal. Neglecting air resistance and any lift he experiences while airborne, find his initial velocity (magnitude and direction) when he left the end of the ramp. Express the direction as an angle relative to the horizontal.

* **40.** The lob in tennis is an effective tactic when your opponent is near the net. It consists of lofting the ball over his head, forcing him to move quickly away from the net (see the drawing). Suppose that you loft the ball with an initial speed of 15.0 m/s, at an angle of 50.0° above the horizontal. At this instant your opponent is 10.0 m away from the ball. He begins moving away from you 0.30 s later, hoping to reach the ball and hit it back at the moment that it is 2.10 m above its launch point. With what minimum average speed

must he move? (Ignore the fact that he can stretch, so that his racket can reach the ball before he does.)

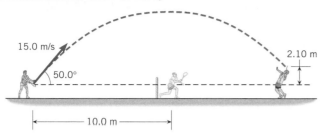

41. ssm The drawing shows an exaggerated view of a rifle that has been "sighted in" for a 91.4-meter target. If the muzzle speed of the bullet is $v_0 = 427$ m/s, what are the two possible angles θ_1 and θ_2 between the rifle barrel and the horizontal such that the bullet will hit the target? One of these angles is so large that it is never used in target shooting. (*Hint: The following trigonometric identity may be useful: $2 \sin \theta \cos \theta = \sin 2\theta$.*)

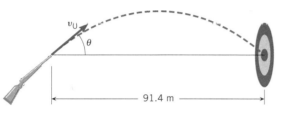

42. A placekicker is about to kick a field goal. The ball is 26.9 m from the goalpost. The ball is kicked with an initial velocity of 19.8 m/s at an angle θ above the ground. Between what two angles, θ_1 and θ_2, will the ball clear the 2.74-m-high crossbar? (*Hint: The following trigonometric identities may be useful: $\sec \theta = 1/(\cos \theta)$ and $\sec^2 \theta = 1 + \tan^2 \theta$.*)

43. ssm From the top of a tall building, a gun is fired. The bullet leaves the gun at a speed of 340 m/s, parallel to the ground. As the drawing shows, the bullet puts a hole in a window of another building and hits the wall that faces the window. Using the data in the drawing, determine the distances D and H, which locate the point where the gun was fired. Assume that the bullet does not slow down as it passes through the window.

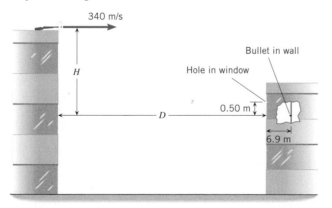

44. A baseball is hit into the air at an initial speed of 36.6 m/s and an angle of 50.0° above the horizontal. At the same time, the center fielder starts running away from the batter and catches the ball 0.914 m above the level at which it was hit. If the center fielder is initially 1.10×10^2 m from home plate, what must be his average speed?

45. Two cannons are mounted as shown in the drawing and rigged to fire simultaneously. They are used in a circus act in which two clowns

serve as human cannonballs. The clowns are fired toward each other and collide at a height of 1.00 m above the muzzles of the cannons. Clown A is launched at a 75.0° angle, with a speed of 9.00 m/s. The horizontal separation between the clowns as they leave the cannons is 6.00 m. Find the launch speed v_{0B} and the launch angle $\theta_B (>45.0°)$ for clown B.

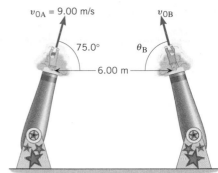

46. A small can is hanging from the ceiling. A rifle is aimed directly at the can, as the figure illustrates. At the instant the gun is fired, the can is released. Ignore air resistance and show that the bullet will always strike the can, regardless of the initial speed of the bullet. Assume that the bullet strikes the can before the can reaches the ground.

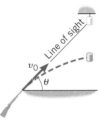

Section 3.4 Relative Velocity

47. ssm Two passenger trains are passing each other on adjacent tracks. Train A is moving east with a speed of 13 m/s, and train B is traveling west with a speed of 28 m/s. (a) What is the velocity (magnitude and direction) of train A as seen by the passengers in train B? (b) What is the velocity (magnitude and direction) of train B as seen by the passengers in train A?

48. Two cars, A and B, are traveling in the same direction, although car A is 186 m behind car B. The speed of A is 24.4 m/s, and the speed of B is 18.6 m/s. How much time does it take for A to catch B?

49. At some airports there are speed ramps to help passengers get from one place to another. A speed ramp is a moving conveyor belt that you can either stand or walk on. Suppose a speed ramp has a length of 105 m and is moving at a speed of 2.0 m/s relative to the ground. In addition, suppose you can cover this distance in 75 s when walking on the ground. If you walk at the same rate with respect to the speed ramp that you walk with respect to the ground, how long does it take for you to travel the 105 m using the speed ramp?

50. The escalator that leads down into a subway station has a length of 30.0 m and a speed of 1.8 m/s relative to the ground. A student is coming out of the station by running in the wrong direction on this escalator. The local record time for this trick is 11 s. Relative to the escalator, what speed must the student exceed in order to beat the record?

51. ssm A swimmer, capable of swimming at a speed of 1.4 m/s in still water (i.e., the swimmer can swim with a speed of 1.4 m/s relative to the water), starts to swim directly across a 2.8-km-wide river. However, the current is 0.91 m/s, and it carries the swimmer downstream. (a) How long does it take the swimmer to cross the river? (b) How far downstream will the swimmer be upon reaching the other side of the river?

52. On a pleasure cruise a boat is traveling relative to the water at a speed of 5.0 m/s due south. Relative to the boat, a passenger walks toward the back of the boat at a speed of 1.5 m/s. (a) What is the magnitude and direction of the passenger's velocity relative to the water? (b) How long does it take for the passenger to walk a distance of 27 m on the boat? (c) How long does it take for the passenger to cover a distance of 27 m on the water?

53. You are in a hot-air balloon that, relative to the ground, has a velocity of 6.0 m/s in a direction due east. You see a hawk moving directly away from the balloon in a direction due north. The speed of the hawk relative to you is 2.0 m/s. What are the magnitude and direction of the hawk's velocity relative to the ground? Express the directional angle relative to due east.

54. The captain of a plane wishes to proceed due west. The cruising speed of the plane is 245 m/s relative to the air. A weather report indicates that a 38.0-m/s wind is blowing from the south to the north. In what direction, measured with respect to due west, should the pilot head the plane relative to the air?

* **55.** As an aid in working this problem, consult **Interactive Solution 3.55** at **www.wiley.com/college/cutnell**. A ferryboat is traveling in a direction 38.0° north of east with a speed of 5.50 m/s relative to the water. A passenger is walking with a velocity of 2.50 m/s due east relative to the boat. What is the velocity (magnitude and direction) of the passenger with respect to the water? Determine the directional angle relative to due east.

* **56.** A person looking out the window of a stationary train notices that raindrops are falling vertically down at a speed of 5.0 m/s relative to the ground. When the train

moves at a constant velocity, the raindrops make an angle of 25° when they move past the window, as the drawing shows. How fast is the train moving?

* **57. ssm** A small aircraft is headed due south with a speed of 57.8 m/s with respect to still air. Then, for 9.00×10^2 s a wind blows the plane so that it moves in a direction 45.0° west of south, even though the plane continues to point due south. The plane travels 81.0 km with respect to the ground in this time. Determine the velocity (magnitude and direction) of the wind with respect to the ground. Determine the directional angle relative to due south.

** **58.** Two boats are heading away from shore. Boat 1 heads due north at a speed of 3.00 m/s relative to the shore. Relative to Boat 1, Boat 2 is moving 30.0° north of east at a speed of 1.60 m/s. A passenger on Boat 2 walks due east across the deck at a speed of 1.20 m/s relative to Boat 2. What is the speed of the passenger relative to the shore?

** **59. ssm www** A Coast Guard ship is traveling at a constant velocity of 4.20 m/s, due east, relative to the water. On his radar screen the navigator detects an object that is moving at a constant velocity. The object is located at a distance of 2310 m with respect to the ship, in a direction 32.0° south of east. Six minutes later, he notes that the object's position relative to the ship has changed to 1120 m, 57.0° south of west. What are the magnitude and direction of the velocity of the object relative to the water? Express the direction as an angle with respect to due west.

Additional Problems

60. A radar antenna is tracking a satellite orbiting the earth. At a certain time, the radar screen shows the satellite to be 162 km away. The radar antenna is pointing upward at an angle of 62.3° from the ground. Find the x and y components (in km) of the position of the satellite.

61. The punter on a football team tries to kick a football so that it stays in the air for a long "hang time." If the ball is kicked with an initial velocity of 25.0 m/s at an angle of 60.0° above the ground, what is the "hang time"?

62. A bullet is fired from a rifle that is held 1.6 m above the ground in a horizontal position. The initial speed of the bullet is 1100 m/s. Find (a) the time it takes for the bullet to strike the ground and (b) the horizontal distance traveled by the bullet.

63. ssm Suppose that the plane in Example 2 is traveling with twice the horizontal velocity—that is, with a velocity of +230 m/s. If all other factors remain the same, determine the time required for the package to hit the ground.

64. On a spacecraft two engines fire for a time of 565 s. One gives the craft an acceleration in the x direction of $a_x = 5.10$ m/s², while the other produces an acceleration in the y direction of $a_y = 7.30$ m/s². At the end of the firing period, the craft has velocity components of $v_x = 3775$ m/s and $v_y = 4816$ m/s. Find the magnitude and direction of the initial velocity. Express the direction as an angle with respect to the +x axis.

65. ssm As preparation for this problem, review Conceptual Example 9. The drawing shows two planes each dropping an empty fuel tank. At the moment of release each plane has the same speed of 135 m/s, and each tank is at the same height of 2.00 km above the ground. Although the speeds are the same, the velocities are different at the instant of release, because one plane is flying at an angle of 15.0° above the horizontal and the other is flying at an angle of 15.0° below the horizontal. Find the magni-

tude and direction of the velocity with which the fuel tank hits the ground if it is from (a) plane A and (b) plane B. In each part, give the directional angles with respect to the horizontal.

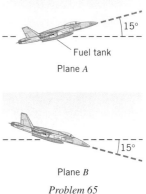

Plane A

Plane B

Problem 65

66. The highest barrier that a projectile can clear is 13.5 m, when the projectile is launched at an angle of 15.0° above the horizontal. What is the projectile's launch speed?

* **67.** Before starting this problem consult **Interactive Solution 3.67** at **www.wiley.com/college/cutnell**. A golfer, standing on a fairway, hits a shot to a green that is elevated 5.50 m above the point where she is standing. If the ball leaves her club with a velocity of 46.0 m/s at an angle of 35.0° above the ground, find the time that the ball is in the air before it hits the green.

* **68.** A diver springs upward from a board that is three meters above the water. At the instant she contacts the water her speed is 8.90 m/s and her body makes an angle of 75.0° with respect to the horizontal surface of the water. Determine her initial velocity, both magnitude and direction.

* **69. ssm** Mario, a hockey player, is skating due south at a speed of 7.0 m/s relative to the ice. A teammate passes the puck to him. The puck has a speed of 11.0 m/s and is moving in a direction of 22° west of south, relative to the ice. What are the magnitude and direction (relative to due south) of the puck's velocity, as observed by Mario?

* **70.** An oceanliner is heading due north with a speed of 8.5 m/s relative to the water. A small sailboat is heading 45° east of north with a speed of 1.0 m/s relative to the water. Find the relative velocity (mag-

nitude and direction) of the sailboat as observed by the passengers on the ocean liner. Determine the directional angle relative to due north.

*71. **ssm** An airplane is flying with a velocity of 240 m/s at an angle of 30.0° with the horizontal, as the drawing shows. When the altitude of the plane is 2.4 km, a flare is released from the plane. The flare hits the target on the ground. What is the angle θ?

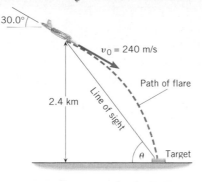

30.0°

v_0 = 240 m/s

Path of flare

2.4 km

Line of sight

θ Target

72. A jetliner can fly 6.00 hours on a full load of fuel. Without any wind it flies at a speed of 2.40×10^2 m/s. The plane is to make a round-trip by heading due west for a certain distance, turning around, and then heading due east for the return trip. During the entire flight, however, the plane encounters a 57.8-m/s wind from the jet stream, which blows from west to east. What is the maximum distance that the plane can travel due west and just be able to return home?

73. Concept Simulation 3.2 at **www.wiley.com/college/cutnell** reviews the general principles in this problem. A projectile is launched from ground level at an angle of 12.0° above the horizontal. It returns to ground level. To what value should the launch angle be adjusted, without changing the launch speed, so that the range doubles?

Concepts & Calculations Group Learning Problems

Note: Each of these problems consists of Concept Questions followed by a related quantitative Problem. They are designed for use by students working alone or in small learning groups. The Concept Questions involve little or no mathematics and are intended to stimulate group discussions. They focus on the concepts with which the problems deal. Recognizing the concepts is the essential initial step in any problem-solving technique.

74. Concept Questions A puck is moving on an air hockey table. Relative to an x, y coordinate system at time $t = 0$ s, the x components of the puck's initial velocity and acceleration are $v_{0x} = +1.0$ m/s and $a_x = +2.0$ m/s². The y components of the initial velocity and acceleration are $v_{0y} = +2.0$ m/s and $a_y = -2.0$ m/s². (a) Is the magnitude of the x component of the velocity increasing or decreasing in time? (b) Is the magnitude of the y component of the velocity increasing or decreasing in time?

Problem Find the magnitude and direction of the puck's velocity at a time of $t = 0.50$ s. Specify the direction relative to the $+x$ axis. Be sure that your calculations are consistent with your answers to the Concept Questions.

75. Concept Questions (a) A projectile is launched at a speed v_0 and at an angle above the horizontal; its initial velocity components are v_{0x} and v_{0y}. In the absence of air resistance, what is the speed of the projectile at the peak of its trajectory? (b) What is its speed just before it lands at the same vertical level from which it was launched? (c) Consider its speed at a point that is at a vertical level between that in (a) and (b). How does the speed at this point compare with the speeds identified in (a) and (b)? In each case, give your reasoning.

Problem A golfer hits a shot to a green that is elevated 3.0 m above the point where the ball is struck. The ball leaves the club at a speed of 14.0 m/s at an angle of 40.0° above the horizontal. It rises to its maximum height and then falls down to the green. Ignoring air resistance, find the speed of the ball just before it lands. Check to see that your answer is consistent with your answer to part (c) of the Concept Questions.

76. Concept Questions (a) When a projectile is launched horizontally from a rooftop at a speed v_{0x}, does its horizontal velocity component ever change in the absence of air resistance? (b) Can you calculate the horizontal distance D traveled after launch simply as $D = v_{0x}t$, where t is the fall time of the projectile? (c) In calculating the fall time, is the vertical part of the motion just like that of a ball dropped from rest? Explain each of your answers.

Problem A criminal is escaping across a rooftop and runs off the roof horizontally at a speed of 5.3 m/s, hoping to land on the roof of an adjacent building. The horizontal distance between the two buildings is D, and the roof of the adjacent building is 2.0 m below the jumping-off point. Find the maximum value for D.

77. Concept Questions (a) When a projectile is thrown at a speed v_0 at an angle above the horizontal, which component of the launch velocity determines the maximum height reached by the projectile? (b) Do the horizontal and vertical components of the launch velocity have larger or smaller magnitudes than v_0? (c) If the projectile is thrown straight upward at a speed v_0, does it attain a greater or lesser height than that in part (a)? In each case, account for your answer.

Problem A ball is thrown upward at a speed v_0 at an angle of 52° above the horizontal. It reaches a maximum height of 7.5 m. How high would this ball go if it were thrown straight upward at speed v_0? Verify that your answer is consistent with your answer to part (c) of the Concept Questions.

78. Concept Questions (a) For a projectile that follows a trajectory like that in Figure 3.12 the range is given by $R = v_{0x}t$, where v_{0x} is the horizontal component of the launch velocity and t is the time of flight. Is v_{0x} proportional to the launch speed v_0? (b) Is the time of flight t proportional to the launch speed v_0? (c) Is the range R proportional to v_0 or v_0^2? Explain each answer.

Problem In the absence of air resistance, a projectile is launched from and returns to ground level. It follows a trajectory similar to that in Figure 3.12 and has a range of 23 m. Suppose the launch speed is doubled, and the projectile is fired at the same angle above the ground. What is the new range?

79. Concept Questions Two friends, Barbara and Neil, are out rollerblading. With respect to the ground, Barbara is skating due south. Neil is in front of her and to her left. With respect to the ground, he is skating due west. (a) Does Barbara see him moving toward the east or toward the west? (b) Does Barbara see him moving toward the north or toward the south? (c) Considering your answers to parts (a) and (b), how does Barbara see Neil moving relative to herself, toward the east and north, toward the east and south, toward the west and north, or toward the west and south? Justify your answers in each case.

Problem With respect to the ground, Barbara is skating due south at a speed of 4.0 m/s. With respect to the ground, Neil is skating due west at a speed of 3.2 m/s. Find Neil's velocity (magnitude and direction relative to due west) as seen by Barbara. Make sure that your answer agrees with your answer to part (c) of the Concept Questions.

Chapter 4

Forces and Newton's Laws of Motion

As this elephant stands on its hind legs to reach some choice morsels, it remains in equilibrium because its weight is balanced by the normal force that the ground exerts on its feet. Equilibrium, weight, and normal force are all concepts discussed in this chapter. (© Age Fotostock America, Inc.)

4.1 The Concepts of Force and Mass

In common usage, a *force* is a push or a pull, as the examples in Figure 4.1 illustrate. In basketball, a player launches a shot by pushing on the ball. The tow bar attached to a speeding boat pulls a water skier. Forces such as those that launch the basketball or pull the skier are called **contact forces,** because they arise from the physical contact between two objects. There are circumstances, however, in which two objects exert forces on one another even though they are not touching. Such forces are referred to as **noncontact forces** or **action-at-a-distance forces.** One example of such a noncontact force occurs when a skydiver is pulled toward the earth because of the force of gravity. The earth exerts this force even when it is not in direct contact with the skydiver. In Figure 4.1, arrows are used to represent the forces. It is appropriate to use arrows, because a force is a vector quantity and has both a magnitude and a direction. The direction of the arrow gives the direction of the force, and the length is proportional to its strength or magnitude.

The word **mass** is just as familiar as the word force. A massive supertanker, for instance, is one that contains an enormous amount of mass. As we will see in the next section, it is difficult to set such a massive object into motion and difficult to bring it to a halt once it is moving. In comparison, a penny does not contain much mass. The emphasis here is on the amount of mass, and the idea of direction is of no concern. Therefore, mass is a scalar quantity.

During the seventeenth century, Isaac Newton, starting with the work of Galileo, developed three important laws that deal with force and mass. Collectively they are called "Newton's laws of motion" and provide the basis for understanding the effect that forces have on an object. Because of the importance of these laws, a separate section will be devoted to each one.

(a)

(b)

4.2 Newton's First Law of Motion

THE FIRST LAW

To gain some insight into Newton's first law, think about the game of ice hockey (Figure 4.2). If a player does not hit a stationary puck, it will remain at rest on the ice. After the puck is struck, however, it coasts on its own across the ice, slowing down only slightly because of friction. Since ice is very slippery, there is only a relatively small amount of friction to slow down the puck. In fact, if it were possible to remove all friction and wind resistance, and if the rink were infinitely large, the puck would coast forever in a straight line at a constant speed. Left on its own, the puck would lose none of the velocity imparted to it at the time it was struck. This is the essence of Newton's first law of motion:

> ■ **NEWTON'S FIRST LAW OF MOTION**
>
> An object continues in a state of rest or in a state of motion at a constant speed along a straight line, unless compelled to change that state by a net force.

In the first law the phrase "net force" is crucial. Often, several forces act simultaneously on a body, and *the net force is the vector sum of all of them.* Individual forces matter only to the extent that they contribute to the total. For instance, if friction and other opposing forces were absent, a car could travel forever at 30 m/s in a straight line, without using any gas after it has come up to speed. In reality gas is needed, but only so that the engine can produce the necessary force to cancel opposing forces such as friction. This cancellation ensures that there is no net force to change the state of motion of the car.

When an object moves at a constant speed along a straight line, its velocity is constant. Newton's first law indicates that a state of rest (zero velocity) and a state of constant velocity are completely equivalent, in the sense that neither one requires the application of a net force to sustain it. *The purpose served when a net force acts on an object is not to sustain the object's velocity, but, rather, to change it.*

(c)

Figure 4.1 The arrow labeled **F** represents the force that acts on the basketball, the water skier, and the skydiver. (*a*. © AFP/Corbis Images, *b*. © Rick Doyle/Corbis Images, *c*. © Taxi/ Getty Images)

Figure 4.2 The game of ice hockey can give some insight into Newton's laws of motion. (© Getty Images News and Sport Services)

The physics of **seat belts.**

INERTIA AND MASS

A greater net force is required to change the velocity of some objects than of others. For instance, a net force that is just enough to cause a bicycle to pick up speed will cause only an imperceptible change in the motion of a freight train. In comparison to the bicycle, the train has a much greater tendency to remain at rest. Accordingly, we say that the train has more *inertia* than the bicycle. Quantitatively, the inertia of an object is measured by its *mass*. The following definition of inertia and mass indicates why Newton's first law is sometimes called the law of inertia:

■ **DEFINITION OF INERTIA AND MASS**

Inertia is the natural tendency of an object to remain at rest or in motion at a constant speed along a straight line. The mass of an object is a quantitative measure of inertia.

SI Unit of Inertia and Mass: kilogram (kg)

The SI unit for mass is the kilogram (kg), whereas the units in the CGS system and the BE system are the gram (g) and the slug (sl), respectively. Conversion factors between these units are given on the page facing the inside of the front cover. Figure 4.3 gives the masses of various objects, ranging from a penny to a supertanker. The larger the mass, the greater is the inertia. Often the words "mass" and "weight" are used interchangeably, but this is incorrect. Mass and weight are different concepts, and Section 4.7 will discuss the distinction between them.

Figure 4.4 shows a useful application of inertia. Automobile seat belts unwind freely when pulled gently, so they can be buckled. But in an accident, they hold you safely in place. One seat-belt mechanism consists of a ratchet wheel, a locking bar, and a pendulum. The belt is wound around a spool mounted on the ratchet wheel. While the car is at rest or moving at a constant velocity, the pendulum hangs straight down, and the locking bar rests horizontally, as the gray part of the drawing shows. Consequently, nothing prevents the ratchet wheel from turning, and the seat belt can be pulled out easily. When the car suddenly slows down in an accident, however, the relatively massive lower part of the pendulum keeps moving forward because of its inertia. The pendulum swings on its pivot into the position shown in color and causes the locking bar to block the rotation of the ratchet wheel, thus preventing the seat belt from unwinding.

AN INERTIAL REFERENCE FRAME

Newton's first law (and also the second law) can appear to be invalid to certain observers. Suppose, for instance, that you are a passenger riding in a friend's car. While the car moves at a constant speed along a straight line, you do not feel the seat pushing against your back to any unusual extent. This experience is consistent with the first law, which indicates that in the absence of a net force you should move with a constant velocity. Suddenly the driver floors the gas pedal. Immediately you feel the seat pressing against your back as the car accelerates. Therefore, you sense that a force is being applied to you. The first law leads you to believe that your motion should change, and, relative to the ground outside, your motion does change. But *relative to the car,* you can see that your motion does *not* change, because you remain stationary with respect to the car. Clearly, Newton's first law does not hold for observers who use the accelerating car as a frame of reference. As a result, such a reference frame is said to be noninertial. All accelerating reference frames are noninertial. In contrast, observers for whom the law of inertia is valid are said to be using *inertial reference frames* for their observations, as defined below:

■ **DEFINITION OF AN INERTIAL REFERENCE FRAME**

An inertial reference frame is one in which Newton's law of inertia is valid.

The acceleration of an inertial reference frame is zero, so it moves with a constant velocity. All of Newton's laws of motion are valid in inertial reference frames, and when we apply these laws, we will be assuming such a reference frame. In particular, the earth itself is a good approximation of an inertial reference frame.

4.3 Newton's Second Law of Motion

Newton's first law indicates that if no net force acts on an object, then the velocity of the object remains unchanged. The second law deals with what happens when a net force does act. Consider a hockey puck once again. When a player strikes a stationary puck, he causes the velocity of the puck to change. In other words, he makes the puck accelerate. The cause of the acceleration is the force that the hockey stick applies. As long as this force acts, the velocity increases, and the puck accelerates. Now, suppose another player strikes the puck and applies twice as much force as the first player does. The greater force produces a greater acceleration. In fact, if the friction between the puck and the ice is negligible, and if there is no wind resistance, the acceleration of the puck is directly proportional to the force. Twice the force produces twice the acceleration. Moreover, the acceleration is a vector quantity, just as the force is, and points in the same direction as the force.

Often, several forces act on an object simultaneously. Friction and wind resistance, for instance, do have some effect on a hockey puck. In such cases, it is the net force, or the vector sum of all the forces acting, that is important. Mathematically, the net force is written as $\Sigma\mathbf{F}$, where the Greek capital letter Σ (sigma) denotes the vector sum. Newton's second law states that the acceleration is proportional to the net force acting on the object.

In Newton's second law, the net force is only one of two factors that determine the acceleration. The other is the inertia or mass of the object. After all, the same net force that imparts an appreciable acceleration to a hockey puck (small mass) will impart very little acceleration to a semitrailer truck (large mass). Newton's second law states that for a given net force, the magnitude of the acceleration is inversely proportional to the mass. Twice the mass means one-half the acceleration, if the same net force acts on both objects. Thus, the second law shows how the acceleration depends on both the net force and the mass, as given in Equation 4.1.

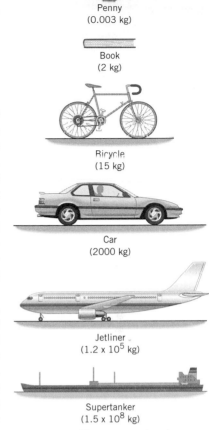

Figure 4.3 The masses of various objects.

Penny
(0.003 kg)

Book
(2 kg)

Bicycle
(15 kg)

Car
(2000 kg)

Jetliner
(1.2×10^5 kg)

Supertanker
(1.5×10^8 kg)

■ **NEWTON'S SECOND LAW OF MOTION**

When a net external force $\Sigma\mathbf{F}$ acts on an object of mass m, the acceleration $\mathbf{a}$ that results is directly proportional to the net force and has a magnitude that is inversely proportional to the mass. The direction of the acceleration is the same as the direction of the net force.

$$\mathbf{a} = \frac{\Sigma\mathbf{F}}{m} \quad \text{or} \quad \Sigma\mathbf{F} = m\mathbf{a} \tag{4.1}$$

SI Unit of Force: $\text{kg}\cdot\text{m/s}^2$ = newton (N)

Note that the net force in Equation 4.1 includes only the forces that the environment exerts on the object of interest. Such forces are called **external forces.** In contrast, **internal forces** are forces that one part of an object exerts on another part of the object and are not included in Equation 4.1.

According to Equation 4.1, the SI unit for force is the unit for mass (kg) times the unit for acceleration (m/s^2), or

$$\text{SI unit for force} = (\text{kg})\left(\frac{\text{m}}{\text{s}^2}\right) = \frac{\text{kg}\cdot\text{m}}{\text{s}^2}$$

The combination of $\text{kg}\cdot\text{m/s}^2$ is called a *newton* (N) and is a derived SI unit, not a base unit; 1 newton = 1 N = 1 $\text{kg}\cdot\text{m/s}^2$.

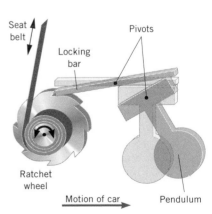

Seat belt

Locking bar

Pivots

Ratchet wheel

Motion of car

Pendulum

Figure 4.4 Inertia plays a central role in one seat-belt mechanism. The gray part of the drawing applies when the car is at rest or moving at a constant velocity. The colored parts show what happens when the car suddenly slows down, as in an accident.

Table 4.1   Units for Mass, Acceleration, and Force

System	Mass	Acceleration	Force
SI	kilogram (kg)	meter/second^2 (m/s^2)	newton (N)
CGS	gram (g)	centimeter/second^2 (cm/s^2)	dyne (dyn)
BE	slug (sl)	foot/second^2 (ft/s^2)	pound (lb)

In the CGS system, the procedure for establishing the unit of force is the same as with SI units, except that mass is expressed in grams (g) and acceleration in cm/s^2. The resulting unit for force is the *dyne*; 1 dyne = 1 g·cm/s^2.

In the BE system, the unit for force is defined to be the pound (lb),* and the unit for acceleration is ft/s^2. With this procedure, Newton's second law can then be used to obtain the unit for mass:

$$\text{BE unit for force} = \text{lb} = (\text{unit for mass})\left(\frac{\text{ft}}{\text{s}^2}\right)$$

$$\text{Unit for mass} = \frac{\text{lb} \cdot \text{s}^2}{\text{ft}}$$

The combination of lb·s^2/ft is the unit for mass in the BE system and is called the *slug* (sl); 1 slug = 1 sl = 1 lb·s^2/ft.

Table 4.1 summarizes the various units for mass, acceleration, and force. Conversion factors between force units from different systems are provided on the page facing the inside of the front cover.

When using the second law to calculate the acceleration, it is necessary to determine the net force that acts on the object. In this determination a *free-body diagram* helps enormously. A free-body diagram is a diagram that represents the object and the forces that act on it. Only the forces that *act on the object* appear in a free-body diagram. Forces that the object exerts on its environment are not included. Example 1 illustrates the use of a free-body diagram.

Problem solving insight
A free-body diagram is very helpful when applying Newton's second law. Always start a problem by drawing the free-body diagram.

Example 1   Pushing a Stalled Car

Two people are pushing a stalled car, as Figure 4.5a indicates. The mass of the car is 1850 kg. One person applies a force of 275 N to the car, while the other applies a force of 395 N. Both forces act in the same direction. A third force of 560 N also acts on the car, but in a direction opposite to that in which the people are pushing. This force arises because of friction and the extent to which the pavement opposes the motion of the tires. Find the acceleration of the car.

Reasoning According to Newton's second law, the acceleration is the net force divided by the mass of the car. To determine the net force, we use the free-body diagram in Figure 4.5b. In this diagram, the car is represented as a dot, and its motion is along the +x axis. The diagram makes it clear that the forces all act along one direction. Therefore, they can be added as colinear vectors to obtain the net force.

Solution The net force is

$$\Sigma F = +275 \text{ N} + 395 \text{ N} - 560 \text{ N} = +110 \text{ N}$$

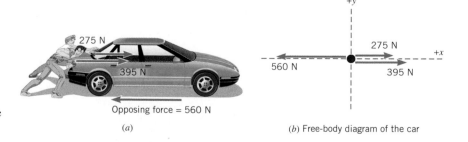

Figure 4.5 (*a*) Two people push a stalled car, in opposition to a force created by friction and the pavement. (*b*) A free-body diagram that shows the horizontal forces acting on the car.

275 N
395 N
Opposing force = 560 N
(*a*)

+y
275 N
560 N
+x
395 N
(*b*) Free-body diagram of the car

* We refer here to the gravitational version of the BE system, in which a force of one pound is defined to be the pull of the earth on a certain standard body at a location where the acceleration due to gravity is 32.174 ft/s^2.

The acceleration can now be obtained:

$$a = \frac{\Sigma F}{m} = \frac{+110 \text{ N}}{1850 \text{ kg}} = \boxed{+0.059 \text{ m/s}^2} \qquad (4.1)$$

Problem solving insight
The direction of the acceleration is always the same as the direction of the net force.

The plus sign indicates that the acceleration points along the $+x$ axis, in the same direction as the net force.

4.4 The Vector Nature of Newton's Second Law of Motion

When a football player throws a pass, the direction of the force he applies to the ball is important. Both the force and the resulting acceleration of the ball are vector quantities, as are all forces and accelerations. The directions of these vectors can be taken into account in two dimensions by using x and y components. The net force $\Sigma\mathbf{F}$ in Newton's second law has components ΣF_x and ΣF_y, while the acceleration $\mathbf{a}$ has components a_x and a_y. Consequently, Newton's second law, as expressed in Equation 4.1, can be written in an equivalent form as two equations, one for the x components and one for the y components:

$$\Sigma F_x = ma_x \qquad (4.2a)$$

$$\Sigma F_y = ma_y \qquad (4.2b)$$

Problem solving insight
Applications of Newton's second law always involve the net external force, which is the vector sum of all the external forces that act on an object. Each component of the net force leads to a corresponding component of the acceleration.

This procedure is similar to that employed in Chapter 3 for the equations of two-dimensional kinematics (see Table 3.1). The components in Equations 4.2a and 4.2b are scalar components and will be either positive or negative numbers, depending on whether they point along the positive or negative x or y axis. The remainder of this section deals with examples that show how these equations are used.

Example 2 Applying Newton's Second Law Using Components

A man is stranded on a raft (mass of man and raft = 1300 kg), as shown in Figure 4.6a. By paddling, he causes an average force **P** of 17 N to be applied to the raft in a direction due east

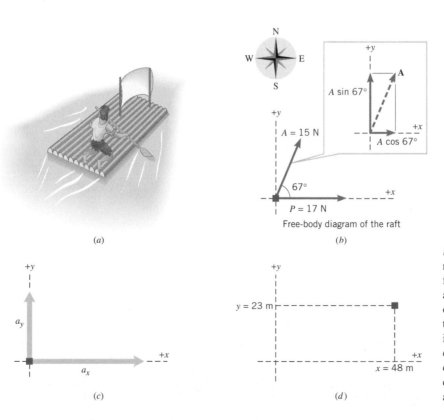

(a)

(b) Free-body diagram of the raft

(c)

(d)

Figure 4.6 (a) A man is paddling a raft, as in Examples 2 and 3. (b) The free-body diagram shows the forces **P** and **A** that act on the raft. Forces acting on the raft in a direction perpendicular to the surface of the water play no role in the examples and are omitted for clarity. (c) The raft's acceleration components a_x and a_y. (d) In 65 s, the components of the raft's displacement are $x = 48$ m and $y = 23$ m.

(the $+x$ direction). The wind also exerts a force **A** on the raft. This force has a magnitude of 15 N and points 67° north of east. Ignoring any resistance from the water, find the x and y components of the raft's acceleration.

Reasoning Since the mass of the man and the raft is known, Newton's second law can be used to determine the acceleration components from the given forces. According to the form of the second law in Equations 4.2a and 4.2b, the acceleration component in a given direction is the component of the net force in that direction divided by the mass. As an aid in determining the components ΣF_x and ΣF_y of the net force, we use the free-body diagram in Figure 4.6b. In this diagram, the direction due east is the $+x$ direction.

Solution Figure 4.6b shows the force components:

Force	x Component	y Component
P	$+17$ N	0 N
A	$+(15\text{ N})\cos 67° = +6$ N	$+(15\text{ N})\sin 67° = +14$ N
	$\Sigma F_x = +17\text{ N} + 6\text{ N} = +23$ N	$\Sigma F_y = +14$ N

The plus signs indicate that ΣF_x points in the direction of the $+x$ axis and ΣF_y points in the direction of the $+y$ axis. The x and y components of the acceleration point in the directions of ΣF_x and ΣF_y, respectively, and can now be calculated:

$$a_x = \frac{\Sigma F_x}{m} = \frac{+23\text{ N}}{1300\text{ kg}} = \boxed{+0.018\text{ m/s}^2} \tag{4.2a}$$

$$a_y = \frac{\Sigma F_y}{m} = \frac{+14\text{ N}}{1300\text{ kg}} = \boxed{+0.011\text{ m/s}^2} \tag{4.2b}$$

These acceleration components are shown in Figure 4.6c.

Example 3 The Displacement of a Raft

At the moment the forces **P** and **A** begin acting on the raft in Example 2, the velocity of the raft is 0.15 m/s, in a direction due east (the $+x$ direction). Assuming that the forces are maintained for 65 s, find the x and y components of the raft's displacement during this time interval.

Reasoning Once the net force acting on an object and the object's mass have been used in Newton's second law to determine the acceleration, it becomes possible to use the equations of kinematics to describe the resulting motion. We know from Example 2 that the acceleration components are $a_x = +0.018$ m/s^2 and $a_y = +0.011$ m/s^2, and it is given here that the initial velocity components are $v_{0x} = +0.15$ m/s and $v_{0y} = 0$ m/s. Thus, Equation 3.5a ($x = v_{0x}t + \frac{1}{2}a_x t^2$) and Equation 3.5b ($y = v_{0y}t + \frac{1}{2}a_y t^2$) can be used with $t = 65$ s to determine the x and y components of the raft's displacement.

Solution According to Equations 3.5a and 3.5b, the x and y components of the displacement are

$$x = v_{0x}t + \tfrac{1}{2}a_x t^2 = (0.15\text{ m/s})(65\text{ s}) + \tfrac{1}{2}(0.018\text{ m/s}^2)(65\text{ s})^2 = \boxed{48\text{ m}}$$

$$y = v_{0y}t + \tfrac{1}{2}a_y t^2 = (0\text{ m/s})(65\text{ s}) + \tfrac{1}{2}(0.011\text{ m/s}^2)(65\text{ s})^2 = \boxed{23\text{ m}}$$

Figure 4.6d shows the final location of the raft.

Need more practice?

Interactive LearningWare 4.1
A catapult on an aircraft carrier is capable of accelerating a 13 300-kg plane from 0 to 56.0 m/s in a distance of 80.0 m. Find the net force, assumed constant, that the jet's engine and the catapult exert on the plane.

Related Homework: Problems 6, 96

Go to
www.wiley.com/college/cutnell
for an interactive solution.

✔ **Check Your Understanding 1**

All of the following, except one, cause the acceleration of an object to double. Which one is it? (a) All forces acting on the object double. (b) The net force acting on the object doubles. (c) Both the net force acting on the object and the mass of the object double. (d) The mass of the object is reduced by a factor of two. *(The answer is given at the end of the book.)*

Background: This problem depends on the concepts of force, net force, mass, and acceleration, because Newton's second law of motion deals with them.

For similar questions (including calculational counterparts), consult Self-Assessment Test 4.1. This test is described at the end of Section 4.5.

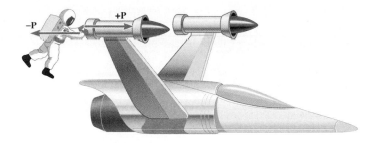

Figure 4.7 The astronaut pushes on the spacecraft with a force $+\mathbf{P}$. According to Newton's third law, the spacecraft simultaneously pushes back on the astronaut with a force $-\mathbf{P}$.

4.5 Newton's Third Law of Motion

Imagine you are in a football game. You line up facing your opponent, the ball is snapped, and the two of you crash together. No doubt, you feel a force. But think about your opponent. He too feels something, for while he is applying a force to you, you are applying a force to him. In other words, there isn't just one force on the line of scrimmage; there is a pair of forces. Newton was the first to realize that all forces occur in pairs and there is no such thing as an isolated force, existing all by itself. His third law of motion deals with this fundamental characteristic of forces.

■ **NEWTON'S THIRD LAW OF MOTION**

Whenever one body exerts a force on a second body, the second body exerts an oppositely directed force of equal magnitude on the first body.

These two polar bears exert action and reaction forces on each other. (© Norbert Rosing/National Geographic/Getty Images)

The third law is often called the "action–reaction" law, because it is sometimes quoted as follows: "For every action (force) there is an equal, but opposite, reaction."

Figure 4.7 illustrates how the third law applies to an astronaut who is drifting just outside a spacecraft and who pushes on the spacecraft with a force $\mathbf{P}$. According to the third law, the spacecraft pushes back on the astronaut with a force $-\mathbf{P}$ that is equal in magnitude but opposite in direction. In Example 4, we examine the accelerations produced by each of these forces.

Example 4 The Accelerations Produced by Action and Reaction Forces

Suppose that the mass of the spacecraft in Figure 4.7 is $m_S = 11\ 000$ kg and that the mass of the astronaut is $m_A = 92$ kg. In addition, assume that the astronaut exerts a force of $\mathbf{P} = +36$ N on the spacecraft. Find the accelerations of the spacecraft and the astronaut.

Reasoning According to Newton's third law, when the astronaut applies the force $\mathbf{P} = +36$ N to the spacecraft, the spacecraft applies a reaction force $-\mathbf{P} = -36$ N to the astronaut. As a result, the spacecraft and the astronaut accelerate in opposite directions. Although the action and reaction forces have the same magnitude, they do not create accelerations of the same magnitude, because the spacecraft and the astronaut have different masses. According to Newton's second law, the astronaut, having a much smaller mass, will experience a much larger acceleration. In applying the second law, we note that the net force acting on the spacecraft is $\Sigma\mathbf{F} = \mathbf{P}$, while the net force acting on the astronaut is $\Sigma\mathbf{F} = -\mathbf{P}$.

Solution Using the second law, we find that the acceleration of the spacecraft is

$$\mathbf{a}_S = \frac{\mathbf{P}}{m_S} = \frac{+36\ \text{N}}{11\ 000\ \text{kg}} = \boxed{+0.0033\ \text{m/s}^2}$$

The acceleration of the astronaut is

$$\mathbf{a}_A = \frac{-\mathbf{P}}{m_A} = \frac{-36\ \text{N}}{92\ \text{kg}} = \boxed{-0.39\ \text{m/s}^2}$$

Problem solving insight
Even though the magnitudes of the action and reaction forces are always equal, these forces do not necessarily produce accelerations that have equal magnitudes, since each force acts on a different object that may have a different mass.

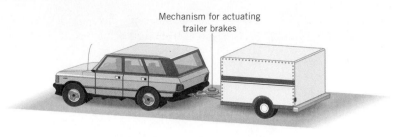

Mechanism for actuating
trailer brakes

Figure 4.8 Some rental trailers
include an automatic brake-actuating
mechanism.

**The physics of
automatic trailer brakes.**

There is a clever application of Newton's third law in some rental trailers. As Figure 4.8 illustrates, the tow bar connecting the trailer to the rear bumper of a car contains a mechanism that can automatically actuate brakes on the trailer wheels. This mechanism works without the need for electrical connections between the car and the trailer. When the driver applies the car brakes, the car slows down. Because of inertia, however, the trailer continues to roll forward and begins pushing against the bumper. In reaction, the bumper pushes back on the tow bar. The reaction force is used by the mechanism in the tow bar to "push the brake pedal" for the trailer.

Self-Assessment Test 4.1

Test your understanding of the material in Sections 4.1–4.5:

• Newton's First Law   • Newton's Second Law   • Newton's Third Law

Go to **www.wiley.com/college/cutnell**

4.6 Types of Forces: An Overview

▶ CONCEPTS AT A GLANCE   Newton's three laws of motion make it clear that forces play a central role in determining the motion of an object. In the next four sections some common forces will be discussed: the gravitational force (Section 4.7), the normal force (Section 4.8), frictional forces (Section 4.9), and the tension force (Section 4.10). In later chapters, we will encounter still others, such as electric and magnetic forces. It is important to realize that Newton's second law is always valid, regardless of which of these forces may act on an object. One does not have a different law for every type of common force. Thus, we need only to determine what forces are acting on an object, add them together to form the net force, and then use Newton's second law to determine the object's acceleration. The Concepts-at-a-Glance chart in Figure 4.9 illustrates this important idea. ◀

Figure 4.9 CONCEPTS AT A GLANCE
When any of the external forces listed here act on an object, they are included as part of the net force $\Sigma\mathbf{F}$ in any application of Newton's second law. In this event of the *World's Strongest Man Competition,* each of the four external forces acts on the man and the truck.
(© Bill Greenblatt/Getty Images Sport Services)

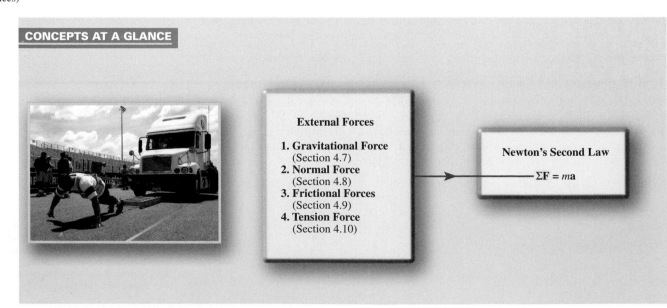

CONCEPTS AT A GLANCE

External Forces

1. **Gravitational Force**
   (Section 4.7)
2. **Normal Force**
   (Section 4.8)
3. **Frictional Forces**
   (Section 4.9)
4. **Tension Force**
   (Section 4.10)

Newton's Second Law

$\Sigma\mathbf{F} = m\mathbf{a}$

In nature there are two general types of forces, fundamental and nonfundamental. Fundamental forces are the ones that are truly unique, in the sense that all other forces can be explained in terms of them. Only three fundamental forces have been discovered:

1. Gravitational force
2. Strong nuclear force
3. Electroweak force

The gravitational force is discussed in the next section. The strong nuclear force plays a primary role in the stability of the nucleus of the atom (see Section 31.2). The electroweak force is a single force that manifests itself in two ways (see Section 32.6). One manifestation is the electromagnetic force that electrically charged particles exert on one another (see Sections 18.5, 21.2, and 21.8). The other manifestation is the so-called weak nuclear force that plays a role in the radioactive disintegration of certain nuclei (see Section 31.5).

Except for the gravitational force, all of the forces discussed in this chapter are non-fundamental, because they are related to the electromagnetic force. They arise from the interactions between the electrically charged particles that comprise atoms and molecules. Our understanding of which forces are fundamental, however, is continually evolving. For instance, in the 1860s and 1870s James Clerk Maxwell showed that the electric force and the magnetic force could be explained as manifestations of a single electromagnetic force. Then, in the 1970s, Sheldon Glashow (1932–), Abdus Salam (1926–1996), and Steven Weinberg (1933–) presented the theory that explains how the electromagnetic force and the weak nuclear force are related to the electroweak force. They received a Nobel prize in 1979 for their achievement. Today, efforts continue that have the goal of reducing further the number of fundamental forces.

4.7 The Gravitational Force
NEWTON'S LAW OF UNIVERSAL GRAVITATION

Objects fall downward because of gravity, and Chapters 2 and 3 discuss how to describe the effects of gravity by using a value of $g = 9.80 \text{ m/s}^2$ for the downward acceleration it causes. However, nothing has been said about why g is 9.80 m/s^2. The reason is fascinating, as we will now see.

The acceleration due to gravity is like any other acceleration, and Newton's second law indicates that it must be caused by a net force. In addition to his famous three laws of motion, Newton also provided a coherent understanding of the *gravitational force*. His "law of universal gravitation" is stated as follows:

■ **NEWTON'S LAW OF UNIVERSAL GRAVITATION**

Every particle in the universe exerts an attractive force on every other particle. A particle is a piece of matter, small enough in size to be regarded as a mathematical point. For two particles that have masses m_1 and m_2 and are separated by a distance r, the force that each exerts on the other is directed along the line joining the particles (see Figure 4.10) and has a magnitude given by

$$F = G\frac{m_1 m_2}{r^2} \tag{4.3}$$

The symbol G denotes the universal gravitational constant, whose value is found experimentally to be

$$G = 6.673 \times 10^{-11} \text{ N·m}^2/\text{kg}^2$$

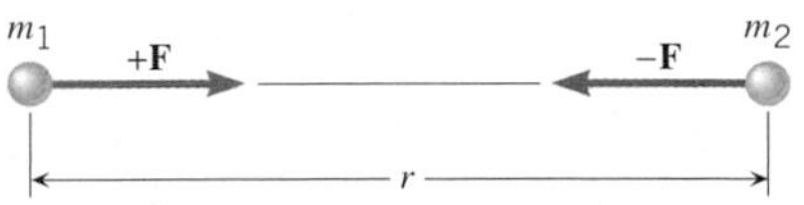

Figure 4.10 The two particles, whose masses are m_1 and m_2, are attracted by gravitational forces $+\mathbf{F}$ and $-\mathbf{F}$.

The constant G that appears in Equation 4.3 is called the ***universal gravitational constant,*** because it has the same value for all pairs of particles anywhere in the universe, no matter what their separation. The value for G was first measured in an experiment by the English scientist Henry Cavendish (1731–1810), more than a century after Newton proposed his law of universal gravitation.

To see the main features of Newton's law of universal gravitation, look at the two particles in Figure 4.10. They have masses m_1 and m_2 and are separated by a distance r. In the picture, it is assumed that a force pointing to the right is positive. The gravitational forces point along the line joining the particles and are

$+\mathbf{F}$, the gravitational force exerted on particle 1 by particle 2

$-\mathbf{F}$, the gravitational force exerted on particle 2 by particle 1

These two forces have equal magnitudes and opposite directions. They act on different bodies, causing them to be mutually attracted. In fact, these forces are an action–reaction pair, as required by Newton's third law. Example 5 shows that the magnitude of the gravitational force is extremely small for ordinary values of the masses and the distance between them.

Example 5   Gravitational Attraction

What is the magnitude of the gravitational force that acts on each particle in Figure 4.10, assuming $m_1 = 12$ kg (approximately the mass of a bicycle), $m_2 = 25$ kg, and $r = 1.2$ m?

Reasoning and Solution The magnitude of the gravitational force can be found using Equation 4.3:

$$F = G\,\frac{m_1 m_2}{r^2} = (6.67 \times 10^{-11}\ \text{N} \cdot \text{m}^2/\text{kg}^2)\,\frac{(12\ \text{kg})(25\ \text{kg})}{(1.2\ \text{m})^2} = \boxed{1.4 \times 10^{-8}\ \text{N}}$$

For comparison, you exert a force of about 1 N when pushing a doorbell, so that the gravitational force is exceedingly small in circumstances such as those here. This result is due to the fact that G itself is very small. However, if one of the bodies has a large mass, like that of the earth (5.98×10^{24} kg), the gravitational force can be large.

As expressed by Equation 4.3, Newton's law of gravitation applies only to particles. However, most familiar objects are too large to be considered particles. Nevertheless, the law of universal gravitation can be applied to such objects with the aid of calculus. Newton was able to prove that an object of finite size can be considered to be a particle for purposes of using the gravitation law, provided the mass of the object is distributed with spherical symmetry about its center. Thus, Equation 4.3 can be applied when each object is a sphere whose mass is spread uniformly over its entire volume. Figure 4.11 shows this kind of application, assuming that the earth and the moon are such uniform spheres of matter. In this case, r is the distance *between the centers of the spheres* and not the distance between the outer surfaces. The gravitational forces that the spheres exert on each other are the same as if the entire mass of each were concentrated at its center. Even if the objects are not uniform spheres, Equation 4.3 can be used to a good degree of approximation if the sizes of the objects are small relative to the distance of separation r.

Problem solving insight
When applying Newton's gravitation law to uniform spheres of matter, remember that the distance r is between the centers of the spheres, not between the surfaces.

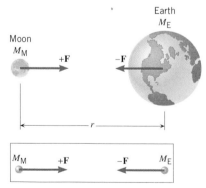

Figure 4.11 The gravitational force that each uniform sphere of matter exerts on the other is the same as if each sphere were a particle with its mass concentrated at its center. The earth (mass M_E) and the moon (mass M_M) approximate such uniform spheres.

WEIGHT

The weight of an object arises because of the gravitational pull of the earth.

■ **DEFINITION OF WEIGHT**

The weight of an object on or above the earth is the gravitational force that the earth exerts on the object. The weight always acts downward, toward the center of the earth. On or above another astronomical body, the weight is the gravitational force exerted on the object by that body.

SI Unit of Weight: newton (N)

Using W for the magnitude of the weight,* m for the mass of the object, and M_E for the mass of the earth, it follows from Equation 4.3 that

$$W = G\frac{M_E m}{r^2} \qquad (4.4)$$

Equation 4.4 and Figure 4.12 both emphasize that an object has weight whether or not it is resting on the earth's surface, because the gravitational force is acting even when the distance r is not equal to the radius R_E of the earth. However, the gravitational force becomes weaker as r increases, since r is in the denominator of Equation 4.4. Figure 4.13, for example, shows how the weight of the Hubble Space Telescope becomes smaller as the distance r from the center of the earth increases. In Example 6 the telescope's weight is determined when it is on earth and in orbit.

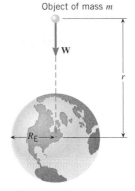

Object of mass m

W

R_E

Mass of earth = M_E

Figure 4.12 On or above the earth, the weight **W** of an object is the gravitational force exerted on the object by the earth.

Example 6 The Hubble Space Telescope

The mass of the Hubble Space Telescope is 11 600 kg. Determine the weight of the telescope (a) when it was resting on the earth and (b) as it is in its orbit 598 km above the earth's surface.

Reasoning The weight of the Hubble Space Telescope is the gravitational force exerted on it by the earth. According to Equation 4.4, the weight varies inversely as the square of the radial distance r. Thus, we expect the telescope's weight on the earth's surface (r smaller) to be greater than its weight in orbit (r larger).

Solution

(a) On the earth's surface, the weight is given by Equation 4.4 with $r = 6.38 \times 10^6$ m (the earth's radius):

$$W = G\frac{M_E m}{r^2} = \frac{(6.67 \times 10^{-11}\ \text{N}\cdot\text{m}^2/\text{kg}^2)(5.98 \times 10^{24}\ \text{kg})(11\ 600\ \text{kg})}{(6.38 \times 10^6\ \text{m})^2}$$

$$\boxed{W = 1.14 \times 10^5\ \text{N}}$$

(b) When the telescope is 598 km above the surface, its distance from the center of the earth is

$$r = 6.38 \times 10^6\ \text{m} + 598 \times 10^3\ \text{m} = 6.98 \times 10^6\ \text{m}$$

The weight now can be calculated as in part (a), except the new value of r must be used: $\boxed{W = 0.950 \times 10^5\ \text{N}}$. As expected, the weight is less in orbit.

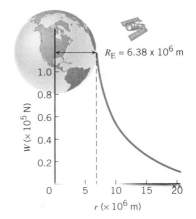

$R_E = 6.38 \times 10^6$ m

Figure 4.13 The weight of the Hubble Space Telescope decreases as it gets farther from the earth. The distance from the center of the earth to the telescope is r.

The space age has forced us to broaden our ideas about weight. For instance, an astronaut weighs only about one-sixth as much on the moon as on the earth. To obtain his weight on the moon from Equation 4.4, it is only necessary to replace M_E by M_M (the mass of the moon) and let $r = R_M$ (the radius of the moon).

RELATION BETWEEN MASS AND WEIGHT

Although massive objects weigh a lot on the earth, mass and weight are not the same quantity. As Section 4.2 discusses, mass is a quantitative measure of inertia. As such, mass is an intrinsic property of matter and does not change as an object is moved from one location to another. Weight, in contrast, is the gravitational force acting on the object and can vary, depending on how far the object is above the earth's surface or whether it is located near another body such as the moon.

---

* Often, the word "weight" and the phrase "magnitude of the weight" are used interchangeably, even though weight is a vector. Generally, the context makes it clear when the direction of the weight vector must be taken into account.

Problem solving insight
Mass and weight are different quantities.
They cannot be interchanged when solving
problems.

The relation between weight W and mass m can be written in one of two ways:

$$W = \boxed{G\frac{M_E\, m}{r^2}} \tag{4.4}$$

$$W = m\,\boxed{g} \tag{4.5}$$

Equation 4.4 is Newton's law of universal gravitation, and Equation 4.5 is Newton's second law (net force equals mass times acceleration) incorporating the acceleration g due to gravity. These expressions make the distinction between mass and weight stand out. The weight of an object whose mass is m depends on the values for the universal gravitational constant G, the mass M_E of the earth, and the distance r. These three parameters together determine the acceleration g due to gravity. The specific value of $g = 9.80 \text{ m/s}^2$ applies only when r equals the radius R_E of the earth. For larger values of r, as would be the case on top of a mountain, the effective value of g is less than 9.80 m/s^2. The fact that g decreases as the distance r increases means that the weight likewise decreases. The mass of the object, however, does not depend on these effects and does not change. Conceptual Example 7 further explores the difference between mass and weight.

Conceptual Example 7   **Mass Versus Weight**

A vehicle is being designed for use in exploring the moon's surface and is being tested on earth, where it weighs roughly six times more than it will on the moon. In one test, the acceleration of the vehicle along the ground is measured. To achieve the same acceleration on the moon, will the net force acting on the vehicle be greater than, less than, or the same as that required on earth?

Reasoning and Solution The net force $\Sigma\mathbf{F}$ required to accelerate the vehicle is specified by Newton's second law as $\Sigma\mathbf{F} = m\mathbf{a}$, where m is the vehicle's mass and $\mathbf{a}$ is the acceleration along the ground. For a given acceleration, the net force depends only on the mass. But the mass is an intrinsic property of the vehicle and is the same on the moon as it is on the earth. Therefore, the same net force would be required for a given acceleration on the moon as on the earth. Do not be misled by the fact that the vehicle weighs more on earth. The greater weight occurs only because the earth's mass and radius are different than the moon's. In any event, *in Newton's second law, the net force is proportional to the vehicle's mass, not its weight.*

Related Homework: *Problems 21, 22*

The lunar exploration vehicle that this astronaut is driving on the moon has the same mass that it has on the earth. However, its weight is different on the moon than on the earth, as Conceptual Example 7 discusses. (© World Perspectives/Stone/Getty Images)

✓ **Check Your Understanding 2**

One object has a mass m_1, and a second object has a mass m_2, which is greater than m_1. The two are separated by a distance $2d$. A third object has a mass m_3. All three objects are located on the same straight line. The net gravitational force acting on the third object is zero. Which of the drawings correctly represents the locations of the objects? *The answer is given at the end of the book.)*

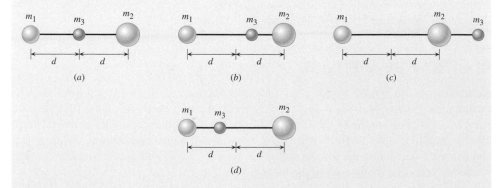

Background: The gravitational force and Newton's law of universal gravitation are the focus of this problem.

For similar questions (including calculational counterparts), consult Self-Assessment Test 4.2. This test is described at the end of Section 4.10.

4.8 The Normal Force

THE NORMAL FORCE AND NEWTON'S THIRD LAW

In many situations, an object is in contact with a surface, such as a tabletop. Because of the contact, there is a force acting on the object. The present section discusses only one component of this force, the component that acts perpendicular to the surface. The next section discusses the component that acts parallel to the surface. The perpendicular component is called the ***normal force.***

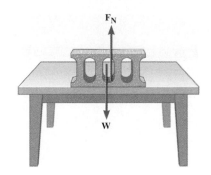

Figure 4.14 Two forces act on the block, its weight **W** and the normal force **F**$_N$ exerted by the surface of the table.

■ **DEFINITION OF THE NORMAL FORCE**

The normal force **F**$_N$ is one component of the force that a surface exerts on an object with which it is in contact—namely, the component that is perpendicular to the surface.

Figure 4.14 shows a block resting on a horizontal table and identifies the two forces that act on the block, the weight **W** and the normal force **F**$_N$. To understand how an inanimate object, such as a tabletop, can exert a normal force, think about what happens when you sit on a mattress. Your weight causes the springs in the mattress to compress. As a result, the compressed springs exert an upward force (the normal force) on you. In a similar manner, the weight of the block causes invisible "atomic springs" in the surface of the table to compress, thus producing a normal force on the block.

Newton's third law plays an important role in connection with the normal force. In Figure 4.14, for instance, the block exerts a force on the table by pressing down on it. Consistent with the third law, the table exerts an oppositely directed force of equal magnitude on the block. This reaction force is the normal force. The magnitude of the normal force indicates how hard the two objects press against each other.

If an object is resting on a horizontal surface and there are no vertically acting forces except the object's weight and the normal force, the magnitudes of these two forces are equal; that is, $F_N = W$. This is the situation in Figure 4.14. The weight must be balanced by the normal force for the object to remain at rest on the table. If the magnitudes of these forces were not equal, there would be a net force acting on the block, and the block would accelerate either upward or downward, in accord with Newton's second law.

If other forces in addition to **W** and **F**$_N$ act in the vertical direction, the magnitudes of the normal force and the weight are no longer equal. In Figure 4.15a, for instance, a box whose weight is 15 N is being pushed downward against a table. The pushing force has a magnitude of 11 N. Thus, the total downward force exerted on the box is 26 N, and this must be balanced by the upwardacting normal force if the box is to remain at rest. In this situation, then, the normal force is 26 N, which is considerably larger than the weight of the box.

Figure 4.15b illustrates a different situation. Here, the box is being pulled upward by a rope that applies a force of 11 N. The net force acting on the box due to its weight and the rope is only 4 N, downward. To balance this force, the normal force needs to be only 4 N. It is not hard to imagine what would happen if the force applied by the rope were increased to 15 N—exactly equal to the weight of the box. In this situation, the normal force would become zero. In fact, the table could be removed, since the block would be supported entirely by the rope. The situations in Figure 4.15 are consistent with the idea that the magnitude of the normal force indicates how hard two objects press against each other. Clearly, the box and the table press against each other harder in part *a* of the picture than in part *b*.

Like the box and the table in Figure 4.15, various parts of the human body press against one another and exert normal forces. Example 8 illustrates the remarkable ability of the human skeleton to withstand a wide range of normal forces.

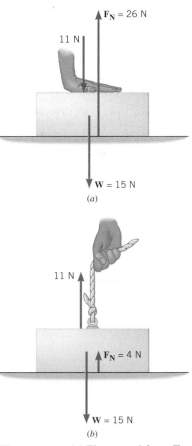

Figure 4.15 (*a*) The normal force **F**$_N$ is greater than the weight of the box, because the box is being pressed downward with an 11-N force. (*b*) The normal force is smaller than the weight, because the rope supplies an upward force of 11 N that partially supports the box.

(a)

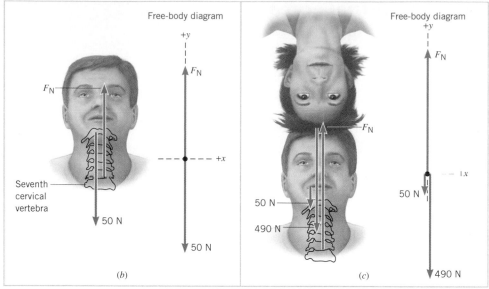

Free-body diagram

(b)

Free-body diagram

(c)

Figure 4.16 (a) A balancing act and free-body diagrams for the man's body above the shoulders (b) before the act and (c) during the act. For convenience, the scales used for the vectors in parts b and c are different. (© Lester Sloan/ Woodfin Camp & Associates)

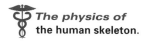
The physics of the human skeleton.

Example 8　A Balancing Act

In a circus balancing act, a woman performs a headstand on top of a man's head, as Figure 4.16a illustrates. The woman weighs 490 N, and the man's head and neck weigh 50 N. It is primarily the seventh cervical vertebra in the spine that supports all the weight above the shoulders. What is the normal force that this vertebra exerts on the neck and head of the man (a) before the act and (b) during the act?

Reasoning To begin, we draw a free-body diagram for the neck and head of the man. Before the act, there are only two forces, the weight of the man's head and neck, and the normal force. During the act, an additional force is present due to the woman's weight. In both cases, the upward and downward forces must balance for the head and neck to remain at rest. This condition of balance will lead us to values for the normal force.

Solution

(a) Figure 4.16b shows the free-body diagram for the man's head and neck before the act. The only forces acting are the normal force $\mathbf{F_N}$ and the 50-N weight. These two forces must balance for the man's head and neck to remain at rest. Therefore, the seventh cervical vertebra exerts a normal force of $\boxed{F_N = 50 \text{ N}}$.

(b) Figure 4.16c shows the free-body diagram that applies during the act. Now, the total downward force exerted on the man's head and neck is 50 N + 490 N = 540 N, which must be balanced by the upward normal force, so that $\boxed{F_N = 540 \text{ N}}$.

In summary, the normal force does not necessarily have the same magnitude as the weight of the object. The value of the normal force depends on what other forces are present. It also depends on whether the objects in contact are accelerating. In one situation that involves accelerating objects, the magnitude of the normal force can be regarded as a kind of "apparent weight," as we will now see.

APPARENT WEIGHT

Usually, the weight of an object can be determined with the aid of a scale. However, even though a scale is working properly, there are situations in which it does not give the correct weight. In such situations, the reading on the scale gives only the "apparent" weight, rather than the gravitational force or "true" weight. The apparent weight is the force that the object exerts on the scale with which it is in contact.

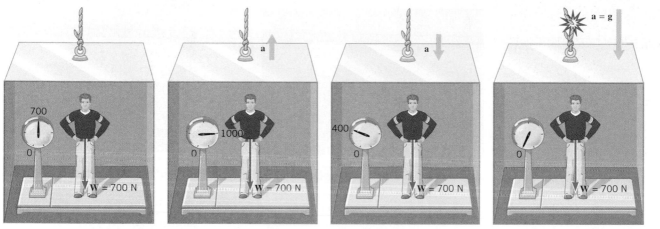

(a) No acceleration (**v** = constant)   (b) Upward acceleration   (c) Downward acceleration   (d) Free-fall

Figure 4.17 (a) When the elevator is not accelerating, the scale registers the true weight ($W = 700$ N) of the person. (b) When the elevator accelerates upward, the apparent weight (1000 N) exceeds the true weight. (c) When the elevator accelerates downward, the apparent weight (400 N) is less than the true weight. (d) The apparent weight is zero if the elevator falls freely—that is, if it falls with the acceleration due to gravity.

To see the discrepancies that can arise between true weight and apparent weight, consider the scale in the elevator in Figure 4.17. The reasons for the discrepancies will be explained shortly. A person whose true weight is 700 N steps on the scale. If the elevator is at rest or moving with a constant velocity (either upward or downward), the scale registers the true weight, as Figure 4.17a illustrates.

If the elevator is accelerating, the apparent weight and the true weight are not equal. When the elevator accelerates upward, the apparent weight is greater than the true weight, as Figure 4.17b shows. Conversely, if the elevator accelerates downward, as in part c, the apparent weight is less than the true weight. In fact, if the elevator falls freely, so its acceleration is equal to the acceleration due to gravity, the apparent weight becomes zero, as part d indicates. In a situation such as this, where the apparent weight is zero, the person is said to be "weightless." The apparent weight, then, does not equal the true weight if the scale and the person on it are accelerating.

The discrepancies between true weight and apparent weight can be understood with the aid of Newton's second law. Figure 4.18 shows a free-body diagram of the person in the elevator. The two forces that act on him are the true weight $\mathbf{W} = m\mathbf{g}$ and the normal force $\mathbf{F_N}$ exerted by the platform of the scale. Applying Newton's second law in the vertical direction gives

$$\Sigma F_y = +F_N - mg = ma$$

where a is the acceleration of the elevator and person. In this result, the symbol g stands for the magnitude of the acceleration due to gravity and can never be a negative quantity. However, the acceleration a may be either positive or negative, depending on whether the elevator is accelerating upward ($+$) or downward ($-$). Solving for the normal force F_N shows that

$$F_N = \underbrace{mg}_{\substack{\text{Apparent} \\ \text{weight}}} + \underbrace{ma}_{\substack{\text{True} \\ \text{weight}}} \qquad (4.6)$$

In Equation 4.6, F_N is the magnitude of the normal force exerted on the person by the scale. But in accord with Newton's third law, F_N is also the magnitude of the downward force that the person exerts on the scale—namely, the apparent weight.

Equation 4.6 contains all the features shown in Figure 4.17. If the elevator is not accelerating, $a = 0$ m/s², and the apparent weight equals the true weight. If the elevator accelerates upward, a is positive, and the equation shows that the apparent weight is greater than the true weight. If the elevator accelerates downward, a is negative, and the apparent weight is less than the true weight. If the elevator falls freely, $a = -g$, and the apparent weight is zero. The apparent weight is zero because when both the person and the scale fall freely, they cannot push against one another. In this text, when the weight is given, it is assumed to be the true weight, unless stated otherwise.

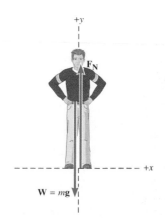

Figure 4.18 A free-body diagram showing the forces acting on the person riding in the elevator of Figure 4.17. **W** is the true weight, and $\mathbf{F_N}$ is the normal force exerted on the person by the platform of the scale.

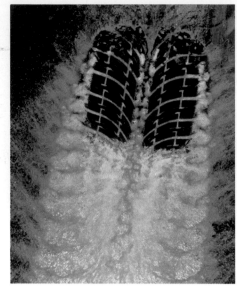

Figure 4.19 This photo, shot from underneath a transparent surface, shows two tires rolling under wet conditions. The channels in the tires collect and divert the water away from the regions where the tires contact the surface, thus providing better traction. (Courtesy Goodyear Tire & Rubber Co.)

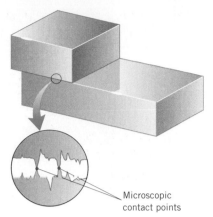

Microscopic contact points

Figure 4.20 Even when two highly polished surfaces are in contact, they touch only at a relatively few points.

4.9 *Static and Kinetic Frictional Forces*

When an object is in contact with a surface, there is a force acting on the object. The previous section discusses the component of this force that is perpendicular to the surface, which is called the normal force. When the object moves or attempts to move along the surface, there is also a component of the force that is parallel to the surface. This parallel force component is called the ***frictional force,*** or simply ***friction.***

In many situations considerable engineering effort is expended trying to reduce friction. For example, oil is used to reduce the friction that causes wear and tear in the pistons and cylinder walls of an automobile engine. Sometimes, however, friction is absolutely necessary. Without friction, car tires could not provide the traction needed to move the car. In fact, the raised tread on a tire is designed to maintain friction. On a wet road, the spaces in the tread pattern (see Figure 4.19) provide channels for the water to collect and be diverted away. Thus, these channels largely prevent the water from coming between the tire surface and the road surface, where it would reduce friction and allow the tire to skid.

Surfaces that appear to be highly polished can actually look quite rough when examined under a microscope. Such an examination reveals that two surfaces in contact touch only at relatively few spots, as Figure 4.20 illustrates. The microscopic area of contact for these spots is substantially less than the apparent macroscopic area of contact between the surfaces—perhaps thousands of times less. At these contact points the molecules of the different bodies are close enough together to exert strong attractive intermolecular forces on one another, leading to what are known as "cold welds." Frictional forces are associated with these welded spots, but the exact details of how frictional forces arise are not well understood. However, some empirical relations have been developed that make it possible to account for the effects of friction.

Figure 4.21 helps to explain the main features of the type of friction known as ***static friction.*** The block in this drawing is initially at rest on a table, and as long as there is no attempt to move the block, there is no static frictional force. Then, a horizontal force **F** is applied to the block by means of a rope. If **F** is small, as in part *a*, experience tells us that the block still does not move. Why? It does not move because the static frictional force f_s exactly cancels the effect of the applied force. The direction of f_s is opposite to that of **F**, and the magnitude of f_s equals the magnitude of the applied force, $f_s = F$. Increasing the applied force in Figure 4.21 by a small amount still does not cause the block to move. There is no movement because the static frictional force also increases by an amount that cancels out the increase in the applied force (see part *b* of the drawing). If the applied

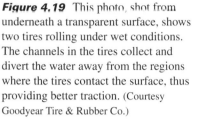

No movement
(*a*)

No movement
(*b*)

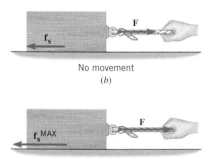

Just when movement begins
(*c*)

Figure 4.21 Applying a small force **F** to the block, as in parts *a* and *b*, produces no movement, because the static frictional force f_s exactly balances the applied force. (*c*) The block just begins to move when the applied force is slightly greater than the maximum static frictional force f_s^{MAX}.

force continues to increase, however, there comes a point when the block finally "breaks away" and begins to slide. The force just before breakaway represents the *maximum static frictional force* $\mathbf{f_s}^{MAX}$ that the table can exert on the block (see part *c* of the drawing). Any applied force that is greater than $\mathbf{f_s}^{MAX}$ cannot be balanced by static friction, and the resulting net force accelerates the block to the right.

Experimental evidence shows that, to a good degree of approximation, the maximum static frictional force between a pair of dry, unlubricated surfaces has two main characteristics. It is independent of the apparent macroscopic area of contact between the objects, provided that the surfaces are hard or nondeformable. For instance, in Figure 4.22 the maximum static frictional force that the surface of the table can exert on a block is the same, whether the block is resting on its largest or its smallest side. The other main characteristic of $\mathbf{f_s}^{MAX}$ is that its magnitude is proportional to the magnitude of the normal force $\mathbf{F_N}$. As Section 4.8 points out, the magnitude of the normal force indicates how hard two surfaces are being pressed together. The harder they are pressed, the larger is f_s^{MAX}, presumably because the number of "cold-welded," microscopic contact points is increased. Equation 4.7 expresses the proportionality between f_s^{MAX} and F_N with the aid of a proportionality constant μ_s, which is called the ***coefficient of static friction.***

Figure 4.22 The maximum static frictional force $\mathbf{f_s}^{MAX}$ would be the same, no matter which side of the block is in contact with the table.

■ **STATIC FRICTIONAL FORCE**

The magnitude f_s of the static frictional force can have any value from zero up to a maximum value of f_s^{MAX}, depending on the applied force. In other words, $f_s \leq f_s^{MAX}$, where the symbol "$\leq$" is read as "less than or equal to." The equality holds only when f_s attains its maximum value, which is

$$f_s^{MAX} = \mu_s F_N \qquad (4.7)$$

In Equation 4.7, μ_s is the coefficient of static friction, and F_N is the magnitude of the normal force.

It should be emphasized that Equation 4.7 relates only the magnitudes of $\mathbf{f_s}^{MAX}$ and $\mathbf{F_N}$, *not the vectors themselves.* This equation does not imply that the directions of the vectors are the same. In fact, $\mathbf{f_s}^{MAX}$ is parallel to the surface, while $\mathbf{F_N}$ is perpendicular to it.

The coefficient of static friction, being the ratio of the magnitudes of two forces ($\mu_s = f_s^{MAX}/F_N$), has no units. Also, it depends on the type of material from which each surface is made (steel on wood, rubber on concrete, etc.), the condition of the surfaces (polished, rough, etc.), and other variables such as temperature. Typical values for μ_s range from about 0.01 for smooth surfaces to about 1.5 for rough surfaces. Example 9 illustrates the use of Equation 4.7 for determining the maximum static frictional force.

Concept Simulation 4.2

Static friction will hold a block in place on an inclined surface, provided the angle of the incline is not too steep. Here you can adjust the angle to see when the block begins to slide. From this angle the coefficient of static friction can be determined. You can also adjust the block's mass and explore its effect on the motion.

Related Homework: Problem 74

Go to
www.wiley.com/college/cutnell

Example 9 The Force Needed to Start a Sled Moving

A sled is resting on a horizontal patch of snow, and the coefficient of static friction is $\mu_s = 0.350$. The sled and its rider have a total mass of 38.0 kg. What is the magnitude of the maximum horizontal force that can be applied to the sled before it just begins to move?

Reasoning The maximum horizontal force occurs when its magnitude equals that of the maximum force of static friction, as indicated in Figure 4.21c. Equation 4.7 ($f_s^{MAX} = \mu_s F_N$) specifies how the magnitude f_s^{MAX} of the maximum static frictional force is related to the magnitude F_N of the normal force. We can determine F_N by noting that the sled does not accelerate in the vertical direction. Thus, the net force acting vertically on the sled must be zero. Consequently, the normal force and the weight of the sled and its rider must balance, so that $F_N = mg$.

Solution The magnitude of the maximum horizontal force is

$$f_s^{MAX} = \mu_s F_N = \mu_s mg = (0.350)(38.0 \text{ kg})(9.80 \text{ m/s}^2) = \boxed{130 \text{ N}}$$

Need more practice?

Interactive LearningWare 4.2
A sofa rests on the horizontal bed of a moving van. The coefficient of static friction between the sofa and van bed is 0.30. The van starts from rest and accelerates for a time of 5.1 s. What is the maximum distance that the van can travel in this time period without having the sofa slide?

Related Homework: Problem 82

Go to
www.wiley.com/college/cutnell
for an interactive solution.

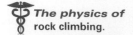

The physics of rock climbing.

Static friction is often very useful, as it is to the rock climber in Figure 4.23, for instance. He presses outward against the walls of the rock formation with his hands and feet to create sufficiently large normal forces, so that the static frictional forces can support his weight.

Once two surfaces begin sliding over one another, the static frictional force is no longer of any concern. Instead, a type of friction known as *kinetic* friction* comes into play. The kinetic frictional force opposes the relative sliding motion. If you have ever pushed an object across a floor, you may have noticed that it takes less force to keep the object sliding than it takes to get it going in the first place. In other words, the kinetic frictional force is usually less than the static frictional force.

Experimental evidence indicates that the kinetic frictional force f_k has three main characteristics, to a good degree of approximation. It is independent of the apparent area of contact between the surfaces (see Figure 4.22). It is independent of the speed of the sliding motion, if the speed is small. And lastly, the magnitude of the kinetic frictional force is proportional to the magnitude of the normal force. Equation 4.8 expresses this proportionality with the aid of a proportionality constant μ_k, which is called the *coefficient of kinetic friction.*

Figure 4.23 In maneuvering his way up Devil's Tower in Wyoming, this rock climber uses the static frictional forces between his hands and feet and the vertical rock walls to support his weight. (© Brian Bailey/Image State)

■ **KINETIC FRICTIONAL FORCE**

The magnitude f_k of the kinetic frictional force is given by

$$f_k = \mu_k F_N \qquad (4.8)$$

In Equation 4.8, μ_k is the coefficient of kinetic friction, and F_N is the magnitude of the normal force.

Equation 4.8, like Equation 4.7, is a relationship between only the magnitudes of the frictional and normal forces. The directions of these forces are perpendicular to each other. Moreover, like the coefficient of static friction, the coefficient of kinetic friction is a number without units and depends on the type and condition of the two surfaces that are in contact. Values for μ_k are typically less than those for μ_s, reflecting the fact that kinetic friction is generally less than static friction. The next example illustrates the effect of kinetic friction.

Example 10 Sled Riding

A sled is traveling at 4.00 m/s along a horizontal stretch of snow, as Figure 4.24a illustrates. The coefficient of kinetic friction is $\mu_k = 0.0500$. How far does the sled go before stopping?

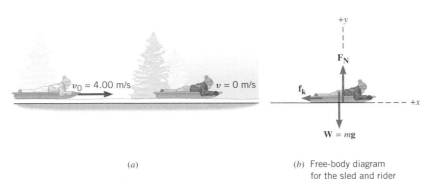

(a)

(b) Free-body diagram for the sled and rider

Figure 4.24 (a) The moving sled decelerates because of the kinetic frictional force. (b) Three forces act on the moving sled, the weight **W** of the sled and its rider, the normal force **F$_N$**, and the kinetic frictional force **f$_k$**. The free-body diagram for the sled shows these forces.

---

* The word "kinetic" is derived from the Greek word *kinetikos,* meaning "of motion."

Reasoning The sled comes to a halt because the kinetic frictional force opposes the motion and causes the sled to slow down. Therefore, we will determine the kinetic frictional force and use it in Newton's second law to find the acceleration of the sled. Knowing the acceleration, we can determine the stopping distance by employing the appropriate equation of kinematics, as discussed in Chapter 3.

Solution To determine the magnitude f_k of the kinetic frictional force, it is necessary to know the magnitude F_N of the normal force, since $f_k = \mu_k F_N$. Part *b* of Figure 4.24 shows the free-body diagram. Since the sled does not accelerate in the vertical direction, there can be no net force acting vertically on the sled. As a result, the normal force and the weight **W** must balance, so the magnitude of the normal force is $F_N = mg$. The magnitude of the kinetic frictional force is

$$f_k = \mu_k F_N = \mu_k mg \tag{4.8}$$

The kinetic frictional force is the only force acting on the sled in the *x* direction, so it is the net force. Newton's second law then gives the acceleration as

$$a_x = \frac{-f_k}{m} = \frac{-\mu_k mg}{m} = -\mu_k g \tag{4.2a}$$

The minus sign before f_k arises because the kinetic frictional force opposes the sliding motion of the sled and is directed to the left, or along the $-x$ axis, in Figure 4.24*b*. Therefore, the acceleration also points along the $-x$ axis. Notice that the acceleration does not depend on the mass of the sled and rider, since the mass is eliminated algebraically. The stopping distance *x* can be obtained with the aid of Equation 3.6a from the equations of kinematics ($v_x^2 = v_{0x}^2 + 2a_x x$) as

$$x = \frac{v_x^2 - v_{0x}^2}{2a_x}$$

where v_x and v_{0x} are the final and initial velocities, respectively. Using the fact that $a_x = -\mu_k g$, we find

$$x = \frac{v_x^2 - v_{0x}^2}{-2\,\mu_k g} = \frac{(0 \text{ m/s})^2 - (4.00 \text{ m/s})^2}{-2(0.0500)(9.80 \text{ m/s}^2)} = \boxed{16.3 \text{ m}}$$

Sliding down a snow-covered slope is fun, but kinetic friction limits how fast you can go. (© Jose Azel/Aurora & Quanta Productions)

Static friction opposes the impending relative motion between two objects, while kinetic friction opposes the relative sliding motion that actually does occur. In either case, *relative motion* is opposed. However, this opposition to relative motion does not mean that friction prevents or works against the motion of *all* objects. For instance, the foot of a person walking exerts a force on the earth, and the earth exerts a reaction force on the foot. This reaction force is a static frictional force, and it opposes the impending backward motion of the foot, propelling the person forward in the process. Kinetic friction can also cause an object to move, all the while opposing relative motion, as it does in Example 10. In this example the kinetic frictional force acts on the sled and opposes the relative motion of the sled and the earth. Newton's third law indicates, however, that since the earth exerts the kinetic frictional force on the sled, the sled must exert a reaction force on the earth. In response, the earth accelerates, but because of the earth's huge mass, the motion is too slight to be noticed.

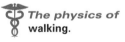
The physics of walking.

Concept Simulation 4.3

A driver slams on the brakes, and her car slides to a halt. In this simulation you can adjust the car's initial speed, the driver's reaction time, and the coefficient of kinetic friction. The effect of changing these parameters is displayed in graphs of velocity and position versus time.

Related Homework: Problem 100

Go to
www.wiley.com/college/cutnell

✔ **Check Your Understanding 3**

A box has a weight of 150 N and is being pulled across a horizontal floor by a force that has a magnitude of 110 N. The pulling force can be applied to the box in two ways. It can point horizontally, or it can point above the horizontal at an angle θ. When the pulling force is applied horizontally, the kinetic frictional force acting on the box is twice as large as when it is applied at an angle θ. What is the value of θ? *(The answer is given at the end of the book.)*

Background: Vector components, the normal force, and the kinetic frictional force play roles in this question. See Section 1.7 to review vector components.

For similar questions (including conceptual counterparts), consult Self-Assessment Test 4.2. This test is described at the end of Section 4.10.

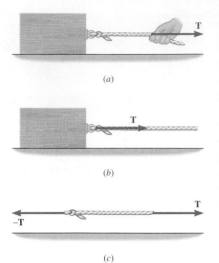

(a)

(b)

(c)

Figure 4.25 (*a*) A force **T** is being applied to the right end of a rope. (*b*) The force is transmitted to the box. (*c*) Forces are applied to both ends of the rope. These forces have equal magnitudes and opposite directions.

4.10 The Tension Force

Forces are often applied by means of cables or ropes that are used to pull an object. For instance, Figure 4.25*a* shows a force **T** being applied to the right end of a rope attached to a box. Each particle in the rope in turn applies a force to its neighbor. As a result, the force is applied to the box, as part *b* of the drawing shows.

In situations such as that in Figure 4.25, we say that "the force **T** is applied to the box because of the tension in the rope," meaning that the tension and the force applied to the box have the same magnitude. However, the word "tension" is commonly used to mean the tendency of the rope to be pulled apart. To see the relationship between these two uses of the word "tension," consider the left end of the rope, which applies the force **T** to the box. In accordance with Newton's third law, the box applies a reaction force to the rope. The reaction force has the same magnitude as **T** but is oppositely directed. In other words, a force −**T** acts on the left end of the rope. Thus, forces of equal magnitude act on opposite ends of the rope, as in Figure 4.25*c*, and tend to pull it apart.

In the previous discussion, we have used the concept of a "massless" rope (*m* = 0 kg) without saying so. In reality, a massless rope does not exist, but it is useful as an idealization when applying Newton's second law. According to the second law, a net force is required to accelerate an object that has mass. In contrast, no net force is needed to accelerate a massless rope, since $\Sigma \mathbf{F} = m\mathbf{a}$ and *m* = 0 kg. Thus, when a force **T** is applied to one end of a massless rope, none of the force is needed to accelerate the rope. As a result, the force **T** is also applied undiminished to the object attached at the other end, as we assumed in Figure 4.25.* If the rope had mass, however, some of the force **T** would have to be used to accelerate the rope. The force applied to the box would then be less than **T**, and the tension would be different at different locations along the rope. In this text we will assume that a rope connecting one object to another is massless, unless stated otherwise. The ability of a massless rope to transmit tension undiminished from one end to the other is not affected when the rope passes around objects such as the pulley in Figure 4.26 (provided the pulley itself is massless and frictionless).

Self-Assessment Test 4.2

Test your understanding of the material in Sections 4.6–4.10:

• Gravitational Force   • Normal Force   • Frictional Forces   • Tension Force

Go to **www.wiley.com/college/cutnell**

Figure 4.26 The force **T** applied at one end of a massless rope is transmitted undiminished to the other end, even when the rope bends around a pulley, provided the pulley is also massless and friction is absent.

4.11 Equilibrium Applications of Newton's Laws of Motion

Have you ever been so upset that it took days to recover your "equilibrium?" In this context, the word "equilibrium" refers to a balanced state of mind, one that is not changing wildly. In physics, the word "equilibrium" also refers to a lack of change, but in the sense that the velocity of an object isn't changing. If its velocity doesn't change, an object is not accelerating. Our definition of equilibrium, then, is as follows:

■ **DEFINITION OF EQUILIBRIUM†**

An object is in equilibrium when it has zero acceleration.

▶ **CONCEPTS AT A GLANCE**   The concept of equilibrium arises directly from Newton's second law. The Concepts-at-a-Glance chart in Figure 4.27, which is an enhanced version of the chart in Figure 4.9, illustrates this important point. When the acceleration of an ob-

* If a rope is not accelerating, **a** is zero in the second law, and $\Sigma \mathbf{F} = m\mathbf{a} = 0$, regardless of the mass of the rope. Then, the rope can be ignored, no matter what mass it has.
† In this discussion of equilibrium we ignore rotational motion, which is discussed in Chapters 8 and 9. In Section 9.2 a more complete treatment of the equilibrium of a rigid object is presented and takes into account the concept of torque and the fact that objects can rotate.

CONCEPTS AT A GLANCE

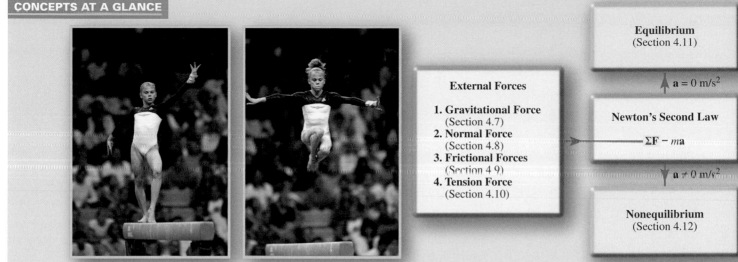

External Forces

1. **Gravitational Force** (Section 4.7)
2. **Normal Force** (Section 4.8)
3. **Frictional Forces** (Section 4.9)
4. **Tension Force** (Section 4.10)

Equilibrium (Section 4.11)

$a = 0$ m/s²

Newton's Second Law

$\Sigma\mathbf{F} - m a$

$a \neq 0$ m/s²

Nonequilibrium (Section 4.12)

Equilibrium   Nonequilibrium

Figure 4.27 CONCEPTS AT A GLANCE Both equilibrium and nonequilibrium problems can be solved with the aid of Newton's second law. For equilibrium situations, such as the gymnast on the balance beam in the left photograph, the acceleration is zero ($a = 0$ m/s²). For nonequilibrium situations, such as the freely falling gymnast in the right photograph, the acceleration is not zero ($a \neq 0$ m/s²). (© Doug Pensinger/Getty Images Sport Services)

ject is zero ($a = 0$ m/s²), the object is in equilibrium, as the upper-right part of the chart indicates. This section presents several examples involving equilibrium situations. On the other hand, when the acceleration is not zero ($a \neq 0$ m/s²), we have a nonequilibrium situation, as the lower-right part of the chart suggests. Section 4.12 deals with nonequilibrium applications of Newton's second law. ◄

Since the acceleration is zero for an object in equilibrium, all of the acceleration components are also zero. In two dimensions, this means that $a_x = 0$ m/s² and $a_y = 0$ m/s². Substituting these values into the second law ($\Sigma F_x = ma_x$ and $\Sigma F_y = ma_y$) shows that the x component and the y component of the net force must each be zero. Thus, in two dimensions, the equilibrium condition is expressed by two equations:

$$\Sigma F_x = 0 \qquad (4.9a)$$

$$\Sigma F_y = 0 \qquad (4.9b)$$

In other words, the forces acting on an object in equilibrium must balance.

In using Equations 4.9a and 4.9b to solve equilibrium problems, we will use the following five-step reasoning strategy:

Reasoning Strategy

Analyzing Equilibrium Situations

1. Select the object (often called the "system") to which Equations 4.9a and 4.9b are to be applied. It may be that two or more objects are connected by means of a rope or a cable. If so, it may be necessary to treat each object separately according to the following steps.

2. Draw a free-body diagram for each object chosen above. Be sure to include only forces that act on the object. *Do not include forces that the object exerts on its environment.*

3. Choose a set of x, y axes for each object and resolve all forces in the free-body diagram into components that point along these axes. Select the axes so that as many forces as possible point along the x axis or the y axis. Such a choice minimizes the calculations needed to determine the force components.

4. Apply Equations 4.9a and 4.9b by setting the sum of the x components and the sum of the y components of the forces each equal to zero.

5. Solve the two equations obtained in Step 4 for the desired unknown quantities, remembering that two equations can yield answers for only two unknowns at most.

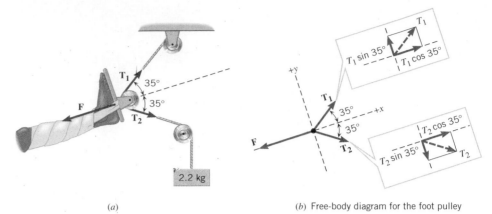

Figure 4.28 (*a*) A traction device for the foot. (*b*) The free-body diagram for the pulley on the foot.

(*a*)

(*b*) Free-body diagram for the foot pulley

Example 11 illustrates how these steps are followed. It deals with a traction device in which three forces act together to bring about the equilibrium.

Example 11   Traction for the Foot

Figure 4.28*a* shows a traction device used with a foot injury. The weight of the 2.2-kg object creates a tension in the rope that passes around the pulleys. Therefore, tension forces $\mathbf{T_1}$ and $\mathbf{T_2}$ are applied to the pulley on the foot. It may seem surprising that the rope applies a force to either side of the foot pulley. A similar effect occurs when you place a finger inside a rubber band and push downward. You can feel each side of the rubber band pulling upward on the finger. The foot pulley is kept in equilibrium because the foot also applies a force $\mathbf{F}$ to it. This force arises in reaction (Newton's third law) to the pulling effect of the forces $\mathbf{T_1}$ and $\mathbf{T_2}$. Ignoring the weight of the foot, find the magnitude of $\mathbf{F}$.

Reasoning The forces $\mathbf{T_1}$, $\mathbf{T_2}$, and $\mathbf{F}$ keep the pulley on the foot at rest. The pulley, therefore, has no acceleration and is in equilibrium. As a result, the sum of the x components and the sum of the y components of the three forces must each be zero. Figure 4.28*b* shows the free-body diagram of the pulley on the foot. The x axis is chosen to be along the direction of force $\mathbf{F}$, and the components of the forces are indicated in the drawing. (See Section 1.7 for a review of vector components.)

Problem solving insight
Choose the orientation of the x, y axes for convenience. In Example 11, the axes have been rotated so the force $\mathbf{F}$ points along the x axis. Since $\mathbf{F}$ does not have a component along the y axis, the analysis is simplified.

Solution Since the sum of the y components of the forces is zero, it follows that

$$\Sigma F_y = +T_1 \sin 35° - T_2 \sin 35° = 0 \qquad (4.9b)$$

or $T_1 = T_2$. In other words, the magnitudes of the tension forces are equal. In addition, the sum of the x components of the forces is zero, so we have that

$$\Sigma F_x = +T_1 \cos 35° + T_2 \cos 35° - F = 0 \qquad (4.9a)$$

Solving for F and letting $T_1 = T_2 = T$, we find that $F = 2T \cos 35°$. However, the tension T in the rope is determined by the weight of the 2.2-kg object: $T = mg$, where m is its mass and g is the acceleration due to gravity. Therefore, the magnitude of $\mathbf{F}$ is

$$F = 2T \cos 35° = 2mg \cos 35° = 2(2.2 \text{ kg})(9.80 \text{ m/s}^2) \cos 35° = \boxed{35 \text{ N}}$$

Example 12 presents another situation in which three forces are responsible for the equilibrium of an object. However, in this example all the forces have different magnitudes.

Example 12   Replacing an Engine

An automobile engine has a weight $\mathbf{W}$, whose magnitude is $W = 3150$ N. This engine is being positioned above an engine compartment, as Figure 4.29*a* illustrates. To position the engine, a worker is using a rope. Find the tension $\mathbf{T_1}$ in the supporting cable and the tension $\mathbf{T_2}$ in the positioning rope.

Reasoning Under the influence of the forces $\mathbf{W}$, $\mathbf{T_1}$, and $\mathbf{T_2}$ the ring in Figure 4.29*a* is at rest and, therefore, in equilibrium. Consequently, the sum of the x components and the sum of the

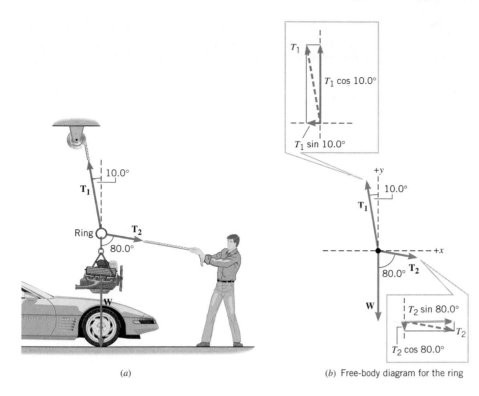

Figure 4.29 (*a*) The ring is in equilibrium because of the three forces **T**$_1$ (the tension force in the supporting cable), **T**$_2$ (the tension force in the positioning rope), and **W** (the weight of the engine). (*b*) The free-body diagram for the ring.

(*a*)

(*b*) Free-body diagram for the ring

y components of these forces must each be zero; $\Sigma F_x = 0$ and $\Sigma F_y = 0$. By using these relations, we can find T_1 and T_2. Figure 4.29*b* shows the free-body diagram of the ring and the force components for a suitable *x*, *y* axis system.

Solution The free-body diagram shows the components for each of the three forces, and the components are listed in the following table:

Force	*x* Component	*y* Component
T₁	$-T_1 \sin 10.0°$	$+T_1 \cos 10.0°$
T₂	$+T_2 \sin 80.0°$	$-T_2 \cos 80.0°$
W	0	$-W$

The plus signs in the table denote components that point along the positive axes, and the minus signs denote components that point along the negative axes. Setting the sum of the *x* components and the sum of the *y* components equal to zero leads to the following two equations:

$$\Sigma F_x = -T_1 \sin 10.0° + T_2 \sin 80.0° = 0 \qquad (4.9a)$$

$$\Sigma F_y = +T_1 \cos 10.0° - T_2 \cos 80.0° - W = 0 \qquad (4.9b)$$

Solving the first of these equations for T_1 shows that

$$T_1 = \left(\frac{\sin 80.0°}{\sin 10.0°}\right) T_2$$

Substituting this expression for T_1 into the second equation gives

$$\left(\frac{\sin 80.0°}{\sin 10.0°}\right) T_2 \cos 10.0° - T_2 \cos 80.0° - W = 0$$

$$T_2 = \frac{W}{\left(\dfrac{\sin 80.0°}{\sin 10.0°}\right) \cos 10.0° - \cos 80.0°}$$

Setting $W = 3150$ N in this result yields $\boxed{T_2 = 582 \text{ N}}$.

Since $T_1 = \left(\dfrac{\sin 80.0°}{\sin 10.0°}\right) T_2$ and $T_2 = 582$ N, it follows that $\boxed{T_1 = 3.30 \times 10^3 \text{ N}}$.

Problem solving insight
When an object is in equilibrium, as here in Example 12, the net force is zero, $\Sigma F = 0$. This does not mean that each individual force is zero. It means that the vector sum of all the forces is zero.

An object can be moving and still be in equilibrium, provided there is no acceleration. Example 13 illustrates such a case, and the solution is again obtained using the five-step reasoning strategy summarized at the beginning of the section.

Example 13 Equilibrium at Constant Velocity

A jet plane is flying with a constant speed along a straight line, at an angle of 30.0° above the horizontal, as Figure 4.30a indicates. The plane has a weight **W** whose magnitude is $W = 86\,500$ N, and its engines provide a forward thrust **T** of magnitude $T = 103\,000$ N. In addition, the lift force **L** (directed perpendicular to the wings) and the force **R** of air resistance (directed opposite to the motion) act on the plane. Find **L** and **R**.

Reasoning Figure 4.30b shows the free-body diagram of the plane, including the forces **W**, **L**, **T**, and **R**. Since the plane is not accelerating, it is in equilibrium, and the sum of the x components and the sum of the y components of these forces must be zero. The lift force **L** and the force **R** of air resistance can be obtained from these equilibrium conditions. To calculate the components, we have chosen axes in the free-body diagram that are rotated by 30.0° from their usual horizontal–vertical positions. This has been done purely for convenience, since the weight **W** is then the only force that does not lie along either axis.

Solution When determining the components of the weight, it is necessary to realize that the angle β in Figure 4.30a is 30.0°. Part c of the drawing focuses attention on the geometry that is responsible for this fact. There it can be seen that $\alpha + \beta = 90°$ and $\alpha + 30° = 90°$, with the result that $\beta = 30°$. The table below lists the components of the forces that act on the jet.

Force	x Component	y Component
W	$-W \sin 30.0°$	$-W \cos 30.0°$
L	0	$+L$
T	$+T$	0
R	$-R$	0

Setting the sum of the x component of the forces to zero gives

$$\Sigma F_x = -W \sin 30.0° + T - R = 0 \tag{4.9a}$$

$$R = T - W \sin 30.0° = 103\,000 \text{ N} - (86\,500 \text{ N}) \sin 30.0° = \boxed{59\,800 \text{ N}}$$

Setting the sum of the y component of the forces to zero gives

$$\Sigma F_y = -W \cos 30.0° + L = 0 \tag{4.9b}$$

$$L = W \cos 30.0° = (86\,500 \text{ N}) \cos 30.0° = \boxed{74\,900 \text{ N}}$$

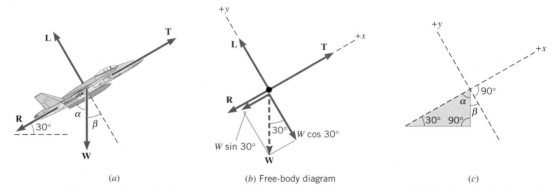

Figure 4.30 (a) A plane moves with a constant velocity at an angle of 30.0° above the horizontal due to the action of four forces, the weight **W**, the lift **L**, the engine thrust **T**, and the air resistance **R**. (b) The free-body diagram for the plane. (c) This geometry occurs often in physics.

4.12 Nonequilibrium Applications of Newton's Laws of Motion

When an object is accelerating, it is not in equilibrium, as indicated in Figure 4.27. The forces acting on it are not balanced, so the net force is not zero in Newton's second law. However, with one exception, the reasoning strategy followed in solving nonequilibrium problems is identical to that used in equilibrium situations. The exception occurs in Step 4 of the five steps outlined at the beginning of the last section. Since the object is now accelerating, the representation of Newton's second law in Equations 4.2a and 4.2b applies instead of Equations 4.9a and 4.9b:

$$\Sigma F_x = ma_x \quad (4.2a) \quad \text{and} \quad \Sigma F_y = ma_y \quad (4.2b)$$

Example 14 uses these equations in a situation where the forces are applied in directions similar to those in Example 11, except that now an acceleration is present.

Example 14 Towing a Supertanker

A supertanker of mass $m = 1.50 \times 10^8$ kg is being towed by two tugboats, as in Figure 4.31a. The tensions in the towing cables apply the forces $\mathbf{T}_1$ and $\mathbf{T}_2$ at equal angles of 30.0° with respect to the tanker's axis. In addition, the tanker's engines produce a forward drive force $\mathbf{D}$, whose magnitude is $D = 75.0 \times 10^3$ N. Moreover, the water applies an opposing force $\mathbf{R}$, whose magnitude is $R = 40.0 \times 10^3$ N. The tanker moves forward with an acceleration that points along the tanker's axis and has a magnitude of 2.00×10^{-3} m/s². Find the magnitudes of the tensions $\mathbf{T}_1$ and $\mathbf{T}_2$.

Figure 4.31 (*a*) Four forces act on a supertanker: $\mathbf{T}_1$ and $\mathbf{T}_2$ are the tension forces due to the towing cables, $\mathbf{D}$ is the forward drive force produced by the tanker's engines, and $\mathbf{R}$ is the force with which the water opposes the tanker's motion. (*b*) The free-body diagram for the tanker.

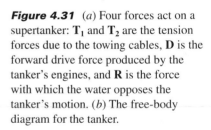

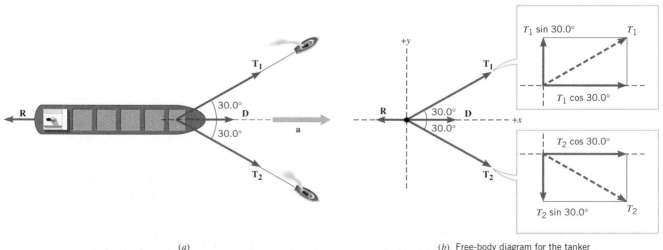

(*a*)

(*b*) Free-body diagram for the tanker

Reasoning The unknown forces $\mathbf{T}_1$ and $\mathbf{T}_2$ contribute to the net force that accelerates the tanker. To determine $\mathbf{T}_1$ and $\mathbf{T}_2$, therefore, we analyze the net force, which we will do using components. The various force components can be found by referring to the free-body diagram for the tanker in Figure 4.31*b*, where the ship's axis is chosen as the *x* axis. We will then use Newton's second law in its component form, $\Sigma F_x = ma_x$ and $\Sigma F_y = ma_y$, to obtain the magnitudes of $\mathbf{T}_1$ and $\mathbf{T}_2$.

Solution The individual force components are summarized as follows:

Force	*x* Component	*y* Component
$\mathbf{T}_1$	$+T_1 \cos 30.0°$	$+T_1 \sin 30.0°$
$\mathbf{T}_2$	$+T_2 \cos 30.0°$	$-T_2 \sin 30.0°$
$\mathbf{D}$	$+D$	0
$\mathbf{R}$	$-R$	0

Since the acceleration points along the *x* axis, there is no *y* component of the acceleration ($a_y = 0$ m/s²). Consequently, the sum of the *y* components of the forces must be zero:

$$\Sigma F_y = +T_1 \sin 30.0° - T_2 \sin 30.0° = 0$$

This result shows that the magnitudes of the tensions in the cables are equal, $T_1 = T_2$. Since the ship accelerates along the *x* direction, the sum of the *x* components of the forces is not zero. The second law indicates that

$$\Sigma F_x = T_1 \cos 30.0° + T_2 \cos 30.0° + D - R = ma_x$$

Since $T_1 = T_2$, we can replace the two separate tension symbols by a single symbol T, the magnitude of the tension. Solving for T gives

$$T = \frac{ma_x + R - D}{2 \cos 30.0°}$$

$$= \frac{(1.50 \times 10^8 \text{ kg})(2.00 \times 10^{-3} \text{ m/s}^2) + 40.0 \times 10^3 \text{ N} - 75.0 \times 10^3 \text{ N}}{2 \cos 30.0°}$$

$$= \boxed{1.53 \times 10^5 \text{ N}}$$

It often happens that two objects are connected somehow, perhaps by a drawbar like that used when a truck pulls a trailer. If the tension in the connecting device is of no interest, the objects can be treated as a single composite object when applying Newton's second law. However, if it is necessary to find the tension, as in the next example, then the second law must be applied separately to at least one of the objects.

Example 15   Hauling a Trailer

A truck is hauling a trailer along a level road, as Figure 4.32*a* illustrates. The mass of the truck is $m_1 = 8500$ kg and that of the trailer is $m_2 = 27\,000$ kg. The two move along the *x* axis with an acceleration of $a_x = 0.78$ m/s². Ignoring the retarding forces of friction and air resistance, determine (a) the tension $\mathbf{T}$ in the horizontal drawbar between the trailer and the truck and (b) the force $\mathbf{D}$ that propels the truck forward.

Reasoning Since the truck and the trailer accelerate along the horizontal direction and friction is being ignored, only forces that have components in the horizontal direction are of interest. Therefore, Figure 4.32 omits the weight and the normal force, which act vertically. To determine the tension force $\mathbf{T}$ in the drawbar, we draw the free-body diagram for the trailer and apply Newton's second law, $\Sigma F_x = ma_x$. Similarly, we can determine the propulsion force $\mathbf{D}$ by drawing the free-body diagram for the truck and applying Newton's second law.

Solution

(a) The free-body diagram for the trailer is shown in Figure 4.32*b*. There is only one horizontal force acting on the trailer, the tension force $\mathbf{T}$ due to the drawbar. Therefore, it is straightforward to obtain the tension from $\Sigma F_x = m_2 a_x$, since the mass of the trailer and the acceleration are known:

$$\Sigma F_x = T = m_2 a_x = (27\,000 \text{ kg})(0.78 \text{ m/s}^2) = \boxed{21\,000 \text{ N}}$$

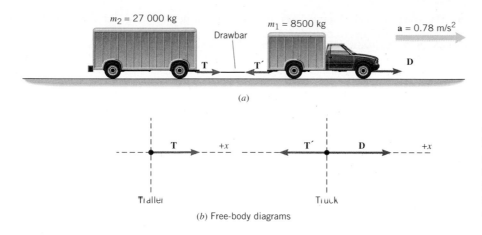

(a)

(b) Free-body diagrams

Figure 4.32 (a) The force **D** acts on the truck and propels it forward. The drawbar exerts the tension force **T′** on the truck and the tension force **T** on the trailer. (b) The free-body diagrams for the truck and the trailer, ignoring the vertical forces.

(**b**) Two horizontal forces act on the truck, as the free-body diagram in Figure 4.32b shows. One is the desired force **D**. The other is the force **T′**. According to Newton's third law, **T′** is the force with which the trailer pulls back on the truck, in reaction to the truck pulling forward. If the drawbar has negligible mass, the magnitude of **T′** is equal to the magnitude of **T**—namely, 21 000 N. Since the magnitude of **T′**, the mass of the truck, and the acceleration are known, $\Sigma F_x = m_1 a_x$ can be used to determine the drive force:

$$\Sigma F_x = +D - T' = m_1 a_x$$

$$D = m_1 a_x + T' = (8500 \text{ kg})(0.78 \text{ m/s}^2) + 21\,000 \text{ N} = \boxed{28\,000 \text{ N}}$$

In Section 4.11 we examined situations where the net force acting on an object is zero, and in this section we have considered two examples where the net force is not zero. Conceptual Example 16 illustrates a common situation where the net force is zero at certain times but is not zero at other times.

Conceptual Example 16   The Motion of a Water Skier

Figure 4.33 shows a water skier at four different moments:

(**a**) The skier is floating motionless in the water.

(**b**) The skier is being pulled out of the water and up onto the skis.

(**c**) The skier is moving at a constant speed along a straight line.

(**d**) The skier has let go of the tow rope and is slowing down.

For each moment, explain whether the net force acting on the skier is zero.

Reasoning and Solution According to Newton's second law, if an object has zero acceleration, the net force acting on it is zero. In such a case, the object is in equilibrium. In contrast, if the object has an acceleration, the net force acting on it is not zero. Such an object is not in equilibrium. We will apply this criterion to each of the four phases of the motion to decide whether the net force is zero.

Figure 4.33 A water skier (a) floating in water, (b) being pulled up by the boat, (c) moving at a constant velocity, and (d) slowing down.

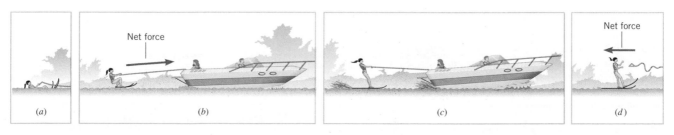

(a) The skier is floating motionless in the water, so her velocity and acceleration are both zero. Therefore, the net force acting on her is zero, and she is in equilibrium.

(b) As the skier is being pulled up and out of the water, her velocity is increasing. Thus, she is accelerating, and the net force acting on her is not zero. The skier is not in equilibrium. The direction of the net force is shown in Figure 4.33*b*.

(c) The skier is now moving at a constant speed along a straight line, so her velocity is constant. Since her velocity is constant, her acceleration is zero. Thus, the net force acting on her is zero, and she is again in equilibrium, even though she is moving.

(d) After the skier lets go of the tow rope, her speed decreases, so she is decelerating. Thus, the net force acting on her is not zero, and she is not in equilibrium. The direction of the net force is shown in Figure 4.33*d*, and it is opposite to that in (*b*).

Related Homework: *Problem 92*

The force of gravity is often present among the forces that affect the acceleration of an object. Examples 17–19 deal with typical situations.

Example 17　Hauling a Crate

A flatbed truck is carrying a crate up a 10.0° hill, as Figure 4.34*a* illustrates. The coefficient of static friction between the truck bed and the crate is $\mu_s = 0.350$. Find the maximum acceleration that the truck can attain before the crate begins to slip backward relative to the truck.

Reasoning The crate will not slip as long as it has the same acceleration as the truck. Therefore, a net force must act on the crate to accelerate it, and the static frictional force $\mathbf{f_s}$ contributes in a major way to this net force. Since the crate tends to slip backward, the static frictional force must be directed forward, up the hill. As the acceleration of the truck increases, $\mathbf{f_s}$ must also increase to produce a corresponding increase in the acceleration of the crate. However, the static frictional force can increase only until its maximum magnitude $f_s^{\text{MAX}} = \mu_s F_N$ is reached, at which point the crate and the truck have the maximum acceleration $\mathbf{a}^{\text{MAX}}$. If the acceleration increases even more, the crate will slip. To find $\mathbf{a}^{\text{MAX}}$, we focus our attention on the crate, and part *b* of the drawing shows its free-body diagram. The three forces acting on the crate at the instant slipping begins are its weight $\mathbf{W} = m\mathbf{g}$, the normal force $\mathbf{F_N}$ exerted by the truck bed, and the maximum static frictional force $\mathbf{f_s^{\text{MAX}}}$.

Solution Using the *x* components of the forces (see the free-body diagram for the magnitudes of the components) and Newton's second law ($\Sigma F_x = ma_x$), we have

$$\Sigma F_x = -mg \sin 10.0° + \mu_s F_N = ma^{\text{MAX}}$$

$$a^{\text{MAX}} = \frac{-mg \sin 10.0° + \mu_s F_N}{m}$$

Before this equation can be used to calculate a^{MAX}, however, a value is needed for F_N. This can be obtained by considering the force components along the *y* axis. Since the crate does not accelerate along this axis, the sum of the *y* components must be zero according to Newton's

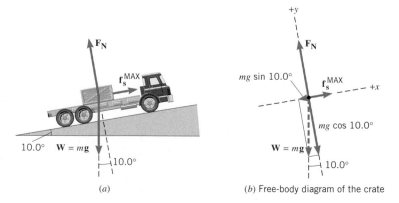

Figure 4.34　(*a*) A crate on a truck is kept from slipping by the static frictional force $\mathbf{f_s^{\text{MAX}}}$. The other forces that act on the crate are its weight $\mathbf{W}$ and the normal force $\mathbf{F_N}$. (*b*) The free-body diagram of the crate.

(*a*)

(*b*) Free-body diagram of the crate

second law ($\Sigma F_y = ma_y = 0$):

$$\Sigma F_y = -mg \cos 10.0° + F_N = 0 \quad \text{or} \quad F_N = mg \cos 10.0°$$

Problem solving insight
The magnitude F_N of the normal force is not necessarily equal to the weight of the object.

Substituting this expression for F_N into the equation for a^{MAX} and algebraically eliminating the mass m reveals that

$$a^{\text{MAX}} = -g \sin 10.0° + \mu_s g \cos 10.0°$$
$$= -(9.80 \text{ m/s}^2) \sin 10.0° + (0.350)(9.80 \text{ m/s}^2) \cos 10.0° = \boxed{1.68 \text{ m/s}^2}$$

Example 18   Accelerating Blocks

Block 1 (mass $m_1 = 8.00$ kg) is moving on a frictionless 30.0° incline. This block is connected to block 2 (mass $m_2 = 22.0$ kg) by a massless cord that passes over a massless and frictionless pulley (see Figure 4.35a). Find the acceleration of each block and the tension in the cord.

Reasoning Since both blocks accelerate, there must be a net force acting on each one. The key to this problem is to realize that Newton's second law can be used separately for each block to relate the net force and the acceleration. Note also that both blocks have accelerations of the same magnitude a, since they move as a unit. We assume that block 1 accelerates up the incline and choose this direction to be the $+x$ axis. If block 1 in reality accelerates down the incline, then the value obtained for the acceleration will be a negative number.

Solution Three forces act on block 1: (1) $\mathbf{W_1}$ is its weight [$W_1 = m_1 g = (8.00 \text{ kg}) \times (9.80 \text{ m/s}^2) = 78.4 \text{ N}$], (2) $\mathbf{T}$ is the force applied because of the tension in the cord, and (3) $\mathbf{F_N}$ is the normal force that the incline exerts. Figure 4.35b shows the free-body diagram for block 1. The weight is the only force that does not point along the x, y axes, and its x and y components are given in the diagram. Applying Newton's second law ($\Sigma F_x = m_1 a_x$) to block 1 shows that

$$\Sigma F_x = -W_1 \sin 30.0° + T = m_1 a$$

where we have set $a_x = a$. This equation cannot be solved as it stands, since both T and a are unknown quantities. To complete the solution, we next consider block 2.

Two forces act on block 2, as the free-body diagram in Figure 4.35b indicates: (1) $\mathbf{W_2}$ is its weight [$W_2 = m_2 g = (22.0 \text{ kg})(9.80 \text{ m/s}^2) = 216 \text{ N}$] and (2) $\mathbf{T'}$ is exerted as a result of block 1 pulling back on the connecting cord. Since the cord and the frictionless pulley are massless, the magnitudes of $\mathbf{T'}$ and $\mathbf{T}$ are the same: $T' = T$. Applying Newton's second law ($\Sigma F_y = m_2 a_y$) to block 2 reveals that

$$\Sigma F_y = T - W_2 = m_2(-a)$$

The acceleration a_y has been set equal to $-a$ since block 2 moves downward along the $-y$ axis in the free-body diagram, consistent with the assumption that block 1 moves up the incline. Now there are two equations in two unknowns, and they may be solved simultaneously (see Appendix C) to give T and a:

$$\boxed{T = 86.3 \text{ N}} \quad \text{and} \quad \boxed{a = 5.89 \text{ m/s}^2}$$

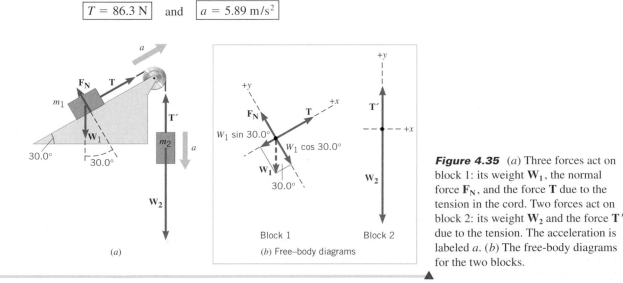

Figure 4.35 (a) Three forces act on block 1: its weight $\mathbf{W_1}$, the normal force $\mathbf{F_N}$, and the force $\mathbf{T}$ due to the tension in the cord. Two forces act on block 2: its weight $\mathbf{W_2}$ and the force $\mathbf{T'}$ due to the tension. The acceleration is labeled a. (b) The free-body diagrams for the two blocks.

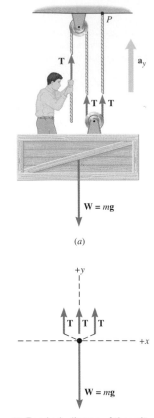

W = mg

(a)

(b) Free-body diagram of the unit

Figure 4.36 (a) A window washer pulls down on the rope to hoist the scaffold up the side of a building. The force **T** results from the effort of the window washer and acts on him and the scaffold in three places, as discussed in Example 19. (b) The free-body diagram of the unit comprising the man and the scaffold.

Example 19 Hoisting a Scaffold

A window washer on a scaffold is hoisting the scaffold up the side of a building by pulling downward on a rope, as in Figure 4.36a. The magnitude of the pulling force is 540 N, and the combined mass of the worker and the scaffold is 155 kg. Find the upward acceleration of the unit.

Reasoning The worker and the scaffold form a single unit, on which the rope exerts a force in three places. The left end of the rope exerts an upward force **T** on the worker's hands. This force arises because he pulls downward with a 540-N force, and the rope exerts an oppositely directed force of equal magnitude on him, in accord with Newton's third law. Thus, the magnitude T of the upward force is $T = 540$ N and is the magnitude of the tension in the rope. If the masses of the rope and each pulley are negligible and if the pulleys are friction-free, the tension is transmitted undiminished along the rope. Then, a 540-N tension force **T** acts upward on the left side of the scaffold pulley (see part a of the drawing). A tension force is also applied to the point P, where the rope attaches to the roof. The roof pulls back on the rope in accord with the third law, and this pull leads to the 540-N tension force **T** that acts on the right side of the scaffold pulley. In addition to the three upward forces, the weight of the unit must be taken into account [$W = mg = (155 \text{ kg})(9.80 \text{ m/s}^2) = 1520$ N]. Part b of the drawing shows the free-body diagram.

Solution Newton's second law ($\Sigma F_y = ma_y$) can be applied to calculate the acceleration a_y:

$$\Sigma F_y = +T + T + T - W = ma_y$$

$$a_y = \frac{3T - W}{m} = \frac{3(540 \text{ N}) - 1520 \text{ N}}{155 \text{ kg}} = \boxed{0.65 \text{ m/s}^2}$$

✓ Check Your Understanding 5

Two boxes have masses m_1 and m_2, and m_2 is greater than m_1. The boxes are being pushed across a frictionless horizontal surface. As the drawing shows, there are two possible arrangements, and the pushing force is the same in each. In which arrangement does the force that the left box applies to the right box have a greater magnitude, or is the magnitude the same in both cases? *(The answer is given at the end of the book.)*

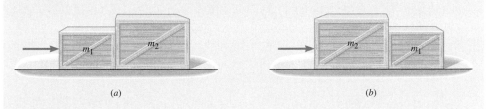

(a)   (b)

Background: This question deals with net force, acceleration, and Newton's second and third laws of motion.

For similar questions (including calculational counterparts), consult Self-Assessment Test 4.3. The test is described next.

🖱 Self-Assessment Test 4.3

Test your understanding of the material in Sections 4.11 and 4.12:

• Equilibrium Applications of Newton's Laws of Motion
• Nonequilibrium Applications of Newton's Laws of Motion

Go to **www.wiley.com/college/cutnell**

4.13 Concepts & Calculations

Newton's three laws of motion provide the basis for understanding the effect of forces on the motion of an object, as we have seen. The second law is especially important, because it provides the quantitative relationship between force and acceleration. The examples in this section serve as a review of the essential features of this relationship.

Concepts & Calculations Example 20
Velocity, Acceleration, and Newton's Second Law of Motion

Figure 4.37 shows two forces, $\mathbf{F}_1 = +3000$ N and $\mathbf{F}_2 = +5000$ N acting on a spacecraft, where the plus signs indicate that the forces are directed along the $+x$ axis. A third force $\mathbf{F}_3$ also acts on the spacecraft but is not shown in the drawing. The craft is moving with a constant velocity of $+850$ m/s. Find the magnitude and direction of $\mathbf{F}_3$.

Figure 4.37 Two horizontal forces, $\mathbf{F}_1$ and $\mathbf{F}_2$, act on the spacecraft. A third force $\mathbf{F}_3$ also acts but is not shown.

Concept Questions and Answers Suppose the spacecraft were stationary. What would be the direction of $\mathbf{F}_3$?

Answer If the spacecraft is stationary, its acceleration is zero. According to Newton's second law, the acceleration of an object is proportional to the net force acting on it. Thus, the net force must also be zero. But the net force is the vector sum of the three forces in this case. Therefore, the force $\mathbf{F}_3$ must have a direction such that it balances to zero the forces $\mathbf{F}_1$ and $\mathbf{F}_2$. Since $\mathbf{F}_1$ and $\mathbf{F}_2$ point along the $+x$ axis in Figure 4.37, $\mathbf{F}_3$ must then point along the $-x$ axis.

When the spacecraft is moving at a constant velocity of $+850$ m/s, what is the direction of $\mathbf{F}_3$?

Answer Since the velocity is constant, the acceleration is still zero. As a result, everything we said in the stationary case applies again here. The net force is zero, and the force $\mathbf{F}_3$ must point along the $-x$ axis in Figure 4.37.

Solution Since the velocity is constant, the acceleration is zero. The net force must also be zero, so that

$$\Sigma F_x = F_1 + F_2 + F_3 = 0$$

Solving for F_3 yields

$$F_3 = -(F_1 + F_2) = -(3000 \text{ N} + 5000 \text{ N}) = \boxed{-8000 \text{ N}}$$

The minus sign in the answer means that $\mathbf{F}_3$ points opposite to the sum of $\mathbf{F}_1$ and $\mathbf{F}_2$, or along the $-x$ axis in Figure 4.37. The force $\mathbf{F}_3$ has a magnitude of 8000 N, which is the magnitude of the sum of the forces $\mathbf{F}_1$ and $\mathbf{F}_2$. The answer is independent of the velocity of the spacecraft, as long as that velocity remains constant.

Concepts & Calculations Example 21   The Importance of Mass

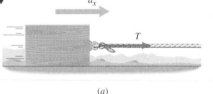

On earth a block has a weight of 88 N. This block is sliding on a horizontal surface on the moon, where the acceleration due to gravity is 1.60 m/s². As Figure 4.38a shows, the block is being pulled by a horizontal rope in which the tension is $T = 24$ N. The coefficient of kinetic friction between the block and the surface is $\mu_k = 0.20$. Determine the acceleration of the block.

(a)

Concept Questions and Answers Which of Newton's laws of motion provides a way to determine the acceleration of the block?

Answer Newton's second law allows us to calculate the acceleration as $a_x = \Sigma F_x/m$, where ΣF_x is the net force acting in the horizontal direction and m is the mass of the block.

This problem deals with a situation on the moon, but the block's mass on the moon is not given. Instead, the block's earth weight is given. Why can the earth weight be used to obtain a value for the block's mass that applies on the moon?

Answer Since the block's earth weight W_{earth} is related to the block's mass according to $W_{earth} = mg_{earth}$, we can use $W_{earth} = 88$ N and $g_{earth} = 9.80$ m/s² to obtain m. But mass is an intrinsic property of the block and does not depend on whether it is on the earth or on the moon. Therefore, the value obtained for m applies on the moon as well as on the earth.

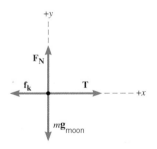

(b) Free-body diagram for the block

Figure 4.38 *(a)* A block is sliding on a horizontal surface on the moon. The tension in the rope is $\mathbf{T}$. *(b)* The free-body diagram for the block, including a kinetic frictional force $\mathbf{f}_k$.

Does the net force ΣF_x equal the tension T?

Answer No. The net force ΣF_x is the vector sum of all the external forces acting in the horizontal direction. It includes the kinetic frictional force f_k as well as the tension T.

Solution Figure 4.38b shows the free-body diagram for the block. The net force along the x axis is $\Sigma F_x = +T - f_k$, where T is the magnitude of the tension in the rope and f_k is the magnitude of the kinetic frictional force. According to Equation 4.8, f_k is related to the magnitude F_N of the normal force by $f_k = \mu_k F_N$, where μ_k is the coefficient of kinetic friction. The acceleration a_x of the block is given by Newton's second law as

$$a_x = \frac{\Sigma F_x}{m} = \frac{+T - \mu_k F_N}{m}$$

We can obtain an expression for F_N by noting that the block does not move in the y direction, so $a_y = 0 \text{ m/s}^2$. Therefore, the net force ΣF_y along the y direction must also be zero. An examination of the free-body diagram reveals that $\Sigma F_y = +F_N - mg_{moon} = 0$, so that $F_N = mg_{moon}$. The acceleration in the x direction becomes

$$a_x = \frac{+T - \mu_k mg_{moon}}{m}$$

Using the earth weight of the block to determine its mass, we find

$$W_{earth} = mg_{earth} \quad \text{or} \quad m = \frac{W_{earth}}{g_{earth}} = \frac{88 \text{ N}}{9.80 \text{ m/s}^2} = 9.0 \text{ kg}$$

The acceleration of the block is, then,

$$a_x = \frac{+T - \mu_k mg_{moon}}{m} = \frac{24 \text{ N} - (0.20)(9.0 \text{ kg})(1.60 \text{ m/s}^2)}{9.0 \text{ kg}} = \boxed{+2.3 \text{ m/s}^2}$$

▲

At the end of the problem set for this chapter, you will find homework problems that contain both conceptual and quantitative parts. These problems are grouped under the heading *Concepts & Calculations, Group Learning Problems*. They are designed for use by students working alone or in small learning groups. The conceptual part of each problem provides a convenient focus for group discussions.

Concept Summary

This summary presents an abridged version of the chapter, including the important equations and all available learning aids. For convenient reference, the learning aids (including the text's examples) are placed next to or immediately after the relevant equation or discussion. The following learning aids may be found on-line at **www.wiley.com/college/cutnell**:

Interactive LearningWare examples are solved according to a five-step interactive format that is designed to help you develop problem-solving skills.	**Concept Simulations** are animated versions of text figures or animations that illustrate important concepts. You can control parameters that affect the display, and we encourage you to experiment.
Interactive Solutions offer specific models for certain types of problems in the chapter homework. The calculations are carried out interactively.	**Self-Assessment Tests** include both qualitative and quantitative questions. Extensive feedback is provided for both incorrect and correct answers, to help you evaluate your understanding of the material.

Topic	*Discussion*	*Learning Aids*

4.1 The Concepts of Force and Mass

Contact and noncontact forces

A force is a push or a pull and is a vector quantity. Contact forces arise from the physical contact between two objects. Noncontact forces are also called action-at-a-distance forces, because they arise without physical contact between two objects.

Mass

Mass is a property of matter that determines how difficult it is to accelerate or decelerate an object. Mass is a scalar quantity.

Topic	*Discussion*	*Learning Aids*

4.2 Newton's First Law of Motion

Newton's first law

Newton's first law of motion, sometimes called the law of inertia, states that an object continues in a state of rest or in a state of motion at a constant speed along a straight line unless compelled to change that state by a net force.

Inertia
Mass

Inertia is the natural tendency of an object to remain at rest or in motion at a constant speed along a straight line. The mass of a body is a quantitative measure of inertia and is measured in an SI unit called the kilogram (kg). An iner-

Inertial reference frame

tial reference frame is one in which Newton's law of inertia is valid.

4.3 Newton's Second Law of Motion

4.4 The Vector Nature of Newton's Second Law of Motion

Newton's second law of motion states that when a net force $\Sigma\mathbf{F}$ acts on an object of mass m, the acceleration $\mathbf{a}$ of the object can be obtained from the following equation:

Newton's second law (vector form)

$$\Sigma\mathbf{F} = m\mathbf{a} \tag{4.1}$$

Concept Simulation 4.1

This is a vector equation and, for motion in two dimensions, is equivalent to the following two equations:

Examples 1–3, 20, 21

Newton's second law (component form)

$$\Sigma F_x = ma_x \tag{4.2a}$$

Interactive LearningWare 4.1

$$\Sigma F_y = ma_y \tag{4.2b}$$

Interactive Solution 4.11

In these equations the x and y subscripts refer to the scalar components of the force and acceleration vectors. The SI unit of force is the Newton (N).

Free-body diagram

When determining the net force, a free-body diagram is helpful. A free-body diagram is a diagram that represents the object and the forces acting on it.

4.5 Newton's Third Law of Motion

Newton's third law of motion

Newton's third law of motion, often called the action–reaction law, states that whenever one object exerts a force on a second object, the second object exerts an oppositely directed force of equal magnitude on the first object.

Example 4

Use Self-Assessment Test 4.1 to evaluate your understanding of Sections 4.1–4.5.

4.6 Types of Forces: An Overview

Fundamental forces

Only three fundamental forces have been discovered: the gravitational force, the strong nuclear force, and the electroweak force. The electroweak force manifests itself as either the electromagnetic force or the weak nuclear force.

4.7 The Gravitational Force

Newton's law of universal gravitation states that every particle in the universe exerts an attractive force on every other particle. For two particles that are separated by a distance r and have masses m_1 and m_2, the law states that the magnitude of this attractive force is

Newton's law of universal gravitation

$$F = G\frac{m_1 m_2}{r^2} \tag{4.3}$$

Examples 5, 6

The direction of this force lies along the line between the particles. The constant G has a value of $G = 6.673 \times 10^{-11}\ \text{N}\cdot\text{m}^2/\text{kg}^2$ and is called the universal gravitational constant.

The weight W of an object on or above the earth is the gravitational force that the earth exerts on the object and can be calculated from the mass m of the object and the acceleration g due to the earth's gravity according to

Weight and mass

$$W = mg \tag{4.5}$$

Example 7

4.8 The Normal Force

Normal force

The normal force $\mathbf{F}_N$ is one component of the force that a surface exerts on an object with which it is in contact—namely, the component that is perpendicular to the surface.

Example 8

Topic	Discussion	Learning Aids

The apparent weight is the force that an object exerts on the platform of a scale and may be larger or smaller than the true weight mg if the object and the scale have an acceleration a ($+$ if upward, $-$ if downward). The apparent weight is

Apparent weight

$$\text{Apparent weight} = mg + ma \qquad (4.6)$$

4.9 Static and Kinetic Frictional Forces

Friction

A surface exerts a force on an object with which it is in contact. The component of the force perpendicular to the surface is called the normal force. The component parallel to the surface is called friction.

The force of static friction between two surfaces opposes any impending relative motion of the surfaces. The magnitude of the static frictional force depends on the magnitude of the applied force and can assume any value up to a maximum of

Example 9

Maximum static frictional force

$$f_s^{\text{MAX}} = \mu_s F_N \qquad (4.7)$$

Concept Simulations 4.2, 4.4

where μ_s is the coefficient of static friction and F_N is the magnitude of the normal force.

Interactive LearningWare 4.2

The force of kinetic friction between two surfaces sliding against one another opposes the relative motion of the surfaces. This force has a magnitude given by

Example 10

Concept Simulation 4.3

Kinetic frictional force

$$f_k = \mu_k F_N \qquad (4.8)$$

Interactive Solutions 4.41, 4.101

where μ_k is the coefficient of kinetic friction.

4.10 The Tension Force

The word "tension" is commonly used to mean the tendency of a rope to be pulled apart due to forces that are applied at each end. Because of tension, a rope transmits a force from one end to the other. When a rope is accelerating, the force is transmitted undiminished only if the rope is massless.

Use Self-Assessment Test 4.2 to evaluate your understanding of Sections 4.6–4.10.

4.11 Equilibrium Applications of Newton's Laws of Motion

Definition of equilibrium

An object is in equilibrium when the object has zero acceleration, or, in other words, when it moves at a constant velocity (which may be zero). The sum of the forces that act on an object in equilibrium is zero. Under equilibrium conditions in two dimensions, the separate sums of the force components in the x direction and in the y direction must each be zero:

The equilibrium condition

$$\Sigma F_x = 0 \qquad (4.9\text{a})$$

Examples 11, 12, 13

$$\Sigma F_y = 0 \qquad (4.9\text{b})$$

Interactive LearningWare 4.3

4.12 Nonequilibrium Applications of Newton's Laws of Motion

If an object is not in equilibrium, then Newton's second law must be used to account for the acceleration:

Examples 14–19

$$\Sigma F_x = ma_x \qquad (4.2\text{a})$$

Interactive LearningWare 4.4

$$\Sigma F_y = ma_y \qquad (4.2\text{b})$$

Interactive Solution 4.77

Use Self-Assessment Test 4.3 to evaluate your understanding of Sections 4.11 and 4.12.

Conceptual Questions

1. Why do you lunge forward when your car suddenly comes to a halt? Why are you pressed backward against the seat when your car rapidly accelerates? In your explanation, refer to the most appropriate one of Newton's three laws of motion.

2. A bird feeder of large mass is hung from a tree limb, as the drawing shows. A cord attached to the bottom of the feeder has been left dangling free. Curiosity gets the best of a child, who pulls on the dangling cord in an attempt to see what's in the feeder. The dangling cord is cut from the same source as the cord attached to the limb. Is the cord between the feeder and the limb more likely to snap with a slow continuous pull or a sudden downward pull? Give your reasoning.

3. The net external force acting on an object is zero. Is it possible for the object to be traveling with a velocity that is *not* zero? If your answer is yes, state whether any conditions must be placed on the magnitude and direction of the velocity. If your answer is no, provide a reason for your answer.

4. Is a net force being applied to an object when the object is moving downward (a) with a constant acceleration of 9.80 m/s² and (b) with a constant velocity of 9.80 m/s? Explain.

5. Newton's second law indicates that when a net force acts on an object, it must accelerate. Does this mean that when two or more forces are applied to an object simultaneously, it must accelerate? Explain.

6. A father and his seven-year-old daughter are facing each other on ice skates. With their hands, they push off against one another. (a) Compare the magnitudes of the pushing forces that they experience. (b) Which one, if either, experiences the larger acceleration? Account for your answers.

7. A gymnast is bouncing on a trampoline. After a high bounce the gymnast comes down and hits the elastic surface of the trampoline. In so doing the gymnast applies a force to the trampoline. (a) Describe the effect this force has on the elastic surface. (b) The surface applies a reaction force to the gymnast. Describe the effect that this reaction force has on the gymnast.

8. According to Newton's third law, when you push on an object, the object pushes back on you with an oppositely directed force of equal magnitude. If the object is a massive crate resting on the floor, it will probably not move. Some people think that the reason the crate does not move is that the two oppositely directed pushing forces cancel. Explain why this logic is faulty and why the crate does not move.

9. Three particles have identical masses. Each particle experiences only the gravitational forces due to the other two particles. How should the particles be arranged so each one experiences a net gravitational force that has the same magnitude? Give your reasoning.

10. When a body is moved from sea level to the top of a mountain, what changes—the body's mass, its weight, or both? Explain.

11. The force of air resistance acts to oppose the motion of an object moving through the air. A ball is thrown upward and eventually returns to the ground. (a) As the ball moves upward, is the net force

that acts on the ball greater than, less than, or equal to its weight? Justify your answer. (b) Repeat part (a) for the downward motion of the ball.

12. Object A weighs twice as much as object B at the same spot on the earth. Would the same be true at a given spot on Mars? Account for your answer.

13. Does the acceleration of a freely falling object depend to any extent on the location—that is, whether the object is on top of Mt. Everest or in Death Valley, California? Explain.

14. A "bottle rocket" is a type of fireworks that has a long thin tail that you insert into an empty bottle, to provide a launch platform. One of these rockets is fired with the bottle pointing vertically upward. An identical rocket is fired with the bottle lying on its side, pointing horizontally. In which case does the rocket leave the bottle with the greater acceleration? Explain, ignoring air resistance and friction.

15. A 10-kg suitcase is placed on a scale that is in an elevator. Is the elevator accelerating up or down when the scale reads (a) 75 N and (b) 120 N? Justify your answers.

16. A stack of books whose true weight is 165 N is placed on a scale in an elevator. The scale reads 165 N. Can you tell from this information whether the elevator is moving with a constant velocity of 2 m/s upward or 2 m/s downward or whether the elevator is at rest? Explain.

17. Suppose you are in an elevator that is moving upward with a constant velocity. A scale inside the elevator shows your weight to be 600 N. (a) Does the scale register a value that is greater than, less than, or equal to 600 N during the time when the elevator slows down as it comes to a stop? (b) What is the reading when the elevator is stopped? (c) How does the value registered on the scale compare to 600 N during the time when the elevator picks up speed again on its way back down? Give your reasoning in each case.

18. A person has a choice of either pushing or pulling a sled at a constant velocity, as the drawing illustrates. Friction is present. If the angle θ is the same in both cases, does it require less force to push or to pull? Account for your answer.

19. Suppose that the coefficients of static and kinetic friction have values such that $\mu_s = 2.0\mu_k$ for a crate in contact with a cement floor. Does this mean that the magnitude of the static frictional force acting on the crate at rest would always be twice the magnitude of the kinetic frictional force acting on the moving crate? Give your reasoning.

20. A box rests on the floor of an elevator. Because of static friction, a force is required to start the box sliding across the floor when the elevator is (a) stationary, (b) accelerating upward, and (c) accelerating downward. Rank the forces required in these three situations in ascending order—that is, smallest first. Explain.

21. A rope is used in a tug-of-war between two teams of five people each. Both teams are equally strong, so neither team wins. An identical rope is tied to a tree, and the same ten people pull just as hard on the loose end as they did in the contest. In both cases, the people pull steadily with no jerking. Which rope, if either, is more likely to break? Justify your answer.

22. A stone is thrown from the top of a cliff. As the stone falls, is it in equilibrium? Explain, ignoring air resistance.

23. Can an object ever be in equilibrium if the object is acted on by only (a) a single nonzero force, (b) two forces that point in mutually perpendicular directions, and (c) two forces that point in directions that are not perpendicular? Account for your answers.

24. A circus performer hangs stationary from a rope. She then begins to climb upward by pulling herself up, hand over hand. When she starts climbing, is the tension in the rope less than, equal to, or greater than it is when she hangs stationary? Explain.

25. During the final stages of descent, a sky diver with an open parachute approaches the ground with a constant velocity. The wind does not blow him from side to side. Is the sky diver in equilibrium and, if so, what forces are responsible for the equilibrium?

26. A weight hangs from a ring at the middle of a rope, as the drawing illustrates. Can the person who is pulling on the right end of the rope ever make the rope perfectly horizontal? Explain your answer in terms of the forces that act on the ring.

27. A freight train is accelerating on a level track. Other things being equal, would the tension in the coupling between the engine and the first car change if some of the cargo in the last car were transferred to any one of the other cars? Account for your answer.

Problems

ssm   Solution is in the Student Solutions Manual.   **www**   Solution is available on the World Wide Web at www.wiley.com/college/cutnell
☤   This icon represents a biomedical application.

Section 4.3 Newton's Second Law of Motion

1. An airplane has a mass of 3.1×10^4 kg and takes off under the influence of a constant net force of 3.7×10^4 N. What is the net force that acts on the plane's 78-kg pilot?

2. Concept Simulation 4.1 at **www.wiley.com/college/cutnell** reviews the central idea in this problem. A boat has a mass of 6800 kg. Its engines generate a drive force of 4100 N, due west, while the wind exerts a force of 800 N, due east, and the water exerts a resistive force of 1200 N due east. What is the magnitude and direction of the boat's acceleration?

3. In the amusement park ride known as Magic Mountain Superman, powerful magnets accelerate a car and its riders from rest to 45 m/s (about 100 mi/h) in a time of 7.0 s. The mass of the car and riders is 5.5×10^3 kg. Find the average net force exerted on the car and riders by the magnets.

4. Scientists are experimenting with a kind of gun that may eventually be used to fire payloads directly into orbit. In one test, this gun accelerates a 5.0-kg projectile from rest to a speed of 4.0×10^3 m/s. The net force accelerating the projectile is 4.9×10^5 N. How much time is required for the projectile to come up to speed?

5. ssm When a 58-g tennis ball is served, it accelerates from rest to a speed of 45 m/s. The impact with the racket gives the ball a constant acceleration over a distance of 44 cm. What is the magnitude of the net force acting on the ball?

6. Interactive LearningWare 4.1 at **www.wiley.com/college/cutnell** reviews the approach taken in problems such as this one. A 1580-kg car is traveling with a speed of 15.0 m/s. What is the magnitude of the horizontal net force that is required to bring the car to a halt in a distance of 50.0 m?

7. ssm A person with a black belt in karate has a fist that has a mass of 0.70 kg. Starting from rest, this fist attains a velocity of 8.0 m/s in 0.15 s. What is the magnitude of the average net force applied to the fist to achieve this level of performance?

*8. An arrow, starting from rest, leaves the bow with a speed of 25.0 m/s. If the average force exerted on the arrow by the bow were doubled, all else remaining the same, with what speed would the arrow leave the bow?

*9. ssm www Two forces $\mathbf{F_A}$ and $\mathbf{F_B}$ are applied to an object whose mass is 8.0 kg. The larger force is $\mathbf{F_A}$. When both forces point due east, the object's acceleration has a magnitude of 0.50 m/s². However, when $\mathbf{F_A}$ points due east and $\mathbf{F_B}$ points due west, the acceleration is 0.40 m/s², due east. Find (a) the magnitude of $\mathbf{F_A}$ and (b) the magnitude of $\mathbf{F_B}$.

Section 4.4 The Vector Nature of Newton's Second Law of Motion, Section 4.5 Newton's Third Law of Motion

10. A force vector has a magnitude of 720 N and a direction of 38° north of east. Determine the magnitude and direction of the components of the force that point along the north–south line and along the east–west line.

11. Review **Interactive Solution 4.11** at **www.wiley.com/college/cutnell** before starting this problem. Two forces, $\mathbf{F_1}$ and $\mathbf{F_2}$, act on the 7.00-kg block shown in the drawing. The magnitudes of the forces are $F_1 = 59.0$ N and $F_2 = 33.0$ N. What is the horizontal acceleration (magnitude and direction) of the block?

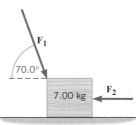

12. A 350-kg sailboat has an acceleration of 0.62 m/s² at an angle of 64° north of east. Find the magnitude and direction of the net force that acts on the sailboat.

13. ssm Only two forces act on an object (mass = 3.00 kg), as in the drawing. Find the magnitude and direction (relative to the x axis) of the acceleration of the object.

14. Airplane flight recorders must be able to survive catastrophic crashes. Therefore, they are typically encased in crash-resistant steel or titanium boxes that are subjected to rigorous testing. One of the tests is an impact shock test, in which the box must sur-

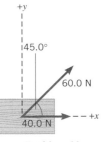

Problem 13

vive being thrown at high speeds against a barrier. A 41-kg box is thrown at a speed of 220 m/s and is brought to a halt in a collision that lasts for a time of 6.5 ms. What is the magnitude of the average net force that acts on the box during the collision?

* **15.** A duck has a mass of 2.5 kg. As the duck paddles, a force of 0.10 N acts on it in a direction due east. In addition, the current of the water exerts a force of 0.20 N in a direction of 52° south of east. When these forces begin to act, the velocity of the duck is 0.11 m/s in a direction due east. Find the magnitude and direction (relative to due east) of the displacement that the duck undergoes in 3.0 s while the forces are acting.

** **16.** At a time when mining asteroids has become feasible, astronauts have connected a line between their 3500-kg space tug and a 6200-kg asteroid. Using their ship's engine, they pull on the asteroid with a force of 490 N. Initially the tug and the asteroid are at rest, 450 m apart. How much time does it take for the ship and the asteroid to meet?

** **17. ssm www** A 325-kg boat is sailing 15.0° north of east at a speed of 2.00 m/s. Thirty seconds later, it is sailing 35.0° north of east at a speed of 4.00 m/s. During this time, three forces act on the boat: a 31.0-N force directed 15.0° north of east (due to an auxiliary engine), a 23.0-N force directed 15.0° south of west (resistance due to the water), and $\mathbf{F_W}$ (due to the wind). Find the magnitude and direction of the force $\mathbf{F_W}$. Express the direction as an angle with respect to due east.

Section 4.7 The Gravitational Force

18. A bowling ball (mass = 7.2 kg, radius = 0.11 m) and a billiard ball (mass = 0.38 kg, radius = 0.028 m) may each be treated as uniform spheres. What is the magnitude of the maximum gravitational force that each can exert on the other?

19. On earth, two parts of a space probe weigh 11 000 N and 3400 N. These parts are separated by a center-to-center distance of 12 m and may be treated as uniform spherical objects. Find the magnitude of the gravitational force that each part exerts on the other out in space, far from any other objects.

20. A rock of mass 45 kg accidentally breaks loose from the edge of a cliff and falls straight down. The magnitude of the air resistance that opposes its downward motion is 250 N. What is the magnitude of the acceleration of the rock?

21. ssm In preparation for this problem, review Conceptual Example 7. A space traveler whose mass is 115 kg leaves earth. What are his weight and mass (a) on earth and (b) in interplanetary space where there are no nearby planetary objects?

22. Review Conceptual Example 7 in preparation for this problem. In tests on earth a lunar surface exploration vehicle (mass = 5.90 × 10³ kg) achieves a forward acceleration of 0.220 m/s². To achieve this same acceleration on the moon, the vehicle's engines must produce a drive force of 1.43 × 10³ N. What is the magnitude of the frictional force that acts on the vehicle on the moon?

23. ssm Synchronous communications satellites are placed in a circular orbit that is 3.59 × 10⁷ m above the surface of the earth. What is the magnitude of the acceleration due to gravity at this distance?

24. The drawing (not to scale) shows one alignment of the sun, earth, and moon. The gravitational force $\mathbf{F_{SM}}$ that the sun exerts on the moon is perpendicular to the force $\mathbf{F_{EM}}$ that the earth exerts on the moon. The masses are: mass of sun = 1.99 × 10³⁰ kg, mass of earth = 5.98 × 10²⁴ kg, mass of moon = 7.35 × 10²² kg. The distances shown in the drawing are r_{SM} = 1.50 × 10¹¹ m and r_{EM} = 3.85 × 10⁸ m. Determine the magnitude of the net gravitational force on the moon.

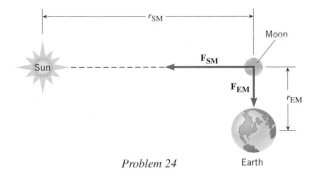

Problem 24

25. The mass of a robot is 5450 kg. This robot weighs 3620 N more on planet A than it does on planet B. Both planets have the same radius of 1.33 × 10⁷ m. What is the difference $M_A − M_B$ in the masses of these planets?

26. The weight of an object is the same on two different planets. The mass of planet A is only sixty percent that of planet B. Find r_A/r_B, which is the ratio of the radii of the planets.

27. ssm Mars has a mass of 6.46 × 10²³ kg and a radius of 3.39 × 10⁶ m. (a) What is the acceleration due to gravity on Mars? (b) How much would a 65-kg person weigh on this planet?

* **28.** Three uniform spheres are located at the corners of an equilateral triangle. Each side of the triangle has a length of 1.20 m. Two of the spheres have a mass of 2.80 kg each. The third sphere (mass unknown) is released from rest. Considering only the gravitational forces that the spheres exert on each other, what is the magnitude of the initial acceleration of the third sphere?

* **29. ssm www** Several people are riding in a hot-air balloon. The combined mass of the people and balloon is 310 kg. The balloon is motionless in the air, because the downward-acting weight of the people and balloon is balanced by an upward-acting "buoyant" force. If the buoyant force remains constant, how much mass should be dropped overboard so the balloon acquires an upward acceleration of 0.15 m/s²?

* **30.** The sun is more massive than the moon, but the sun is farther from the earth. Which one exerts a greater gravitational force on a person standing on the earth? Give your answer by determining the ratio F_{sun}/F_{moon} of the magnitudes of the gravitational forces. Use the data on the inside of the front cover.

* **31.** At a distance H above the surface of a planet, the true weight of a remote probe is one percent less than its true weight on the surface. The radius of the planet is R. Find the ratio H/R.

* **32.** Jupiter is the largest planet in our solar system, having a mass and radius that are, respectively, 318 and 11.2 times that of earth. Suppose that an object falls from rest near the surface of each planet and that the acceleration due to gravity remains constant during the fall. Each object falls the same distance before striking the ground. Determine the ratio of the time of fall on Jupiter to that on earth.

** **33.** Two particles are located on the x axis. Particle 1 has a mass m and is at the origin. Particle 2 has a mass $2m$ and is at $x = +L$. A third particle is placed between particles 1 and 2. Where on the x axis should the third particle be located so that the magnitude of the gravitational force on *both* particle 1 and particle 2 doubles? Express your answer in terms of L.

Section 4.8 The Normal Force, Section 4.9 Static and Kinetic Frictional Forces

34. A 35-kg crate rests on a horizontal floor, and a 65-kg person is standing on the crate. Determine the magnitude of the normal force that (a) the floor exerts on the crate and (b) the crate exerts on the person.

35. ssm A rocket blasts off from rest and attains a speed of 45 m/s in 15 s. An astronaut has a mass of 57 kg. What is the astronaut's apparent weight during takeoff?

36. A woman stands on a scale in a moving elevator. Her mass is 60.0 kg, and the combined mass of the elevator and scale is an additional 815 kg. Starting from rest, the elevator accelerates upward. During the acceleration, the hoisting cable applies a force of 9410 N. What does the scale read during the acceleration?

37. A block whose weight is 45.0 N rests on a horizontal table. A horizontal force of 36.0 N is applied to the block. The coefficients of static and kinetic friction are 0.650 and 0.420, respectively. Will the block move under the influence of the force, and, if so, what will be the block's acceleration? Explain your reasoning.

38. A cup of coffee is sitting on a table in an airplane that is flying at a constant altitude and a constant velocity. The coefficient of static friction between the cup and the table is 0.30. Suddenly, the plane accelerates, its altitude remaining constant. What is the maximum acceleration that the plane can have without the cup sliding backward on the table?

39. ssm A 20.0-kg sled is being pulled across a horizontal surface at a constant velocity. The pulling force has a magnitude of 80.0 N and is directed at an angle of 30.0° above the horizontal. Determine the coefficient of kinetic friction.

40. A 6.00-kg box is sliding across the horizontal floor of an elevator. The coefficient of kinetic friction between the box and the floor is 0.360. Determine the kinetic frictional force that acts on the box when the elevator is (a) stationary, (b) accelerating upward with an acceleration whose magnitude is 1.20 m/s², and (c) accelerating downward with an acceleration whose magnitude is 1.20 m/s².

41. Review **Interactive Solution 4.41** at **www.wiley.com/college/cutnell** in preparation for this problem. An 81-kg baseball player slides into second base. The coefficient of kinetic friction between the player and the ground is 0.49. (a) What is the magnitude of the frictional force? (b) If the player comes to rest after 1.6 s, what was his initial velocity?

* **42.** One block rests upon a horizontal surface. A second identical block rests upon the first one. The coefficient of static friction between the blocks is the same as the coefficient of static friction between the lower block and the horizontal surface. A horizontal force is applied to the upper block, and its magnitude is slowly increased. When the force reaches 47.0 N, the upper block just begins to slide. The force is then removed from the upper block, and the blocks are returned to their original configuration. What is the magnitude of the horizontal force that should be applied to the lower block, so that it just begins to slide out from under the upper block?

* **43. ssm** A skater with an initial speed of 7.60 m/s is gliding across the ice. Air resistance is negligible. (a) The coefficient of kinetic friction between the ice and the skate blades is 0.100. Find the deceleration caused by kinetic friction. (b) How far will the skater travel before coming to rest?

* **44.** Refer to **Concept Simulation 4.4** at **www.wiley.com/college/cutnell** for background relating to this problem. The drawing shows a large cube (mass = 25 kg) being accelerated across a horizontal frictionless surface by a horizontal force **P**. A small cube (mass = 4.0 kg) is in contact with the front surface of the large cube and will slide downward unless **P** is sufficiently large. The coefficient of static friction between the cubes is 0.71. What is the smallest magnitude that **P** can have in order to keep the small cube from sliding downward?

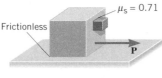

μ_s = 0.71
Frictionless
P

** **45.** While moving in, a new homeowner is pushing a box across the floor at a constant velocity. The coefficient of kinetic friction between the box and the floor is 0.41. The pushing force is directed downward at an angle θ below the horizontal. When θ is greater than a certain value, it is not possible to move the box, no matter how large the pushing force is. Find that value of θ.

Section 4.10 The Tension Force, Section 4.11 Equilibrium Applications of Newton's Laws of Motion

46. Part *a* of the drawing shows a bucket of water suspended from the pulley of a well; the tension in the rope is 92.0 N. Part *b* shows the same bucket of water being pulled up from the well at a constant velocity. What is the tension in the rope in part *b*?

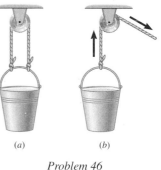

(a)   (b)

Problem 46

47. A supertanker (mass = 1.70 × 10⁸ kg) is moving with a constant velocity. Its engines generate a forward thrust of 7.40 × 10⁵ N. Determine (a) the magnitude of the resistive force exerted on the tanker by the water and (b) the magnitude of the upward buoyant force exerted on the tanker by the water.

48. Review **Interactive LearningWare 4.3** at **www.wiley.com/college/cutnell** in preparation for this problem. The helicopter in the drawing is moving horizontally to the right at a constant velocity. The weight of the helicopter is W = 53 800 N. The lift force **L** generated by the rotating blade makes an angle of 21.0° with respect to the vertical. (a) What is the magnitude of the lift force? (b) Determine the magnitude of the air resistance **R** that opposes the motion.

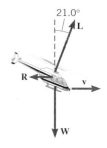

21.0°
L
R
V
W

49. ssm A 1.40-kg bottle of vintage wine is lying horizontally in the rack shown in the drawing. The two surfaces on which the bottle rests are 90.0° apart, and the right surface makes an angle of 45.0° with respect to the ground. Each surface exerts a force on the bottle that is perpendicular to the surface. What is the magnitude of each of these forces?

90.0°

50. The steel I-beam in the drawing has a weight of 8.00 kN and is being lifted at a constant velocity. What is the tension in each cable attached to its ends?

51. A stuntman is being pulled along a rough road at a constant velocity, by a cable attached to a moving truck. The cable is parallel to the ground. The mass of the stuntman is 109 kg, and the coefficient of kinetic friction between the road and him is 0.870. Find the tension in the cable.

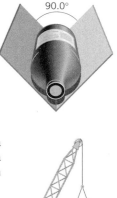

70.0° 70.0°
I-beam

Problem 50

52. As preparation for this problem, review Example 13. Suppose that the pilot suddenly jettisons 2800 N of fuel. If the plane is to continue moving with the same velocity under the influence of the same air resistance **R**, by how much does the pilot have to reduce (a) the thrust and (b) the lift?

53. ssm Three forces act on a moving object. One force has a magnitude of 80.0 N and is directed due north. Another has a magnitude of 60.0 N and is directed due west. What must be the magnitude and direction of the third force, such that the object continues to move with a constant velocity?

54. The drawing shows a circus clown who weighs 890 N. The coefficient of static friction between the clown's feet and the ground is 0.53. He pulls vertically downward on a rope that passes around three pulleys and is tied around his feet. What is the minimum pulling force that the clown must exert to yank his feet out from under himself?

* 55. The drawing shows box 1 resting on a table, with box 2 resting on top of box 1. A massless rope passes over a massless, frictionless pulley. One end of the rope is connected to box 2 and the other end is connected to box 3. The weights of the three boxes are $W_1 = 55$ N, $W_2 = 35$ N, and $W_3 = 28$ N. Determine the magnitude of the normal force that the table exerts on box 1.

* 56. Interactive LearningWare 4.3 at **www.wiley.com/college/cutnell** reviews the principles that play a role in this problem. During a storm, a tree limb breaks off and comes to rest across a barbed wire fence at a point that is not in the middle between two fence posts. The limb exerts a downward force of 151 N on the wire. The left section of the wire makes an angle of 14.0° relative to the horizontal and sustains a tension of 447 N. Find the magnitude and direction of the tension that the right section of the wire sustains.

Problem 55

* 57. ssm A person is trying to judge whether a picture (mass = 1.10 kg) is properly positioned by temporarily pressing it against a wall. The pressing force is perpendicular to the wall. The coefficient of static friction between the picture and the wall is 0.660. What is the minimum amount of pressing force that must be used?

* 58. A mountain climber, in the process of crossing between two cliffs by a rope, pauses to rest. She weighs 535 N. As the drawing shows, she is closer to the left cliff than to the right cliff, with the result that the tensions in the left and right sides of the rope are not the same. Find the tensions in the rope to the left and to the right of the mountain climber.

* 59. A skier is pulled up a slope at a constant velocity by a tow bar. The slope is inclined at 25.0° with respect to the horizontal. The force applied to the skier by the tow bar is parallel to the slope. The skier's mass is 55.0 kg, and the coefficient of kinetic friction between the skis and the snow is 0.120. Find the magnitude of the force that the tow bar exerts on the skier.

** 60. The weight of the block in the drawing is 88.9 N. The coefficient of static friction between the block and the vertical wall is 0.560. (a) What minimum force **F** is required to prevent the block from sliding down the wall? *(Hint: The static frictional force exerted*

on the block is directed upward, parallel to the wall.) (b) What minimum force is required to start the block moving up the wall? *(Hint: The static frictional force is now directed down the wall.)*

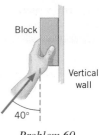

Problem 60

** 61. ssm A bicyclist is coasting straight down a hill at a constant speed. The mass of the rider and bicycle is 80.0 kg, and the hill is inclined at 15.0° with respect to the horizontal. Air resistance opposes the motion of the cyclist. Later, the bicyclist climbs the same hill at the same constant speed. How much force (directed parallel to the hill) must be applied to the bicycle in order for the bicyclist to climb the hill?

** 62. A damp washcloth is hung over the edge of a table to dry. Thus, part (mass = m_{on}) of the washcloth rests on the table and part (mass = m_{off}) does not. The coefficient of static friction between the table and the washcloth is 0.40. Determine the maximum fraction [$m_{off}/(m_{on} + m_{off})$] that can hang over the edge without causing the whole washcloth to slide off the table.

Section 4.12 Nonequilibrium Applications of Newton's Laws of Motion

63. Only two forces act on an object (mass = 4.00 kg), as in the drawing. Find the magnitude and direction (relative to the x axis) of the acceleration of the object.

64. Concept Simulation 4.1 at **www.wiley.com/college/cutnell** reviews the concepts that are important in this problem. The speed of a bobsled is increasing because it has an acceleration of 2.4 m/s². At a given instant in time, the forces resisting the motion, including kinetic friction and air resistance, total 450 N. The mass of the bobsled and its riders is 270 kg. (a) What is the magnitude of the force propelling the bobsled forward? (b) What is the magnitude of the net force that acts on the bobsled?

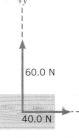

Problem 63

65. ssm A 1380-kg car is moving due east with an initial speed of 27.0 m/s. After 8.00 s the car has slowed down to 17.0 m/s. Find the magnitude and direction of the net force that produces the deceleration.

66. A falling skydiver has a mass of 110 kg. (a) What is the magnitude of the skydiver's acceleration when the upward force of air resistance has a magnitude that is equal to one-third of his weight? (b) After the parachute opens, the skydiver descends at a constant velocity. What is the force of air resistance (magnitude and direction) that acts on the skydiver?

67. In the drawing, the weight of the block on the table is 422 N and that of the hanging block is 185 N. Ignoring all frictional effects and assuming the pulley to be massless, find (a) the acceleration of the two blocks and (b) the tension in the cord.

68. A 292-kg motorcycle is accelerating up along a ramp that is inclined 30.0° above the horizontal. The propulsion force pushing the motorcycle up the ramp is 3150 N, and air resistance produces a force of 250 N that opposes the motion. Find the magnitude of the motorcycle's acceleration.

69. ssm www A student is skateboarding down a ramp that is 6.0 m long and inclined at 18° with respect to the horizontal. The initial speed of the skateboarder at the top of the ramp is 2.6 m/s. Neglect friction and find the speed at the bottom of the ramp.

70. A rescue helicopter is lifting a man (weight = 822 N) from a capsized boat by means of a cable and harness. (a) What is the tension in the cable when the man is given an initial upward acceleration of 1.10 m/s²? (b) What is the tension during the remainder of the rescue when he is pulled upward at a constant velocity?

71. ssm In a supermarket parking lot, an employee is pushing ten empty shopping carts, lined up in a straight line. The acceleration of the carts is 0.050 m/s². The ground is level, and each cart has a mass of 26 kg. (a) What is the net force acting on any one of the carts? (b) Assuming friction is negligible, what is the force exerted by the fifth cart on the sixth cart?

72. Interactive LearningWare 4.4 at **www.wiley.com/college/cutnell** provides a review of the concepts that are important in this problem. A rocket of mass 4.50×10^5 kg is in flight. Its thrust is directed at an angle of 55.0° above the horizontal and has a magnitude of 7.50×10^6 N. Find the magnitude and direction of the rocket's acceleration. Give the direction as an angle above the horizontal.

73. A 1.14×10^4-kg lunar landing craft is about to touch down on the surface of the moon, where the acceleration due to gravity is 1.60 m/s². At an altitude of 165 m the craft's downward velocity is 18.0 m/s. To slow down the craft, a retrorocket is firing to provide an upward thrust. Assuming the descent is vertical, find the magnitude of the thrust needed to reduce the velocity to zero at the instant when the craft touches the lunar surface.

* **74.** Consult **Concept Simulation 4.2** at **www.wiley.com/college/cutnell** in preparation for this problem. A crate is resting on a ramp that is inclined at an angle θ above the horizontal. As θ is increased, the crate remains in place until θ reaches a value of 38.0°. Then the crate begins to slide down the slope. (a) Determine the coefficient of static friction between the crate and the ramp surface. (b) The coefficient of kinetic friction between the crate and the ramp surface is 0.600. Find the acceleration of the moving crate.

* **75. ssm** To hoist himself into a tree, a 72.0-kg man ties one end of a nylon rope around his waist and throws the other end over a branch of the tree. He then pulls downward on the free end of the rope with a force of 358 N. Neglect any friction between the rope and the branch, and determine the man's upward acceleration.

* **76.** A 205-kg log is pulled up a ramp by means of a rope that is parallel to the surface of the ramp. The ramp is inclined at 30.0° with respect to the horizontal. The coefficient of kinetic friction between the log and the ramp is 0.900, and the log has an acceleration of 0.800 m/s². Find the tension in the rope.

* **77.** Review **Interactive Solution 4.77** at **www.wiley.com/college/cutnell** before staring this problem. The drawing shows Robin Hood (mass = 77.0 kg) about to escape from a dangerous situation. With one hand, he is gripping the rope that holds up a chandelier (mass = 195 kg). When he cuts the rope where it is tied to the floor, the chandelier will fall, and he will be pulled up toward a balcony above. Ignore the friction between the rope and the beams over which it slides, and find (a) the ac-

celeration with which Robin is pulled upward and (b) the tension in the rope while Robin escapes.

* **78.** A train consists of 50 cars, each of which has a mass of 6.8×10^3 kg. The train has an acceleration of $+8.0 \times 10^{-2}$ m/s². Ignore friction and determine the tension in the coupling (a) between the 30th and 31st cars and (b) between the 49th and 50th cars.

* **79. ssm** A box is sliding up an incline that makes an angle of 15.0° with respect to the horizontal. The coefficient of kinetic friction between the box and the surface of the incline is 0.180. The initial speed of the box at the bottom of the incline is 1.50 m/s. How far does the box travel along the incline before coming to rest?

* **80.** A girl is sledding down a slope that is inclined at 30.0° with respect to the horizontal. A moderate wind is aiding the motion by providing a steady force of 105 N that is parallel to the motion of the sled. The combined mass of the girl and sled is 65.0 kg, and the coefficient of kinetic friction between the runners of the sled and the snow is 0.150. How much time is required for the sled to travel down a 175-m slope, starting from rest?

* **81.** At an airport, luggage is unloaded from a plane into the three cars of a luggage carrier, as the drawing shows. The acceleration of the carrier is 0.12 m/s², and friction is negligible. The coupling bars have negligible mass. By how much would the tension in *each* of the coupling bars A, B, and C change if 39 kg of luggage were removed from car 2 and placed in (a) car 1 and (b) car 3? If the tension changes, specify whether it increases or decreases.

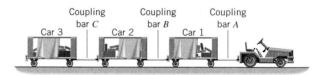

* **82.** Consult **Interactive LearningWare 4.2** at **www.wiley.com/college/cutnell** before beginning this problem. A truck is traveling at a speed of 25.0 m/s along a level road. A crate is resting on the bed of the truck, and the coefficient of static friction between the crate and the truck bed is 0.650. Determine the shortest distance in which the truck can come to a halt without causing the crate to slip forward relative to the truck.

** **83. ssm** A penguin slides at a constant velocity of 1.4 m/s down an icy incline. The incline slopes above the horizontal at an angle of 6.9°. At the bottom of the incline, the penguin slides onto a horizontal patch of ice. The coefficient of kinetic friction between the penguin and the ice is the same for the incline as for the horizontal patch. How much time is required for the penguin to slide to a halt after entering the horizontal patch of ice?

** **84.** As part *a* of the drawing shows, two blocks are connected by a rope that passes over a set of pulleys. One block has a weight of 412 N, and the other has a weight of 908 N. The rope and the pulleys are massless and there is no friction. (a) What is the accelera-

(a)   (b)

tion of the lighter block? (b) Suppose that the heavier block is removed, and a downward force of 908 N is provided by someone pulling on the rope, as part *b* of the drawing shows. Find the acceleration of the remaining block. (c) Explain why the answers in (a) and (b) are different.

** **85.** In the drawing, the rope and the pulleys are massless, and there is no friction. Find (a) the tension in the rope and (b) the acceleration of the 10.0-kg block. *(Hint: The larger mass moves twice as far as the smaller mass.)*

** **86.** A 5.00-kg block is placed on top of a 12.0-kg block that rests on a frictionless table. The coefficient of static friction between the two blocks is 0.600. What is the maximum horizontal force that can be applied before the 5.00-kg block begins to slip relative to the 12.0-kg block, if the force is applied to (a) the more massive block and (b) the less massive block?

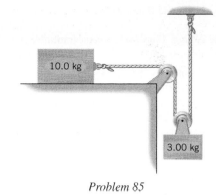

Problem 85

Additional Problems

87. ssm Saturn has an equatorial radius of 6.00×10^7 m and a mass of 5.67×10^{26} kg. (a) Compute the acceleration of gravity at the equator of Saturn. (b) What is the ratio of a person's weight on Saturn to that on earth?

88. A 75-kg water skier is being pulled by a horizontal force of 520 N and has an acceleration of 2.4 m/s². Assuming that the total resistive force exerted on the skier by the water and the wind is constant, what force is needed to pull the skier at a constant velocity?

89. ssm A student presses a book between his hands, as the drawing indicates. The forces that he exerts on the front and back covers of the book are perpendicular to the book and are horizontal. The book weighs 31 N. The coefficient of static friction between his hands and the book is 0.40. To keep the book from falling, what is the magnitude of the minimum pressing force that each hand must exert?

90. The space probe *Deep Space 1* was launched on October 24, 1998. Its mass was 474 kg. The goal of the mission was to test a new kind of engine called an ion propulsion drive. This engine generates only a weak thrust, but it can do so over long periods of time with the consumption of only small amounts of fuel. The mission has been spectacularly successful. At a thrust of 56 mN how many days are required for the probe to attain a velocity of 805 m/s (1800 mi/h), assuming that the probe starts from rest and that the mass remains nearly constant?

91. ssm A 60.0-kg crate rests on a level floor at a shipping dock. The coefficients of static and kinetic friction are 0.760 and 0.410, respectively. What horizontal pushing force is required to (a) just start the crate moving and (b) slide the crate across the dock at a constant speed?

92. Review Conceptual Example 16 as background for this problem. The water skier there has a mass of 73 kg. Find the magnitude of the net force acting on the skier when (a) she is accelerated from rest to a speed of 11 m/s in 8.0 s and (b) she lets go of the tow rope and glides to a halt in 21 s.

93. A cable is lifting a construction worker and a crate, as the drawing shows. The weights of the worker and crate are 965 and 1510 N, respectively. The acceleration of the cable is 0.620 m/s², upward. What is the tension in the cable (a) below the worker and (b) above the worker?

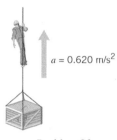

$a = 0.620$ m/s²

Problem 93

94. A 95.0-kg person stands on a scale in an elevator. What is the apparent weight when the elevator is (a) accelerating upward with an acceleration of 1.80 m/s², (b) moving upward at a constant speed, and (c) accelerating downward with an acceleration of 1.30 m/s²?

95. ssm A 15-g bullet is fired from a rifle. It takes 2.50×10^{-3} s for the bullet to travel the length of the barrel, and it exits the barrel with a speed of 715 m/s. Assuming that the acceleration of the bullet is constant, find the average net force exerted on the bullet.

96. Review **Interactive LearningWare 4.1** at **www.wiley.com/college/cutnell** in preparation for this problem. During a circus performance, a 72-kg human cannonball is shot out of an 18-m-long cannon. If the human cannonball spends 0.95 s in the cannon, determine the average net force exerted on him in the barrel of the cannon.

97. (a) Calculate the magnitude of the gravitational force exerted on a 425-kg satellite that is a distance of two earth radii from the center of the earth. (b) What is the magnitude of the gravitational force exerted on the earth by the satellite? (c) Determine the magnitude of the satellite's acceleration. (d) What is the magnitude of the earth's acceleration?

98. The drawing shows three particles far away from any other objects and located on a straight line. The masses of these particles are $m_A = 363$ kg, $m_B = 517$ kg, and $m_C = 154$ kg. Find the magnitude and direction of the net gravitational force acting on (a) particle A, (b) particle B, and (c) particle C.

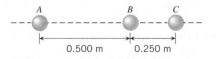

* **99. ssm** Two objects (45.0 and 21.0 kg) are connected by a massless string that passes over a massless, frictionless pulley. The pulley hangs from the ceiling. Find (a) the acceleration of the objects and (b) the tension in the string.

* **100. Concept Simulation 4.3** at **www.wiley.com/college/cutnell** reviews the concepts that play a role in this problem. Traveling at a speed of 16.1 m/s, the driver of an automobile suddenly locks the wheels by slamming on the brakes. The coefficient of kinetic friction between the tires and the road is 0.720. How much time does it take for the car to come to a halt? Ignore the effects of air resistance.

* **101.** Consult **Interactive Solution 4.101** at **www.wiley.com/college/cutnell** before beginning this problem. A toboggan slides down a hill and has a constant velocity. The angle of the hill is 8.00° with respect to the horizontal. What is the coefficient of kinetic friction between the surface of the hill and the toboggan?

* **102.** A bicyclist is coasting down an 8.0° hill at a constant velocity of +8.9 m/s. The mass of the bicycle and rider is 85 kg. A force due to air resistance is directed opposite to the motion and has a magnitude f_{air} that is proportional to the cyclist's speed v, so that $f_{air} = cv$, where c is a constant. Using the fact that the cyclist is coasting at a constant velocity, determine the numerical value (including units) for the constant c.

* **103. ssm** A person whose weight is 5.20×10^2 N is being pulled up vertically by a rope from the bottom of a cave that is 35.1 m deep. The maximum tension that the rope can withstand without breaking is 569 N. What is the shortest time, starting from rest, in which the person can be brought out of the cave?

* **104.** A space probe has two engines. Each generates the same amount of force when fired, and the directions of these forces can be independently adjusted. When the engines are fired simultaneously and each applies its force in the same direction, the probe, starting from rest, takes 28 s to travel a certain distance. How long does it take to travel the same distance, again starting from rest, if the engines are fired simultaneously and the forces that they apply to the probe are perpendicular?

* **105.** A sports car is accelerating up a hill that rises 18° above the horizontal. The coefficient of static friction between the wheels and the road is $\mu_s = 0.88$. It is the static frictional force that propels the car forward. (a) What is the magnitude of the maximum acceleration that the car can have? (b) What is the magnitude of the maximum acceleration if the car is being driven down the hill?

* **106.** A neutron star has a mass of 2.0×10^{30} kg (about the mass of our sun) and a radius of 5.0×10^3 m (about the height of a good-sized mountain). Suppose an object falls from rest near the surface of such a star. How fast would it be moving after it had fallen a distance of 0.010 m? (Assume that the gravitational force is constant over the distance of the fall, and that the star is not rotating.)

* **107. ssm** A 44-kg chandelier is suspended 1.5 m below a ceiling by three wires, each of which has the same tension and the same length of 2.0 m (see the drawing). Find the tension in each wire.

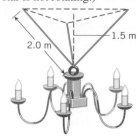

Problem 107

** **108.** A small sphere is hung by a string from the ceiling of a van. When the van is stationary, the sphere hangs vertically. However, when the van accelerates, the sphere swings backward so that the string makes an angle of θ with respect to the vertical. (a) Derive an expression for the magnitude a of the acceleration of the van in terms of the angle θ and the magnitude g of the acceleration due to gravity. (b) Find the acceleration of the van when $\theta = 10.0°$. (c) What is the angle θ when the van moves with a constant velocity?

** **109.** The drawing shows three objects. They are connected by strings that pass over massless and friction-free pulleys. The objects move, and the coefficient of kinetic friction between the middle object and the surface of the table is 0.100. (a) What is the acceleration of the three objects? (b) Find the tension in each of the two strings.

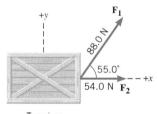

** **110.** The drawing shows a 25.0-kg crate that is initially at rest. Note that the view is one looking down on the top of the crate. Two forces, $\mathbf{F_1}$ and $\mathbf{F_2}$, are applied to the crate, and it begins to move. The coefficient of kinetic friction between the crate and the floor is $\mu_k = 0.350$. Determine the magnitude and direction (relative to the x axis) of the acceleration of the crate.

** **111. ssm** A 225-kg crate rests on a surface that is inclined above the horizontal at an angle of 20.0°. A horizontal force (magnitude = 535 N and parallel to the ground, not the incline) is required to start the crate moving down the incline. What is the coefficient of static friction between the crate and the incline?

Concepts & Calculations Group Learning Problems

Note: Each of these problems consists of Concept Questions followed by a related quantitative Problem. They are designed for use by students working alone or in small learning groups. The Concept Questions involve little or no mathematics and are intended to stimulate group discussions. They focus on the concepts with which the problems deal. Recognizing the concepts is the essential initial step in any problem-solving technique.

112. Concept Questions Two skaters, a man and a woman, are standing on ice. Neglect any friction between the skate blades and the ice. The woman pushes on the man with a certain force that is parallel to the ground. (a) Must the man accelerate under the action of this force? If so, what three factors determine the magnitude and direction of his acceleration? (b) Is there a corresponding force exerted on the woman? If so, where does it originate? Is this force related to the magnitude and direction of the force the woman exerts on the man? If so, how?

Problem The mass of the man is 82 kg and that of the woman is 48 kg. The woman pushes on the man with a force of 45 N due east. Determine the acceleration (magnitude and direction) of (a) the man and (b) the woman.

113. Concept Questions A person is attempting to push a refrigerator across a room. He exerts a horizontal force on the refrigerator, but it does not move. (a) What other horizontal force must be acting on the refrigerator? (b) How are the magnitude and direction of this force related to the force that the person exerts? (c) Suppose that the person applies the force such that it is the largest possible force before the refrigerator begins to move. What factors determine the magnitude of this force?

Problem A person pushes on a 57-kg refrigerator with a horizontal force of -267 N; the minus sign indicates that the force is directed along the $-x$ direction. The coefficient of static friction is 0.65. (a) If the refrigerator does not move, what is the magnitude and direction of the static frictional force that the floor exerts on the refrigerator? (b) What is the magnitude of the largest pushing force that can be applied to the refrigerator before it just begins to move?

114. Concept Questions A large rock and a small pebble are held at the same height above the ground. (a) Is the gravitational force exerted on the rock greater than, less than, or equal to that exerted on the pebble? Justify your answer. (b) When the rock and the pebble are released, is the downward acceleration of the rock greater than, less than, or equal to that of the pebble? Why?

Problem A 5.0-kg rock and a 3.0×10^{-4}-kg pebble are held near the surface of the earth. (a) Determine the magnitude of the gravitational force exerted on each by the earth. (b) Calculate the acceleration of each object when released. Verify that your answers are consistent with your answers to the Concept Questions.

115. Concept Question The earth exerts a gravitational force on a falling raindrop, pulling it down. Does the raindrop exert a gravitational force on the earth, pulling it up? If so, is this force greater than, less than, or equal to the force that the earth exerts on the raindrop? Provide a reason for your answer.

Problem A raindrop has a mass of 5.2×10^{-7} kg, and is falling near the surface of the earth. Calculate the magnitude of the gravitational force exerted (a) on the raindrop by the earth and (b) on the earth by the raindrop.

116. Concept Questions Two horizontal forces, F_1 and F_2, are acting on a box, but only F_1 is shown (see the drawing). F_2 can point either to the right or to the left. The box moves only along the x axis. There is no friction between the box and the surface. What is the direction of F_2 and how does its magnitude compare to the magnitude of F_1 when the acceleration of the box is (a) positive, (b) negative, and (c) zero?

Problem Suppose that $F_1 = +9.0$ N and the mass of the box is 3.0 kg. Find the magnitude and direction of F_2 when the acceleration of the box is (a) $+5.0$ m/s^2, (b) -5.0 m/s^2, and

(c) 0 m/s^2. Be sure that your answers are consistent with your answers to the Concept Questions.

117. Concept Questions A car is driving up a hill. (a) Is the magnitude of the normal force exerted on the car equal to the magnitude of its weight? Why or why not? (b) If the car drives up a steeper hill,

does the normal force increase, decrease, or remain the same? Justify your answer. (c) Does the magnitude of the normal force depend on whether the car is traveling up the hill or down the hill? Give your reasoning.

Problem A car is traveling up a hill that is inclined at an angle of θ above the horizontal. Determine the ratio of the magnitude of the normal force to the weight of the car when (a) $\theta = 15°$ and (b) $\theta = 35°$. Check to see that your answers are consistent with your answers to the Concept Questions.

* **118. Concept Questions** Two blocks are sliding to the right across a horizontal surface, as the drawing shows. In Case A the masses of both blocks are 3.0 kg. In Case B the mass of block 1, the block behind, is 6.0 kg, and the mass of block 2 is 3.0 kg. No frictional force acts on block 1 in either Case A or Case B. However, a kinetic frictional force does act on block 2 in both cases and opposes the motion. (a) Identify the forces that contribute to the horizontal net force acting on block 1. (b) Identify the forces that contribute to the horizontal net force acting on block 2. (c) In which case, if either, do the blocks push against each other with greater forces? Explain. (d) Are the blocks accelerating or decelerating, and in which case, if either, is the magnitude of the acceleration greater?

Problem The magnitude of the kinetic frictional force acting on block 2 in the drawing is 5.8 N. For both Case A and Case B determine (a) the magnitude of the forces with which the blocks push against each other and (b) the acceleration of the blocks. Check to see that your answers are consistent with your answers to the Concept Questions.

Case A   Case B

* **119. Concept Questions** A block is pressed against a vertical wall by a force P, as the drawing shows. This force can either push the block upward at a constant velocity or allow it to slide downward at a constant velocity, the magnitude of the force being different in the two cases, while the directional angle θ is the same. Kinetic friction exists between the block and the wall. (a) Is the block in equilibrium in each case? Explain. (b) In each case what is the direction of the kinetic frictional force that acts on the block? Why? (c) In each case is the magnitude of the frictional force the same or different? Justify your answer. (d) In which case is the magnitude of the force P greater? Provide a reason for your answer.

Problem The weight of the block is 39.0 N, and the coefficient of kinetic friction between the block and the wall is 0.250. The direction of the force P is $\theta = 30.0°$. Determine the magnitude of P when the block slides up the wall and when it slides down the wall. Check to see that your answers are consistent with your answers to the Concept Questions.

Chapter 5

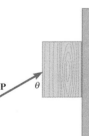

Dynamics of Uniform Circular Motion

When traveling in a circular turn at a constant speed, each power boat experiences a net force and, hence, an acceleration that both point toward the center of the circle. (© AFP/ Corbis Images)

5.1 Uniform Circular Motion

There are many examples of motion on a circular path. Of the many possibilities, we single out those that satisfy the following definition:

■ **DEFINITION OF UNIFORM CIRCULAR MOTION**

Uniform circular motion is the motion of an object traveling at a constant (uniform) speed on a circular path.

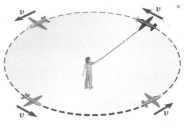

Figure 5.1 The motion of an airplane flying at a constant speed on a horizontal circular path is an example of uniform circular motion.

As an example of uniform circular motion, Figure 5.1 shows a model airplane on a guideline. The speed of the plane is the magnitude of the velocity vector **v**, and since the speed is constant, the vectors in the drawing have the same magnitude at all points on the circle.

Sometimes it is more convenient to describe uniform circular motion by specifying the period of the motion, rather than the speed. The **period T** is the time required to travel once around the circle—that is, to make one complete revolution. There is a relationship between period and speed, since speed v is the distance traveled (circumference of the circle = $2\pi r$) divided by the time T:

$$v = \frac{2\pi r}{T} \tag{5.1}$$

If the radius is known, as in Example 1, the speed can be calculated from the period or vice versa.

Example 1   A Tire-Balancing Machine

The wheel of a car has a radius of $r = 0.29$ m and is being rotated at 830 revolutions per minute (rpm) on a tire-balancing machine. Determine the speed (in m/s) at which the outer edge of the wheel is moving.

Reasoning The speed v can be obtained directly from $v = 2\pi r/T$, but first the period T is needed. The period is the time for one revolution, and it must be expressed in seconds, because the problem asks for the speed in meters per second.

Solution Since the tire makes 830 revolutions in one minute, the number of minutes required for a single revolution is

$$\frac{1}{830 \text{ revolutions/min}} = 1.2 \times 10^{-3} \text{ min/revolution}$$

Therefore, the period is $T = 1.2 \times 10^{-3}$ min, which corresponds to 0.072 s. Equation 5.1 can now be used to find the speed:

$$v = \frac{2\pi r}{T} = \frac{2\pi(0.29 \text{ m})}{0.072 \text{ s}} = \boxed{25 \text{ m/s}}$$

The definition of uniform circular motion emphasizes that the speed, or the magnitude of the velocity vector, is constant. It is equally significant that the direction of the vector is *not constant*. In Figure 5.1, for instance, the velocity vector changes direction as the plane moves around the circle. Any change in the velocity vector, even if it is only a change in direction, means that an acceleration is occurring. This particular acceleration is called "centripetal acceleration," because it points toward the center of the circle, as the next section explains.

5.2 Centripetal Acceleration

In this section we determine how the magnitude a_c of the centripetal acceleration depends on the speed v of the object and the radius r of the circular path. We will see that $a_c = v^2/r$.

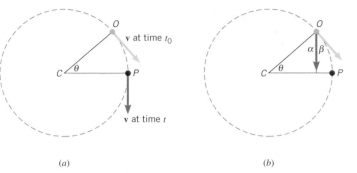

Figure 5.2 (a) For an object (●) in uniform circular motion, the velocity **v** has different directions at different places on the circle. (b) The velocity vector has been removed from point P, shifted parallel to itself, and redrawn with its tail at point O.

(a)　　　　　　(b)

In Figure 5.2a an object (symbolized by a dot ●) is in uniform circular motion. At time t_0 the velocity is tangent to the circle at point O, and at a later time t the velocity is tangent at point P. As the object moves from O to P, the radius traces out the angle θ, and the velocity vector changes direction. To emphasize the change, part b of the picture shows the velocity vector removed from point P, shifted parallel to itself, and redrawn with its tail at point O. The angle β between the two vectors indicates the change in direction. Since the radii CO and CP are perpendicular to the tangents at points O and P, respectively, it follows that $\alpha + \beta = 90°$ and $\alpha + \theta = 90°$. Therefore, angle β and angle θ are equal.

As always, acceleration is the change $\Delta\mathbf{v}$ in velocity divided by the elapsed time Δt, or $\mathbf{a} = \Delta\mathbf{v}/\Delta t$. Figure 5.3a shows the two velocity vectors oriented at the angle θ with respect to one another, together with the vector $\Delta\mathbf{v}$ that represents the change in velocity. The change $\Delta\mathbf{v}$ is the increment that must be added to the velocity at time t_0, so that the resultant velocity has the new direction after an elapsed time $\Delta t = t - t_0$. Figure 5.3b shows the sector of the circle COP. In the limit that Δt is very small, the arc length OP is approximately a straight line whose length is the distance $v\Delta t$ traveled by the object. In this limit, COP is an isosceles triangle, as is the triangle in part a of the drawing. Since both triangles have equal apex angles θ, they are similar, so that

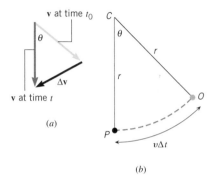

Figure 5.3 (a) The directions of the velocity vector at times t and t_0 differ by the angle θ. (b) When the object moves along the circle from O to P, the radius r traces out the same angle θ. Here, the sector COP has been rotated clockwise by 90° relative to its orientation in Figure 5.2.

$$\frac{\Delta v}{v} = \frac{v\,\Delta t}{r}$$

This equation can be solved for $\Delta v/\Delta t$, to show that the magnitude a_c of the centripetal acceleration is given by $a_c = v^2/r$.

Centripetal acceleration is a vector quantity and, therefore, has a direction as well as a magnitude. The direction is toward the center of the circle, and Conceptual Example 2 helps us to set the stage for explaining this important fact.

Conceptual Example 2　Which Way Will the Object Go?

In Figure 5.4 an object, such as a model airplane on a guideline, is in uniform circular motion. The object is symbolized by a dot (●), and at point O it is released suddenly from its circular path. For instance, the guideline for a model plane is cut suddenly. Does the object move along the straight tangent line between points O and A or along the circular arc between points O and P?

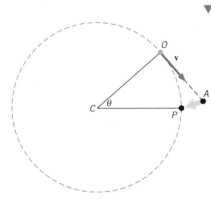

Figure 5.4 If an object (●) moving on a circular path were released from its path at point O, it would move along the straight tangent line OA in the absence of a net force.

Reasoning and Solution Newton's first law of motion guides our reasoning. An object continues in a state of rest or in a state of motion at a constant speed along a straight line unless compelled to change that state by a net force. When the object is suddenly released from its circular path, there is no longer a net force being applied to the object. In the case of a model airplane, the guideline cannot apply a force, since it is cut. Gravity certainly acts on the plane, but the wings provide a lift force that balances the weight of the plane. In the absence of a net force, then, the plane or any object would continue to move at a constant speed along a straight line in the direction it had at the time of release. This speed and direction are given in Figure 5.4 by the velocity vector **v**. As a result, *the object would move along the straight line between points O and A, not on the circular arc between points O and P.*

Related Homework: *Problems 4, 45*

As Example 2 discusses, the object in Figure 5.4 would travel on a tangent line if it were released from its circular path suddenly at point *O*. It would move in a straight line to point *A* in the time it would have taken to travel on the circle to point *P*. It is as if the object drops through the distance *AP* in the process of remaining on the circle, and *AP* is directed toward the center of the circle in the limit that the angle θ is small. Thus, the object accelerates toward the center of the circle at every moment. The acceleration is called *centripetal acceleration,* because the word "centripetal" means "center-seeking."

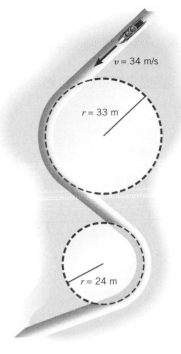

Figure 5.5 This bobsled travels at the same speed around two curves with different radii. For the turn with the larger radius, the sled has a smaller centripetal acceleration.

■ **CENTRIPETAL ACCELERATION**

Magnitude: The centripetal acceleration of an object moving with a speed v on a circular path of radius r has a magnitude a_c given by

$$a_c = \frac{v^2}{r} \tag{5.2}$$

Direction: The centripetal acceleration vector always points toward the center of the circle and continually changes direction as the object moves.

The following example illustrates the effect of the radius r on the centripetal acceleration.

Example 3 The Effect of Radius on Centripetal Acceleration

The bobsled track at the 1994 Olympics in Lillehammer, Norway, contained turns with radii of 33 m and 24 m, as Figure 5.5 illustrates. Find the centripetal acceleration at each turn for a speed of 34 m/s, a speed that was achieved in the two-man event. Express the answers as multiples of $g = 9.8$ m/s^2.

Reasoning In each case, the magnitude of the centripetal acceleration can be obtained from $a_c = v^2/r$. Since the radius r is in the denominator on the right side of this expression, we expect the acceleration to be smaller when r is larger.

Solution From $a_c = v^2/r$ it follows that

Radius = 33 m $a_c = \dfrac{(34 \text{ m/s})^2}{33 \text{ m}} = 35 \text{ m/s}^2 = \boxed{3.6 \, g}$

Radius = 24 m $a_c = \dfrac{(34 \text{ m/s})^2}{24 \text{ m}} = 48 \text{ m/s}^2 = \boxed{4.9 \, g}$

The physics of a bobsled track.

The centripetal acceleration is indeed smaller when the radius is larger. In fact, with r in the denominator on the right of $a_c = v^2/r$, the acceleration approaches zero when the radius becomes very large. Uniform circular motion along the arc of an infinitely large circle entails no acceleration, because it is just like motion at a constant speed along a straight line.

In Section 4.11 we learned that an object is in equilibrium when it has zero acceleration. Conceptual Example 4 discusses whether an object undergoing uniform circular motion can ever be at equilibrium.

Conceptual Example 4 Uniform Circular Motion and Equilibrium

A car moves at a constant speed, and there are three parts to the motion. It moves along a straight line toward a circular turn, goes around the turn, and then moves away along a straight line. In each of these parts, is the car in equilibrium?

Reasoning and Solution An object in equilibrium has no acceleration, according to the definition given in Section 4.11. As the car approaches the turn, both the speed and direction of the motion are constant. Thus, the velocity vector does not change, and there is no acceleration. The same is true as the car moves away from the turn. For these parts of the motion, then, the car is in equilibrium. As the car goes around the turn, however, the direction of travel changes, so the car has a centripetal acceleration that is characteristic of uniform circular mo-

Need more practice?

Interactive LearningWare 5.1
Car A negotiates a curve at a speed of 32 m/s and experiences a centripetal acceleration of 6.4 m/s^2. Car B negotiates the same curve at a speed of 16 m/s. What centripetal acceleration does it experience?

Related Homework: *Problem 56*

Go to
www.wiley.com/college/cutnell
for an interactive solution.

tion. Because of this acceleration, the car is not in equilibrium during the turn. In general, ***an object that is in uniform circular motion can never be in equilibrium.***

Related Homework: *Problem 43*

We have seen that going around tight turns (smaller *r*) and gentle turns (larger *r*) at the same speed entails different centripetal accelerations. And most drivers know that such turns "feel" different. This feeling is associated with the force that is present in uniform circular motion, and we now turn to this topic.

> **✔ Check Your Understanding 1**
>
> The car in the drawing is moving clockwise around a circular section of road at a constant speed. What are the directions of its velocity and acceleration at (a) position 1 and (b) position 2? Specify your responses as north, east, south, or west. *(The answers are given at the end of the book.)*
>
>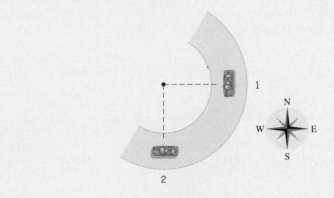
>
> *Background:* Centripetal acceleration lies at the heart of this question. In particular, the direction of the object's velocity and the direction of its centripetal acceleration are different.
>
> *For similar questions (including calculational counterparts), consult Self-Assessment Test 5.1. This test is described at the end of Section 5.3.*

Figure 5.6 CONCEPTS AT A GLANCE
Uniform circular motion entails a centripetal acceleration $a_c = v^2/r$. The net force required to produce this acceleration is called the centripetal force and is given by Newton's second law as $F_c = mv^2/r$. In the death spiral in pairs figure skating, the woman is spun around in a circle by her partner, and the centripetal force is largely provided by a component of the force applied to her by his hand. (© AFP/ Corbis Images)

5.3 *Centripetal Force*

▶ CONCEPTS AT A GLANCE Newton's second law indicates that whenever an object accelerates, there must be a net force to create the acceleration. Thus, in uniform circular motion there must be a net force to produce the centripetal acceleration. As the Concept-at-a-Glance chart in Figure 5.6 indicates, the second law gives this net force as the prod-

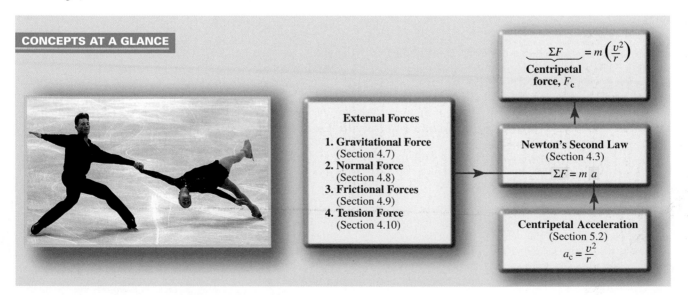

CONCEPTS AT A GLANCE

$$\underbrace{\Sigma F}_{\substack{\text{Centripetal} \\ \text{force, } F_c}} = m\left(\frac{v^2}{r}\right)$$

External Forces

1. **Gravitational Force**
   (Section 4.7)
2. **Normal Force**
   (Section 4.8)
3. **Frictional Forces**
   (Section 4.9)
4. **Tension Force**
   (Section 4.10)

Newton's Second Law
(Section 4.3)

$\Sigma F = m\,a$

Centripetal Acceleration
(Section 5.2)

$a_c = \dfrac{v^2}{r}$

uct of the object's mass m and its acceleration v^2/r. This chart is an expanded version of the chart shown previously in Figure 4.9. The net force causing the centripetal acceleration is called the ***centripetal force*** F_c and points in the same direction as the acceleration—that is, toward the center of the circle. ◄

■ **CENTRIPETAL FORCE**

Magnitude: The centripetal force is the name given to the net force required to keep an object of mass m, moving at a speed v, on a circular path of radius r, and it has a magnitude of

$$F_c = \frac{mv^2}{r} \tag{5.3}$$

Direction: The centripetal force always points toward the center of the circle and continually changes direction as the object moves.

The phrase "centripetal force" does not denote a new and separate force created by nature. The phrase merely labels the net force pointing toward the center of the circular path, and this net force is the vector sum of all the force components that point along the radial direction.

In some cases, it is easy to identify the source of the centripetal force, as when a model airplane on a guideline flies in a horizontal circle. The only force pulling the plane inward is the tension in the line, so this force alone (or a component of it) is the centripetal force. Example 5 illustrates the fact that higher speeds require greater tensions.

Example 5 The Effect of Speed on Centripetal Force

The model airplane in Figure 5.7 has a mass of 0.90 kg and moves at a constant speed on a circle that is parallel to the ground. The path of the airplane and its guideline lie in the same horizontal plane, because the weight of the plane is balanced by the lift generated by its wings. Find the tension T in the guideline (length = 17 m) for speeds of 19 and 38 m/s.

Reasoning Since the plane flies on a circular path, it experiences a centripetal acceleration that is directed toward the center of the circle. According to Newton's second law of motion, this acceleration is produced by a net force that acts on the plane and this net force is called the centripetal force. The centripetal force is also directed toward the center of the circle. Since the tension T in the guideline is the only force pulling the plane inward, it must be the centripetal force.

Solution Equation 5.3 gives the tension directly: $F_c = T = mv^2/r$.

Speed = 19 m/s $T = \dfrac{(0.90 \text{ kg})(19 \text{ m/s})^2}{17 \text{ m}} = \boxed{19 \text{ N}}$

Speed = 38 m/s $T = \dfrac{(0.90 \text{ kg})(38 \text{ m/s})^2}{17 \text{ m}} = \boxed{76 \text{ N}}$

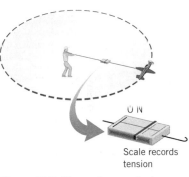

Figure 5.7 The scale records the tension in the guideline. See Example 5.

Need more practice?

🖱**Interactive LearningWare 5.2**
A 0.60-kg and a 1.2-kg model airplane fly in horizontal circles on guidelines that are parallel to the ground. The speeds of the planes are the same, and the same type of cord is used for each guideline. The smallest circle on which the 0.60-kg plane can fly without breaking its guideline has a radius of 3.5 m. What is the radius of the smallest circle on which the 1.2-kg plane can fly?

Related Homework: *Problem 58*

Go to
www.wiley.com/college/cutnell for an interactive solution.

Conceptual Example 6 deals with another case where it is easy to identify the source of the centripetal force.

🖱**Concept Simulation 5.1**

This simulation illustrates the model airplane discussed in Example 5. The speed and mass of the plane are under your control, so you can explore how these quantities affect the centripetal force. For each setting, the simulation calculates the magnitude of the centripetal force and displays the velocity and centripetal acceleration vectors associated with the moving plane.

Related Homework: *Problem 49*

Go to
www.wiley.com/college/cutnell

Conceptual Example 6   A Trapeze Act

In a circus, a man hangs upside down from a trapeze, legs bent over the bar and arms downward, holding his partner (see Figure 5.8). Is it harder for the man to hold his partner when the partner hangs straight down and is stationary or when the partner is swinging through the straight-down position?

Reasoning and Solution When the man and his partner are stationary, the man's arms must support his partner's weight. When the two are swinging, however, the man's arms must do an additional job. Then, the partner is moving on a circular arc and has a centripetal acceleration. The man's arms must exert an additional pull so that there will be sufficient centripetal force to produce this acceleration. Because of the additional pull, *it is harder for the man to hold his partner while swinging than while stationary.*

Related Homework: *Problem 46*

Figure 5.8 Does the man hanging upside down from the trapeze have a harder job holding his partner when the team is swinging through the straight-down position than when they are hanging straight down and stationary? (Courtesy Ringling Brothers and Barnum & Bailey Circus Combined Shows, Inc.)

When a car moves at a steady speed around an unbanked curve, the centripetal force keeping the car on the curve comes from the static friction between the road and the tires, as Figure 5.9 indicates. It is static, rather than kinetic friction, because the tires are not slipping with respect to the radial direction. If the static frictional force is insufficient, given the speed and the radius of the turn, the car will skid off the road. Example 7 shows how an icy road can limit safe driving.

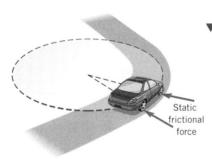

Static frictional force

Figure 5.9 When the car moves without skidding around a curve, static friction between the road and the tires provides the centripetal force to keep the car on the road.

Problem solving insight
When using an equation to obtain a numerical answer, algebraically solve for the unknown variable in terms of the known variables. Then substitute in the numbers for the known variables, as this example shows.

Example 7   Centripetal Force and Safe Driving

Compare the maximum speeds at which a car can safely negotiate an unbanked turn (radius = 50.0 m) in dry weather (coefficient of static friction = 0.900) and icy weather (coefficient of static friction = 0.100).

Reasoning At the maximum speed, the maximum centripetal force acts on the tires, and static friction must provide it. The magnitude of the maximum force of static friction is specified by Equation 4.7 as $f_s^{MAX} = \mu_s F_N$, where μ_s is the coefficient of static friction and F_N is the magnitude of the normal force. Our strategy, then, is to find the normal force, substitute it into the expression for the maximum force of static friction, and then equate the result to mv^2/r. Experience indicates that the maximum speed should be greater for the dry road than for the icy road.

Solution Since the car does not accelerate in the vertical direction, the weight mg of the car is balanced by the normal force, so $F_N = mg$. From Equations 4.7 and 5.3 it follows that

$$F_c = \mu_s F_N = \mu_s mg = \frac{mv^2}{r}$$

Consequently, $\mu_s g = v^2/r$, and

$$v = \sqrt{\mu_s g r}$$

The mass m of the car has been eliminated algebraically from this result. All cars, heavy or light, have the same maximum speed:

Dry road (μ_s = 0.900) $v = \sqrt{(0.900)(9.80 \text{ m/s}^2)(50.0 \text{ m})} = \boxed{21.0 \text{ m/s}}$

Icy road (μ_s = 0.100) $v = \sqrt{(0.100)(9.80 \text{ m/s}^2)(50.0 \text{ m})} = \boxed{7.00 \text{ m/s}}$

As expected, the dry road allows the greater maximum speed.

The passenger in Figure 5.9 must also experience a centripetal force to remain on the circular path. However, if the upholstery is very slippery, there may not be enough static friction to keep him in place as the car makes a tight turn at high speed. Then, when viewed from inside the car, he appears to be thrown toward the outside of the curve. What really happens is that the passenger slides off on a tangent to the circle, until he encounters a source of centripetal force to keep him in place while the car turns. This occurs when the passenger bumps into the side of the car, which pushes on him with the necessary force.

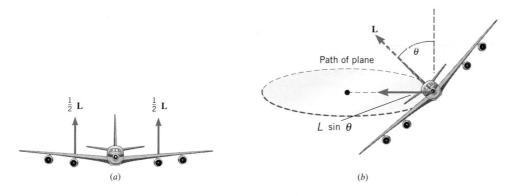

(a)

(b)

Figure 5.10 (*a*) The air exerts an upward lifting force $\frac{1}{2}\mathbf{L}$ on each wing. (*b*) When a plane executes a circular turn, the plane banks at an angle θ. The lift component $L \sin \theta$ is directed toward the center of the circle and provides the centripetal force.

Sometimes the source of the centripetal force is not obvious. A pilot making a turn, for instance, banks or tilts the plane at an angle to create the centripetal force. As a plane flies, the air pushes upward on the wing surfaces with a net lifting force **L** that is perpendicular to the wing surfaces, as Figure 5.10*a* shows. Part *b* of the drawing illustrates that when the plane is banked at an angle θ, a component $L \sin \theta$ of the lifting force is directed toward the center of the turn. It is this component that provides the centripetal force. Greater speeds and/or tighter turns require greater centripetal forces. In such situations, the pilot must bank the plane at a larger angle, so that a larger component of the lift points toward the center of the turn. The technique of banking into a turn also has an application in the construction of high-speed roadways, where the road itself is banked to achieve a similar effect, as the next section discusses.

The physics of
flying an airplane in a banked turn.

✔ **Check Your Understanding 2**

A car is traveling in uniform circular motion on a section of road whose radius is *r* (see the drawing). The road is slippery, and the car is just on the verge of sliding. (a) If the car's speed were doubled, what would have to be the smallest radius in order that the car does not slide? Express your answer in terms of *r* (b) What would be your answer to part (a) if the car were replaced by one that weighed twice as much? *(The answers are given at the end of the book.)*

r

Background: The concepts of centripetal force, the maximum force of static friction, and the normal force are featured in this problem. The relationship between weight and mass also plays a role.

For similar questions (including calculational counterparts), consult Self-Assessment Test 5.1. The test is described next.

🖱 **Concept Simulation 5.2**

A car is entering a circular section of the road, and you can simulate the road conditions—from icy to dry by changing the coefficient of static friction μ_s. The speed of the car is also under your control. The centripetal force needed to negotiate the turn is provided by the static frictional force between the tires and the road. If the static frictional force is sufficiently large, the car will stay on the road; if not, the car will skid off and crash.

Related Homework: *Conceptual Question 16, Problem 15*

Go to
www.wiley.com/college/cutnell

🖱 **Self-Assessment Test 5.1**

Several important ideas were presented in Sections 5.1–5.3:

• Uniform Circular Motion   • Centripetal Acceleration   • Centripetal Force

Use Self-Assessment Test 5.1 to check your understanding of these concepts.

Go to **www.wiley.com/college/cutnell**

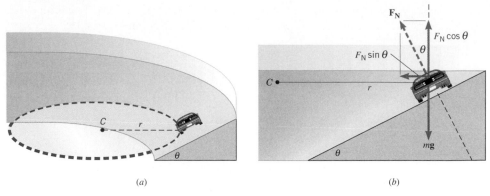

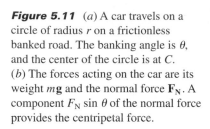

Figure 5.11 (*a*) A car travels on a circle of radius *r* on a frictionless banked road. The banking angle is θ, and the center of the circle is at *C*. (*b*) The forces acting on the car are its weight $m\mathbf{g}$ and the normal force $\mathbf{F_N}$. A component $F_N \sin \theta$ of the normal force provides the centripetal force.

5.4 Banked Curves

When a car travels without skidding around an unbanked curve, the static frictional force between the tires and the road provides the centripetal force. The reliance on friction can be eliminated completely for a given speed, however, if the curve is banked at an angle relative to the horizontal, much in the same way that a plane is banked while making a turn.

Figure 5.11*a* shows a car going around a friction-free banked curve. The radius of the curve is *r*, where *r* is measured parallel to the horizontal and not to the slanted surface. Part *b* shows the normal force $\mathbf{F_N}$ that the road applies to the car, the normal force being perpendicular to the road. Because the roadbed makes an angle θ with respect to the horizontal, the normal force has a component $F_N \sin \theta$ that points toward the center *C* of the circle and provides the centripetal force:

$$F_c = F_N \sin \theta = \frac{mv^2}{r}$$

The vertical component of the normal force is $F_N \cos \theta$ and, since the car does not accelerate in the vertical direction, this component must balance the weight mg of the car. Therefore, $F_N \cos \theta = mg$. Dividing this equation into the previous one shows that

$$\frac{F_N \sin \theta}{F_N \cos \theta} = \frac{mv^2/r}{mg}$$

$$\tan \theta = \frac{v^2}{rg} \tag{5.4}$$

Equation 5.4 indicates that, for a given speed *v*, the centripetal force needed for a turn of radius *r* can be obtained from the normal force by banking the turn at an angle θ, independent of the mass of the vehicle. Greater speeds and smaller radii require more steeply banked curves—that is, larger values of θ. At a speed that is too small for a given θ, a car would slide down a frictionless banked curve; at a speed that is too large, a car would slide off the top. The next example deals with a famous banked curve.

Example 8   The Daytona 500

The physics of the Daytona International Speedway.

The Daytona 500 is the major event of the NASCAR (National Association for Stock Car Auto Racing) season. It is held at the Daytona International Speedway in Daytona, Florida. The turns in this oval track have a maximum radius (at the top) of $r = 316$ m and are banked steeply, with $\theta = 31°$ (see Figure 5.11). Suppose these maximum-radius turns were frictionless. At what speed would the cars have to travel around them?

Reasoning In the absence of friction, the horizontal component of the normal force that the track exerts on the car must provide the centripetal force. Therefore, the speed of the car is given by Equation 5.4.

Solution From Equation 5.4, it follows that

$$v = \sqrt{rg \tan \theta} = \sqrt{(316 \text{ m})(9.80 \text{ m/s}^2) \tan 31°} = \boxed{43 \text{ m/s } (96 \text{ mph})}$$

Drivers actually negotiate the turns at speeds up to 195 mph, however, which requires a greater centripetal force than that implied by Equation 5.4 for frictionless turns. Static friction provides the additional force.

5.5 Satellites in Circular Orbits

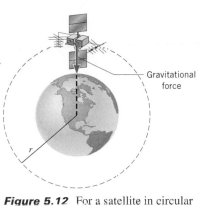

Today there are many satellites in orbit about the earth. The ones in circular orbits are examples of uniform circular motion. Like a model airplane on a guideline, each satellite is kept on its circular path by a centripetal force. The gravitational pull of the earth provides the centripetal force and acts like an invisible guideline for the satellite.

There is only one speed that a satellite can have if the satellite is to remain in an orbit with a fixed radius. To see how this fundamental characteristic arises, consider the gravitational force acting on the satellite of mass m in Figure 5.12. Since the gravitational force is the only force acting on the satellite in the radial direction, it alone provides the centripetal force. Therefore, using Newton's law of gravitation (Equation 4.3), we have

$$F_c = G\frac{mM_E}{r^2} = \frac{mv^2}{r}$$

Figure 5.12 For a satellite in circular orbit around the earth, the gravitational force provides the centripetal force.

where G is the universal gravitational constant, M_E is the mass of the earth, and r is the distance from the center of the earth to the satellite. Solving for the speed v of the satellite gives

$$v = \sqrt{\frac{GM_E}{r}} \tag{5.5}$$

If the satellite is to remain in an orbit of radius r, the speed must have precisely this value. Note that the radius r of the orbit is in the denominator in Equation 5.5. This means that the closer the satellite is to the earth, the smaller is the value for r and the greater the orbital speed must be.

The mass m of the satellite does not appear in Equation 5.5, having been eliminated algebraically. *Consequently, for a given orbit, a satellite with a large mass has exactly the same orbital speed as a satellite with a small mass.* However, more effort is certainly required to lift the larger-mass satellite into orbit. The orbital speed of one famous artificial satellite is determined in the following example.

Figure 5.13 The Hubble Space Telescope orbits the earth, after being released from the Space Shuttle *Discovery*. (Courtesy NASA)

Example 9 Orbital Speed of the Hubble Space Telescope

Determine the speed of the Hubble Space Telescope (see Figure 5.13) orbiting at a height of 598 km above the earth's surface.

Reasoning Before Equation 5.5 can be applied, the orbital radius r must be determined *relative to the center of the earth*. Since the radius of the earth is approximately 6.38×10^6 m, and the height of the telescope above the earth's surface is 0.598×10^6 m, the orbital radius is $r = 6.98 \times 10^6$ m.

The physics of the Hubble Space Telescope.

Solution The orbital speed is

$$v = \sqrt{\frac{GM_E}{r}} = \sqrt{\frac{(6.67 \times 10^{-11}\ \text{N}\cdot\text{m}^2/\text{kg}^2)(5.98 \times 10^{24}\ \text{kg})}{6.98 \times 10^6\ \text{m}}}$$

$$v = \boxed{7.56 \times 10^3\ \text{m/s (16 900 mi/h)}}$$

Problem solving insight
The orbital radius r that appears in the relation $v = \sqrt{GM_E/r}$ is the distance from the satellite to the center of the earth (not to the surface of the earth).

Many applications of satellite technology affect our lives. One is a network of 24 satellites called the Global Positioning System (GPS), which can be used to determine the position of an object to within 15 m or less. Figure 5.14 illustrates how the system works. Each GPS satellite carries a highly accurate atomic clock, whose time is transmitted to the ground continually by means of radio waves. In the drawing, a car carries a computerized GPS receiver that can detect the waves and is synchronized to the satellite clock. The receiver can, therefore, determine the distance between the car and a satellite from a knowledge of the travel time of the waves and the speed at which they move. This speed,

The physics of the Global Positioning System.

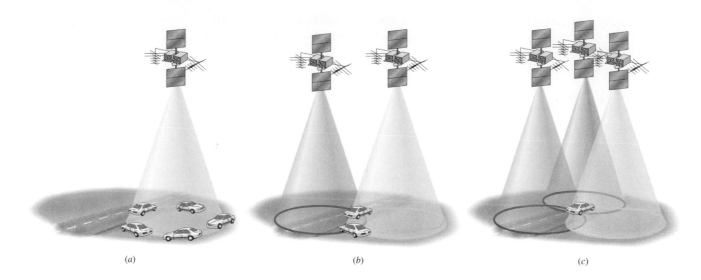

(a)　　　　　　　　　　　　(b)　　　　　　　　　　　　(c)

Figure 5.14 The Navstar Global Positioning System (GPS) of satellites can be used with a GPS receiver to locate an object, such as a car, on the earth. (*a*) One satellite identifies the car as being somewhere on a circle. (*b*) A second places it on another circle, which identifies two possibilities for the exact spot. (*c*) A third provides the means for deciding where the car is.

as we will see in Chapter 24, is the speed of light and is known with great precision. A measurement using a single satellite locates the car somewhere on a circle, as Figure 5.14*a* shows, while a measurement using a second satellite locates the car on another circle. The intersection of the circles reveals two possible positions for the car, as in Figure 5.14*b*. With the aid of a third satellite, a third circle can be established, which intersects the other two and identifies the car's exact position, as in Figure 5.14*c*. The use of ground-based radio beacons to provide additional reference points leads to a system called Differential GPS, which can locate objects even more accurately than the satellite-based system alone. Navigational systems for automobiles and portable systems that tell hikers and people with visual impairments where they are located are two of the many uses for the GPS technique. GPS applications are so numerous that they have developed into a multibillion dollar industry.

Equation 5.5 applies to man-made earth satellites or to natural satellites like the moon. It also applies to circular orbits about any astronomical object, provided M_E is replaced by the mass of the object on which the orbit is centered. Example 10, for instance, shows how scientists have applied this equation to conclude that a supermassive black hole is probably located at the center of the galaxy known as M87. This galaxy is located at a distance of about 50 million light-years away from the earth. (One light-year is the distance that light travels in a year, or 9.5×10^{15} m.)

The physics of locating a black hole. (See below.)

Figure 5.15 This image of the ionized gas (yellow) at the heart of galaxy M87 was obtained by the Hubble Space Telescope. The circle identifies the center of the galaxy, at which a black hole is thought to exist. (Courtesy NASA and Space Telescope Science Institute)

Example 10　**A Supermassive Black Hole**

The Hubble telescope has detected the light being emitted from different regions of galaxy M87, which is shown in Figure 5.15. The black circle identifies the center of the galaxy. From the characteristics of this light, astronomers have determined an orbiting speed of 7.5×10^5 m/s for matter located at a distance of 5.7×10^{17} m from the center. Find the mass M of the object located at the galactic center.

Reasoning and Solution Replacing M_E in Equation 5.5 with M gives $v = \sqrt{GM/r}$, which can be solved to show that

$$M = \frac{v^2 r}{G} = \frac{(7.5 \times 10^5 \text{ m/s})^2 (5.7 \times 10^{17} \text{ m})}{6.67 \times 10^{-11} \text{ N} \cdot \text{m}^2/\text{kg}^2} = \boxed{4.8 \times 10^{39} \text{ kg}}$$

The ratio of this incredibly large mass to the mass of our sun is $(4.8 \times 10^{39}$ kg$)/(2.0 \times 10^{30}$ kg$) = 2.4 \times 10^9$. Thus, matter equivalent to 2.4 billion suns is located at the center of galaxy M87. Considering that the volume of space in which this matter is located contains relatively few visible stars, researchers have concluded that the data provide strong evidence for the existence of a supermassive black hole. The term "black hole" is used because the tremendous mass prevents even light from escaping. The light that forms the image in Figure 5.15 comes not from the black hole itself, but from matter that surrounds it.

The period T of a satellite is the time required for one orbital revolution. As in any uniform circular motion, the period is related to the speed of the motion by $v = 2\pi r/T$. Substituting v from Equation 5.5 shows that

$$\sqrt{\frac{GM_E}{r}} = \frac{2\pi r}{T}$$

Solving this expression for the period T gives

$$T = \frac{2\pi r^{3/2}}{\sqrt{GM_E}} \tag{5.6}$$

Although derived for earth orbits, Equation 5.6 can also be used for calculating the periods of those planets in nearly circular orbits about the sun, if M_E is replaced by the mass M_S of the sun and r is interpreted as the distance between the center of the planet and the center of the sun. The fact that the period is proportional to the three-halves power of the orbital radius is known as Kepler's third law, and it is one of the laws discovered by Johannes Kepler (1571–1630) during his studies of planetary motion. Kepler's third law also holds for elliptical orbits, which will be discussed in Chapter 9.

An important application of Equation 5.6 occurs in the field of communications, where "synchronous satellites" are put into a circular orbit that is in the plane of the equator, as Figure 5.16 shows. The orbital period is chosen to be one day, which is also the time it takes for the earth to turn once about its axis. Therefore, these satellites move around their orbits in a way that is synchronized with the rotation of the earth. For earth-based observers, synchronous satellites have the useful characteristic of appearing in fixed positions in the sky and can serve as "stationary" relay stations for communication signals sent up from the earth's surface. This is exactly what is done in the digital satellite systems that are a popular alternative to cable TV. As the blowup in Figure 5.16 indicates, a small "dish" antenna on your house picks up the digital TV signals relayed to earth by the satellite. After being decoded, these signals are delivered to your TV set. All synchronous satellites are in orbit at the same height above the earth's surface, as Example 11 shows.

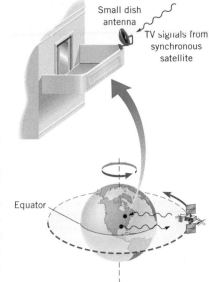

Figure 5.16 A synchronous satellite orbits the earth once per day on a circular path that lies in the plane of the equator. Digital satellite system television uses such satellites as relay stations for TV signals that are sent up from the earth's surface and then rebroadcast down toward your own small dish antenna.

The physics of
digital satellite system TV.

Example 11 The Orbital Radius for Synchronous Satellites

What is the height H above the earth's surface at which all synchronous satellites (regardless of mass) must be placed in orbit?

Reasoning The period T of a synchronous satellite is one day, so we can use Equation 5.6 to find the distance r from the center of the earth. To find the height H of the satellite above the earth's surface we will have to take into account the fact that the earth itself has a radius of 6.38×10^6 m.

Solution A period of one day* corresponds to $T = 8.64 \times 10^4$ s. In using this value it is convenient to rearrange the equation $T = 2\pi r^{3/2}/\sqrt{GM_E}$ as follows:

$$r^{3/2} = \frac{T\sqrt{GM_E}}{2\pi} = \frac{(8.64 \times 10^4 \text{ s}) \sqrt{(6.67 \times 10^{-11} \text{ N} \cdot \text{m}^2/\text{kg}^2)(5.98 \times 10^{24} \text{ kg})}}{2\pi}$$

By squaring and then taking the cube root, we find that $r = 4.23 \times 10^7$ m. Since the radius of the earth is approximately 6.38×10^6 m, the height of the satellite above the earth's surface is

$$H = 4.23 \times 10^7 \text{ m} - 0.64 \times 10^7 \text{ m} = \boxed{3.59 \times 10^7 \text{ m (22 300 mi)}}$$

---

* Successive appearances of the sun define the solar day of 24 h or 8.64×10^4 s. The sun moves against the background of the stars, however, and the time required for the earth to turn once on its axis relative to the fixed stars is 23 h 56 min, which is called the sidereal day. The sidereal day should be used in Example 11, but the neglect of this effect introduces an error of less than 0.4% in the answer.

✔ **Check Your Understanding 3**

Two satellites are placed in orbit, one about Mars and the other about Jupiter, such that the orbital speeds are the same. Mars has the smaller mass. Is the radius of the satellite in orbit about Mars less than, greater than, or equal to the radius of the satellite orbiting Jupiter? *(The answer is given at the end of the book.)*

Background: Understanding the concepts of centripetal force and the gravitational force is the key to answering this question.

For similar questions (including calculational counterparts), consult Self-Assessment Test 5.2. This test is described at the end of Section 5.7.

5.6 *Apparent Weightlessness and Artificial Gravity*

The physics of apparent weightlessness.

The idea of life on board an orbiting satellite conjures up visions of astronauts floating around in a state of "weightlessness," as in Figure 5.17. Actually, this state should be called "apparent weightlessness," because it is similar to the condition of zero apparent weight that occurs in an elevator during free-fall. Conceptual Example 12 explores this similarity.

Figure 5.17 As she orbits the earth, astronaut Janet Kavandi floats around in a state of apparent weightlessness. (Courtesy NASA)

Conceptual Example 12 Apparent Weightlessness and Free-Fall

Figure 5.18 shows a person on a scale in a freely falling elevator and in a satellite in a circular orbit. In each case, what apparent weight is recorded by the scale?

Reasoning and Solution As Section 4.8 discusses, apparent weight is the force that an object exerts on the platform of a scale. In the freely falling elevator in Figure 5.18a, the apparent weight is zero. The reason is that both the scale and the person on it fall together and, therefore, cannot push against one another. In the orbiting satellite in Figure 5.18b, both the person and the scale are in uniform circular motion. Objects in uniform circular motion continually accelerate or "fall" toward the center of the circle, in order to remain on the circular path. Consequently, the scale in Figure 5.18b and the person on it both "fall" with the same acceleration toward the center of the orbit and cannot push against one another. Thus, *the apparent weight in the satellite is zero, just as it is in the freely falling elevator.* The only difference between the satellite and the elevator is that the satellite moves on a circle, so that its "falling" does not bring it closer to the earth. In contrast to the apparent weight, the true

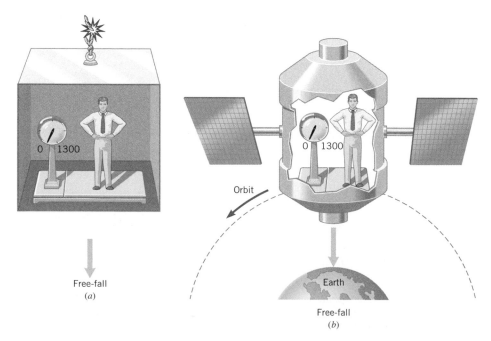

Free-fall
(a)

Orbit

Earth

Free-fall
(b)

Figure 5.18 (a) During free-fall, the elevator accelerates downward with the acceleration due to gravity, and the apparent weight of the person is zero. (b) The orbiting space station is also in free-fall toward the center of the earth.

weight is the gravitational force ($F = GmM_E/r^2$) that the earth exerts on an object and is not zero in a freely falling elevator or aboard an orbiting satellite.

The physiological effects of prolonged apparent weightlessness are only partially known. To minimize such effects, it is likely that artificial gravity will be provided in large space stations of the future. To help explain artificial gravity, Figure 5.19 shows a space station rotating about an axis. Because of the rotational motion, any object located at a point P on the interior surface of the station experiences a centripetal force directed toward the axis. The surface of the station provides this force by pushing on the feet of an astronaut, for instance. The centripetal force can be adjusted to match the astronaut's earth weight by properly selecting the rotational speed of the space station, as Examples 13 and 14 illustrate.

The physics of artificial gravity.

Example 13 Artificial Gravity

At what speed must the surface of the space station ($r = 1700$ m) move in Figure 5.19, so that the astronaut at point P experiences a push on his feet that equals his earth weight?

Reasoning The floor of the rotating space station exerts a normal force on the feet of the astronaut. This is the centripetal force ($F_c = mv^2/r$) that keeps the astronaut moving in a circular path. Since the magnitude of the normal force equals the astronaut's earth weight, we can determine the speed v of the space station.

Solution The earth weight of the astronaut (mass $= m$) is mg, and with this substitution, Equation 5.3 can be used to determine the required speed: $F_c = mg = mv^2/r$. Solving this equation for the speed, we find that

$$v = \sqrt{rg} = \sqrt{(1700 \text{ m})(9.80 \text{ m/s}^2)} = \boxed{130 \text{ m/s}}$$

Figure 5.19 The surface of the rotating space station pushes on an object with which it is in contact and thereby provides the centripetal force that keeps the object moving on a circular path.

Example 14 A Rotating Space Laboratory

A space laboratory is rotating to create artificial gravity, as Figure 5.20 indicates. Its period of rotation is chosen so the outer ring ($r_O = 2150$ m) simulates the acceleration due to gravity on earth (9.80 m/s^2). What should be the radius r_I of the inner ring, so it simulates the acceleration due to gravity on the surface of Mars (3.72 m/s^2)?

Reasoning The value given for either acceleration corresponds to the centripetal acceleration $a_c = v^2/r$ in the corresponding ring. Moreover, the speed v and radius r are related according to $v = 2\pi r/T$ (Equation 5.1), where T is the period of the motion. Although no value is given for T, we do know that the laboratory is rigid. This is an important observation, because all points on a rigid object make one revolution in the same time. Therefore, both rings have the same period, a fact that will make a solution possible.

Solution Substituting Equation 5.1 into the expression for the centripetal acceleration gives

$$a_c = \frac{v^2}{r} = \frac{\left(\dfrac{2\pi r}{T}\right)^2}{r} = \frac{4\pi^2 r}{T^2}$$

Applying this result to both rings shows that

$$\underbrace{9.80 \text{ m/s}^2 = \frac{4\pi^2 (2150 \text{ m})}{T^2}}_{\text{Outer ring}} \quad \text{and} \quad \underbrace{3.72 \text{ m/s}^2 = \frac{4\pi^2 r_I}{T^2}}_{\text{Inner ring}}$$

Dividing the inner-ring expression by the outer-ring expression and eliminating the common algebraic factors of $4\pi^2$ and T^2 gives

$$\frac{3.72 \text{ m/s}^2}{9.80 \text{ m/s}^2} = \frac{r_I}{2150 \text{ m}} \quad \text{or} \quad r_I = \boxed{816 \text{ m}}$$

Problem solving insight
All points on a rotating rigid body have the same value for the period T of the motion.

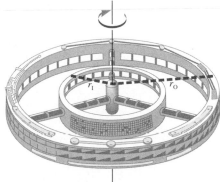

Figure 5.20 The outer ring (radius $= r_O$) of this rotating space laboratory simulates gravity on earth, while the inner ring (radius $= r_I$) simulates gravity on Mars.

✔ **Check Your Understanding 4**

The acceleration due to gravity on the moon is one-sixth that on earth. (a) Is the true weight of a person on the moon less than, greater than, or equal to the true weight of the same person on the earth? (b) Is the apparent weight of a person in orbit about the moon less than, greater than, or equal to the apparent weight of the same person in orbit about the earth? *(The answers are given at the end of the book.)*

Background: The solution to this question involves an understanding of the difference between the concepts of true weight and apparent weight. In particular, pay attention to the apparent weight of a person in orbit.

For similar questions (including calculational counterparts), consult Self-Assessment Test 5.2. This test is described at the end of Section 5.7.

*5.7 Vertical Circular Motion

Motorcycle stunt drivers perform a feat in which they drive their cycles around a vertical circular track, as in Figure 5.21*a*. Usually, the speed varies in this stunt. When the speed of travel on a circular path changes from moment to moment, the motion is said to be "nonuniform." Nonetheless, we can use the concepts that apply to uniform circular motion to gain considerable insight into the motion that occurs on a vertical circle.

There are four points on a vertical circle where the centripetal force can be identified easily, as Figure 5.21*b* indicates. As you look at Figure 5.21*b*, keep in mind that the centripetal force is not a new and separate force of nature. Instead, at each point the centripetal force is the net sum of all the force components oriented along the radial direction and points toward the center of the circle. The drawing shows only the weight of the cycle plus rider (magnitude $= mg$) and the normal force pushing on the cycle (magnitude $= F_N$). The propulsion and braking forces are omitted for simplicity, because they do not act in the radial direction. The magnitude of the centripetal force at each of the four points is given as follows in terms of mg and F_N:

Problem solving insight
Centripetal force $\mathbf{F}_c$ is the name given to the net force that points toward the center of a circular path. As shown here, there may be several forces that contribute to this net force.

(1) $\underbrace{F_{N1} - mg}_{= F_{c1}} = \dfrac{mv_1^2}{r}$ (3) $\underbrace{F_{N3} + mg}_{= F_{c3}} = \dfrac{mv_3^2}{r}$

(2) $\underbrace{F_{N2}}_{= F_{c2}} = \dfrac{mv_2^2}{r}$ (4) $\underbrace{F_{N4}}_{= F_{c4}} = \dfrac{mv_4^2}{r}$

As the cycle goes around, the magnitude of the normal force changes. It changes because the speed changes and because the weight does not have the same effect at every point. At the bottom, the normal force and the weight oppose one another, giving a centripetal force of magnitude $F_{N1} - mg$. At the top, the normal force and the weight reinforce each other to provide a centripetal force whose magnitude is $F_{N3} + mg$. At points 2 and 4 on either side, only F_{N2} and F_{N4} provide the centripetal force. The weight is tangent to the circle at points 2 and 4 and has no component pointing toward the center. If the speed at each of the four places is known, along with the mass and the radius, the normal forces can be determined.

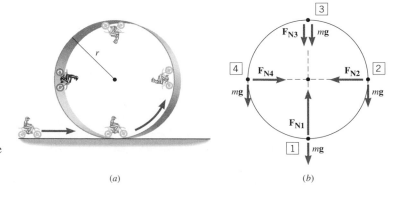

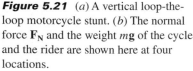

Figure 5.21 (*a*) A vertical loop-the-loop motorcycle stunt. (*b*) The normal force **F**$_N$ and the weight mg of the cycle and the rider are shown here at four locations.

(*a*)   (*b*)

Riders who perform the loop-the-loop trick know that they must have at least a minimum speed at the top of the circle to remain on the track. This speed can be determined by considering the centripetal force at point 3. The speed v_3 in the equation $F_{N3} + mg - mv_3{}^2/r$ is a minimum when F_{N3} is zero. Then, the speed is given by $v_3 = \sqrt{rg}$. At this speed, the track does not exert a normal force to keep the cycle on the circle, because the weight mg provides all the centripetal force. Under these conditions, the rider experiences an apparent weightlessness like that discussed in Section 5.6, because for an instant the rider and the cycle are falling freely toward the center of the circle.

The physics of
the loop-the-loop motorcycle stunt.

Self-Assessment Test 5.2

Sections 5.4–5.7 discuss a number of applications of centripetal acceleration and centripetal force:

• Banked Curves • Satellites in Circular Orbits • Apparent Weightlessness and Artificial Gravity • Vertical Circular Motion

Self-Assessment Test 5.2 can be used to check your understanding of the physics concepts on which these applications are based.

Go to **www.wiley.com/college/cutnell**

5.8 *Concepts & Calculations*

In uniform circular motion the concepts of acceleration and force play central roles. Example 15 deals with acceleration, which we first discussed in Chapter 2, particularly in connection with the equations of kinematics. This example emphasizes the difference between the acceleration that is used in the equations of kinematics and the acceleration that arises in uniform circular motion.

Concepts & Calculations Example 15 Acceleration

At time $t = 0$ s, automobile A is traveling at a speed of 18 m/s along a straight road and is picking up speed with an acceleration that has a magnitude of 3.5 m/s² (Figure 5.22a). At time $t = 0$ s, automobile B is traveling at a speed of 18 m/s in uniform circular motion as it negotiates a turn (Figure 5.22b). It has a centripetal acceleration whose magnitude is also 3.5 m/s². Determine the speed of each automobile when $t = 2.0$ s.

Concept Questions and Answers Which automobile has a constant acceleration?

Answer Acceleration is a vector, and for it to be constant, both its magnitude and direction must be constant. Automobile A has a constant acceleration, because its acceleration has a constant magnitude of 3.5 m/s² and its direction always points forward along the straight road. Automobile B has an acceleration with a constant magnitude of 3.5 m/s², but the acceleration does not have a constant direction. Automobile B is in uniform circular motion, so it has a centripetal acceleration, which points toward the center of the circle at every instant. Therefore, the direction of the acceleration vector continually changes, and the vector is not constant.

For which automobile do the equations of kinematics apply?

Answer The equations of kinematics apply for automobile A, because it has a constant acceleration, which must be the case when you use these equations. They do not apply to automobile B, because it does not have a constant acceleration.

Solution To determine the speed of automobile A at $t = 2.0$ s, we use Equation 2.4 from the equations of kinematics:

$$v = v_0 + at = 18 \text{ m/s} + (3.5 \text{ m/s}^2)(2.0 \text{ s}) = \boxed{25 \text{ m/s}}$$

To determine the speed of automobile B we note that this car is in uniform circular motion and, therefore, has a constant speed as it goes around the turn. At a time of $t = 2.0$ s its speed is the same as it was at $t = 0$ s—namely, $v = \boxed{18 \text{ m/s}}$.

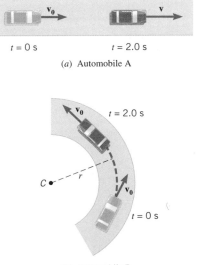

(a) Automobile A

(b) Automobile B

Figure 5.22 (a) Automobile A accelerates to the right along a straight road. (b) Automobile B travels at a constant speed of v_0 along a circular turn. It is also accelerating, but the acceleration is a centripetal acceleration.

The next example deals with force, and it stresses that the centripetal force is the net force along the radial direction. As we will see, this means that all forces along the radial direction must be taken into account when identifying the centripetal force. In addition, the example is a good review of the tension force, which we first encountered in Chapter 4.

Concepts & Calculations Example 16   Centripetal Force

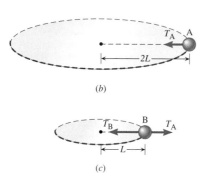

Ball A is attached to one end of a rigid massless rod, while an identical ball B is attached to the center of the rod, as Figure 5.23a illustrates. Each ball has a mass of $m = 0.50$ kg, and the length of each half of the rod is $L = 0.40$ m. This arrangement is held by the empty end and is whirled around in a horizontal circle at a constant rate, so each ball is in uniform circular motion. Ball A travels at a constant speed of $v_A = 5.0$ m/s. Find the tension in each half of the rod.

Concept Questions and Answers How many tension forces contribute to the centripetal force that acts on ball A?

Answer As Figure 5.23b illustrates, only a single tension force of magnitude T_A acts on ball A. It points to the left in the drawing and is due to the tension in the rod between the two balls. This force alone provides the centripetal force keeping ball A on its circular path of radius 2L.

How many tension forces contribute to the centripetal force that acts on ball B?

Answer As Figure 5.23c shows, two tension forces act on ball B. One has a magnitude T_B and points to the left in the drawing, which is the positive direction. It is due to the tension in the left half of the rod. The other has a magnitude T_A and points to the right. It is due to the tension in the right half of the rod. The centripetal force acting on ball B points toward the center of the circle and is the vector sum of these two forces, or $T_B - T_A$.

Is the speed of ball B the same as the speed of ball A?

Answer No, it is not. The reason is that ball A travels farther than ball B in the same time. Consider a time of one period, which is the same for either ball, since the arrangement is a rigid unit. It is the time for either ball to travel once around its circular path. In this time ball A travels a distance equal to the circumference of its path, which is $2\pi(2L)$. In contrast, the circumference for ball B is only $2\pi L$. Therefore, the speed of ball B is one-half the speed of ball A, or $v_B = 2.5$ m/s.

Solution Applying Equation 5.3 to each ball, we have

Ball A
$$\underbrace{T_A}_{\text{Centripetal force, } F_c} = \frac{mv_A^2}{2L}$$

Ball B
$$\underbrace{T_B - T_A}_{\text{Centripetal force, } F_c} = \frac{mv_B^2}{L}$$

The tension in the right half of the rod follows directly from the first of these equations:

$$T_A = \frac{mv_A^2}{2L} = \frac{(0.50 \text{ kg})(5.0 \text{ m/s})^2}{2(0.40 \text{ m})} = \boxed{16 \text{ N}}$$

Adding the equations for ball A and ball B gives the following result for the tension in the left half of the rod:

$$T_B = \frac{mv_B^2}{L} + \frac{mv_A^2}{2L} = \frac{(0.50 \text{ kg})(2.5 \text{ m/s})^2}{0.40 \text{ m}} + \frac{(0.50 \text{ kg})(5.0 \text{ m/s})^2}{2(0.40 \text{ m})} = \boxed{23 \text{ N}}$$

Figure 5.23 Two identical balls attached to a rigid massless rod are whirled around on horizontal circles, as Example 16 explains.

At the end of the problem set for this chapter, you will find homework problems that contain both conceptual and quantitative parts. These problems are grouped under the heading *Concepts & Calculations, Group Learning Problems*. They are designed for use by students working alone or in small learning groups. The conceptual part of each problem provides a convenient focus for group discussions.

Concept Summary

This summary presents an abridged version of the chapter, including the important equations and all available learning aids. For convenient reference, the learning aids (including the text's examples) are placed next to or immediately after the relevant equation or discussion. The following learning aids may be found on-line at **www.wiley.com/college/cutnell:**

Interactive LearningWare examples are solved according to a five-step interactive format that is designed to help you develop problem-solving skills.	**Concept Simulations** are animated versions of text figures or animations that illustrate important concepts. You can control parameters that affect the display, and we encourage you to experiment.
Interactive Solutions offer specific models for certain types of problems in the chapter homework. The calculations are carried out interactively.	**Self-Assessment Tests** include both qualitative and quantitative questions. Extensive feedback is provided for both incorrect and correct answers, to help you evaluate your understanding of the material.

Topic	Discussion	Learning Aids

5.1 Uniform Circular Motion

Uniform circular motion is the motion of an object traveling at a constant (uniform) speed on a circular path.

Period of the motion — The period T is the time required for the object to travel once around the circle. The speed v of the object is related to the period and the radius r of the circle by

$$v = \frac{2\pi r}{T} \qquad (5.1)$$

Learning Aids: Example 1

5.2 Centripetal Acceleration

An object in uniform circular motion experiences an acceleration, known as centripetal acceleration. The magnitude a_c of the centripetal acceleration is

$$a_c = \frac{v^2}{r} \qquad (5.2)$$

Learning Aids: Examples 2, 3, 4, 15; Interactive LearningWare 5.1; Interactive Solution 5.3

Centripetal acceleration (magnitude)

where v is the speed of the object and r is the radius of the circle. The direction of the centripetal acceleration vector always points toward the center of the circle and continually changes as the object moves.

Centripetal acceleration (direction)

5.3 Centripetal Force

To produce a centripetal acceleration, a net force pointing toward the center of the circle is required. This net force is called the centripetal force, and its magnitude F_c is

$$F_c = \frac{mv^2}{r} \qquad (5.3)$$

Learning Aids: Examples 5, 6, 7, 16; Interactive LearningWare 5.2; Concept Simulations 5.1, 5.2; Interactive Solution 5.19

Centripetal force (magnitude)

where m and v are the mass and speed of the object, and r is the radius of the circle. The direction of the centripetal force vector, like that of the centripetal acceleration vector, always points toward the center of the circle.

Centripetal force (direction)

 Use Self-Assessment Test 5.1 to evaluate your understanding of Sections 5.1–5.3.

5.4 Banked Curves

A vehicle can negotiate a circular turn without relying on static friction to provide the centripetal force, provided the turn is banked at an angle relative to the horizontal. The angle θ at which a friction-free curve must be banked is related to the speed v of the vehicle, the radius r of the curve, and the magnitude g of the acceleration due to gravity by

$$\tan\theta = \frac{v^2}{rg} \qquad (5.4)$$

Learning Aids: Example 8; Interactive Solution 5.23

Angle of a banked turn

5.5 Satellites in Circular Orbits

When a satellite orbits the earth, the gravitational force provides the centripetal force that keeps the satellite moving in a circular orbit. The speed v and period T of a satellite depend on the mass M_E of the earth and the radius r of the orbit according to

Topic	Discussion	Learning Aids
Orbital speed	$$v = \sqrt{\dfrac{GM_{\mathrm{E}}}{r}}$$ (5.5)	Examples 9, 10, 11
Orbital period	$$T = \dfrac{2\pi r^{3/2}}{\sqrt{GM_{\mathrm{E}}}}$$ (5.6)	Interactive Solution 5.31

where G is the universal gravitational constant.

5.6 Apparent Weightlessness and Artificial Gravity

The apparent weight of an object is the force that it exerts on a scale with which it is in contact. All objects, including people, on board an orbiting satellite are in free-fall, since they experience negligible air resistance and they have an acceleration that is equal to the acceleration due to gravity. When a person is in free-fall, his or her apparent weight is zero, because both the person and the scale fall freely and cannot push against one another.

Examples 12, 13, 14

5.7 Vertical Circular Motion

Vertical circular motion occurs when an object, such as a motorcycle, moves on a vertical circular path. The speed of the object often varies from moment to moment, and so do the magnitudes of the centripetal acceleration and centripetal force.

Interactive Solution 5.41

 Use Self-Assessment Test 5.2 to evaluate your understanding of Sections 5.4–5.7.

Conceptual Questions

1. The speedometer of your car shows that you are traveling at a constant speed of 35 m/s. Is it possible that your car is accelerating? If so, explain how this could happen.

2. Consider two people, one on the earth's surface at the equator and the other at the north pole. Which has the larger centripetal acceleration? Explain.

3. The equations of kinematics (Equations 3.3–3.6) describe the motion of an object that has a constant acceleration. These equations cannot be applied to uniform circular motion. Why not?

4. Is it possible for an object to have an acceleration when the velocity of the object is constant? When the speed of the object is constant? In each case, give your reasoning.

5. A car is traveling at a constant speed along the road ABCDE shown in the drawing. Sections AB and DE are straight. Rank the accelerations in each of the four sections according to magnitude, listing the smallest first. Justify your ranking.

6. Other things being equal, would it be easier to drive at high speed around an unbanked horizontal curve on the moon than to drive around the same curve on the earth? Explain.

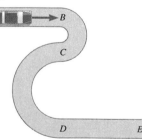

Question 5

7. A bug lands on a windshield wiper. Explain why the bug is more likely to be dislodged when the wipers are turned on at the high rather than the low setting.

8. What is the chance of a light car safely rounding an unbanked curve on an icy road as compared to that of a heavy car: worse, the

same, or better? Assume that both cars have the same speed and are equipped with identical tires. Account for your answer.

9. A container is filled with water and attached to a rope. Holding the free end of the rope, you whirl the container around in a horizontal circle at a constant speed. There is a hole in the container, so that as you whirl it, water continually leaks out. Explain what, if anything, you feel as the water leaks out.

10. A propeller, operating under test conditions, is being made to rotate at ever faster speeds. Explain what is likely to happen, and why, when the maximum rated speed of the propeller is exceeded.

11. A penny is placed on a rotating turntable. Where on the turntable does the penny require the largest centripetal force to remain in place? Give your reasoning.

12. Explain why a real airplane must bank as it flies in a circle, but a model airplane on a guideline can fly in a circle without banking.

13. Would a change in the earth's mass affect (a) the banking of airplanes as they turn, (b) the banking of roadbeds, (c) the speeds with which satellites are put into circular orbits, and (d) the performance of the loop-the-loop motorcycle stunt? In each case, give your reasoning.

14. A stone is tied to a string and whirled around in a circle at a constant speed. Is the string more likely to break when the circle is horizontal or when it is vertical? Account for your answer, assuming the constant speed is the same in each case.

15. A fighter plane is diving toward the earth. At the last instant, the pilot pulls out of the dive on a vertical circle and begins climbing upward. In such maneuvers, the pilot can black out if too much blood drains from his head. Why does blood drain from his head?

16. Go to **Concept Simulation 5.2** at **www.wiley.com/college/cutnell** to review the concepts involved in this question. Two cars are identical, except for the type of tread design on their tires. The cars are driven at the same speed and enter the same unbanked horizontal turn. Car A cannot negotiate the turn, but car B can. Which tread design yields a larger coefficient of static friction between the tires and the road? Give your reasoning in terms of centripetal force.

Problems

Section 5.1 Uniform Circular Motion, Section 5.2 Centripetal Acceleration

1. ssm How long does it take a plane, traveling at a constant speed of 110 m/s, to fly once around a circle whose radius is 2850 m?

2. A car travels at a constant speed around a circular track whose radius is 2.6 km. The car goes once around the track in 360 s. What is the magnitude of the centripetal acceleration of the car?

3. Interactive Solution 5.3 at www.wiley.com/college/cutnell presents a method for modeling this problem. The blade of a windshield wiper moves through an angle of 90.0° in 0.40 s. The tip of the blade moves on the arc of a circle that has a radius of 0.45 m. What is the magnitude of the centripetal acceleration of the tip of the blade?

4. Review Conceptual Example 2 in preparation for this problem. In Figure 5.4, an object, after being released from its circular path, travels the distance *OA* in the same time it would have moved from *O* to *P* on the circle. The speed of the object on and off the circle remains constant at the same value. Suppose that the radius of the circle in Figure 5.4 is 3.6 m and the angle θ is 25°. What is the distance *OA*?

5. ssm Computer-controlled display screens provide drivers in the Indianapolis 500 with a variety of information about how their cars are performing. For instance, as a car is going through a turn, a speed of 221 mi/h (98.8 m/s) and a centripetal acceleration of $3.00g$ (three times the acceleration due to gravity) are displayed. Determine the radius of the turn (in meters).

6. There is a clever kitchen gadget for drying lettuce leaves after you wash them. It consists of a cylindrical container mounted so that it can be rotated about its axis by turning a hand crank. The outer wall of the cylinder is perforated with small holes. You put the wet leaves in the container and turn the crank to spin off the water. The radius of the container is 12 cm. When the cylinder is rotating at 2.0 revolutions per second, what is the magnitude of the centripetal acceleration at the outer wall?

7. A bicycle chain is wrapped around a rear sprocket ($r = 0.039$ m) and a front sprocket ($r = 0.10$ m). The chain moves with a speed of 1.4 m/s around the sprockets, while the bike moves at a constant velocity. Find the magnitude of the acceleration of a chain link that is in contact with (a) the rear sprocket, (b) neither sprocket, and (c) the front sprocket.

* 8. Each of the space shuttle's main engines is fed liquid hydrogen by a high-pressure pump. Turbine blades inside the pump rotate at 617 rev/s. A point on one of the blades traces out a circle with a radius of 0.020 m as the blade rotates. (a) What is the magnitude of the centripetal acceleration that the blade must sustain at this point? (b) Express this acceleration as a multiple of $g = 9.80 \text{ m/s}^2$.

* 9. ssm The large blade of a helicopter is rotating in a horizontal circle. The length of the blade is 6.7 m, measured from its tip to the center of the circle. Find the ratio of the centripetal acceleration at the end of the blade to that which exists at a point located 3.0 m from the center of the circle.

* 10. ⚕ A centrifuge is a device in which a small container of material is rotated at a high speed on a circular path. Such a device is used in medical laboratories, for instance, to cause the more dense red blood cells to settle through the less dense blood serum and collect at the bottom of the container. Suppose the centripetal acceleration of the sample is 6.25×10^3 times as large as the acceleration due to gravity. How many revolutions per minute is the sample making, if it is located at a radius of 5.00 cm from the axis of rotation?

Section 5.3 Centripetal Force

11. ssm A 0.015-kg ball is shot from the plunger of a pinball machine. Because of a centripetal force of 0.028 N, the ball follows a circular arc whose radius is 0.25 m. What is the speed of the ball?

12. In a skating stunt known as "crack-the-whip," a number of skaters hold hands and form a straight line. They try to skate so that the line rotates about the skater at one end, who acts as the pivot. The skater farthest out has a mass of 80.0 kg and is 6.10 m from the pivot. He is skating at a speed of 6.80 m/s. Determine the magnitude of the centripetal force that acts on him.

13. Concept Simulation 5.1 at www.wiley.com/college/cutnell reviews the concepts that are involved in this problem. A child is twirling a 0.0120-kg ball on a string in a horizontal circle whose radius is 0.100 m. The ball travels once around the circle in 0.500 s. (a) Determine the centripetal force acting on the ball. (b) If the speed is doubled, does the centripetal force double? If not, by what factor does the centripetal force increase?

14. At an amusement park there is a ride in which cylindrically shaped chambers spin around a central axis. People sit in seats facing the axis, their backs against the outer wall. At one instant the outer wall moves at a speed of 3.2 m/s, and an 83-kg person feels a 560-N force pressing against his back. What is the radius of a chamber?

15. Concept Simulation 5.2 at www.wiley.com/college/cutnell reviews the concepts that play a role in this problem. A car is safely negotiating an unbanked circular turn at a speed of 21 m/s. The maximum static frictional force acts on the tires. Suddenly a wet patch in the road reduces the maximum static frictional force by a factor of three. If the car is to continue safely around the curve, to what speed must the driver slow the car?

16. A stone has a mass of 6.0×10^{-3} kg and is wedged into the tread of an automobile tire, as the drawing shows. The coefficient of static friction between the stone and each side of the tread channel is 0.90. When the tire surface is rotating at 13 m/s, the stone flies out of the tread. The magnitude F_N of the normal force that each side of the tread channel exerts on the stone is 1.8 N. Assume that only static friction supplies the centripetal force, and determine the radius r of the tire.

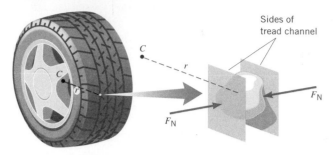

Sides of tread channel

*17. ssm www** The hammer throw is a track-and-field event in which a 7.3-kg ball (the "hammer") is whirled around in a circle several times and released. It then moves upward on the familiar curving path of projectile motion and eventually returns to earth some distance away. The world record for this distance is 86.75 m, achieved in 1986 by Yuriy Sedykh. Ignore air resistance and the fact that the ball is released above the ground rather than at ground level. Furthermore, assume that the ball is whirled on a circle that has a radius of 1.8 m and that its velocity at the instant of release is directed 41° above the horizontal. Find the magnitude of the centripetal force acting on the ball just prior to the moment of release.

*18.** A block is hung by a string from the inside roof of a van. When the van goes straight ahead at a speed of 28 m/s, the block hangs vertically down. But when the van maintains this same speed around an unbanked curve (radius = 150 m), the block swings toward the outside of the curve. Then the string makes an angle θ with the vertical. Find θ.

*19. Interactive Solution 5.19** at **www.wiley.com/college/cutnell** illustrates a method for modeling this problem. A "swing" ride at a carnival consists of chairs that are swung in a circle by 15.0-m cables attached to a vertical rotating pole, as the drawing shows. Suppose the total mass of a chair and its occupant is 179 kg. (a) Determine the tension in the cable attached to the chair. (b) Find the speed of the chair.

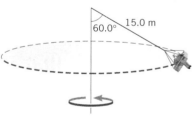

60.0° 15.0 m

Section 5.4 Banked Curves

20. At what angle should a curve of radius 150 m be banked, so cars can travel safely at 25 m/s without relying on friction?

21. ssm A curve of radius 120 m is banked at an angle of 18°. At what speed can it be negotiated under icy conditions where friction is negligible?

22. On a banked race track, the smallest circular path on which cars can move has a radius of 112 m, while the largest has a radius of 165 m, as the drawing illustrates. The height of the outer wall is 18 m. Find (a) the smallest and (b) the largest speed at which cars can move on this track without relying on friction.

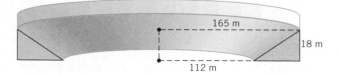

165 m

18 m

112 m

23. As an aid in working this problem, consult **Interactive Solution 5.23** at **www.wiley.com/college/cutnell.** A racetrack has the shape of an inverted cone, as the drawing shows. On this surface the cars race in circles that are parallel to the ground. For a speed of 34.0 m/s, at what value of the distance d should a driver locate his car if he wishes to stay on a circular path without depending on friction?

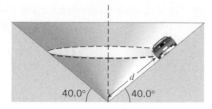

40.0°   40.0°

*24.** Before attempting this problem, review Examples 7 and 8. Two curves on a highway have the same radii. However, one is unbanked and the other is banked at an angle θ. A car can safely travel along the unbanked curve at a maximum speed v_0 under conditions when the coefficient of static friction between the tires and the road is $\mu_s = 0.81$. The banked curve is frictionless, and the car can negotiate it at the same maximum speed v_0. Find the angle θ of the banked curve.

*25. ssm www** A jet ($m = 2.00 \times 10^5$ kg), flying at 123 m/s, banks to make a horizontal circular turn. The radius of the turn is 3810 m. Calculate the necessary lifting force.

26. The drawing shows a baggage carousel at an airport. Your suitcase has not slid all the way down the slope and is going around at a constant speed on a circle ($r = 11.0$ m) as the carousel turns. The coefficient of static friction between the suitcase and the carousel is 0.760, and the angle θ in the drawing is 36.0°. How much time is required for your suitcase to go around once?

Section 5.5 Satellites in Circular Orbits, Section 5.6 Apparent Weightlessness and Artificial Gravity

27. ssm www A satellite is in a circular orbit around an unknown planet. The satellite has a speed of 1.70×10^4 m/s, and the radius of the orbit is 5.25×10^6 m. A second satellite also has a circular orbit around this same planet. The orbit of this second satellite has a radius of 8.60×10^6 m. What is the orbital speed of the second satellite?

28. A rocket is used to place a synchronous satellite in orbit about the earth. What is the speed of the satellite in orbit?

29. A satellite is placed in orbit 6.00×10^5 m above the surface of Jupiter. Jupiter has a mass of 1.90×10^{27} kg and a radius of 7.14×10^7 m. Find the orbital speed of the satellite.

30. The moon orbits the earth at a distance of 3.85×10^8 m. Assume that this distance is between the centers of the earth and the moon and that the mass of the earth is 5.98×10^{24} kg. Find the period for the moon's motion around the earth. Express the answer in days and compare it to the length of a month.

31. Review **Interactive Solution 5.31** at **www.wiley.com/college/cutnell** before beginning this problem. Two satellites, A and B, are in different circular orbits about the earth. The orbital speed of satellite A is three times that of satellite B. Find the ratio (T_A/T_B) of the periods of the satellites.

*32.** A satellite is in a circular earth orbit that has a radius of 6.7×10^6 m. A model airplane is flying on a 15-m guideline in a horizontal circle. The guideline is parallel to the ground. Find the speed of the plane such that the plane and the satellite have the same centripetal acceleration.

* **33. ssm** A satellite has a mass of 5850 kg and is in a circular orbit 4.1×10^5 m above the surface of a planet. The period of the orbit is two hours. The radius of the planet is 4.15×10^6 m. What is the true weight of the satellite when it is at rest on the planet's surface?

* **34.** The earth orbits the sun once a year at a distance of 1.50×10^{11} m. Venus orbits the sun at a distance of 1.08×10^{11} m. These distances are between the centers of the planets and the sun. How long (in earth days) does it take for Venus to make one orbit around the sun?

** **35.** To create artificial gravity, the space station shown in the drawing is rotating at a rate of 1.00 rpm. The radii of the cylindrically shaped chambers have the ratio $r_A/r_B = 4.00$. Each chamber A simulates an acceleration due to gravity of 10.0 m/s². Find values for (a) r_A, (b) r_B, and (c) the acceleration due to gravity that is simulated in chamber B.

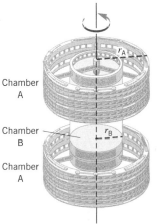

Problem 35

Section 5.7 Vertical Circular Motion

36. A roller coaster at an amusement park has a dip that bottoms out in a vertical circle of radius r. A passenger feels the seat of the car pushing upward on her with a force equal to twice her weight as she goes through the dip. If $r = 20.0$ m, how fast is the roller coaster traveling at the bottom of the dip?

37. ssm A motorcycle has a constant speed of 25.0 m/s as it passes over the top of a hill whose radius of curvature is 126 m. The mass of the motorcycle and driver is 342 kg. Find the magnitude of (a) the centripetal force and (b) the normal force that acts on the cycle.

38. For the normal force in Figure 5.21 to have the same magnitude at all points on the vertical track, the stunt driver must adjust the speed to be different at different points. Suppose, for example, that the track has a radius of 3.0 m and that the driver goes past point 1 at the bottom with a speed of 15 m/s. What speed must she have at point 3, so that the normal force at the top has the same magnitude as it did at the bottom?

39. ssm The condition of apparent weightlessness for the passengers can be created for a brief instant when a plane flies over the top of a vertical circle. At a speed of 215 m/s, what is the radius of the vertical circle that the pilot must use?

* **40.** A motorcycle is traveling up one side of a hill and down the other side. The crest is a circular arc with a radius of 45.0 m. Determine the maximum speed that the cycle can have while moving over the crest without losing contact with the road.

* **41.** Reviewing **Interactive Solution 5.41** at **www.wiley.com/college/cutnell** will help in solving this problem. A stone is tied to a string (length = 1.10 m) and whirled in a circle at the same constant speed in two different ways. First, the circle is horizontal and the string is nearly parallel to the ground. Next, the circle is vertical. In the vertical case the maximum tension in the string is 15.0% larger than the tension that exists when the circle is horizontal. Determine the speed of the stone.

** **42.** In an automatic clothes dryer, a hollow cylinder moves the clothes on a vertical circle (radius $r = 0.32$ m), as the drawing shows. The appliance is designed so that the clothes tumble gently as they dry. This means that when a piece of clothing reaches an angle of θ above the horizontal, it loses contact with the wall of the cylinder and falls onto the clothes below. How many revolutions per second should the cylinder make in order that the clothes lose contact with the wall when $\theta = 70.0°$?

Additional Problems

43. ssm Review Example 3, which deals with the bobsled in Figure 5.5. Also review Conceptual Example 4. The mass of the sled and its two riders in Figure 5.5 is 350 kg. Find the magnitude of the centripetal force that acts on the sled during the turn with a radius of (a) 33 m and (b) 24 m.

44. Suppose the surface (radius = r) of the space station in Figure 5.19 is rotating at 35.8 m/s. What must be the value of r for the astronauts to weigh one-half of their earth weight?

45. Review Conceptual Example 2 as background for this problem. One kind of slingshot consists of a pocket that holds a pebble and is whirled on a circle of radius r. The pebble is released from the circle at the angle θ, so that it will hit the target. The distance to the target from the center of the circle is d. (See the drawing, which is not to scale.) The circular path is parallel to the ground, and the target lies in the plane of the circle. The distance d is ten times the ra-

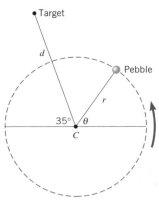

dius r. Ignore the effect of gravity in pulling the stone downward after it is released and find the angle θ.

46. For background pertinent to this problem, review Conceptual Example 6. In Figure 5.8 the man hanging upside down is holding a partner who weighs 475 N. Assume that the partner moves on a circle that has a radius of 6.50 m. At a swinging speed of 4.00 m/s, what force must the man apply to his partner in the straight-down position?

47. ssm Speedboat A negotiates a curve whose radius is 120 m. Speedboat B negotiates a curve whose radius is 240 m. Each boat experiences the same centripetal acceleration. What is the ratio v_A/v_B of the speeds of the boats?

48. A 9.5-kg monkey is hanging by one arm from a branch and is swinging on a vertical circle. As an approximation, assume a radial distance of 85 cm between the branch and the point where the monkey's mass is located. As the monkey swings through the lowest point on the circle, it has a speed of 2.8 m/s. Find (a) the magnitude of the centripetal force acting on the monkey and (b) the magnitude of the tension in the monkey's arm.

49. ssm A 2100-kg demolition ball swings at the end of a 15-m cable on the arc of a vertical circle. At the lowest point of the swing, the ball is moving at a speed of 7.6 m/s. Determine the tension in the cable. (*Hint: The tension serves the same purpose as the normal force at point 1 in Figure 5.21.*)

50. A satellite circles the earth in an orbit whose radius is twice the earth's radius. The earth's mass is 5.98×10^{24} kg, and its radius is 6.38×10^6 m. What is the period of the satellite?

* **51.** A computer is reading data from a rotating CD-ROM. At a point that is 0.030 m from the center of the disc, the centripetal acceleration is 120 m/s². What is the centripetal acceleration at a point that is 0.050 m from the center of the disc?

* **52.** The earth rotates once per day about an axis passing through the north and south poles, an axis that is perpendicular to the plane of the equator. Assuming the earth is a sphere with a radius of 6.38×10^6 m, determine the speed and centripetal acceleration of a person situated (a) at the equator and (b) at a latitude of 30.0° north of the equator.

* **53.** A rigid massless rod is rotated about one end in a horizontal circle. There is a mass m_1 attached to the center of the rod and a mass

m_2 attached to the outer end of the rod. The inner section of the rod sustains three times as much tension as the outer section. Find the ratio m_2/m_1.

** **54.** At amusement parks, there is a popular ride where the floor of a rotating cylindrical room falls away, leaving the backs of the riders "plastered" against the wall. Suppose the radius of the room is 3.30 m and the speed of the wall is 10.0 m/s when the floor falls away. (a) What is the source of the centripetal force acting on the riders? (b) How much centripetal force acts on a 55.0-kg rider? (c) What is the minimum coefficient of static friction that must exist between a rider's back and the wall, if the rider is to remain in place when the floor drops away?

** **55.** ssm www Redo Example 5, assuming that there is no upward lift on the plane due to its wings. Without such lift, the guideline slopes downward due to the weight of the plane. For purposes of significant figures, use 0.900 kg for the mass of the plane, 17.0 m for the length of the guideline, and 19.0 and 38.0 m/s for the speeds.

Concepts & Calculations Group Learning Problems

Note: Each of these problems consists of Concept Questions followed by a related quantitative Problem. They are designed for use by students working alone or in small learning groups. The Concept Questions involve little or no mathematics and are intended to stimulate group discussions. They focus on the concepts with which the problems deal. Recognizing the concepts is the essential initial step in any problem-solving technique.

56. Concept Questions Interactive LearningWare 5.1 at **www.wiley.com/college/cutnell** explores how to solve problems that have concepts similar to this one. The second hand and the minute hand on one type of clock are the same length. (a) What is the period T of the motion for the second hand and for the minute hand? (b) We have seen that the centripetal acceleration is given by $a_c = v^2/r$, where v is the speed and r is the radius. How is the centripetal acceleration related to the period? In each case, explain your reasoning.

Problem Find the ratio $(a_{c, \text{ second}}/a_{c, \text{ minute}})$ of the centripetal accelerations for the tips of the second hand and the minute hand.

57. Concept Questions The following table lists data for the speed and the radius in three examples of uniform circular motion.

	Radius	Speed
Example 1	0.50 m	12 m/s
Example 2	Infinitely large	35 m/s
Example 3	1.8 m	2.3 m/s

(a) Without doing any calculations, identify the example with the smallest centripetal acceleration. (b) Similarly identify the example with the greatest centripetal acceleration. In each case, justify your answer.

Problem Find the value for the centripetal acceleration for each example. Verify that your answers are consistent with your answers to the Concept Questions.

58. Concept Questions Interactive LearningWare 5.2 at **www.wiley.com/college/cutnell** illustrates good problem-solving techniques for this type of problem. Two cars are traveling at the same speed of 27 m/s on a curve that has a radius of 120 m. Car A has a mass of 1100 kg, and car B has a mass of 1600 kg. Without doing any calcu-

lations, decide (a) which car, if either, has the greatest centripetal acceleration and (b) which car, if either, experiences the greatest centripetal force. Justify your answers.

Problem Find the centripetal acceleration and the centripetal force for each car. Be sure that your answers are consistent with your answers to the Concept Questions.

59. Concept Questions A penny is placed at the outer edge of a $33\frac{1}{3}$-rpm record (radius = 0.150 m). (a) What is producing the centripetal force that enables the penny to rotate along with the record—kinetic or static friction? (b) The centripetal force needed depends on the speed at which the penny is moving. How can the speed be determined from the data given? Account for your answer in each case.

Problem Find the minimum coefficient of friction necessary to allow the penny to rotate along with the record.

60. Concept Questions Two satellites are in circular orbits around the earth. The orbit for satellite A is at a height of 360 km above the earth's surface, while that for satellite B is at a height of 720 km. (a) Which satellite has the greatest orbital speed? (b) The speed of a satellite in a circular orbit is given by Equation 5.5 ($v = \sqrt{GM_E/r}$). In this expression, do you substitute the heights of 360×10^3 m and 720×10^3 m for the term r? Explain your answers.

Problem Find the orbital speed for each satellite. Check to see that your answers are consistent with your answers to the Concept Questions.

61. Concept Questions A 0.20-kg ball on a string is whirled on a vertical circle at a constant speed. When the ball is at the three o'clock position, the tension is 16 N. (a) Is the centripetal force acting on the ball the same at each point on the circle? (b) When the ball is at the three o'clock position, what is providing the centripetal force—gravity, the tension in the string, or a combination of both? (c) Is the tension in the string greater when the ball is at the twelve o'clock position or when it is at the six o'clock position? Justify your answers.

Problem Find the tensions in the string when the ball is at the twelve o'clock and at the six o'clock positions. Verify that your answers are consistent with your answers to the Concept Questions.

Chapter 6

Work and Energy

As the car of a roller coaster moves up and down its track, kinetic energy and gravitational potential energy are constantly being interconverted.
(©Kurita Kaku/Gamma Presse, Inc.)

Figure 6.1 Work is done when a force **F** pushes a car through a displacement **s**.

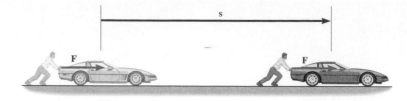

6.1　Work Done by a Constant Force

Work is a familiar concept. For example, it takes work to push a stalled car. In fact, more work is done when the pushing force is greater or when the displacement of the car is greater. Force and displacement are, in fact, the two essential elements of work, as Figure 6.1 illustrates. The drawing shows a constant pushing force **F** that points in the same direction as the resulting displacement **s**.* In such a case, the work W is defined as the magnitude F of the force times the magnitude s of the displacement: $W = Fs$. The work done to push a car is the same whether the car is moved north to south or east to west, provided that the amount of force used and the distance moved are the same. Since work does not convey directional information, it is a scalar quantity.

The equation $W = Fs$ indicates that the unit of work is the unit of force times the unit of distance, or the newton·meter in SI units. One newton·meter is referred to as a *joule* (J) (rhymes with "cool"), in honor of James Joule (1818–1889) and his research into the nature of work, energy, and heat. Table 6.1 summarizes the units for work in several systems of measurement.

The definition of work as $W = Fs$ does have one surprising feature: If the distance s is zero, the work is zero, even if a force is applied. Pushing on an immovable object, such as a brick wall, may tire your muscles, but there is no work done of the type we are discussing. In physics, the idea of work is intimately tied up with the idea of motion. If the object does not move, the force acting on the object does no work.

Often, the force and displacement do not point in the same direction. For instance, Figure 6.2*a* shows a suitcase-on-wheels being pulled to the right by a force that is applied along the handle. The force is directed at an angle θ relative to the displacement. In such a case, only the component of the force along the displacement is used in defining work. As part *b* of the drawing illustrates, this component is $F \cos \theta$, and it appears in the general definition of work as follows:

■ **DEFINITION OF WORK DONE BY A CONSTANT[†] FORCE**

The work done on an object by a constant force **F** is

$$W = (F \cos \theta)s \qquad (6.1)$$

where F is the magnitude of the force, s is the magnitude of the displacement, and θ is the angle between the force and the displacement.

SI Unit of Work: newton·meter = joule (J)

Table 6.1　*Units of Measurement for Work*

System	Force	×	Distance	=	Work
SI	newton (N)		meter (m)		joule (J)
CGS	dyne (dyn)		centimeter (cm)		erg
BE	pound (lb)		foot (ft)		foot·pound (ft·lb)

* When discussing work, it is customary to use the symbol **s** for the displacement, rather than **x** or **y**.
† Section 6.9 considers the work done by a variable force.

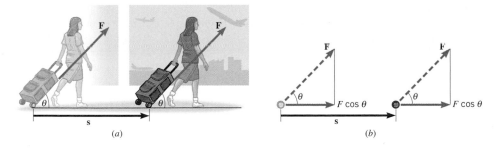

Figure 6.2 (*a*) Work can be done by a force **F** that points at an angle θ relative to the displacement **s**. (*b*) The force component that points along the displacement is $F \cos \theta$.

When the force points in the same direction as the displacement, then $\theta = 0°$, and Equation 6.1 reduces to $W = Fs$. The following example illustrates how Equation 6.1 is used to calculate work.

Example 1 Pulling a Suitcase-on-Wheels

Find the work done by a 45.0-N force in pulling the suitcase in Figure 6.2*a* at an angle $\theta = 50.0°$ for a distance $s = 75.0$ m.

Reasoning The pulling force causes the suitcase to move a distance of 75.0 m and does work. However, the force makes an angle of 50.0° with the displacement, and we must take this angle into account by using the definition of work given by Equation 6.1.

Solution The work done by the 45.0-N force is

$$W = (F \cos \theta)s = [(45.0 \text{ N}) \cos 50.0°](75.0 \text{ m}) = \boxed{2170 \text{ J}}$$

The answer is expressed in newton·meters or joules (J).

The definition of work in Equation 6.1 takes into account only the component of the force in the direction of the displacement. The force component perpendicular to the displacement does no work. To do work, there must be a force *and* a displacement, and since there is no displacement in the perpendicular direction, there is no work done by the perpendicular component of the force. If the entire force is perpendicular to the displacement, the angle θ in Equation 6.1 is 90°, and the force does no work at all.

Work can be either positive or negative, depending on whether a component of the force points in the same direction as the displacement or in the opposite direction. Example 2 illustrates how positive and negative work arise.

Example 2 Bench-Pressing

The weight lifter in Figure 6.3*a* is bench-pressing a barbell whose weight is 710 N. In part *b* of the figure, he raises the barbell a distance of 0.65 m above his chest, and in part *c* he lowers it the same distance. The weight is raised and lowered at a constant velocity. Determine the work done on the barbell by the weight lifter during (a) the lifting phase and (b) the lowering phase.

Reasoning To calculate the work, it is necessary to know the force exerted by the weight lifter. The barbell is raised and lowered at a constant velocity and, therefore, is in equilibrium.

Figure 6.3 (*a*) In the bench press, work is done during both the lifting and lowering phases. (©Patrick Bennet/Corbis Images) (*b*) During the lifting phase, the force **F** does positive work. (*c*) During the lowering phase, the force does negative work.

(*a*)

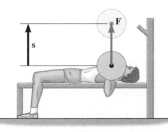

(*b*)

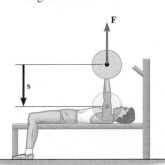

(*c*)

Consequently, the force **F** exerted by the weight lifter must balance the weight of the barbell, so $F = 710$ N. During the lifting phase, the force **F** and displacement **s** are in the same direction, as Figure 6.3*b* shows. The angle between them is $\theta = 0°$. When the barbell is lowered, however, the force and displacement are in opposite directions, as in Figure 6.3*c*. The angle between the force and the displacement is now $\theta = 180°$. With these observations, we can find the work.

Solution

(a) During the lifting phase, the work done by the force **F** is given by Equation 6.1 as

$$W = (F \cos \theta)s = [(710 \text{ N}) \cos 0°](0.65 \text{ m}) = 460 \text{ J}$$

(b) The work done during the lowering phase is

$$W - (F \cos \theta)s$$
$$= [(710 \text{ N}) \cos 180°](0.65 \text{ m}) = \boxed{-460 \text{ J}}$$

since $\cos 180° = -1$. The work is negative, because the force is opposite to the displacement. Weight lifters call each complete up-and-down movement of the barbell a repetition, or "rep." The lifting of the weight is referred to as the positive part of the rep, and the lowering is known as the negative part.

The physics of
positive and negative "reps" in
weight lifting.

Example 3 deals with the work done by a static frictional force when it acts on a crate that is resting on the bed of an accelerating truck.

Example 3 Accelerating a Crate

Figure 6.4*a* shows a 120-kg crate on the flatbed of a truck that is moving with an acceleration of $a = +1.5$ m/s² along the positive x axis. The crate does not slip with respect to the truck, as the truck undergoes a displacement whose magnitude is $s = 65$ m. What is the total work done on the crate by all of the forces acting on it?

Reasoning The free-body diagram in Figure 6.4*b* shows the forces that act on the crate: (1) the weight **W** of the crate, (2) the normal force $\mathbf{F_N}$ exerted by the flatbed, and (3) the static frictional force $\mathbf{f_s}$, which is exerted by the flatbed in the forward direction and keeps the crate from slipping backward. The weight and the normal force are perpendicular to the displacement, so they do no work. Only the static frictional force does work, since it acts in the x direction. To determine the frictional force, we note that the crate does not slip and, therefore, must have the same acceleration of $a = +1.5$ m/s² as does the truck. The force creating this acceleration is the static frictional force, and knowing the mass of the crate and its acceleration, we can use Newton's second law to obtain its magnitude. Then, knowing the frictional force and the displacement, we can determine the total work done on the crate.

Solution From Newton's second law, we find that the magnitude f_s of the static frictional force is

$$f_s = ma = (120 \text{ kg})(1.5 \text{ m/s}^2) = 180 \text{ N}$$

The total work is that done by the static frictional force and is

$$W = (f_s \cos \theta)s = (180 \text{ N})(\cos 0°)(65 \text{ m}) = \boxed{1.2 \times 10^4 \text{ J}} \qquad (6.1)$$

The work is positive, because the frictional force is in the same direction as the displacement ($\theta = 0°$).

Need more practice?

Interactive LearningWare 6.1
A cyclist coasts down a hill, losing 20.0 m in height, and then coasts up an identical hill to the same initial vertical height. The bicycle and rider have a mass of 73.5 kg. Two forces act on the bicycle/rider system: the weight of the bicycle and rider and the normal force exerted on the tires by the ground. Determine the work done by each of these forces during (a) the downward part of the motion and (b) the upward part of the motion.

Related Homework: *Problem 6*

Go to
www.wiley.com/college/cutnell
for an interactive solution.

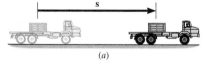

(*a*)

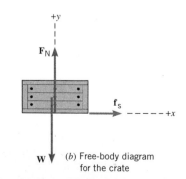

(*b*) Free-body diagram
for the crate

Figure 6.4 (*a*) The truck and crate are accelerating to the right for a distance of $s = 65$ m. (*b*) The free-body diagram for the crate.

✔ **Check Your Understanding 1**

A suitcase is hanging straight down from your hand as you ride an escalator. Your hand exerts a force on the suitcase, and this force does work. Which one of the following statements is correct? (a) The work is negative when you ride up the escalator and positive when you ride down the escalator. (b) The work is positive when you ride up the escalator and negative when you ride down the escalator. (c) The work is positive irrespective of whether you ride up or down the escalator. (d) The work is negative irrespective of whether you ride up or down the escalator. *(The answer is given at the end of the book.)*

Background: This question deals with work and its relationship to force and displacement. Pay close attention to the directions of the force and displacement vectors.

For similar questions (including calculational counterparts), consult Self-Assessment Test 6.1. This test is described at the end of Section 6.2.

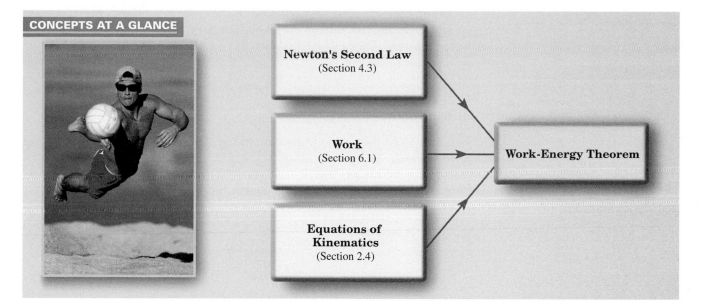

CONCEPTS AT A GLANCE

Newton's Second Law
(Section 4.3)

Work
(Section 6.1)

Equations of
Kinematics
(Section 2.4)

Work-Energy Theorem

Figure 6.5 CONCEPTS AT A GLANCE In this section Newton's second law, the definition of work, and the equations of kinematics are brought together to produce the work–energy theorem. In hitting a volleyball, a player does work on it. This work causes the ball's kinetic energy to change in accordance with the work–energy theorem. (©Yellow Dog Productions/The Image Bank)

6.2 The Work–Energy Theorem and Kinetic Energy

▶ CONCEPTS AT A GLANCE Most people expect that if you do work, you get something as a result. In physics, when a net force performs work on an object, there is always a result from the effort. The result is a change in the *kinetic energy* of the object. As we will now see, the relationship that relates work to the change in kinetic energy is known as the *work–energy theorem.* This theorem is obtained by bringing together three basic concepts that we've already learned about, as the Concepts-at-a-Glance chart in Figure 6.5 illustrates. First we'll apply Newton's second law of motion, $\Sigma F = ma$, which relates the net force ΣF to the acceleration a of an object. Then, we'll determine the work done by the net force when the object moves through a certain distance. Finally, we'll use Equation 2.9, one of the equations of kinematics, to relate the distance and acceleration to the initial and final speeds of the object. The result of this approach will be the work–energy theorem. ◀

To gain some insight into the idea of kinetic energy and the work–energy theorem, look at Figure 6.6, where a constant net external force $\Sigma \mathbf{F}$ acts on an airplane of mass m. This net force is the vector sum of all the external forces acting on the plane, and, for simplicity, it is assumed to have the same direction as the displacement $\mathbf{s}$. According to Newton's second law, the net force produces an acceleration a, given by $a = \Sigma F/m$. Consequently, the speed of the plane changes from an initial value of v_0 to a final value of v_f.* Multiplying both sides of $\Sigma F = ma$ by the distance s gives

$$\underbrace{(\Sigma F)s}_{\substack{\text{Work done by} \\ \text{net ext. force}}} = mas$$

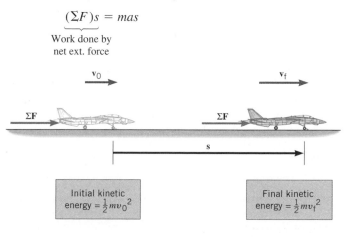

Figure 6.6 A constant net external force $\Sigma \mathbf{F}$ acts over a displacement $\mathbf{s}$ and does work on the plane. As a result of the work done, the plane's kinetic energy changes.

*For extra emphasis, the final speed is now represented by the symbol v_f, rather than v.

The left side of this equation is the work done by the net external force. The term as on the right side can be related to v_0 and v_f by using Equation 2.9 ($v_f^2 = v_0^2 + 2as$) from the equations of kinematics. Solving this equation to give $as = \frac{1}{2}(v_f^2 - v_0^2)$ and substituting into $(\Sigma F)s = mas$ shows that

$$\underbrace{(\Sigma F)s}_{\substack{\text{Work done by} \\ \text{net ext. force}}} = \underbrace{\tfrac{1}{2}mv_f^2}_{\substack{\text{Final} \\ \text{KE}}} - \underbrace{\tfrac{1}{2}mv_0^2}_{\substack{\text{Initial} \\ \text{KE}}}$$

This expression is the work–energy theorem. Its left side is the work W done by the net external force, while its right side involves the difference between two terms, each of which has the form $\frac{1}{2}(\text{mass})(\text{speed})^2$. The quantity $\frac{1}{2}(\text{mass})(\text{speed})^2$ is called kinetic energy (KE) and plays a significant role in physics, as we will soon see.

■ **DEFINITION OF KINETIC ENERGY**

The kinetic energy KE of an object with mass m and speed v is given by

$$\text{KE} = \tfrac{1}{2}mv^2 \tag{6.2}$$

SI Unit of Kinetic Energy: joule (J)

The SI unit of kinetic energy is the same as the unit for work, the joule. Kinetic energy, like work, is a scalar quantity. These are not surprising observations, for work and kinetic energy are closely related, as is clear from the following statement of the work–energy theorem.

■ **THE WORK–ENERGY THEOREM**

When a net external force does work W on an object, the kinetic energy of the object changes from its initial value of KE_0 to a final value of KE_f, the difference between the two values being equal to the work:

$$W = \text{KE}_f - \text{KE}_0 = \tfrac{1}{2}mv_f^2 - \tfrac{1}{2}mv_0^2 \tag{6.3}$$

The work–energy theorem may be derived for any direction of the force relative to the displacement, not just the situation in Figure 6.6. In fact, the force may even vary from point to point along a path that is curved rather than straight, and the theorem remains valid. According to the work–energy theorem, a moving object has kinetic energy, because work was done to accelerate the object from rest to a speed v_f.* Conversely, an object with kinetic energy can perform work, if it is allowed to push or pull on another object. Example 4 illustrates the work–energy theorem and considers a single force that does work to change the kinetic energy of a space probe.

Example 4   Deep Space 1

The physics of an ion propulsion drive.

The space probe *Deep Space 1* was launched October 24, 1998. Its mass was 474 kg. The goal of the mission was to test a new kind of engine called an ion propulsion drive, which generates only a weak thrust, but can do so for long periods of time using only small amounts of fuel. The mission has been spectacularly successful. Consider the probe traveling at an initial speed of $v_0 = 275$ m/s. No forces act on it except the 56.0-mN thrust of its engine. This external force **F** is directed parallel to the displacement s of magnitude 2.42×10^9 m (Figure 6.7). Determine the final speed of the probe, assuming that the mass remains nearly constant.

---

* Strictly speaking, the work–energy theorem, as given by Equation 6.3, applies only to a single particle, which occupies a mathematical point in space. A macroscopic object, however, is a collection or system of particles and is spread out over a region of space. Therefore, when a force is applied to a macroscopic object, the point of application of the force may be anywhere on the object. To take into account this and other factors, a discussion of work and energy is required that is beyond the scope of this text. The interested reader may refer to A. B. Arons, *The Physics Teacher,* October 1989, p. 506.

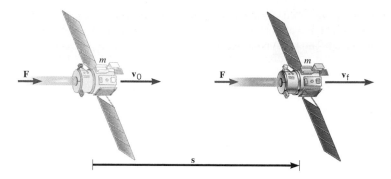

Figure 6.7 The engine of *Deep Space 1* generates a single force **F** that points in the same direction as the displacement **s**. The force performs positive work, causing the probe to gain kinetic energy.

Reasoning Since the force **F** is the only force acting on the probe, it is the net external force, and the work it does causes the kinetic energy of the spacecraft to change. The work can be calculated directly from Equation 6.1 and is positive, because **F** points in the same direction as the displacement **s**. According to the work–energy theorem ($W = \text{KE}_f - \text{KE}_0$), a positive value for W means that the kinetic energy increases. We can use the theorem to find the final kinetic energy and, hence, the final speed. Since the kinetic energy increases, we expect the final speed to be greater than the initial speed.

Solution From the definition of work (Equation 6.1), we have

$$W = (F \cos \theta)s = [(56.0 \times 10^{-3}\ \text{N}) \cos 0°](2.42 \times 10^9\ \text{m}) = 1.36 \times 10^8\ \text{J}$$

where $\theta = 0°$, because the force and displacement are in the same direction (see the drawing). Since $W = \text{KE}_f - \text{KE}_0$ according to the work–energy theorem, the final kinetic energy of the probe is

$$\text{KE}_f = W + \text{KE}_0 = (1.36 \times 10^8\ \text{J}) + \tfrac{1}{2}(474\ \text{kg})(275\ \text{m/s})^2 = 1.54 \times 10^8\ \text{J}$$

The final kinetic energy is $\text{KE}_f = \tfrac{1}{2}mv_f^2$, so the final speed is

$$v_f = \sqrt{\frac{2(\text{KE}_f)}{m}} = \sqrt{\frac{2(1.54 \times 10^8\ \text{J})}{474\ \text{kg}}} = \boxed{806\ \text{m/s}}$$

As expected, the final speed is greater than the initial speed.

Concept Simulation 6.1

This simulation illustrates the work–energy theorem, as applied to a block being pushed along a frictionless horizontal surface by a single horizontal force. The force and the mass, initial velocity, and initial position of the block are under your control. The effect of adjusting these variables can be seen in graphs of the block's kinetic energy, velocity, and position versus time.

Related Homework: Problem 14

Go to
www.wiley.com/college/cutnell

In Example 4 only the force of the engine does work. If several external forces act on an object, they must be added together vectorially to give the net force. The work done by the net force can then be related to the change in the object's kinetic energy by using the work–energy theorem, as in the next example.

Example 5 Downhill Skiing

A 58-kg skier is coasting down a 25° slope, as Figure 6.8*a* shows. A kinetic frictional force of magnitude $f_k = 70$ N opposes her motion. Near the top of the slope, the skier's speed is $v_0 = 3.6$ m/s. Ignoring air resistance, determine the speed v_f at a point that is displaced 57 m downhill.

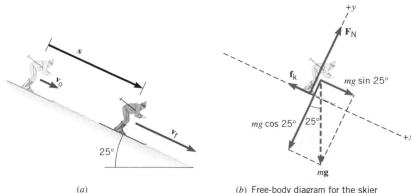

(*a*)

(*b*) Free-body diagram for the skier

Figure 6.8 (*a*) A skier coasting downhill. (*b*) The free-body diagram for the skier.

Need more practice?

Interactive LearningWare 6.2
In screeching to a halt, a car leaves skid marks that are 65 m long. The coefficient of kinetic friction between the tires and the road is μ_k = 0.71. How fast was the car going before the driver applied the brakes?

Related Homework: *Problem 18*

Go to
www.wiley.com/college/cutnell
for an interactive solution.

Reasoning As in Example 4, we will use the work–energy theorem to find the final speed. Note that the work done by the net external force is needed, not just the work done by any single force. The net force is the vector sum of the forces shown in the free-body diagram in Figure 6.8*b*. In this diagram, the normal force $\mathbf{F_N}$ is balanced by the component of the skier's weight ($mg \cos 25°$) perpendicular to the slope, because no acceleration occurs along that direction. Therefore, the net force points along the *x* axis.

Solution The net external force in Figure 6.8*b* points along the *x* axis and is

$$\Sigma F = mg \sin 25° - f_k = (58 \text{ kg})(9.80 \text{ m/s}^2) \sin 25° - 70 \text{ N} = +170 \text{ N}$$

The work done by the net force is

$$W = (\Sigma F \cos \theta)s = [(170 \text{ N}) \cos 0°](57 \text{ m}) = 9700 \text{ J} \qquad (6.1)$$

where $\theta = 0°$ because the net force and the displacement point in the same direction down the slope. From the work–energy theorem ($W = KE_f - KE_0$), it follows that the final kinetic energy of the skier is

$$KE_f = W + KE_0$$
$$= 9700 \text{ J} + \tfrac{1}{2}(58 \text{ kg})(3.6 \text{ m/s})^2 = 10\ 100 \text{ J}$$

Since the final kinetic energy is $KE_f = \tfrac{1}{2}mv_f^2$, the final speed of the skier is

$$v_f = \sqrt{\frac{2(KE_f)}{m}} = \sqrt{\frac{2(10\ 100 \text{ J})}{58 \text{ kg}}} = \boxed{19 \text{ m/s}}$$

✓ **Check Your Understanding 2**

A rocket is at rest on the launch pad. When the rocket is launched, its kinetic energy increases. Is the following statement true or false? "The amount by which the kinetic energy increases is equal to the work done by the force generated by the rocket's engine." *(The answer is given at the end of the book.)*

Background: The concepts of work, kinetic energy, and the work–energy theorem are pertinent to this question. Think about what forces determine the work that appears in the work–energy theorem.

For similar questions (including calculational counterparts), consult Self-Assessment Test 6.1. This test is described at the end of this section.

Problem solving insight

Example 5 emphasizes that *the work–energy theorem deals with the work done by the net external force. The work–energy theorem does not apply to the work done by an individual force,* unless that force happens to be the only one present, in which case it is the net force. If the work done by the net force is *positive,* as in Example 5, the kinetic energy of the object *increases.* If the work done is *negative,* the kinetic energy *decreases.* If the work is zero, the kinetic energy remains the same. Conceptual Example 6 explores these ideas further.

Conceptual Example 6 **Work and Kinetic Energy**

Figure 6.9 illustrates a satellite moving about the earth in a circular orbit and in an elliptical orbit. The only external force that acts on the satellite is the gravitational force. For these two orbits, determine whether the kinetic energy of the satellite changes during the motion.

Reasoning and Solution For the circular orbit in Figure 6.9*a* the gravitational force **F** does no work on the satellite, since the force is perpendicular to the instantaneous displacement **s** at

Figure 6.9 (*a*) In a circular orbit, the gravitational force **F** is always perpendicular to the displacement **s** of the satellite and does no work. (*b*) In an elliptical orbit, there can be a component of the force along the displacement, and, consequently, work can be done.

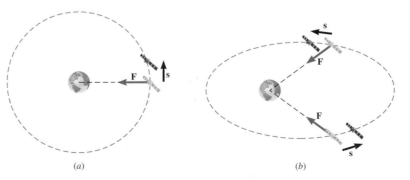

(*a*)   (*b*)

all times. The gravitational force is the only external force acting on the satellite, so it is the net force. Thus, the work done by the net force is zero, and according to the work–energy theorem (Equation 6.3), the kinetic energy of the satellite (and, hence, its speed) remains the same everywhere on the orbit.

In contrast, the gravitational force does do work on a satellite in an elliptical orbit. For example, as the satellite moves toward the earth in the top part of Figure 6.9*b*, there is a component of the gravitational force that points in the same direction as the displacement. Consequently, the gravitational force does positive work during this part of the orbit, and the kinetic energy of the satellite increases. When the satellite moves away from the earth, as in the lower part of Figure 6.9*b*, the gravitational force has a component that points opposite to the displacement. Now, the gravitational force does negative work, and the kinetic energy of the satellite decreases.

Can you identify the two places on an elliptical orbit where the gravitational force does no work?

Related Homework: *Problem 16*

Self-Assessment Test 6.1

Test your understanding of the material in Sections 6.1 and 6.2:

• Work • Kinetic Energy • The Work–Energy Theorem

Go to **www.wiley.com/college/cutnell**

6.3 *Gravitational Potential Energy*

WORK DONE BY THE FORCE OF GRAVITY

The gravitational force is a well-known force that can do positive or negative work, and Figure 6.10 helps to show how the work can be determined. This drawing depicts a basketball of mass m moving vertically downward, the force of gravity mg being the only force acting on the ball. The initial height of the ball is h_0, and the final height is h_f, both distances measured from the earth's surface. The displacement **s** is downward and has a magnitude of $s = h_0 - h_f$. To calculate the work $W_{gravity}$ done on the ball by the force of gravity, we use $W = (F\cos\theta)s$ with $F = mg$ and $\theta = 0°$, since the force and displacement are in the same direction:

$$W_{gravity} = (mg\cos 0°)(h_0 - h_f) = mg(h_0 - h_f) \qquad (6.4)$$

Equation 6.4 is valid for *any path* taken between the initial and final heights, and not just for the straight-down path shown in Figure 6.10. For example, the same expression can be derived for both paths shown in Figure 6.11. Thus, only the *difference in vertical dis-*

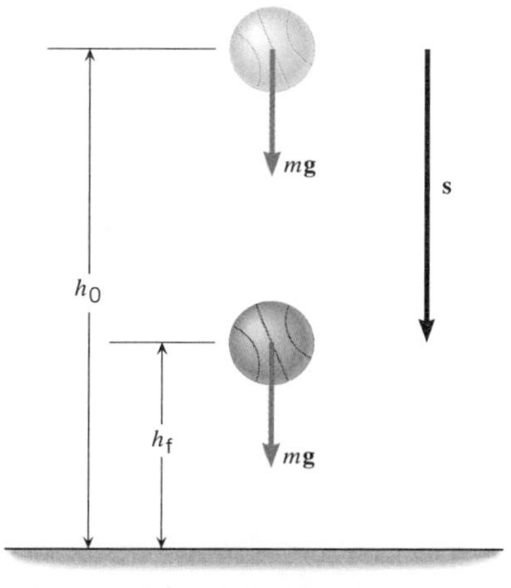

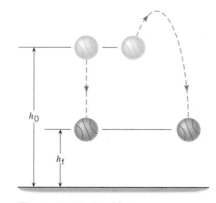

Figure 6.10 Gravity exerts a force mg on the basketball. Work is done by the gravitational force as the basketball falls from a height of h_0 to a height of h_f.

Figure 6.11 An object can move along different paths in going from an initial height of h_0 to a final height of h_f. In each case, the work done by the gravitational force is the same $[W_{gravity} = mg(h_0 - h_f)]$, since the change in vertical distance $(h_0 - h_f)$ is the same.

tances $(h_0 - h_f)$ need be considered when calculating the work done by gravity. Since the difference in the vertical distances is the same for each path in the drawing, the work done by gravity is the same in each case. We are assuming here that the difference in heights is small compared to the radius of the earth, so that the magnitude g of the acceleration due to gravity is the same at every height. Moreover, for positions close to the earth's surface, we can use the value of $g = 9.80$ m/s².

Since only the difference between h_0 and h_f appears in Equation 6.4, the vertical distances themselves need not be measured from the earth. For instance, they could be measured relative to a zero level that is one meter above the ground, and $h_0 - h_f$ would still have the same value. Example 7 illustrates how the work done by gravity is used in conjunction with the work–energy theorem.

Example 7   A Gymnast on a Trampoline

A gymnast springs vertically upward from a trampoline as in Figure 6.12*a*. The gymnast leaves the trampoline at a height of 1.20 m and reaches a maximum height of 4.80 m before falling back down. All heights are measured with respect to the ground. Ignoring air resistance, determine the initial speed v_0 with which the gymnast leaves the trampoline.

Reasoning We can find the initial speed of the gymnast (mass = m) by using the work–energy theorem, provided the work done by the net external force can be determined. Since only the gravitational force acts on the gymnast in the air, it is the net force, and we can evaluate the work by using the relation $W_{gravity} = mg(h_0 - h_f)$.

Solution Figure 6.12*b* shows the gymnast moving upward. The initial and final heights are $h_0 = 1.20$ m and $h_f = 4.80$ m, respectively. The initial speed is v_0 and the final speed is $v_f = 0$ m/s, since the gymnast comes to a momentary halt at the highest point. Since $v_f = 0$ m/s, the final kinetic energy is $KE_f = 0$ J, and the work–energy theorem becomes $W = KE_f - KE_0 = -KE_0$. The work W is that due to gravity, so this theorem reduces to $W_{gravity} = mg(h_0 - h_f) = -\frac{1}{2}mv_0^2$. Solving for v_0 gives

$$v_0 = \sqrt{-2g(h_0 - h_f)} = \sqrt{-2(9.80 \text{ m/s}^2)(1.20 \text{ m} - 4.80 \text{ m})} = \boxed{8.40 \text{ m/s}}$$

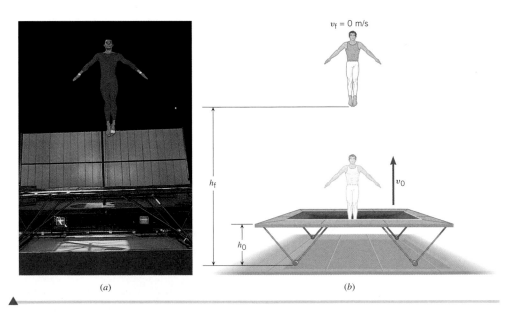

Figure 6.12 (*a*) A gymnast bounces on a trampoline. (©James D. Wilson Liaison/Getty Images) (*b*) The gymnast moves upward with an initial speed v_0 and reaches maximum height with a final speed of zero.

(*a*)   (*b*)

GRAVITATIONAL POTENTIAL ENERGY

We have seen that an object in motion has kinetic energy. There are also other types of energy. For example, an object may possess energy by virtue of its position relative to the earth; such an object is said to have gravitational potential energy. A pile driver, for instance, is used by construction workers to pound "piles," or structural support beams, into the ground. The pile driver contains a massive hammer that is raised to a height h and then dropped (see Figure 6.13). As a result, the hammer has the potential to do the work of driving the pile into the ground. The greater the height of the hammer, the greater is the potential for doing work, and the greater is the gravitational potential energy.

Now, let's obtain an expression for the gravitational potential energy. Our starting point is Equation 6.4 for the work done by the gravitational force as an object moves from an initial height h_0 to a final height h_f:

$$W_{\text{gravity}} = \underbrace{mgh_0}_{\substack{\text{Initial} \\ \text{gravitational} \\ \text{potential energy} \\ \text{PE}_0}} - \underbrace{mgh_f}_{\substack{\text{Final} \\ \text{gravitational} \\ \text{potential energy} \\ \text{PE}_f}} \qquad (6.4)$$

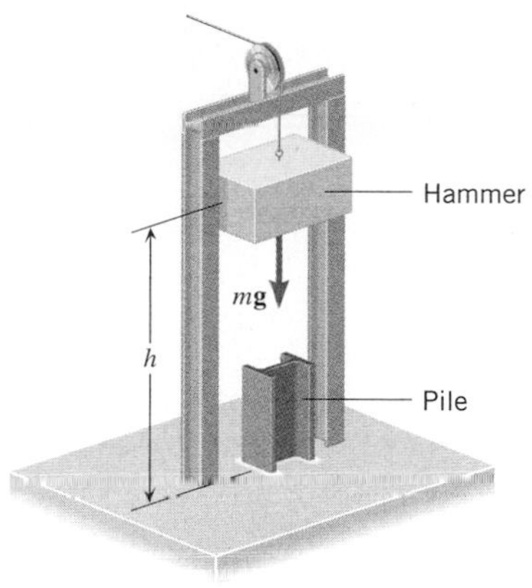

This equation indicates that the work done by the gravitational force is equal to the difference between the initial and final values of the quantity mgh. The value of mgh is larger when the height is larger and smaller when the height is smaller. We are led, then, to identify the quantity mgh as the **gravitational potential energy.** The concept of potential energy is associated only with a type of force known as a "conservative" force, as we will discuss in Section 6.4.

Figure 6.13 In a pile driver, the gravitational potential energy of the hammer relative to the ground is $PE = mgh$.

■ **DEFINITION OF GRAVITATIONAL POTENTIAL ENERGY**

The gravitational potential energy PE is the energy that an object of mass m has by virtue of its position relative to the surface of the earth. That position is measured by the height h of the object relative to an arbitrary zero level:

$$PE = mgh \qquad (6.5)$$

SI Unit of Gravitational Potential Energy: joule (J)

Gravitational potential energy, like work and kinetic energy, is a scalar quantity and has the same SI unit as they do, the joule. It is the *difference* between two potential energies that is related by Equation 6.4 to the work done by the force of gravity. Therefore, the zero level for the heights can be taken anywhere, as long as both h_0 and h_f are measured relative to the same zero level. The gravitational potential energy depends on both the object and the earth (m and g, respectively), as well as the height h. Therefore, the gravitational potential energy belongs to the object and the earth as a system, although one often speaks of the object alone as possessing the gravitational potential energy.

6.4 Conservative Versus Nonconservative Forces

The gravitational force has an interesting property that when an object is moved from one place to another, the work done by the gravitational force does not depend on the choice of path. In Figure 6.11, for instance, an object moves from an initial height h_0 to a final height h_f along two different paths. As Section 6.3 discusses, the work done by gravity depends only on the initial and final heights, and not on the path between these heights. For this reason, the gravitational force is called a **conservative force,** according to version 1 of the following definition:

■ **DEFINITION OF A CONSERVATIVE FORCE**

Version 1 A force is conservative when the work it does on a moving object is independent of the path between the object's initial and final positions.
Version 2 A force is conservative when it does no net work on an object moving around a closed path, starting and finishing at the same point.

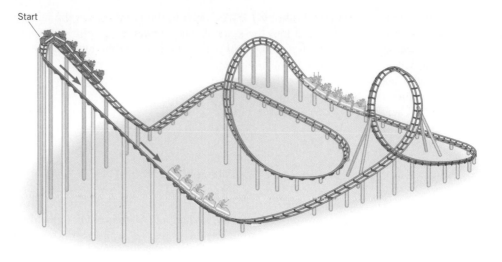

Start

Figure 6.14 A roller coaster track is an example of a closed path.

Table 6.2 *Some Conservative and Nonconservative Forces*

Conservative Forces
  Gravitational force (Ch. 4)
  Elastic spring force (Ch. 10)
  Electric force (Ch. 18, 19)

Nonconservative Forces
  Static and kinetic frictional forces
  Air resistance
  Tension
  Normal force
  Propulsion force of a rocket

The gravitational force is our first example of a conservative force. Later, we will encounter others, such as the elastic force of a spring and the electrical force of electrically charged particles. With each conservative force we will associate a potential energy, as we have already done in the gravitational case (see Equation 6.5). For other conservative forces, however, the algebraic form of the potential energy will differ from that in Equation 6.5.

Figure 6.14 helps us to illustrate version 2 of the definition of a conservative force. The picture shows a roller coaster car racing through dips and double dips, ultimately returning to its starting point. This kind of path, which begins and ends at the same place, is called a *closed* path. Gravity provides the only force that does work on the car, assuming that there is no friction or air resistance. Of course, the track exerts a normal force, but this force is always directed perpendicular to the motion and does no work. On the downward parts of the trip, the gravitational force does positive work, increasing the car's kinetic energy. Conversely, on the upward parts of the motion, the gravitational force does negative work, decreasing the car's kinetic energy. Over the entire trip, the gravitational force does as much positive work as negative work, so the net work is zero, and the car returns to its starting point with the same kinetic energy it had at the start. Therefore, consistent with version 2 of the definition of a conservative force, $W_{\text{gravity}} = 0$ J for a closed path.

Not all forces are conservative forces. A force is nonconservative if the work it does on an object moving between two points depends on the path of the motion between the points. The kinetic frictional force is one example of a nonconservative force. When an object slides over a surface, the kinetic frictional force points opposite to the sliding motion and does negative work. Between any two points, greater amounts of work are done over longer paths between the points, so that the work depends on the choice of path. Thus, the kinetic frictional force is nonconservative. Air resistance is another nonconservative force. The concept of potential energy is not defined for a nonconservative force.

For a closed path, the total work done by a nonconservative force is not zero as it is for a conservative force. In Figure 6.14, for instance, a frictional force would oppose the motion and slow down the car. Unlike gravity, friction would do negative work on the car throughout the entire trip, on *both* the up and down parts of the motion. Assuming that the car makes it back to the starting point, the car would have *less* kinetic energy than it had originally. Table 6.2 gives some examples of conservative and nonconservative forces.

In normal situations, conservative forces (such as gravity) and nonconservative forces (such as friction and air resistance) act simultaneously on an object. Therefore, we write the work W done by the net external force as $W = W_{\text{c}} + W_{\text{nc}}$, where W_{c} is the work done by the conservative forces and W_{nc} is the work done by the nonconservative forces. According to the work–energy theorem, the work done by the net external force is equal to the change in the object's kinetic energy, or $W_c + W_{nc} = \frac{1}{2}mv_f^2 - \frac{1}{2}mv_0^2$. If the only conservative force acting is the gravitational force, then $W_c = W_{gravity} = mg(h_0 - h_f)$, and the work–energy theorem becomes

$$mg(h_0 - h_f) + W_{nc} = \frac{1}{2}mv_f^2 - \frac{1}{2}mv_0^2$$

The work done by the gravitational force can be moved to the right side of this equation, with the result that

$$W_{nc} = (\tfrac{1}{2}mv_f^2 - \tfrac{1}{2}mv_0^2) + (mgh_f - mgh_0) \qquad (6.6)$$

In terms of kinetic and potential energies, we find that

$$\underbrace{W_{nc}}_{\substack{\text{Net work done by} \\ \text{nonconservative} \\ \text{forces}}} = \underbrace{(KE_f - KE_0)}_{\substack{\text{Change in} \\ \text{kinetic energy}}} + \underbrace{(PE_f - PE_0)}_{\substack{\text{Change in} \\ \text{gravitational} \\ \text{potential energy}}} \qquad (6.7a)$$

Equation 6.7a states that the net work W_{nc} done by all the external nonconservative forces equals the change in the object's kinetic energy plus the change in its gravitational potential energy. It is customary to use the delta symbol (Δ) to denote such changes; thus, $\Delta KE = (KE_f - KE_0)$ and $\Delta PE = (PE_f - PE_0)$. With the delta notation, the work–energy theorem takes the form

Problem solving insight
The change in both the kinetic and the potential energy is always the final value minus the initial value: $\Delta KE = KE_f - KE_0$ and $\Delta PE = PE_f - PE_0$.

$$W_{nc} = \Delta KE + \Delta PE \qquad (6.7b)$$

In the next two sections, we will show why the form of the work–energy theorem expressed by Equations 6.7a and 6.7b is useful.

6.5 *The Conservation of Mechanical Energy*

▶ CONCEPTS AT A GLANCE The concept of work and the work–energy theorem have led us to the conclusion that an object can possess two kinds of energy: kinetic energy, KE, and gravitational potential energy, PE. The sum of these two energies is called the ***total mechanical energy*** E, so that $E = KE + PE$. The Concepts-at-a-Glance chart in Figure 6.15 illustrates this summation. The concept of total mechanical energy will be extremely useful in describing the motion of objects in this and other chapters. Later on, in a number of places, we will update this chart to include other forms of energy. ◀

Figure 6.15
CONCEPTS AT A GLANCE
The total mechanical energy E is formed by combining the concepts of kinetic energy and gravitational potential energy. This soccer ball has both types of energy as it sails through the air. (©Nathan Bilow/Allsport/Getty Images)

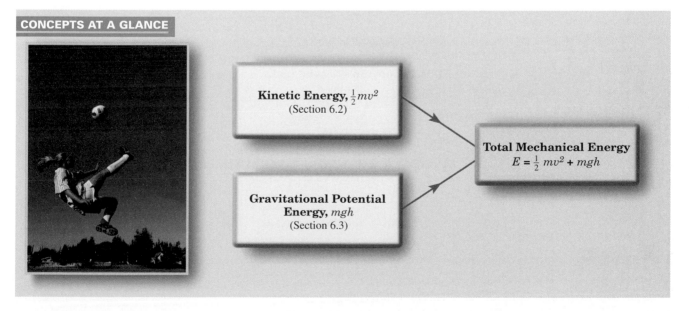

CONCEPTS AT A GLANCE

Kinetic Energy, $\frac{1}{2}mv^2$
(Section 6.2)

Gravitational Potential Energy, mgh
(Section 6.3)

Total Mechanical Energy
$E = \frac{1}{2}mv^2 + mgh$

By rearranging the terms on the right side of Equation 6.7a, the work–energy theorem can be expressed in terms of the total mechanical energy:

$$W_{nc} = (KE_f - KE_0) + (PE_f - PE_0) \tag{6.7a}$$

$$= \underbrace{(KE_f + PE_f)}_{E_f} - \underbrace{(KE_0 + PE_0)}_{E_0}$$

or

$$W_{nc} = E_f - E_0 \tag{6.8}$$

Remember: Equation 6.8 is just another form of the work–energy theorem. It states that W_{nc}, the net work done by external nonconservative forces, changes the total mechanical energy from an initial value of E_0 to a final value of E_f.

The conciseness of the work–energy theorem in the form $W_{nc} = E_f - E_0$ allows an important basic principle of physics to stand out. This principle is known as the conservation of mechanical energy. Suppose that the net work W_{nc} done by external nonconservative forces is zero, so $W_{nc} = 0$ J. Then, Equation 6.8 reduces to

$$E_f = E_0 \tag{6.9a}$$

$$\underbrace{\tfrac{1}{2}mv_f^2 + mgh_f}_{E_f} = \underbrace{\tfrac{1}{2}mv_0^2 + mgh_0}_{E_0} \tag{6.9b}$$

Equation 6.9a indicates that the final mechanical energy is equal to the initial mechanical energy. Consequently, the total mechanical energy *remains constant all along the path between the initial and final points,* never varying from the initial value of E_0. A quantity that remains constant throughout the motion is said to be "conserved." The fact that the total mechanical energy is conserved when $W_{nc} = 0$ J is called the ***principle of conservation of mechanical energy.***

Figure 6.16

CONCEPTS AT A GLANCE
The work–energy theorem leads to the principle of conservation of mechanical energy under circumstances in which $W_{nc} = 0$ J, where W_{nc} is the net work done by external nonconservative forces. To the extent that air resistance, a nonconservative force, can be ignored, the total mechanical energy of a skydiver is conserved as he or she falls toward the earth. (©Peter Mason/The Image Bank/Getty Images)

> ■ **THE PRINCIPLE OF CONSERVATION OF MECHANICAL ENERGY**
>
> The total mechanical energy ($E = KE + PE$) of an object remains constant as the object moves, provided that the net work done by external nonconservative forces is zero, $W_{nc} = 0$ J.

▶ **CONCEPTS AT A GLANCE** The Concepts-at-a-Glance chart in Figure 6.16 outlines how the conservation of mechanical energy arises naturally from concepts that we have encountered earlier. You might recognize that this chart is an expanded version of the one

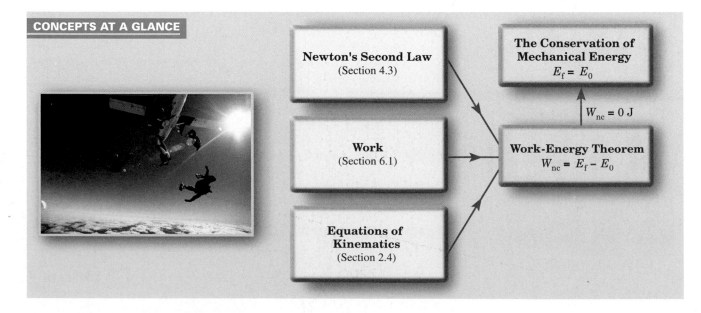

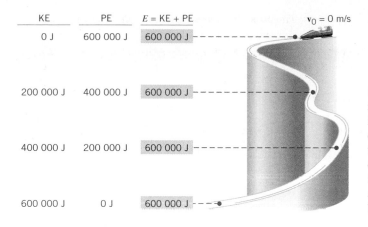

KE	PE	$E = KE + PE$	
			$v_0 = 0$ m/s
0 J	600 000 J	600 000 J	
200 000 J	400 000 J	600 000 J	
400 000 J	200 000 J	600 000 J	
600 000 J	0 J	600 000 J	

Figure 6.17 If friction and wind resistance are ignored, a bobsled run illustrates how kinetic and potential energy can be interconverted, while the total mechanical energy remains constant. The total mechanical energy is 600 000 J, being all potential energy at the top and all kinetic energy at the bottom.

in Figure 6.5. In the middle are the three concepts that we used to develop the work–energy theorem: Newton's second law, work, and the equations of kinematics. On the right side of the chart is the work–energy theorem, stated in the form $W_{nc} = E_f - E_0$. If, as the chart shows, $W_{nc} = 0$ J, then the final and initial total mechanical energies are equal. In other words, $E_f = E_0$, which is the mathematical statement of the conservation of mechanical energy. ◄

The principle of conservation of mechanical energy offers keen insight into the way in which the physical universe operates. While the sum of the kinetic and potential energies at any point is conserved, the two forms may be interconverted or transformed into one another. Kinetic energy of motion is converted into potential energy of position, for instance, when a moving object coasts up a hill. Conversely, potential energy is converted into kinetic energy when an object is allowed to fall. Figure 6.17 illustrates such transformations of energy for a bobsled run, assuming that nonconservative forces, such as friction and wind resistance, can be ignored. The normal force, being directed perpendicular to the path, does no work. Only the force of gravity does work, so the total mechanical energy E remains constant at all points along the run. The conservation principle is well known for the ease with which it can be applied, as in the following examples.

Example 8 A Daredevil Motorcyclist

A motorcyclist is trying to leap across the canyon shown in Figure 6.18 by driving horizontally off the cliff at a speed of 38.0 m/s. Ignoring air resistance, find the speed with which the cycle strikes the ground on the other side.

Reasoning Once the cycle leaves the cliff, no forces other than gravity act on the cycle, since air resistance is being ignored. Thus, the work done by external nonconservative forces is zero, $W_{nc} = 0$ J. Accordingly, the principle of conservation of mechanical energy holds, so the total mechanical energy is the same at the final and initial positions of the motorcycle. We will use this important observation to determine the final speed of the cyclist.

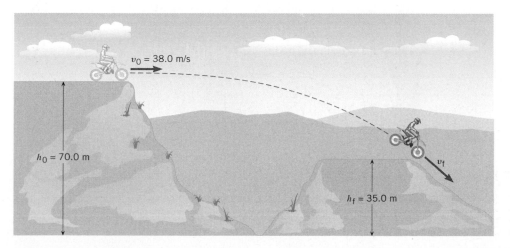

$v_0 = 38.0$ m/s

$h_0 = 70.0$ m

$h_f = 35.0$ m

v_f

Figure 6.18 A daredevil jumping a canyon.

Solution The principle of conservation of mechanical energy is written as

$$\underbrace{\tfrac{1}{2}mv_f^2 + mgh_f}_{E_f} = \underbrace{\tfrac{1}{2}mv_0^2 + mgh_0}_{E_0} \qquad (6.9b)$$

Problem solving insight
Be on the alert for factors, such as the mass *m* here in Example 8, that sometimes can be eliminated algebraically when using the conservation of mechanical energy.

The mass *m* of the rider and cycle can be eliminated algebraically from this equation, since *m* appears as a factor in every term. Solving for v_f gives

$$v_f = \sqrt{v_0^2 + 2g(h_0 - h_f)}$$

$$v_f = \sqrt{(38.0 \text{ m/s})^2 + 2(9.80 \text{ m/s}^2)(70.0 \text{ m} - 35.0 \text{ m})} = \boxed{46.2 \text{ m/s}}$$

Examples 9 and 10 emphasize that the principle of conservation of mechanical energy can be applied even when forces act perpendicular to the path of a moving object.

✔ **Check Your Understanding 3**

Some of the following situations are consistent with the principle of conservation of mechanical energy, and some are not. Which ones are consistent with the principle? (a) An object moves uphill with an increasing speed. (b) An object moves uphill with a decreasing speed. (c) An object moves uphill with a constant speed. (d) An object moves downhill with an increasing speed. (e) An object moves downhill with a decreasing speed. (f) An object moves downhill with a constant speed. *(The answers are given at the end of the book.)*

Background: Kinetic energy, gravitational potential energy, and the principle of conservation of mechanical energy play roles in this question. Consider what the conservation principle implies about the way in which kinetic and potential energies change.

For similar questions (including calculational counterparts), consult Self-Assessment Test 6.2. This test is described at the end of this section.

Conceptual Example 9   The Favorite Swimming Hole

A rope is tied to a tree limb and used by a swimmer to swing into the water below. The person starts from rest with the rope held in the horizontal position, as in Figure 6.19, swings downward, and then lets go of the rope. Three forces act on him: his weight, the tension in the rope, and the force due to air resistance. His initial height h_0 and final height h_f are known. Considering the nature of these forces, conservative versus nonconservative, can we use the principle of conservation of mechanical energy to find his speed v_f at the point where he lets go of the rope?

Reasoning and Solution The principle of conservation of mechanical energy can be used only if the net work W_{nc} done by nonconservative forces is zero, $W_{nc} = 0$ J. The tension and the force due to air resistance are nonconservative forces (see Table 6.2), so we need to inquire whether the net work done by these forces is zero. The tension **T** is always perpendicular to the circular path of the motion, as shown in the drawing. Thus, the angle θ between the tension and the displacement is 90°. According to Equation 6.1, the work depends on the cosine of this angle, or cos 90°, which is zero. Thus, the work done by the tension is zero. On the other hand, the force due to air resistance is directed opposite to the motion of the swinging person.

Problem solving insight
When nonconservative forces are perpendicular to the motion, we can still use the principle of conservation of mechanical energy, because such "perpendicular" forces do no work.

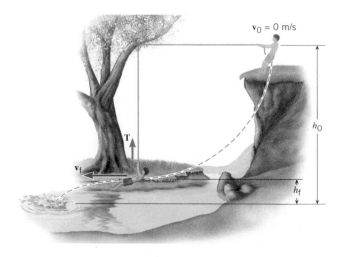

Figure 6.19 During the downward swing, the tension **T** in the rope acts perpendicular to the circular arc and, hence, does no work on the person.

The angle θ between this force and the displacement is 180°. The work done by the air resistance is not zero, because the cosine of 180° is not zero. Since the net work done by the two nonconservative forces is not zero, $W_{\text{net}} \neq 0$ J, we cannot use the principle of conservation of mechanical energy to determine the final speed. On the other hand, if the force due to air resistance is very small, then the work done by this force is negligible. In this case, the net work done by the nonconservative forces is effectively zero, and the principle of conservation of mechanical energy can be used to determine the final speed of the swimmer.

Related Homework: *Problem 43*

Need more practice?

Interactive LearningWare 6.3
A person is standing on the edge of a cliff that is 8.10 m above a lake. Using a rope tied to a tree limb, he swings downward, lets go of the rope, and subsequently splashes into the water. He can let go of the rope either as it is swinging downward or a little later as it is swinging upward. Ignore air resistance and, for each case, determine the speed of the person just before entering the water.

Related Homework: *Problem 32*

Go to **www.wiley.com/college/cutnell** for an interactive solution.

The next example illustrates how the conservation of mechanical energy is applied to the breathtaking drop of a roller coaster.

Example 10   The Steel Dragon

The tallest and fastest roller coaster in the world is now the Steel Dragon in Mie, Japan (Figure 6.20). The ride includes a vertical drop of 93.5 m. The coaster has a speed of 3.0 m/s at the top of the drop. Neglect friction and find the speed of the riders at the bottom.

The physics of a giant roller coaster.

Reasoning Since we are neglecting friction, we may set the work done by the frictional force equal to zero. A normal force from the seat acts on each rider, but this force is perpendicular to the motion, so it does not do any work. Thus, the work done by external nonconservative forces is zero, $W_{\text{nc}} = 0$ J, and we may use the principle of conservation of mechanical energy to find the speed of the riders at the bottom.

Solution The principle of conservation of mechanical energy states that

$$\underbrace{\tfrac{1}{2}mv_{\text{f}}^2 + mgh_{\text{f}}}_{E_{\text{f}}} = \underbrace{\tfrac{1}{2}mv_0^2 + mgh_0}_{E_0} \tag{6.9b}$$

The mass m of the rider appears as a factor in every term in this equation and can be eliminated algebraically. Solving for the final speed gives

$$v_{\text{f}} = \sqrt{v_0^2 + 2g(h_0 - h_{\text{f}})}$$

$$v_{\text{f}} = \sqrt{(3.0 \text{ m/s})^2 + 2(9.80 \text{ m/s})(93.5 \text{ m})} = \boxed{42.9 \text{ m/s (about 96 mi/h)}}$$

where the vertical drop is $h_0 - h_{\text{f}} = 93.5$ m.

When applying the principle of conservation of mechanical energy to solving problems, we have been using the following reasoning strategy:

Reasoning Strategy

Applying the Principle of Conservation of Mechanical Energy

1. Identify the external conservative and nonconservative forces that act on the object. For this principle to apply, the total work done by nonconservative forces must be zero, $W_{\text{nc}} = 0$ J. A nonconservative force that is perpendicular to the displacement of the object does no work, for example.

Figure 6.20 The Steel Dragon roller coaster in Mie, Japan, is a giant. It includes a vertical drop of 93.5 m. (©AFP/Corbis Images)

2. Choose the location where the gravitational potential energy is taken to be zero. This location is arbitrary but must not be changed during the course of solving a problem.

3. Set the final total mechanical energy of the object equal to the initial total mechanical energy, as in Equations 6.9a and 6.9b. The total mechanical energy is the sum of the kinetic and potential energies.

Self-Assessment Test 6.2

Test your understanding of the material in Sections 6.3–6.5:

• Gravitational Potential Energy   • Conservative and Nonconservative Forces
• Conservation of Mechanical Energy

Go to **www.wiley.com/college/cutnell**

6.6 *Nonconservative Forces and the Work–Energy Theorem*

Most moving objects experience nonconservative forces, such as friction, air resistance, and propulsive forces, and the work W_{nc} done by the net external nonconservative force is not zero. In these situations, the difference between the final and initial total mechanical energies is equal to W_{nc}, according to $W_{nc} = E_f - E_0$ (Equation 6.8). Consequently, the total mechanical energy is not conserved. The next two examples illustrate how Equation 6.8 is used when nonconservative forces are present and do work.

Example 11 **The Steel Dragon, Revisited**

In Example 10, we ignored nonconservative forces, such as friction. In reality, however, such forces are present when the roller coaster descends. The actual speed of the riders at the bottom is 41.0 m/s, which is less than that determined in Example 10. Assuming again that the coaster has a speed of 3.0 m/s at the top, find the work done by nonconservative forces on a 55.0-kg rider during the descent from a height h_0 to a height h_f, where $h_0 - h_f = 93.5$ m.

Reasoning Since the speed at the top, the final speed, and the vertical drop are given, we can determine the initial and final total mechanical energies of the rider. The work–energy theorem, $W_{nc} = E_f - E_0$, can then be used to determine the work W_{nc} done by the nonconservative forces.

Solution The work–energy theorem is

$$W_{nc} = \underbrace{(\tfrac{1}{2}mv_f^2 + mgh_f)}_{E_f} - \underbrace{(\tfrac{1}{2}mv_0^2 + mgh_0)}_{E_0} \qquad (6.8)$$

Problem solving insight
As illustrated here and in Example 3, a nonconservative force such as friction can do negative or positive work. It does negative work when it has a component opposite to the displacement and slows down the object. It does positive work when it has a component in the direction of the displacement and speeds up the object.

Rearranging the terms on the right side of this equation gives

$$W_{nc} = \tfrac{1}{2}m(v_f^2 - v_0^2) - mg(h_0 - h_f)$$

$$W_{nc} = \tfrac{1}{2}(55.0\text{ kg})[(41.0\text{ m/s})^2 - (3.0\text{ m/s})^2] - (55.0\text{ kg})(9.80\text{ m/s}^2)(93.5\text{ m}) = \boxed{-4400\text{ J}}$$

Example 12 **Fireworks**

A 0.20-kg rocket in a fireworks display is launched from rest and follows an erratic flight path to reach the point *P*, as Figure 6.21 shows. Point *P* is 29 m above the starting point. In the process, 425 J of work is done on the rocket by the nonconservative force generated by the burning propellant. Ignoring air resistance and the mass lost due to the burning propellant, find the speed v_f of the rocket at the point *P*.

Reasoning The only nonconservative force acting on the rocket is the force generated by the burning propellant, and the work done by this force is $W_{nc} = 425$ J. Because work is done by a nonconservative force, we use the work–energy theorem in the form $W_{nc} = E_f - E_0$ to find the final speed v_f of the rocket.

Solution From the work–energy theorem we have

$$W_{nc} = (\tfrac{1}{2}mv_f^2 + mgh_f) - (\tfrac{1}{2}mv_0^2 + mgh_0) \tag{6.8}$$

Solving this expression for the final speed of the rocket, we get

$$v_f = \sqrt{\frac{2[W_{nc} + \tfrac{1}{2}mv_0^2 - mg(h_f - h_0)]}{m}}$$

$$v_f = \sqrt{\frac{2[425 \text{ J} + \tfrac{1}{2}(0.20 \text{ kg})(0 \text{ m/s})^2 - (0.20 \text{ kg})(9.80 \text{ m/s}^2)(29 \text{ m})]}{0.20 \text{ kg}}} = \boxed{61 \text{ m/s}}$$

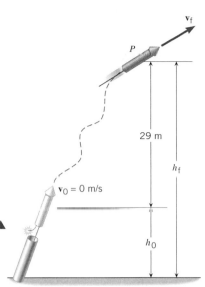

Figure 6.21 A fireworks rocket, moving along an erratic flight path, reaches a point P that is 29 m above the launch point.

6.7 Power

In many situations, the time it takes to do work is just as important as the amount of work that is done. Consider two automobiles that are identical in all respects (e.g., same mass), except that one has a "souped-up" engine. The car with the "souped-up" engine can go from 0 to 27 m/s (60 mph) in 4 seconds, while the other car requires 8 seconds to achieve the same speed. Each engine does work in accelerating its car, but one does it more quickly. Where cars are concerned, we associate the quicker performance with an engine that has a larger horsepower rating. A large horsepower rating means that the engine can do a large amount of work in a short time. In physics, the horsepower rating is just one way to measure an engine's ability to generate power. The idea of *power* incorporates both the concepts of work and time, for power is work done per unit time.

■ **DEFINITION OF AVERAGE POWER**

Average power $\overline{P}$ is the average rate at which work W is done, and it is obtained by dividing W by the time t required to perform the work:

$$\overline{P} = \frac{\text{Work}}{\text{Time}} = \frac{W}{t} \tag{6.10a}$$

SI Unit of Power: joule/s = watt (W)

✔ **Check Your Understanding 4**

Engine A has a greater power rating than engine B. Which one of the following statements correctly describes the abilities of these engines to do work? (a) Engines A and B can do the same amount of work in the same amount of time. (b) In the same amount of time, engine B can do more work than engine A. (c) Engines A and B can do the same amount of work, but engine A can do it more quickly. *(The answer is given at the end of the book.)*

Background: Work, time, and power are the key concepts. Remember that work and power are not the same thing.

For similar questions (including calculational counterparts), consult Self-Assessment Test 6.3. This test is described at the end of Section 6.9.

The definition of average power presented in Equation 6.10a involves the work that is done by the net force. However, the work–energy theorem relates this work to the change in the energy of the object (see, for example, Equations 6.3 and 6.8). Therefore, we can also define average power as the rate at which the energy is changing, or as the change in

Table 6.3 *Units of Measurement for Power*

System	Work	÷	Time	=	Power
SI	joule (J)		second (s)		watt (W)
CGS	erg		second (s)		erg per second (erg/s)
BE	foot · pound (ft · lb)		second (s)		foot · pound per second (ft · lb/s)

energy divided by the time during which the change occurs:

$$\overline{P} = \frac{\text{Change in energy}}{\text{Time}} \qquad (6.10b)$$

Since work, energy, and time are scalar quantities, power is also a scalar quantity. The unit in which power is expressed is that of work divided by time, or a joule per second in SI units. One joule per second is called a watt (W), in honor of James Watt (1736–1819), the developer of the steam engine. The unit of power in the BE system is the foot-pound per second (ft·lb/s), although the familiar horsepower (hp) unit is used frequently for specifying the power generated by electric motors and internal combustion engines:

$$1 \text{ horsepower} = 550 \text{ foot·pounds/second} = 745.7 \text{ watts}$$

Table 6.3 summarizes the units for power in the various systems of measurement.

Equation 6.10b provides the basis for understanding the production of power in the human body. In this context the "Change in energy" on the right-hand side of the equation refers to the energy produced by metabolic processes, which, in turn, is derived from the food we eat. Table 6.4 gives typical metabolic rates of energy production needed to sustain various activities. Running at 15 km/h (9.3 mi/h), for example, requires metabolic power sufficient to operate eighteen 75-watt light bulbs, and the metabolic power used in sleeping would operate a single 75-watt bulb.

An alternative expression for power can be obtained from Equation 6.1, which indicates that the work W done when a constant net force of magnitude F points in the same direction as the displacement is $W = (F \cos 0°)s = Fs$. Dividing both sides of this equation by the time t it takes for the force to move the object through the distance s, we obtain

$$\frac{W}{t} = \frac{Fs}{t}$$

But W/t is the average power $\overline{P}$, and s/t is the average speed $\overline{v}$, so that

$$\overline{P} = F\overline{v} \qquad (6.11)$$

The next example illustrates the use of Equation 6.11.

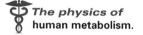

The physics of human metabolism.

Table 6.4 *Human Metabolic Rates[a]*

Activity	Rate (watts)
Running (15 km/h)	1340 W
Skiing	1050 W
Biking	530 W
Walking (5 km/h)	280 W
Sleeping	77 W

[a]For a young 70-kg male.

Example 13 The Power to Accelerate a Car

A 1.10×10^3-kg car, starting from rest, accelerates for 5.00 s. The magnitude of the acceleration is $a = 4.60 \text{ m/s}^2$. Determine the average power generated by the net force that accelerates the vehicle.

Reasoning We can find the average power by using the relation $\overline{P} = F\overline{v}$, provided the magnitude F of the net force acting on the car and the average speed $\overline{v}$ of the car can be determined. The net force can be obtained from Newton's second law, and the average speed can be calculated from the equations of kinematics.

Solution According to Newton's second law, the magnitude of the net force is

$$F = ma = (1.10 \times 10^3 \text{ kg})(4.60 \text{ m/s}^2) = 5060 \text{ N}$$

Since the car starts from rest ($v_0 = 0$ m/s) and has a constant acceleration, the average speed $\bar{v}$ of the car is one-half of its final speed $\bar{v}_f$:

$$v = \tfrac{1}{2}(v_0 + v_f) = \tfrac{1}{2}v_f \qquad (2.6)$$

Because the initial speed of the car is zero, the final speed of the car after 5.00 s is the product of its acceleration and time:

$$v_f = v_0 + at = (0 \text{ m/s}) + (4.60 \text{ m/s}^2)(5.00 \text{ s}) = 23.0 \text{ m/s} \qquad (2.4)$$

Thus, the average speed is $\bar{v} = 11.5$ m/s, and the average power is

$$\overline{P} = F\bar{v} = (5060 \text{ N})(11.5 \text{ m/s}) = \boxed{5.82 \times 10^4 \text{ W } (78.0 \text{ hp})}$$

6.8 Other Forms of Energy and the Conservation of Energy

Up to now, we have considered only two types of energy, kinetic energy and gravitational potential energy. There are many other types, however. Electrical energy is used to run electrical appliances. Energy in the form of heat is utilized in cooking food. Moreover, the work done by the kinetic frictional force often appears as heat, as you can experience by rubbing your hands back and forth. When gasoline is burned, some of the stored chemical energy is released and does the work of moving cars, airplanes, and boats. The chemical energy stored in food provides the energy needed for metabolic processes.

One of the most controversial forms of energy is nuclear energy. The research of many scientists, most notably Albert Einstein, led to the discovery that mass itself is one manifestation of energy. Einstein's famous equation, $E_0 = mc^2$, describes how mass m and energy E_0 are related, where c is the speed of light in a vacuum and has a value of 3.00×10^8 m/s. Because the speed of light is so large, this equation implies that very small masses are equivalent to large amounts of energy. The relationship between mass and energy will be discussed further in Chapter 28.

We have seen that kinetic energy can be converted into gravitational potential energy and vice versa. In general, energy of all types can be converted from one form to another. Part of the chemical energy stored in food, for example, is transformed into gravitational potential energy when a hiker climbs a mountain. Suppose a 65-kg hiker eats a 250-Calorie* snack, which contains 1.0×10^6 J of chemical energy. If this were 100% converted into potential energy $mg(h_f - h_0)$, the change in height would be

The physics of transforming chemical energy in food into mechanical energy.

$$h_f - h_0 = \frac{1.0 \times 10^6 \text{ J}}{(65 \text{ kg})(9.8 \text{ m/s}^2)} = 1600 \text{ m}$$

At a more realistic conversion efficiency of 50%, the change in height would be 800 m. Similarly, in a moving car the chemical energy of gasoline is converted into kinetic energy, as well as into electrical energy and heat.

Whenever energy is transformed from one form to another, it is found that no energy is gained or lost in the process; the total of all the energies before the process is equal to the total of the energies after the process. This observation leads to the following important principle:

> ■ **THE PRINCIPLE OF CONSERVATION OF ENERGY**
>
> Energy can neither be created nor destroyed, but can only be converted from one form to another.

---

* Energy content in food is typically given in units called Calories, which we will discuss in Section 12.7.

Learning how to convert energy from one form to another more efficiently is one of the main goals of modern science and technology.

6.9 Work Done by a Variable Force

The physics of the compound bow.

The work W done by a constant force (constant in both magnitude and direction) is given by Equation 6.1 as $W = (F\cos\theta)s$. Quite often, situations arise in which the force is not constant but changes with the displacement of the object. For instance, Figure 6.22a shows an archer using a high-tech compound bow. This type of bow consists of a series of pulleys and strings that produce a force-versus-displacement graph like that in Figure 6.22b. One of the key features of the compound bow is that the force rises to a maximum as the string is drawn back, and then falls to 60% of this maximum value when the string is fully drawn. The reduced force at $s = 0.500$ m makes it much easier for the archer to hold the fully drawn bow while aiming the arrow.

When the force varies with the displacement, as in Figure 6.22b, we cannot use the relation $W = (F\cos\theta)s$ to find the work, because this equation is valid only when the force is constant. However, we can use a graphical method. In this method we divide the total displacement into very small segments, Δs_1, Δs_2, and so on (see Figure 6.23a). For each segment, the *average value* of the force component is indicated by a short horizontal line. For example, the short horizontal line for segment Δs_1 is labeled $(F\cos\theta)_1$ in Figure 6.23a. We can then use this average value as the constant-force component in Equation 6.1 and determine an approximate value for the work ΔW_1 done during the first segment: $\Delta W_1 = (F\cos\theta)_1\Delta s_1$. But this work is just the area of the colored rectangle in the drawing. The word "area" here refers to the area of a rectangle that has a width of Δs_1 and a height of $(F\cos\theta)_1$; it does not mean an area in square meters, such as the area of a parcel of land. In a like manner, we can calculate an approximate value for the work for each segment. Then we add the results for the segments to get, approximately, the work W done by the variable force:

$$W \approx (F\cos\theta)_1\Delta s_1 + (F\cos\theta)_2\Delta s_2 + \cdots$$

The symbol "$\approx$" means "approximately equal to." The right side of this equation is the sum of all the rectangular areas in Figure 6.23a and is an approximate value for the area shaded in color under the graph in Figure 6.23b. If the rectangles are made narrower and narrower by decreasing each Δs, the right side of this equation eventually becomes equal to the area under the graph. Thus, we define the work done by a variable force as follows: *The work done by a variable force in moving an object is equal to the area under the graph of $F\cos\theta$ versus s.* Example 14 illustrates how to use this graphical method to determine the approximate work done when a high-tech compound bow is drawn.

Example 14 Work and the Compound Bow

Find the work that the archer must do in drawing back the string of the compound bow in Figure 6.22 from 0 to 0.500 m.

Reasoning The work is equal to the colored area under the curved line in Figure 6.22b. For convenience, this area is divided into a number of small squares, each having an area of

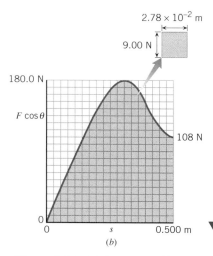

Figure 6.22 (a) A compound bow. (©Ron Chappel) (b) A plot of $F\cos\theta$ versus s as the bowstring is drawn back.

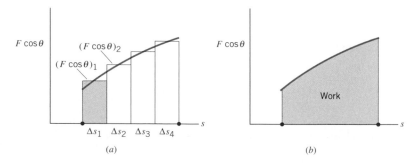

Figure 6.23 (a) The work done by the average-force component $(F\cos\theta)_1$ during the small displacement Δs_1 is $(F\cos\theta)_1\Delta s_1$, which is the area of the colored rectangle. (b) The work done by a variable force is equal to the colored area under the $F\cos\theta$-versus-s curve.

$(9.00 \text{ N})(2.78 \times 10^{-2} \text{ m}) = 0.250 \text{ J}$. The area can be found by counting the number of squares under the curve and multiplying by the area per square.

Solution We estimate that there are 242 colored squares in the drawing. Since each square represents 0.250 J of work, the total work done is

$$W = (242 \text{ squares})\left(0.250 \frac{\text{J}}{\text{square}}\right) = \boxed{60.5 \text{ J}}$$

When the arrow is fired, part of this work is imparted to it as kinetic energy.

Self-Assessment Test 6.3

Test your understanding of the material in Sections 6.6–6.9:

• Nonconservative Forces and the Work–Energy Theorem   • Power
• Work Done by a Variable Force

Go to **www.wiley.com/college/cutnell**

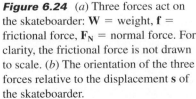

6.10 *Concepts & Calculations*

This section contains examples that discuss one or more conceptual questions, followed by a related quantitative problem. Example 15 reviews the important concept of work and illustrates how forces can give rise to positive, negative, and zero work. Example 16 examines the all-important conservation of mechanical energy and the work–energy theorem.

Concepts & Calculations Example 15   **Skateboarding and Work**

The skateboarder in Figure 6.24*a* is coasting down a ramp, and there are three forces acting on her: her weight **W** (magnitude = 675 N), a frictional force **f** (magnitude = 125 N) that opposes her motion, and a normal force **F**$_N$ (magnitude = 612 N). Determine the net work done by the three forces when she coasts for a distance of 9.2 m.

Concept Questions and Answers Figure 6.24*b* shows each force, along with the displacement **s** of the skateboarder. By examining these diagrams and without doing any numerical calculations, determine whether the work done by each force is positive, negative, or zero. Provide a reason for each answer.

Answer The work done by a force is positive if the force has a component that points in the same direction as the displacement. The work is negative if there is a force component pointing opposite to the displacement. The work done by the weight **W** is positive, be-

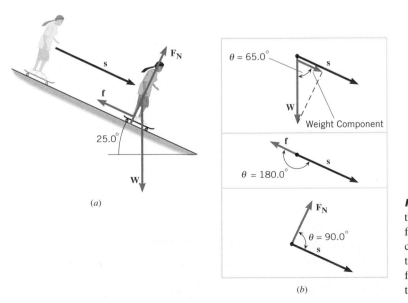

(a)

(b)

Figure 6.24 (*a*) Three forces act on the skateboarder: **W** = weight, **f** = frictional force, **F**$_N$ = normal force. For clarity, the frictional force is not drawn to scale. (*b*) The orientation of the three forces relative to the displacement **s** of the skateboarder.

cause the weight has a component that points in the same direction as the displacement. The top drawing in Figure 6.24*b* shows this weight component. The work done by the frictional force **f** is negative, because it points opposite to the direction of the displacement ($\theta = 180.0°$ in the middle drawing). The work done by the normal force $\mathbf{F_N}$ is zero, because the normal force is perpendicular to the displacement ($\theta = 90.0°$ in the bottom drawing) and does not have a component along the displacement.

Solution The work W done by a force is given by Equation 6.1 as $W = (F \cos \theta)s$, where F is the magnitude of the force, s is the magnitude of the displacement, and θ is the angle between the force and the displacement. Figure 6.24*b* shows this angle for each of the three forces. The work done by each force is computed in the table below. The net work is the algebraic sum of the three values.

Force	Magnitude F of the Force	Angle, θ	Magnitude s of the Displacement	Work Done by the Force $W = (F \cos \theta)\,s$
W	675 N	65.0°	9.2 m	$W = (675 \text{ N})(\cos 65.0°)$ $\times (9.2 \text{ m}) = +2620 \text{ J}$
f	125 N	180.0°	9.2 m	$W = (125 \text{ N})(\cos 180.0°)$ $\times (9.2 \text{ m}) = -1150 \text{ J}$
$\mathbf{F_N}$	612 N	90.0°	9.2 m	$W = (612 \text{ N})(\cos 90.0°)$ $\times (9.2 \text{ m}) = 0 \text{ J}$

The net work done by the three forces is

$$+2620 \text{ J} + (-1150 \text{ J}) + 0 \text{ J} = \boxed{+1470 \text{ J}}$$

Concepts & Calculations Example 16
Conservation of Mechanical Energy and the Work–Energy Theorem

Figure 6.25 shows a 0.41-kg block sliding from A to B along a frictionless surface. When the block reaches B, it continues to slide along the horizontal surface BC where a kinetic frictional force acts. As a result, the block slows down, coming to rest at C. The kinetic energy of the block at A is 37 J, and the heights of A and B are 12.0 and 7.0 m above the ground, respectively. (a) What is the kinetic energy of the block when it reaches B? (b) How much work does the kinetic frictional force do during the BC segment of the trip?

Concept Questions and Answers Is the total mechanical energy of the block conserved as the block goes from A to B? Why or why not?

Answer The total mechanical energy of the block is conserved if we can show that the net work done by the nonconservative forces is zero, or $W_{nc} = 0$ J (see Section 6.5). Only two forces act on the block during its trip from A to B: its weight and the normal force. The block's weight, as we have seen in Section 6.4, is a conservative force. The normal force, on the other hand, is a nonconservative force. However, it is always perpendicular to the displacement of the block, so it does no work. Thus, we conclude that $W_{nc} = 0$ J, with the result that the total mechanical energy is conserved during the AB part of the trip.

When the block reaches point B, has its kinetic energy increased, decreased, or remained the same relative to what it had at A? Provide a reason for your answer.

Answer As we have seen, the total mechanical energy is the sum of the kinetic and gravitational potential energies and remains constant during the trip. Therefore, as one type of energy decreases, the other must increase for the sum to remain constant. Since B

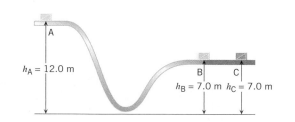

Figure 6.25 The block slides on a frictionless surface from A to B. From B to C, a kinetic frictional force slows down the block until it comes to rest at C.

is lower than A, the gravitational potential energy at B is less than that at A. As a result, the kinetic energy at B must be greater than that at A.

Is the total mechanical energy of the block conserved as the block goes from B to C? Justify your answer.

Answer During this part of the trip, a kinetic frictional force acts on the block. This force is nonconservative, and it does work on the block, just as the frictional force does work on the skateboarder in Example 15. Consequently, the net work W_{nc} done by nonconservative forces is not zero ($W_{nc} \neq 0$ J), so the total mechanical energy is not conserved during this part of the trip.

Solution

(a) Since the total mechanical energy is conserved during the AB segment, we can set the total mechanical energy at B equal to that at A.

$$\underbrace{KE_B + mgh_B}_{\substack{\text{Total mechanical} \\ \text{energy at B}}} = \underbrace{KE_A + mgh_A}_{\substack{\text{Total mechanical} \\ \text{energy at A}}}$$

Solving this equation for the kinetic energy at B gives

$$KE_B = KE_A + mg(h_A - h_B)$$
$$= 37 \text{ J} + (0.41 \text{ kg})(9.80 \text{ m/s}^2)(12.0 \text{ m} - 7.0 \text{ m}) = \boxed{57 \text{ J}}$$

As expected, the kinetic energy at B is greater than that at A.

(b) During the BC part of the trip, the total mechanical energy is not conserved because a kinetic frictional force is present. The work W_{nc} done by this nonconservative force is given by the work–energy theorem (see Equation 6.8) as the final total mechanical energy minus the initial total mechanical energy:

$$W_{nc} = \underbrace{KE_C + mgh_C}_{\substack{\text{Total mechanical} \\ \text{energy at C}}} - \underbrace{(KE_B + mgh_B)}_{\substack{\text{Total mechanical} \\ \text{energy at B}}}$$

Rearranging this equation gives

$$W_{nc} = \underbrace{KE_C}_{=0 \text{ J}} - KE_B + mg\underbrace{(h_C - h_B)}_{=0 \text{ m}}$$

In the preceding equation we have noted that the kinetic energy KE_C at C is equal to zero, because the block comes to rest at this point. The term ($h_C - h_B$) is also zero, because the two heights are the same. Thus, the work W_{nc} done by the kinetic frictional force during the BC part of the trip is

$$W_{nc} = -KE_B = \boxed{-57 \text{ J}}$$

The work done by the nonconservative frictional force is negative, because this force points opposite to the displacement of the block.

▲

At the end of the problem set for this chapter, you will find homework problems that contain both conceptual and quantitative parts. These problems are grouped under the heading *Concepts & Calculations, Group Learning Problems.* They are designed for use by students working alone or in small learning groups. The conceptual part of each problem provides a convenient focus for group discussions.

Concept Summary

This summary presents an abridged version of the chapter, including the important equations and all available learning aids. For convenient reference, the learning aids (including the text's examples) are placed next to or immediately after the relevant equation or discussion. The following learning aids may be found on-line at **www.wiley.com/college/cutnell**:

Interactive LearningWare examples are solved according to a five-step interactive format that is designed to help you develop problem-solving skills.	**Concept Simulations** are animated versions of text figures or animations that illustrate important concepts. You can control parameters that affect the display, and we encourage you to experiment.
Interactive Solutions offer specific models for certain types of problems in the chapter homework. The calculations are carried out interactively.	**Self-Assessment Tests** include both qualitative and quantitative questions. Extensive feedback is provided for both incorrect and correct answers, to help you evaluate your understanding of the material.

Topic	Discussion	Learning Aids
	6.1 Work Done by a Constant Force The work W done by a constant force acting on an object is	
Work done by constant force	$$W = (F \cos\theta)s \qquad (6.1)$$	Example 1
	where F is the magnitude of the force, s is the magnitude of the displacement, and θ is the angle between the force and the displacement vectors. Work is a scalar quantity and can be positive or negative, depending on whether the force has a component that points, respectively, in the same direction as the displacement or in the opposite direction. The work is zero if the force is perpendicular ($\theta = 90°$) to the displacement.	Examples 2, 3 Interactive LearningWare 6.1 Example 15
	6.2 The Work–Energy Theorem and Kinetic Energy The kinetic energy KE of an object of mass m and speed v is	
Kinetic energy	$$\mathrm{KE} = \tfrac{1}{2}mv^2 \qquad (6.2)$$	
	The work–energy theorem states that the work W done by the net external force acting on an object equals the difference between the object's final kinetic energy $\mathrm{KE_f}$ and initial kinetic energy $\mathrm{KE_0}$:	Examples 4, 5, 6 Concept Simulation 6.1
Work–energy theorem	$$W = \mathrm{KE_f} - \mathrm{KE_0} \qquad (6.3)$$	Interactive LearningWare 6.2
	The kinetic energy increases when the net force does positive work and decreases when the net force does negative work.	

Use Self-Assessment Test 6.1 to evaluate your understanding of Sections 6.1 and 6.2.

Topic	Discussion	Learning Aids
	6.3 Gravitational Potential Energy The work done by the force of gravity on an object of mass m is	
Work done by force of gravity	$$W_{\mathrm{gravity}} = mg(h_0 - h_f) \qquad (6.4)$$	Example 7
	where h_0 and h_f are the initial and final heights of the object, respectively. Gravitational potential energy PE is the energy that an object has by virtue of its position. For an object near the surface of the earth, the gravitational potential energy is given by	
Gravitational potential energy	$$\mathrm{PE} = mgh \qquad (6.5)$$	
	where h is the height of the object relative to an arbitrary zero level.	
	6.4 Conservative versus Nonconservative Forces	
Conservative force	A conservative force is one that does the same work in moving an object between two points, independent of the path taken between the points. Alternatively, a force is conservative if the work it does in moving an object around	
Nonconservative force	any closed path is zero. A force is nonconservative if the work it does on an object moving between two points depends on the path of the motion between the points.	
	6.5 The Conservation of Mechanical Energy The total mechanical energy E is the sum of the kinetic energy and potential energy:	
Total mechanical energy	$$E = \mathrm{KE} + \mathrm{PE}$$	

Topic	Discussion	Learning Aids
Alternate form for work–energy theorem	The work–energy theorem can be expressed in an alternate form as $$W_{nc} = E_f - E_0 \qquad (6.8)$$ where W_{nc} is the net work done by the external nonconservative forces, and E_f and E_0 are the final and initial total mechanical energies, respectively.	
Principle of conservation of mechanical energy	The principle of conservation of mechanical energy states that the total mechanical energy E remains constant along the path of an object, provided that the net work done by external nonconservative forces is zero. Whereas E is constant, KE and PE may be transformed into one another.	**Examples 8, 9, 10** **Interactive LearningWare 6.3** **Interactive Solutions 6.33, 6.39**

Use Self-Assessment Test 6.2 to evaluate your understanding of Sections 6.3–6.5.

6.6 Nonconservative Forces and the Work–Energy Theorem

Examples 11, 12, 16
Interactive Solutions 6.51, 6.75

6.7 Power

| **Average power** | Average power $\overline{P}$ is the work done per unit time or the rate at which work is done: $$\overline{P} = \frac{\text{Work}}{\text{Time}} \qquad (6.10a)$$ It is also the rate at which energy changes: $$\overline{P} = \frac{\text{Change in energy}}{\text{Time}} \qquad (6.10b)$$ When a force of magnitude F acts on an object moving with an average speed $\overline{v}$, the average power is given by $$\overline{P} = F\overline{v} \qquad (6.11)$$ | **Example 13** **Interactive Solution 6.55** |

6.8 Other Forms of Energy and the Conservation of Energy

| **Principle of conservation of energy** | The principle of conservation of energy states that energy cannot be created or destroyed but can only be transformed from one form to another. | |

6.9 Work Done by a Variable Force

| **Work done by a variable force** | The work done by a variable force of magnitude F in moving an object through a displacement of magnitude s is equal to the area under the graph of $F \cos\theta$ versus s. The angle θ is the angle between the force and displacement vectors. | **Example 14** |

Use Self-Assessment Test 6.3 **to evaluate your understanding of Sections 6.6–6.9.**

Conceptual Questions

1. Two forces $\mathbf{F}_1$ and $\mathbf{F}_2$ are acting on the box shown in the drawing, causing the box to move across the floor. The two force vectors are drawn to scale. Which force does more work? Justify your answer.

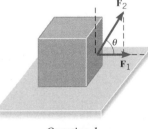

Question 1

2. A box is being moved with a velocity $\mathbf{v}$ by a force $\mathbf{P}$ (parallel to $\mathbf{v}$) along a level horizontal floor. The normal force is $\mathbf{F}_N$, the kinetic frictional force is $\mathbf{f}_k$, and the weight of the box is $m\mathbf{g}$. Decide which forces do positive, zero, or negative work. Provide a reason for each of your answers.

3. A force does positive work on a particle that has a displacement pointing in the $+x$ direction. This same force does negative work on a particle that has a displacement pointing in the $+y$ direction. In what quadrant does the force lie? Account for your answer.

4. A sailboat is moving at a constant velocity. (a) Is work being done by a net external force acting on the boat? Explain. (b) Recognizing that the wind propels the boat forward and the water resists the boat's motion, what does your answer in part (a) imply about the work done by the wind's force compared to the work done by the water's resistive force?

5. A ball has a speed of 15 m/s. Only one external force acts on the ball. After this force acts, the speed of the ball is 7 m/s. Has the force done positive or negative work? Explain.

6. A slow-moving car may have more kinetic energy than a fast-moving motorcycle. How is this possible?

7. A net external force acts on a particle. This net force is not zero. Is this sufficient information to conclude that (a) the velocity of the particle changes, (b) the kinetic energy of the particle changes, and (c) the speed of the particle changes? Give your reasoning in each case.

8. The speed of a particle doubles and then doubles again because a net external force acts on it. Does the net force do more work during the first or the second doubling? Justify your answer.

9. A shopping bag is hanging straight down from your hand as you walk across a horizontal floor at a constant velocity. (a) Does the force that your hand exerts on the bag's handle do any work? Explain. (b) Does this force do any work while you are riding up an escalator at a constant velocity? Give a reason for your answer.

10. In a simulation on earth, an astronaut in his space suit climbs up a vertical ladder. On the moon, the same astronaut makes the same climb. In which case does the gravitational potential energy of the astronaut change by a greater amount? Account for your answer.

11. A net external nonconservative force does positive work on a particle, and both its kinetic and potential energies change. What, if anything, can you conclude about (a) the change in the particle's total mechanical energy and (b) the individual changes in the kinetic and potential energies? Justify your answers.

12. Suppose the total mechanical energy of an object is conserved. (a) If the kinetic energy decreases, what must be true about the gravitational potential energy? (b) If the potential energy decreases, what must be true about the kinetic energy? (c) If the kinetic energy does not change, what must be true about the potential energy?

13. In Example 10 the Steel Dragon starts with a speed of 3.0 m/s at the top of the drop and attains a speed of 42.9 m/s when it reaches the bottom. If the roller coaster were to then start up an identical hill, would its speed be 3.0 m/s at the top of this hill? Assume that friction is negligible. Explain your answer in terms of energy concepts.

14. A person is riding on a Ferris wheel. When the wheel makes one complete turn, is the net work done by the gravitational force positive, negative, or zero? Justify your answer.

15. Consider the following two situations in which the retarding effects of friction and air resistance are negligible. Car A approaches a hill. The driver turns off the engine at the bottom of the hill, and the car coasts up the hill. Car B, its engine running, is driven up the hill at a constant speed. Which situation is an example of the principle of conservation of mechanical energy? Provide a reason for your answer.

16. A trapeze artist, starting from rest, swings downward on the bar, lets go at the bottom of the swing, and falls freely to the net. An assistant, standing on the same platform as the trapeze artist, jumps from rest straight downward. Friction and air resistance are negligible. (a) On which person, if either, does gravity do the greatest amount of work? Explain. (b) Who, if either, strikes the net with a greater speed? Why?

17. The drawing shows an empty fuel tank being released by three different jet planes. At the moment of release, each plane has the same speed and each tank is at the same height above the ground. However, the directions of travel are different. In the absence of air resistance, do the tanks have different speeds when they hit the ground? If so, which tank has the largest speed and which has the smallest speed? Explain.

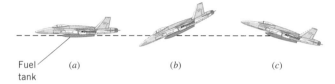

Fuel
tank   (a)   (b)   (c)

18. Is it correct to conclude that one engine is doing twice the work of another just because it is generating twice the power? Explain, neglecting friction and taking into account the time of operation of the engines.

Problems

Problems that are not marked with a star are considered the easiest to solve. Problems that are marked with a single star (*) are more difficult, while those marked with a double star (**) are the most difficult.

ssm Solution is in the Student Solutions Manual.   **www** Solution is available on the World Wide Web at www.wiley.com/college/cutnell
⚕ **This icon represents a biomedical application.**

Section 6.1 Work Done by a Constant Force

1. ssm The brakes of a truck cause it to slow down by applying a retarding force of 3.0×10^3 N to the truck over a distance of 850 m. What is the work done by this force on the truck? Is the work positive or negative? Why?

2. The drawing shows a boat being pulled by two locomotives through a canal of length 2.00 km. The tension in each cable is 5.00×10^3 N, and $\theta = 20.0°$. What is the net work done on the boat by the two locomotives?

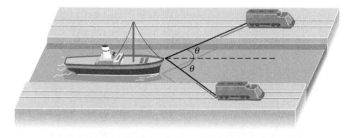

3. A person pulls a toboggan for a distance of 35.0 m along the snow with a rope directed 25.0° above the snow. The tension in the rope is 94.0 N. (a) How much work is done on the toboggan by the

tension force? (b) How much work is done if the same tension is directed parallel to the snow?

4. A 75.0-kg man is riding an escalator in a shopping mall. The escalator moves the man at a constant velocity from ground level to the floor above, a vertical height of 4.60 m. What is the work done on the man by (a) the gravitational force and (b) the escalator?

5. ssm Suppose in Figure 6.2 that $+1.10 \times 10^3$ J of work are done by the force **F** (magnitude = 30.0 N) in moving the suitcase a distance of 50.0 m. At what angle θ is the force oriented with respect to the ground?

6. Consult **Interactive LearningWare 6.1** at **www.wiley.com/college/cutnell** for background pertinent to this problem. The drawing shows a plane diving toward the ground and then climbing back upward. During each of these motions, the lift force **L** acts perpendicular to the displacement **s**, which has the same magnitude of 1.7×10^3 m in each case. The engines of the plane exert a thrust **T**, which points in the direction of the displacement and has the same magnitude during the dive and the climb. The weight **W** of the plane has a magnitude of 5.9×10^4 N. In both motions, net work is performed due to the combined action of the forces **L**, **T**, and **W**. (a) Is more net work done during the dive or the climb? Explain. (b) Find the difference between the net work done during the dive and the climb.

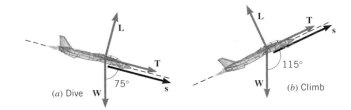

(a) Dive   (b) Climb

7. A person pushes a 16.0-kg shopping cart at a constant velocity for a distance of 22.0 m. She pushes in a direction 29.0° below the horizontal. A 48.0-N frictional force opposes the motion of the cart. (a) What is the magnitude of the force that the shopper exerts? Determine the work done by (b) the pushing force, (c) the frictional force, and (d) the gravitational force.

* **8.** A 55-kg box is being pushed a distance of 7.0 m across the floor by a force **P** whose magnitude is 150 N. The force **P** is parallel to the displacement of the box. The coefficient of kinetic friction is 0.25. Determine the work done on the box by each of the *four* forces that act on the box. Be sure to include the proper plus or minus sign for the work done by each force.

* **9. ssm** A husband and wife take turns pulling their child in a wagon along a horizontal sidewalk. Each exerts a constant force and pulls the wagon through the same displacement. They do the same amount of work, but the husband's pulling force is directed 58° above the horizontal, and the wife's pulling force is directed 38° above the horizontal. The husband pulls with a force whose magnitude is 67 N. What is the magnitude of the pulling force exerted by his wife?

* **10.** A 1.00×10^2-kg crate is being pushed across a horizontal floor by a force **P** that makes an angle of 30.0° below the horizontal. The coefficient of kinetic friction is 0.200. What should be the magnitude of **P**, so that the net work done by it and the kinetic frictional force is zero?

** **11.** A 1200-kg car is being driven up a 5.0° hill. The frictional force is directed opposite to the motion of the car and has a magnitude of $f = 524$ N. A force **F** is applied to the car by the road and propels the car forward. In addition to these two forces, two other forces act on the car: its weight **W** and the normal force $\mathbf{F}_N$ directed perpendicular to the road surface. The length of the road up the hill is 290 m. What should be the magnitude of **F**, so that the net work done by all the forces acting on the car is +150 kJ?

Section 6.2 The Work–Energy Theorem and Kinetic Energy

12. A fighter jet is launched from an aircraft carrier with the aid of its own engines and a steam-powered catapult. The thrust of its engines is 2.3×10^5 N. In being launched from rest it moves through a distance of 87 m and has a kinetic energy of 4.5×10^7 J at lift-off. What is the work done on the jet by the catapult?

13. ssm The hammer throw is a track-and-field event in which a 7.3-kg ball (the "hammer"), starting from rest, is whirled around in a circle several times and released. It then moves upward on the familiar curving path of projectile motion. In one throw, the hammer is given a speed of 29 m/s. For comparison, a .22 caliber bullet has a mass of 2.6 g and, starting from rest, exits the barrel of a gun with a speed of 410 m/s. Determine the work done to launch the motion of (a) the hammer and (b) the bullet.

14. Refer to Concept Simulation 6.1 at **www.wiley.com/college/cutnell** for a review of the concepts with which this problem deals. A 0.075-kg arrow is fired horizontally. The bowstring exerts an average force of 65 N on the arrow over a distance of 0.90 m. With what speed does the arrow leave the bow?

15. When a 0.045-kg golf ball takes off after being hit, its speed is 41 m/s. (a) How much work is done on the ball by the club? (b) Assume that the force of the golf club acts parallel to the motion of the ball and that the club is in contact with the ball for a distance of 0.010 m. Ignore the weight of the ball and determine the average force applied to the ball by the club.

16. As background for this problem, review Conceptual Example 6. A 7420-kg satellite has an elliptical orbit, as in Figure 6.9*b*. The point on the orbit that is farthest from the earth is called the *apogee* and is at the far right side of the drawing. The point on the orbit that is closest to the earth is called the *perigee* and is at the left side of the drawing. Suppose that the speed of the satellite is 2820 m/s at the apogee and 8450 m/s at the perigee. Find the work done by the gravitational force when the satellite moves from (a) the apogee to the perigee and (b) the perigee to the apogee.

17. ssm www Two cars, A and B, are traveling with the same speed of 40.0 m/s, each having started from rest. Car A has a mass of 1.20×10^3 kg, and car B has a mass of 2.00×10^3 kg. Compared to the work required to bring car A up to speed, how much *additional* work is required to bring car B up to speed?

18. Interactive LearningWare 6.2 at **www.wiley.com/college/cutnell** provides a useful review of the concepts that play a role in this problem. A 5.0×10^4-kg space probe is traveling at a speed of 11 000 m/s through deep space. Retrorockets are fired along the line of motion to reduce the probe's speed. The retrorockets generate a force of 4.0×10^5 N over a distance of 2500 km. What is the final speed of the probe?

* **19. ssm** A sled is being pulled across a horizontal patch of snow. Friction is negligible. The pulling force points in the same direction as the sled's displacement, which is along the $+x$ axis. As a result, the kinetic energy of the sled increases by 38%. By what percentage would the sled's kinetic energy have increased if this force had pointed 62° above the $+x$ axis?

* **20.** A 16-kg sled is being pulled along the horizontal snow-covered ground by a horizontal force of 24 N. Starting from rest, the sled attains a speed of 2.0 m/s in 8.0 m. Find the coefficient of kinetic friction between the runners of the sled and the snow.

* **21. ssm www** A 6200-kg satellite is in a circular earth orbit that has a radius of 3.3×10^7 m. A net external force must act on the satellite to make it change to a circular orbit that has a radius of 7.0×10^6 m. What work must the net external force do?

* **22.** A rescue helicopter lifts a 79-kg person straight up by means of a cable. The person has an upward acceleration of 0.70 m/s² and is lifted from rest through a distance of 11 m. (a) What is the tension in the cable? How much work is done by (b) the tension in the cable and (c) the person's weight? (d) Use the work–energy theorem and find the final speed of the person.

* **23.** An extreme skier, starting from rest, coasts down a mountain that makes an angle of 25.0° with the horizontal. The coefficient of kinetic friction between her skis and the snow is 0.200. She coasts for a distance of 10.4 m before coming to the edge of a cliff. Without slowing down, she skis off the cliff and lands downhill at a point whose vertical distance is 3.50 m below the edge. How fast is she going just before she lands?

** **24.** The model airplane in Figure 5.7 is flying at a speed of 22 m/s on a horizontal circle of radius 16 m. The mass of the plane is 0.90 kg. The person holding the guideline pulls it in until the radius of the circle becomes 14 m. The plane speeds up, and the tension in the guideline becomes four times greater. What is the net work done on the plane?

Section 6.3 Gravitational Potential Energy

Section 6.4 Conservative versus Nonconservative Forces

25. ssm A bicyclist rides 5.0 km due east, while the resistive force from the air has a magnitude of 3.0 N and points due west. The rider then turns around and rides 5.0 km due west, back to her starting point. The resistive force from the air on the return trip has a magnitude of 3.0 N and points due east. (a) Find the work done by the resistive force during the round trip. (b) Based on your answer to part (a), is the resistive force a conservative force? Explain.

26. A 0.60-kg basketball is dropped out of a window that is 6.1 m above the ground. The ball is caught by a person whose hands are 1.5 m above the ground. (a) How much work is done on the ball by its weight? What is the gravitational potential energy of the basketball, relative to the ground, when it is (b) released and (c) caught? (d) How is the change ($PE_f - PE_0$) in the ball's gravitational potential energy related to the work done by its weight?

27. Relative to the ground, what is the gravitational potential energy of a 55.0-kg person who is at the top of the Sears Tower, a height of 443 m above the ground?

28. A shot-putter puts a shot (weight = 71.1 N) that leaves his hand at a distance of 1.52 m above the ground. (a) Find the work done by the gravitational force when the shot has risen to a height of 2.13 m above the ground. Include the correct sign for the work. (b) Determine the change ($\Delta PE = PE_f - PE_0$) in the gravitational potential energy of the shot.

29. A 75.0-kg skier rides a 2830-m-long lift to the top of a mountain. The lift makes an angle of 14.6° with the horizontal. What is the change in the skier's gravitational potential energy?

30. When an 81.0-kg adult uses a spiral staircase to climb to the second floor of his house, his gravitational potential energy increases by 2.00×10^3 J. By how much does the potential energy of an 18.0-kg child increase when the child climbs a normal staircase to the second floor?

31. ssm "Rocket man" has a propulsion unit strapped to his back. He starts from rest on the ground, fires the unit, and is propelled straight upward. At a height of 16 m, his speed is 5.0 m/s. His mass, including the propulsion unit, has the approximately constant value of 136 kg. Find the work done by the force generated by the propulsion unit.

Section 6.5 The Conservation of Mechanical Energy

32. Consult **Interactive Learn-ingWare 6.3** at **www.wiley. com/college/cutnell** for a review of the concepts on which this problem is based. A gymnast is swinging on a high bar. The distance between his waist and the bar is 1.1 m, as the drawing shows. At the top of the swing his speed is momentarily zero. Ignoring friction and treating the gymnast as if all of his mass is located at his waist, find his speed at the bottom of the swing.

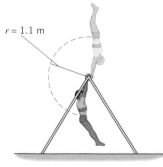

$r = 1.1$ m

33. Interactive Solution 6.33 at **www.wiley.com/college/cutnell** presents a model for solving this problem. A slingshot fires a pebble from the top of a building at a speed of 14.0 m/s. The

building is 31.0 m tall. Ignoring air resistance, find the speed with which the pebble strikes the ground when the pebble is fired (a) horizontally, (b) vertically straight up, and (c) vertically straight down.

34. A water-skier lets go of the tow rope upon leaving the end of a jump ramp at a speed of 14.0 m/s. As the drawing indicates, the skier has a speed of 13.0 m/s at the highest point of the jump. Ignoring air resistance, determine the skier's height H above the *top of the ramp* at the highest point.

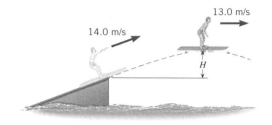

13.0 m/s

14.0 m/s

H

35. ssm A pole-vaulter approaches the takeoff point at a speed of 9.00 m/s. Assuming that only this speed determines the height to which he can rise, find the maximum height at which the vaulter can clear the bar.

36. A 47.0-g golf ball is driven from the tee with an initial speed of 52.0 m/s and rises to a height of 24.6 m. (a) Neglect air resistance and determine the kinetic energy of the ball at its highest point. (b) What is its speed when it is 8.0 m below its highest point?

37. A cyclist approaches the bottom of a gradual hill at a speed of 11 m/s. The hill is 5.0 m high, and the cyclist estimates that she is going fast enough to coast up and over it without pedaling. Ignoring air resistance and friction, find the speed at which the cyclist crests the hill.

* **38.** A particle, starting from point A in the drawing, is projected down the curved runway. Upon leaving the runway at point B, the particle is traveling straight upward and reaches a height of 4.00 m above the floor before falling back down. Ignoring friction and air resistance, find the speed of the particle at point A.

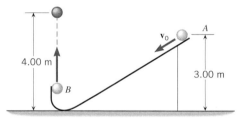

4.00 m

v_0

A

3.00 m

B

* **39.** Review **Interactive Solution 6.39** at **www.wiley.com/college/cutnell** for background on this problem. A wrecking ball swings at the end of a 12.0-m cable on a vertical circular arc. The crane operator manages to give the ball a speed of 5.00 m/s as the ball passes through the lowest point of its swing and then gives the ball no further assistance. Friction and air resistance are negligible. What speed v_f does the ball have when the cable makes an angle of 20.0° with respect to the vertical?

* **40.** The drawing shows a skateboarder moving at 5.4 m/s along a horizontal section of a track that is slanted upward by 48° above the horizontal at its end, which is 0.40 m above the ground. When she leaves the track, she follows the characteristic path of projectile motion. Ignoring friction and air resistance, find the maximum height H to which she rises above the end of the track.

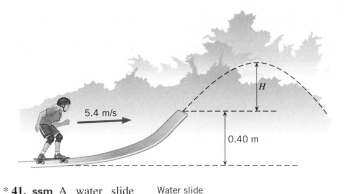

41. ssm A water slide

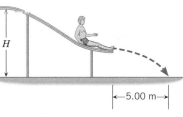

Water slide

is constructed so that swimmers, starting from rest at the top of the slide, leave the end of the slide traveling horizontally. As the drawing shows, one person hits the water 5.00 m from the end of the slide in a time of 0.500 s after leaving the slide. Ignoring friction and air resistance, find the height H in the drawing.

* **42.** A skier starts from rest at the top of a hill. The skier coasts down the hill and up a second hill, as the drawing illustrates. The crest of the second hill is circular, with a radius of $r = 36$ m. Neglect friction and air resistance. What must be the height h of the first hill so that the skier just loses contact with the snow at the crest of the second hill?

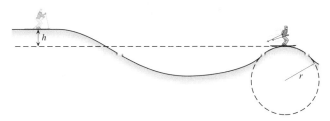

** **43.** Conceptual Example 9 provides background for this problem. A swing is made from a rope that will tolerate a maximum tension of 8.00×10^2 N without breaking. Initially, the swing hangs vertically. The swing is then pulled back at an angle of $60.0°$ with respect to the vertical and released from rest. What is the mass of the heaviest person who can ride the swing?

** **44.** The drawing shows a version of the loop-the-loop trick for a small car. If the car is given an initial speed of 4.0 m/s, what is the largest value that the radius r can have if the car is to remain in contact with the circular track at all times?

Section 6.6 Nonconservative Forces and the Work–Energy Theorem

45. ssm A roller coaster (375 kg) moves from A (5.00 m above the ground) to B (20.0 m above the ground). Two nonconservative forces are present: friction does -2.00×10^4 J of work on the car, and a chain mechanism does $+3.00 \times 10^4$ J of work to help the car up a long climb. What is the change in the car's kinetic energy, $\Delta KE = KE_f - KE_0$, from A to B?

46. A basketball player makes a jump shot. The 0.600-kg ball is released at a height of 2.00 m above the floor with a speed of 7.20 m/s. The ball goes through the net 3.10 m above the floor at a speed of 4.20 m/s. What is the work done on the ball by air resistance, a nonconservative force?

47. A 5.00×10^2-kg hot-air balloon takes off from rest at the surface of the earth. The nonconservative wind and lift forces take the balloon up, doing $+9.70 \times 10^4$ J of work on the balloon in the process. At what height above the surface of the earth does the balloon have a speed of 8.00 m/s?

48. A projectile of mass 0.750 kg is shot straight up with an initial speed of 18.0 m/s. (a) How high would it go if there were no air friction? (b) If the projectile rises to a maximum height of only 11.8 m, determine the magnitude of the average force due to air resistance.

49. ssm The (nonconservative) force propelling a 1.50×10^3-kg car up a mountain road does 4.70×10^6 J of work on the car. The car starts from rest at sea level and has a speed of 27.0 m/s at an altitude of 2.00×10^2 m above sea level. Obtain the work done on the car by the combined forces of friction and air resistance, both of which are nonconservative forces.

* **50.** A pitcher throws a 0.140-kg baseball, and it approaches the bat at a speed of 40.0 m/s. The bat does $W_{nc} = 70.0$ J of work on the ball in hitting it. Ignoring air resistance, determine the speed of the ball after the ball leaves the bat and is 25.0 m above the point of impact.

* **51. Interactive Solution 6.51** at **www.wiley.com/college/cutnell** offers help in modeling this problem. A basketball of mass 0.60 kg is dropped from rest from a height of 1.05 m. It rebounds to a height of 0.57 m. (a) How much mechanical energy was lost during the collision with the floor? (b) A basketball player dribbles the ball from a height of 1.05 m by exerting a constant downward force on it for a distance of 0.080 m. In dribbling, the player compensates for the mechanical energy lost during each bounce. If the ball now returns to a height of 1.05 m, what is the magnitude of the force?

* **52.** In attempting to pass the puck to a teammate, a hockey player gives it an initial speed of 1.7 m/s. However, this speed is inadequate to compensate for the kinetic friction between the puck and the ice. As a result, the puck travels only one-half the distance between the players before sliding to a halt. What minimum initial speed should the puck have been given so that it reached the teammate, assuming that the same force of kinetic friction acted on the puck everywhere between the two players?

* **53. ssm** At a carnival, you can try to ring a bell by striking a target with a 9.00-kg hammer. In response, a 0.400-kg metal piece is sent upward toward the bell, which is 5.00 m above. Suppose that 25.0% of the hammer's kinetic energy is used to do the work of sending the metal piece upward. How fast must the hammer be moving when it strikes the target so that the bell just barely rings?

** **54.** A 3.00-kg model rocket is launched straight up. It reaches a maximum height of 1.00×10^2 m above where its engine cuts out, even though air resistance performs -8.00×10^2 J of work on the rocket. What would have been this height if there were no air resistance?

Section 6.7 Power

55. Interactive Solution 6.55 at **www.wiley.com/college/cutnell** offers a model for solving this problem. A car accelerates uniformly from rest to 20.0 m/s in 5.6 s along a level stretch of road. Ignoring friction, determine the average power required to accelerate the car if (a) the weight of the car is 9.0×10^3 N, and (b) the weight of the car is 1.4×10^4 N.

56. A person is making homemade ice cream. She exerts a force of magnitude 22 N on the free end of the crank handle, and this end moves in a circular path of radius 0.28 m. The force is always applied parallel to the motion of the handle. If the handle is turned once every 1.3 s, what is the average power being expended?

57. ssm One kilowatt·hour (kWh) is the amount of work or energy generated when one kilowatt of power is supplied for a time of one hour. A kilowatt·hour is the unit of energy used by power companies when figuring your electric bill. Determine the number of joules of energy in one kilowatt·hour.

58. A 3.00×10^2-kg piano is being lifted at a steady speed from ground level straight up to an apartment 10.0 m above the ground. The crane that is doing the lifting produces a steady power of 4.00×10^2 W. How much time does it take to lift the piano?

* 59. In 2.0 minutes, a ski lift raises four skiers at constant speed to a height of 140 m. The average mass of each skier is 65 kg. What is the average power provided by the tension in the cable pulling the lift?

* 60. A motorcycle (mass of cycle plus rider = 2.50×10^2 kg) is traveling at a steady speed of 20.0 m/s. The force of air resistance acting on the cycle and rider is 2.00×10^2 N. Find the power necessary to sustain this speed if (a) the road is level and (b) the road is sloped upward at 37.0° with respect to the horizontal.

** 61. ssm The motor of a ski boat generates an average power of 7.50×10^4 W when the boat is moving at a constant speed of 12 m/s. When the boat is pulling a skier at the same speed, the engine must generate an average power of 8.30×10^4 W. What is the tension in the tow rope that is pulling the skier?

** 62. A 1900-kg car experiences a combined force of air resistance and friction that has the same magnitude whether the car goes up or down a hill at 27 m/s. Going up a hill, the car's engine needs to produce 47 hp more power to sustain the constant velocity than it does going down the same hill. At what angle is the hill inclined above the horizontal?

Section *6.9 Work Done by a Variable Force

63. The graph shows the net external force component $F \cos \theta$ along the displacement as a function of the magnitude s of the displacement. The graph applies to a 65-kg ice skater. How much work does the net force component do on the skater from (a) 0 to 3.0 m and (b) 3.0 m to 6.0 m? (c) If the initial speed of the skater is 1.5 m/s when $s = 0$ m, what is the speed when $s = 6.0$ m?

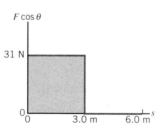

64. The graph shows how the force component $F \cos \theta$ along the displacement varies with the magnitude s of the displacement. Find the work done by the force. (*Hint: Recall how the area of a triangle is related to the triangle's base and height.*)

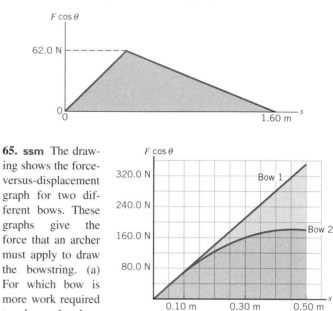

65. ssm The drawing shows the force-versus-displacement graph for two different bows. These graphs give the force that an archer must apply to draw the bowstring. (a) For which bow is more work required to draw the bow fully from $s = 0$ to $s = 0.50$ m? Give your reasoning. (b) Estimate the additional work required for the bow identified in part (a) compared to the other bow.

66. Review Example 14, in which the work done in drawing the bowstring in Figure 6.22 from $s = 0$ to $s = 0.500$ m is determined. In part b of the figure, the force component $F \cos \theta$ reaches a maximum at $s = 0.306$ m. Find the percentage of the total work that is done when the bowstring is moved (a) from $s = 0$ to 0.306 m and (b) from $s = 0.306$ to 0.500 m.

67. A net external force is applied to a 6.00-kg object that is initially at rest. The net force component along the displacement of the object varies with the magnitude of the displacement as shown in the drawing. (a) How much work is done by the net force? (b) What is the speed of the object at $s = 20.0$ m?

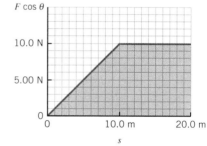

Additional Problems

68. A cable lifts a 1200-kg elevator at a constant velocity for a distance of 35 m. What is the work done by (a) the tension in the cable and (b) the elevator's weight?

69. ssm A 2.00-kg rock is released from rest at a height of 20.0 m. Ignore air resistance and determine the kinetic energy, gravitational potential energy, and total mechanical energy at each of the following heights: 20.0, 10.0, and 0 m.

70. A softball pitcher has a "windmill" windup in which a 0.25-kg ball moves on a vertical arc of radius $r = 0.51$ m. To accelerate the

ball, she exerts a 28-N force parallel to the ball's motion along the circular arc. The speed of the ball is 12 m/s at the top of the arc. With what speed is the ball released one-half a revolution later at the bottom of the arc?

71. ssm During a tug-of-war, team A pulls on team B by applying a force of 1100 N to the rope between them. How much work does team A do if they pull team B toward them a distance of 2.0 m?

72. A 55.0-kg skateboarder starts out with a speed of 1.80 m/s. He does +80.0 J of work on himself by pushing with his feet against the ground. In addition, friction does −265 J of work on him. In both cases, the forces doing the work are nonconservative. The final speed of the skateboarder is 6.00 m/s. (a) Calculate the change (ΔPE = PE_f − PE_0) in the gravitational potential energy. (b) How much has the vertical height of the skater changed, and is the skater above or below the starting point?

* **73.** A 2.40×10^2-N force is pulling an 85.0-kg refrigerator across a horizontal surface. The force acts at an angle of 20.0° above the surface. The coefficient of kinetic friction is 0.200, and the refrigerator moves a distance of 8.00 m. Find (a) the work done by the pulling force, and (b) the work done by the kinetic frictional force.

* **74.** The cheetah is one of the fastest accelerating animals, because it can go from rest to 27 m/s (about 60 mi/h) in 4.0 s. If its mass is 110 kg, determine the average power developed by the cheetah during the acceleration phase of its motion. Express your answer in (a) watts and (b) horsepower.

* **75.** Refer to **Interactive Solution 6.75** for a review of the approach taken in problems such as this one. A 67.0-kg person jumps from rest off a 3.00-m-high tower straight down into the water. Neglect air resistance during the descent. She comes to rest 1.10 m under the surface of the water. Determine the magnitude of the average force that the water exerts on the diver. This force is nonconservative.

* **76.** A 63-kg skier coasts up a snow-covered hill that makes an angle of 25° with the horizontal. The initial speed of the skier is 6.6 m/s. After coasting a distance of 1.9 m up the slope, the speed of the skier is 4.4 m/s. (a) Find the work done by the kinetic frictional force that acts on the skis. (b) What is the magnitude of the kinetic frictional force?

* **77. ssm www** Two pole-vaulters just clear the bar at the same height. The first lands at a speed of 8.90 m/s, and the second lands at a speed of 9.00 m/s. The first vaulter clears the bar at a speed of 1.00 m/s. Ignore air resistance and friction and determine the speed at which the second vaulter clears the bar.

** **78.** A person starts from rest at the top of a large frictionless spherical surface, and slides into the water below (see the drawing). At what angle θ does the person leave the surface? *(Hint: When the person leaves the surface, the normal force is zero.)*

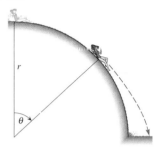

** **79. ssm www** A truck is traveling at 11.1 m/s down a hill when the brakes on all four wheels lock. The hill makes an angle of 15.0° with respect to the horizontal. The coefficient of kinetic friction between the tires and the road is 0.750. How far does the truck skid before coming to a stop?

Concepts & Calculations Group Learning Problems

Note: Each of these problems consists of Concept Questions followed by a related quantitative Problem. They are designed for use by students working alone or in small learning groups. The Concept Questions involve little or no mathematics and are intended to stimulate group discussions. They focus on the concepts with which the problems deal. Recognizing the concepts is the essential initial step in any problem-solving technique.

80. Concept Questions You are moving into an apartment and take the elevator to the 6th floor. Does the force exerted on you by the elevator do positive or negative work when the elevator (a) goes up and (b) goes down? Explain your answers.

Problem Suppose your weight is 685 N and that of your belongings is 915 N. (a) Determine the work done by the elevator in lifting you and your belongings up to the 6th floor (15.2 m) at a constant velocity. (b) How much work does the elevator do on you alone (without belongings) on the downward trip, which is also made at a constant velocity? Check to see that your answers are consistent with your answers to the Concept Questions.

81. Concept Questions An asteroid is moving along a straight line. A force acts along the displacement of the asteroid and slows it down. (a) Is the direction of the force the same as or opposite to the direction of the displacement of the asteroid? Why? (b) Does the force do positive, negative, or zero work? Justify your answer. (c) What type of energy is changing as the object slows down? (d) What is the relationship between the work done by this force and the change in the object's energy?

Problem The asteroid has a mass of 4.5×10^4 kg, and the force causes its speed to change from 7100 to 5500 m/s. (a) What is the work done by the force? (b) If the asteroid slows down over a distance of 1.8×10^6 m, determine the magnitude of the force. Verify that your answers are consistent with the answers to the Concept Questions.

82. Concept Questions The drawing shows two boxes resting on frictionless ramps. One box is relatively light and sits on a steep ramp. The other box is heavier and rests on a ramp that is less steep. The boxes are released from rest at A and allowed to slide down the ramps. Which box, if either, has (a) the greatest speed and (b) the greatest kinetic energy at B? Provide a reason for each answer.

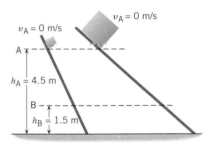

Problem The two boxes have masses of 11 and 44 kg. If A and B are 4.5 and 1.5 m, respectively, above the ground, determine the speed of (a) the lighter box and (b) the heavier box when each reaches B. (c) What is the ratio of the kinetic energy of the heavier box to that of the lighter box at B? Be sure that your answers are consistent with your answers to the Concept Questions.

83. Concept Questions A student, starting from rest, slides down a water slide. On the way down, a kinetic frictional force (a nonconservative force) acts on her. (a) Does the kinetic frictional force do positive, negative, or zero work? Provide a reason for your answer. (b) Does the total mechanical energy of the student increase, decrease, or remain the same as she descends the slide? Why? (c) If the kinetic frictional force does work, how is this work related to the change in the total mechanical energy of the student?

Problem The student has a mass of 83.0 kg and the height of the water slide is 11.8 m. If the kinetic frictional force does -6.50×10^3 J of work, how fast is the student going at the bottom of the slide?

84. Concept Questions A helicopter, starting from rest, accelerates straight up from the roof of a hospital. The lifting force does work in raising the helicopter. (a) What two types of energy are changing? Is each type increasing or decreasing? Why? (b) How are the two types of energy related to the work done by the lifting force? (c) If you want to determine the average power generated by the lifting force, what other variable besides the work must be known?

Problem An 810-kg helicopter rises from rest to a speed of 7.0 m/s in a time of 3.5 s. During this time it climbs to a height of 8.2 m. What is the average power generated by the lifting force?

85. Concept Questions A water-skier is being pulled by a towrope attached to a boat. As the driver pushes the throttle forward, the skier accelerates. (a) What type of energy is changing? (b) Is the work being done by the net external force acting on the skier positive or negative? Why? (c) Explain how this work is related to the change in the energy of the skier.

Problem A 70.3-kg water-skier has an initial speed of 6.10 m/s. Later, the speed increases to 11.3 m/s. Determine the work done by the net external force acting on the skier.

* **86. Concept Questions** Under the influence of its drive force, a snowmobile is moving at a constant velocity along a horizontal patch of snow. When the drive force is shut off, the snowmobile coasts to a halt. (a) The work–energy theorem is given by Equation 6.3 as $W = \frac{1}{2}mv_\text{f}^2 - \frac{1}{2}mv_0^2$, where W is the work done by the net external force acting on the moving object. During the coasting phase, what is the net external force acting on the snowmobile, and how does W depend on the magnitude and direction of the net force and on the displacement of the snowmobile? (b) What is the net force acting on the snowmobile during the constant-velocity phase? Explain. (c) By using only the work–energy theorem, can you determine the time t it takes for the snowmobile to coast to a halt? If not, what other information is needed to determine t?

Problem The snowmobile and its rider have a mass of 136 kg. Under the influence of a drive force of 205 N, it is moving at a constant velocity, whose magnitude is 5.50 m/s. Its drive force is shut off.

Find (a) the distance in which the snowmobile coasts to a halt and (b) the time required to do so.

* **87. Concept Questions** The drawing shows two frictionless inclines that begin at ground level ($h = 0$ m) and slope upward at the same angle θ. One track is longer than the other, however. Identical blocks are projected up each track with the same initial speed v_0. On the longer track the block slides upward until it reaches a maximum height H above the ground. On the shorter track the block slides upward, flies off the end of the track at a height H_1 above the ground, and then follows the familiar parabolic trajectory of projectile motion. At the highest point of this trajectory, the block is a height H_2 above the end of the track. The initial total mechanical energy of each block is the same and is all kinetic energy. (a) When the block on the longer track reaches its maximum height, is its final total mechanical energy all kinetic energy, all potential energy, or some of each? Explain. (b) When the block on the shorter track reaches the top of its trajectory after leaving the track, is its final total mechanical energy all kinetic energy, all potential energy, or some of each? Justify your answer. (c) Which is the greater height above the gound: H or $H_1 + H_2$? Why?

Problem The initial speed of each block is $v_0 = 7.00$ m/s, and each incline slopes upward at an angle of $\theta = 50.0°$. The block on the shorter track leaves the track at a height of $H_1 = 1.25$ m above the ground. Find the height H for the block on the longer track and the total height $H_1 + H_2$ for the block on the shorter track. Verify that your answers are consistent with your answers to the Concept Questions.

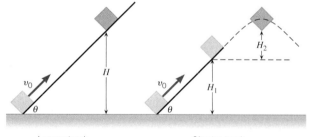

Longer track   Shorter track

Chapter 7

Impulse and Momentum

The concepts of impulse and momentum discussed in this chapter aid us in understanding collisions such as this one between a bowling ball and the pins. (© Alan Thornton/Stone/Getty Images)

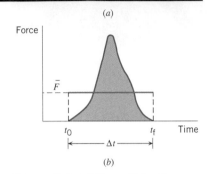

(a)

(b)

Figure 7.1 (*a*) The collision time between a bat and a ball is very short, often less than a millisecond, but the force can be quite large. (© Chuck Savage/Corbis Images) (*b*) When the bat strikes the ball, the magnitude of the force exerted on the ball rises to a maximum value and then returns to zero when the ball leaves the bat. The time interval during which the force acts is Δt, and the magnitude of the average force is $\overline{F}$.

7.1 The Impulse–Momentum Theorem

There are many situations in which the force acting on an object is not constant, but varies with time. For instance, Figure 7.1a shows a baseball being hit, and part *b* of the figure illustrates approximately how the force applied to the ball by the bat changes during the time of contact. The magnitude of the force is zero at the instant t_0 just before the bat touches the ball. During contact, the force rises to a maximum and then returns to zero at the time t_f when the ball leaves the bat. The time interval $\Delta t = t_f - t_0$ during which the bat and ball are in contact is quite short, being only a few-thousandths of a second, although the maximum force can be very large, often exceeding thousands of newtons. For comparison, the graph also shows the magnitude $\overline{F}$ of the average force exerted on the ball during the time of contact. Figure 7.2 depicts other situations in which a time-varying force is applied to a ball.

▶ **CONCEPTS AT A GLANCE** To describe how a time-varying force affects the motion of an object, we will introduce two new ideas: the impulse of a force and the linear momentum of an object. As the Concepts-at-a-Glance chart in Figure 7.3 illustrates, these ideas will be used with Newton's second law of motion to produce an important result known as the impulse–momentum theorem. This theorem plays a central role in describing collisions, such as that between a ball and a bat. Later on, we will see also that the theorem leads in a natural way to one of the most fundamental laws in physics, the conservation of momentum. ◀

If a baseball is to be hit well, both the magnitude of the force and the time of contact are important. When a large average force acts on the ball for a long enough time, the ball is hit solidly. To describe such situations, we bring together the average force and the time of contact, calling the product of the two the *impulse* of the force.

■ **DEFINITION OF IMPULSE**

The impulse $\mathbf{J}$ of a force is the product of the average force $\overline{\mathbf{F}}$ and the time interval Δt during which the force acts:

$$\mathbf{J} = \overline{\mathbf{F}}\,\Delta t \qquad (7.1)$$

Impulse is a vector quantity and has the same direction as the average force.

SI Unit of Impulse: newton · second (N · s)

When a ball is hit, it responds to the value of the impulse. A large impulse produces a large response; that is, the ball departs from the bat with a large velocity. However, we know from experience that the more massive the ball, the less velocity it has after leaving the bat. Both mass and velocity play a role in how an object responds to a given impulse, and the effect of each of them is included in the concept of *linear momentum*, which is defined as follows:

Figure 7.2 In each of these situations, the force applied to the ball varies with time. The time of contact is small, but the maximum force can be large. (*left,* © Tommy Hindley/Aurora Photos; *right,* © Allsport/Getty Images)

CONCEPTS AT A GLANCE

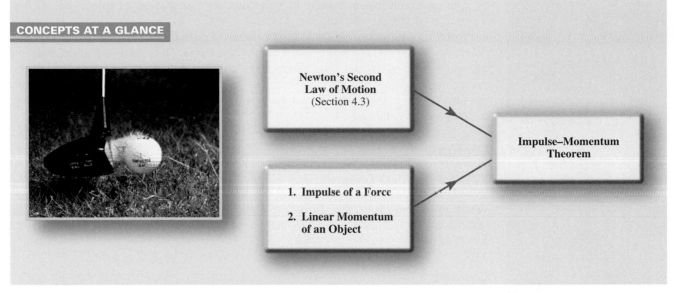

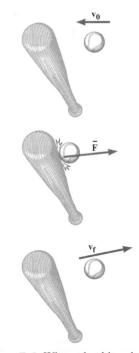

Figure 7.3 **CONCEPTS AT A GLANCE**
Two new concepts, impulse and momentum, will be combined with Newton's second law of motion to produce the impulse–momentum theorem, a theorem used for describing collisions. When the golf club collides with the ball in this photograph, an impulse is imparted to the ball. In response, the ball's momentum changes, in accord with the impulse–momentum theorem. (© Stephen Dalton/Photo Researchers)

■ **DEFINITION OF LINEAR MOMENTUM**

The linear momentum **p** of an object is the product of the object's mass m and velocity **v**:

$$\mathbf{p} = m\mathbf{v} \qquad (7.2)$$

Linear momentum is a vector quantity that points in the same direction as the velocity.

SI Unit of Linear Momentum: kilogram · meter/second (kg · m/s)

Newton's second law of motion can now be used to reveal a relationship between impulse and momentum. Figure 7.4 shows a ball with an initial velocity of $\mathbf{v_0}$ approaching a bat, being struck by the bat, and then departing with a final velocity of $\mathbf{v_f}$. When the velocity of an object changes from $\mathbf{v_0}$ to $\mathbf{v_f}$ during a time interval Δt, the average acceleration $\bar{\mathbf{a}}$ is given by Equation 2.4 as

$$\bar{\mathbf{a}} = \frac{\mathbf{v_f} - \mathbf{v_0}}{\Delta t}$$

According to Newton's second law, $\Sigma \bar{\mathbf{F}} = m\bar{\mathbf{a}}$, the average acceleration is produced by the net average force $\Sigma \bar{\mathbf{F}}$. Here $\Sigma \bar{\mathbf{F}}$ represents the vector sum of all the average forces that act on the object. Thus,

$$\Sigma \bar{\mathbf{F}} = m \left(\frac{\mathbf{v_f} - \mathbf{v_0}}{\Delta t} \right) = \frac{m\mathbf{v_f} - m\mathbf{v_0}}{\Delta t} \qquad (7.3)$$

In this result, the numerator on the far right is the final momentum minus the initial momentum, which is the change in momentum. Thus, the net average force is given by the change in momentum per unit of time.* Multiplying both sides of Equation 7.3 by Δt yields Equation 7.4, which is known as the *impulse–momentum theorem.*

■ **IMPULSE–MOMENTUM THEOREM**

When a net force acts on an object, the impulse of this force is equal to the change in momentum of the object:

$$\underbrace{(\Sigma \bar{\mathbf{F}}) \Delta t}_{\text{Impulse}} = \underbrace{m\mathbf{v_f}}_{\substack{\text{Final} \\ \text{momentum}}} - \underbrace{m\mathbf{v_0}}_{\substack{\text{Initial} \\ \text{momentum}}} \qquad (7.4)$$

Impulse = Change in momentum

*The equality between the net force and the change in momentum per unit time is the version of the second law of motion presented originally by Newton.

Figure 7.4 When a bat hits a ball, an average force $\bar{\mathbf{F}}$ is applied to the ball by the bat. As a result, the ball's velocity changes from an initial value of $\mathbf{v_0}$ (top drawing) to a final value of $\mathbf{v_f}$ (bottom drawing).

This karate expert applies an impulse to the stack of tiles. As a result, the momentum of the stack changes as the tiles break. (© Terje Rakke/The Image Bank/Getty Images)

During a collision, it is often difficult to measure the net average force $\Sigma\overline{\mathbf{F}}$, so it is not easy to determine the impulse, $(\Sigma\overline{\mathbf{F}})\,\Delta t$, directly. On the other hand, it is usually straightforward to measure the mass and velocity of an object, so that its momentum just after the collision $m\mathbf{v_f}$ and just before it $m\mathbf{v_0}$ can be found. Thus, the impulse–momentum theorem allows us to gain information about the impulse indirectly by measuring the change in momentum that the impulse causes. Then, armed with a knowledge of the contact time Δt, we can evaluate the net average force. Examples 1 and 2 illustrate how the theorem is used in this way.

Example 1   A Well-Hit Ball

A baseball ($m = 0.14$ kg) has an initial velocity of $\mathbf{v_0} = -38$ m/s as it approaches a bat. We have chosen the direction of approach as the negative direction. The bat applies an average force $\overline{\mathbf{F}}$ that is much larger than the weight of the ball, and the ball departs from the bat with a final velocity of $\mathbf{v_f} = +58$ m/s. (a) Determine the impulse applied to the ball by the bat. (b) Assuming that the time of contact is $\Delta t = 1.6 \times 10^{-3}$ s, find the average force exerted on the ball by the bat.

Reasoning Two forces act on the ball during impact, and together they constitute the net average force: the average force $\overline{\mathbf{F}}$ exerted by the bat, and the weight of the ball. Since $\overline{\mathbf{F}}$ is much greater than the weight of the ball, we neglect the weight. Thus, the net average force is equal to $\overline{\mathbf{F}}$, or $\Sigma\overline{\mathbf{F}} = \overline{\mathbf{F}}$. In hitting the ball, the bat imparts an impulse to it. We cannot use Equation 7.1 ($\mathbf{J} = \overline{\mathbf{F}}\,\Delta t$) to determine the impulse $\mathbf{J}$ directly, since $\overline{\mathbf{F}}$ is not known. We can find the impulse indirectly, however, by turning to the impulse–momentum theorem, which states that the impulse is equal to the ball's final momentum minus its initial momentum. With values for the impulse and the time of contact, Equation 7.1 can be used to determine the average force applied to the ball by the bat.

Solution

(a) According to the impulse–momentum theorem, the impulse $\mathbf{J}$ applied to the ball is

$$\mathbf{J} = m\mathbf{v_f} - m\mathbf{v_0}$$
$$= \underbrace{(0.14 \text{ kg})(+58 \text{ m/s})}_{\text{Final momentum}} - \underbrace{(0.14 \text{ kg})(-38 \text{ m/s})}_{\text{Initial momentum}} = \boxed{+13.4 \text{ kg} \cdot \text{m/s}}$$

(b) Now that the impulse is known, the contact time can be used in Equation 7.1 to find the average force $\overline{\mathbf{F}}$ exerted by the bat on the ball:

$$\overline{\mathbf{F}} = \frac{\mathbf{J}}{\Delta t} = \frac{+13.4 \text{ kg} \cdot \text{m/s}}{1.6 \times 10^{-3} \text{ s}} = \boxed{+8400 \text{ N}}$$

The force is positive, indicating that it points opposite to the velocity of the approaching ball. A force of 8400 N corresponds to 1900 lb, such a large value being necessary to change the ball's momentum during the brief contact time.

Example 2   A Rain Storm

During a storm, rain comes straight down with a velocity of $\mathbf{v_0} = -15$ m/s and hits the roof of a car perpendicularly (see Figure 7.5). The mass of rain per second that strikes the car roof is 0.060 kg/s. Assuming that the rain comes to rest upon striking the car ($\mathbf{v_f} = 0$ m/s), find the average force exerted by the rain on the roof.

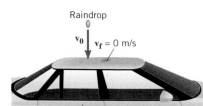

Figure 7.5 A raindrop falling on a car roof has an initial velocity of $\mathbf{v_0}$ just before striking the roof. The final velocity of the raindrop is $\mathbf{v_f} = 0$ m/s, because it comes to rest on the roof.

Reasoning This example differs from Example 1 in an important way. Example 1 gives information about the ball and asks for the force applied to the ball. In contrast, the present example gives information about the rain but asks for the force acting on the roof. However, the force exerted on the roof by the rain and the force exerted on the rain by the roof have equal magnitudes and opposite directions, according to Newton's law of action and reaction (see Section 4.5). Thus, we will find the force exerted on the rain and then apply the law of action and reaction to obtain the force on the roof. Two forces act on the rain while it impacts with the roof: the average force $\overline{\mathbf{F}}$ exerted by the roof, and the weight of the rain. These two forces constitute the net average force. By comparison, however, the force $\overline{\mathbf{F}}$ is much greater than the weight, so we may neglect the weight. Thus, the net average force becomes equal to $\overline{\mathbf{F}}$, or $\Sigma\overline{\mathbf{F}} = \overline{\mathbf{F}}$. The value of $\overline{\mathbf{F}}$ can be obtained by applying the impulse–momentum theorem to the rain.

Solution The average force $\overline{\mathbf{F}}$ needed to reduce the rain's velocity from $\mathbf{v_0} = -15$ m/s to $\mathbf{v_f} = 0$ m/s is given by Equation 7.4 as

$$\overline{\mathbf{F}} = \frac{m\mathbf{v_f} - m\mathbf{v_0}}{\Delta t} = -\left(\frac{m}{\Delta t}\right)\mathbf{v_0}$$

The term $m/\Delta t$ is the mass of rain per second that strikes the roof, so that $m/\Delta t = 0.060$ kg/s. Thus, the average force exerted on the rain by the roof is

$$\overline{\mathbf{F}} = -(0.060 \text{ kg/s})(-15 \text{ m/s}) = +0.90 \text{ N}$$

This force is in the positive or upward direction, which is reasonable since the roof exerts an upward force on each falling drop in order to bring it to rest. According to the action–reaction law, the force exerted on the roof by the rain also has a magnitude of 0.90 N but points downward: Force on roof = $\boxed{-0.90 \text{ N}}$.

As you reason through problems such as those in Examples 1 and 2, take advantage of the impulse–momentum theorem. It is a powerful statement that can lead to significant insights. The following Conceptual Example further illustrates its use.

Conceptual Example 3 Hailstones Versus Raindrops

In Example 2 rain is falling on the roof of a car and exerts a force on it. Instead of rain, suppose hail is falling. The hail comes straight down at a mass rate of $m/\Delta t = 0.060$ kg/s and an initial velocity of $\mathbf{v_0} = -15$ m/s and strikes the roof perpendicularly, just as the rain does in Example 2. However, unlike rain, hail usually does not come to rest after striking a surface. Instead, the hailstones bounce off the roof of the car. If hail fell instead of rain, would the force on the roof be smaller than, equal to, or greater than that calculated in Example 2?

Reasoning and Solution The raindrops and the hailstones fall in exactly the same way. That is, they both fall with the same initial velocity and mass rate, and they both strike the roof perpendicularly. However, there is an important difference: the raindrops come to rest (see Figure 7.5), while the hailstones bounce upward after striking the roof (see Figure 7.6). According to the impulse–momentum theorem, the impulse that acts on an object is given by the change in the momentum of the object. This change is $m\mathbf{v_f} - m\mathbf{v_0} = m \Delta\mathbf{v}$ and is proportional to the change in velocity $\Delta\mathbf{v}$. For a hailstone, the change in velocity is from $\mathbf{v_0}$ (downward) to $\mathbf{v_f}$ (upward). This is a larger change than for the raindrop, for which the change is only from $\mathbf{v_0}$ (downward) to zero. Therefore, the change in momentum is greater for the hailstone, and, correspondingly, a greater impulse must act on it. But an impulse is the product of the average force and the time interval Δt. Since the same amount of mass falls in the same time interval in either case, Δt is the same for hailstones as for raindrops. The greater impulse acting on the hailstones, then, means that the car roof must exert a greater force on the hailstones than on the raindrops. Conversely, according to Newton's action–reaction law, *the car roof experiences a greater force from the hailstones than from the raindrops.*

Related Homework: *Problems 6, 50*

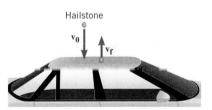

Figure 7.6 Hailstones have a downward velocity of $\mathbf{v_0}$ just before striking this car roof. They rebound with an upward velocity of $\mathbf{v_f}$.

✓ **Check Your Understanding 1**

Suppose you are standing on the edge of a dock and jump straight down. If you land on sand your stopping time is much shorter than if you land on water. Using the impulse–momentum theorem as a guide, determine which one of the following statements is correct. *(The answer is given at the end of the book.)*

a. In bringing you to a halt, the sand exerts a greater impulse on you than does the water.
b. In bringing you to a halt, the sand and the water exert the same impulse on you, but the sand exerts a greater average force.
c. In bringing you to a halt, the sand and the water exert the same impulse on you, but the sand exerts a smaller average force.

Background: The concept of impulse and its relation to the change in an object's momentum are pertinent to this question. Think carefully about the definition of impulse.

For similar questions (including calculational counterparts), consult Self-Assessment Test 7.1. This test is described at the end of Section 7.2.

Need more practice?

🖱️ **Interactive LearningWare 7.1**
A 1200-kg car has an initial velocity of 13 m/s, eastward. The driver applies the brakes and brings the car to rest in 6.0 s. Determine the net average force (magnitude and direction) exerted on the car by using (a) the impulse–momentum theorem and (b) kinematics and Newton's second law.

Related Homework: *Problem 2*

Go to
www.wiley.com/college/cutnell
for an interactive solution.

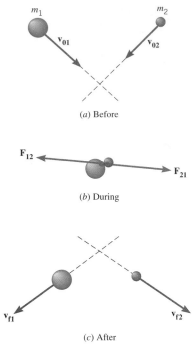

Figure 7.7 (*a*) The velocities of the two objects before the collision are $\mathbf{v}_{01}$ and $\mathbf{v}_{02}$. (*b*) During the collision, each object exerts a force on the other. These forces are $\mathbf{F}_{12}$ and $\mathbf{F}_{21}$. (*c*) The velocities after the collision are $\mathbf{v}_{f1}$ and $\mathbf{v}_{f2}$.

7.2 The Principle of Conservation of Linear Momentum

It is worthwhile to compare the impulse–momentum theorem to the work–energy theorem discussed in Chapter 6. The impulse–momentum theorem states that the impulse produced by a net force is equal to the change in the object's momentum, while the work–energy theorem states that the work done by a net force is equal to the change in the object's kinetic energy. The work–energy theorem leads directly to the principle of conservation of mechanical energy (see Figure 6.16), and, as we will see, the impulse–momentum theorem also leads to a conservation principle, known as the conservation of linear momentum.

We begin by applying the impulse–momentum theorem to a midair collision between two objects. The two objects (masses m_1 and m_2) are approaching each other with initial velocities $\mathbf{v}_{01}$ and $\mathbf{v}_{02}$, as Figure 7.7a shows. The collection of objects being studied is referred to as the "system." In this case, the system contains only the two objects. They interact during the collision in part *b* and then depart with the final velocities $\mathbf{v}_{f1}$ and $\mathbf{v}_{f2}$ shown in part *c*. Because of the collision, the initial and final velocities are not the same.

Two types of forces act on the system:

1. *Internal forces*—Forces that the objects within the system exert on each other.

2. *External forces*—Forces exerted on the objects by agents external to the system.

During the collision in Figure 7.7b, $\mathbf{F}_{12}$ is the force exerted on object 1 by object 2, while $\mathbf{F}_{21}$ is the force exerted on object 2 by object 1. These forces are action–reaction forces that are equal in magnitude but opposite in direction, so $\mathbf{F}_{12} = -\mathbf{F}_{21}$. They are also internal forces, since they are forces that the two objects within the system exert on each other. The force of gravity also acts on the objects, their weights being $\mathbf{W}_1$ and $\mathbf{W}_2$. These weights, however, are external forces, because they are applied by the earth, which is outside the system. Friction and air resistance would also be considered external forces, although these forces are ignored here for the sake of simplicity. The impulse–momentum theorem, as applied to each object, gives the following results:

Object 1 $(\underbrace{\mathbf{W}_1}_{\substack{\text{External} \\ \text{force}}} + \underbrace{\overline{\mathbf{F}}_{12}}_{\substack{\text{Internal} \\ \text{force}}}) \Delta t = m_1 \mathbf{v}_{f1} - m_1 \mathbf{v}_{01}$

Object 2 $(\underbrace{\mathbf{W}_2}_{\substack{\text{External} \\ \text{force}}} + \underbrace{\overline{\mathbf{F}}_{21}}_{\substack{\text{Internal} \\ \text{force}}}) \Delta t = m_2 \mathbf{v}_{f2} - m_2 \mathbf{v}_{02}$

Adding these equations produces a single result for the system as a whole:

$$(\underbrace{\mathbf{W}_1 + \mathbf{W}_2}_{\substack{\text{External} \\ \text{forces}}} + \underbrace{\overline{\mathbf{F}}_{12} + \overline{\mathbf{F}}_{21}}_{\substack{\text{Internal} \\ \text{forces}}}) \Delta t = \underbrace{(m_1 \mathbf{v}_{f1} + m_2 \mathbf{v}_{f2})}_{\substack{\text{Total final} \\ \text{momentum } \mathbf{P}_f}} - \underbrace{(m_1 \mathbf{v}_{01} + m_2 \mathbf{v}_{02})}_{\substack{\text{Total initial} \\ \text{momentum } \mathbf{P}_0}}$$

On the right side of this equation, the quantity $m_1 \mathbf{v}_{f1} + m_2 \mathbf{v}_{f2}$ is the vector sum of the final momenta for each object, or the total final momentum $\mathbf{P}_f$ of the system. Likewise, $m_1 \mathbf{v}_{01} + m_2 \mathbf{v}_{02}$ is the total initial momentum $\mathbf{P}_0$. Therefore, the result above can be rewritten as

$$\left(\begin{array}{c} \textbf{Sum of average} \\ \textbf{external forces} \end{array} + \begin{array}{c} \textbf{Sum of average} \\ \textbf{internal forces} \end{array} \right) \Delta t = \mathbf{P}_f - \mathbf{P}_0 \qquad (7.5)$$

The advantage of the internal/external force classification is that the internal forces always add together to give zero, as a consequence of Newton's law of action–reaction; $\overline{\mathbf{F}}_{12} = -\overline{\mathbf{F}}_{21}$ so that $\overline{\mathbf{F}}_{12} + \overline{\mathbf{F}}_{21} = 0$. Cancellation of the internal forces occurs no matter how many parts there are to the system and allows us to ignore the internal forces, as Equation 7.6 indicates:

$$\textbf{(Sum of average external forces)} \, \Delta t = \mathbf{P}_f - \mathbf{P}_0 \qquad (7.6)$$

We developed this result with gravity as the only external force. But, in general, the sum of the external forces on the left includes *all* external forces.

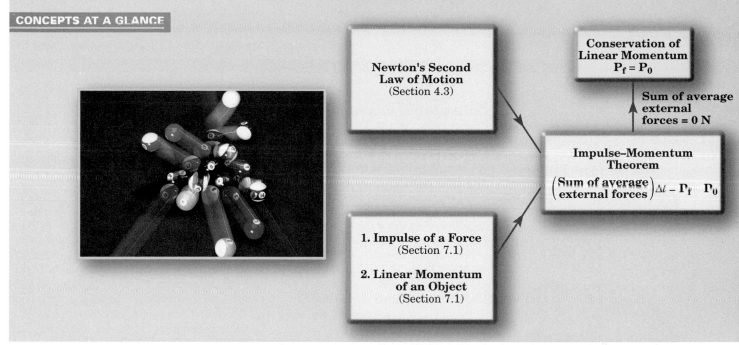

Figure 7.8 CONCEPTS AT A GLANCE
The impulse–momentum theorem leads to the principle of conservation of linear momentum when the sum of the average external forces acting on a system is zero. In the collision in the photograph, the conservation of linear momentum means that the total linear momentum of the cue ball and the other balls before the collision is equal to the total linear momentum after the collision. (© David Leah/Stone/Getty Images)

With the aid of Equation 7.6, it is possible to see how the conservation of linear momentum arises. Suppose that the sum of the external forces is zero. A system for which this is true is called an *isolated system.* Then Equation 7.6 indicates that

$$0 = \mathbf{P_f} - \mathbf{P_0} \quad \text{or} \quad \mathbf{P_f} = \mathbf{P_0} \tag{7.7a}$$

In other words, the final total momentum of the isolated system after the objects in Figure 7.7 collide is the same as the initial total momentum.* Explicitly writing out the final and initial momenta for the two-body collision, we obtain for Equation 7.7a that

$$\underbrace{m_1\mathbf{v_{f1}} + m_2\mathbf{v_{f2}}}_{\mathbf{P_f}} = \underbrace{m_1\mathbf{v_{01}} + m_2\mathbf{v_{02}}}_{\mathbf{P_0}} \tag{7.7b}$$

This result is an example of a general principle known as the **principle of conservation of linear momentum.**

■ PRINCIPLE OF CONSERVATION OF LINEAR MOMENTUM

The total linear momentum of an isolated system remains constant (is conserved). An isolated system is one for which the vector sum of the average external forces acting on the system is zero.

▶ CONCEPTS AT A GLANCE The Concepts-at-a-Glance chart in Figure 7.8 presents an overview of how concepts that we have studied lead to the conservation of linear momentum. The chart is an extension of that shown earlier in Figure 7.3. In the middle, it shows that the origin of the impulse–momentum theorem lies in Newton's second law and the ideas of the impulse of a force and the linear momentum of an object. The theorem is stated at the right of the chart in the form (**Sum of average external forces**) $\Delta t = \mathbf{P_f} - \mathbf{P_0}$. If the sum of the average external forces is zero, then the final and initial total momenta are equal, or $\mathbf{P_f} = \mathbf{P_0}$, which is the mathematical statement of the conservation of linear momentum. ◀

*Technically, the initial and final momenta are equal when the impulse of the sum of the external forces is zero—that is, when the left-hand side of Equation 7.6 is zero. Somtimes, however, the initial and final momenta are very nearly equal even when the sum of the external forces is not zero. This occurs when the time Δt during which the forces act is so short that it is effectively zero. Then, the left-hand side of Equation 7.6 is approximately zero.

This principle applies to a system containing any number of objects, regardless of the internal forces, provided the system is isolated. Whether the system is isolated depends on whether the vector sum of the external forces is zero. Judging whether a force is internal or external depends on which objects are included in the system, as Conceptual Example 4 illustrates.

Conceptual Example 4   Is the Total Momentum Conserved?

Imagine two balls colliding on a billiard table that is friction-free. Use the momentum conservation principle in answering the following questions. (a) Is the total momentum of the two-ball system the same before and after the collision? (b) Answer part (a) for a system that contains only one of the two colliding balls.

Reasoning and Solution

(a) Figure 7.9a shows the balls colliding, and the dashed outline emphasizes that both balls are part of the system. The collision forces are not shown, because they are internal forces and cannot cause the total momentum of the system to change. The external forces include the weights $\mathbf{W}_1$ and $\mathbf{W}_2$ of the balls. There are two additional external forces, however: the upward-pointing normal force that the table exerts on each ball. These normal forces are $\mathbf{F}_{N1}$ and $\mathbf{F}_{N2}$. Since the balls do not accelerate in the vertical direction, the normal forces must balance the weights, so that the vector sum of the four external forces in Figure 7.9a is zero. Furthermore, the table is friction-free. Thus, there is no net external force to change the total momentum of the two-ball system, and it is the same before and after the collision. *The total momentum of this two-ball system is conserved.*

(b) In Figure 7.9b only one ball is included in the system, as the dashed outline indicates. The forces acting on this system are all external and include the weight $\mathbf{W}_1$ of the ball and the normal force $\mathbf{F}_{N1}$. As in part a of the drawing, these two forces balance. However, there is a third external force to consider. Ball 2 is outside the system, so the force $\mathbf{F}_{12}$ that it applies to the system during the collision is now an external force. As a result, the vector sum of the three external forces is not zero, and the net external force causes the total momentum of the one-ball system to be different after the collision than it was before the collision. *The total momentum of this one-ball system is not conserved.*

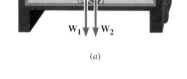
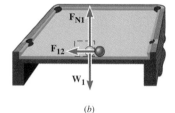

Figure 7.9 (a) Two billiard balls collide. They have weights $\mathbf{W}_1$ and $\mathbf{W}_2$. The dashed box emphasizes that both balls are included in the system. $\mathbf{F}_{N1}$ and $\mathbf{F}_{N2}$ are the normal forces that the table applies to the balls. (b) Now only ball 1 is included in the system.

Next, we apply the principle of conservation of linear momentum to the problem of assembling a freight train.

Example 5   Assembling a Freight Train

A freight train is being assembled in a switching yard, and Figure 7.10 shows two boxcars. Car 1 has a mass of $m_1 = 65 \times 10^3$ kg and moves at a velocity of $v_{01} = +0.80$ m/s. Car 2, with a mass of $m_2 = 92 \times 10^3$ kg and a velocity of $v_{02} = +1.3$ m/s, overtakes car 1 and couples to it. Neglecting friction, find the common velocity v_f of the cars after they become coupled.

Reasoning The two boxcars constitute the system. The sum of the external forces acting on the system is zero, because the weight of each car is balanced by a corresponding normal force, and friction is being neglected. Thus, the system is isolated, and the principle of conservation of linear momentum applies. The coupling forces that each car exerts on the other are internal forces and do not affect the applicability of this principle.

Solution Momentum conservation indicates that

$$\underbrace{(m_1 + m_2)v_f}_{\substack{\text{Total momentum} \\ \text{after collision}}} = \underbrace{m_1 v_{01} + m_2 v_{02}}_{\substack{\text{Total momentum} \\ \text{before collision}}}$$

Problem solving insight
The conservation of linear momentum is applicable only when the net external force acting on the system is zero. Therefore, the first step in applying momentum conservation is to be sure that the net external force is zero.

Figure 7.10 (a) The boxcar on the left eventually catches up with the other boxcar and (b) couples to it. The coupled cars move together with a common velocity after the collision.

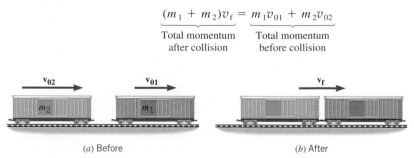

(a) Before   (b) After

This equation can be solved for v_f, the common velocity of the two cars after the collision:

$$v_f = \frac{m_1 v_{01} + m_2 v_{02}}{m_1 + m_2}$$

$$= \frac{(65 \times 10^3 \text{ kg})(0.80 \text{ m/s}) + (92 \times 10^3 \text{ kg})(1.3 \text{ m/s})}{(65 \times 10^3 \text{ kg} + 92 \times 10^3 \text{ kg})} = \boxed{+1.1 \text{ m/s}}$$

In the previous example it can be seen that the velocity of car 1 increases, while the velocity of car 2 decreases as a result of the collision. The acceleration and deceleration arise at the moment the cars become coupled, because the cars exert internal forces on each other. The powerful feature of the momentum conservation principle is that it allows us to determine the changes in velocity without knowing what the internal forces are. Example 6 further illustrates this feature.

Example 6   Ice Skaters

Starting from rest, two skaters "push off" against each other on smooth level ice, where friction is negligible. As Figure 7.11*a* shows, one is a woman ($m_1 = 54$ kg), and one is a man ($m_2 = 88$ kg). Part *b* of the drawing shows that the woman moves away with a velocity of $v_{f1} = +2.5$ m/s. Find the "recoil" velocity v_{f2} of the man.

Reasoning For a system consisting of the two skaters on level ice, the sum of the external forces is zero. This is because the weight of each skater is balanced by a corresponding normal force and friction is negligible. The skaters, then, constitute an isolated system, and the principle of conservation of linear momentum applies. We expect the man to have a smaller recoil speed for the following reason. The internal forces that the man and woman exert on each other during pushoff have equal magnitudes but opposite directions, according to Newton's action–reaction law. The man, having the larger mass, experiences a smaller acceleration according to Newton's second law. Hence, he acquires a smaller recoil speed.

Solution The total momentum of the skaters before they push on each other is zero, since they are at rest. Momentum conservation requires that the total momentum remains zero after the skaters have separated, as in part *b* of the drawing:

$$\underbrace{m_1 v_{f1} + m_2 v_{f2}}_{\substack{\text{Total momentum} \\ \text{after pushing}}} = \underbrace{0}_{\substack{\text{Total momentum} \\ \text{before pushing}}}$$

Solving for the recoil velocity of the man gives

$$v_{f2} = \frac{-m_1 v_{f1}}{m_2} = \frac{-(54 \text{ kg})(+2.5 \text{ m/s})}{88 \text{ kg}} = \boxed{-1.5 \text{ m/s}}$$

The minus sign indicates that the man moves to the left in the drawing. After the skaters separate, the total momentum of the system remains zero, because momentum is a vector quantity, and the momenta of the man and the woman have equal magnitudes but opposite directions.

(a) Before

v_{f2} v_{f1}

(b) After

Figure 7.11 (*a*) In the absence of friction, two skaters pushing on each other constitute an isolated system. (*b*) As the skaters move away, the total linear momentum of the system remains zero, which is what it was initially.

Concept Simulation 7.1

This simulation illustrates the two ice skaters in Figure 7.11, who are initially at rest and then push off against one another. In the absence of friction, the total linear momentum of the two-skater system is conserved, so it is the same before and after the push-off. The mass of one of the skaters is under your control, so you can explore how the speeds of the two recoiling skaters change as the mass changes.

Related Homework: Conceptual Question 10, Problems 17, 18

Go to **www.wiley.com/college/cutnell**

It is important to realize that the total linear momentum may be conserved even when the kinetic energies of the individual parts of a system change. In Example 6, for instance, the initial kinetic energy is zero, since the skaters are stationary. But after they push off, the skaters are moving, so each has kinetic energy. The kinetic energy changes

Problem solving insight

Need more practice?

Interactive LearningWare 7.2
An astronaut is motionless in outer space. Upon command, the propulsion unit strapped to his back ejects some gas with a velocity of +32 m/s, and the astronaut recoils with a velocity of −0.30 m/s. After the gas is ejected, the mass of the astronaut is 160 kg. What is the mass of the ejected gas?

Related Homework: *Problem 15*

Go to
www.wiley.com/college/cutnell
for an interactive solution to the problem.

because work is done by the internal force that each skater exerts on the other. However, internal forces cannot change the total linear momentum of a system, since the total linear momentum of an isolated system is conserved in the presence of such forces.

When applying the principle of conservation of linear momentum, we have been following a definite reasoning strategy that is summarized below.

Reasoning Strategy

Applying the Principle of Conservation of Linear Momentum

1. Decide which objects are included in the system.
2. Relative to the system that you have chosen, identify the internal forces and the external forces.
3. Verify that the system is isolated. In other words, verify that the sum of the external forces applied to the system is zero. Only if this sum is zero can the conservation principle be applied. If the sum of the average external forces is not zero, consider a different system for analysis.
4. Set the total final momentum of the isolated system equal to the total initial momentum. Remember that linear momentum is a vector. If necessary, apply the conservation principle separately to the various vector components.

✓ Check Your Understanding 2

A canoe with two people aboard is coasting with an initial momentum of +110 kg·m/s. Then, one of the people (person 1) dives off the back of the canoe. During this time, the net average external force acting on the system (canoe and the two people) is zero. The table lists four possibilities for the final momentum of person 1 and the final momentum of person 2 plus the canoe, immediately after person 1 leaves the canoe. Only one possibility could be correct. Which is it? *(The answer is given at the end of the book.)*

	Final Momenta	
	Person 1	Person 2 and Canoe
a.	−60 kg·m/s	+170 kg·m/s
b.	−30 kg·m/s	+110 kg·m/s
c.	−40 kg·m/s	−70 kg·m/s
d.	+80 kg·m/s	−30 kg·m/s

Background: Momentum and the principle of conservation of linear momentum play a role in this question. Remember that momentum is a vector quantity.

For similar questions (including conceptual counterparts), consult Self-Assessment Test 7.1. The test is described next.

Self-Assessment Test 7.1

Four important ideas are discussed in Sections 7.1 and 7.2:

• The Impulse of a Force   • The Linear Momentum of an Object   • The Impulse–Momentum Theorem   • The Conservation of Linear Momentum

Use Self-Assessment Test 7.1 to check your understanding of these concepts.

Go to **www.wiley.com/college/cutnell**

7.3 Collisions in One Dimension

As discussed in the last section, the total linear momentum is conserved when two objects collide, provided they constitute an isolated system. When the objects are atoms or subatomic particles, the total kinetic energy of the system is often conserved also. In other words, the total kinetic energy of the particles before the collision equals the total kinetic energy of the particles after the collision, so that kinetic energy gained by one particle is lost by another.

In contrast, when two macroscopic objects collide, such as two cars, the total kinetic energy after the collision is generally less than that before the collision. Kinetic energy is lost mainly in two ways: (1) it can be converted into heat because of friction, and (2) it is spent in creating permanent distortion or damage, as in an automobile collision. With very hard objects, such as a solid steel ball and a marble floor, the permanent distortion suffered upon collision is much smaller than with softer objects and, consequently, less kinetic energy is lost.

Collisions are often classified according to whether the total kinetic energy changes during the collision:

1. *Elastic collision*—One in which the total kinetic energy of the system after the collision is equal to the total kinetic energy before the collision.

2. *Inelastic collision*—One in which the total kinetic energy of the system is *not* the same before and after the collision; if the objects stick together after colliding, the collision is said to be completely inelastic.

The boxcars coupling together in Figure 7.10 are an example of a completely inelastic collision. When a collision is completely inelastic, the greatest amount of kinetic energy is lost. Example 7 shows how one particular elastic collision is described using the conservation of linear momentum and the fact that no kinetic energy is lost.

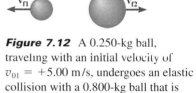

Figure 7.12 A 0.250-kg ball, traveling with an initial velocity of $v_{01} = +5.00$ m/s, undergoes an elastic collision with a 0.800-kg ball that is initially at rest.

Example 7 A Collision in One Dimension

As in Figure 7.12, a ball of mass $m_1 = 0.250$ kg and velocity $v_{01} = +5.00$ m/s collides head-on with a ball of mass $m_2 = 0.800$ kg that is initially at rest ($v_{02} = 0$ m/s). No external forces act on the balls. If the collision is elastic, what are the velocities of the balls after the collision?

Reasoning The total linear momentum of the two-ball system is conserved, because no external forces act on the system. Momentum conservation applies whether or not the collision is elastic:

$$\underbrace{m_1 v_{f1} + m_2 v_{f2}}_{\substack{\text{Total momentum} \\ \text{after collision}}} = \underbrace{m_1 v_{01} + 0}_{\substack{\text{Total momentum} \\ \text{before collision}}}$$

Before and after an elastic collision, the total kinetic energy is the same.

$$\underbrace{\tfrac{1}{2} m_1 v_{f1}^2 + \tfrac{1}{2} m_2 v_{f2}^2}_{\substack{\text{Total kinetic energy} \\ \text{after collision}}} = \underbrace{\tfrac{1}{2} m_1 v_{01}^2 + 0}_{\substack{\text{Total kinetic energy} \\ \text{before collision}}}$$

We expect that ball 1, having the smaller mass, will rebound to the left after striking ball 2, which is more massive. Ball 2 will be driven to the right in the process. One solution to the equations above, then, should give a value for v_{f1} that is negative (the left direction in Figure 7.12) and a value for v_{f2} that is positive.

Solution The equations above are simultaneous equations containing the two unknown quantities v_{f1} and v_{f2}. To solve them, we begin by rearranging the equation expressing momentum conservation to show that $v_{f2} = m_1(v_{01} - v_{f1})/m_2$. Substituting this result into the equation expressing the conservation of kinetic energy leads to the following equation for v_{f1}:

$$v_{f1} = \left(\frac{m_1 - m_2}{m_1 + m_2} \right) v_{01} \tag{7.8a}$$

The expression for v_{f1} in Equation 7.8a can be substituted into either of the equations obtained in the Reasoning to show that

$$v_{f2} = \left(\frac{2m_1}{m_1 + m_2} \right) v_{01} \tag{7.8b}$$

With the given values for m_1, m_2, and v_{01}, Equations 7.8 yield the following values for v_{f1} and v_{f2}:

$$\boxed{v_{f1} = -2.62 \text{ m/s}} \quad \text{and} \quad \boxed{v_{f2} = +2.38 \text{ m/s}}$$

The negative value for v_{f1} indicates that ball 1 rebounds to the left after the collision in Figure 7.12, while the positive value for v_{f2} indicates that ball 2 moves to the right, as expected.

Problem solving insight
As long as the net external force is zero, the conservation of linear momentum applies to any type of collision. This is true whether the collision is elastic or inelastic.

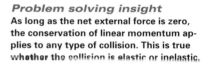
Concept Simulation 7.2

In this simulation two balls experience a head-on collision that can be either elastic or completely inelastic. When the collision is elastic, the total linear momentum and the total kinetic energy of the two-ball system are conserved. You can alter the masses and initial velocities of the objects. The simulation illustrates the collision itself, and it displays the momentum and kinetic energy of each ball, both before and after the collision.

Related Homework: Conceptual Question 17, Problem 38

Go to
www.wiley.com/college/cutnell

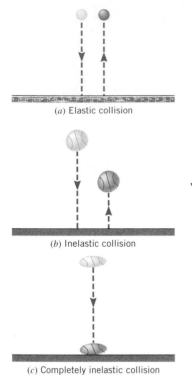

(a) Elastic collision

(b) Inelastic collision

(c) Completely inelastic collision

Figure 7.13 (a) A hard steel ball would rebound to its original height after striking a hard marble surface if the collision were elastic. (b) A partially deflated basketball has little bounce on a soft asphalt surface. (c) A deflated basketball has no bounce at all.

We can get a feel for an elastic collision by dropping a steel ball onto a hard surface, such as a marble floor. If the collision is elastic, the ball will rebound to its original height, as Figure 7.13a illustrates. In contrast, a partially deflated basketball exhibits little rebound from a relatively soft asphalt surface, as in part b, indicating that a fraction of the ball's kinetic energy is dissipated during the inelastic collision. The completely deflated basketball in part c has no bounce at all, and a maximum amount of kinetic energy is lost during the completely inelastic collision.

The next example illustrates a completely inelastic collision in a device called a "ballistic pendulum." This device can be used to measure the speed of a bullet.

Example 8   A Ballistic Pendulum

A ballistic pendulum can be used to measure the speed of a projectile, such as a bullet. The ballistic pendulum shown in Figure 7.14a consists of a block of wood (mass $m_2 = 2.50$ kg) suspended by a wire of negligible mass. A bullet (mass $m_1 = 0.0100$ kg) is fired with a speed v_{01}. Just after the bullet collides with it, the block (with the bullet in it) has a speed v_f and then swings to a maximum height of 0.650 m above the initial position (see part b of the drawing). Find the speed v_{01} of the bullet, assuming that air resistance is negligible.

Reasoning The physics of the ballistic pendulum can be divided into two parts. The first is the completely inelastic collision between the bullet and the block. The second is the resulting motion of the block and bullet as they swing upward. The total momentum of the system (block plus bullet) is conserved during the collision, because the suspension wire supports the system's weight, which means that the sum of the external forces acting on the system is nearly zero. Furthermore, as the system swings upward, the principle of conservation of mechanical energy applies, because nonconservative forces do no work. The tension force in the wire does no work because it acts perpendicular to the motion. Since air resistance is negligible, we can ignore the work it does.

Solution Applying the momentum conservation principle gives

$$\underbrace{(m_1 + m_2)v_f}_{\substack{\text{Total momentum} \\ \text{after collision}}} = \underbrace{m_1 v_{01}}_{\substack{\text{Total momentum} \\ \text{before collision}}}$$

This equation can be solved for the initial speed v_{01} of the bullet:

$$v_{01} = \frac{m_1 + m_2}{m_1} v_f$$

A value is now needed for the speed v_f immediately after the collision. It can be obtained from the maximum height to which the system swings, by using the principle of conservation of mechanical energy:

$$\underbrace{(m_1 + m_2)gh_f}_{\substack{\text{Total mechanical energy} \\ \text{at the top of the swing,} \\ \text{all potential}}} = \underbrace{\tfrac{1}{2}(m_1 + m_2)v_f^2}_{\substack{\text{Total mechanical energy} \\ \text{at the bottom of the} \\ \text{swing, all kinetic}}}$$

The physics of measuring the speed of a bullet.

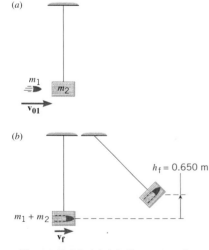

(a)

(b)

$h_f = 0.650$ m

$m_1 + m_2$

v_f

Figure 7.14 (a) A bullet approaches a ballistic pendulum. (b) The block and bullet swing upward after the collision.

Solving this expression for v_f gives $v_f = \sqrt{2gh_f}$. Substituting this result into the equation for v_{01}, we find that

$$v_{01} = \left(\frac{m_1 + m_2}{m_1}\right)\sqrt{2gh_f} = \left(\frac{0.0100 \text{ kg} + 2.50 \text{ kg}}{0.0100 \text{ kg}}\right)\sqrt{2(9.80 \text{ m/s}^2)(0.650 \text{ m})}$$

$$= \boxed{+896 \text{ m/s}}$$

✔ Check Your Understanding 3

Two balls collide in a one-dimensional, elastic collision. The two balls constitute a system, and the net external force acting on them is zero. The table shows four possible sets of values for the initial and final momenta of the two balls, as well as their initial and final kinetic energies. Only one set of values could be correct. Which is it? (*The answer is given at the end of the book.*)

		Initial (Before Collision)		Final (After Collision)	
		Momentum	Kinetic Energy	Momentum	Kinetic Energy
a.	Ball 1:	+4 kg·m/s	12 J	−5 kg·m/s	10 J
	Ball 2:	−3 kg·m/s	5 J	−1 kg·m/s	7 J
b.	Ball 1:	+7 kg·m/s	22 J	+5 kg·m/s	18 J
	Ball 2:	+2 kg·m/s	8 J	+4 kg·m/s	15 J
c.	Ball 1:	−5 kg·m/s	12 J	−6 kg·m/s	15 J
	Ball 2:	−8 kg·m/s	31 J	−9 kg·m/s	25 J
d.	Ball 1:	+9 kg·m/s	25 J	+6 kg·m/s	18 J
	Ball 2:	+4 kg·m/s	15 J	+7 kg·m/s	22 J

Background: Momentum, kinetic energy, the principle of conservation of linear momentum, and the notion of an elastic collision are involved in this question. Remember that momentum is a vector quantity, and kinetic energy is a scalar quantity.

For similar questions (including conceptual counterparts) consult Self-Assessment Test 7.2. This test is described at the end of Section 7.5.

7.4 *Collisions in Two Dimensions*

The collisions discussed so far have been "head-on," or one-dimensional, because the velocities of the objects all point along a single line before and after contact. Collisions often occur, however, in two or three dimensions. Figure 7.15 shows a two-dimensional case in which two balls collide on a horizontal frictionless table.

For the system consisting of the two balls, the external forces include the weights of the balls and the corresponding normal forces produced by the table. Since each weight is balanced by a normal force, the sum of the external forces is zero, and the total momentum of the system is conserved, as Equation 7.7b indicates. Momentum is a vector quantity, however, and in two dimensions the *x* and *y* components of the total momentum are conserved separately. In other words, Equation 7.7b is equivalent to the following two equations:

x Component

$$\underbrace{m_1 v_{f1x} + m_2 v_{f2x}}_{\mathbf{P}_{fx}} = \underbrace{m_1 v_{01x} + m_2 v_{02x}}_{\mathbf{P}_{0x}} \tag{7.9a}$$

y Component

$$\underbrace{m_1 v_{f1y} + m_2 v_{f2y}}_{\mathbf{P}_{fy}} = \underbrace{m_1 v_{01y} + m_2 v_{02y}}_{\mathbf{P}_{0y}} \tag{7.9b}$$

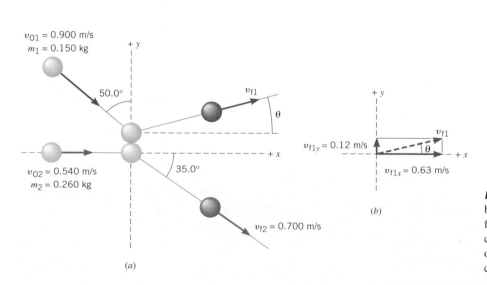

(a)

(b)

Figure 7.15 (*a*) Top view of two balls colliding on a horizontal frictionless table. (*b*) This part of the drawing shows the *x* and *y* components of the velocity of ball 1 after the collision.

These equations are written for a system that contains two objects. If a system contains more than two objects, a mass-times-velocity term must be included for each additional object on either side of Equations 7.9a and 7.9b. Example 9 shows how to deal with a two-dimensional collision when the total linear momentum of the system is conserved.

Example 9   A Collision in Two Dimensions

For the data given in Figure 7.15, use momentum conservation to determine the magnitude and direction of the final velocity of ball 1 after the collision.

Reasoning The magnitude and direction of the final velocity of ball 1 can be obtained once the components v_{f1x} and v_{f1y} of this velocity are known. We will begin by determining these components with the aid of the momentum conservation principle.

Solution Applying momentum conservation in the x-direction (Equation 7.9a), we find that

x Component

$$\underbrace{(0.150 \text{ kg})(v_{f1x})}_{\text{Ball 1, after}} + \underbrace{(0.260 \text{ kg})(0.700 \text{ m/s})(\cos 35.0°)}_{\text{Ball 2, after}}$$

$$= \underbrace{(0.150 \text{ kg})(0.900 \text{ m/s})(\sin 50.0°)}_{\text{Ball 1, before}} + \underbrace{(0.260 \text{ kg})(0.540 \text{ m/s})}_{\text{Ball 2, before}}$$

This equation can be solved to show that $v_{f1x} = +0.63$ m/s. Applying momentum conservation in the y-direction (Equation 7.9b), we find that

y Component

$$\underbrace{(0.150 \text{ kg})(v_{f1y})}_{\text{Ball 1, after}} + \underbrace{(0.260 \text{ kg})[-(0.700 \text{ m/s})(\sin 35.0°)]}_{\text{Ball 2, after}}$$

$$= \underbrace{(0.150 \text{ kg})[-(0.900 \text{ m/s})(\cos 50.0°)]}_{\text{Ball 1, before}} + \underbrace{0}_{\text{Ball 2, before}}$$

The solution to this equation reveals that $v_{f1y} = +0.12$ m/s.

Figure 7.15b shows the x and y components of the final velocity of ball 1. The magnitude of the velocity is

$$v_{f1} = \sqrt{(0.63 \text{ m/s})^2 + (0.12 \text{ m/s})^2} = \boxed{0.64 \text{ m/s}}$$

The direction of the velocity is given by the angle θ:

$$\theta = \tan^{-1}\left(\frac{0.12 \text{ m/s}}{0.63 \text{ m/s}}\right) = \boxed{11°}$$

7.5   Center of Mass

In previous sections, we have encountered situations in which objects interact with one another, such as the two skaters pushing off in Example 6. In these situations, the mass of the system is located in several places, and the various objects move relative to each other before, after, and even during the interaction. It is possible, however, to speak of a kind of average location for the total mass by introducing a concept known as the *center of mass* (abbreviated as "cm"). With the aid of this concept, we will be able to gain additional insight into the principle of conservation of linear momentum.

The center of mass is a point that represents the average location for the total mass of a system. Figure 7.16, for example, shows two particles of mass m_1 and m_2 that are located on the x axis at the positions x_1 and x_2, respectively. The position x_{cm} of the center-of-mass point from the origin is defined to be

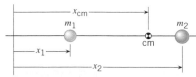

Figure 7.16 The center of mass cm of the two particles is located on a line between them and lies closer to the more massive one.

Center of mass $$x_{cm} = \frac{m_1 x_1 + m_2 x_2}{m_1 + m_2}$$ (7.10)

Each term in the numerator of this equation is the product of a particle's mass and position, while the denominator is the total mass of the system. If the two masses are equal, we expect the average location of the total mass to be midway between the particles. With $m_1 = m_2 = m$, Equation 7.10 becomes $x_{cm} = (mx_1 + mx_2)/(m + m) = \frac{1}{2}(x_1 + x_2)$, which indeed corresponds to the point midway between the particles. Alternatively, suppose that $m_1 = 5.0$ kg and $x_1 = 2.0$ m, while $m_2 = 12$ kg and $x_2 = 6.0$ m. Then we expect the average location of the total mass to be located closer to particle 2, since it is more massive. Equation 7.10 is also consistent with this expectation, for it gives

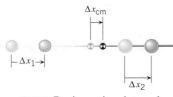

Figure 7.17 During a time interval Δt, the displacements of the particles are Δx_1 and Δx_2, while the displacement of the center of mass is Δx_{cm}.

$$x_{cm} = \frac{(5.0 \text{ kg})(2.0 \text{ m}) + (12 \text{ kg})(6.0 \text{ m})}{5.0 \text{ kg} + 12 \text{ kg}} = 4.8 \text{ m}$$

If a system contains more than two particles, the center-of-mass point can be determined by generalizing Equation 7.10. For three particles, for instance, the numerator would contain a third term m_3x_3, and the total mass in the denominator would be $m_1 + m_2 + m_3$. For a macroscopic object, which contains many, many particles, the center-of-mass point is located at the geometric center of the object, provided that the mass is distributed symmetrically about the center. Such would be the case for a billiard ball. For objects such as a golf club, the mass is not distributed symmetrically and the center-of-mass point is not located at the geometric center of the club. The driver used to launch a golf ball from the tee, for instance, has more mass in the club head than in the handle, so the center-of-mass point is closer to the head than to the handle.

To see how the center-of-mass concept is related to momentum conservation, suppose that the two particles in a system are moving, as they would be during a collision. With the aid of Equation 7.10, we can determine the velocity v_{cm} of the center-of-mass point. During a time interval Δt, the particles experience displacements of Δx_1 and Δx_2, as Figure 7.17 shows. They have different displacements during this time because they have different velocities. Equation 7.10 can be used to find the displacement Δx_{cm} of the center of mass by replacing x_{cm} by Δx_{cm}, x_1 by Δx_1, and x_2 by Δx_2:

$$\Delta x_{cm} = \frac{m_1 \Delta x_1 + m_2 \Delta x_2}{m_1 + m_2}$$

Now we divide both sides of this equation by the time interval Δt. In the limit as Δt becomes infinitesimally small, the ratio $\Delta x_{cm}/\Delta t$ becomes equal to the instantaneous velocity v_{cm} of the center of mass. (See Section 2.2 for a review of instantaneous velocity.) Likewise, the ratios $\Delta x_1/\Delta t$ and $\Delta x_2/\Delta t$ become equal to the instantaneous velocities v_1 and v_2, respectively. Thus, we have

Velocity of center of mass
$$v_{cm} = \frac{m_1 v_1 + m_2 v_2}{m_1 + m_2} \tag{7.11}$$

The numerator $(m_1v_1 + m_2v_2)$ on the right-hand side in Equation 7.11 is the momentum of particle 1 (mv_1) plus the momentum of particle 2 (m_2v_2), which is the total linear momentum of the system. In an isolated system, the total linear momentum does not change because of an interaction such as a collision. Therefore, Equation 7.11 indicates that the velocity v_{cm} of the center of mass does not change either. To emphasize this important point, consider the collision discussed in Example 7. With the data from that example, we can apply Equation 7.11 to determine the velocity of the center of mass before and after the collision:

Before collision
$$v_{cm} = \frac{(0.250 \text{ kg})(+5.00 \text{ m/s}) + (0.800 \text{ kg})(0 \text{ m/s})}{0.250 \text{ kg} + 0.800 \text{ kg}} = +1.19 \text{ m/s}$$

After collision
$$v_{cm} = \frac{(0.250 \text{ kg})(-2.62 \text{ m/s}) + (0.800 \text{ kg})(+2.38 \text{ m/s})}{0.250 \text{ kg} + 0.800 \text{ kg}} = +1.19 \text{ m/s}$$

Thus, the velocity of the center of mass is the same before and after the objects interact during a collision in which the total linear momentum is conserved.

✔ **Check Your Understanding 4**

Water, dripping at a constant rate from a faucet, falls to the ground. At any instant, there are many drops in the air between the faucet and the ground. Where does the center of mass of the drops lie relative to the halfway point between the faucet and the ground: above it, below it, or exactly at the halfway point? *(The answer is given at the end of the book.)*

Background: The solution to this question involves an understanding of the concepts of accelerated motion (the falling drops) and the center of mass.

For similar questions (including calculational counterparts), consult Self-Assessment Test 7.2. The test is described next.

🖱 Self-Assessment Test 7.2

Several important ideas are discussed in Sections 7.3–7.5:

• Collisions in One and Two Dimensions
• The Center of Mass of a System of Particles

Use Self-Assessment Test 7.2 to check your understanding of these concepts.

Go to **www.wiley.com/college/cutnell**

7.6 *Concepts* & *Calculations*

Momentum and energy are two of the most fundamental concepts in physics. As we have seen in this chapter, momentum is a vector and, like all vectors, has a magnitude and a direction. In contrast, energy is a scalar quantity, as Chapter 6 discusses, and does not have a direction associated with it. Example 10 provides the opportunity to review how the vector nature of momentum and the scalar nature of kinetic energy influence calculations using these quantities.

Concepts & Calculations Example 10   A Scalar and a Vector

Two joggers, Jim and Tom, are both running at a speed of 4.00 m/s. Jim has a mass of 90.0 kg, and Tom has a mass of 55.0 kg. Find the kinetic energy and momentum of the two-jogger system when (a) Jim and Tom are both running due north (Figure 7.18a) and (b) Jim is running due north and Tom is running due south (Figure 7.18b).

Concept Questions and Answers Does the total kinetic energy have a smaller value in case (a) or (b), or is it the same in both cases?

Answer   Everything is the same in both cases, except that both joggers are running due north in (a), but one is running due north and one is running due south in (b). Kinetic energy is a scalar quantity and does not depend on directional differences such as these. Therefore, the kinetic energy of the two-jogger system is the same in both cases.

Does the total momentum have a smaller magnitude in case (a) or (b), or is it the same in both cases?

Answer   Momentum is a vector. As a result, when we add Jim's momentum to Tom's momentum to get the total for the two-jogger system, we must take the directions into account. For example, we can use positive to denote north and negative to denote south. In case (a) two positive momentum vectors are added, while in case (b) one positive and one negative vector are added. Due to the partial cancellation in case (b), the resulting total momentum in case (b) will have a smaller magnitude than in case (a).

Solution

(a) The kinetic energy of the two-jogger system can be obtained by applying the definition of kinetic energy as $\frac{1}{2}mv^2$ (see Equation 6.2) to each jogger:

$$\mathrm{KE_{system}} = \tfrac{1}{2}m_{\mathrm{Jim}}v_{\mathrm{Jim}}^2 + \tfrac{1}{2}m_{\mathrm{Tom}}v_{\mathrm{Tom}}^2$$
$$= \tfrac{1}{2}(90.0\ \mathrm{kg})(4.00\ \mathrm{m/s})^2 + \tfrac{1}{2}(55.0\ \mathrm{kg})(4.00\ \mathrm{m/s})^2 = \boxed{1160\ \mathrm{J}}$$

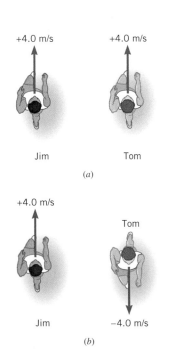

Figure 7.18 (a) Two joggers are running due north with the same velocities. (b) One jogger is running due north and the other due south, with velocities that have the same magnitudes but different directions.

The momentum of each jogger is his mass times his velocity, according to the definition given in Equation 7.2. Using P_{system} to denote the momentum of the two-jogger system and denoting due north as the positive direction, we have

$$P_{\text{system}} = m_{\text{Jim}}v_{\text{Jim}} + m_{\text{Tom}}v_{\text{Tom}}$$
$$= (90.0 \text{ kg})(+4.00 \text{ m/s}) + (55.0 \text{ kg})(+4.00 \text{ m/s}) = \boxed{580 \text{ kg} \cdot \text{m/s}}$$

(b) In these calculations, we note that Tom's velocity is in the due south direction, so it is -4.00 m/s, not $+4.00$ m/s. Proceeding as in part (a), we find

$$\text{KE}_{\text{system}} = \tfrac{1}{2}m_{\text{Jim}}v_{\text{Jim}}^2 + \tfrac{1}{2}m_{\text{Tom}}v_{\text{Tom}}^2$$
$$= \tfrac{1}{2}(90.0 \text{ kg})(4.00 \text{ m/s})^2 + \tfrac{1}{2}(55.0 \text{ kg})(-4.00 \text{ m/s})^2 = \boxed{1160 \text{ J}}$$
$$P_{\text{system}} = m_{\text{Jim}}v_{\text{Jim}} + m_{\text{Tom}}v_{\text{Tom}}$$
$$= (90.0 \text{ kg})(+4.00 \text{ m/s}) + (55.0 \text{ kg})(-4.00 \text{ m/s}) = \boxed{140 \text{ kg} \cdot \text{m/s}}$$

▲

Momentum and kinetic energy are not the same concepts, and the next example explores further some of the differences between them.

Concepts & Calculations Example 11
Momentum and Kinetic Energy

The following table gives mass and speed data for the two objects in Figure 7.19. Find the magnitude of the momentum and the kinetic energy for each object.

	Mass	Speed
Object A	2.0 kg	6.0 m/s
Object B	6.0 kg	2.0 m/s

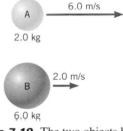

Figure 7.19 The two objects have different masses and speeds. Are their momenta and kinetic energies the same?

Concept Questions and Answers Is it possible for two objects to have different speeds when their momenta have the same magnitude?

Answer Yes. The magnitude of an object's momentum is the product of its mass and its speed, according to Equation 7.2. Both mass and speed play a role. The speed can be reduced, for instance, but if the mass is increased proportionally, the magnitude of the momentum will be the same.

If two objects have the same momentum, do they necessarily have the same kinetic energy?

Answer No. Kinetic energy is $\text{KE} = \tfrac{1}{2}mv^2$. Momentum is $p = mv$, which means that $v = p/m$. Substituting this relation into the kinetic energy expression shows that $\text{KE} = \tfrac{1}{2}m(p/m)^2 = p^2/(2m)$. Thus, two objects can have the same momentum p, but if each has different mass m, their kinetic energies will be different.

Solution Using Equation 7.2 for the momentum and Equation 6.2 for the kinetic energy, we find that

Momenta
$$p_A = m_A v_A = (2.0 \text{ kg})(6.0 \text{ m/s}) = \boxed{12 \text{ kg} \cdot \text{m/s}}$$
$$p_B = m_B v_B = (6.0 \text{ kg})(2.0 \text{ m/s}) = \boxed{12 \text{ kg} \cdot \text{m/s}}$$

Kinetic energies
$$\text{KE}_A = \tfrac{1}{2}m_A v_A^2 = \tfrac{1}{2}(2.0 \text{ kg})(6.0 \text{ m/s})^2 = \boxed{36 \text{ J}}$$
$$\text{KE}_B = \tfrac{1}{2}m_B v_B^2 = \tfrac{1}{2}(6.0 \text{ kg})(2.0 \text{ m/s})^2 = \boxed{12 \text{ J}}$$

▲

At the end of the problem set for this chapter, you will find homework problems that contain both conceptual and quantitative parts. These problems are grouped under the heading *Concepts & Calculations, Group Learning Problems*. They are designed for use by students working alone or in small learning groups. The conceptual part of each problem provides a convenient focus for group discussions.

Concept Summary

This summary presents an abridged version of the chapter, including the important equations and all available learning aids. For convenient reference, the learning aids (including the text's examples) are placed next to or immediately after the relevant equation or discussion. The following learning aids may be found on-line at **www.wiley.com/college/cutnell:**

Interactive LearningWare examples are solved according to a five-step interactive format that is designed to help you develop problem-solving skills.	**Concept Simulations** are animated versions of text figures or animations that illustrate important concepts. You can control parameters that affect the display, and we encourage you to experiment.
Interactive Solutions offer specific models for certain types of problems in the chapter homework. The calculations are carried out interactively.	**Self-Assessment Tests** include both qualitative and quantitative questions. Extensive feedback is provided for both incorrect and correct answers, to help you evaluate your understanding of the material.

Topic	Discussion	Learning Aids
	7.1 The Impulse–Momentum Theorem	
	The impulse $\mathbf{J}$ of a force is the product of the average force $\overline{\mathbf{F}}$ and the time interval Δt during which the force acts:	
Impulse	$$\mathbf{J} = \overline{\mathbf{F}}\,\Delta t \qquad (7.1)$$	
	Impulse is a vector that points in the same direction as the average force.	
	The linear momentum $\mathbf{p}$ of an object is the product of the object's mass m and velocity $\mathbf{v}$:	**Examples 10, 11**
Linear momentum	$$\mathbf{p} = m\mathbf{v} \qquad (7.2)$$	
	Linear momentum is a vector that points in the same direction as the velocity. The total linear momentum of a system of objects is the vector sum of the momenta of the individual objects.	
	The impulse–momentum theorem states that when a net force $\Sigma\overline{\mathbf{F}}$ acts on an object, the impulse of the net force is equal to the change in momentum of the object:	**Examples 1, 2, 3**
		Interactive LearningWare 7.1
Impulse–momentum theorem	$$(\Sigma\overline{\mathbf{F}})\,\Delta t = m\mathbf{v_f} - m\mathbf{v_0} \qquad (7.4)$$	**Interactive Solution 7.13**
	7.2 The Principle of Conservation of Linear Momentum	
	External forces are those that agents external to the system exert on objects within the system. An isolated system is one for which the vector sum of the average external forces acting on the system is zero.	
	The principle of conservation of linear momentum states that the total linear momentum of an isolated system remains constant. For a two-body collision, the conservation of linear momentum can be written as	
Conservation of linear momentum	$$\underbrace{m_1\mathbf{v_{f1}} + m_2\mathbf{v_{f2}}}_{\substack{\text{Final total linear}\\\text{momentum}}} = \underbrace{m_1\mathbf{v_{01}} + m_2\mathbf{v_{02}}}_{\substack{\text{Initial total linear}\\\text{momentum}}} \qquad (7.7b)$$	**Examples 4, 5, 6**
		Concept Simulation 7.1
		Interactive LearningWare 7.2
	where m_1 and m_2 are the masses, $\mathbf{v_{f1}}$ and $\mathbf{v_{f2}}$ are the final velocities, and $\mathbf{v_{01}}$ and $\mathbf{v_{02}}$ are the initial velocities of the objects.	**Interactive Solution 7.19**

Use *Self-Assessment Test 7.1* to evaluate your understanding of Sections 7.1 and 7.2.

Topic	Discussion	Learning Aids
	7.3 Collisions in One Dimension	
Elastic collision	An elastic collision is one in which the total kinetic energy of the system after the collision is equal to the total kinetic energy of the system before the collision.	**Examples 7, 8**
		Concept Simulation 7.2
Inelastic collision	An inelastic collision is one in which the total kinetic energy of the system is not the same before and after the collision. If the objects stick together after the collision, the collision is said to be completely inelastic.	**Interactive Solutions 7.33, 7.55**
	7.4 Collisions in Two Dimensions	
	When the total linear momentum is conserved in a two-dimensional collision, the x and y components of the total linear momentum are conserved separately.	

Topic	Discussion	Learning Aids

For a collision between two objects, the conservation of total linear momentum can be written as

Conservation of linear momentum

$$\underbrace{m_1 v_{f1x} + m_2 v_{f2x}}_{\substack{x \text{ component of final} \\ \text{total linear momentum}}} = \underbrace{m_1 v_{01x} + m_2 v_{02x}}_{\substack{x \text{ component of initial} \\ \text{total linear momentum}}}$$

(7.9a) **Example 9**

Interactive LearningWare 7.3

$$\underbrace{m_1 v_{f1y} + m_2 v_{f2y}}_{\substack{y \text{ component of final} \\ \text{total linear momentum}}} = \underbrace{m_1 v_{01y} + m_2 v_{02y}}_{\substack{y \text{ component of initial} \\ \text{total linear momentum}}}$$

(7.9b)

7.5 Center of Mass

The location of the center of mass of two particles lying on the x axis is given by

Location of center of mass

$$x_{cm} = \frac{m_1 x_1 + m_2 x_2}{m_1 + m_2}$$

(7.10) **Interactive Solution 7.43**

where m_1 and m_2 are the masses of the particles and x_1 and x_2 are their positions relative to the coordinate origin. If the particles move with velocities v_1 and v_2, the velocity v_{cm} of the center of mass is

Velocity of center of mass

$$v_{cm} = \frac{m_1 v_1 + m_2 v_2}{m_1 + m_2}$$

(7.11)

If the total linear momentum of a system of particles remains constant during an interaction such as a collision, the velocity of the center of mass also remains constant.

Use Self-Assessment Test 7.2 to evaluate your understanding of Sections 7.3–7.5.

Conceptual Questions

1. Two identical automobiles have the same speed, one traveling east and one traveling west. Do these cars have the same momentum? Explain.

2. In Times Square in New York City, people celebrate on New Year's Eve. Some just stand around, but many move about randomly. Consider a system consisting of all of these people. Approximately, what is the total linear momentum of this system at any given instant? Justify your answer.

3. Two objects have the same momentum. Do the velocities of these objects necessarily have (a) the same directions and (b) the same magnitudes? Give your reasoning in each case.

4. (a) Can a single object have kinetic energy but no momentum? (b) Can a system of two or more objects have a total kinetic energy that is not zero but a total momentum that is zero? Account for your answers.

5. An airplane is flying horizontally with a constant momentum during a time interval Δt. (a) With the aid of Equation 7.4, decide whether a net impulse is acting on the plane during this time interval. (b) In the horizontal direction, both the thrust generated by the engines and air resistance act on the plane. What does the answer in part (a) imply about the impulse of the thrust compared to the impulse of the resistive force?

6. You have a choice. You may get hit head-on either by an adult moving slowly on a bicycle or by a child that is moving twice as fast on a bicycle. The mass of the child and bicycle is one-half that of the adult and bicycle. Considering only the issues of mass and velocity, which collision do you prefer? Or doesn't it matter? Account for your answer.

7. An object slides along the surface of the earth and slows down because of kinetic friction. If the object alone is considered as the system, the kinetic frictional force must be identified as an external force that, according to Equation 7.4, decreases the momentum of the system. (a) If *both* the object and the earth are considered to be the system, is the force of kinetic friction still an external force? (b) Can the friction force change the total linear momentum of the two-body system? Give your reasoning for both answers.

8. When you're driving a golf ball, a good "follow-through" helps to increase the distance of the drive. A good follow-through means that the club head is kept in contact with the ball as long as possible. Using the impulse–momentum theorem, explain why this technique allows you to hit the ball farther.

9. The drawing shows a garden sprinkler that whirls around a vertical axis. From each of the three arms of the sprinkler, water exits through a tapered nozzle. Because of this nozzle, the water leaves each nozzle with a speed that is greater than the speed inside the arm. (a) Apply the impulse–momentum theorem to deduce the direction of the force applied to the water. (b) Then, with the aid of Newton's third law, explain how the water causes the whirling motion.

10. Concept Simulation 7.1 at **www.wiley.com/college/cutnell** reviews the concepts that are pertinent in this question. In movies, Superman hovers in midair, grabs a villain by the neck, and throws him forward. Superman, however, remains stationary. Using the conservation of linear momentum, explain what is wrong with this scene.

11. A satellite explodes in outer space, far from any other body, sending thousands of pieces in all directions. How does the linear momentum of the satellite before the explosion compare with the total linear momentum of all the pieces after the explosion? Account for your answer.

12. You are a passenger on a jetliner that is flying at a constant velocity. You get up from your seat and walk toward the front of the plane. Because of this action, your forward momentum increases. What, if anything, happens to the forward momentum of the plane? Give your reasoning.

13. An ice boat is coasting along on a frozen lake. Friction between the ice and the boat is negligible, and so is air resistance. Nothing is propelling the boat. From a bridge someone jumps straight down and lands in the boat, which continues to coast straight ahead. (a) Does the horizontal momentum of the boat change? (b) Does the speed of the boat increase, decrease, or remain the same? Explain your answers.

14. On a distant asteroid, a large catapult is used to throw chunks of stone into space. Could such a device be used as a propulsion system to move the asteroid closer to the earth? Explain.

15. A collision occurs between three moving billiard balls such that no net external force acts on the three-ball system. Is the momentum of *each* ball conserved during the collision? If so, explain why. If not, what quantity is conserved?

16. In an elastic collision, is the kinetic energy of *each* object the same before and after the collision? Explain.

17. Concept Simulation 7.2 at **www.wiley.com/college/cutnell** illustrates the concepts that are involved in this question. Review Example 7. Now, suppose both objects have the same mass, $m_1 = m_2$. Describe what happens to the velocities of both objects as a result of the collision, using Equations 7.8a and 7.8b to justify your answers.

18. Where would you expect the center of mass of a doughnut to be located? Why?

19. Would you expect the center of mass of a baseball bat to be located halfway between the ends of the bat, nearer the lighter end or nearer the heavier end? Provide a reason for your answer.

20. A sunbather is lying on a floating raft that is stationary. She then gets up and walks to one end of the raft. Consider the sunbather and raft as an isolated system. (a) What is the velocity of the center of mass of this system while she is walking? Why? (b) Does the raft itself move while she is walking? If so, what is the direction of the raft's velocity relative to that of the sunbather? Provide a reason for your answer.

Problems

ssm Solution is in the Student Solutions Manual.   www Solution is available on the World Wide Web at www.wiley.com/college/cutnell
⚕ This icon represents a biomedical application.

Section 7.1 The Impulse–Momentum Theorem

1. ssm A volleyball is spiked so that its incoming velocity of $+4.0$ m/s is changed to an outgoing velocity of -21 m/s. The mass of the volleyball is 0.35 kg. What impulse does the player apply to the ball?

2. Interactive LearningWare 7.1 at **www.wiley.com/college/cutnell** provides a review of the concepts that are involved in this problem. A 62.0-kg person, standing on a diving board, dives straight down into the water. Just before striking the water, her speed is 5.50 m/s. At a time of 1.65 s after she enters the water, her speed is reduced to 1.10 m/s. What is the net average force (magnitude and direction) that acts on her when she is in the water?

3. A golfer, driving a golf ball off the tee, gives the ball a velocity of $+38$ m/s. The mass of the ball is 0.045 kg, and the duration of the impact with the golf club is 3.0×10^{-3} s. (a) What is the change in momentum of the ball? (b) Determine the average force applied to the ball by the club.

4. A baseball ($m = 149$ g) approaches a bat horizontally at a speed of 40.2 m/s (90 mi/h) and is hit straight back at a speed of 45.6 m/s (102 mi/h). If the ball is in contact with the bat for a time of 1.10 ms, what is the average force exerted on the ball by the bat? Neglect the weight of the ball, since it is so much less than the force of the bat. Choose the direction of the incoming ball as the positive direction.

5. ssm A soccer player kicks a ball with an average force of $+1400$ N, and her foot remains in contact with the ball for a time of 7.9×10^{-3} s. What is the impulse (magnitude and direction) of this force?

6. Before starting this problem, review Conceptual Example 3. Suppose that the hail described there bounces off the roof of the car with a velocity of $+15$ m/s. Ignoring the weight of the hailstones, calculate the force exerted by the hail on the roof. Compare your answer to that obtained in Example 2 for the rain, and verify that your answer is consistent with the conclusion reached in Conceptual Example 3.

7. A 46-kg skater is standing still in front of a wall. By pushing against the wall she propels herself backward with a velocity of -1.2 m/s. Her hands are in contact with the wall for 0.80 s. Ignore friction and wind resistance. Find the magnitude and direction of the average force she exerts on the wall (which has the same magnitude, but opposite direction, as the force that the wall applies to her).

8. ⚕ When jumping straight down, you can be seriously injured if you land stiff-legged. One way to avoid injury is to bend your knees upon landing to reduce the force of the impact. A 75-kg man just before contact with the ground has a speed of 6.4 m/s. (a) In a stiff-legged landing he comes to a halt in 2.0 ms. Find the average net force that acts on him during this time. (b) When he bends his knees, he comes to a halt in 0.10 s. Find the average net force now. (c) During the landing, the force of the ground on the man points upward, while the force due to gravity points downward. The average net force acting on the man includes both of these forces. Taking into account the directions of the forces, find the force of the ground on the man in parts (a) and (b).

***9. ssm www** A golf ball strikes a hard, smooth floor at an angle of 30.0° and, as the drawing shows, rebounds at the same angle. The mass of the ball is 0.047 kg, and its speed is 45 m/s just before and after striking the floor. What is the magnitude of the impulse applied to the golf ball by the floor? (*Hint: Note that only the vertical component of the ball's momentum changes during impact with the floor, and ignore the weight of the ball.*)

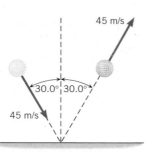

*10. A student ($m = 63$ kg) falls freely from rest and strikes the ground. During the collision with the ground, he comes to rest in a time of 0.010 s. The average force exerted on him by the ground is $+18\,000$ N, where the upward direction is taken to be the positive direction. From what height did the student fall? Assume that the only force acting on him during the collision is that due to the ground.

*11. A 0.500-kg ball is dropped from rest at a point 1.20 m above the floor. The ball rebounds straight upward to a height of 0.700 m. What are the magnitude and direction of the impulse of the net force applied to the ball during the collision with the floor?

*12. An 85-kg jogger is heading due east at a speed of 2.0 m/s. A 55-kg jogger is heading 32° north of east at a speed of 3.0 m/s. Find the magnitude and direction of the sum of the momenta of the two joggers.

*13. Consult **Interactive Solution 7.13** at **www.wiley.com/college/cutnell** for a review of problem-solving skills that are involved in this problem. A stream of water strikes a stationary turbine blade horizontally, as the drawing illustrates. The incident water stream has a velocity of $+16.0$ m/s, while the exiting water stream has a velocity of -16.0 m/s. The mass of water per second that strikes the blade is 30.0 kg/s. Find the magnitude of the average force exerted on the water by the blade.

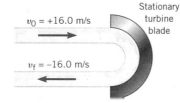

**14. A dump truck is being filled with sand. The sand falls straight downward from rest from a height of 2.00 m above the truck bed, and the mass of sand that hits the truck per second is 55.0 kg/s. The truck is parked on the platform of a weight scale. By how much does the scale reading exceed the weight of the truck and sand?

Section 7.2 The Principle of Conservation of Linear Momentum

15. **Interactive LearningWare 7.2** at **www.wiley.com/college/cutnell** provides a review of the concepts that are important in this problem. For tests using a *ballistocardiograph,* a patient lies on a horizontal platform that is supported on jets of air. Because of the air jets, the friction impeding the horizontal motion of the platform is negligible. Each time the heart beats, blood is pushed out of the heart in a direction that is nearly parallel to the platform. Since momentum must be conserved, the body and the platform recoil, and this recoil can be detected to provide information about the heart. For each beat, suppose that 0.050 kg of blood is pushed out of the heart with a velocity of $+0.25$ m/s and that the mass of the patient and platform is 85 kg. Assuming that the patient does not slip with respect to the platform, and that the patient and platform start from rest, determine the recoil velocity.

16. A 55-kg swimmer is standing on a stationary 210-kg floating raft. The swimmer then runs off the raft horizontally with a velocity of $+4.6$ m/s relative to the shore. Find the recoil velocity that the raft would have if there were no friction and resistance due to the water.

17. **ssm Concept Simulation 7.1** at **www.wiley.com/college/cutnell** illustrates the physics principles in this problem. In a science fiction novel two enemies, Bonzo and Ender, are fighting in outer space. From stationary positions they push against each other. Bonzo flies off with a velocity of $+1.5$ m/s, while Ender recoils with a velocity of -2.5 m/s. (a) Without doing any calculations, decide which person has the greater mass. Give your reasoning. (b) Determine the ratio of the masses ($m_{\text{Bonzo}}/m_{\text{Ender}}$) of these two people.

18. Consult **Concept Simulation 7.1** at **www.wiley.com/college/cutnell** in preparation for this problem. Two friends, Al and Jo, have a combined mass of 168 kg. At an ice skating rink they stand close together on skates, at rest and facing each other, with a compressed spring between them. The spring is kept from pushing them apart because they are holding each other. When they release their arms, Al moves off in one direction at a speed of 0.90 m/s, while Jo moves off in the opposite direction at a speed of 1.2 m/s. Assuming that friction is negligible, find Al's mass.

*19. To view an interactive solution to a problem that is very similar to this one, go to **www.wiley.com/college/cutnell** and select **Interactive Solution 7.19**. A fireworks rocket is moving at a speed of 45.0 m/s. The rocket suddenly breaks into two pieces of equal mass, which fly off with velocities $\mathbf{v}_1$ and $\mathbf{v}_2$, as shown in the drawing. What is the magnitude of (a) $\mathbf{v}_1$ and (b) $\mathbf{v}_2$?

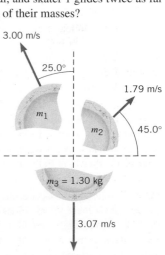

Problem 19

*20. Two ice skaters have masses m_1 and m_2 and are initially stationary. Their skates are identical. They push against one another, as in Figure 7.11, and move in opposite directions with different speeds. While they are pushing against each other, any kinetic frictional forces acting on their skates can be ignored. However, once the skaters separate, kinetic frictional forces eventually bring them to a halt. As they glide to a halt, the magnitudes of their accelerations are equal, and skater 1 glides twice as far as skater 2. What is the ratio m_1/m_2 of their masses?

*21. **ssm** By accident, a large plate is dropped and breaks into three pieces. The pieces fly apart parallel to the floor. As the plate falls, its momentum has only a vertical component, and no component parallel to the floor. After the collision, the component of the total momentum parallel to the floor must remain zero, since the net external force acting on the plate has no component parallel to the floor. Using the data shown in the drawing, find the masses of pieces 1 and 2.

* **22.** The lead female character in the movie *Diamonds Are Forever* is standing at the edge of an offshore oil rig. As she fires a gun, she is driven back over the edge and into the sea. Suppose the mass of a bullet is 0.010 kg and its velocity is +720 m/s. Her mass (including the gun) is 51 kg. (a) What recoil velocity does she acquire in response to a single shot from a stationary position, assuming that no external force keeps her in place? (b) Under the same assumption, what would be her recoil velocity if, instead, she shoots a blank cartridge that ejects a mass of 5.0×10^{-4} kg at a velocity of +720 m/s?

** **23. ssm www** A cannon of mass 5.80×10^3 kg is rigidly bolted to the earth so it can recoil only by a negligible amount. The cannon fires an 85.0-kg shell horizontally with an initial velocity of +551 m/s. Suppose the cannon is then unbolted from the earth, and no external force hinders its recoil. What would be the velocity of a shell fired by this loose cannon? *(Hint: In both cases assume that the burning gunpowder imparts the same kinetic energy to the system.)*

** **24.** A wagon is coasting at a speed v_A along a straight and level road. When ten percent of the wagon's mass is thrown off the wagon, parallel to the ground and in the forward direction, the wagon is brought to a halt. If the direction in which this mass is thrown is exactly reversed, but the speed of this mass relative to the wagon remains the same, the wagon accelerates to a new speed v_B. Calculate the ratio v_B/v_A.

Section 7.3 Collisions in One Dimension, Section 7.4 Collisions in Two Dimensions

25. In a football game, a receiver is standing still, having just caught a pass. Before he can move, a tackler, running at a velocity of +4.5 m/s, grabs him. The tackler holds onto the receiver, and the two move off together with a velocity of +2.6 m/s. The mass of the tackler is 115 kg. Assuming that momentum is conserved, find the mass of the receiver.

26. A 1055-kg van, stopped at a traffic light, is hit directly in the rear by a 715-kg car traveling with a velocity of +2.25 m/s. Assume that the transmission of the van is in neutral, the brakes are not being applied, and the collision is elastic. What is the final velocity of (a) the car and (b) the van?

27. ssm A golf ball bounces down a flight of steel stairs, striking each stair once on the way down. The ball starts at the top step with a vertical velocity component of zero. If all the collisions with the stairs are elastic, and if the vertical height of the staircase is 3.00 m, determine the bounce height when the ball reaches the bottom of the stairs. Neglect air resistance.

28. A 2.50-g bullet, traveling at a speed of 425 m/s, strikes the wooden block of a ballistic pendulum, such as that in Figure 7.14. The block has a mass of 215 g. (a) Find the speed of the bullet/block combination immediately after the collision. (b) How high does the combination rise above its initial position?

29. A cue ball (mass = 0.165 kg) is at rest on a frictionless pool table. The ball is hit dead center by a pool stick, which applies an impulse of +1.50 N·s to the ball. The ball then slides along the table and makes an elastic head-on collision with a second ball of equal mass that is initially at rest. Find the velocity of the second ball just after it is struck.

30. The drawing shows a collision between two pucks on an air-hockey table. Puck A has a mass of 0.025 kg and is moving along the x axis with a velocity of +5.5 m/s. It makes a collision with puck B, which has a mass of 0.050 kg and is initially at rest. The collision is not head-on. After the collision, the two pucks fly apart

with the angles shown in the drawing. Find the final speed of (a) puck A and (b) puck B.

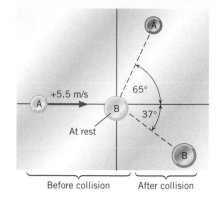

Before collision　　After collision

31. ssm www A 5.00-kg ball, moving to the right at a velocity of +2.00 m/s on a frictionless table, collides head-on with a stationary 7.50-kg ball. Find the final velocities of the balls if the collision is (a) elastic and (b) completely inelastic.

32. A 0.150-kg projectile is fired with a velocity of +715 m/s at a 2.00-kg wooden block that rests on a frictionless table. The velocity of the block, immediately after the projectile passes through it, is +40.0 m/s. Find the velocity with which the projectile exits from the block.

* **33. Interactive Solution 7.33** at **www.wiley.com/college/cutnell** illustrates how to model a problem similar to this one. An automobile has a mass of 2100 kg and a velocity of +17 m/s. It makes a rear-end collision with a stationary car whose mass is 1900 kg. The cars lock bumpers and skid off together with the wheels locked. (a) What is the velocity of the two cars just after the collision? (b) Find the impulse (magnitude and direction) that acts on the skidding cars from just after the collision until they come to a halt. (c) If the coefficient of kinetic friction between the wheels of the cars and the pavement is $\mu_k = 0.68$, determine how far the cars skid before coming to rest.

* **34.** A mine car, whose mass is 440 kg, rolls at a speed of 0.50 m/s on a horizontal track, as the drawing shows. A 150-kg chunk of coal has a speed of 0.80 m/s when it leaves the chute. Determine the velocity of the car/coal system after the coal has come to rest in the car.

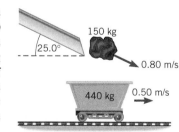

* **35. ssm** A 50.0-kg skater is traveling due east at a speed of 3.00 m/s. A 70.0-kg skater is moving due south at a speed of 7.00 m/s. They collide and hold on to each other after the collision, managing to move off at an angle θ south of east, with a speed of v_f. Find (a) the angle θ and (b) the speed v_f, assuming that friction can be ignored.

* **36.** An electron collides elastically with a stationary hydrogen atom. The mass of the hydrogen atom is 1837 times that of the electron. Assume that all motion, before and after the collision, occurs along the same straight line. What is the ratio of the kinetic energy of the hydrogen atom after the collision to that of the electron before the collision?

* **37.** A 60.0-kg person, running horizontally with a velocity of +3.80 m/s, jumps onto a 12.0-kg sled that is initially at rest. (a) Ignoring the effects of friction during the collision, find the velocity of the sled and person as they move away. (b) The sled and person coast

30.0 m on level snow before coming to rest. What is the coefficient of kinetic friction between the sled and the snow?

** 38. Concept Simulation 7.2 at **www.wiley.com/college/cutnell** provides a view of this elastic collision. Two identical balls are traveling toward each other with velocities of −4.0 and +7.0 m/s, and they experience an elastic head-on collision. Obtain the velocities (magnitude and direction) of each ball after the collision.

** 39. ssm Starting with an initial speed of 5.00 m/s at a height of 0.300 m, a 1.50-kg ball swings downward and strikes a 4.60-kg ball that is at rest, as the drawing shows. (a) Using the principle of conservation of mechanical energy, find the speed of the 1.50-kg ball just before impact. (b) Assuming that the collision is elastic, find the velocities (magnitude and direction) of both balls just after the collision. (c) How high does each ball swing after the collision, ignoring air resistance?

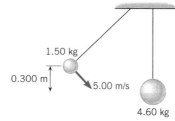

** 40. A ball is dropped from rest at the top of a 6.10-m-tall building, falls straight downward, collides inelastically with the ground, and bounces back. The ball loses 10.0% of its kinetic energy every time it collides with the ground. How many bounces can the ball make and still reach a windowsill that is 2.44 m above the ground?

Section 7.5 Center of Mass

41. ssm The earth and moon are separated by a center-to-center distance of 3.85×10^8 m. The mass of the earth is 5.98×10^{24} kg and that of the moon is 7.35×10^{22} kg. How far does the center of mass lie from the center of the earth?

42. Consider the two moving boxcars in Example 5. Determine the velocity of their center of mass (a) before and (b) after the collision. (c) Should your answer in part (b) be less than, greater than, or equal to the common velocity v_f of the two coupled cars after the collision? Justify your answer.

43. Interactive Solution 7.43 at **www.wiley.com/college/cutnell** presents a method for modeling this problem. The carbon monoxide molecule (CO) consists of a carbon atom and an oxygen atom separated by a distance of 1.13×10^{-10} m. The mass m_C of the carbon atom is 0.750 times the mass m_O of the oxygen atom: $m_C = 0.750\, m_O$. Determine the location of the center of mass of this molecule relative to the carbon atom.

* 44. The drawing shows the bond lengths and angles in the nitric acid (HNO_3) molecule, which is planar. The masses of the atoms are $m_H = 1.67 \times 10^{-27}$ kg, $m_N = 23.3 \times 10^{-27}$ kg, and $m_O = 26.6 \times 10^{-27}$ kg. Locate the center of mass of this molecule relative to the hydrogen atom. *(Hint: The oxygen atoms are located symmetrically on either side of the H−O−N line.)*

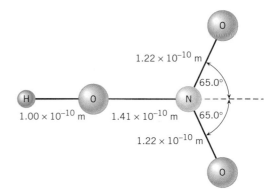

Additional Problems

45. ssm www A lumberjack (mass = 98 kg) is standing at rest on one end of a floating log (mass = 230 kg) that is also at rest. The lumberjack runs to the other end of the log, attaining a velocity of +3.6 m/s relative to the shore, and then hops onto an identical floating log that is initially at rest. Neglect any friction and resistance between the logs and the water. (a) What is the velocity of the first log just before the lumberjack jumps off? (b) Determine the velocity of the second log if the lumberjack comes to rest on it.

46. With the engines off, a spaceship is coasting at a velocity of +230 m/s through outer space. The ship carries rockets that are mounted in firing tubes, the back ends of which are closed. It fires a rocket straight ahead at an enemy vessel. The mass of the rocket is 1300 kg, and the mass of the spaceship (not including the rocket) is 4.0×10^6 kg. The firing of the rocket brings the spaceship to a halt. What is the velocity of the rocket?

47. ssm Batman (mass = 91 kg) jumps straight down from a bridge into a boat (mass = 510 kg) in which a criminal is fleeing. The velocity of the boat is initially +11 m/s. What is the velocity of the boat after Batman lands in it?

48. Two men pushing a stalled car generate a net force of +680 N for 7.2 s. What is the final momentum of the car?

49. Kevin has a mass of 87 kg and is skating with in-line skates. He sees his 22-kg younger brother up ahead standing on the sidewalk, with his back turned. Coming up from behind, he grabs his brother and rolls off at a speed of 2.4 m/s. Ignoring friction, find Kevin's speed just before he grabbed his brother.

50. Refer to Conceptual Example 3 as an aid in understanding this problem. A hockey goalie is standing on ice. Another player fires a puck ($m = 0.17$ kg) at the goalie with a velocity of +65 m/s. (a) If the goalie catches the puck with his glove in a time of 5.0×10^{-3} s, what is the average force (magnitude and direction) exerted on the goalie by the puck? (b) Instead of catching the puck, the goalie slaps it with his stick and returns the puck straight back to the player with a velocity of −65 m/s. The puck and stick are in contact for a time of 5.0×10^{-3} s. Now what is the average force exerted on the goalie by the puck? Verify that your answers to parts (a) and (b) are consistent with the conclusion of Conceptual Example 3.

51. ssm A two-stage rocket moves in space at a constant velocity of 4900 m/s. The two stages are then separated by a small explosive charge placed between them. Immediately after the explosion the velocity of the 1200-kg upper stage is 5700 m/s in the same direction as before the explosion. What is the velocity (magnitude and direction) of the 2400-kg lower stage after the explosion?

52. Two balls are approaching each other head-on. Their velocities are +9.70 and −11.8 m/s. Determine the velocity of the center of mass of the two balls if (a) they have the same mass and (b) if the mass of one ball ($v = 9.70$ m/s) is twice the mass of the other ball ($v = −11.8$ m/s).

* 53. ssm During July 1994 the comet Shoemaker–Levy 9 smashed into Jupiter in a spectacular fashion. The comet actually consisted of 21 distinct pieces, the largest of which had a mass of about 4.0×10^{12} kg and a speed of 6.0×10^4 m/s. Jupiter, the largest planet in the solar system, has a mass of 1.9×10^{27} kg and an orbital speed of 1.3×10^4 m/s. If this piece of the comet had hit Jupiter head-on, what would have been the *change* (magnitude only) in Jupiter's orbital speed (not its final speed)?

* **54.** A person stands in a stationary canoe and throws a 5.00-kg stone with a velocity of 8.00 m/s at an angle of 30.0° above the horizontal. The person and canoe have a combined mass of 105 kg. Ignoring air resistance and effects of the water, find the horizontal recoil velocity (magnitude and direction) of the canoe.

* **55.** To view an interactive solution to a problem that is similar to this one, go to **www.wiley.com/college/cutnell** and select **Interactive Solution 7.55.** A 0.015-kg bullet is fired straight up at a falling wooden block that has a mass of 1.8 kg. The bullet has a speed of 810 m/s when it strikes the block. The block originally was dropped from rest from the top of a building and had been falling for a time t when the collision with the bullet occurred. As a result of the collision, the block (with the bullet in it) reverses direction, rises, and comes to a momentary halt at the top of the building. Find the time t.

* **56.** The drawing shows a sulfur dioxide molecule. It consists of two oxygen atoms and a sulfur atom. A sulfur atom is twice as massive as an oxygen atom. Using this information and the data provided in the drawing, find (a) the x coordinate

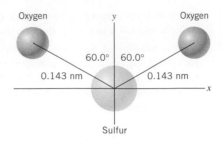

and (b) the y coordinate of the center of mass of the sulfur dioxide molecule. Express your answers in nanometers (1 nm = 10^{-9} m).

** **57. ssm** Two people are standing on a 2.0-m-long platform, one at each end. The platform floats parallel to the ground on a cushion of air, like a hovercraft. One person throws a 6.0-kg ball to the other, who catches it. The ball travels nearly horizontally. Excluding the ball, the total mass of the platform and people is 118 kg. Because of the throw, this 118-kg mass recoils. How far does it move before coming to rest again?

Concepts & Calculations *Group Learning Problems*

Note: Each of these problems consists of Concept Questions followed by a related quantitative Problem. They are designed for use by students working alone or in small learning groups. The Concept Questions involve little or no mathematics and are intended to stimulate group discussions. They focus on the concepts with which the problems deal. Recognizing the concepts is the essential initial step in any problem-solving technique.

58. Concept Questions In a performance test two cars take the same time to accelerate from rest up to the same speed. Car A has a mass of 1400 kg, and car B has a mass of 1900 kg. During the test, which car (a) has the greatest change in momentum, (b) experiences the greatest impulse, and (c) is acted upon by the greatest net average force? In each case, give your reasoning.

Problem Each car takes 9.0 s to accelerate from rest to 27 m/s. Find the net average force that acts on each car during the test. Verify that your answers are consistent with your answers to the Concept Questions.

59. Concept Questions As the drawing illustrates, two disks with masses m_1 and m_2 are moving horizontally to the right at a speed v_0. They are on an air-hockey table, which supports them with an essentially frictionless cushion of air. They move as a unit, with a compressed spring between them, which has a negligible mass. (a) When the spring is released and allowed to push outward, what are the directions of the forces that act on disk 1 and disk 2? (b) After the spring is released, is the speed of each disk larger than, smaller than, or the same as the speed v_0? Explain.

Problem Consider the situation where disk 1 comes to a momentary halt shortly after the spring is released. Assuming that $m_1 = 1.2$ kg, $m_2 = 2.4$ kg, and $v_0 = 5.0$ m/s, find the speed of disk 2 at that moment. Verify that your answer is consistent with your answers to the Concept Questions.

60. Concept Questions A wagon is rolling forward on level ground. Friction is negligible. The person sitting in the wagon throws a rock. Does the momentum of the wagon increase, decrease,

or remain the same (a) when the rock is thrown directly forward and (b) when the rock is thrown directly backward? (c) In which case does the wagon have the greatest speed after the rock is thrown?

Problem The total mass of the wagon, rider, and rock is 95.0 kg. The mass of the rock is 0.300 kg. Initially the wagon is rolling forward at a speed of 0.500 m/s. Then the rock is thrown with a speed of 16.0 m/s. Both speeds are relative to the ground. Find the speed of the wagon after the rock is thrown directly forward in one case and directly backward in another. Check to see that your answers are consistent with your answers to the Concept Questions.

61. Concept Questions One object is at rest and another is moving. Furthermore, one object is more massive than the other. The two collide in a one-dimensional completely inelastic collision. In other words, they stick together after the collision and move off with a common velocity. Momentum is conserved. (a) Consider the final momentum of the two-object system after the collision. Is it greater when the large-mass object is moving initially or when the small-mass object is moving initially? (b) Is the final speed after the collision greater when the large-mass or the small-mass object is moving initially?

Problem The speed of the object that is moving initially is 25 m/s. The masses of the two objects are 3.0 and 8.0 kg. Determine the final speed of the two-object system after the collision for the case when the large-mass object is moving initially and the case when the small-mass object is moving initially. Be sure that your answers are consistent with your answers to the Concept Questions.

62. Concept Questions Object A is moving due east, while object B is moving due north. They collide and stick together in a completely inelastic collision. Momentum is conserved. (a) Is it possible that the two-object system has a final total momentum of zero after the collision? (b) Roughly, what is the direction of the final total momentum of the two-object system after the collision?

Problem Object A has a mass of $m_A = 17.0$ kg and an initial velocity of $\mathbf{v}_{0A} = 8.00$ m/s, due east. Object B has a mass of $m_B = 29.0$ kg and an initial velocity of $\mathbf{v}_{0B} = 5.00$ m/s, due north. Find the magnitude and direction of the total momentum of the two-object system after the collision. Make sure that your answers are consistent with your answers to the Concept Questions.

63. Concept Questions (a) John has a larger mass than Barbara has. He is standing on the x axis at $x_J = +9.0$ m, while she is standing on the x axis at $x_B = +2.0$ m. Is their center-of-mass point closer to the 9.0-m point or the 2.0-m point? (b) They switch positions. Is their center-of-mass point now closer to the 9.0-m point or the 2.0-m point? (c) In which direction, toward or away from the origin, does their center of mass move as a result of the switch?

Problem John's mass is 86 kg, and Barbara's is 55 kg. How far and in which direction does their center of mass move as a result of the switch? Verify that your answer is consistent with your answers to the Concept Questions.

* **64. Concept Questions** A ball is attached to one end of a wire, the other end being fastened to the ceiling. The wire is held horizontal, and the ball is released from rest (see the drawing). It swings downward and strikes a block initially at rest on a horizontal frictionless surface. Air resistance is negligible, and the collision is elastic. (a) During the downward motion of the ball, are any of the following conserved: its momentum, its kinetic energy, its total mechanical energy? (b) During the collision with the block, are any of the following conserved: the horizontal component of the total momentum of the ball/block system, the total kinetic energy of the system? Provide reasons for your choices.

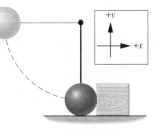

Problem The masses of the ball and block are, respectively, 1.60 kg and 2.40 kg, and the length of the wire is 1.20 m. Find the velocity (magnitude and direction) of the ball (a) just before the collision, and (b) just after the collision.

* **65. Concept Questions** Part *a* of the drawing shows a bullet approaching two blocks resting on a horizontal frictionless surface. Air resistance is negligible. The bullet passes completely through the first block and embeds itself in the second one, as indicated in part *b*. Note that both blocks are moving after the collision with the bullet. (a) Can the conservation of linear momentum be applied to this three-object system, even though the second collision occurs a bit later than the first one? Justify your answer. Neglect any mass removed from the first block by the bullet. (b) Is the total kinetic energy of this three-body system conserved? If not, would the total kinetic energy after the collisions be greater than or smaller than that before the collisions? Justify your answer.

Problem A 4.00-g bullet is moving horizontally with a velocity of +355 m/s, where the + sign indicates that it is moving to the right. The mass of the first block is 1150 g, and its velocity is +0.550 m/s after the bullet passes through it. The mass of the second block is 1530 g. (a) What is the velocity of the second block after the bullet imbeds itself? (b) Find the ratio of the total kinetic energy after the collision to that before the collision. Be sure your answer is consistent with that in part (b) of the Concept Questions.

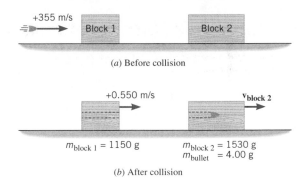

(*a*) Before collision

(*b*) After collision

Chapter 8

Rotational Kinematics

The rotating wheels of this bicycle illustrate the central theme of this chapter, rotational kinematics, which deals with the concepts needed to describe rotational motion. These concepts are angular displacement, angular velocity, and angular acceleration. They are analogous to those used in Chapters 2 and 3 for describing linear motion. (© Adam Pretty/Getty Images News and Sport Services)

8.1 Rotational Motion and Angular Displacement

In the simplest kind of rotation, points on a rigid object move on circular paths. In Figure 8.1, for example, we see the circular paths for points *A*, *B*, and *C* on a spinning skater. The centers of all such circular paths define a line, called the ***axis of rotation.***

The angle through which a rigid object rotates about a fixed axis is called the ***angular displacement.*** Figure 8.2 shows how the angular displacement is measured for a rotating compact disc (CD). Here, the axis of rotation passes through the center of the disc and is perpendicular to its surface. On the surface of the CD we draw a radial line, which is a line that intersects the axis of rotation perpendicularly. As the CD turns, we observe the angle through which this line moves relative to a convenient reference line that does not rotate. The radial line moves from its initial orientation at angle θ_0 to a final orientation at angle θ (Greek letter theta). In the process, the line sweeps out the angle $\theta - \theta_0$. As with other differences that we have encountered ($\Delta x = x - x_0$, $\Delta v = v - v_0$, and $\Delta t = t - t_0$), it is customary to denote the difference between the final and initial angles by the notation $\Delta\theta$ (read as "delta theta"): $\Delta\theta = \theta - \theta_0$. The angle $\Delta\theta$ is the angular displacement. A rotating object may rotate either counterclockwise or clockwise, and standard convention calls a counterclockwise displacement positive and a clockwise displacement negative.

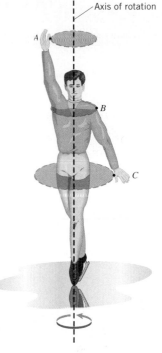

Figure 8.1 When an object rotates, points on the object, such as *A*, *B*, or *C*, move on circular paths. The centers of the circles form a line that is the axis of rotation.

■ **DEFINITION OF ANGULAR DISPLACEMENT**

When a rigid body rotates about a fixed axis, the angular displacement is the angle $\Delta\theta$ swept out by a line passing through any point on the body and intersecting the axis of rotation perpendicularly. By convention, the angular displacement is positive if it is counterclockwise and negative if it is clockwise.

SI Unit of Angular Displacement: radian (rad)*

Angular displacement is often expressed in one of three units. The first is the familiar ***degree,*** and it is well known that there are 360 degrees in a circle. The second unit is the ***revolution (rev),*** one revolution representing one complete turn of 360°. The most useful unit from a scientific viewpoint is the SI unit called the ***radian (rad).*** Figure 8.3 shows how the radian is defined, again using a CD as an example. The picture focuses attention on a point *P* on the disc. This point starts out on the stationary reference line, so that $\theta_0 = 0$ rad, and the angular displacement is $\Delta\theta = \theta - \theta_0 = \theta$. As the disc rotates, the point traces out an arc of length *s*, which is measured along a circle of radius *r*. Equation 8.1 defines the angle θ in radians:

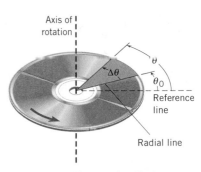

Figure 8.2 The angular displacement of a CD is the angle $\Delta\theta$ swept out by a radial line as the disc turns about its axis of rotation.

$$\theta\ (\text{in radians}) = \frac{\text{Arc length}}{\text{Radius}} = \frac{s}{r} \tag{8.1}$$

According to this definition, an angle in radians is the ratio of two lengths, for example, meters/meters. In calculations, therefore, the radian is treated as a number without units and has no effect on other units that it multiplies or divides.

To convert between degrees and radians, it is only necessary to remember that the arc length of an entire circle of radius *r* is the circumference $2\pi r$. Therefore, according to Equation 8.1, *the number of radians that corresponds to 360°, or one revolution, is*

$$\theta = \frac{2\pi r}{r} = 2\pi\ \text{rad}$$

Since 2π rad corresponds to 360°, the number of degrees in one radian is

$$1\ \text{rad} = \frac{360°}{2\pi} = 57.3°$$

Figure 8.3 In radian measure, the angle θ is defined to be the arc length *s* divided by the radius *r*.

*The radian is neither a base SI unit nor a derived one. It is regarded as a supplementary SI unit.

It is useful to express an angle θ in radians, because then the arc length s subtended at any radius r can be calculated by multiplying θ by r. Example 1 illustrates this point and also shows how to convert between degrees and radians.

Example 1   Adjacent Synchronous Satellites

The physics of communications satellites.

Synchronous or "stationary" communications satellites are put into an orbit whose radius is $r = 4.23 \times 10^7$ m. The orbit is in the plane of the equator, and two adjacent satellites have an angular separation of $\theta = 2.00°$, as Figure 8.4 illustrates. Find the arc length s (see the drawing) that separates the satellites.

Reasoning Since the radius r and the angle θ are known, we may find the arc length s by using the relation θ (in radians) $= s/r$. But first, the angle must be converted to radians from degrees.

Solution To convert 2.00° into radians, we use the fact that 2π radians is equivalent to 360°:

$$2.00° = (2.00 \; \cancel{degrees}) \left(\frac{2\pi \text{ radians}}{360 \; \cancel{degrees}} \right) = 0.0349 \text{ radians}$$

From Equation 8.1, it follows that the arc length between the satellites is

$$s = r\theta = (4.23 \times 10^7 \text{ m})(0.0349 \text{ rad}) = \boxed{1.48 \times 10^6 \text{ m (920 miles)}}$$

The radian, being a unitless quantity, is dropped from the final result, leaving the answer expressed in meters.

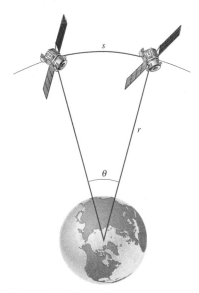

Figure 8.4 Two adjacent synchronous satellites have an angular separation of $\theta = 2.00°$. The distances and angles have been exaggerated for clarity.

Conceptual Example 2 takes advantage of the radian as a unit for measuring angles and explains the spectacular phenomenon of a total solar eclipse.

Conceptual Example 2   A Total Eclipse of the Sun

The diameter of the sun is about 400 times greater than that of the moon. By coincidence, the sun is also about 400 times farther from the earth than is the moon. For an observer on earth, compare the angle subtended by the moon to the angle subtended by the sun, and explain why this result leads to a total solar eclipse.

The physics of a total solar eclipse.

Reasoning and Solution Figure 8.5a shows a person on earth viewing the moon and the sun. The angle θ_{moon} subtended by the moon is given by Equation 8.1 as the arc length s_{moon} divided by the distance r_{moon} from the earth to the moon: $\theta_{\text{moon}} = s_{\text{moon}}/r_{\text{moon}}$. The earth and moon are sufficiently far apart that the arc length s_{moon} is very nearly equal to the moon's diameter. Similar reasoning applies for the sun, so that the angle subtended by the sun is $\theta_{\text{sun}} = s_{\text{sun}}/r_{\text{sun}}$. But since $s_{\text{sun}} \approx 400 \, s_{\text{moon}}$ (where the symbol "$\approx$" means "approximately equal to") and $r_{\text{sun}} \approx 400 \, r_{\text{moon}}$, the two angles are approximately equal: $\theta_{\text{moon}} \approx \theta_{\text{sun}}$. Figure 8.5b shows what happens when the moon comes between the sun and the earth. ***Since the angle subtended by the moon is nearly equal to the angle subtended by the sun, the moon blocks most of the sun's light from reaching the observer's eyes,*** and a total solar eclipse like that in Figure 8.5c occurs.

Related Homework: *Problem 14*

✔ **Check Your Understanding 1**

Three objects are visible in the night sky. They have the following diameters (in multiples of d) and subtend the following angles (in multiples of θ_0) at the eye of the observer. Object A has a diameter of $4d$ and subtends an angle of $2\,\theta_0$. Object B has a diameter of $3d$ and subtends an angle of $\theta_0/2$. Object C has a diameter of $d/2$ and subtends an angle of $\theta_0/8$. Rank them in descending order (greatest first) according to their distance from the observer. *(The answer is given at the end of the book.)*

Background: The radian is a unit for measuring angular displacement, and it lies at the heart of this problem.

For similar questions (including calculational counterparts), consult Self-Assessment Test 8.1. This test is described at the end of Section 8.3.

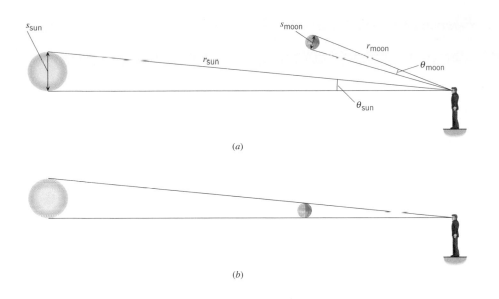

(a)

(b)

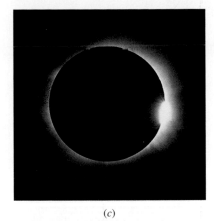

(c)

Figure 8.5 (*a*) The angles subtended by the moon and sun at the eyes of the observer are θ_{moon} and θ_{sun}. (The distances and angles are exaggerated for the sake of clarity.) (*b*) Since the moon and sun subtend approximately the same angle, the moon blocks nearly all the sun's light from reaching the observer's eyes. (*c*) The result is a total solar eclipse. (© Roger Ressmeyer/Corbis Images)

8.2 *Angular Velocity and Angular Acceleration*
ANGULAR VELOCITY

In Section 2.2 we introduced the idea of linear velocity to describe how fast an object moves and the direction of its motion. We now introduce the analogous idea of angular velocity to describe the motion of a rigid object rotating about an axis.

▶ CONCEPTS AT A GLANCE To define angular velocity, we use two concepts previously encountered. As the Concepts-at-a-Glance chart in Figure 8.6 illustrates, the angular velocity is obtained by combining the angular displacement and the time during which the displacement occurs. A comparison of this chart with that in Figure 2.2 shows that angular velocity is defined in a manner analogous to that used for linear velocity. Taking advantage of this analogy between the two types of velocities will help us understand rotational motion. ◀

According to Equation 2.2 ($\overline{v} = \Delta x/\Delta t$), the average linear velocity is the linear displacement of the object divided by the time required for the displacement to occur. For rotational motion about a fixed axis, the ***average angular velocity*** $\overline{\omega}$ (Greek letter omega) is obtained in an analogous way, as the angular displacement divided by the elapsed time during which the displacement occurs.

Figure 8.6 CONCEPTS AT A GLANCE Angular displacement and time are combined to produce the concept of angular velocity. This Ferris wheel has an angular velocity that depends on how fast the wheel is spinning as well as the direction of its spin. (© Carolyn Brown/The Image Bank/Getty Images)

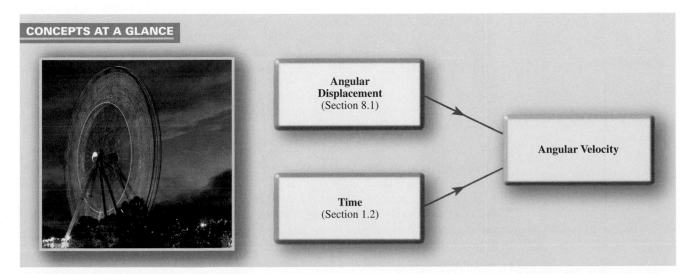

CONCEPTS AT A GLANCE

Angular Displacement
(Section 8.1)

Time
(Section 1.2)

Angular Velocity

■ **DEFINITION OF AVERAGE ANGULAR VELOCITY**

$$\frac{\text{Average angular}}{\text{velocity}} = \frac{\text{Angular displacement}}{\text{Elapsed time}}$$

$$\overline{\omega} = \frac{\theta - \theta_0}{t - t_0} = \frac{\Delta\theta}{\Delta t} \tag{8.2}$$

SI Unit of Angular Velocity: radian per second (rad/s)

The SI unit for angular velocity is the radian per second (rad/s), although other units such as revolutions per minute (rev/min or rpm) are also used. In agreement with the sign convention adopted for angular displacement, angular velocity is positive when the rotation is counterclockwise and negative when it is clockwise. Example 3 shows how the concept of average angular velocity is applied to a gymnast.

Example 3   Gymnast on a High Bar

A gymnast on a high bar swings through two revolutions in a time of 1.90s, as Figure 8.7 suggests. Find the average angular velocity (in rad/s) of the gymnast.

Reasoning The average angular velocity of the gymnast in rad/s is the angular displacement in radians divided by the elapsed time. However, the angular displacement is given as two revolutions, so we begin by converting this value into radian measure.

Solution The angular displacement (in radians) of the gymnast is

$$\Delta\theta = -2.00 \text{ revolutions} \left(\frac{2\pi \text{ radians}}{1 \text{ revolution}}\right) = -12.6 \text{ radians}$$

where the minus sign denotes that the gymnast rotates clockwise (see the drawing). The average angular velocity is

$$\overline{\omega} = \frac{\Delta\theta}{\Delta t} = \frac{-12.6 \text{ rad}}{1.90 \text{ s}} = \boxed{-6.63 \text{ rad/s}} \tag{8.2}$$

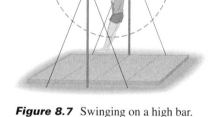

Figure 8.7 Swinging on a high bar.

The **instantaneous angular velocity** ω is the angular velocity that exists at any given instant. To measure it, we follow the same procedure used in Chapter 2 for the instantaneous linear velocity. In this procedure, a small angular displacement $\Delta\theta$ occurs during a small time interval Δt. The time interval is so small that it approaches zero ($\Delta t \rightarrow 0$), and in this limit, the measured average angular velocity, $\overline{\omega} = \Delta\theta/\Delta t$, becomes the instantaneous angular velocity ω:

$$\omega = \lim_{\Delta t \rightarrow 0} \overline{\omega} = \lim_{\Delta t \rightarrow 0} \frac{\Delta\theta}{\Delta t} \tag{8.3}$$

The magnitude of the instantaneous angular velocity, without reference to whether it is a positive or negative quantity, is called the **instantaneous angular speed.** If a rotating object has a constant angular velocity, the instantaneous value and the average value are the same.

ANGULAR ACCELERATION

In linear motion, a changing velocity means that an acceleration is occurring. Such is also the case in rotational motion; a changing angular velocity means that an **angular acceleration** is occurring. There are many examples of angular acceleration. For instance, as a compact disc recording is played, the disc turns with an angular velocity that is continually decreasing. And when the push buttons of an electric blender are changed from a lower setting to a higher setting, the angular velocity of the blades increases.

▶ **CONCEPTS AT A GLANCE** The idea of angular acceleration describes how rapidly or slowly the angular velocity changes during a given time interval. The Concepts-at-a-

CONCEPTS AT A GLANCE

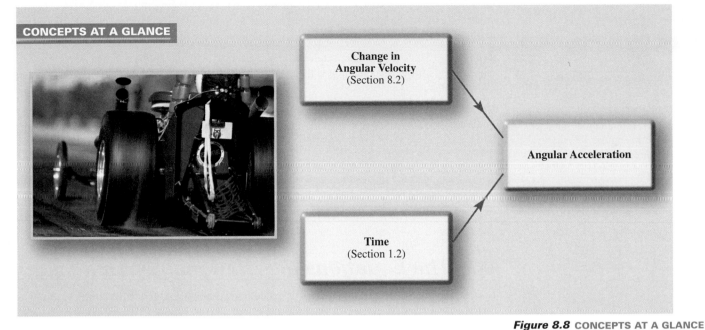

Figure 8.8 CONCEPTS AT A GLANCE To define the concept of angular acceleration, we bring together the change in angular velocity and the time required for the change to occur. Each wheel on this drag racer has an angular acceleration because the wheel's angular velocity changes as the racer accelerates from the starting line. (© Philip Bailey/Corbis Images)

Glance chart in Figure 8.8 illustrates that the notion of angular acceleration is formulated by bringing together the change in the angular velocity with the time required for the change to occur. A comparison of this chart with that in Figure 2.5 reveals the analogy between angular acceleration and linear acceleration. This analogy, like that between angular velocity and linear velocity, will be useful in helping us to understand rotational motion. ◄

When the linear velocity of an object changes, Equation 2.4 ($\overline{\mathbf{a}} = \Delta\mathbf{v}/\Delta t$) defines the average linear acceleration as the change in velocity per unit time. When the angular velocity changes from an initial value of ω_0 at time t_0 to a final value of ω at time t, the average angular acceleration $\overline{\alpha}$ (Greek letter alpha) is defined similarly:

■ **DEFINITION OF AVERAGE ANGULAR ACCELERATION**

$$\frac{\text{Average angular}}{\text{acceleration}} = \frac{\text{Change in angular velocity}}{\text{Elapsed time}}$$

$$\overline{\alpha} = \frac{\omega - \omega_0}{t - t_0} = \frac{\Delta\omega}{\Delta t} \tag{8.4}$$

SI Unit of Average Angular Acceleration: radian per second squared (rad/s^2)

The SI unit for average angular acceleration is the unit for angular velocity divided by the unit for time, or (rad/s)/s = rad/s^2. An angular acceleration of +5 rad/s^2, for example, means that the angular velocity of the rotating object increases by +5 radians per second during each second of acceleration.

The *instantaneous angular acceleration* α is the angular acceleration at a given instant. In discussing linear motion, we assumed a condition of constant acceleration, so that the average and instantaneous accelerations were identical ($\overline{\mathbf{a}} = \mathbf{a}$). Similarly, we assume that the angular acceleration is constant, so that the instantaneous angular acceleration α and the average angular acceleration $\overline{\alpha}$ are the same ($\overline{\alpha} = \alpha$). The next example illustrates the concept of angular acceleration.

Example 4 A Jet Revving Its Engines

A jet awaiting clearance for takeoff is momentarily stopped on the runway. As seen from the front of one engine, the fan blades are rotating with an angular velocity of −110 rad/s, where

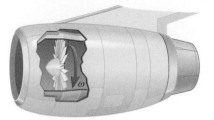

Figure 8.9 The fan blades of a jet engine have an angular velocity in a clockwise direction.

the negative sign indicates a clockwise rotation (see Figure 8.9). As the plane takes off, the angular velocity of the blades reaches -330 rad/s in a time of 14 s. Find the angular acceleration, assuming it to be constant.

Reasoning Since the angular acceleration is constant, it is equal to the average angular acceleration. The average acceleration is the change in the angular velocity, $\omega - \omega_0$, divided by the elapsed time, $t - t_0$.

Solution Applying the definition of average angular acceleration given in Equation 8.4, we find that

$$\overline{\alpha} = \frac{\omega - \omega_0}{t - t_0} = \frac{(-330 \text{ rad/s}) - (-110 \text{ rad/s})}{14 \text{ s}} = \boxed{-16 \text{ rad/s}^2}$$

Thus, the magnitude of the angular velocity increases by 16 rad/s during each second that the blades are accelerating. The negative sign in the answer indicates that the direction of the angular acceleration is also in the clockwise direction.

8.3 The Equations of Rotational Kinematics

In Chapters 2 and 3 the concepts of displacement, velocity, and acceleration were introduced. As illustrated in the Concepts-at-a-Glance chart in Figure 2.10, we combined these concepts and developed a set of equations called the equations of kinematics for constant acceleration (see Tables 2.1 and 3.1). These equations are a great aid in solving problems involving linear motion in one and two dimensions.

▶ CONCEPTS AT A GLANCE We now take a similar approach for rotational motion. The Concepts-at-a-Glance chart in Figure 8.10, which is analogous to that in Figure 2.10, shows that we will bring together the ideas of angular displacement, angular velocity, and angular acceleration to produce a set of equations called the equations of kinematics for constant angular acceleration. These equations, like those developed in Chapters 2 and 3, will prove very useful in solving problems that involve rotational motion. ◀

A complete description of rotational motion requires values for the angular displacement $\Delta\theta$, the angular acceleration α, the final angular velocity ω, the initial angular velocity ω_0, and the elapsed time Δt. In Example 4, for instance, only the angular displacement of the fan blades during the 14-s interval is missing. Such missing information can be cal-

Figure 8.10 CONCEPTS AT A GLANCE The equations of rotational kinematics for constant angular acceleration are obtained by bringing together the notions of angular displacement, angular velocity, and angular acceleration. The rotational motion of the blades on this wind turbine can be described by using these equations, provided that the angular acceleration is constant. (© John W. Warden/Age Fotostock America, Inc.)

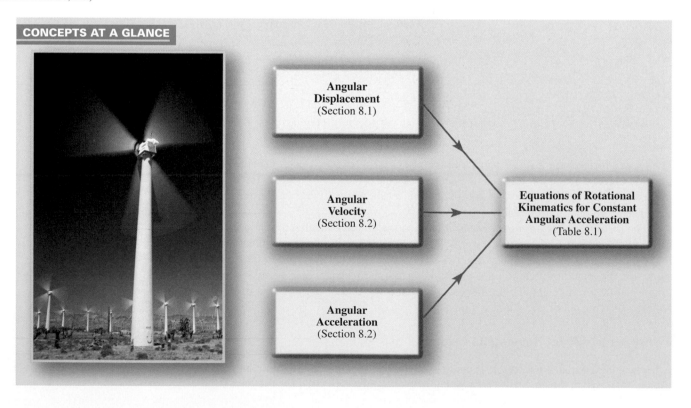

CONCEPTS AT A GLANCE

Angular Displacement (Section 8.1)

Angular Velocity (Section 8.2)

Angular Acceleration (Section 8.2)

Equations of Rotational Kinematics for Constant Angular Acceleration (Table 8.1)

Table 8.1 *The Equations of Kinematics for Rotational and Linear Motion*

Rotational Motion (α = constant)		Linear Motion (a = constant)	
$\omega = \omega_0 + \alpha t$	(8.4)	$v = v_0 + at$	(2.4)
$\theta = \frac{1}{2}(\omega_0 + \omega)t$	(8.6)	$x = \frac{1}{2}(v_0 + v)t$	(2.7)
$\theta = \omega_0 t + \frac{1}{2}\alpha t^2$	(8.7)	$x = v_0 t + \frac{1}{2}at^2$	(2.8)
$\omega^2 = \omega_0^2 + 2\alpha\theta$	(8.8)	$v^2 = v_0^2 + 2ax$	(2.9)

culated, however. For convenience in the calculations, we assume that the orientation of the rotating object is given by $\theta_0 = 0$ rad at time $t_0 = 0$ s. Then, the angular displacement becomes $\Delta\theta = \theta - \theta_0 = \theta$, and the time interval becomes $\Delta t = t - t_0 = t$.

In Example 4, the angular velocity of the fan blades changes at a constant rate from an initial value of $\omega_0 = -110$ rad/s to a final value of $\omega = -330$ rad/s. Therefore, the average angular velocity is midway between the initial and final values:

$$\overline{\omega} = \tfrac{1}{2}[(-110 \text{ rad/s}) + (-330 \text{ rad/s})] = -220 \text{ rad/s}$$

In other words, when the angular acceleration is constant, the average angular velocity is given by

$$\overline{\omega} = \tfrac{1}{2}(\omega_0 + \omega) \tag{8.5}$$

With a value for the average angular velocity, Equation 8.2 can be used to obtain the angular displacement of the fan blades:

$$\theta = \overline{\omega}t = (-220 \text{ rad/s})(14 \text{ s}) = -3100 \text{ rad}$$

In general, when the angular acceleration is constant, the angular displacement can be obtained from

$$\theta = \overline{\omega}t = \tfrac{1}{2}(\omega_0 + \omega)t \tag{8.6}$$

This equation and Equation 8.4 provide a complete description of rotational motion under the condition of constant angular acceleration. Equation 8.4 (with $t_0 = 0$ s) and Equation 8.6 are compared with the analogous results for linear motion in the first two rows of Table 8.1. The purpose of this comparison is to emphasize that the mathematical forms of Equations 8.4 and 2.4 are identical, as are the forms of Equations 8.6 and 2.7. Of course, the symbols used for the rotational variables are different from those used for the linear variables, as Table 8.2 indicates.

In Chapter 2, Equations 2.4 and 2.7 are used to derive the remaining two equations of kinematics (Equations 2.8 and 2.9). These additional equations convey no new information but are convenient to have when solving problems. Similar derivations can be carried out here. The results are listed as Equations 8.7 and 8.8 below and in Table 8.1; they can be inferred directly from their counterparts in linear motion by making the substitution of symbols indicated in Table 8.2:

$$\theta = \omega_0 t + \tfrac{1}{2}\alpha t^2 \tag{8.7}$$

$$\omega^2 = \omega_0^2 + 2\alpha\theta \tag{8.8}$$

The four equations in the left column of Table 8.1 are called the ***equations of rotational kinematics for constant angular acceleration.*** The following example illustrates that they are used in the same fashion as the equations of linear kinematics.

Example 5 Blending with a Blender

The blades of an electric blender are whirling with an angular velocity of +375 rad/s while the "puree" button is pushed in, as Figure 8.11 shows. When the "blend" button is pressed, the blades accelerate and reach a greater angular velocity after the blades have rotated through an angular displacement of +44.0 rad (seven revolutions). The angular acceleration has a constant value of +1740 rad/s². Find the final angular velocity of the blades.

Table 8.2 *Symbols Used in Rotational and Linear Kinematics*

Rotational Motion	Quantity	Linear Motion
θ	Displacement	x
ω_0	Initial velocity	v_0
ω	Final velocity	v
α	Acceleration	a
t	Time	t

Figure 8.11 The angular velocity of the blades in an electric blender changes each time a different push button is chosen.

Reasoning The three known variables are listed in the table below, along with a question mark indicating that a value for the final angular velocity ω is being sought.

θ	α	ω	ω_0	t
$+44.0$ rad	$+1740$ rad/s^2	?	$+375$ rad/s	

We can use Equation 8.8, because it relates the angular variables θ, α, ω, and ω_0.

Solution From Equation 8.8 ($\omega^2 = \omega_0{}^2 + 2\alpha\theta$) it follows that

$$\omega = +\sqrt{\omega_0{}^2 + 2\alpha\theta} = \sqrt{(375 \text{ rad/s})^2 + 2(1740 \text{ rad/s}^2)(44.0 \text{ rad})} = \boxed{+542 \text{ rad/s}}$$

The negative root is disregarded, since the blades do not reverse their direction of rotation.

Problem solving insight

The equations of rotational kinematics can be used with any self-consistent set of units for θ, α, ω, ω_0, *and* t. Radians are used in Example 5 only because data are given in terms of radians. Had the data for θ, α, and ω_0 been provided in rev, rev/s^2, and rev/s, respectively, then Equation 8.8 could have been used to determine the answer for ω directly in rev/s. In any case, the reasoning strategy for applying the kinematics equations is summarized as follows.

Reasoning Strategy

Applying the Equations of Rotational Kinematics

1. Make a drawing to represent the situation being studied, showing the direction of rotation.

2. Decide which direction of rotation is to be called positive ($+$) and which direction is to be called negative ($-$). In this text we choose the counterclockwise direction to be positive and the clockwise direction to be negative, but this is arbitrary. However, do not change your decision during the course of a calculation.

3. In an organized way, write down the values (with appropriate $+$ and $-$ signs) that are given for any of the five rotational kinematic variables (θ, α, ω, ω_0, and t). Be on the alert for implied data, such as the phrase "starts from rest," which means that the value of the initial angular velocity is $\omega_0 = 0$ rad/s. The data box in Example 5 is a good way to keep track of this information. In addition, identify the variable(s) that you are being asked to determine.

4. Before attempting to solve a problem, verify that the given information contains values for at least three of the five kinematic variables. Once the three variables are identified for which values are known, the appropriate relation from Table 8.1 can be selected.

5. When the rotational motion is divided into segments, the final angular velocity of one segment is the initial angular velocity for the next segment.

6. Keep in mind that there may be two possible answers to a kinematics problem. Try to visualize the different physical situations to which the answers correspond.

✓ **Check Your Understanding 2**

The blades of a ceiling fan start from rest and, after two revolutions, have an angular speed of 0.50 rev/s. The angular acceleration of the blades is constant. What is the angular speed after eight revolutions? *(The answer is given at the end of the book.)*

Background: Since the angular acceleration is constant, the equations of kinematics for rotational motion apply. These equations are given in Table 8.1.

For similar questions (including conceptual counterparts), consult Self-Assessment Test 8.1. The test is described next.

Self-Assessment Test 8.1

Test your understanding of the material in Sections 8.1–8.3:

• Angular Displacement   • Angular Velocity   • Angular Acceleration
• The Equations of Rotational Kinematics

Go to **www.wiley.com/college/cutnell**

8.4 Angular Variables and Tangential Variables

In the familiar ice-skating stunt known as "crack-the-whip," a number of skaters attempt to maintain a straight line as they skate around the one person (the pivot) who remains in place. Figure 8.12 shows each skater moving on a circular arc and includes the corresponding velocity vector at the instant portrayed in the picture. For every individual skater, the vector is drawn tangent to the appropriate circle and, therefore, is called the *tangential velocity* $\mathbf{v}_T$. The magnitude of the tangential velocity is referred to as the *tangential speed.*

Of all the skaters involved in the stunt, the one farthest from the pivot has the hardest job. Why? Because, in keeping the line straight, this skater covers more distance than anyone else. To accomplish this, he must skate faster than anyone else and, thus, must have the largest tangential speed. In fact, the line remains straight only if each person skates with the correct tangential speed. The skaters closer to the pivot must move with smaller tangential speeds than those farther out, as indicated by the magnitudes of the tangential velocity vectors in Figure 8.12.

With the aid of Figure 8.13, it is possible to show that the tangential speed of any skater is directly proportional to his distance r from the pivot, assuming a given angular speed for the rotating line. When the line rotates as a rigid unit for a time t, it sweeps out the angle θ shown in the drawing. The distance s through which a skater moves along a circular arc can be calculated from Equation 8.1, $s = r\theta$, provided θ is measured in radians. Dividing both sides of this equation by t gives $s/t = r(\theta/t)$. The term s/t is the tangential speed v_T (e.g., in meters/second) of the skater, while θ/t is the angular speed ω (in radians/second) of the line:

$$v_T = r\omega \qquad (\omega \text{ in rad/s}) \qquad (8.9)$$

In this expression, the terms v_T and ω refer to the magnitudes of the tangential and angular velocities, respectively, and are numbers without algebraic signs.

It is important to emphasize that the angular speed ω in Equation 8.9 must be expressed in radian measure (e.g., in rad/s); no other units, such as revolutions per second, are acceptable. This restriction arises because the equation was derived by using the definition of radian measure, $s = r\theta$.

The real challenge for the "crack-the-whip" skaters is to keep the line straight while making it pick up angular speed—that is, while giving it an angular acceleration. To make the angular speed of the line increase, each skater must increase his tangential speed, since the two speeds are related according to $v_T = r\omega$. Of course, the fact that a skater must skate faster and faster means that he must accelerate, and his tangential acceleration a_T can be related to the angular acceleration α of the line. If time is measured relative to $t_0 = 0$ s, the definition of linear acceleration is given by Equation 2.4 as $a_T = (v_T - v_{T0})/t$, where v_T and v_{T0} are the final and initial tangential speeds, respectively. Substituting $v_T = r\omega$ for the tangential speed shows that

$$a_T = \frac{v_T - v_{T0}}{t} = \frac{(r\omega) - (r\omega_0)}{t} = r\left(\frac{\omega - \omega_0}{t}\right)$$

Since $\alpha = (\omega - \omega_0)/t$ according to Equation 8.4, it follows that

$$a_T = r\alpha \qquad (\alpha \text{ in rad/s}^2) \qquad (8.10)$$

The physics of "crack-the-whip."

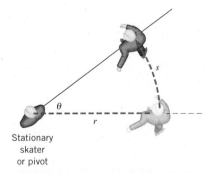

Figure 8.12 When doing a stunt known as "crack-the-whip," each skater along the radial line moves on a circular arc. The tangential velocity $\mathbf{v}_T$ of each skater is represented by an arrow that is tangent to each arc.

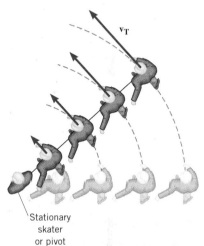

Figure 8.13 During a time t, the line of skaters sweeps through an angle θ. An individual skater, located at a distance r from the stationary skater, moves through a distance s on a circular arc.

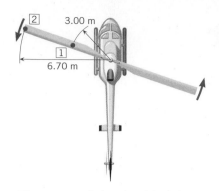

Figure 8.14 Points 1 and 2 on the rotating blade of the helicopter have the same angular speed and acceleration, but they have *different* tangential speeds and accelerations.

This result shows that, for a given value of α, the tangential acceleration a_T is proportional to the radius r, so the skater farthest from the pivot must have the largest tangential acceleration. In this expression, the terms a_T and α refer to the magnitudes of the numbers involved, without reference to any algebraic sign. Moreover, as is the case for ω in $v_T = r\omega$, only radian measure can be used for α in Equation 8.10.

There is an advantage to using the angular velocity ω and the angular acceleration α to describe the rotational motion of a rigid object. The advantage is that these angular quantities describe the motion of the *entire object*. In contrast, the tangential quantities v_T and a_T describe only the motion of a single point on the object, and Equations 8.9 and 8.10 indicate that different points located at different distances r have different tangential velocities and accelerations. Example 6 stresses this advantage.

Example 6   A Helicopter Blade

A helicopter blade has an angular speed of $\omega = 6.50$ rev/s and an angular acceleration of $\alpha = 1.30$ rev/s^2. For points 1 and 2 on the blade in Figure 8.14, find the magnitudes of (a) the tangential speeds and (b) the tangential accelerations.

Reasoning Since the radius r for each point and the angular speed ω of the helicopter blade are known, we can find the tangential speed v_T for each point by using the relation $v_T = r\omega$. However, since this equation can be used only with radian measure, the angular speed ω must be converted to rad/s from rev/s. In a similar manner, the tangential acceleration a_T for points 1 and 2 can be found using $a_T = r\alpha$, provided the angular acceleration α is expressed in rad/s^2 rather than in rev/s^2.

Solution

(a) Converting the angular speed ω to rad/s from rev/s, we obtain

$$\omega = \left(6.50 \, \frac{\text{rev}}{\text{s}}\right)\left(\frac{2\pi \, \text{rad}}{1 \, \text{rev}}\right) = 40.8 \, \frac{\text{rad}}{\text{s}}$$

The tangential speed for each point is

Point 1 $v_T = r\omega = (3.00 \text{ m})(40.8 \text{ rad/s}) = \boxed{122 \text{ m/s (273 mph)}}$ (8.9)

Point 2 $v_T = r\omega = (6.70 \text{ m})(40.8 \text{ rad/s}) = \boxed{273 \text{ m/s (611 mph)}}$ (8.9)

The rad unit, being dimensionless, does not appear in the final answers.

(b) Converting the angular acceleration α to rad/s^2 from rev/s^2, we find

$$\alpha = \left(1.30 \, \frac{\text{rev}}{\text{s}^2}\right)\left(\frac{2\pi \, \text{rad}}{1 \, \text{rev}}\right) = 8.17 \, \frac{\text{rad}}{\text{s}^2}$$

The tangential accelerations can now be determined:

Point 1 $a_T = r\alpha = (3.00 \text{ m})(8.17 \text{ rad/s}^2) = \boxed{24.5 \text{ m/s}^2}$ (8.10)

Point 2 $a_T = r\alpha = (6.70 \text{ m})(8.17 \text{ rad/s}^2) = \boxed{54.7 \text{ m/s}^2}$ (8.10)

✓ **Check Your Understanding 3**

The blade of a lawn mower is rotating at an angular speed of 17 rev/s. The tangential speed of the outer edge of the blade is 32 m/s. What is the radius of the blade? *(The answer is given at the end of the book.)*

Background: Angular variables, tangential variables, and the relation between them are the focus of this problem.

For similar questions (including conceptual counterparts), consult Self-Assessment Test 8.2. This test is described at the end of Section 8.6.

8.5   Centripetal Acceleration and Tangential Acceleration

When an object picks up speed as it moves around a circle, it has a tangential acceleration, as discussed in the last section. In addition, the object also has a centripetal acceleration, as emphasized in Chapter 5. That chapter deals with **uniform circular motion,** in which a particle moves at a constant tangential speed on a circular path. The tangential speed v_T is the magnitude of the tangential velocity vector. Even when the magnitude of the tangential velocity is constant, an acceleration is present, since the direction of the velocity changes continually. Because the resulting acceleration points toward the center of the circle, it is called the centripetal acceleration. Figure 8.15a shows the centripetal acceleration $\mathbf{a}_c$ for a model airplane flying in uniform circular motion on a guide wire. The magnitude of $\mathbf{a}_c$ is

$$a_c = \frac{v_T^2}{r} \qquad (5.2)$$

The subscript "T" has now been included in this equation as a reminder that it is the tangential speed that appears in the numerator.

The centripetal acceleration can be expressed in terms of the angular speed ω by using $v_T = r\omega$ (Equation 8.9):

$$a_c = \frac{v_T^2}{r} = \frac{(r\omega)^2}{r} = r\omega^2 \qquad (\omega \text{ in rad/s}) \qquad (8.11)$$

Only radian measure, such as rad/s, can be used for ω in this result, since the relation $v_T = r\omega$ presumes radian measure.

While considering uniform circular motion in Chapter 5, we ignored the details of how the motion is established in the first place. In Figure 8.15b, for instance, the engine of the plane produces a thrust in the tangential direction, and this force leads to a tangential acceleration. In response, the tangential speed of the plane increases from moment to moment, until the situation shown in the drawing results. While the tangential speed is changing, the motion is called **nonuniform circular motion.**

Figure 8.15b illustrates an important feature of nonuniform circular motion. Since the direction and the magnitude of the tangential velocity are both changing, the airplane experiences two acceleration components simultaneously. The changing direction means that there is a centripetal acceleration $\mathbf{a}_c$. The magnitude of $\mathbf{a}_c$ at any moment can be calculated using the value of the instantaneous angular speed and the radius: $a_c = r\omega^2$. The fact that the magnitude of the tangential velocity is changing means that there is also a tangential acceleration $\mathbf{a}_T$. The magnitude of $\mathbf{a}_T$ can be determined from the angular acceleration α according to $a_T = r\alpha$, as the previous section explains. If the magnitude F_T of the net tangential force and the mass m are known, a_T also can be calculated using Newton's second law, $F_T = ma_T$. Figure 8.15b shows the two acceleration components. The total acceleration is given by the vector sum of $\mathbf{a}_c$ and $\mathbf{a}_T$. Since $\mathbf{a}_c$ and $\mathbf{a}_T$ are perpendicular, the magnitude of the total acceleration $\mathbf{a}$ can be obtained from the Pythagorean theorem as $a = \sqrt{a_c^2 + a_T^2}$, while the angle ϕ in the drawing can be determined from $\tan \phi = a_T/a_c$. The next example applies these concepts to a discus thrower.

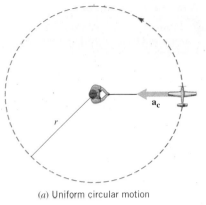

(a) Uniform circular motion

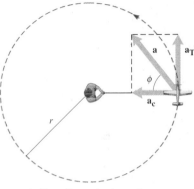

(b) Nonuniform circular motion

Figure 8.15 (a) If a model airplane flying on a guide wire has a constant tangential speed, the motion is uniform circular motion, and the plane experiences only a centripetal acceleration $\mathbf{a}_c$. (b) Nonuniform circular motion occurs when the tangential speed changes. Then there is a tangential acceleration $\mathbf{a}_T$ in addition to the centripetal acceleration.

Example 7 A Discus Thrower

Discus throwers often warm up by standing with both feet flat on the ground and throwing the discus with a twisting motion of their bodies. Figure 8.16a illustrates a top view of such a warm-up throw. Starting from rest, the thrower accelerates the discus to a final angular velocity of $+15.0$ rad/s in a time of 0.270 s before releasing it. During the acceleration, the discus moves on a circular arc of radius 0.810 m. Find (a) the magnitude a of the total acceleration of the discus just before it is released and (b) the angle ϕ that the total acceleration makes with the radius at this moment.

Reasoning Since the tangential speed of the discus is increasing, the discus simultaneously experiences a tangential acceleration $\mathbf{a}_T$ and a centripetal acceleration $\mathbf{a}_c$ that are oriented at right angles to each other. The magnitude of the total acceleration is $a = \sqrt{a_c^2 + a_T^2}$, where a_c and a_T are the magnitudes of the centripetal and tangential accelerations. The angle ϕ in Figure 8.16b is given by $\phi = \tan^{-1}(a_T/a_c)$. The magnitude of the centripetal acceleration can be evaluated from $a_c = r\omega^2$. The magnitude of the tangential acceleration follows from $a_T = r\alpha$, where α can be found from the definition of angular acceleration in Equation 8.4.

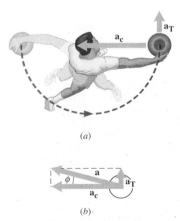

(a)

(b)

Figure 8.16 (a) A discus thrower and the centripetal acceleration $\mathbf{a}_c$ and tangential acceleration $\mathbf{a}_T$ that act on the discus. (b) The total acceleration $\mathbf{a}$ of the discus just before the discus is released is the vector sum of $\mathbf{a}_c$ and $\mathbf{a}_T$.

Solution

(a) Just before the moment of release, the magnitude of the total acceleration of the discus is $a = \sqrt{a_c^2 + a_T^2}$. According to Equation 8.11, the magnitude of the centripetal acceleration is $a_c = r\omega^2$, where the radius is $r = 0.810$ m and the final angular velocity is $\omega = +15.0$ rad/s. According to Equation 8.10, the magnitude of the tangential acceleration is $a_T = r\alpha$, where the angular acceleration α is not given. However, it can be obtained from the definition of angular acceleration in Equation 8.4 as $\alpha = (\omega - \omega_0)/t$, where the time is $t = 0.270$ s and

Need more practice?

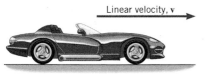

Interactive LearningWare 8.2
A thin rigid rod is rotating with a constant angular acceleration about an axis that passes perpendicularly through one of its ends. At one instant, the total acceleration vector (centripetal plus tangential) at the other end of the rod makes a 60.0° angle with respect to the rod and has a magnitude of 15.0 m/s². The rod has an angular speed of 2.00 rad/s at this instant. What is the rod's length?

Related Homework: Problems 42, 60

Go to
www.wiley.com/college/cutnell
for an interactive solution.

the initial angular velocity is $\omega_0 = 0$ rad/s, since the discus starts from rest. With these substitutions we find that

$$a = \sqrt{a_c{}^2 + a_T{}^2} = \sqrt{(r\omega^2)^2 + (r\alpha)^2} = r\sqrt{\omega^4 + \alpha^2}$$

$$= r\sqrt{\omega^4 + \frac{(\omega - \omega_0)^2}{t^2}} = (0.810 \text{ m})\sqrt{(15.0 \text{ rad/s})^4 + \left(\frac{15.0 \text{ rad/s} - 0 \text{ rad/s}}{0.270 \text{ s}}\right)^2}$$

$$= \boxed{188 \text{ m/s}^2}$$

(b) The angle ϕ in Figure 8.16b is given by

$$\phi = \tan^{-1}\left(\frac{a_T}{a_c}\right) = \tan^{-1}\left(\frac{r\alpha}{r\omega^2}\right) = \tan^{-1}\left(\frac{\alpha}{\omega^2}\right)$$

$$= \tan^{-1}\left[\frac{(\omega - \omega_0)/t}{\omega^2}\right] = \tan^{-1}\left[\frac{(15.0 \text{ rad/s} - 0 \text{ rad/s})/(0.270 \text{ s})}{(15.0 \text{ rad/s})^2}\right] = \boxed{13.9°}$$

▲

✔ **Check Your Understanding 4**

A rotating object starts from rest and has a constant angular acceleration. Three seconds later the centripetal acceleration of a point on the object has a magnitude of 2.0 m/s². What is the magnitude of the centripetal acceleration of this point six seconds after the motion begins? *(The answer is given at the end of the book.)*

Background: The concept of centripetal acceleration is the jumping-off point for this question. Furthermore, since the angular acceleration is constant, the equations of kinematics for rotational motion apply (see Table 8.1).

For similar questions (including conceptual counterparts), consult Self-Assessment Test 8.2. This test is described at the end of Section 8.6.

8.6 Rolling Motion

Rolling motion is a familiar situation that involves rotation, as Figure 8.17 illustrates for the case of an automobile tire. The essence of rolling motion is that there is *no slipping* at the point of contact where the tire touches the ground. To a good approximation, the tires on a normally moving automobile roll and do not slip. On the other hand, the squealing tires that accompany the start of a drag race are rotating, but they are not rolling while they rapidly spin and slip against the ground.

When the tires in Figure 8.17 roll, there is a relationship between the angular speed at which the tires rotate and the linear speed (assumed constant) at which the car moves forward. In part *b* of the drawing, consider the points labeled *A* and *B* on the left tire. Between these points we apply a coat of red paint to the tread of the tire; the length of this circular arc of paint is *s*. The tire then rolls to the right until point *B* comes in contact with the ground. As the tire rolls, all the paint comes off the tire and sticks to the ground, leaving behind the horizontal red line shown in the drawing. The axle of the wheel moves through a linear distance *d*, which is equal to the length of the horizontal strip of paint. Since the tire does not slip, the distance *d* must be equal to the circular arc length *s*, measured along the outer edge of the tire: $d = s$. Dividing both sides of this equation by the elapsed time *t* shows that $d/t = s/t$. The term d/t is the speed at which the axle moves parallel to the ground—namely, the linear speed *v* of the car. The term s/t is the tangential speed v_T at which a point on the outer edge of the tire moves relative to the axle. In addition, v_T is related to the angular speed ω about the axle according to $v_T = r\omega$ (Equation 8.9). Therefore, it follows that

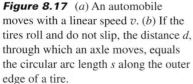

Linear velocity, **v**

(a)

B   *A*

r   *s*

A   *B*

|← $d = s$ →|

(b)

Figure 8.17 (*a*) An automobile moves with a linear speed *v*. (*b*) If the tires roll and do not slip, the distance *d*, through which an axle moves, equals the circular arc length *s* along the outer edge of a tire.

$$\underbrace{v}_{\substack{\text{Linear} \\ \text{speed}}} = \underbrace{r\omega}_{\substack{\text{Tangential} \\ \text{speed}, v_T}} \qquad (\omega \text{ in rad/s}) \qquad (8.12)$$

If the car in Figure 8.17 has a linear acceleration **a** parallel to the ground, a point on the tire's outer edge experiences a tangential acceleration **a**$_T$ relative to the axle. The same kind of reasoning used in the last paragraph reveals that the magnitudes of these accelerations are the same and that they are related to the angular acceleration α of the wheel relative to the axle:

$$\underbrace{a}_{\substack{\text{Linear} \\ \text{acceleration}}} = \underbrace{r\alpha}_{\substack{\text{Tangential} \\ \text{acceleration, } a_T}} \qquad (\alpha \text{ in rad/s}^2) \qquad (8.13)$$

Equations 8.12 and 8.13 may be applied to any rolling motion, because the object does not slip against the surface on which it is rolling. Example 8 illustrates the basic features of rolling motion.

Example 8 An Accelerating Car

An automobile starts from rest and for 20.0 s has a constant linear acceleration of 0.800 m/s^2 to the right, as in Figure 8.17. During this period, the tires do not slip. The radius of the tires is 0.330 m. At the end of the 20.0-s interval, what is the angle through which each wheel has rotated?

Reasoning As the car accelerates, the tires rotate faster and faster. Thus, each tire has an angular acceleration, which must be taken into account when we determine the angular displacement of each wheel. Since the tires roll and do not slip, the magnitude α of the angular acceleration is related to the magnitude a of the linear acceleration of the car by $a = r\alpha$ (Equation 8.13). Therefore, we find that

$$\alpha = \frac{a}{r} = \frac{0.800 \text{ m/s}^2}{0.330 \text{ m}} = 2.42 \text{ rad/s}^2$$

Since the tires rotate faster and faster in the clockwise, or negative, direction as the car moves to the right (see Figure 8.17), the angular acceleration is negative. Taking into account the negative value of the acceleration, the angular data are as follows:

θ	α	ω	ω_0	t
?	-2.42 rad/s^2		0 rad/s	20.0 s

The angular displacement θ is given in terms of α, ω_0, and t by Equation 8.7.

Solution From Equation 8.7, we find that

$$\theta = \omega_0 t + \tfrac{1}{2}\alpha t^2 = (0 \text{ rad/s})(20.0 \text{ s}) + \tfrac{1}{2}(-2.42 \text{ rad/s}^2)(20.0 \text{ s})^2 = \boxed{-484 \text{ rad}}$$

The angular displacement θ is negative, because the wheels rotate in the clockwise direction.

*8.7 The Vector Nature of Angular Variables

We have presented angular velocity and angular acceleration by taking advantage of the analogy between angular variables and linear variables. Like the linear velocity and the linear acceleration, the angular quantities are also vectors and have a direction as well as a magnitude. As yet, however, we have not discussed the directions of these vectors.

Figure 8.18 The angular velocity vector ω of a rotating object points along the axis of rotation. The direction along the axis depends on the sense of the rotation and can be determined with the aid of a right-hand rule (see text).

When a rigid object rotates about a fixed axis, it is the axis that identifies the motion, and the angular velocity vector points along this axis. Figure 8.18 shows how to determine the direction using a *right-hand rule:*

> **Right-Hand Rule**   Grasp the axis of rotation with your right hand, so that your fingers circle the axis in the same sense as the rotation. Your extended thumb points along the axis in the direction of the angular velocity vector.

No part of the rotating object moves in the direction of the angular velocity vector.

Angular acceleration arises when the angular velocity changes, and the acceleration vector also points along the axis of rotation. The acceleration vector has the same direction as the *change* in the angular velocity. That is, when the angular velocity is increasing, the angular acceleration vector points in the same direction as the angular velocity. Conversely, when the angular velocity is decreasing, the angular acceleration vector points in the direction opposite to the angular velocity.

8.8   Concepts & Calculations

In this chapter we have studied the concepts of angular displacement, angular velocity, and angular acceleration. We conclude with several examples that review some of the important features of these ideas. Example 9 illustrates that the angular acceleration and the angular velocity can have the same or the opposite direction, depending on whether the angular speed is increasing or decreasing.

Concepts & Calculations Example 9   Riding a Mountain Bike

A rider on a mountain bike is traveling to the left in Figure 8.19. Each wheel has an angular velocity of $+21.7$ rad/s, where, as usual, the plus sign indicates that the wheel is rotating in the counterclockwise direction. (a) To pass another cyclist, the rider pumps harder, and the angular velocity of the wheels increases from $+21.7$ to $+28.5$ rad/s in a time of 3.50 s. (b) After passing the cyclist, the rider begins to coast, and the angular velocity of the wheels decreases from $+28.5$ to $+15.3$ rad/s in a time of 10.7 s. In both instances, determine the magnitude and direction of the angular acceleration (assumed constant) of the wheels.

Concept Questions and Answers Is the angular acceleration positive or negative when the rider is passing the cyclist and the angular speed of the wheels is increasing?

> *Answer*   Since the angular speed is increasing, the angular acceleration has the same direction as the angular velocity, which is the counterclockwise, or positive, direction (see Figure 8.19*a*).

Is the angular acceleration positive or negative when the rider is coasting and the angular speed of the wheels is decreasing?

> *Answer*   Since the angular speed is decreasing during the coasting phase, the direction of the angular acceleration is opposite to that of the angular velocity. The angular velocity is in the counterclockwise (positive) direction, so the angular acceleration must be in the clockwise (negative) direction (see Figure 8.19*b*).

Solution

(a) The angular acceleration α is the change in the angular velocity, $\omega - \omega_0$, divided by the elapsed time t:

$$\alpha = \frac{\omega - \omega_0}{t} = \frac{+28.5 \text{ rad/s} - (+21.7 \text{ rad/s})}{3.50 \text{ s}} = \boxed{+1.9 \text{ rad/s}^2} \qquad (8.4)$$

As expected, the angular acceleration is positive (counterclockwise).

(b) Now the wheels are slowing down, but still rotating in the positive (counterclockwise) direction. The angular acceleration for this part of the motion is

$$\alpha = \frac{\omega - \omega_0}{t} = \frac{+15.3 \text{ rad/s} - (+28.5 \text{ rad/s})}{10.7 \text{ s}} = \boxed{-1.23 \text{ rad/s}^2} \qquad (8.4)$$

Now, as anticipated, the angular acceleration is negative (clockwise).

(a) Angular speed increasing   (b) Angular speed decreasing

Figure 8.19 (*a*) When the angular speed of the wheel is increasing, the angular velocity ω and the angular acceleration α point in the same direction (counterclockwise in this drawing). (*b*) When the angular speed is decreasing, the angular velocity and the angular acceleration point in opposite directions.

Example 10 reviews the two different types of acceleration, centripetal and tangential, that a car can have when it travels on a circular road.

Concepts & Calculations Example 10
A Circular Roadway and the Acceleration of Your Car

Suppose you are driving a car in a counterclockwise direction on a circular road whose radius is $r = 390$ m (see Figure 8.20). You look at the speedometer and it reads a steady 32 m/s (about 72 mi/h). (a) What is the angular speed of the car? (b) Determine the acceleration (magnitude and direction) of the car. (c) To avoid a rear-end collision with a vehicle ahead, you apply the brakes and reduce your angular speed to 4.9×10^{-2} rad/s in a time of 4.0 s. What is the tangential acceleration (magnitude and direction) of the car?

Concept Questions and Answers Does an object traveling at a constant tangential speed (for example, $v_T = 32$ m/s) along a circular path have an acceleration?

Answer Yes. Recall that an object has an acceleration if its velocity is changing in time. The velocity has two attributes, a magnitude (or speed) and a direction. In this instance the speed is not changing, since it is steady at 32 m/s. However, the direction of the velocity is changing continually as the car moves on the circular road. As Sections 5.2 and 8.5 discuss, this change in direction gives rise to an acceleration, called the centripetal acceleration $\mathbf{a}_c$. The centripetal acceleration is directed toward the center of the circle (see Figure 8.20*a*).

Is there a tangential acceleration $\mathbf{a}_T$ when the angular speed of an object changes (e.g., when the car's angular speed decreases to 4.9×10^{-2} rad/s)?

Answer Yes. When the car's angular speed ω decreases, for example, its tangential speed v_T also decreases. This is because they are related by $v_T = r\omega$ (Equation 8.9), where r is the radius of the circular road. A decreasing tangential speed v_T, in turn, means that the car has a tangential acceleration $\mathbf{a}_T$. The direction of the tangential acceleration $\mathbf{a}_T$ must be *opposite* to that of the tangential velocity $\mathbf{v}_T$, because the tangential speed is decreasing. (If the two vectors were in the same direction, the tangential speed would increase.) Figure 8.20*b* shows these two vectors.

Solution

(a) The angular speed ω of the car is equal to its tangential speed v_T (the speed indicated by the speedometer) divided by the radius of the circular road:

$$\omega = \frac{v_T}{r} = \frac{32 \text{ m/s}}{390 \text{ m}} = \boxed{8.2 \times 10^{-2} \text{ rad/s}} \tag{8.9}$$

(b) The acceleration is the centripetal acceleration and arises because the tangential velocity is changing direction as the car travels around the circular path. The magnitude of the centripetal acceleration is

$$a_c = r\omega^2 = (390 \text{ m})(8.2 \times 10^{-2} \text{ rad/s})^2 = \boxed{2.6 \text{ m/s}^2} \tag{8.11}$$

As always, the centripetal acceleration is directed toward the center of the circle.

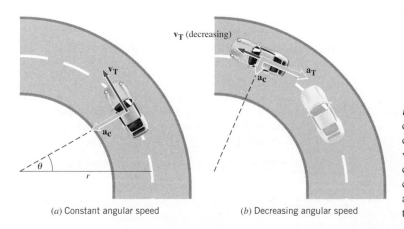

(*a*) Constant angular speed   (*b*) Decreasing angular speed

Figure 8.20 (*a*) A car is traveling in a counterclockwise direction at a constant angular speed. The tangential velocity of the car is $\mathbf{v}_T$ and its centripetal acceleration is $\mathbf{a}_c$. (*b*) The car is slowing down, so the tangential acceleration $\mathbf{a}_T$ points opposite to the tangential velocity $\mathbf{v}_T$.

(c) The tangential acceleration arises because the tangential speed of the car is changing in time. The magnitude of the tangential acceleration is $a_T = r\alpha$ (Equation 8.10), where α is the magnitude of the angular acceleration. The angular acceleration is given by Equation 8.4 as

$$\alpha = \frac{\omega - \omega_0}{t} = \frac{4.9 \times 10^{-2}\,\text{rad/s} - 8.2 \times 10^{-2}\,\text{rad/s}}{4.0\,\text{s}} = -8.3 \times 10^{-3}\,\text{rad/s}^2$$

The magnitude of the angular acceleration is $8.3 \times 10^{-3}\,\text{rad/s}^2$. Therefore, the magnitude of the tangential acceleration is

$$a_T = r\alpha = (390\,\text{m})(8.3 \times 10^{-3}\,\text{rad/s}^2) = \boxed{3.2\,\text{m/s}^2} \qquad (8.10)$$

Since the car is slowing down, the tangential acceleration is directed opposite to the direction of the tangential velocity, as shown in Figure 8.20b.

▲

At the end of the problem set for this chapter, you will find homework problems that contain both conceptual and quantitative parts. These problems are grouped under the heading *Concepts & Calculations, Group Learning Problems.* They are designed for use by students working alone or in small learning groups. The conceptual part of each problem provides a convenient focus for group discussions.

Concept Summary

This summary presents an abridged version of the chapter, including the important equations and all available learning aids. For convenient reference, the learning aids (including the text's examples) are placed next to or immediately after the relevant equation or discussion. The following learning aids may be found on-line at **www.wiley.com/college/cutnell**:

Interactive LearningWare examples are solved according to a five-step interactive format that is designed to help you develop problem-solving skills.	**Concept Simulations** are animated versions of text figures or animations that illustrate important concepts. You can control parameters that affect the display, and we encourage you to experiment.
Interactive Solutions offer specific models for certain types of problems in the chapter homework. The calculations are carried out interactively.	**Self-Assessment Tests** include both qualitative and quantitative questions. Extensive feedback is provided for both incorrect and correct answers, to help you evaluate your understanding of the material.

Topic	**Discussion**	**Learning Aids**

8.1 Rotational Motion and Angular Displacement

Angular displacement   When a rigid body rotates about a fixed axis, the angular displacement is the angle swept out by a line passing through any point on the body and intersecting the axis of rotation perpendicularly. By convention, the angular displacement is positive if it is counterclockwise and negative if it is clockwise.

Radian   The radian (rad) is the SI unit of angular displacement. In radians, the angle θ is defined as the circular arc length s traveled by a point on the rotating body divided by the radial distance r of the point from the axis:

$$\theta\,(\text{in radians}) = \frac{s}{r} \qquad (8.1) \quad \textbf{Examples 1, 2}$$

8.2 Angular Velocity and Angular Acceleration

The average angular velocity $\overline{\omega}$ is the angular displacement $\Delta\theta$ divided by the elapsed time Δt:

Average angular velocity

$$\overline{\omega} = \frac{\Delta\theta}{\Delta t} \qquad (8.2) \quad \textbf{Example 3}$$

Instantaneous angular velocity   As Δt approaches zero, the average angular velocity becomes equal to the instantaneous angular velocity ω. The magnitude of the instantaneous angular velocity is called the instantaneous angular speed.

The average angular acceleration $\overline{\alpha}$ is the change $\Delta\omega$ in the angular velocity divided by the elapsed time Δt:

Average angular acceleration

$$\overline{\alpha} = \frac{\Delta\omega}{\Delta t} \qquad (8.4) \quad \textbf{Example 4}$$

Topic	Discussion	Learning Aids
Instantaneous angular acceleration	As Δt approaches zero, the average angular acceleration becomes equal to the instantaneous angular acceleration α.	

8.3 The Equations of Rotational Kinematics

The equations of rotational kinematics apply when a rigid body rotates with a constant angular acceleration about a fixed axis. These equations relate the angular displacement $\theta - \theta_0$, the angular acceleration α, the final angular velocity ω, the initial angular velocity ω_0, and the elapsed time $t - t_0$. Assuming that $\theta_0 = 0$ rad at $t_0 = 0$ s, the equations of rotational kinematics are

Equations of rotational kinematics	$\omega = \omega_0 + \alpha t$	(8.4)	**Examples 5, 9**
	$\theta = \frac{1}{2}(\omega + \omega_0)t$	(8.6)	
			Interactive LearningWare 8.1
	$\theta = \omega_0 t + \frac{1}{2}\alpha t^2$	(8.7)	
	$\omega^2 = \omega_0{}^2 + 2\alpha\theta$	(8.8)	**Interactive Solution 8.25**

These equations may be used with any self-consistent set of units and are not restricted to radian measure.

Use *Self-Assessment Test 8.1* to evaluate your understanding of Sections 8.1–8.3.

8.4 Angular Variables and Tangential Variables

When a rigid body rotates through an angle θ about a fixed axis, any point on the body moves on a circular arc of length s and radius r. Such a point has a tangential velocity (magnitude $= v_{\mathrm{T}}$) and, possibly, a tangential acceleration (magnitude $= a_{\mathrm{T}}$). The angular and tangential variables are related by the following equations.

Relations between angular and tangential variables	$s = r\theta$	(θ in rad)	(8.1)	**Example 6**
	$v_{\mathrm{T}} = r\omega$	(ω in rad/s)	(8.9)	
	$a_{\mathrm{T}} = r\alpha$	(α in rad/s^2)	(8.10)	**Interactive Solutions 8.31, 8.61**

These equations refer to the magnitudes of the variables involved, without reference to positive or negative signs, and only radian measure can be used when applying them.

8.5 Centripetal Acceleration and Tangential Acceleration

The magnitude a_{c} of the centripetal acceleration of a point on an object rotating with uniform or nonuniform circular motion can be expressed in terms of the radial distance r of the point from the axis and the angular speed ω:

Examples 7, 10

Centripetal acceleration	$a_{\mathrm{c}} = r\omega^2$	(ω in rad/s)	(8.11)	**Interactive LearningWare 8.2**

Total acceleration This point experiences a total acceleration $\mathbf{a}$ that is the vector sum of two perpendicular acceleration components, the centripetal acceleration $\mathbf{a_{\mathrm{c}}}$ and the tangential acceleration $\mathbf{a_{\mathrm{T}}}$; $\mathbf{a} = \mathbf{a_{\mathrm{c}}} + \mathbf{a_{\mathrm{T}}}$.

8.6 Rolling Motion

The essence of rolling motion is that there is no slipping at the point where the object touches the surface upon which it is rolling. As a result, the tangential speed v_{T} of a point on the outer edge of a rolling object, measured relative to the axis through the center of the object, is equal to the linear speed v with which the object moves parallel to the surface. In other words, we have

Concept Simulation 8.1

$$v = v_{\mathrm{T}} = r\omega \qquad (\omega \text{ in rad/s}) \qquad (8.12)$$

The magnitudes of the tangential acceleration a_{T} and the linear acceleration a of a rolling object are similarly related:

Interactive Solution 8.49

	$a = a_{\mathrm{T}} = r\alpha$	(α in rad/s^2)	(8.13)	**Example 8**

Topic	Discussion	Learning Aids

Use Self-Assessment Test 8.2 to evaluate your understanding of Sections 8.4–8.6.

8.7 The Vector Nature of Angular Variables

The direction of the angular velocity vector is given by a right-hand rule. Grasp the axis of rotation with your right hand, so that your fingers circle the axis in the same sense as the rotation. Your extended thumb points along the axis in the direction of the angular velocity vector.

The angular acceleration vector has the same direction as the change in the angular velocity.

Conceptual Questions

1. In the drawing, the flat triangular sheet *ABC* is lying in the plane of the paper. This sheet is going to rotate about an axis that also lies in the plane of the paper and passes through point *A*. Draw two such axes that are oriented so that points *B* and *C* will move on circular paths having the same radii.

2. A pair of scissors is being used to cut a string. Does each blade of the scissors have the same angular velocity (both magnitude and direction) at a given instant? Give your reasoning.

3. An electric clock is hanging on the wall in the living room. The clock is unplugged, and the second hand comes to a halt over a brief period of time. During this period, what is the direction of the angular acceleration of the second hand? Why?

4. The earth rotates once per day about its axis. Where on the earth's surface should you stand in order to have the smallest possible tangential speed? Justify your answer.

5. A thin rod rotates at a constant angular speed. Consider the tangential speed of each point on the rod for the case when the axis of rotation is perpendicular to the rod (a) at its center and (b) at one end. Explain for each case whether there are any points on the rod that have the same tangential speeds.

6. A car is up on a hydraulic lift at a garage. The wheels are free to rotate, and the drive wheels are rotating with a constant angular velocity. Does a point on the rim of a wheel have (a) a tangential acceleration and (b) a centripetal acceleration? In each case, give your reasoning.

7. Two points are located on a rigid wheel that is rotating with an increasing angular velocity about a fixed axis. The axis is perpendicular to the wheel at its center. Point 1 is located on the rim, and point 2 is halfway between the rim and the axis. At any given instant, which point (if either) has the greater (a) angular velocity, (b) angular acceleration, (c) tangential speed, (d) tangential acceleration, and (e) centripetal acceleration? Provide a reason for each of your answers.

8. A building is located on the earth's equator. Which has the greatest tangential speed due to the earth's rotation, the top floor, the bottom floor, or neither? Justify your answer.

9. Section 5.6 discusses how the uniform circular motion of a space station can be used to create "artificial" gravity for the astronauts. This can be done by adjusting the angular speed of the space station, so the centripetal acceleration at the astronaut's feet equals *g*, the magnitude of the acceleration due to gravity (see Figure 5.19). If such an adjustment is made, will the acceleration due to the "artificial" gravity be greater than, equal to, or less than *g* at the astronaut's head? Account for your answer.

10. It is possible, but not very practical, to build a clock in which the tips of the second hand, the minute hand, and the hour hand move with the same tangential speed. Explain why not.

11. Explain why a point on the rim of a tire has an acceleration when the tire is on a car that is moving at a constant linear velocity.

12. A bicycle is turned upside down, the front wheel is spinning (see the drawing), and there is an angular acceleration. At the instant shown, there are six points on the wheel that have arrows associated with them. Which of the following quantities could the arrows represent: (a) tangential velocity, (b) tangential acceleration, (c) centripetal acceleration? In each case, answer why the arrows do or do not represent the quantity.

13. Suppose that the speedometer of a truck is set to read the linear speed of the truck, but uses a device that actually measures the angular speed of the tires. If larger diameter tires are mounted on the truck, will the reading on the speedometer be correct? If not, will the reading be greater than or less than the true linear speed of the truck? Why?

14. The blades of a fan rotate more and more slowly after the fan is shut off. Eventually they stop rotating altogether. In such a situation, we sometimes assume that the angular acceleration of the blades is constant and apply the equations of rotational kinematics as an approximation. Explain why the angular acceleration can never really be constant in this kind of situation.

15. Rolling motion is one example that involves rotation about an axis that is not fixed. Give three other examples. In each case, identify the axis of rotation and explain why it is not fixed.

Problems

ssm   Solution is in the Student Solutions Manual.   www   Solution is available on the World Wide Web at www.wiley.com/college/cutnell
⚕   This icon represents a biomedical application.

Section 8.1 Rotational Motion and Angular Displacement, Section 8.2 Angular Velocity and Angular Acceleration

1. ssm A diver completes $3\frac{1}{2}$ somersaults in 1.7 s. What is the average angular speed (in rad/s) of the diver?

2. A pitcher throws a curveball that reaches the catcher in 0.60 s. The ball curves because it is spinning at an average angular velocity of 330 rev/min (assumed constant) on its way to the catcher's mitt. What is the angular displacement of the baseball (in radians) as it travels from the pitcher to the catcher?

3. ssm In Europe, surveyors often measure angles in *grads*. There are 100 grads in one-quarter of a circle. How many grads are there in one radian?

4. A pulsar is a rapidly rotating neutron star that continuously emits a beam of radio waves in a searchlight manner. Each time the pulsar makes one revolution, the rotating beam sweeps across the earth, and the earth receives a pulse of radio waves. For one particular pulsar, the time between two successive pulses is 0.033 s. Determine the average angular speed (in rad/s) of this pulsar.

5. A CD has a playing time of 74 minutes. When the music starts, the CD is rotating at an angular speed of 480 revolutions per minute (rpm). At the end of the music, the CD is rotating at 210 rpm. Find the magnitude of the average angular acceleration of the CD. Express your answer in rad/s^2.

6. A Ferris wheel rotates at an angular velocity of 0.24 rad/s. Starting from rest, it reaches its operating speed with an average angular acceleration of 0.030 rad/s^2. How long does it take the wheel to come up to operating speed?

7. ssm An electric circular saw is designed to reach its final angular speed, starting from rest, in 1.50 s. Its average angular acceleration is 328 rad/s^2. Obtain its final angular speed.

*8. A floor polisher has a rotating disk of radius 15 cm. The disk rotates at a constant angular velocity of 1.4 rev/s and is covered with a soft material that does the polishing. An operator holds the polisher in one place for 45 s, in order to buff an especially scuffed area of the floor. How far (in meters) does a spot on the outer edge of the disk move during this time?

*9. A space station consists of two donut-shaped living chambers, A and B, that have the radii shown in the drawing. As the station rotates, an astronaut in chamber A is moved 2.40×10^2 m along a circular arc. How far along a circular arc is an astronaut in chamber B moved during the same time?

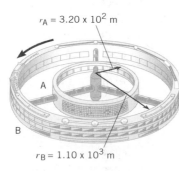

$r_A = 3.20 \times 10^2$ m

$r_B = 1.10 \times 10^3$ m

*10. The drawing shows a graph of the angular velocity of a rotating wheel as a function of time. Although not shown in the

graph, the angular velocity continues to increase at the same rate until $t = 8.0$ s. What is the angular displacement of the wheel from 0 to 8.0 s?

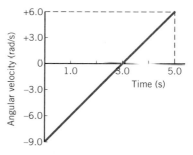

*11. ssm The drawing shows a device that can be used to measure the speed of a bullet. The device consists of two rotating disks, separated by a distance of $d = 0.850$ m, and rotating with an angular speed of 95.0 rad/s. The bullet first passes through the left disk and then through the right disk. It is found that the angular displacement between the two bullet holes is $\theta = 0.240$ rad. From these data, determine the speed of the bullet.

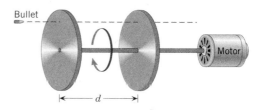

*12. A stroboscope is a light that flashes on and off at a constant rate. It can be used to illuminate a rotating object, and if the flashing rate is adjusted properly, the object can be made to appear stationary. (a) What is the shortest time between flashes of light that will make a three-bladed propeller appear stationary when it is rotating with an angular speed of 16.7 rev/s? (b) What is the next shortest time?

*13. A baton twirler throws a spinning baton directly upward. As it goes up and returns to the twirler's hand, the baton turns through four revolutions. Ignoring air resistance and assuming that the average angular speed of the baton is 1.80 rev/s, determine the height to which the center of the baton travels above the point of release.

**14. Review Conceptual Example 2 before attempting this problem. The moon has a diameter of 3.48×10^6 m and is a distance of 3.85×10^8 m from the earth. The sun has a diameter of 1.39×10^9 m and is 1.50×10^{11} m from the earth. (a) Determine (in radians) the angles subtended by the moon and the sun, as measured by a person standing on the earth. (b) Based on your answers to part (a), decide whether a total eclipse of the sun is really "total." Give your reasoning. (c) Determine the ratio, expressed as a percentage, of the apparent circular area of the moon to the apparent circular area of the sun.

**15. A quarterback throws a pass that is a perfect spiral. In other words, the football does not wobble, but spins smoothly about an axis passing through each end of the ball. Suppose the ball spins at 7.7 rev/s. In addition, the ball is thrown with a linear speed of 19 m/s at an angle of 55° with respect to the ground. If the ball is caught at the same height at which it left the quarterback's hand, how many revolutions has the ball made while in the air?

Section 8.3 The Equations of Rotational Kinematics

16. A gymnast is performing a floor routine. In a tumbling run she spins through the air, increasing her angular velocity from 3.00 to 5.00 rev/s while rotating through one-half of a revolution. How much time does this maneuver take?

17. ssm An electric fan is running on HIGH. After the LOW button is pressed, the angular speed of the fan decreases to 83.8 rad/s in 1.75 s. The deceleration is 42.0 rad/s². Determine the initial angular speed of the fan.

18. A basketball player is balancing a spinning basketball on the tip of his finger. The angular velocity of the ball slows down from 18.5 to 14.1 rad/s. During the slow-down, the angular displacement is 85.1 rad. Determine the time it takes for the ball to slow down.

19. ssm www A flywheel has a constant angular deceleration of 2.0 rad/s². (a) Find the angle through which the flywheel turns as it comes to rest from an angular speed of 220 rad/s. (b) Find the time required for the flywheel to come to rest.

20. The angular speed of the rotor in a centrifuge increases from 420 to 1420 rad/s in a time of 5.00 s. (a) Obtain the angle through which the rotor turns. (b) What is the magnitude of the angular acceleration?

* **21.** A top is a toy that is made to spin on its pointed end by pulling on a string wrapped around the body of the top. The string has a length of 64 cm and is wrapped around the top at a place where its radius is 2.0 cm. The thickness of the string is negligible. The top is initially at rest. Someone pulls the free end of the string, thereby unwinding it and giving the top an angular acceleration of +12 rad/s². What is the final angular velocity of the top when the string is completely unwound?

* **22. Interactive LearningWare 8.1** at **www.wiley.com/college/cutnell** reviews the approach that is necessary for solving problems such as this one. A motorcyclist is traveling along a road and accelerates for 4.50 s to pass another cyclist. The angular acceleration of each wheel is +6.70 rad/s², and, just after passing, the angular velocity of each is +74.5 rad/s, where the plus signs indicate counterclockwise directions. What is the angular displacement of each wheel during this time?

* **23. ssm** A spinning wheel on a fireworks display is initially rotating in a counterclockwise direction. The wheel has an angular acceleration of −4.00 rad/s². Because of this acceleration, the angular velocity of the wheel changes from its initial value to a final value of −25.0 rad/s. While this change occurs, the angular displacement of the wheel is zero. (Note the similarity to that of a ball being thrown vertically upward, coming to a momentary halt, and then falling downward to its initial position.) Find the time required for the change in the angular velocity to occur.

* **24.** A dentist causes the bit of a high-speed drill to accelerate from an angular speed of 1.05 × 10⁴ rad/s to an angular speed of 3.14 × 10⁴ rad/s. In the process, the bit turns through 1.88 × 10⁴ rad. Assuming a constant angular acceleration, how long would it take the bit to reach its maximum speed of 7.85 × 10⁴ rad/s, starting from rest?

* **25. Interactive Solution 8.25** at **www.wiley.com/college/cutnell** offers a model for solving this problem. The drive propeller of a ship starts from rest and accelerates at 2.90 × 10⁻³ rad/s² for 2.10 × 10³ s. For the next 1.40 × 10³ s the propeller rotates at a constant angular speed. Then it decelerates at 2.30 × 10⁻³ rad/s² until it slows (without reversing direction) to an angular speed of 4.00 rad/s. Find the total angular displacement of the propeller.

* **26.** At the local swimming hole, a favorite trick is to run horizontally off a cliff that is 8.3 m above the water. One diver runs off the edge of the cliff, tucks into a "ball," and rotates on the way down with an average angular speed of 1.6 rev/s. Ignore air resistance and determine the number of revolutions she makes while on the way down.

** **27. ssm www** A child, hunting for his favorite wooden horse, is running on the ground around the edge of a stationary merry-go-round. The angular speed of the child has a constant value of 0.250 rad/s. At the instant the child spots the horse, one-quarter of a turn away, the merry-go-round begins to move (in the direction the child is running) with a constant angular acceleration of 0.0100 rad/s². What is the shortest time it takes for the child to catch up with the horse?

Section 8.4 Angular Variables and Tangential Variables

28. Our sun rotates in a circular orbit about the center of the Milky Way galaxy. The radius of the orbit is 2.2 × 10²⁰ m, and the angular speed of the sun is 1.2 × 10⁻¹⁵ rad/s. (a) What is the tangential speed of the sun? (b) How long (in years) does it take for the sun to make one revolution around the center?

29. ssm A disk (radius = 2.00 mm) is attached to a high-speed drill at a dentist's office and is turning at 7.85 × 10⁴ rad/s. Determine the tangential speed of a point on the outer edge of this disk.

30. A string trimmer is a tool for cutting grass and weeds; it utilizes a length of nylon "string" that rotates about an axis perpendicular to one end of the string. The string rotates at an angular speed of 47 rev/s, and its tip has a tangential speed of 54 m/s. What is the length of the rotating string?

31. Interactive Solution 8.31 at **www.wiley.com/college/cutnell** offers one approach to solving this problem. The drawing shows the blade of a chain saw. The rotating sprocket tip at the end of the guide bar has a radius of 4.0 × 10⁻² m. The linear speed of a chain link at point A is 5.6 m/s. Find the angular speed of the sprocket tip in rev/s.

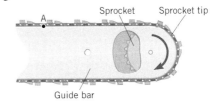

32. An auto race is held on a circular track. A car completes one lap in a time of 18.9 s, with an average tangential speed of 42.6 m/s. Find (a) the average angular speed and (b) the radius of the track.

33. The earth has a radius of 6.38 × 10⁶ m and turns on its axis once every 23.9 h. (a) What is the tangential speed (in m/s) of a person living in Ecuador, a country that lies on the equator? (b) At what latitude (i.e., the angle θ in the drawing) is the tangential speed one-third that of a person living in Ecuador?

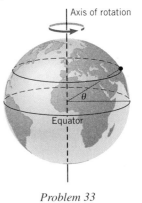

Problem 33

* **34.** A baseball pitcher throws a baseball horizontally at a linear speed of 42.5 m/s (about 95 mi/h). Before being caught, the baseball travels a horizontal distance of 16.5 m and rotates through an angle of 49.0 rad. The baseball has a radius of 3.67 cm and is rotating about an axis as it travels, much like the earth does. What is the tangential speed of a point on the "equator" of the baseball?

* **35. ssm** A compact disc (CD) contains music on a spiral track. Music is put onto a CD with the assumption that, during playback, the music will be detected at a *constant tangential speed* at any point. Since $v_T = r\omega$, a CD rotates at a smaller angular speed for

music near the outer edge and a larger angular speed for music near the inner part of the disc. For music at the outer edge ($r = 0.0568$ m), the angular speed is 3.50 rev/s. Find (a) the constant tangential speed at which music is detected and (b) the angular speed (in rev/s) for music at a distance of 0.0249 m from the center of a CD.

* **36.** A thin rod (length = 1.50 m) is oriented vertically, with its bottom end attached to the floor by means of a frictionless hinge. The mass of the rod may be ignored, compared to the mass of the object fixed to the top of the rod. The rod, starting from rest, tips over and rotates downward. (a) What is the angular speed of the rod just before it strikes the floor? *(Hint: Consider using the principle of conservation of mechanical energy.)* (b) What is the magnitude of the angular acceleration of the rod just before it strikes the floor?

** **37.** One type of slingshot can be made from a length of rope and a leather pocket for holding the stone. The stone can be thrown by whirling it rapidly in a horizontal circle and releasing it at the right moment. Such a slingshot is used to throw a stone from the edge of a cliff, the point of release being 20.0 m above the base of the cliff. The stone lands on the ground below the cliff at a point *X*. The horizontal distance of point *X* from the base of the cliff (directly beneath the point of release) is thirty times the radius of the circle on which the stone is whirled. Determine the angular speed of the stone at the moment of release.

Section 8.5 Centripetal Acceleration and Tangential Acceleration

38. A ceiling fan has two different angular speed settings: $\omega_1 = 440$ rev/min and $\omega_2 = 110$ rev/min. What is the ratio a_1/a_2 of the centripetal accelerations of a given point on a fan blade?

39. ssm A race car travels with a constant tangential speed of 75.0 m/s around a circular track of radius 625 m. Find (a) the magnitude of the car's total acceleration and (b) the direction of its total acceleration relative to the radial direction.

40. The earth orbits the sun once a year (3.16×10^7 s) in a nearly circular orbit of radius 1.50×10^{11} m. With respect to the sun, determine (a) the angular speed of the earth, (b) the tangential speed of the earth, and (c) the magnitude and direction of the earth's centripetal acceleration.

41. ssm A 220-kg speedboat is negotiating a circular turn (radius = 32 m) around a buoy. During the turn, the engine causes a net tangential force of magnitude 550 N to be applied to the boat. The initial tangential speed of the boat going into the turn is 5.0 m/s. (a) Find the tangential acceleration. (b) After the boat is 2.0 s into the turn, find the centripetal acceleration.

* **42. Interactive LearningWare 8.2** at **www.wiley.com/college/cutnell** provides a review of the concepts that are important in this problem. A race car, starting from rest, travels around a circular turn of radius 23.5 m. At a certain instant, the car is still accelerating, and its angular speed is 0.571 rad/s. At this time, the total acceleration (centripetal plus tangential) makes an angle of 35.0° with respect to the radius. (The situation is similar to that in Figure 8.15*b*.) What is the magnitude of the total acceleration?

* **43.** A rectangular plate is rotating with a constant angular acceleration about an axis that passes perpendicularly through one corner, as the drawing shows. The tangential acceleration measured at corner *A* has twice the magnitude of that measured at corner *B*. What is the ratio L_1/L_2 of the lengths of the sides of the rectangle?

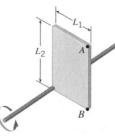

* **44.** A centrifuge is used for training astronauts to withstand large accelerations. It consists of a chamber (in which the astronaut sits)

that is fixed to the end of a long horizontal and rigid pole. The arrangement is rotated about an axis perpendicular to the pole's free end. Such a centrifuge starts from rest and has an angular acceleration of 0.25 rad/s². The chamber is 3.0 m from the axis of rotation. Through what angle has the device rotated when the centripetal acceleration experienced by an astronaut in the chamber is four times the acceleration due to the earth's gravity?

** **45. ssm** An electric drill starts from rest and rotates with a constant angular acceleration. After the drill has rotated through a certain angle, the magnitude of the centripetal acceleration of a point on the drill is twice the magnitude of the tangential acceleration. What is the angle?

Section 8.6 Rolling Motion

Note: All problems in this section assume that there is no slipping of the surfaces in contact during the rolling motion.

46. An automobile tire has a radius of 0.330 m, and its center moves forward with a linear speed of $v = 15.0$ m/s. (a) Determine the angular speed of the wheel. (b) Relative to the axle, what is the tangential speed of a point located 0.175 m from the axle?

47. ssm www A motorcycle accelerates uniformly from rest and reaches a linear speed of 22.0 m/s in a time of 9.00 s. The radius of each tire is 0.280 m. What is the magnitude of the angular acceleration of each tire?

48. Suppose you are riding a stationary exercise bicycle, and the electronic meter indicates that the wheel is rotating at 9.1 rad/s. The wheel has a radius of 0.45 m. If you ride the bike for 35 min, how far would you have gone if the bike could move?

49. Refer to **Interactive Solution 8.49** at **www.wiley.com/college/cutnell** in preparation for this problem. A car is traveling with a speed of 20.0 m/s along a straight horizontal road. The wheels have a radius of 0.300 m. If the car speeds up with a linear acceleration of 1.50 m/s² for 8.00 s, find the angular displacement of each wheel during this period.

50. Concept Simulation 8.1 at **www.wiley.com/college/cutnell** reviews the concept that plays the central role in this problem. The warranty on a new tire says that an automobile can travel for a distance of 96 000 km before the tire wears out. The radius of the tire is 0.31 m. How many revolutions does the tire make before wearing out?

* **51. ssm www** The drawing shows a view (from beneath) of the platter on a belt-drive turntable. The platter has an angular speed of 3.49 rad/s. The pulley on the motor shaft has a radius of 1.27 cm. Assuming that the belt does not slip, determine the angular speed of the motor shaft.

* **52.** The penny-farthing is a bicycle that was popular between 1870 and 1890. As the drawing shows, this type of bicycle has a large front wheel and a small rear wheel. On a Sunday ride in the park the front wheel (radius = 1.20 m) makes 276 revolutions. How many revolutions does the rear wheel (radius = 0.340 m) make?

* **53.** A ball of radius 0.200 m rolls along a horizontal table top with a constant linear speed of 3.60 m/s. The ball rolls off the edge and falls a vertical distance of 2.10 m before hitting the floor. What is the angular displacement of the ball while the ball is in the air?

* **54.** The two-gear combination shown in the drawing is being used to hoist the load L with a constant upward speed of 2.50 m/s. The rope attached to the load is being wound onto a cylinder behind the big gear. The depth of the teeth of the gears is negligible compared to

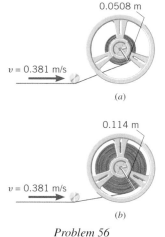

0.300 m

0.170 m

$v = 2.50$ m/s

L

the radii. Determine the angular velocity (magnitude and direction) of (a) the larger gear and (b) the smaller gear.

** **55.** A bicycle is rolling down a circular hill that has a radius of 9.00 m. As the drawing illustrates, the angular displacement of the bicycle is 0.960 rad. The radius of each wheel is 0.400 m. What is the angle (in radians) through which each tire rotates?

0.960 rad

Additional Problems

56. On an open-reel tape deck, the tape is being pulled past the playback head at a constant linear speed of 0.381 m/s. (a) Using the data in part *a* of the drawing, find the angular speed of the take-up reel. (b) After 2.40×10^3 s, the take-up reel is almost full, as part *b* of the drawing indicates. Find the average angular acceleration of the reel and specify whether the acceleration indicates an increasing or decreasing angular velocity.

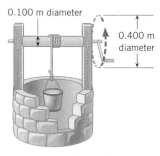

0.0508 m

$v = 0.381$ m/s

(a)

0.114 m

$v = 0.381$ m/s

(b)

Problem 56

57. ssm During a tennis serve, a racket is given an angular acceleration of magnitude 160 rad/s². At the top of the serve, the racket has an angular speed of 14 rad/s. If the distance between the top of the racket and the shoulder is 1.5 m, find the magnitude of the total acceleration of the top of the racket.

58. The shaft of a pump starts from rest and has an angular acceleration of 3.00 rad/s² for 18.0 s. At the end of this interval, what is (a) the shaft's angular speed and (b) the angle through which the shaft has turned?

59. ssm The drill bit of a variable-speed electric drill has a constant angular acceleration of 2.50 rad/s². The initial angular speed of the bit is 5.00 rad/s. After 4.00 s, (a) what angle has the bit turned through and (b) what is the bit's angular speed?

60. Review **Interactive LearningWare 8.2** at **www.wiley.com/college/cutnell** as an aid in solving this problem. A train is rounding a circular curve whose radius is 2.00×10^2 m. At one instant, the train has an angular acceleration of 1.50×10^{-3} rad/s² and an angular speed of 0.0500 rad/s. (a) Find the magnitude of the total acceleration (centripetal plus tangential) of the train. (b) Determine the angle of the total acceleration relative to the radial direction.

61. Refer to **Interactive Solution 8.61** at **www.wiley.com/college/cutnell** to review a model for solving this problem. The take-up reel of a cassette tape has an average radius of 1.4 cm. Find the length of tape (in meters) that passes around the reel in 13 s when the reel rotates at an average angular speed of 3.4 rad/s.

* **62.** The sun has a mass of 1.99×10^{30} kg and is moving in a circular orbit about the center of our galaxy, the Milky Way. The radius of the orbit is 2.3×10^4 light-years (1 light-year $= 9.5 \times 10^{15}$ m), and the angular speed of the sun is 1.1×10^{-15} rad/s. (a) Determine the tangential speed of the sun. (b) What is the magnitude of the net force that acts on the sun to keep it moving around the center of the Milky Way?

* **63. ssm** A person lowers a bucket into a well by turning the hand crank, as the drawing illustrates. The crank handle moves with a

constant tangential speed of 1.20 m/s on its circular path. Find the linear speed with which the bucket moves down the well.

* **64.** Two people start at the same place and walk around a circular lake in opposite directions. One has an angular speed of 1.7×10^{-3} rad/s, while the other has an angular speed of 3.4×10^{-3} rad/s. How long will it be before they meet?

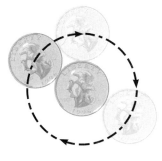

0.100 m diameter

0.400 m diameter

Problem 63

* **65.** A fan blade, whose angular acceleration is a constant 2.00 rad/s², rotates through an angle of 285 radians in 11.0 s. How long did it take the blade, starting from rest, to reach the *beginning* of the 11.0-s interval?

* **66.** The front and rear sprockets on a bicycle have radii of 9.00 and 5.10 cm, respectively. The angular speed of the front sprocket is 9.40 rad/s. Determine (a) the linear speed (in cm/s) of the chain as it moves between the sprockets and (b) the centripetal acceleration (in cm/s²) of the chain as it passes around the rear sprocket.

** **67. ssm** Consult **Concept Simulation 8.1** at **www.wiley.com/college/cutnell** for help in understanding the concepts that are important in this problem. Take two quarters and lay them on a table. Press down on one quarter so it cannot move. Then, starting at the 12:00 position, roll the other quarter along the edge of the stationary quarter, as the drawing suggests. How many revolutions does the rolling quarter make when it travels once around the circumference of the stationary quarter? Surprisingly, the answer is *not* one revolution. (*Hint: Review the paragraph just before Equation 8.12 that discusses how the distance traveled by the axle of a wheel is related to the circular arc length along the outer edge of the wheel.*)

** **68.** The drawing shows a golf ball passing through a windmill at a miniature golf course. The windmill has 8 blades and rotates at an angular speed of 1.25 rad/s. The opening between successive blades is equal to the width of a blade. A golf ball of diameter 4.50×10^{-2} m is just passing by one of the rotating blades. What must be the *minimum* speed of the ball so that it will not be hit by the next blade?

Golf ball

Concepts & Calculations *Group Learning Problems*

Note: Each of these problems consists of Concept Questions followed by a related quantitative Problem. They are designed for use by students working alone or in small learning groups. The Concept Questions involve little or no mathematics and are intended to stimulate group discussions. They focus on the concepts with which the problems deal. Recognizing the concepts is the essential initial step in any problem-solving technique.

69. Concept Question In general, is the direction of an object's average angular velocity the same as its initial angle θ_0, its final angle θ, or the difference $\theta - \theta_0$ between its final and initial angles? The table that follows lists four pairs of initial and final angles of a wheel on a moving car. Decide which pairs give a positive average angular velocity and which give a negative average angular velocity. Provide reasons for your answers.

	Initial angle θ_0	Final angle θ
(a)	0.45 rad	0.75 rad
(b)	0.94 rad	0.54 rad
(c)	5.4 rad	4.2 rad
(d)	3.0 rad	3.8 rad

Problem The elapsed time for each pair of angles is 2.0 s. Review the concept of average angular velocity in Section 8.2 and then determine the average angular velocity (magnitude and direction) for each of the four pairs of angles in the table. Check to see that the directions (positive or negative) of the angular velocities agree with the directions found in the Concept Question.

70. Concept Question In general, does the average angular acceleration of a rotating object have the same direction as its initial angular velocity ω_0, its final angular velocity ω, or the difference $\omega - \omega_0$ between its final and initial angular velocities? The table that follows lists four pairs of initial and final angular velocities for a rotating fan blade. Determine the direction (positive or negative) of the average angular acceleration for each pair. Provide reasons for your answers.

	Initial angular velocity ω_0	Final angular velocity ω
(a)	+2.0 rad/s	+5.0 rad/s
(b)	+5.0 rad/s	+2.0 rad/s
(c)	−7.0 rad/s	−3.0 rad/s
(d)	+4.0 rad/s	−4.0 rad/s

Problem The elapsed time for each of the four pairs of angular velocities is 4.0 s. Find the average angular acceleration (magnitude and direction) for each of the four pairs. Be sure that your directions agree with those found in the Concept Question.

71. Concept Question In the table are listed the initial angular velocity ω_0 and the angular acceleration α of four rotating objects at a given instant in time.

	Initial angular velocity ω_0	Angular acceleration α
(a)	+12 rad/s	+3.0 rad/s^2
(b)	+12 rad/s	−3.0 rad/s^2
(c)	−12 rad/s	+3.0 rad/s^2
(d)	−12 rad/s	−3.0 rad/s^2

In each case, state whether the angular *speed* of the object is increasing or decreasing in time. Account for your answers.

Problem For each of the four pairs in the table, determine the final angular speed of the object if the elapsed time is 2.0 s. Compare your final angular speeds with the initial angular speeds and make sure that your answers are consistent with your answers to the Concept Question.

72. Concept Questions Does the tip of a rotating fan blade have a tangential acceleration when the blade is rotating (a) at a constant angular velocity and (b) at a constant angular acceleration? Provide reasons for your answers.

Problem A fan blade is rotating with a constant angular acceleration of +12.0 rad/s^2. At what point on the blade, measured from the axis of rotation, does the magnitude of the tangential acceleration equal that of the acceleration due to gravity?

73. Concept Questions During a time t, a wheel has a constant angular acceleration, so its angular velocity increases from an initial value of ω_0 to a final value of ω. (a) During this time, is the angular displacement less than, greater than, or equal to (a) $\omega_0 t$ or (b) ωt? In each case, justify your answer. (c) If you conclude that the angular displacement does not equal $\omega_0 t$ or ωt, then how does it depend on ω_0, ω, and t?

Problem A car is traveling along a road, and the engine is turning over with an angular velocity of +220 rad/s. The driver steps on the accelerator, and in a time of 10.0 s the angular velocity increases to +280 rad/s. (a) What would have been the angular displacement of the engine if its angular velocity had remained constant at the initial value of +220 rad/s during the 10.0 s? (b) What would have been the angular displacement if the angular velocity had been equal to its final value of +280 rad/s during the 10.0 s? (c) Determine the actual value of the angular displacement during this period. Check your answers to see that they are consistent with those to the Concept Questions.

74. Concept Questions Two identical dragsters, starting from rest, accelerate side-by-side along a straight track. The wheels on one of the cars roll without slipping, while the wheels on the other slip during part of the time. (a) For which car, the winner or the loser, do the wheels roll without slipping? Why? For the dragster whose wheels roll without slipping, is there (b) a relationship between its linear speed and the angular speed of its wheels, and (c) a relationship between the magnitude of its linear acceleration and the magnitude of the angular acceleration of its wheels? If a relationship exists in either case, what is it?

Problem A dragster starts from rest and accelerates down the track. Each tire has a radius of 0.320 m and rolls without slipping. At a distance of 384 m, the angular speed of the wheels is 288 rad/s. Determine, if possible, (a) the linear speed of the dragster and (b) the magnitude of the angular acceleration of its wheels.

75. Concept Questions A propeller is rotating about an axis perpendicular to its center, as the drawing shows. The axis is parallel to the ground. An arrow is fired at the propeller, travels parallel to the axis, and passes through one of the open spaces between the propeller blades. The vertical drop of the arrow may be ignored. There is a maximum value for the angular speed ω of the propeller beyond which the arrow cannot pass through an open space without being struck by one of the blades. (a) If the arrow is to pass through an open space, does it matter if the arrow is aimed closer to or farther away from the axis (see points A and B in the drawing, for example)? Explain. (b) Does the maximum value of ω increase or de-

crease with increasing arrow speed v? Why? (c) Does the maximum value of ω increase or decrease with increasing arrow length L? Justify your answer.

Problem The angular open spaces between the three propeller blades are each 60.0°. Find the maximum value of the angular speed ω when the arrow has the lengths L and speeds v shown in the following table. Check to see that your answers agree with your answers to the Concept Questions.

	L	v
(a)	0.71 m	75.0 m/s
(b)	0.71 m	91.0 m/s
(c)	0.81 m	91.0 m/s

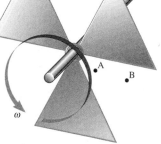

Problem 75

* **76. Concept Questions** At a county fair there is a betting game that involves a spinning wheel. As the drawing shows, the wheel is set into rotational motion with the beginning of the angular section labeled "1" at the marker at the top of the wheel. The wheel then decelerates and eventually comes to a halt on one of the numbered angular sections. (a) Given the initial angular velocity and the magnitude of the angular deceleration, which one of the equations of rotational kinematics would you use to calculate the angular displacement of the wheel? (b) When using the equations of rotational kinematics, can the initial angular velocity and the angular deceleration be expressed in rev/s and rev/s², respectively, or is it necessary to use radian measure? Explain. (c) Does a greater initial angular velocity necessarily mean that the wheel comes to a halt on an angular section labeled with a greater number (8 versus 6, for example)? Provide a reason for your answer.

Problem The wheel in the drawing is divided into twelve angular sections, each of which is 30.0°. Determine the numbered section on which the wheel comes to a halt when the deceleration of the wheel has a magnitude of 0.200 rev/s² and the initial angular velocity is (a) 1.20 rev/s and (b) 1.47 rev/s. Check to see that your answers are consistent with your answers to the Concept Questions.

Chapter 9

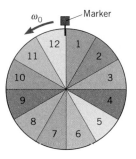

Rotational Dynamics

This climber is (hopefully) in equilibrium. The location of any force acting on the climber, as well as the force itself, determines a quantity called *torque*. We will see that the net torque acting on a body in equilibrium must be zero.

(© H. Saxgren/Peter Arnold)

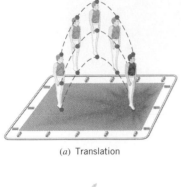

(a) Translation

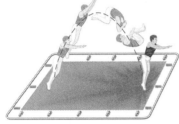

(b) Combined translation and rotation

Figure 9.1 An example of (a) translational motion and (b) combined translational and rotational motions.

9.1 The Action of Forces and Torques on Rigid Objects

The mass of most rigid objects, such as a propeller or a wheel, is spread out and not concentrated at a single point. These objects can move in a number of ways. Figure 9.1a illustrates one possibility called translational motion, in which all points on the body travel on parallel paths (not necessarily straight lines). In pure translation there is no rotation of any line in the body. Because translational motion can occur along a curved line, it is often called curvilinear motion or linear motion. Another possibility is rotational motion, which may occur in combination with translational motion, as is the case for the somersaulting gymnast in Figure 9.1b.

We have seen many examples of how a net force affects linear motion by causing an object to accelerate. We now need to take into account the possibility that a rigid object can also have an angular acceleration. A net external force causes linear motion to change, but what causes rotational motion to change? For example, something causes the rotational velocity of a speedboat's propeller to change when the boat accelerates. Is it simply the net force? As it turns out, it is not the net external force, but rather the net external torque that causes the rotational velocity to change. Just as greater net forces cause greater linear accelerations, greater net torques cause greater rotational or angular accelerations.

Figure 9.2 helps to explain the idea of torque. When you push on a door with a force **F**, as in part a, the door opens more quickly when the force is larger. Other things being equal, a larger force generates a larger torque. However, the door does not open as quickly if you apply the same force at a point closer to the hinge, as in part b, because the force now produces less torque. Furthermore, if your push is directed nearly at the hinge, as in part c, you will have a hard time opening the door at all, because the torque is nearly zero. In summary, the torque depends on the magnitude of the force, on the point where the force is applied relative to the axis of rotation (the hinge in Figure 9.2), and on the direction of the force.

For simplicity, we deal with situations in which the force lies in a plane that is perpendicular to the axis of rotation. In Figure 9.3, for instance, the axis is perpendicular to the page and the force lies in the plane of the paper. The drawing shows the line of action and the lever arm of the force, two concepts that are important in the definition of torque. The *line of action* is an extended line drawn colinear with the force. The *lever arm* is the distance ℓ between the line of action and the axis of rotation, measured on a line that is perpendicular to both. The torque is represented by the symbol τ (Greek letter *tau*), and its magnitude is defined as the magnitude of the force times the lever arm:

Figure 9.2 It is easier to open a door with a force of a given magnitude by (a) pushing at the door's outer edge than by (b) pushing closer to the axis of rotation (the hinge). (c) Pushing nearly into the hinge makes it very difficult to open the door.

■ **DEFINITION OF TORQUE**

$$\text{Torque} = (\text{Magnitude of the force}) \times (\text{Lever arm})$$

$$\tau = F\ell \tag{9.1}$$

Direction: The torque is positive when the force tends to produce a counterclockwise rotation about the axis, and negative when the force tends to produce a clockwise rotation.

SI Unit of Torque: newton · meter (N · m)

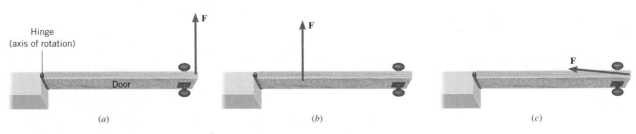

(a) (b) (c)

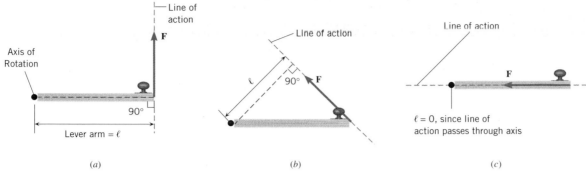

(a)   (b)   (c)

Figure 9.3 In this top view, the hinges of a door appear as a black dot (●) and define the axis of rotation. The line of action and lever arm ℓ are illustrated for a force applied to the door (a) perpendicularly and (b) at an angle. (c) The lever arm is zero because the line of action passes through the axis of rotation.

Equation 9.1 indicates that forces of the same magnitude can produce *different* torques, depending on the value of the lever arm, and Example 1 illustrates this important feature.

Example 1 Different Lever Arms, Different Torques

In Figure 9.3 a force whose magnitude is 55 N is applied to a door. However, the lever arms are different in the three parts of the drawing: (a) $\ell = 0.80$ m, (b) $\ell = 0.60$ m, and (c) $\ell = 0$ m. Find the magnitude of the torque in each case.

Reasoning In each case the lever arm is the perpendicular distance between the axis of rotation and the line of action of the force. In part *a* this perpendicular distance is equal to the width of the door. In parts *b* and *c*, however, the lever arm is less than the width. Because the lever arm is different in each case, the torque is different, even though the magnitude of the applied force is the same.

Solution Equation 9.1 gives the following values for the torques:

(a) $\tau = F\ell = (55 \text{ N})(0.80 \text{ m}) = \boxed{44 \text{ N} \cdot \text{m}}$

(b) $\tau = F\ell = (55 \text{ N})(0.60 \text{ m}) = \boxed{33 \text{ N} \cdot \text{m}}$

(c) $\tau = F\ell = (55 \text{ N})(0 \text{ m}) = \boxed{0 \text{ N} \cdot \text{m}}$

In part *c* the line of action of *F* passes through the axis of rotation (the hinge). Hence, the lever arm is zero, and the torque is zero.

In our bodies, muscles and tendons produce torques about various joints. Example 2 illustrates how the Achilles tendon produces a torque about the ankle joint.

Example 2 The Achilles Tendon

Figure 9.4*a* shows the ankle joint and the Achilles tendon attached to the heel at point *P*. The tendon exerts a force of magnitude $F = 720$ N, as Figure 9.4*b* indicates. Determine the torque (magnitude and direction) of this force about the ankle joint, which is located 3.6×10^{-2} m away from point *P*.

Reasoning To calculate the magnitude of the torque, it is necessary to have a value for the lever arm ℓ. However, the lever arm is not the given distance of 3.6×10^{-2} m. Instead, the lever arm is the perpendicular distance between the axis of rotation at the ankle joint and the line of action of the force **F**. In Figure 9.4*b* this distance is indicated by the dashed red line.

Solution From the drawing, it can be seen that the lever arm is $\ell = (3.6 \times 10^{-2} \text{ m}) \cos 55°$. The magnitude of the torque is

$$\tau = F\ell = (720 \text{ N})(3.6 \times 10^{-2} \text{ m}) \cos 55° = 15 \text{ N} \cdot \text{m} \qquad (9.1)$$

The force **F** tends to produce a clockwise rotation about the ankle joint, so the torque is negative: $\boxed{\tau = -15 \text{ N} \cdot \text{m}}$.

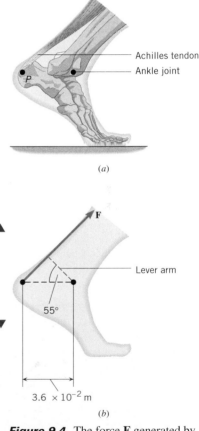

(a)

(b)

Figure 9.4 The force **F** generated by the Achilles tendon produces a clockwise (negative) torque about the ankle joint.

The physics of the Achilles tendon.

Need more practice?

Interactive LearningWare 9.1
A thin rod is mounted on an axle that passes perpendicularly through the center of the rod and allows the rod to rotate in the vertical plane. The weight of the rod is negligible. The rod is held at an angle of 33° above the horizontal by the action of two forces. These two forces produce a total torque of zero with respect to the axle. One force has a magnitude F and is applied perpendicularly to one end of the rod. The other force is due to a 49-N weight that hangs vertically from the other end of the rod. Find F.

Related Homework: *Problem 8*

Go to
www.wiley.com/college/cutnell
for an interactive solution.

✓ **Check Your Understanding 1**

The drawing shows an overhead view of a horizontal bar that is free to rotate about an axis perpendicular to the page. Two forces act on the bar, and they have the same magnitude. However, one force is perpendicular to the bar, and the other makes an angle ϕ with respect to it. The angle ϕ can be 90°, 45°, or 0°. Rank the values of ϕ according to the magnitude of the net torque (the sum of the torques) that the two forces produce, largest net torque first. *(The answer is given at the end of the book.)*

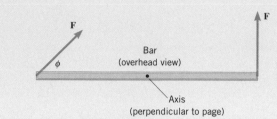

Background: This question deals with the concept of torque and how it depends on the lever arm, the magnitude of the force, and the direction of the rotation that the force causes.

For similar questions (including calculational counterparts), consult Self-Assessment Test 9.1. This test is described at the end of Section 9.3.

9.2 Rigid Objects in Equilibrium

▶ CONCEPTS AT A GLANCE If a rigid body is in equilibrium, neither its linear motion nor its rotational motion changes. This lack of change leads to certain equations that apply for rigid-body equilibrium, as the Concepts-at-a-Glance chart in Figure 9.5 emphasizes. For instance, an object whose linear motion is not changing has no acceleration **a**. Therefore, the net force $\Sigma\mathbf{F}$ applied to the object must be zero, since $\Sigma\mathbf{F} = m\mathbf{a}$ and $\mathbf{a} = 0$. For two-dimensional motion the x and y components of the net force are separately zero: $\Sigma F_x = 0$ and $\Sigma F_y = 0$ (Equations 4.9a and 4.9b). In calculating the net force, we include only forces from external agents, or *external forces.** In addition to linear motion, we must consider rotational motion, which also does not change under equilibrium conditions. This means that the net external torque acting on the object must be zero, because it is what causes rotational motion to change. Using the symbol $\Sigma\tau$ to represent the net external torque (the sum of all positive and negative torques), we have

$$\Sigma\tau = 0 \qquad (9.2)$$

We define rigid-body equilibrium, then, in the following way. ◀

Figure 9.5 CONCEPTS AT A GLANCE
A body is in equilibrium when the sum of the external forces is zero and when the sum of the external torques is zero. These two conditions apply to each frog, since each is nicely balanced on the limb and is in equilibrium.
(© Renee Lynn/Stone/Getty Images)

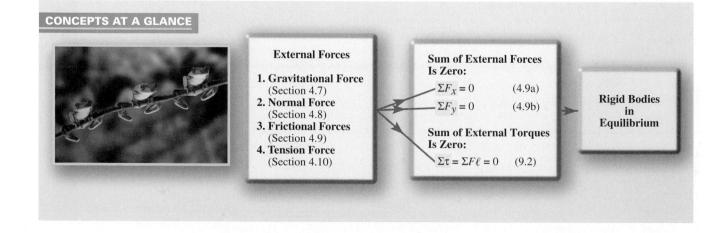

CONCEPTS AT A GLANCE

External Forces

1. **Gravitational Force** (Section 4.7)
2. **Normal Force** (Section 4.8)
3. **Frictional Forces** (Section 4.9)
4. **Tension Force** (Section 4.10)

Sum of External Forces Is Zero:
$\Sigma F_x = 0$ (4.9a)
$\Sigma F_y = 0$ (4.9b)

Sum of External Torques Is Zero:
$\Sigma\tau = \Sigma F\ell = 0$ (9.2)

Rigid Bodies in Equilibrium

---

* We ignore internal forces that one part of an object exerts on another part, because they occur in action–reaction pairs, each of which consists of oppositely directed forces of equal magnitude. The effect of one force cancels the effect of the other, as far as the acceleration of the entire object is concerned.

■ **EQUILIBRIUM OF A RIGID BODY**

A rigid body is in equilibrium if it has zero translational acceleration and zero angular acceleration. In equilibrium, the sum of the externally applied forces is zero, and the sum of the externally applied torques is zero:

$$\Sigma F_x = 0 \quad \text{and} \quad \Sigma F_y = 0 \qquad (4.9a \text{ and } 4.9b)$$

$$\Sigma \tau = 0 \qquad (9.2)$$

The reasoning strategy for analyzing the forces and torques acting on a body in equilibrium is given below. The first four steps of the strategy are essentially the same as those outlined in Section 4.11, where only forces are considered. Steps 5 and 6 have been added to account for any external torques that may be present. Example 3 illustrates how this reasoning strategy is applied to a diving board.

Reasoning Strategy

Applying the Conditions of Equilibrium to a Rigid Body

1. Select the object to which the equations for equilibrium are to be applied.

2. Draw a free-body diagram that shows all the external forces acting on the object.

3. Choose a convenient set of x, y axes and resolve all forces into components that lie along these axes.

4. Apply the equations that specify the balance of forces at equilibrium: $\Sigma F_x = 0$ and $\Sigma F_y = 0$.

5. Select a convenient axis of rotation. Identify the point where each external force acts on the object, and calculate the torque produced by each force about the axis of rotation. Set the sum of the torques about this axis equal to zero: $\Sigma \tau = 0$.

6. Solve the equations in steps 4 and 5 for the desired unknown quantities.

Example 3 A Diving Board

A woman whose weight is 530 N is poised at the right end of a diving board with a length of 3.90 m. The board has negligible weight and is bolted down at the left end, while being supported 1.40 m away by a fulcrum, as Figure 9.6a shows. Find the forces $\mathbf{F}_1$ and $\mathbf{F}_2$ that the bolt and the fulcrum, respectively, exert on the board.

Reasoning Part b of the figure shows the free-body diagram of the diving board. Three forces act on the board: $\mathbf{F}_1$, $\mathbf{F}_2$, and the force due to the diver's weight $\mathbf{W}$. In choosing the directions of $\mathbf{F}_1$ and $\mathbf{F}_2$ we have used our intuition: $\mathbf{F}_1$ points downward, because the bolt must pull in that direction to counteract the tendency of the board to rotate clockwise about the fulcrum; $\mathbf{F}_2$ points upward, because the board pushes downward against the fulcrum, which, in reaction, pushes upward on the board. Since the board is stationary, it is in equilibrium.

Solution Since the board is in equilibrium, the sum of the vertical forces must be zero:

$$\Sigma F_y = -F_1 + F_2 - W = 0 \qquad (4.9b)$$

Similarly, the sum of the torques must be zero, $\Sigma \tau = 0$. For calculating torques, we select an axis that passes through the left end of the board and is perpendicular to the page. (We will see shortly that this choice is arbitrary.) The force $\mathbf{F}_1$ produces no torque since it passes through the axis and has a zero lever arm, while $\mathbf{F}_2$ creates a counterclockwise (positive) torque, and $\mathbf{W}$ produces a clockwise (negative) torque. The free-body diagram shows the lever arms for the torques:

$$\Sigma \tau = +F_2 \ell_2 - W\ell_w = 0 \qquad (9.2)$$

Solving this equation for F_2 yields

$$F_2 = \frac{W\ell_w}{\ell_2} = \frac{(530 \text{ N})(3.90 \text{ m})}{1.40 \text{ m}} = \boxed{1480 \text{ N}}$$

This value for F_2, along with $W = 530$ N, can be substituted into Equation 4.9b above to show that $\boxed{F_1 = 950 \text{ N}}$.

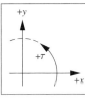

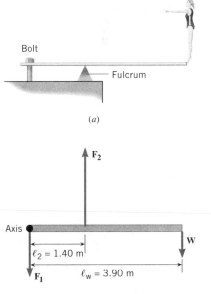

(a)

(b) Free-body diagram of the diving board

Figure 9.6 (a) A diver stands at the end of a diving board. (b) The free-body diagram for the diving board. The box at the upper left shows the positive x and y directions for the forces, as well as the positive (counterclockwise) direction for the torques.

Problem solving insight

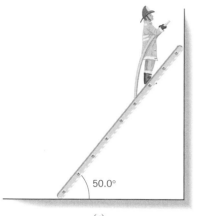

(a)

In Example 3 the sum of the external torques is calculated using an axis that passes through the left end of the diving board. *However, the choice of the axis is completely arbitrary, because if an object is in equilibrium, it is in equilibrium with respect to any axis whatsoever.* Thus, the sum of the external torques is zero, no matter where the axis is placed. One usually chooses the location so that the lines of action of one or more of the unknown forces pass through the axis. Such a choice simplifies the torque equation, because the torques produced by these forces are zero. For instance, in Example 3 the torque due to the force F_1 does not appear in Equation 9.2, because the lever arm of this force is zero.

Need more practice?

Interactive LearningWare 9.2
A woman whose weight is 530 N is poised at the right end of a diving board with a length of 3.90 m. The board has negligible weight and is bolted down at the left end, while being supported 1.40 m away by a fulcrum, as Figure 9.6a shows. Find the forces F_1 and F_2 that the bolt and the fulcrum, respectively, exert on the board.

This Interactive LearningWare tutorial is identical to Example 3. However, the solution is carried out differently to emphasize that it does not matter where you choose the axis of rotation. Go to **www.wiley.com/college/cutnell** for an interactive solution.

In a calculation of torque, the lever arm of the force must be determined relative to the axis of rotation. In Example 3 the lever arms are obvious, but sometimes a little care is needed in determining them, as in the next example.

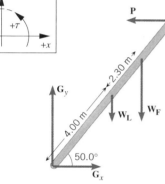

(b) Free-body diagram of the ladder

Example 4 Fighting a Fire

In Figure 9.7a an 8.00-m ladder of weight $W_L = 355$ N leans against a smooth vertical wall. The term "smooth" means that the wall can exert only a normal force directed perpendicular to the wall and cannot exert a frictional force parallel to it. A firefighter, whose weight is $W_F = 875$ N, stands 6.30 m from the bottom of the ladder. Assume that the ladder's weight acts at the ladder's center and neglect the hose's weight. Find the forces that the wall and the ground exert on the ladder.

Reasoning Part *b* of the figure shows the free-body diagram of the ladder. The following forces act on the ladder:

1. Its weight W_L

2. A force due to the weight W_F of the firefighter

3. The force **P** applied to the top of the ladder by the wall and directed perpendicular to the wall

4. The forces G_x and G_y, which are the horizontal and vertical components of the force exerted by the ground on the bottom of the ladder

The ground, unlike the wall, is not smooth, so that the force G_x is produced by static friction and prevents the ladder from slipping. The force G_y is the normal force applied to the ladder by the ground. The ladder is in equilibrium, so the sum of these forces and the sum of the torques produced by them must be zero.

Solution Since the net force acting on the ladder is zero, we have

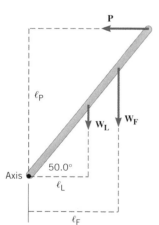

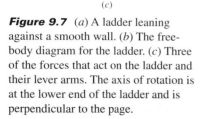

(c)

Figure 9.7 (*a*) A ladder leaning against a smooth wall. (*b*) The free-body diagram for the ladder. (*c*) Three of the forces that act on the ladder and their lever arms. The axis of rotation is at the lower end of the ladder and is perpendicular to the page.

$$\Sigma F_x = G_x - P = 0 \qquad (4.9a)$$

$$\Sigma F_y = G_y - W_L - W_F = 0 \qquad (4.9b)$$

$$\text{or} \quad G_y = W_L + W_F$$

$$= 355\text{ N} + 875\text{ N} = \boxed{1230\text{ N}}$$

Equation 4.9a cannot be solved as it stands, because it contains two unknown variables. However, another equation can be obtained from the fact that the net torque acting on an object in equilibrium is zero. In calculating torques, it is convenient to use an axis at the left end of the ladder, directed perpendicular to the page, as Figure 9.7c indicates. This axis is convenient, because G_x and G_y produce no torques about it, their lever arms being zero. Consequently, these forces will not appear in the equation representing the balance of torques. The lever arms for the remaining forces are shown in part c as red dashed lines. The following list summarizes these forces, their lever arms, and the torques:

Force	Lever Arm	Torque
$W_L = 355 \text{ N}$	$\ell_L = (4.00 \text{ m}) \cos 50.0°$	$-W_L\ell_L$
$W_F = 875 \text{ N}$	$\ell_F = (6.30 \text{ m}) \cos 50.0°$	$-W_F\ell_F$
P	$\ell_P = (8.00 \text{ m}) \sin 50.0°$	$+P\ell_P$

Setting the sum of the torques equal to zero gives

$$\Sigma\tau = -W_L\ell_L - W_F\ell_F + P\ell_P = 0 \qquad (9.2)$$

Solving this equation for P gives

$$P = \frac{W_L\ell_L + W_F\ell_F}{\ell_P}$$

$$= \frac{(355 \text{ N})(4.00 \text{ m}) \cos 50.0° + (875 \text{ N})(6.30 \text{ m}) \cos 50.0°}{(8.00 \text{ m}) \sin 50.0°} = \boxed{727 \text{ N}}$$

Substituting $P = 727$ N into Equation 4.9a indicates that $G_x = P = \boxed{727 \text{ N}}$.

To a large extent the directions of the forces acting on an object in equilibrium can be deduced using intuition. Sometimes, however, the direction of an unknown force is not obvious, and it is inadvertently drawn reversed in the free-body diagram. This kind of mistake causes no difficulty. ***Choosing the direction of an unknown force backward in the free-body diagram simply means that the value determined for the force will be a negative number,*** as the next example illustrates.

Problem solving insight

The physics of bodybuilding.

Example 5 Bodybuilding

A bodybuilder holds a dumbbell of weight $\mathbf{W_d}$ as in Figure 9.8a. His arm is horizontal and weighs $W_a = 31.0$ N. The deltoid muscle is assumed to be the only muscle acting and is attached to the arm as shown. The maximum force $\mathbf{M}$ that the deltoid muscle can supply has a magnitude of 1840 N. Figure 9.8b shows the distances that locate where the various forces act on the arm. What is the weight of the heaviest dumbbell that can be held, and what are the horizontal and vertical force components, $\mathbf{S_x}$ and $\mathbf{S_y}$, that the shoulder joint applies to the left end of the arm?

Figure 9.8 (a) The fully extended, horizontal arm of a bodybuilder supports a dumbbell. (b) The free-body diagram for the arm. (c) Three of the forces that act on the arm and their lever arms. The axis of rotation at the left end of the arm is perpendicular to the page. Force vectors are not to scale.

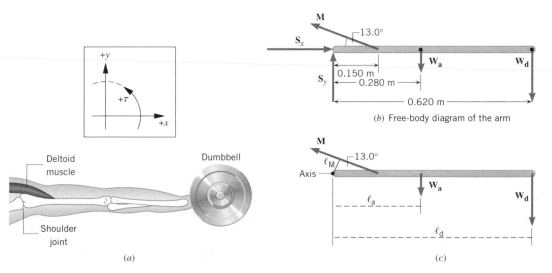

(b) Free-body diagram of the arm

(a)

(c)

Reasoning Figure 9.8b is the free-body diagram for the arm. Note that $\mathbf{S}_x$ is directed to the right, because the deltoid muscle pulls the arm in toward the shoulder joint, and the joint pushes back in accordance with Newton's third law. The direction of the force $\mathbf{S}_y$, however, is less obvious, and we are alert for the possibility that the direction chosen in the free-body diagram is backward. If so, the value obtained for $\mathbf{S}_y$ will be negative.

Solution The arm is in equilibrium, so the net force acting on it is zero:

$$\Sigma F_x = S_x - M \cos 13.0° = 0 \tag{4.9a}$$

$$\text{or} \quad S_x = M \cos 13.0° = (1840 \text{ N}) \cos 13.0° = \boxed{1790 \text{ N}}$$

$$\Sigma F_y = S_y + M \sin 13.0° - W_a - W_d = 0 \tag{4.9b}$$

Equation 4.9b cannot be solved at this point, because it contains two unknowns, S_y and W_d. However, since the arm is in equilibrium, the torques acting on the arm must balance, and this fact provides another equation. To calculate torques, we choose an axis through the left end of the arm and perpendicular to the page. With this axis, the torques due to $\mathbf{S}_x$ and $\mathbf{S}_y$ are zero, because the line of action of each force passes through the axis and the lever arm of each force is zero. The list below summarizes the remaining forces, their lever arms (see Figure 9.8c), and the torques.

Force	Lever Arm	Torque
$W_a = 31.0 \text{ N}$	$\ell_a = 0.280 \text{ m}$	$-W_a \ell_a$
W_d	$\ell_d = 0.620 \text{ m}$	$-W_d \ell_d$
$M = 1840 \text{ N}$	$\ell_M = (0.150 \text{ m}) \sin 13.0°$	$+M \ell_M$

The condition specifying a zero net torque is

$$\Sigma \tau = -W_a \ell_a - W_d \ell_d + M \ell_M = 0 \tag{9.2}$$

Solving this equation for W_d yields

$$W_d = \frac{-W_a \ell_a + M \ell_M}{\ell_d}$$

$$= \frac{-(31.0 \text{ N})(0.280 \text{ m}) + (1840 \text{ N})(0.150 \text{ m}) \sin 13.0°}{0.620 \text{ m}} = \boxed{86.1 \text{ N}}$$

Problem solving insight
When a force is negative, such as $S_y = -297$ N in this example, it means that the direction of the force is opposite to that chosen originally.

Substituting this value for W_d into Equation 4.9b above and solving for S_y gives us $\boxed{S_y = -297 \text{ N}}$. The minus sign indicates that the choice of direction for S_y in the free-body diagram is wrong. In reality, S_y has a magnitude of 297 N but is directed downward, not upward.

▲

✓ Check Your Understanding 2

Three forces act on each of the thin, square sheets shown in the drawing. In parts A and B of the drawing, the force labeled 2F acts at the center of the sheet. The forces can have different magnitudes (F or 2F) and can be applied at different points on an object. In which drawing is (a) the translational acceleration equal to zero, but the angular acceleration is not equal to zero, (b) the translational acceleration is not equal to zero, but the angular acceleration is equal to zero, and (c) the object in equilibrium? *(The answers are given at the end of the book.)*

Concept Simulation 9.1

This simulation illustrates two boxes resting on a beam that is held horizontal by two supports under the beam. The distance between the supports, the weights of the boxes and the beam, and the position of each box along the beam are under your control. For each setting of these variables, the simulation uses the conditions of equilibrium and calculates the upward force that each support exerts on the beam.

Related Homework: *Problem 14*

Go to
www.wiley.com/college/cutnell

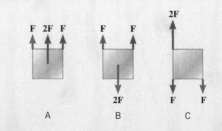

A   B   C

Background: Understanding the concept of equilibrium is essential for answering this question. Think about what must be true concerning the net force and the net torque acting on an object in equilibrium.

For similar questions (including calculational counterparts), consult Self-Assessment Test 9.1. This test is described at the end of Section 9.3.

9.3 Center of Gravity

Often, it is important to know the torque produced by the weight of an *extended* body. In Examples 4 and 5, for instance, it is necessary to determine the torques caused by the weight of the ladder and the arm, respectively. In both cases the weight is considered to act at a definite point for the purpose of calculating the torque. This point is called the *center of gravity* (abbreviated "cg").

■ DEFINITION OF CENTER OF GRAVITY

The center of gravity of a rigid body is the point at which its weight can be considered to act when the torque due to the weight is being calculated.

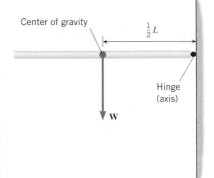

Figure 9.9 A thin, uniform, horizontal rod of length L is attached to a vertical wall by a hinge. The center of gravity of the rod is at its geometrical center.

When an object has a symmetrical shape and its weight is distributed uniformly, the center of gravity lies at its geometrical center. For instance, Figure 9.9 shows a thin, uniform, horizontal rod of length L attached to a vertical wall by a hinge. The center of gravity of the rod is located at the geometrical center. The lever arm for the weight **W** is $L/2$, and the magnitude of the torque is $\tau = W(L/2)$. In a similar fashion, the center of gravity of any symmetrically shaped and uniform object, such as a sphere, disk, cube, or cylinder, is located at its geometrical center. However, this does not mean that the center of gravity must lie within the object itself. The center of gravity of a compact disc recording, for instance, lies at the center of the hole in the disc and is, therefore, "outside" the object.

Suppose we have a group of objects, with known weights and centers of gravity, and it is necessary to know the center of gravity for the group as a whole. As an example, Figure 9.10a shows a group composed of two parts: a horizontal uniform board (weight $\mathbf{W_1}$) and a uniform box (weight $\mathbf{W_2}$) near the left end of the board. The center of gravity can be determined by calculating the net torque created by the board and box about an axis that is picked arbitrarily to be at the right end of the board. Part *a* of the figure shows the weights $\mathbf{W_1}$ and $\mathbf{W_2}$ and their corresponding lever arms x_1 and x_2. The net torque is $\Sigma\tau = W_1 x_1 + W_2 x_2$. It is also possible to calculate the net torque by treating the total weight $\mathbf{W_1} + \mathbf{W_2}$ as if it were located at the center of gravity and had the lever arm x_{cg}, as part *b* of the drawing indicates: $\Sigma\tau = (W_1 + W_2)x_{cg}$. The two values for the net torque must be the same, so that

$$W_1 x_1 + W_2 x_2 = (W_1 + W_2)x_{cg}$$

This expression can be solved for x_{cg}, which locates the center of gravity relative to the axis:

Center of gravity
$$x_{cg} = \frac{W_1 x_1 + W_2 x_2 + \cdots}{W_1 + W_2 + \cdots} \tag{9.3}$$

The notation "$+ \cdots$" indicates that Equation 9.3 can be extended to account for any number of weights distributed along a horizontal line. Figure 9.10c illustrates that the

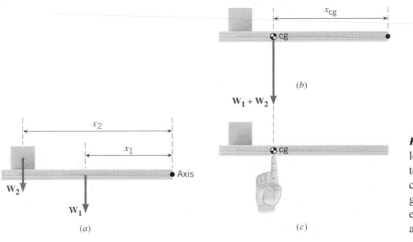

Figure 9.10 (*a*) A box rests near the left end of a horizontal board. (*b*) The total weight ($\mathbf{W_1} + \mathbf{W_2}$) acts at the center of gravity of the group. (*c*) The group can be balanced by applying an external force (due to the index finger) at the center of gravity.

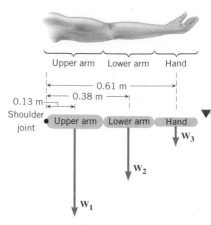

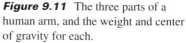

Figure 9.11 The three parts of a human arm, and the weight and center of gravity for each.

group can be balanced by a single external force (due to the index finger), if the line of action of the force passes through the center of gravity, and if the force is equal in magnitude, but opposite in direction, to the weight of the group. Example 6 demonstrates how to calculate the center of gravity for the human arm.

Example 6   The Center of Gravity of an Arm

The horizontal arm in Figure 9.11 is composed of three parts: the upper arm (weight $W_1 = 17$ N), the lower arm ($W_2 = 11$ N), and the hand ($W_3 = 4.2$ N). The drawing shows the center of gravity of each part, measured with respect to the shoulder joint. Find the center of gravity of the entire arm, relative to the shoulder joint.

Reasoning and Solution   The coordinate x_{cg} of the center of gravity is given by

$$x_{cg} = \frac{W_1 x_1 + W_2 x_2 + W_3 x_3}{W_1 + W_2 + W_3} \tag{9.3}$$

$$= \frac{(17 \text{ N})(0.13 \text{ m}) + (11 \text{ N})(0.38 \text{ m}) + (4.2 \text{ N})(0.61 \text{ m})}{17 \text{ N} + 11 \text{ N} + 4.2 \text{ N}} = \boxed{0.28 \text{ m}}$$

The center of gravity plays an important role in determining whether a group of objects remains in equilibrium as the weight distribution within the group changes. A change in the weight distribution causes a change in the position of the center of gravity, and if the change is too great, the group will not remain in equilibrium. Conceptual Example 7 discusses a shift in the center of gravity that led to an embarrassing result.

Figure 9.12 (*a*) This stationary cargo plane is sitting on its tail at Los Angeles International Airport, after being overloaded toward the rear. (© AP/Wide World Photos) (*b*) In a correctly loaded plane, the center of gravity is between the front and the rear landing gear. (*c*) When the plane is overloaded toward the rear, the center of gravity shifts behind the rear landing gear, and the accident in part *a* occurs.

Conceptual Example 7   Overloading a Cargo Plane

Figure 9.12*a* shows a stationary cargo plane with its front landing gear 9 meters off the ground. This accident occurred because the plane was overloaded toward the rear. How did a shift in the center of gravity of the loaded plane cause the accident?

Reasoning and Solution   Figure 9.12*b* shows a drawing of a correctly loaded plane, with the center of gravity located between the front and the rear landing gears. The weight **W** of the plane and cargo acts downward at the center of gravity, and the normal forces $\mathbf{F}_{N1}$ and $\mathbf{F}_{N2}$ act upward at the front and at the rear landing gear, respectively. With respect to an axis at the rear landing gear, the counterclockwise torque due to $\mathbf{F}_{N1}$ balances the clockwise torque due to **W**,

(*a*)

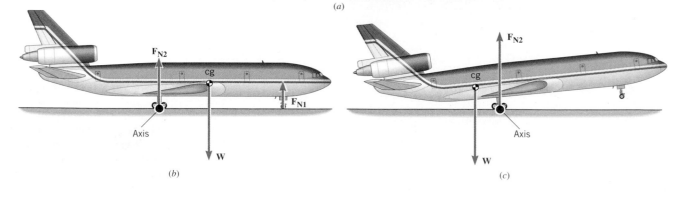

(*b*)   (*c*)

and the plane remains in equilibrium. Figure 9.12c shows the plane with too much cargo loaded toward the rear, just after the plane has begun to rotate counterclockwise. Because of the overloading, the center of gravity has shifted behind the rear landing gear. The torque due to **W** is now counterclockwise and is not balanced by any clockwise torque. Due to the unbalanced counterclockwise torque, the plane rotates until its tail hits the ground, which applies an upward force to the tail. The clockwise torque due to this upward force balances the counterclockwise torque due to **W**, and the plane comes again into an equilibrium state, this time with the front landing gear 9 meters off the ground.

Related Homework: *Problems 12, 19*

The center of gravity of an object with an irregular shape and a nonuniform weight distribution can be found by suspending the object from two different points P_1 and P_2, one at a time. Figure 9.13a shows the object at the moment of release, when its weight **W**, acting at the center of gravity, has a nonzero lever arm ℓ relative to the axis shown in the drawing. At this instant the weight produces a torque about the axis. The tension force **T** applied to the object by the suspension cord produces no torque because its line of action passes through the axis. Hence, in part *a* there is a net torque applied to the object, and the object begins to rotate. Friction eventually brings the object to rest as in part *b*, where the center of gravity lies directly below the point of suspension. In such an orientation, the line of action of the weight passes through the axis, so there is no longer any net torque. In the absence of a net torque the object remains at rest. By suspending the object from a second point P_2 (see Figure 9.13c), a second line through the object can be established, along which the center of gravity must also lie. The center of gravity, then, must be at the intersection of the two lines.

The center of gravity is closely related to the center-of-mass concept discussed in Section 7.5. To see why they are related, let's replace each occurrence of the weight in Equation 9.3 by $W = mg$, where m is the mass of a given object and g is the acceleration due to gravity at the location of the object. Suppose that g has the same value everywhere that the objects are located. Then it can be algebraically canceled from each term on the right side of Equation 9.3. The resulting equation, which contains only masses and distances, is the same as Equation 7.10, which defines the location of the center of mass. Thus, the two points are identical. For ordinary-sized objects, like cars and boats, the center of gravity coincides with the center of mass.

Self-Assessment Test 9.1

Test your understanding of the key ideas presented in Sections 9.1–9.3:

• Forces, Lever Arms, and Torques   • Rigid Bodies in Equilibrium
• Center of Gravity

Go to **www.wiley.com/college/cutnell**

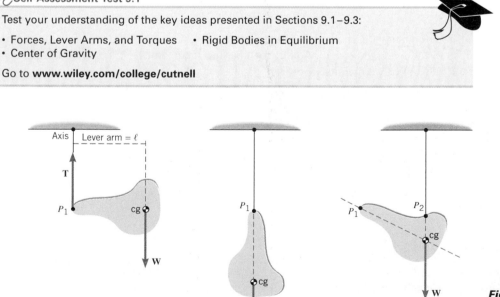

Figure 9.13 The center of gravity (cg) of an object can be located by suspending the object from two different points, P_1 and P_2, one at a time.

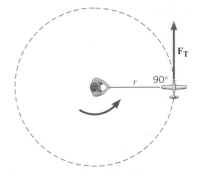

Figure 9.14 A model airplane on a guideline has a mass m and is flying on a circle of radius r (top view). A net tangential force $\mathbf{F_T}$ acts on the plane.

9.4 Newton's Second Law for Rotational Motion About a Fixed Axis

The goal of this section is to put Newton's second law into a form suitable for describing the rotational motion of a rigid object about a fixed axis. We begin by considering a particle moving on a circular path. Figure 9.14 presents a good approximation of this situation by using a small model plane on a guideline of negligible mass. The plane's engine produces a net external tangential force F_T that gives the plane a tangential acceleration a_T. In accord with Newton's second law, it follows that $F_T = ma_T$. The torque τ produced by this force is $\tau = F_T r$, where the radius r of the circular path is also the lever arm. As a result, the torque is $\tau = ma_T r$. But the tangential acceleration is related to the angular acceleration α according to $a_T = r\alpha$ (Equation 8.10), where α must be expressed in rad/s^2. With this substitution for a_T, the torque becomes

$$\tau = \underbrace{(mr^2)}_{\substack{\text{Moment} \\ \text{of inertia } I}}\alpha \qquad (9.4)$$

Equation 9.4 is the form of Newton's second law we have been seeking. It indicates that the net external torque τ is directly proportional to the angular acceleration α. The constant of proportionality is $I = mr^2$, which is called the ***moment of inertia of the particle.*** The SI unit for moment of inertia is kg · m^2.

If all objects were single particles, it would be just as convenient to use the second law in the form $F_T = ma_T$ as in the form $\tau = I\alpha$. The advantage in using $\tau = I\alpha$ is that it can be applied to any rigid body rotating about a fixed axis, and not just to a particle. To illustrate how this advantage arises, Figure 9.15a shows a flat sheet of material that rotates about an axis perpendicular to the sheet. The sheet is composed of a number of mass particles, $m_1, m_2, \ldots, m_N$, where N is very large. Only four particles are shown for the sake of clarity. Each particle behaves in the same way as the model airplane in Figure 9.14 and obeys the relation $\tau = (mr^2)\alpha$:

$$\tau_1 = (m_1 r_1{}^2)\alpha$$
$$\tau_2 = (m_2 r_2{}^2)\alpha$$
$$\vdots$$
$$\tau_N = (m_N r_N{}^2)\alpha$$

In these equations each particle has the same angular acceleration α, since the rotating object is assumed to be rigid. Adding together the N equations and factoring out the common value of α, we find that

$$\underbrace{\Sigma\tau}_{\substack{\text{Net} \\ \text{external torque}}} = \underbrace{(\Sigma mr^2)}_{\substack{\text{Moment} \\ \text{of inertia}}}\alpha \qquad (9.5)$$

Figure 9.15 (a) A rigid body consists of a large number of particles, four of which are shown. (b) The internal forces that particles 3 and 4 exert on each other obey Newton's law of action and reaction.

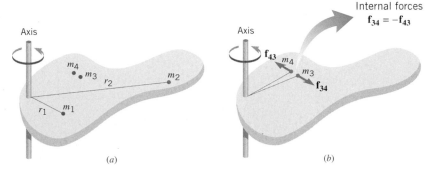

where $\Sigma\tau = \tau_1 + \tau_2 + \cdots + \tau_N$ is the sum of the external torques, and $\Sigma mr^2 = m_1 r_1^2 + m_2 r_2^2 + \cdots + m_N r_N^2$ represents the sum of the individual moments of inertia. The latter quantity is the ***moment of inertia I of the body:***

Moment of inertia of a body

$$I = \Sigma mr^2 \qquad (9.6)$$

In this equation, r is the perpendicular radial distance of each particle from the axis of rotation. Combining Equation 9.6 with Equation 9.5 gives the following result:

■ **ROTATIONAL ANALOG OF NEWTON'S SECOND LAW FOR A RIGID BODY ROTATING ABOUT A FIXED AXIS**

$$\text{Net external torque} = \left(\begin{array}{c}\text{Moment of}\\\text{inertia}\end{array}\right) \times \left(\begin{array}{c}\text{Angular}\\\text{acceleration}\end{array}\right)$$

$$\Sigma\tau = I\alpha \qquad (9.7)$$

Requirement: α must be expressed in rad/s^2.

The form of the second law for rotational motion, $\Sigma\tau = I\alpha$, is similar to that for translational (linear) motion, $\Sigma F = ma$, and is valid only in an inertial frame. The moment of inertia I plays the same role for rotational motion that the mass m does for translational motion. Thus, I is a measure of the rotational inertia of a body. When using Equation 9.7, α must be expressed in rad/s^2, because the relation $a_T = r\alpha$ (which requires radian measure) was used in the derivation.

When calculating the sum of torques in Equation 9.7, it is necessary to include only the *external torques,* those applied by agents outside the body. The torques produced by internal forces need not be considered, because they always combine to produce a net torque of zero. Internal forces are those that one particle within the body exerts on another particle. They always occur in pairs of oppositely directed forces of equal magnitude, in accord with Newton's third law (see m_3 and m_4 in Figure 9.15b). The forces in such a pair have the same line of action, so they have identical lever arms and produce torques of equal magnitudes. One torque is counterclockwise, while the other is clockwise, the net torque from the pair being zero.

It can be seen from Equation 9.6 that the moment of inertia depends on both the mass of each particle and its distance from the axis of rotation. The farther a particle is from the axis, the greater is its contribution to the moment of inertia. Therefore, although a rigid object possesses a unique total mass, it does not have a unique moment of inertia, for ***the moment of inertia depends on the location and orientation of the axis relative to the particles that make up the object.*** Example 8 shows how the moment of inertia can change when the axis of rotation changes.

Example 8 The Moment of Inertia Depends on Where the Axis Is

Two particles each have a mass m and are fixed to the ends of a thin rigid rod, whose mass can be ignored. The length of the rod is L. Find the moment of inertia when this object rotates relative to an axis that is perpendicular to the rod at (a) one end and (b) the center. (See Figure 9.16.)

Reasoning When the axis of rotation changes, the distance r between the axis and each particle changes. In determining the moment of inertia using $I = \Sigma mr^2$, we must be careful to use the distances that apply for each axis.

Solution

(a) Particle 1 lies on the axis, as part a of the drawing shows, and has a zero radial distance: $r_1 = 0$. In contrast, particle 2 moves on a circle whose radius is $r_2 = L$. Noting that $m_1 = m_2 = m$, we find that the moment of inertia is

$$I = \Sigma mr^2 = m_1 r_1^2 + m_2 r_2^2 = \boxed{mL^2} \qquad (9.6)$$

Concept Simulation 9.3

One or two forces can be applied to a rod that is free to rotate about its center. Each force produces a torque about the axis. The user can change the net torque by altering the magnitude and lever arms of the forces. The direction of one of the forces can also be changed. The simulation calculates the net torque and graphically illustrates the resulting angular displacement θ and velocity ω of the rod as a function of time.

Go to
www.wiley.com/college/cutnell

Problem solving insight

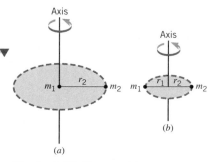

Figure 9.16 Two particles, masses m_1 and m_2, are attached to the ends of a massless rigid rod. The moment of inertia of this object is different, depending on whether the rod rotates about an axis through (*a*) the end or (*b*) the center of the rod.

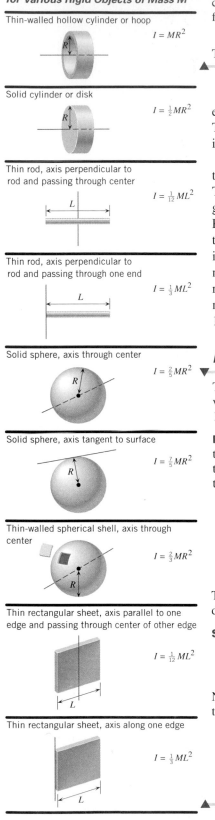

Table 9.1 Moments of inertia
for Various Rigid Objects of Mass M

Thin-walled hollow cylinder or hoop

$I = MR^2$

Solid cylinder or disk

$I = \frac{1}{2}MR^2$

Thin rod, axis perpendicular to
rod and passing through center

$I = \frac{1}{12}ML^2$

Thin rod, axis perpendicular to
rod and passing through one end

$I = \frac{1}{3}ML^2$

Solid sphere, axis through center

$I = \frac{2}{5}MR^2$

Solid sphere, axis tangent to surface

$I = \frac{7}{5}MR^2$

Thin-walled spherical shell, axis through
center

$I = \frac{2}{3}MR^2$

Thin rectangular sheet, axis parallel to one
edge and passing through center of other edge

$I = \frac{1}{12}ML^2$

Thin rectangular sheet, axis along one edge

$I = \frac{1}{3}ML^2$

(b) Part *b* of the drawing shows that particle 1 no longer lies on the axis but now moves on a circle of radius $r_1 = L/2$. Particle 2 moves on a circle with the same radius, $r_2 = L/2$. Therefore,

$$I = \Sigma mr^2 = m_1 r_1^2 + m_2 r_2^2 = m(L/2)^2 + m(L/2)^2 = \boxed{\frac{1}{2}mL^2}$$

This value differs from that in part (a) because the axis of rotation is different.

The procedure illustrated in Example 8 can be extended using integral calculus to evaluate the moment of inertia of a rigid object with a continuous mass distribution, and Table 9.1 gives some typical results. These results depend on the total mass of the object, its shape, and the location and orientation of the axis.

▶ **CONCEPTS AT A GLANCE** When forces act on a rigid object, they can affect its motion in two ways. They can produce a translational acceleration *a* (components a_x and a_y). The forces can also produce torques, which can produce an angular acceleration α. In general, we can deal with the resulting combined motion by using Newton's second law. For the translational motion, we use the law in the form $\Sigma F = ma$. For the rotational motion, we use the law in the form $\Sigma\tau = I\alpha$. The Concepts-at-a-Glance chart in Figure 9.17 illustrates the essence of this joint usage of Newton's second law. When *a* (both components) and α are zero, there is no acceleration of any kind, and the object is in equilibrium. This is the situation already discussed in Section 9.2. If any component of *a* or α is nonzero, we have accelerated motion, and the object is not in equilibrium. Examples 9, 10, and 11 deal with this type of situation. ◀

Example 9 The Torque of an Electric Saw Motor

The motor in an electric saw brings the circular blade from rest up to the rated angular velocity of 80.0 rev/s in 240.0 rev. One type of blade has a moment of inertia of 1.41×10^{-3} kg·m². What net torque (assumed constant) must the motor apply to the blade?

Reasoning Newton's second law for rotational motion (Equation 9.7, $\Sigma\tau = I\alpha$) can be used to find the net torque $\Sigma\tau$, once the angular acceleration α is determined. The angular acceleration can be calculated from the data in the following table and the appropriate equation of rotational kinematics:

θ	α	ω	ω_0	t
1508 rad (240.0 rev)	?	503 rad/s (80.0 rev/s)	0 rad/s	

The data for the angular displacement θ and the angular velocity ω have been converted to radian measure because $\Sigma\tau = I\alpha$ requires that α be expressed in radian measure.

Solution From Equation 8.8 ($\omega^2 = \omega_0^2 + 2\alpha\theta$) it follows that

$$\alpha = \frac{\omega^2 - \omega_0^2}{2\theta}$$

Newton's second law for rotational motion (Equation 9.7) can now be used to obtain the net torque:

$$\Sigma\tau = I\alpha = I\left(\frac{\omega^2 - \omega_0^2}{2\theta}\right)$$

$$= (1.41 \times 10^{-3} \text{ kg·m}^2)\left[\frac{(503 \text{ rad/s})^2 - (0 \text{ rad/s})^2}{2(1508 \text{ rad})}\right] = \boxed{0.118 \text{ N·m}}$$

*The physics of
wheelchairs.*

To accelerate a wheelchair, the rider applies a force to a handrail attached to each wheel. The torque generated by this force is the product of the magnitude of the force and the lever arm. As Figure 9.18 illustrates, the lever arm is just the radius of the circular rail, which is designed to be as large as possible. Thus, a relatively large torque can be generated for a given force, allowing the rider to accelerate quickly.

CONCEPTS AT A GLANCE

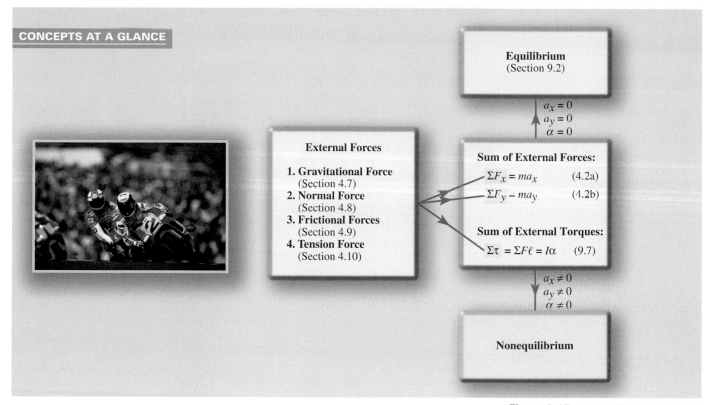

External Forces

1. **Gravitational Force** (Section 4.7)
2. **Normal Force** (Section 4.8)
3. **Frictional Forces** (Section 4.9)
4. **Tension Force** (Section 4.10)

Equilibrium (Section 9.2)

$a_x = 0$
$a_y = 0$
$\alpha = 0$

Sum of External Forces:

$\Sigma F_x = ma_x$ (4.2a)

$\Sigma F_y = ma_y$ (4.2b)

Sum of External Torques:

$\Sigma \tau = \Sigma F\ell = I\alpha$ (9.7)

$a_x \neq 0$
$a_y \neq 0$
$\alpha \neq 0$

Nonequilibrium

Figure 9.17 CONCEPTS AT A GLANCE An object is in equilibrium when its translational acceleration components, a_x and a_y, and its angular acceleration α are zero. If a_x, a_y, or α is not zero, the object has an acceleration and is not in equilibrium. As these motorcyclists round the turn, they have translational and angular accelerations, so they are not in equilibrium. (© Pascal Rondeau/ Stone/Getty Images)

Example 9 shows how Newton's second law for rotational motion is used when design considerations demand an adequately large angular acceleration. There are also situations when it is desirable to have as little angular acceleration as possible, and Conceptual Example 10 deals with one of them.

Conceptual Example 10 **Archery and Bow Stabilizers**

Archers can shoot with amazing accuracy, especially using modern bows such as the one in Figure 9.19. Notice the bow stabilizer, a long, thin rod that extends from the front of the bow and has a relatively massive cylinder at the tip. Advertisements claim that the stabilizer helps to steady the archer's aim. Could there be any truth to this claim? Explain.

Reasoning and Solution To help explain why the stabilizer works, we have added to the photograph an axis for rotation that passes through the archer's shoulder and is perpendicular to the plane of the paper. Any angular acceleration α about this axis will lead to a rotation of the bow that will degrade the archer's aim. The acceleration will be created by any unbalanced

Figure 9.18 A rider applies a force **F** to the circular handrail. The torque produced by this force is the product of its magnitude and the lever arm ℓ about the axis of rotation. (© Dan Coffee/The Image Bank/Getty Images)

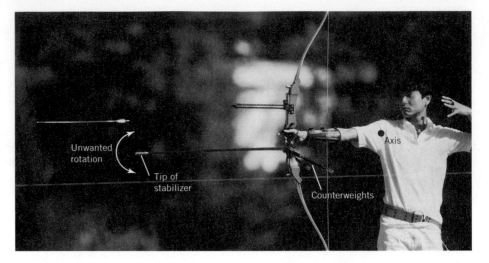

Figure 9.19 The stabilizer helps to steady the archer's aim, as Conceptual Example 10 discusses. Relative to an axis through the archer's shoulder, the moment of inertia of the bow is larger with the stabilizer than without it. The counterweights below the archer's hand help to keep the center of gravity of the bow near the hand. (© Amwell/Stone/Getty Images)

The physics of archery and bow stabilizers.

torques that occur while the archer's tensed muscles try to hold the drawn bow. Newton's second law for rotation indicates, however, that the angular acceleration is $\alpha = (\Sigma \tau)/I$. The moment of inertia I is in the denominator on the right side of this equation. Therefore, to the extent that I is larger, a given net torque $\Sigma \tau$ will create a smaller angular acceleration and less disturbance of the aim. It is to increase the moment of inertia of the bow that the stabilizer has been added. The relatively massive cylinder is particularly effective in increasing the moment of inertia, because it is placed at the tip of the stabilizer, far from the axis of rotation (a large value of r in the equation $I = \Sigma mr^2$).

Rotational motion and translational motion sometimes occur together. The next example deals with an interesting situation in which both angular acceleration and translational acceleration must be considered.

Example 11   Hoisting a Crate

A crate that weighs 4420 N is being lifted by the mechanism shown in Figure 9.20a. The two cables are wrapped around their respective pulleys, which have radii of 0.600 and 0.200 m. The pulleys are fastened together to form a dual pulley and turn as a single unit about the center axle, relative to which the combined moment of inertia is $I = 50.0 \text{ kg} \cdot \text{m}^2$. A tension of

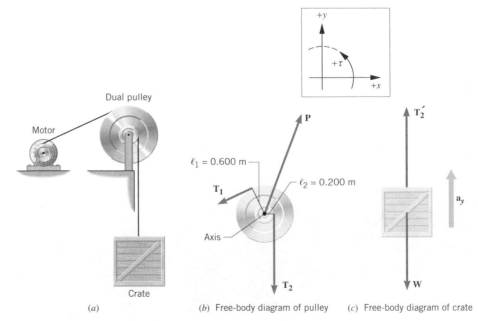

Figure 9.20 (a) The crate is lifted upward by the motor and pulley arrangement. The free-body diagram for (b) the dual pulley and (c) the crate.

(a)

(b) Free-body diagram of pulley

(c) Free-body diagram of crate

magnitude $T_1 = 2150$ N is maintained in the cable attached to the motor. Find the angular acceleration of the dual pulley and the tension in the cable connected to the crate.

Reasoning To determine the angular acceleration of the dual pulley and the tension in the cable attached to the crate, we will apply Newton's second law to the pulley and the crate separately. Three external forces act on the dual pulley, as its free-body diagram shows (Figure 9.20*b*). These are (1) the tension $\mathbf{T}_1$ in the cable connected to the motor, (2) the tension $\mathbf{T}_2$ in the cable attached to the crate, and (3) the reaction force $\mathbf{P}$ exerted on the dual pulley by the axle. The force $\mathbf{P}$ arises because the two cables pull the pulley down and to the left into the axle, and the axle pushes back, thus keeping the pulley in place. The net torque that results from these forces obeys Newton's second law for rotational motion (Equation 9.7). Two external forces act on the crate, as its free-body diagram indicates (Figure 9.20*c*). These are (1) the cable tension $\mathbf{T}_2'$ and (2) the weight $\mathbf{W}$ of the crate. The net force that results from these forces obeys Newton's second law for translational motion (Equation 4.2b).

Solution Using the lever arms ℓ_1 and ℓ_2 shown in part *b* of the figure, we can apply the second law to the rotational motion of the pulley. We note that the force $\mathbf{P}$ has a zero lever arm, since the line of action of $\mathbf{P}$ passes directly through the axle:

$$\Sigma\tau = T_1\ell_1 - T_2\ell_2 = I\alpha \tag{9.7}$$

$$(2150\text{ N})(0.600\text{ m}) - T_2(0.200\text{ m}) = (50.0\text{ kg}\cdot\text{m}^2)\alpha$$

This equation contains two unknown quantities, so a second equation is needed and may be obtained by applying Newton's second law to the upward translational motion of the crate. In doing so, we note that the magnitude of the tension in the cable attached to the crate is $T_2' = T_2$ and that the mass of the crate is $m = (4420\text{ N})/(9.80\text{ m/s}^2) = 451$ kg:

$$\Sigma F_y = ma_y \tag{4.2b}$$

$$T_2 - (4420\text{ N}) = (451\text{ kg})\,a_y$$

Because the cable attached to the crate rolls on the pulley without slipping, the linear acceleration a_y of the crate is related to the angular acceleration α of the pulley via Equation 8.13: $a_y = r\alpha = (0.200\text{ m})\alpha$. With this substitution for a_y, Equation 4.2b becomes

$$T_2 - (4420\text{ N}) = (451\text{ kg})(0.200\text{ m})\alpha$$

This result and Equation 9.7 can be solved simultaneously to yield

$$\boxed{T_2 = 4960\text{ N}} \quad \text{and} \quad \boxed{\alpha = 6.0\text{ rad/s}^2}$$

▲

We have seen that Newton's second law for rotational motion, $\Sigma\tau = I\alpha$, has the same form as that for translational motion, $\Sigma F = ma$, so each rotational variable has a translational analog: torque τ and force F are analogous quantities, as are moment of inertia I and mass m, and angular acceleration α and linear acceleration a. The other physical concepts developed for studying translational motion, such as kinetic energy and momentum, also have rotational analogs. For future reference, Table 9.2 itemizes these concepts and their rotational analogs.

Concept Simulation 9.4

A block is hanging from one end of a string. The other end wraps around a pulley that is fastened to the ceiling above the block. When the block is released it accelerates downward. You can adjust the radius and mass of the pulley, as well as the mass of the block. You can also select from three different pulley designs. The simulation determines the linear acceleration of the block, the angular acceleration of the pulley, and the tension in the string. With a calculation analogous to that in Example 11, check to see whether your values agree with those arrived at by the simulation.

Related Homework: Problem 42

Go to
www.wiley.com/college/cutnell

Table 9.2 *Analogies Between Rotational and Translational Concepts*

Physical Concept	Rotational	Translational
Displacement	θ	s
Velocity	ω	v
Acceleration	α	a
The cause of acceleration	Torque τ	Force F
Inertia	Moment of inertia I	Mass m
Newton's second law	$\Sigma\tau = I\alpha$	$\Sigma F = ma$
Work	$\tau\theta$	Fs
Kinetic energy	$\frac{1}{2}I\omega^2$	$\frac{1}{2}mv^2$
Momentum	$L = I\omega$	$p = mv$

Check Your Understanding 3

Three massless rods are free to rotate about an axis at their left end (see the drawing). The same force **F** is applied to the right end of each rod. Objects with different masses are attached to the rods, but the total mass (3m) of the objects is the same for each rod. Rank the angular acceleration of the rods, largest to smallest. *(The answer is given at the end of the book.)*

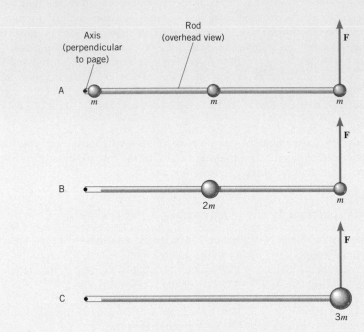

Background: Newton's second law for rotational motion holds the key here. However, it is also necessary to understand the related concepts of net torque, moment of inertia, and angular acceleration.

For similar questions (including calculational counterparts), consult Self-Assessment Test 9.2. This test is described at the end of Section 9.6.

9.5 Rotational Work and Energy

Work and energy are among the most fundamental and useful concepts in physics. Chapter 6 discusses their application to translational motion. These concepts are equally useful for rotational motion, provided they are expressed in terms of angular variables.

The work W done by a constant force that points in the same direction as the displacement is $W = Fs$ (Equation 6.1), where F and s are the magnitudes of the force and displacement, respectively. To see how this expression can be rewritten using angular variables, consider Figure 9.21. Here a rope is wrapped around a wheel and is under a constant tension F. If the rope is pulled out a distance s, the wheel rotates through an angle $\theta = s/r$ (Equation 8.1), where r is the radius of the wheel and θ is in radians. Thus, $s = r\theta$, and the work done by the tension force in turning the wheel is $W = Fs = Fr\theta$. But Fr is the torque τ applied to the wheel by the tension, so the rotational work can be written as follows:

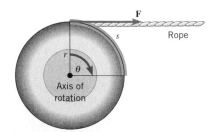

Figure 9.21 The force **F** does work in rotating the wheel through the angle θ.

■ **DEFINITION OF ROTATIONAL WORK**

The rotational work W_R done by a constant torque τ in turning an object through an angle θ is

$$W_R = \tau\theta \tag{9.8}$$

Requirement: θ must be expressed in radians.

SI Unit of Rotational Work: joule (J)

Section 6.2 discusses the work–energy theorem and kinetic energy. There we saw that the work done on an object by a net external force causes the translational kinetic energy ($\frac{1}{2}mv^2$) of the object to change. In an analogous manner, the rotational work done by a net external torque causes the rotational kinetic energy to change. A rotating body possesses kinetic energy, because its constituent particles are moving. If the body is rotating with an angular speed ω, the tangential speed v_T of a particle at a distance r from the axis is $v_T = r\omega$ (Equation 8.9). Figure 9.22 shows two such particles. If a particle's mass is m, its kinetic energy is $\frac{1}{2}mv_T^2 = \frac{1}{2}mr^2\omega^2$. The kinetic energy of the entire rotating body, then, is the sum of the kinetic energies of the particles:

$$\text{Rotational KE} = \Sigma(\tfrac{1}{2}mr^2\omega^2) = \tfrac{1}{2}\underbrace{(\Sigma mr^2)}_{\substack{\text{Moment of}\\\text{inertia, } I}}\omega^2$$

Figure 9.22 The rotating wheel is composed of many particles, two of which are shown.

In this result, the angular speed ω is the same for all particles in a rigid body and, therefore, has been factored outside the summation. According to Equation 9.6, the term in parentheses is the moment of inertia, $I = \Sigma mr^2$, so the rotational kinetic energy takes the following form:

■ **DEFINITION OF ROTATIONAL KINETIC ENERGY**

The rotational kinetic energy KE_R of a rigid object rotating with an angular speed ω about a fixed axis and having a moment of inertia I is

$$\text{KE}_R = \tfrac{1}{2}I\omega^2 \qquad (9.9)$$

Requirement: ω must be expressed in rad/s.

SI Unit of Rotational Kinetic Energy: joule (J)

► **CONCEPTS AT A GLANCE** Kinetic energy is one part of an object's total mechanical energy. The total mechanical energy is the sum of the kinetic and potential energies and obeys the principle of conservation of mechanical energy (see Section 6.5). The Concepts-at-a-Glance chart in Figure 9.23, which is an expanded version of that in Figure 6.15, shows how rotational kinetic energy is incorporated into this principle. Specifically, we need to remember that translational and rotational motion can occur simultaneously.

Figure 9.23 CONCEPTS AT A GLANCE The principle of conservation of mechanical energy can be applied to a rigid object that has both translational and rotational motion, provided the rotational kinetic energy is included in the total mechanical energy and provided that W_{nc}, the net work done by external nonconservative forces and torques, is zero. The principle can be applied to the translational and rotational motion of these divers, to the extent that their shapes are constant and air resistance is negligible. (© Darren England/Getty Images News and Sport Services)

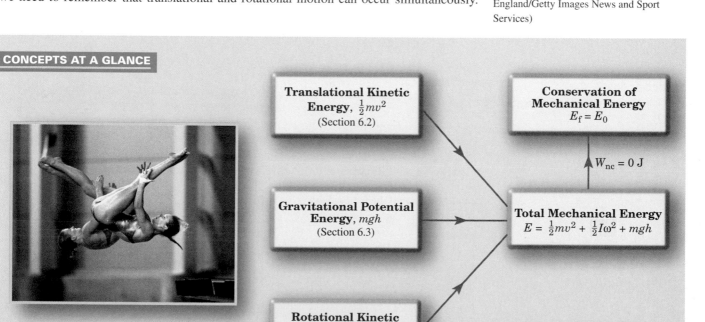

CONCEPTS AT A GLANCE

Translational Kinetic Energy, $\frac{1}{2}mv^2$ (Section 6.2)

Gravitational Potential Energy, mgh (Section 6.3)

Rotational Kinetic Energy, $\frac{1}{2}I\omega^2$ (Section 9.5)

Total Mechanical Energy $E = \frac{1}{2}mv^2 + \frac{1}{2}I\omega^2 + mgh$

$W_{nc} = 0$ J

Conservation of Mechanical Energy $E_f = E_0$

Need more practice?

Interactive LearningWare 9.3
A car is moving with a speed of 27.0 m/s. Each wheel has a radius of 0.300 m and a moment of inertia of 0.850 kg·m². The car has a total mass (including the wheels) of 1.20 × 10³ kg. Find the total kinetic energy of the car, translational plus rotational.

Go to
www.wiley.com/college/cutnell
for an interactive solution.

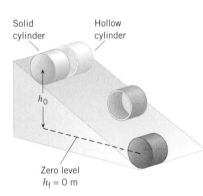

Figure 9.24 A hollow cylinder and a solid cylinder start from rest and roll down the incline plane. The conservation of mechanical energy can be used to show that the solid cylinder, having the greater translational speed, reaches the bottom first.

When a bicycle coasts down a hill, for instance, its tires are both translating and rotating. An object such as a rolling bicycle tire has both translational and rotational kinetic energies, so that the total mechanical energy is

$$E \;=\; \underbrace{\tfrac{1}{2}mv^2}_{\substack{\text{Translational}\\\text{kinetic energy}}} \;+\; \underbrace{\tfrac{1}{2}I\omega^2}_{\substack{\text{Rotational}\\\text{kinetic energy}}} \;+\; \underbrace{mgh}_{\substack{\text{Gravitational}\\\text{potential energy}}}$$

$$\underbrace{}_{\substack{\text{Total}\\\text{mechanical}\\\text{energy}}}$$

Here m is the mass of the object, v is the translational speed of its center of mass, I is its moment of inertia about an axis through the center of mass, ω is its angular speed, and h is the height of the object's center of mass relative to an arbitrary zero level. Mechanical energy is conserved if W_{nc}, the net work done by external nonconservative forces and torques, is zero. If the total mechanical energy is conserved as an object moves, its final total mechanical energy E_f equals its initial total mechanical energy E_0: $E_f = E_0$. ◀

Example 12, like Interactive LearningWare 9.3, also illustrates the effect of combined translational and rotational motion, but in the context of how the total mechanical energy of a cylinder is conserved as it rolls down an incline.

Example 12 Rolling Cylinders

A thin-walled hollow cylinder (mass $= m_h$, radius $= r_h$) and a solid cylinder (mass $= m_s$, radius $= r_s$) start from rest at the top of an incline (Figure 9.24). Both cylinders start at the same vertical height h_0. All heights are measured relative to an arbitrarily chosen zero level that passes through the center of mass of a cylinder when it is at the bottom of the incline (see the drawing). Ignoring energy losses due to retarding forces, determine which cylinder has the greatest translational speed upon reaching the bottom.

Reasoning Only the conservative force of gravity does work on the cylinders, so the total mechanical energy is conserved as they roll down. The total mechanical energy E at any height h above the zero level is the sum of the translational kinetic energy ($\tfrac{1}{2}mv^2$), the rotational kinetic energy ($\tfrac{1}{2}I\omega^2$), and the gravitational potential energy (mgh):

$$E = \tfrac{1}{2}mv^2 + \tfrac{1}{2}I\omega^2 + mgh$$

As the cylinders roll down, potential energy is converted into kinetic energy, but the kinetic energy is shared between the translational form ($\tfrac{1}{2}mv^2$) and the rotational form ($\tfrac{1}{2}I\omega^2$). The object with more of its kinetic energy in the translational form will have the greater translational speed at the bottom of the incline. We expect the solid cylinder to have the greater translational speed, because more of its mass is located near the rotational axis and, thus, possesses less rotational kinetic energy.

Solution The total mechanical energy E_f at the bottom ($h_f = 0$ m) is the same as the total mechanical energy E_0 at the top ($h = h_0$, $v_0 = 0$ m/s, $\omega_0 = 0$ rad/s):

$$\tfrac{1}{2}mv_f^2 + \tfrac{1}{2}I\omega_f^2 + mgh_f = \tfrac{1}{2}mv_0^2 + \tfrac{1}{2}I\omega_0^2 + mgh_0$$

$$\tfrac{1}{2}mv_f^2 + \tfrac{1}{2}I\omega_f^2 = mgh_0$$

Since each cylinder rolls without slipping, the final rotational speed ω_f and the final translational speed v_f of its center of mass are related according to Equation 8.12, $\omega_f = v_f/r$, where r is the radius of the cylinder. Substituting this expression for ω_f into the energy-conservation equation and solving for v_f yields

$$v_f = \sqrt{\frac{2mgh_0}{m + I/r^2}}$$

Setting $m = m_h$, $r = r_h$ and $I = mr_h^2$ for the hollow cylinder and then setting $m = m_s$, $r = r_s$ and $I = \tfrac{1}{2}mr_s^2$ for the solid cylinder (see Table 9.1), we find that the two cylinders have the following translational speeds at the bottom of the incline:

Hollow cylinder $v_f = \sqrt{gh_0}$

Solid cylinder $v_f = \sqrt{\dfrac{4gh_0}{3}} = 1.15\sqrt{gh_0}$

The solid cylinder, having the greater translational speed, arrives at the bottom first.

✓ **Check Your Understanding 4**

Two solid balls are placed side by side at the top of an incline plane and, starting from rest, are allowed to roll down the incline. Which ball, if either, has the greater translational speed at the bottom if (a) they have the same radii but one is more massive than the other, and (b) they have the same mass but one has a larger radius? *(The answers are given at the end of the book.)*

Background: The conservation of total mechanical energy is the central concept here. Note that the mass of an object is not the same as its moment of inertia.

For similar questions (including calculational counterparts), consult Self-Assessment Test 9.2. This test is described at the end of Section 9.6.

9.6 *Angular Momentum*

In Chapter 7 the linear momentum p of an object is defined as the product of its mass m and linear velocity v; that is, $p = mv$. For rotational motion the analogous concept is called the ***angular momentum L***. The mathematical form of angular momentum is analogous to that of linear momentum, with the mass m and the linear velocity v being replaced with their rotational counterparts, the moment of inertia I and the angular velocity ω.

■ **DEFINITION OF ANGULAR MOMENTUM**

The angular momentum L of a body rotating about a fixed axis is the product of the body's moment of inertia I and its angular velocity ω with respect to that axis:

$$L = I\omega \qquad (9.10)$$

Requirement: ω must be expressed in rad/s.

SI Unit of Angular Momentum: $\text{kg} \cdot \text{m}^2/\text{s}$

▶ **CONCEPTS AT A GLANCE** Linear momentum is an important concept in physics because the total linear momentum of a system is conserved when the sum of the average external forces acting on the system is zero. Then, the final total linear momentum P_f and the initial total linear momentum P_0 are the same: $P_f = P_0$. Figure 7.8 outlines the conceptual development of the conservation of linear momentum. The Concepts-at-a-Glance chart in Figure 9.25 outlines a similar development for angular momentum. In construct-

Figure 9.25 **CONCEPTS AT A GLANCE** The impulse–momentum theorem for rotational motion leads to the principle of conservation of angular momentum when the sum of the external torques is zero. This concept diagram is similar to that in Figure 7.8, which shows how the conservation of linear momentum arises from the impulse–momentum theorem. The angular momentum of this spinning gymnast is conserved while he is in the air, assuming that air resistance is negligible. (© David Madison/Stone/Getty Images)

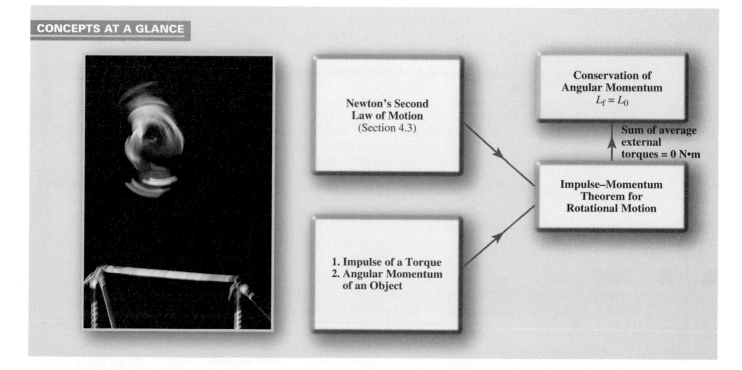

CONCEPTS AT A GLANCE

Newton's Second Law of Motion (Section 4.3)

1. **Impulse of a Torque**
2. **Angular Momentum of an Object**

Impulse–Momentum Theorem for Rotational Motion

Conservation of Angular Momentum $L_f = L_0$

Sum of average external torques = 0 N•m

ing this chart, we recall that each translational variable is analogous to a corresponding rotational variable, as Table 9.2 indicates. Therefore, if we replace "force" by "torque" and "linear momentum" by "angular momentum" in Figure 7.8, we obtain Figure 9.25. This chart indicates that when the sum of the average external torques is zero, the final and initial angular momenta are the same: $L_f = L_0$, which is the **principle of conservation of angular momentum.** ◄

> **■ PRINCIPLE OF CONSERVATION OF ANGULAR MOMENTUM**
>
> The total angular momentum of a system remains constant (is conserved) if the net average external torque acting on the system is zero.

Example 13 illustrates an interesting consequence of the conservation of angular momentum.

The physics of a spinning ice skater.

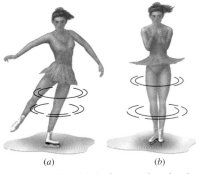

(a)   *(b)*

Figure 9.26 *(a)* A skater spins slowly on one skate, with both arms and one leg outstretched. *(b)* As she pulls her arms and leg in toward the rotational axis, her moment of inertia *I* decreases, and her angular velocity *ω* increases.

Conceptual Example 13 A Spinning Skater

In Figure 9.26*a* an ice skater is spinning with both arms and a leg outstretched. In Figure 9.26*b* she pulls her arms and leg inward. As a result of this maneuver, her spinning motion changes dramatically. Using the principle of conservation of angular momentum, explain how and why it changes.

Reasoning and Solution Choosing the skater as the system, we can apply the conservation principle provided that the net external torque produced by air resistance and by friction between the skates and the ice is negligibly small. We assume that it is. Then the skater in Figure 9.26*a* would spin forever at the same angular velocity, since her angular momentum is conserved in the absence of a net external torque. In Figure 9.26*b* the inward movement of her arms and leg involves internal, not external, torques and, therefore, does not change her angular momentum. But angular momentum is the product of the moment of inertia *I* and angular velocity *ω*. By moving the mass of her arms and leg inward, the skater decreases the distance *r* of the mass from the axis of rotation and, consequently, decreases her moment of inertia *I* ($I = \Sigma mr^2$). If the product of *I* and *ω* is to remain constant, then *ω* must increase. Thus, the consequence of pulling her arms and leg inward is that **she spins with a larger angular velocity.**

Related Homework: *Conceptual Questions 20, 21, and 22, Problem 52*

The next example involves a satellite and illustrates another application of the principle of conservation of angular momentum.

The physics of a satellite in orbit about the earth.

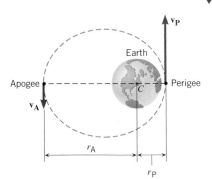

Figure 9.27 A satellite is moving in an elliptical orbit about the earth. The gravitational force exerts no torque on the satellite, so the angular momentum of the satellite is conserved.

Example 14 A Satellite in an Elliptical Orbit

An artificial satellite is placed into an elliptical orbit about the earth, as in Figure 9.27. Telemetry data indicate that its point of closest approach (called the *perigee*) is $r_P = 8.37 \times 10^6$ m from the center of the earth, and its point of greatest distance (called the *apogee*) is $r_A = 25.1 \times 10^6$ m from the center of the earth. The speed of the satellite at the perigee is $v_P = 8450$ m/s. Find its speed v_A at the apogee.

Reasoning The only force of any significance that acts on the satellite is the gravitational force of the earth. However, at any instant, this force is directed toward the center of the earth and passes through the axis about which the satellite instantaneously rotates. Therefore, the gravitational force exerts *no torque* on the satellite (the lever arm is zero). Consequently, the angular momentum of the satellite remains constant at all times.

Solution Since the angular momentum is the same at the apogee (A) and the perigee (P), it follows that $I_A \omega_A = I_P \omega_P$. Furthermore, the orbiting satellite can be considered a point mass, so its moment of inertia is $I = mr^2$ (see Equation 9.4). In addition, the angular speed *ω* of the satellite is related to its tangential speed v_T by $\omega = v_T/r$ (Equation 8.9). If these relations are used at the apogee and perigee, the conservation of angular momentum gives the following result:

$$I_A \omega_A = I_P \omega_P \quad \text{or} \quad (mr_A{}^2)\left(\frac{v_A}{r_A}\right) = (mr_P{}^2)\left(\frac{v_P}{r_P}\right)$$

$$v_A = \frac{r_P v_P}{r_A} = \frac{(8.37 \times 10^6 \text{ m})(8450 \text{ m/s})}{25.1 \times 10^6 \text{ m}} = \boxed{2820 \text{ m/s}}$$

The answer is independent of the mass of the satellite. The satellite behaves just like the skater in Figure 9.26, because its speed is greater at the perigee, where the moment of inertia is smaller.

The result in Example 14 indicates that a satellite does not have a constant speed in an elliptical orbit. The speed changes from a maximum at the perigee to a minimum at the apogee; the closer the satellite comes to the earth, the faster it travels. Planets moving around the sun in elliptical orbits exhibit the same kind of behavior, and Johannes Kepler (1571–1630) formulated his famous second law based on observations of such characteristics of planetary motion. Kepler's second law states that, in a given amount of time, a line joining any planet to the sun sweeps out the same amount of area no matter where the planet is on its elliptical orbit, as Figure 9.28 illustrates. The conservation of angular momentum can be used to show why the law is valid, by means of a calculation similar to that in Example 14.

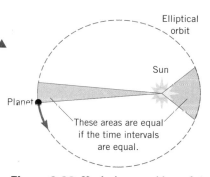

Figure 9.28 Kepler's second law of planetary motion states that a line joining a planet to the sun sweeps out equal areas in equal time intervals.

Self-Assessment Test 9.2

Check your knowledge of the main ideas presented in Sections 9.4–9.6:

- Newton's Second Law for Rotational Motion
- Rotational Work and Energy
- Angular Momentum

Go to **www.wiley.com/college/cutnell**

9.7 *Concepts & Calculations*

In this chapter we have seen that a rotational or angular acceleration results when a net external torque acts on an object. In contrast, when a net external force acts on an object, it leads to a translational or linear acceleration, as Chapter 4 discusses. Torque and force, then, are fundamentally different concepts, and Example 15 focuses on this fact.

Concepts & Calculations Example 15   Torque and Force

Figure 9.29*a* shows a crate resting on a horizontal surface. It has a square cross section and a weight of $W = 580$ N, which is uniformly distributed. At the bottom right edge is a small obstruction that prevents the crate from sliding when a horizontal pushing force **P** is applied to the left side. However, if this force is great enough, the crate will begin to tip or rotate over the obstruction. Determine the minimum pushing force that leads to tipping.

Concept Questions and Answers What causes tipping—the force **P** or the torque that it creates?

Answer Tipping is a rotational or angular motion. Since the crate starts from rest, an angular acceleration is needed, which can only be created by a net external torque. Thus, the torque created by the force **P** causes the tipping.

A given force can create a variety of torques, depending on the lever arm of the force with respect to the rotational axis. In this case, the rotational axis is located at the small obstruction in Figure 9.29*a* and is perpendicular to the page. For this axis, the lever arm of the force **P** is ℓ_P, as the drawing shows. Where should **P** be applied so that a minimum force will give the necessary torque? In other words, should the lever arm be a minimum or a maximum?

Answer The magnitude of the torque is the product of the magnitude of the force and the lever arm. Thus, for a minimum force, the lever arm should be a maximum, and the force **P** should be applied at the upper left corner of the crate, as in Figure 9.29*b*.

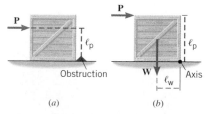

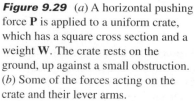

Figure 9.29 (*a*) A horizontal pushing force **P** is applied to a uniform crate, which has a square cross section and a weight **W**. The crate rests on the ground, up against a small obstruction. (*b*) Some of the forces acting on the crate and their lever arms.

Consider the crate just at the instant *before* it begins to rotate. At this instant, the crate is in equilibrium. What must be true about the sum of the external torques acting on the crate?

Answer　Because the crate is in equilibrium, the sum of the external torques must be zero.

Solution　Only the pushing force **P** and the weight **W** of the crate produce external torques with respect to a rotational axis through the lower right corner of the crate. The obstruction also applies a force to the crate, but it creates no torque, since its line of action passes through the axis. Since the sum of the external torques is zero, we refer to Figure 9.29*b* for the lever arms and write

$$\Sigma\tau = -P\ell_P + W\ell_W = 0$$

The lever arm for the force **P** is $\ell_P = L$, where L is the length of the side of the crate. The lever arm for the weight is $\ell_W = L/2$. This is because the crate is uniform, and the center of gravity is at the center of the crate. Substituting these lever arms in the torque equation, we obtain

$$-PL + W(\tfrac{1}{2}L) = 0 \quad \text{or} \quad P = \tfrac{1}{2}W = \tfrac{1}{2}(580\ \text{N}) = \boxed{290\ \text{N}}$$

For rotational motion, the moment of inertia plays a key role. We have seen in this chapter that the moment of inertia and the net external torque determine the angular acceleration of a rotating object, according to Newton's second law. The angular acceleration, in turn, can be used in the equations of rotational kinematics, provided that it remains constant, as Chapter 8 discusses. When applied together in this way, Newton's second law and the equations of rotational kinematics are particularly useful in accounting for a wide variety of rotational motion. The following example reviews this approach in a situation where a rotating object is slowing down.

Concepts & Calculations Example 16
Which Sphere Takes Longer to Stop?

Two spheres are each rotating at an angular speed of 24 rad/s about axes that pass through their centers. Each has a radius of 0.20 m and a mass of 1.5 kg. However, as Figure 9.30 shows, one is solid and the other is a thin-walled spherical shell. Suddenly, a net external torque due to friction (magnitude = 0.12 N·m) begins to act on each sphere and slows the motion down. How long does it take each sphere to come to a halt?

Concept Questions and Answers　Which sphere has the greater moment of inertia and why?

Answer　Referring to Table 9.1, we see that the solid sphere has a moment of inertia of $\tfrac{2}{5}MR^2$, while the shell has a moment of inertia of $\tfrac{2}{3}MR^2$. Since the masses and radii of the spheres are the same, it follows that the shell has the greater moment of inertia. The reason is that more of the mass of the shell is located farther from the rotational axis than is the case for the solid sphere. In the solid sphere, some of the mass is located close to the axis and, therefore, does not contribute as much to the moment of inertia.

Which sphere has the angular acceleration (a deceleration) with the smaller magnitude?

Answer　Newton's second law for rotation (Equation 9.7) specifies that the angular acceleration is $\alpha = (\Sigma\tau)/I$, where $\Sigma\tau$ is the net external torque and I is the moment of inertia. Since the moment of inertia is in the denominator, the angular acceleration is smaller when I is greater. Because it has the greater moment of inertia, the shell has the angular acceleration with the smaller magnitude.

Which sphere takes the longer time to come to a halt?

Answer　Since the angular acceleration of the shell has the smaller magnitude, the shell requires a longer time for the deceleration to reduce its angular velocity to zero.

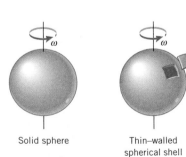

Solid sphere　　　Thin–walled
　　　　　　　　spherical shell

Figure 9.30　The two spheres have identical radii and masses; initially, they also have the same angular velocities. Which one comes to rest first as they slow down due to friction?

Solution　According to Equation 8.4 from the equations of rotational kinematics, the time is given by $t = (\omega - \omega_0)/\alpha$ where ω and ω_0 are, respectively, the final and initial angular velocities. From Newton's second law as given in Equation 9.7 we know that the angular acceleration is $\alpha = (\Sigma\tau)/I$. Substituting this into the expression for the time gives

$$t = \frac{\omega - \omega_0}{(\Sigma\tau)/I} = \frac{I(\omega - \omega_0)}{\Sigma\tau}$$

In applying this result, we arbitrarily choose the direction of the initial rotation to be positive. With this choice, the torque must be negative, since it causes a deceleration. Using the proper moments of inertia, we find the following times for the spheres to come to a halt:

Solid sphere
$$t = \frac{I(\omega - \omega_0)}{\Sigma\tau} = \frac{\frac{2}{5}MR^2(\omega - \omega_0)}{\Sigma\tau}$$

$$t = \frac{\frac{2}{5}(1.5\text{ kg})(0.20\text{ m})^2[(0\text{ rad/s}) - (24\text{ rad/s})]}{-0.12\text{ N}\cdot\text{m}} = \boxed{4.8\text{ s}}$$

Spherical shell
$$t = \frac{I(\omega - \omega_0)}{\Sigma\tau} = \frac{\frac{2}{3}MR^2(\omega - \omega_0)}{\Sigma\tau}$$

$$t = \frac{\frac{2}{3}(1.5\text{ kg})(0.20\text{ m})^2[(0\text{ rad/s}) - (24\text{ rad/s})]}{-0.12\text{ N}\cdot\text{m}} = \boxed{8.0\text{ s}}$$

As expected, the shell requires a longer time to come to a halt.

▲

At the end of the problem set for this chapter, you will find homework problems that contain both conceptual and quantitative parts. These problems are grouped under the heading *Concepts & Calculations, Group Learning Problems*. They are designed for use by students working alone or in small learning groups. The conceptual part of each problem provides a convenient focus for group discussions.

Concept Summary

This summary presents an abridged version of the chapter, including the important equations and all available learning aids. For convenient reference, the learning aids (including the text's examples) are placed next to or immediately after the relevant equation or discussion. The following learning aids may be found on-line at **www.wiley.com/college/cutnell**:

Interactive LearningWare examples are solved according to a five-step interactive format that is designed to help you develop problem-solving skills.	**Concept Simulations** are animated versions of text figures or animations that illustrate important concepts. You can control parameters that affect the display, and we encourage you to experiment.
Interactive Solutions offer specific models for certain types of problems in the chapter homework. The calculations are carried out interactively.	**Self-Assessment Tests** include both qualitative and quantitative questions. Extensive feedback is provided for both incorrect and correct answers, to help you evaluate your understanding of the material.

Topic	Discussion	Learning Aids
	9.1 The Action of Forces and Torques on Rigid Objects	
Line of action Lever arm	The line of action of a force is an extended line that is drawn colinear with the force. The lever arm ℓ is the distance between the line of action and the axis of rotation, measured on a line that is perpendicular to both.	
Torque	The torque of a force has a magnitude τ that is given by the magnitude F of the force times the lever arm ℓ: $$\tau = F\ell \qquad (9.1)$$ The torque is positive when the force tends to produce a counterclockwise rotation about the axis, and negative when the force tends to produce a clockwise rotation.	Examples 1, 2 Interactive LearningWare 9.1 Interactive Solution 9.3
	9.2 Rigid Objects in Equilibrium	
Equilibrium of a rigid body	A rigid body is in equilibrium if it has zero translational acceleration and zero angular acceleration. In equilibrium, the net external force and the net external torque acting on the body are zero: $$\Sigma F_x = 0 \quad \text{and} \quad \Sigma F_y = 0 \qquad (4.9a\text{ and }4.9b)$$ $$\Sigma\tau = 0 \qquad (9.2)$$	Examples 3, 4, 5, 15 Interactive LearningWare 9.2 Concept Simulation 9.1 Interactive Solution 9.17
	9.3 Center of Gravity	
	The center of gravity of a rigid object is the point where its entire weight can be considered to act when calculating the torque due to the weight. For a symmetrical body with uniformly distributed weight, the center of gravity is at the geo-	

Topic	Discussion	Learning Aids

metrical center of the body. When a number of objects whose weights are W_1, W_2, . . . are distributed along the x axis at locations $x_1, x_2, . . .$, the center of gravity x_{cg} is located at

Definition of center of gravity

$$x_{cg} = \frac{W_1 x_1 + W_2 x_2 + \cdots}{W_1 + W_2 + \cdots} \qquad (9.3)$$

Examples 6, 7

Concept Simulation 9.2

The center of gravity is identical to the center of mass, provided the acceleration due to gravity does not vary over the physical extent of the objects.

Use Self-Assessment Test 9.1 to evaluate your understanding of Sections 9.1–9.3.

9.4 Newton's Second Law for Rotational Motion About a Fixed Axis

The moment of inertia I of a body composed of N particles is

Moment of inertia

$$I = m_1 r_1^2 + m_2 r_2^2 + \cdots + m_N r_N^2 = \Sigma m r^2 \qquad (9.6)$$

Example 8

where m is the mass of a particle and r is the perpendicular distance of the particle from the axis of rotation.

For a rigid body rotating about a fixed axis, Newton's second law for rotational motion is

Newton's second law for rotational motion

$$\Sigma \tau = I\alpha \qquad (\alpha \text{ in rad/s}^2) \qquad (9.7)$$

where $\Sigma \tau$ is the net external torque applied to the body, I is the moment of inertia of the body, and α is its angular acceleration.

Examples 9, 10, 11, 16

Concept Simulations 9.3, 9.4

Interactive Solution 9.35

9.5 Rotational Work and Energy

The rotational work W_R done by a constant torque τ in turning a rigid body through an angle θ is

Rotational work

$$W_R = \tau\theta \qquad (\theta \text{ in radians}) \qquad (9.8)$$

The rotational kinetic energy KE_R of a rigid object rotating with an angular speed ω about a fixed axis and having a moment of inertia I is

Rotational kinetic energy

$$KE_R = \tfrac{1}{2} I\omega^2 \qquad (9.9)$$

The total mechanical energy E of a rigid body is the sum of its translational kinetic energy ($\tfrac{1}{2} mv^2$), its rotational kinetic energy ($\tfrac{1}{2} I\omega^2$), and its gravitational potential energy (mgh):

Total mechanical energy

$$E = \tfrac{1}{2} mv^2 + \tfrac{1}{2} I\omega^2 + mgh$$

Interactive LearningWare 9.3

Interactive Solution 9.47

where m is the mass of the object, v is the translational speed of its center of mass, I is its moment of inertia about an axis through the center of mass, ω is its angular speed, and h is the height of the object's center of mass relative to an arbitrary zero level.

Conservation of total mechanical energy

The total mechanical energy is conserved if the net work done by external nonconservative forces and torques is zero. When the total mechanical energy is conserved, the final total mechanical energy E_f equals the initial total mechanical energy E_0: $E_f = E_0$.

Example 12

9.6 Angular Momentum

The angular momentum of a rigid body rotating with an angular velocity ω about a fixed axis and having a moment of inertia I with respect to that axis is

Angular momentum

$$L = I\omega \qquad (\omega \text{ in rad/s}) \qquad (9.10)$$

Conservation of angular momentum

The principle of conservation of angular momentum states that the total angular momentum of a system remains constant (is conserved) if the net average external torque acting on the system is zero. When the total angular momentum is conserved, the final angular momentum L_f equals the initial angular momentum L_0: $L_f = L_0$.

Examples 13, 14

Interactive Solution 9.55

Use Self-Assessment Test 9.2 to evaluate your understanding of Sections 9.4–9.6.

Conceptual Questions

1. Sometimes, even with a wrench, one cannot loosen a nut that is frozen tightly to a bolt. It is often possible to loosen the nut by slipping a long pipe over the wrench handle. The purpose of the pipe is to extend the length of the handle, so that the applied force can be located farther away from the nut. Explain why this trick works.

2. Explain (a) how it is possible for a large force to produce only a small, or even zero, torque, and (b) how it is possible for a small force to produce a large torque.

3. A magnetic tape is being played on a cassette deck. The tension in the tape applies a torque to the supply reel. Assuming the tension remains constant during playback, discuss how this torque varies as the reel becomes empty.

4. A flat rectangular sheet of plywood can rotate about an axis perpendicular to the sheet through one corner. How should a force (acting in the plane of the sheet) be applied to the plywood so as to create the largest possible torque? Give your reasoning.

5. A torque is the product of a force and a distance (lever arm). Work is also the product of a force and a distance. Yet, torque and work *are different*. What is it about the distances that makes torque and work different?

6. Suppose you are standing on a train, both feet together, facing a window. The front of the train is to your left. The train starts moving forward. To keep from falling, you slide your right foot out toward the rear of the train. Explain in terms of torque how this action keeps you from falling over.

7. Starting in the spring, fruit begins to grow on the outer end of a branch on a pear tree. Explain how the center of gravity of the pear-growing branch shifts during the course of the summer.

8. The free-body diagram in the drawing shows the forces that act on a thin rod. The three forces are drawn to scale and lie in the plane of the paper. Are these forces sufficient to keep the rod in equilibrium, or are additional forces necessary? Explain.

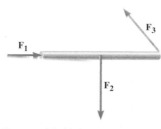

9. An A-shaped stepladder is standing on frictionless ground. The ladder consists of two sections joined at the top and kept from spreading apart by a horizontal crossbar. Draw a free-body diagram showing the forces that keep *one* section of the ladder in equilibrium.

10. The drawing shows a wine rack for a single bottle of wine that seems to defy common sense, as it balances on a table top. Treat the rack and wine bottle as a rigid body and draw the external forces that keep it in equilibrium. In particular, where must the center of gravity of the rigid body be located? Give your reasoning. (*Hint: The physics here is the same as in Figure 9.10c.*)

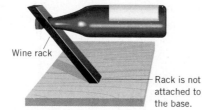

Wine rack

Rack is not attached to the base.

11. A flat triangular sheet of uniform material is shown in the drawing. There are three possible axes of rotation, each perpendicular to the sheet and passing through one corner, *A*, *B*, or *C*. For which axis is the greatest

net external torque required to bring the triangle up to an angular speed of 10.0 rad/s in 10.0 s, starting from rest? Explain, assuming that the net torque is kept constant while it is being applied.

12. An object has an angular velocity. It also has an angular acceleration due to torques that are present. Therefore, the angular velocity is changing. What happens to the angular velocity if (a) additional torques are applied so as to make the net torque suddenly equal to zero and (b) all the torques are suddenly removed?

13. The space probe in the drawing is initially moving with a constant translational velocity and zero angular velocity. (a) When the two engines are fired, each generating a thrust of magnitude *T*, will the translational velocity increase, decrease, or remain the same? Why? (b) Explain what will happen to the angular velocity.

14. Sit-ups are more difficult to do with your hands placed behind your head instead of on your stomach. Why?

15. For purposes of computing the translational kinetic energy of a rigid body, its mass can be considered as concentrated at the center of mass. If one wishes to compute the body's moment of inertia, can the mass be considered as concentrated at the center of mass? If not, why not?

16. A thin sheet of plastic is uniform and has the shape of an equilateral triangle. Consider two axes for rotation. Both are perpendicular to the plane of the triangle, axis *A* passing through the center of the triangle and axis *B* passing through one corner. If the angular speed ω about each axis is the same, for which axis does the triangle have the greater rotational kinetic energy? Explain.

17. Bob and Bill have the same weight and wear identical shoes. Keeping both feet flat on the floor and the body straight, Bob can lean over farther than Bill can before falling. Whose center of gravity is closer to the ground? Account for your answer.

18. A hoop, a solid cylinder, a spherical shell, and a solid sphere are placed at rest at the top of an incline. All the objects have the same radius. They are then released at the same time. What is the order in which they reach the bottom? Justify your answer.

19. A woman is sitting on the spinning seat of a piano stool with her arms folded. What happens to her (a) angular velocity and (b) angular momentum when she extends her arms outward? Justify your answers.

20. Review Conceptual Example 13 as an aid in answering this question. Suppose the ice cap at the South Pole melted and the water was distributed uniformly over the earth's oceans. Would the earth's angular velocity increase, decrease, or remain the same? Explain.

21. The concept behind this question is discussed in Conceptual Example 13. Many rivers, like the Mississippi River, flow from north to south toward the equator. These rivers often carry a large amount of sediment that they deposit when entering the ocean. What effect does this redistribution of the earth's soil have on the angular velocity of the earth? Why?

22. Conceptual Example 13 provides background for this question. A cloud of interstellar gas is rotating. Because the gravitational force pulls the gas particles together, the cloud shrinks, and, under the right conditions, a star may ultimately be formed. Would the angular velocity of the star be less than, equal to, or greater than, the angular velocity of the rotating gas? Justify your answer.

23. A person is hanging motionless from a vertical rope over a swimming pool. She lets go of the rope and drops straight down. After letting go, is it possible for her to curl into a ball and start spinning? Justify your answer.

24. The photograph shows a workman struggling to keep a stack of boxes balanced on a dolly. The man's right foot is on the axle of the dolly. Assuming that the boxes are identical, which one creates the greatest torque with respect to the axle? Why?

(© Paul Miller Photography)

Problems

Section 9.1 The Action of Forces and Torques on Rigid Objects

1. ssm www A square, 0.40 m on a side, is mounted so that it can rotate about an axis that passes through the center of the square. The axis is perpendicular to the plane of the square. A force of 15 N lies in this plane and is applied to the square. What is the magnitude of the maximum torque that such a force could produce?

2. You are installing a new spark plug in your car, and the manual specifies that it be tightened to a torque that has a magnitude of 45 N·m. Using the data in the drawing, determine the magnitude F of the force that you must exert on the wrench.

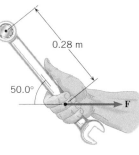

Problem 2

3. Interactive Solution 9.3 at **www.wiley.com/college/cutnell** presents a model for solving this problem. The wheel of a car has a radius of 0.350 m. The engine of the car applies a torque of 295 N·m to this wheel, which does not slip against the road surface. Since the wheel does not slip, the road must be applying a force of static friction to the wheel that produces a countertorque. Moreover, the car has a constant velocity, so this countertorque balances the applied torque. What is the magnitude of the static frictional force?

4. In San Francisco a very simple technique is used to turn around a cable car when it reaches the end of its route. The car rolls onto a turntable, which can rotate about a vertical axis through its center. Then, two people push perpendicularly on the car, one at each end, as in the drawing. The turntable is rotated one-half of a revolution to turn the car around. If the length of the car is 9.20 m and each person pushes with a 185-N force, what is the magnitude of the net torque applied to the car?

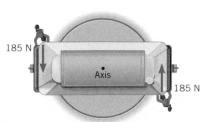

5. ssm The steering wheel of a car has a radius of 0.19 m, while the steering wheel of a truck has a radius of 0.25 m. The same force is applied in the same direction to each. What is the ratio of the torque produced by this force in the truck to the torque produced in the car?

6. The drawing shows a jet engine suspended beneath the wing of an airplane. The weight **W** of the engine is 10 200 N and acts as shown in the drawing. In flight the engine produces a thrust **T** of 62 300 N that is parallel to the ground. The rotational axis in the drawing is perpendicular to the plane of the paper. With respect to this axis, find the magnitude of the torque due to (a) the weight and (b) the thrust.

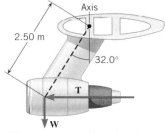

* 7. A pair of forces with equal magnitudes, opposite directions, and different lines of action is called a "couple." When a couple acts on a rigid object, the couple produces a torque that does *not* depend on the location of the axis. The drawing shows a couple acting on a tire wrench, each force being perpendicular to the wrench. Determine an expression for the torque produced by the couple when the axis is perpendicular to the tire and passes through (a) point A, (b) point B, and (c) point C. Express your answers in terms of the magnitude F of the force and the length L of the wrench.

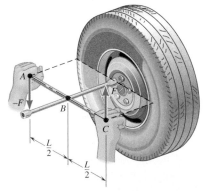

* 8. Interactive LearningWare 9.1 at **www.wiley.com/college/cutnell** reviews the concepts that are important in this problem. One end of a meter stick is pinned to a table, so the stick can rotate freely in a plane parallel to the tabletop (see the drawing). Two forces, both parallel to the tabletop, are applied to the stick in such a way that the net torque is zero. One force has a magnitude of 4.00 N and is applied perpendicular to the stick at the free end. The other force has a magnitude of 6.00 N and acts at a 60.0° angle with respect to the stick. Where along the stick is the 6.00-N force applied? Express this distance with respect to the axis of rotation.

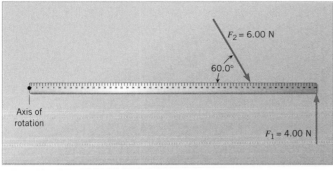

Table (top view)

* **9. ssm** One end of a meter stick is pinned to a table, so the stick can rotate freely in a plane parallel to the tabletop. Two forces, both parallel to the tabletop, are applied to the stick in such a way that the net torque is zero. One force has a magnitude of 2.00 N and is applied perpendicular to the length of the stick at the free end. The other force has a magnitude of 6.00 N and acts at a 30.0° angle with respect to the length of the stick. Where along the stick is the 6.00-N force applied? Express this distance with respect to the end that is pinned.

** **10.** A rotational axis is directed perpendicular to the plane of a square and is located as shown in the drawing. Two forces, $\mathbf{F_1}$ and $\mathbf{F_2}$, are applied to diagonally opposite corners, and act along the sides of the square, first as shown in part *a* and then as shown in part *b* of the drawing. In each case the net torque produced by the forces is zero. The square is one meter on a side, and the magnitude of $\mathbf{F_2}$ is three times that of $\mathbf{F_1}$. Find the distances a and b that locate the axis.

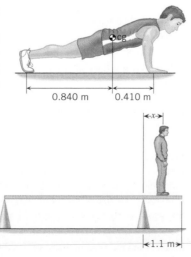

(a)   (b)

Section 9.2 Rigid Objects in Equilibrium, Section 9.3 Center of Gravity

11. The drawing shows a person whose weight is $W = 584$ N doing push-ups. Find the normal force exerted by the floor on *each* hand and *each* foot, assuming that the person holds this position.

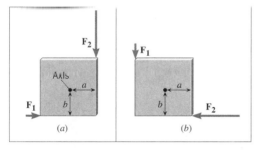

0.840 m   0.410 m

12. Review Conceptual Example 7 before starting this problem. A uniform plank of length 5.0 m and weight 225 N rests horizontally on two supports, with 1.1 m of the plank hanging over the right support (see the drawing). To what

distance *x* can a person who weighs 450 N walk on the overhanging part of the plank before it just begins to tip?

13. ssm Review Example 4 before attempting this problem. What is the minimum value for the coefficient of static friction between the ladder and the ground, so that the ladder does not slip?

14. Concept Simulation 9.1 at **www.wiley.com/college/cutnell** illustrates how the forces can vary in problems of this type. A hiker, who weighs 985 N, is strolling through the woods and crosses a small horizontal bridge. The bridge is uniform, weighs 3610 N, and rests on two concrete supports, one at each end. He stops one-fifth of the way along the bridge. What is the magnitude of the force that a concrete support exerts on the bridge (a) at the near end and (b) at the far end?

15. ssm A uniform door (0.81 m wide and 2.1 m high) weighs 140 N and is hung on two hinges that fasten the long left side of the door to a vertical wall. The hinges are 2.1 m apart. Assume that the lower hinge bears all the weight of the door. Find the magnitude and direction of the horizontal component of the force applied *to the door* by (a) the upper hinge and (b) the lower hinge. Determine the magnitude and direction of the force applied *by the door* to (c) the upper hinge and (d) the lower hinge.

16. A lunch tray is being held in one hand, as the drawing illustrates. The mass of the tray itself is 0.200 kg, and its center of gravity is located at its geometrical center. On the tray is a 1.00-kg plate of food and a 0.250-kg cup of coffee. Obtain the force **T** exerted by the thumb and the force **F** exerted by the four fingers. Both forces act perpendicular to the tray, which is being held parallel to the ground.

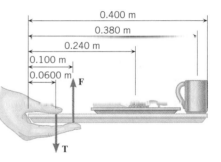

17. Interactive Solution 9.17 at **www.wiley. com/college/cutnell** illustrates how to model this type of problem. A person exerts a horizontal force of 190 N in the test apparatus shown in the drawing. Find the horizontal force **M** (magnitude and direction) that his flexor muscle exerts on his forearm.

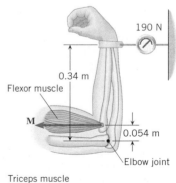

18. In an isometric exercise a person places a hand on a scale and pushes vertically downward, keeping the forearm horizontal. This is possible because the triceps muscle applies an upward force **M** perpendicular to the arm, as the drawing in-

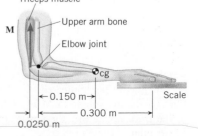

dicates. The forearm weighs 22.0 N and has a center of gravity as indicated. The scale registers 111 N. Determine the magnitude of **M**.

19. Review **Concept Stimulation 9.2** at **www.wiley.com/college/cutnell** and Conceptual Example 7 as background material for this problem. A jet transport has a weight of 1.00×10^6 N and is at rest on the runway. The two rear wheels are 15.0 m behind the front wheel, and the plane's center of gravity is 12.6 m behind the front wheel. Determine the normal force exerted by the ground on (a) the front wheel and on (b) *each* of the two rear wheels.

* **20.** The drawing shows a bicycle wheel resting against a small step whose height is $h = 0.120$ m. The weight and radius of the wheel are $W = 25.0$ N and $r = 0.340$ m. A horizontal force **F** is applied to the axle of the wheel. As the magnitude of **F** increases, there comes a time when the wheel just begins to rise up and loses contact with the ground. What is the magnitude of the force when this happens?

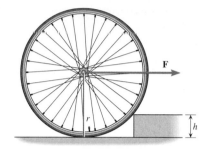

* **21. ssm** A uniform board is leaning against a smooth vertical wall. The board is at an angle θ above the horizontal ground. The coefficient of static friction between the ground and the lower end of the board is 0.650. Find the smallest value for the angle θ, such that the lower end of the board does not slide along the ground.

* **22.** A person is sitting with one leg outstretched so that it makes an angle of $30.0°$ with the horizontal, as the drawing indicates. The weight of the leg below the knee is 44.5 N with the center of gravity located below the knee joint. The leg is being held in this position because of the force **M** applied by the quadriceps muscle, which is attached 0.100 m below the knee joint (see the drawing). Obtain the magnitude of **M**.

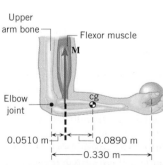

* **23. ssm** A man holds a 178-N ball in his hand, with the forearm horizontal (see the drawing). He can support the ball in this position because of the flexor muscle force **M**, which is applied perpendicular to the forearm. The forearm weighs 22.0 N and has a center of gravity as indicated.

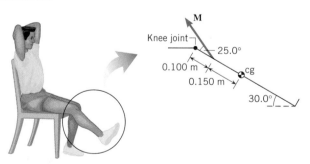

Find (a) the magnitude of **M** and (b) the magnitude and direction of the force applied by the upper arm bone to the forearm at the elbow joint.

* **24.** A woman who weighs 5.00×10^2 N is leaning against a smooth vertical wall, as the drawing shows. Find (a) the force $\mathbf{F_N}$ (directed perpendicular to the wall) exerted on her shoulders by the wall and the (b) horizontal and (c) vertical components of the force exerted on her shoes by the ground.

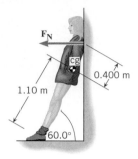

** **25. ssm www** An inverted "V" is made of uniform boards and weighs 356 N. Each side has the same length and makes a $30.0°$ angle with the vertical, as the drawing shows. Find the magnitude of the static frictional force that acts on the lower end of each leg of the "V."

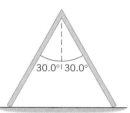

** **26.** The drawing shows an A-shaped ladder. Both sides of the ladder are equal in length. This ladder is standing on a frictionless horizontal surface, and only the crossbar (which has a negligible mass) of the "A" keeps the ladder from collapsing. The ladder is uniform and has a mass of 20.0 kg. Determine the tension in the crossbar of the ladder.

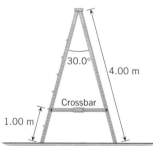

** **27.** Two vertical walls are separated by a distance of 1.5 m, as the drawing shows. Wall 1 is smooth, while wall 2 is not smooth. A uniform board is propped between them. The coefficient of static friction between the board and wall 2 is 0.98. What is the length of the longest board that can be propped between the walls?

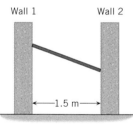

Section 9.4 Newton's Second Law for Rotational Motion About a Fixed Axis

28. A solid circular disk has a mass of 1.2 kg and a radius of 0.16 m. Each of three identical thin rods has a mass of 0.15 kg. The rods are attached perpendicularly to the plane of the disk at its outer edge to form a three-legged stool (see the drawing). Find the moment of inertia of the stool with respect to an axis that is perpendicular to the plane of the disk at its center. (*Hint: When considering the moment of inertia of each rod, note that all of the mass of each rod is located at the same perpendicular distance from the axis.*)

29. ssm A clay vase on a potter's wheel experiences an angular acceleration of 8.00 rad/s^2 due to the application of a 10.0-N · m net torque. Find the total moment of inertia of the vase and potter's wheel.

30. A ceiling fan is turned on and a net torque of 1.8 N·m is applied to the blades. The blades have a total moment of inertia of 0.22 kg·m². What is the angular acceleration of the blades?

31. A uniform solid disk with a mass of 24.3 kg and a radius of 0.314 m is free to rotate about a frictionless axle. Forces of 90.0 and 125 N are applied to the disk, as the drawing illustrates. What is (a) the net torque produced by the two forces and (b) the angular acceleration of the disk?

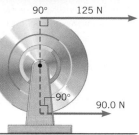

90°   125 N

90°

90.0 N

32. A rotating door is made from four rectangular glass panes, as shown in the drawing. The mass of each pane is 85 kg. A person pushes on the outer edge of one pane with a force of $F = 68$ N that is directed perpendicular to the pane. Determine the magnitude of the door's angular acceleration.

1.2 m

1.2 m

F

Problem 32

33. ssm A bicycle wheel has a radius of 0.330 m and a rim whose mass is 1.20 kg. The wheel has 50 spokes, each with a mass of 0.010 kg. (a) Calculate the moment of inertia of the rim about the axle. (b) Determine the moment of inertia of *any one spoke,* assuming it to be a long, thin rod that can rotate about one end. (c) Find the *total* moment of inertia of the wheel, including the rim and all 50 spokes.

34. A CD has a mass of 17 g and a radius of 6.0 cm. When inserted into a player, the CD starts from rest and accelerates to an angular velocity of 21 rad/s in 0.80 s. Assuming the CD is a uniform solid disk, determine the net torque acting on it.

35. Interactive Solution 9.35 at **www.wiley.com/college/cutnell** presents a method for modeling this problem. A cylinder is rotating about an axis that passes through the center of each circular end piece. The cylinder has a radius of 0.0830 m, an angular speed of 76.0 rad/s, and a moment of inertia of 0.615 kg·m². A brake shoe presses against the surface of the cylinder and applies a tangential frictional force to it. The frictional force reduces the angular speed of the cylinder by a factor of two during a time of 6.40 s. (a) Find the magnitude of the angular deceleration of the cylinder. (b) Find the magnitude of the force of friction applied by the brake shoe.

* **36.** Two thin rectangular sheets (0.20 m × 0.40 m) are identical. In the first sheet the axis of rotation lies along the 0.20-m side, and in the second it lies along the 0.40-m side. The same torque is applied to each sheet. The first sheet, starting from rest, reaches its final angular velocity in 8.0 s. How long does it take for the second sheet, starting from rest, to reach the same angular velocity?

* **37. ssm** A stationary bicycle is raised off the ground, and its front wheel ($m = 1.3$ kg) is rotating at an angular velocity of 13.1 rad/s (see the drawing). The front brake is then applied for 3.0 s, and the wheel slows down to 3.7 rad/s. Assume that all the mass of the wheel is concentrated in the rim, the radius of which is 0.33 m. The coefficient of kinetic friction between each brake pad and the rim is $\mu_k = 0.85$. What is the magnitude of the normal force that *each* brake pad applies to the rim?

Brake pads

0.33 m

* **38.** The drawing shows a model for the motion of the human forearm in throwing a dart. Because of the force **M** applied by the triceps muscle, the forearm can rotate about an axis at the elbow joint. Assume that the forearm has the dimensions shown in the drawing and a moment of inertia of 0.065 kg·m² (including the effect of the dart) relative to the axis at the elbow. Assume also that the force **M** acts perpendicular to the forearm. Ignoring the effect of gravity and any frictional forces, determine the magnitude of the force **M** needed to give the dart a tangential speed of 5.0 m/s in 0.10 s, starting from rest.

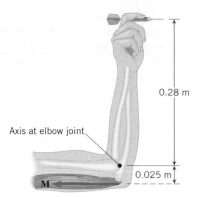

0.28 m

Axis at elbow joint

0.025 m

M

* **39. ssm** A thin, rigid, uniform rod has a mass of 2.00 kg and a length of 2.00 m. (a) Find the moment of inertia of the rod relative to an axis that is perpendicular to the rod at one end. (b) Suppose all the mass of the rod were located at a single point. Determine the perpendicular distance of this point from the axis in part (a), such that this point particle has the same moment of inertia as the rod. This distance is called the *radius of gyration* of the rod.

* **40.** The *parallel axis theorem* provides a useful way to calculate the moment of inertia I about an arbitrary axis. The theorem states that $I = I_{cm} + Mh^2$, where I_{cm} is the moment of inertia of the object relative to an axis that passes through the center of mass and is parallel to the axis of interest, M is the total mass of the object, and h is the perpendicular distance between the two axes. Use this theorem and information to determine an expression for the moment of inertia of a solid cylinder of radius R relative to an axis that lies on the surface of the cylinder and is perpendicular to the circular ends.

* **41.** The drawing shows the top view of two doors. The doors are uniform and identical. Door A rotates about an axis through its left edge, and door B rotates about an axis through the center. The same force **F** is applied perpendicular to each door at its right edge, and the force remains perpendicular as the door turns. Starting from rest, door A rotates through a certain angle in 3.00 s. How long does it take door B to rotate through the same angle?

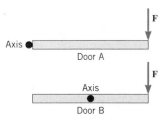

F

Axis

Door A

F

Axis

Door B

** **42.** See **Concept Simulation 9.4** at **www.wiley.com/college/cutnell** to review the principles involved in this problem. By means of a rope whose mass is negligible, two blocks are suspended over a pulley, as the drawing shows. The pulley can be treated as a uniform solid cylindrical disk. The downward acceleration of the 44.0-kg block is observed to be exactly one-half the acceleration due to gravity. Noting that the tension in the rope is not the same on each side of the pulley, find the mass of the pulley.

11.0 kg

44.0 kg

Section 9.5 Rotational Work and Energy

43. ssm Three objects lie in the *x, y* plane. Each rotates about the *z* axis with an angular speed of 6.00 rad/s. The mass *m* of each object and its perpendicular distance *r* from the *z* axis are: (1) $m_1 = 6.00$ kg and $r_1 = 2.00$ m, (2) $m_2 = 4.00$ kg and $r_2 = 1.50$ m,

(3) $m_3 = 3.00$ kg and $r_3 = 3.00$ m. (a) Find the tangential speed of each object. (b) Determine the total kinetic energy of this system using the expression KE $= \frac{1}{2}m_1v_1^2 + \frac{1}{2}m_2v_2^2 + \frac{1}{2}m_3v_3^2$. (c) Obtain the moment of inertia of the system. (d) Find the rotational kinetic energy of the system using the relation $\frac{1}{2}I\omega^2$ to verify that the answer is the same as that in (b).

44. Calculate the kinetic energy that the earth has because of (a) its rotation about its own axis and (b) its motion around the sun. Assume that the earth is a uniform sphere and that its path around the sun is circular. For comparison, the total energy used in the United States in one year is about 9.3×10^{19} J.

45. ssm A flywheel is a solid disk that rotates about an axis that is perpendicular to the disk at its center. Rotating flywheels provide a means for storing energy in the form of rotational kinetic energy and are being considered as a possible alternative to batteries in electric cars. The gasoline burned in a 300-mile trip in a typical midsize car produces about 1.2×10^9 J of energy. How fast would a 13-kg flywheel with a radius of 0.30 m have to rotate to store this much energy? Give your answer in rev/min.

46. A helicopter has two blades (see Figure 8.14), each of which has a mass of 240 kg and can be approximated as a thin rod of length 6.7 m. The blades are rotating at an angular speed of 44 rad/s. (a) What is the total moment of inertia of the two blades about the axis of rotation? (b) Determine the rotational kinetic energy of the spinning blades.

* 47. Interactive Solution 9.47 at **www.wiley.com/college/cutnell** offers a model for solving problems of this type. A solid sphere is rolling on a surface. What fraction of its total kinetic energy is in the form of rotational kinetic energy about the center of mass?

* 48. A thin uniform rod is initially positioned in the vertical direction, with its lower end attached to a frictionless axis that is mounted on the floor. The rod has a length of 2.00 m and is allowed to fall, starting from rest. Find the tangential speed of the free end of the rod, just before the rod hits the floor after rotating through 90°.

* 49. ssm Review Example 12 before attempting this problem. A solid cylinder and a thin-walled hollow cylinder (see Table 9.1) have the same mass and radius. They are rolling horizontally toward the bottom of an incline. The center of mass of each has the same translational speed. The cylinders roll up the incline and reach their highest points. Calculate the ratio of the distances ($s_{\text{solid}}/s_{\text{hollow}}$) along the incline through which each center of mass moves.

* 50. Review Example 12 before attempting this problem. A bowling ball encounters a 0.760-m vertical rise on the way back to the ball rack, as the drawing illustrates. Ignore frictional losses and assume that the mass of the ball is distributed uniformly. The translational speed of the ball is 3.50 m/s at the bottom of the rise. Find the translational speed at the top.

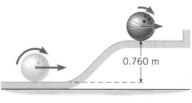

0.760 m

** 51. A tennis ball, starting from rest, rolls down the hill in the drawing. At the end of the hill the ball becomes airborne, leaving at an angle of 35° with respect to the ground. Treat the ball as a thin-walled spherical shell, and determine the range x.

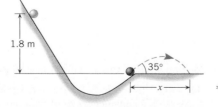

1.8 m

35°

x

Section 9.6 Angular Momentum

52. Before starting this problem, review Conceptual Example 13. A woman stands at the center of a platform. The woman and the platform rotate with an angular speed of 5.00 rad/s. Friction is negligible. Her arms are outstretched, and she is holding a dumbbell in each hand. In this position the total moment of inertia of the rotating system (platform, woman, and dumbbells) is 5.40 kg·m². By pulling in her arms, she reduces the moment of inertia to 3.80 kg·m². Find her new angular speed.

53. ssm www Two disks are rotating about the same axis. Disk A has a moment of inertia of 3.4 kg·m² and an angular velocity of +7.2 rad/s. Disk B is rotating with an angular velocity of −9.8 rad/s. The two disks are then linked together without the aid of any external torques, so that they rotate as a single unit with an angular velocity of −2.4 rad/s. The axis of rotation for this unit is the same as that for the separate disks. What is the moment of inertia of disk B?

54. A solid disk rotates at an angular velocity of 0.067 rad/s with respect to an axis perpendicular to the disk at its center. The moment of inertia of the disk is 0.10 kg·m². From above, sand is dropped straight down onto this rotating disk, so that a thin uniform ring of sand is formed at a distance of 0.40 m from the axis. The sand in the ring has a mass of 0.50 kg. After all the sand is in place, what is the angular velocity of the disk?

55. Interactive Solution 9.55 at **www.wiley.com/college/cutnell** illustrates one way of solving a problem similar to this one. A thin rod has a length of 0.25 m and rotates in a circle on a frictionless tabletop. The axis is perpendicular to the length of the rod at one of its ends. The rod has an angular velocity of 0.32 rad/s and a moment of inertia of 1.1×10^{-3} kg·m². A bug standing on the axis decides to crawl out to the other end of the rod. When the bug (mass $= 4.2 \times 10^{-3}$ kg) gets where it's going, what is the angular velocity of the rod?

56. A flat uniform circular disk (radius = 2.00 m, mass = 1.00×10^2 kg) is initially stationary. The disk is free to rotate in the horizontal plane about a frictionless axis perpendicular to the center of the disk. A 40.0-kg person, standing 1.25 m from the axis, begins to run on the disk in a circular path and has a tangential speed of 2.00 m/s relative to the ground. Find the resulting angular speed of the disk (in rad/s) and describe the direction of the rotation.

* 57. ssm A cylindrically shaped space station is rotating about the axis of the cylinder to create artificial gravity. The radius of the cylinder is 82.5 m. The moment of inertia of the station without people is 3.00×10^9 kg·m². Suppose 500 people, with an average mass of 70.0 kg each, live on this station. As they move radially from the outer surface of the cylinder toward the axis, the angular speed of the station changes. What is the maximum possible percentage change in the station's angular speed due to the radial movement of the people?

* 58. A thin uniform rod is rotating at an angular velocity of 7.0 rad/s about an axis that is perpendicular to the rod at its center. As the drawing indicates, the rod is hinged at two places, one-quarter of the length from each end. Without the aid of external torques, the rod suddenly assumes a "u" shape, with the arms of the "u" parallel to the rotation axis. What is the angular velocity of the rotating "u"?

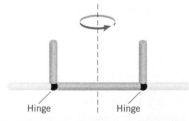

Hinge   Hinge

** 59. A platform is rotating at an angular speed of 2.2 rad/s. A block is resting on this platform at a distance of 0.30 m from the axis. The coefficient of static friction between the block and the platform is 0.75. Without any external torque acting on the system, the block is moved toward the axis. Ignore the moment of inertia of the platform and determine the smallest distance from the axis at which the block can be relocated and still remain in place as the platform rotates.

** 60. A small 0.500-kg object moves on a frictionless horizontal table in a circular path of radius 1.00 m. The angular speed is 6.28 rad/s. The object is attached to a string of negligible mass that passes through a small hole in the table at the center of the circle. Someone under the table begins to pull the string downward to make the circle smaller. If the string will tolerate a tension of no more than 105 N, what is the radius of the smallest possible circle on which the object can move?

Additional Problems

61. ssm A post is driven perpendicularly into the ground and serves as the axis about which a gate rotates. A force of 12 N is applied perpendicular to the gate and acts parallel to the ground. How far from the post should the force be applied to produce a torque with a magnitude of 3.0 N · m?

62. The circular blade on a radial arm saw is turning at 262 rad/s at the instant the motor is turned off. In 18.0 s the speed of the blade is reduced to 85 rad/s. Assume the blade to be a uniform solid disk of radius 0.130 m and mass 0.400 kg. Find the net torque applied to the blade.

63. A baggage carousel at an airport is rotating with an angular speed of 0.20 rad/s when the baggage begins to be loaded onto it. The moment of inertia of the carousel is 1500 kg · m². Ten pieces of baggage with an average mass of 15 kg each are dropped vertically onto the carousel and come to rest at a perpendicular distance of 2.0 m from the axis of rotation. (a) Assuming that no net external torque acts on the system of carousel and baggage, find the final angular speed. (b) In reality, the angular speed of a baggage carousel does not change. Therefore, what can you say qualitatively about the net external torque acting on the system?

64. A person is standing on a level floor. His head, upper torso, arms, and hands together weigh 438 N and have a center of gravity that is 1.28 m above the floor. His upper legs weigh 144 N and have a center of gravity that is 0.760 m above the floor. Finally, his lower legs and feet together weigh 87 N and have a center of gravity that is 0.250 m above the floor. Relative to the floor, find the location of the center of gravity for the entire body.

65. ssm A particle is located at each corner of an imaginary cube. Each edge of the cube is 0.25 m long, and each particle has a mass of 0.12 kg. What is the moment of inertia of these particles with respect to an axis that lies along one edge of the cube?

* 66. A 1220-N uniform beam is attached to a vertical wall at one end and is supported by a cable at the other end. A 1960-N crate hangs from the far end of the beam. Using the data shown in the drawing, find (a) the magnitude of the tension in the wire and (b) the magnitude of the horizontal and vertical components of the force that the wall exerts on the left end of the beam.

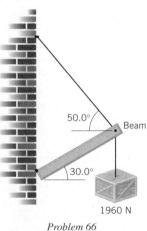

50.0° Beam

30.0°

1960 N

Problem 66

* 67. ssm www A massless, rigid board is placed across two bathroom scales that are separated by a distance of 2.00 m. A person lies on the board. The scale under his head reads 425 N, and the scale under his feet reads 315 N. (a) Find the weight of the person. (b) Locate the center of gravity of the person relative to the scale beneath his head.

* 68. A wrecking ball (weight = 4800 N) is supported by a boom, which may be assumed to be uniform and has a weight of 3600 N. As the drawing shows, a support cable runs from the top of the boom to the tractor. The angle between the support cable and the horizontal is 32°, and the angle between the boom and the horizontal is 48°. Find (a) the tension in the support cable and (b) the magnitude of the force exerted on the lower end of the boom by the hinge at point *P*.

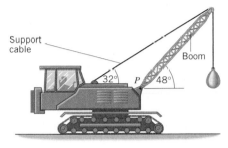

Support cable

Boom

32° *P* 48°

* 69. In outer space two space modules are joined together by a massless cable. These modules are rotating about their center of mass, which is at the center of the cable, because the modules are identical (see the drawing). In each module, the cable is connected to a motor, so that the modules can pull each other together. The initial tangential speed of each module is $v_0 = 17$ m/s. Then they pull together until the distance between them is reduced by a factor of two. Determine the final tangential speed v_i for each module.

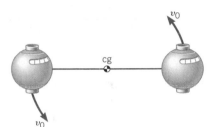

v_0

cg

v_0

* 70. Review Example 12 before attempting this problem. A marble and a cube are placed at the top of a ramp. Starting from rest at the same height, the marble rolls and the cube slides (no kinetic friction) down the ramp. Determine the ratio of the center-of-mass speed of the cube to the center-of-mass speed of the marble at the bottom of the ramp.

** 71. The drawing shows an inverted "A" that is suspended from the ceiling by two vertical ropes. Each leg of the "A" has a length of 2*L* and a weight of 120 N. The horizontal crossbar has a negligible weight. Find the force that the crossbar applies to each leg.

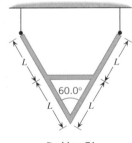

L *L*

60.0°

L *L*

Problem 71

** 72. The crane shown in the drawing is lifting a 180-kg crate upward with an acceleration of 1.2 m/s². The cable from the crate passes over a solid cylindrical pulley at the top of the boom. The pulley has a mass of 130 kg. The cable is then wound onto a hollow cylindrical drum that is mounted on the deck of the crane. The mass of the drum is 150 kg, and its radius is 0.76 m. The engine applies a counterclockwise torque to the drum in order to wind up the cable. What is the magnitude of this torque? Ignore the mass of the cable.

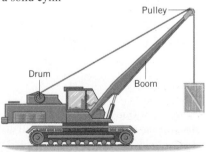

Pulley

Drum

Boom

Concepts & Calculations *Group Learning Problems*

Note: Each of these problems consists of Concept Questions followed by a related quantitative Problem. They are designed for use by students working alone or in small learning groups. The Concept Questions involve little or no mathematics and are intended to stimulate group discussions. They focus on the concepts with which the problems deal. Recognizing the concepts is the essential initial step in any problem-solving technique.

73. Concept Questions The drawing shows a rectangular piece of wood. The forces applied to corners *B* and *D* have the same magnitude and are directed parallel to the long and short sides of the rectangle. An axis of rotation is shown perpendicular to the

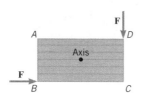

plane of the rectangle at its center. (a) Relative to this axis, which force produces the torque with the greater magnitude? (b) A force is to be applied to corner *A*, directed along the short side of the rectangle. The net torque produced by the three forces is zero. How is the force at corner *A* directed—from *A* toward *B* or away from *B*? Give your reasoning.

Problem The magnitudes of the forces at corners *B* and *D* are each 12 N. The long side of the rectangle is twice as long as the short side. Find the magnitude and direction of the force applied to corner *A*.

74. Concept Questions The wheels, axle, and handles of a wheelbarrow weigh 60.0 N. The load chamber and its contents weigh 525 N. The drawing shows these two forces in two different wheelbarrow designs. To support the wheelbarrow in equilibrium, the man's hands apply a force **F** to the handles that is directed vertically upward. Consider a rotational axis at the point where the tire contacts the ground, directed perpendicular to the plane of the paper. (a) For which design is there a greater total torque from the 60.0- and 525-N forces? (b) For which design does the torque from the man's force need to be greater? (c) For which design does the man's force need to be greater? Account for your answers.

Problem Find the magnitude of the man's force for both designs. Be sure that your answers are consistent with your answers to the Concept Questions.

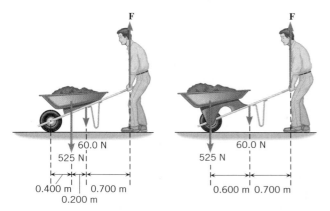

75. Concept Questions Part *a* of the drawing shows a uniform horizontal beam bolted to a vertical wall and supported from below at an angle *θ* by a brace that is attached to a pin. The forces that keep the beam in equilibrium are its weight **W** and the forces applied to the beam by the bolt and the brace. (a) Part *b* of the drawing shows the weight **W** and the force **P** from the brace, while assuming

that the bolt applies a force **V** that is vertical. Can these three forces keep the beam in equilibrium? (b) Part *c* of the drawing is like part *b*, except that it assumes the bolt applies a force **H** that is horizontal rather than vertical. Can the three forces in part *c* keep the beam in equilibrium? (c) Part *d* of the

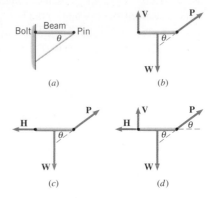

drawing assumes that the bolt applies a force that has both a vertical component **V** and a horizontal component **H**. Can the forces in part *d* keep the beam in equilibrium? Justify your answers.

Problem The brace makes an angle *θ* = 39° with respect to the beam, which has a weight of 340 N. Find the magnitudes of the forces **V**, **H**, and **P**.

76. Concept Questions Two wheels have the same mass and radius. One has the shape of a hoop and the other the shape of a solid disk. Each wheel starts from rest and has a constant angular acceleration with respect to a rotational axis that is perpendicular to the plane of the wheel at its center. Each makes the same number of revolutions in the same time. (a) Which wheel, if either, has the greater angular acceleration? (b) Which, if either, has the greater moment of inertia? (c) To which wheel, if either, is a greater net external torque applied? Explain your answers.

Problem The mass and radius of each wheel are 4.0 kg and 0.35 m, respectively. Each wheel starts from rest and turns through an angle of 13 rad in 8.0 s. Find the net external torque that acts on each wheel. Check to see that your answers are consistent with your answers to the Concept Questions.

77. Concept Questions Two thin rods of length *L* are rotating with the same angular speed *ω* (in rad/s) about axes that pass perpendicularly through one end. Rod A is massless but has a particle of mass 0.66 kg attached to its free end. Rod B has a mass 0.66 kg, which is distributed uniformly along its length. (a) Which has the greater moment of inertia—rod A with its attached particle or rod B? (b) Which has the greater kinetic energy according to Equation 9.9 ($KE_R = \frac{1}{2}I\omega^2$)? Account for your answers.

Problem The length of each rod is 0.75 m, and the angular speed is 4.2 rad/s. Find the kinetic energies of rod A with its attached particle and of rod B. Make sure your answers are consistent with your answers to the Concept Questions.

78. Concept Questions As seen from above, a playground carousel is rotating counterclockwise about its center on frictionless bearings. A person standing still on the ground grabs onto one of the bars on the carousel very close to its outer edge and climbs aboard. Thus, this person begins with an angular speed of zero and ends up with a nonzero angular speed, which means that he underwent a counterclockwise angular acceleration. (a) What applies the force to the person to create the torque causing this acceleration? What is the direction of this force? (b) According to Newton's action–reaction law, what can you say about the direction of the force applied to the carousel by the person and about the nature (clockwise or counterclockwise) of the torque that it creates? (c) Does the torque identified in part (b) increase or decrease the angular speed of the carousel?

Problem The carousel has a radius of 1.50 m, an initial angular speed of 3.14 rad/s, and a moment of inertia of 125 kg·m². The mass of the person is 40 kg. Find the final angular speed of the carousel after the person climbs aboard. Verify that your answer is consistent with your answers to the Concept Questions.

* **79. Concept Questions** The drawing shows two identical systems of objects; each consists of three small balls (masses m_1, m_2, and m_3) connected by massless rods. In both systems the axis is perpendicular to the page, but it is located at a different place, as shown. (a) Although the balls in the two systems have identical masses, do the systems necessarily have the same moments of inertia? If not, why not? (b) The same force of magnitude F is applied to the same ball in each system (see the drawing). Are the torques produced by this force the same? Explain. (c) The two systems start from rest. Will they have the same or different angular velocities at the same later time? Justify your answer.

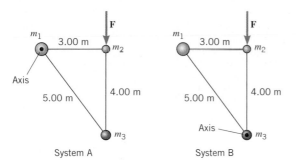

System A   System B

Problem The masses of the balls are $m_1 = 9.00$ kg, $m_2 = 6.00$ kg, and $m_3 = 7.00$ kg. The magnitude of the force is $F = 424$ N. (a) For each of the two systems, determine the moment of inertia about the given axis of rotation. (b) Calculate the torque (magnitude and direction) acting on each system. (c) Both systems start from rest, and the direction of the force moves with the system and always points along the 4.00-m rod. What is the angular velocity of each system after 5.00 s?

* **80. Concept Questions** Two identical wheels are moving on horizontal surfaces. The center of mass of each has the same linear speed. However, one wheel is rolling, while the other is sliding (without rolling) on a frictionless surface. Each wheel then encounters an incline plane. One continues to roll up the incline, while the other continues to slide up. Eventually they come to a momentary halt, because the gravitational force slows them down. (a) As the wheels move on the horizontal surfaces, which, if either, has the greater total kinetic energy? Why? (b) When they come to a momentary halt on the incline, which, if either, has the greater potential energy? Justify your answer. (c) Which wheel, if either, rises to the greater height? Explain. (d) If the mass of the rolling wheel were reduced to one-half that of the sliding wheel, would your conclusion to part (c) still be valid? Give your reasoning.

Problem Each wheel is a disk of mass 2.0 kg. On the horizontal surfaces the center of mass of each moves with a linear speed of 6.0 m/s. (a) What is the total kinetic energy of each wheel? (b) Determine the maximum height reached by each wheel as it moves up the incline. (c) What would be the maximum height reached by the rolling wheel if its mass were halved?

Chapter 10

Simple Harmonic Motion and Elasticity

In this pole-vaulting event the pole is bent sharply into an arc. Because it is made from an elastic material, the pole returns to its original shape; however, pulling the vaulter upward in the process. As this chapter discusses, the elastic behavior of materials leads to an important kind of vibratory motion, known as simple harmonic motion. (© Chris Cole/Stone/Getty Images)

10.1 The Ideal Spring and Simple Harmonic Motion

Springs are familiar objects that have many applications, ranging from push-button switches on electronic components, to automobile suspension systems, to mattresses. They are useful because they can be stretched or compressed. For example, the top drawing in Figure 10.1 shows a spring being stretched. Here a hand applies a pulling force $F_{Applied}$ to the spring. In response, the spring stretches and undergoes a displacement of x from its original, or "unstrained," length. The bottom drawing in Figure 10.1 illustrates the spring being compressed. Now the hand applies a pushing force to the spring, and it again undergoes a displacement from its unstrained length.

Experiment reveals that for relatively small displacements, the force $F_{Applied}$ required to stretch or compress a spring is directly proportional to the displacement x, or $F_{Applied} \propto x$. As is customary, this proportionality may be converted into an equation by introducing a proportionality constant k:

$$F_{Applied} = kx \qquad (10.1)$$

The constant k is called the **spring constant,** and Equation 10.1 shows that it has the dimensions of force per unit length (N/m). A spring that behaves according to $F_{Applied} = kx$ is said to be an **ideal spring.** Example 1 illustrates one application of such a spring.

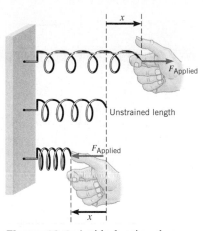

Figure 10.1 An ideal spring obeys the equation $F_{Applied} = kx$, where $F_{Applied}$ is the force applied to the spring, x is the displacement of the spring from its unstrained length, and k is the spring constant.

Example 1   A Tire Pressure Gauge

In a tire pressure gauge, the air in the tire pushes against a plunger attached to a spring when the gauge is pressed against the tire valve, as in Figure 10.2. Suppose the spring constant of the spring is $k = 320$ N/m and the bar indicator of the gauge extends 2.0 cm when the gauge is pressed against the tire valve. What force does the air in the tire apply to the spring?

The physics of a tire pressure gauge.

Reasoning We assume that the spring is an ideal spring, so that the relation $F_{Applied} = kx$ is obeyed. The spring constant k is known, as is the displacement x. Therefore, we can determine the force applied to the spring.

Solution The force needed to compress the spring is given by Equation 10.1 as

$$F_{Applied} = kx = (320 \text{ N/m})(0.020 \text{ m}) = \boxed{6.4 \text{ N}}$$

Thus, the exposed length of the bar indicator indicates the force that the air pressure in the tire exerts on the spring. We will see later that pressure is force per unit area, so force is pressure times area. Since the area of the plunger surface is fixed, the bar indicator can be marked in units of pressure.

Sometimes the spring constant k is referred to as the **stiffness** of the spring, because a large value for k means the spring is "stiff," in the sense that a large force is required to stretch or compress it. Conceptual Example 2 examines what happens to the stiffness of a spring when the spring is cut into two shorter pieces.

Conceptual Example 2   Are Shorter Springs Stiffer Springs?

Figure 10.3a shows a 10-coil spring that has a spring constant k. If this spring is cut in half, so there are two 5-coil springs, what is the spring constant of each of the smaller springs?

Reasoning and Solution It is tempting to say that each 5-coil spring has a spring constant of $\frac{1}{2}k$, or one-half the value for the 10-coil spring. However, the spring constant of each 5-coil spring is, in fact, $2k$. Here's the logic. When a force $F_{Applied}$ is applied to the spring, the displacement of the spring from its unstrained length is x, as Figure 10.3a illustrates. Now consider part b of the drawing, which shows the spring divided in half between the fifth and sixth coil (counting from the right). The spring is in equilibrium, so that the net force acting on the right half (coils 1–5) must be zero. Thus, as part b shows, a force of $-F_{Applied}$ must act on coil 5 in order to balance the force $F_{Applied}$ that acts on coil 1. It is the adjacent coil 6 that exerts the force $-F_{Applied}$ on coil 5, and Newton's action–reaction law now comes into play. It tells us that coil 5, in response, exerts an oppositely directed force of equal magnitude on coil 6. In

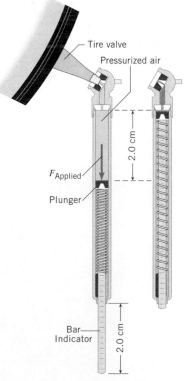

Figure 10.2 In a tire pressure gauge, the pressurized air from the tire exerts a force $F_{Applied}$ that compresses a spring.

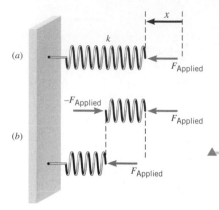

Figure 10.3 (*a*) The 10-coil spring has a spring constant k. The applied force is F_{Applied}, and the displacement of the spring from its unstrained length is x. (*b*) The spring in part *a* is divided in half, so that the forces acting on the two 5-coil springs can be analyzed.

other words, the force F_{Applied} is also exerted on the left half of the spring, as part *b* also indicates. As a result, the left half compresses by an amount that is one-half the displacement x experienced by the 10-coil spring. In summary, we have a 10-coil spring for which the force F_{Applied} causes a displacement x and a 5-coil spring for which the same force causes a displacement $\frac{1}{2}x$. We conclude, then, that the 5-coil spring must be twice as stiff as the 10-coil spring. When the 10-coil spring is cut in half, the spring constant of each resulting 5-coil spring is $2k$. In general, the spring constant is inversely proportional to the number of coils in the spring, so ***shorter springs are stiffer springs,*** all other things being equal.

Related Homework: *Problems 8, 16*

To stretch or compress a spring, a force must be applied to it. In accord with Newton's third law, the spring exerts an oppositely directed force of equal magnitude. This reaction force is applied by the spring to the agent that does the pulling or pushing. In other words, the reaction force is applied to the object attached to the spring. The reaction force is also called a "restoring force," for a reason that will be clarified shortly. The restoring force of an ideal spring is obtained from the relation $F_{\text{Applied}} = kx$ by including the minus sign required by Newton's action–reaction law, as indicated in Equation 10.2.

■ **HOOKE'S LAW* RESTORING FORCE OF AN IDEAL SPRING**

The restoring force of an ideal spring is

$$F = -kx \tag{10.2}$$

where k is the spring constant and x is the displacement of the spring from its unstrained length. The minus sign indicates that the restoring force always points in a direction opposite to the displacement of the spring.

Figure 10.4 **CONCEPTS AT A GLANCE**
The restoring force of a spring may contribute to the net external force $\Sigma\mathbf{F}$ that acts on an object. According to Newton's second law, the resulting acceleration **a** is directly proportional to the net force. The springs on his feet, his weight, and the normal force from the ground provide the net force that causes this fellow's bouncing motion. (© Windsor & Wiehahn/Stone/Getty Images)

▶ **CONCEPTS AT A GLANCE** In Chapter 4 we encountered four types of forces: the gravitational force, the normal force, frictional forces, and the tension force. The Concepts-at-a-Glance chart in Figure 4.9 illustrates that these forces can contribute to the net external force, which Newton's second law relates to the mass and acceleration of an object. The restoring force of a spring can also contribute to the net external force, as the Concepts-at-a-Glance chart in Figure 10.4 indicates. Except for this force, this chart is identical to that in Figure 4.9. Once again, we see the unifying theme of Newton's second law, in that individual forces contribute to the net force, which, in turn, is responsible for the acceleration. Newton's second law plays a central role in describing the motion of objects attached to springs. ◀

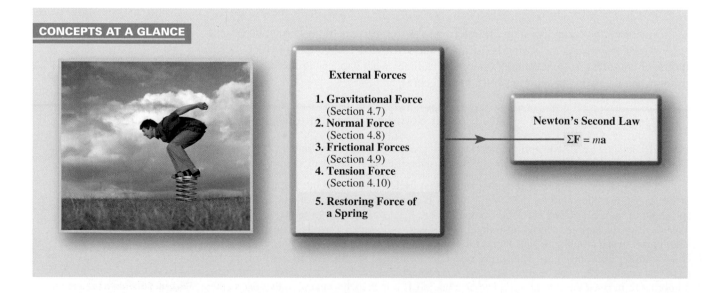

CONCEPTS AT A GLANCE

External Forces

1. **Gravitational Force** (Section 4.7)
2. **Normal Force** (Section 4.8)
3. **Frictional Forces** (Section 4.9)
4. **Tension Force** (Section 4.10)
5. **Restoring Force of a Spring**

Newton's Second Law
$\Sigma\mathbf{F} = m\mathbf{a}$

* As we will see in Section 10.8, Equation 10.2 is similar to a relationship first discovered by Robert Hooke (1635–1703).

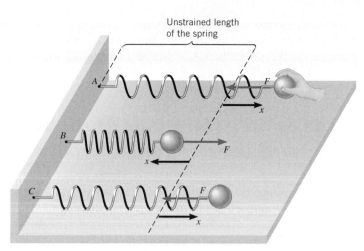

Unstrained length
of the spring

Figure 10.5 The restoring force (see blue arrows) produced by an ideal spring always points opposite to the displacement (see black arrows) of the spring and leads to a back-and-forth motion of the object.

Figure 10.5 helps to explain why the phrase "restoring force" is used. In the picture, an object of mass m is attached to a spring on a frictionless table. In part A, the spring has been stretched to the right, so it exerts the leftward-pointing force F. When the object is released, this force pulls it to the left, restoring it toward its equilibrium position. However, consistent with Newton's first law, the moving object has inertia and coasts beyond the equilibrium position, compressing the spring as in part B. The force exerted by the spring now points to the right and, after bringing the object to a momentary halt, acts to restore the object to its equilibrium position. But the object's inertia again carries it beyond the equilibrium position, this time stretching the spring and leading to the restoring force F shown in part C. The back-and-forth motion illustrated in the drawing then repeats itself, continuing forever, since no friction acts on the object or the spring.

When the restoring force has the mathematical form given by $F = -kx$, the type of friction-free motion illustrated in Figure 10.5 is designated as "simple harmonic motion." By attaching a pen to the object and moving a strip of paper past it at a steady rate, we can record the position of the vibrating object as time passes. Figure 10.6 illustrates the resulting graphical record of simple harmonic motion. The maximum excursion from equilibrium is the ***amplitude*** A of the motion. The shape of this graph is characteristic of simple harmonic motion and is called "sinusoidal," because it has the shape of a trigonometric sine or cosine function.

The restoring force also leads to simple harmonic motion when the object is attached to a vertical spring, just as it does when the spring is horizontal. When the spring is vertical, however, the weight of the object causes the spring to stretch, and the motion occurs

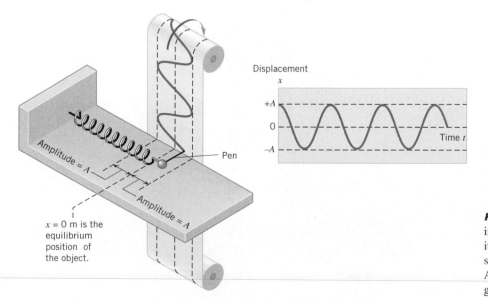

Amplitude = A

Amplitude = A

$x = 0$ m is the equilibrium position of the object.

Pen

Displacement
x

$+A$

0

$-A$

Time t

Figure 10.6 When an object moves in simple harmonic motion, a graph of its position as a function of time has a sinusoidal shape with an amplitude A. A pen attached to the object records the graph.

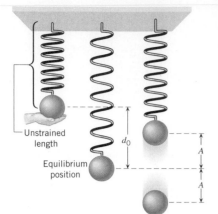

Figure 10.7 The weight of an object on a vertical spring stretches the spring by an amount d_0. Simple harmonic motion of amplitude A occurs with respect to the equilibrium position of the object on the stretched spring.

with respect to the equilibrium position of the object on the stretched spring, as Figure 10.7 indicates. The amount of initial stretching d_0 due to the weight can be calculated by equating the weight to the magnitude of the restoring force that supports it; thus, $mg = kd_0$, which gives $d_0 = mg/k$.

10.2 Simple Harmonic Motion and the Reference Circle

Simple harmonic motion, like any motion, can be described in terms of displacement, velocity, and acceleration, and the model in Figure 10.8 is helpful in explaining these characteristics. The model consists of a small ball attached to the top of a rotating turntable. The ball is moving in uniform circular motion (see Section 5.1) on a path known as the *reference circle.* As the ball moves, its shadow falls on a strip of film, which is moving upward at a steady rate and records where the shadow is. A comparison of the film with the paper in Figure 10.6 reveals the same kind of patterns, suggesting that the shadow of the ball is a good model for simple harmonic motion.

DISPLACEMENT

Figure 10.9 takes a closer look at the reference circle (radius $= A$) and indicates how to determine the displacement of the shadow on the film. The ball starts on the x axis at $x = +A$ and moves through the angle θ in a time t. Since the circular motion is uniform, the ball moves with a constant angular speed ω (in rad/s). Therefore, the angle has a value (in rad) of $\theta = \omega t$. The displacement x of the shadow is just the projection of the radius A onto the x axis:

$$x = A \cos \theta = A \cos \omega t \qquad (10.3)$$

Figure 10.10 shows a graph of this equation. As time passes, the shadow of the ball oscillates between the values of $x = +A$ and $x = -A$, corresponding to the limiting values of $+1$ and -1 for the cosine of an angle. The radius A of the reference circle, then, is the amplitude of the simple harmonic motion.

As the ball moves one revolution or cycle around the reference circle, its shadow executes one cycle of back-and-forth motion. For any object in simple harmonic motion, the time required to complete one cycle is the *period T,* as Figure 10.10 indicates. The value of T depends on the angular speed ω of the ball because the greater the angular speed, the shorter the time it takes to complete one revolution. We can obtain the relationship be-

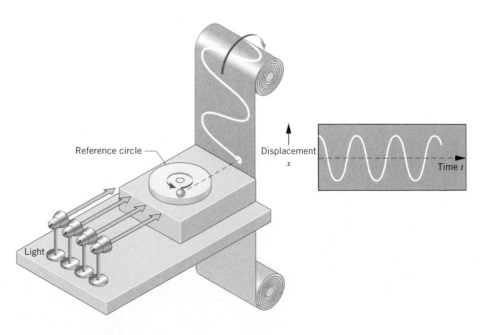

Figure 10.8 The ball mounted on the turntable moves in uniform circular motion, and its shadow, projected on a moving strip of film, executes simple harmonic motion.

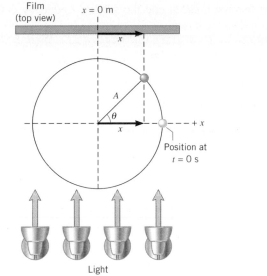

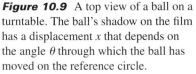

Figure 10.9 A top view of a ball on a turntable. The ball's shadow on the film has a displacement *x* that depends on the angle *θ* through which the ball has moved on the reference circle.

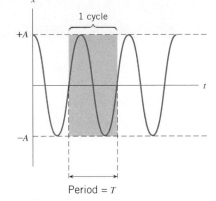

Figure 10.10 For simple harmonic motion, the graph of displacement *x* versus time *t* is a sinusoidal curve. The period *T* is the time required for one complete motional cycle.

tween *ω* and *T* by recalling that $\omega = \Delta\theta/\Delta t$ (Equation 8.2), where $\Delta\theta$ is the angular displacement of the ball and Δt is the time. For one cycle, $\Delta\theta = 2\pi$ rad and $\Delta t = T$, so that

$$\omega = \frac{2\pi}{T} \qquad (\omega \text{ in rad/s}) \qquad (10.4)$$

Often, instead of the period, it is more convenient to speak of the ***frequency f*** of the motion, the frequency being just the number of cycles of the motion per second. For example, if an object on a spring completes 10 cycles in one second, the frequency is $f = 10$ cycles/s. The period *T*, or the time for one cycle, would be $\frac{1}{10}$ s. Thus, frequency and period are related according to

$$f = \frac{1}{T} \qquad (10.5)$$

Usually one cycle per second is referred to as one hertz (Hz), the unit being named after Heinrich Hertz (1857–1894). One thousand cycles per second is called one kilohertz (kHz). Thus, five thousand cycles per second, for instance, can be written as 5 kHz.

Using the relationships $\omega = 2\pi/T$ and $f = 1/T$, we can relate the angular speed *ω* (in rad/s) to the frequency *f* (in cycles/s or Hz):

$$\omega = \frac{2\pi}{T} = 2\pi f \qquad (\omega \text{ in rad/s}) \qquad (10.6)$$

Because *ω* is directly proportional to the frequency *f*, *ω* is often called the ***angular frequency.***

VELOCITY

The reference circle model can also be used to determine the velocity of an object in simple harmonic motion. Figure 10.11 shows the tangential velocity $\mathbf{v}_T$ of the ball on the reference circle. The drawing indicates that the velocity **v** of the shadow is just the *x* component of the vector $\mathbf{v}_T$; that is, $v = -v_T \sin\theta$, where $\theta = \omega t$. The minus sign is necessary, since **v** points to the left, in the direction of the negative *x* axis. Since the tangential speed v_T is related to the angular speed *ω* by $v_T = r\omega$ (Equation 8.9) and since $r = A$, it follows that $v_T = A\omega$. Therefore, the velocity in simple harmonic motion is given by

$$v = -A\omega \sin\theta = -A\omega \sin\omega t \qquad (\omega \text{ in rad/s}) \qquad (10.7)$$

This velocity is *not* constant, but varies between maximum and minimum values as time passes. When the shadow changes direction at either end of the oscillatory motion, the velocity is momentarily zero. When the shadow passes through the $x = 0$ m position, the velocity has a maximum magnitude of $A\omega$, since the sine of an angle is between $+1$ and -1:

$$v_{max} = A\omega \qquad (\omega \text{ in rad/s}) \qquad (10.8)$$

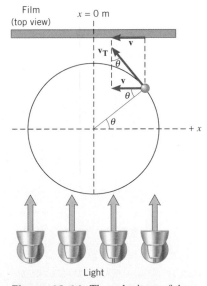

Figure 10.11 The velocity **v** of the ball's shadow is the *x* component of the tangential velocity $\mathbf{v}_T$ of the ball on the reference circle.

Both the amplitude A and the angular frequency ω determine the maximum velocity, as Example 3 emphasizes.

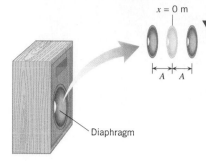

Figure 10.12 The diaphragm of a loudspeaker generates a sound by moving back and forth in simple harmonic motion.

Example 3   The Maximum Speed of a Loudspeaker Diaphragm

The diaphragm of a loudspeaker moves back and forth in simple harmonic motion to create sound, as in Figure 10.12. The frequency of the motion is $f = 1.0$ kHz and the amplitude is $A = 0.20$ mm. (a) What is the maximum speed of the diaphragm? (b) Where in the motion does this maximum speed occur?

Reasoning The maximum speed v_{max} of an object vibrating in simple harmonic motion is $v_{max} = A\omega$ (ω in rad/s), according to Equation 10.8. The angular frequency ω is related to the frequency f by $\omega = 2\pi f$, according to Equation 10.6.

Solution

(a) Using Equations 10.8 and 10.6, we find that the maximum speed of the vibrating diaphragm is

$$v_{max} = A\omega = A(2\pi f) = (0.20 \times 10^{-3} \text{ m})(2\pi)(1.0 \times 10^{3} \text{ Hz}) = \boxed{1.3 \text{ m/s}}$$

(b) The speed of the diaphragm is zero when the diaphragm momentarily comes to rest at either end of its motion: $x = +A$ and $x = -A$. Its maximum speed occurs midway between these two positions, or at $x = 0$ m.

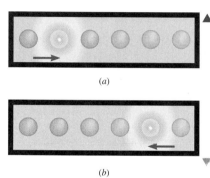

(a)

(b)

Figure 10.13 The motion of a lighted bulb is from (a) left to right and then from (b) right to left.

Simple harmonic motion is not just any kind of vibratory motion. It is a very specific kind and, among other things, must have the velocity given by Equation 10.7. For instance, advertising signs often use a "moving light" display to grab your attention. Conceptual Example 4 examines the back-and-forth motion in one such display, to see whether it is simple harmonic motion.

Conceptual Example 4   Moving Lights

Over the entrance to a restaurant is mounted a strip of equally spaced light bulbs, as Figure 10.13a illustrates. Starting at the left end, each bulb turns on in sequence for one-half second. Thus, a lighted bulb appears to move from left to right. Once the apparent motion of a lighted bulb reaches the right side of the sign, the motion reverses. The lighted bulb then appears to move to the left, as part b of the drawing indicates. Thus, the lighted bulb appears to oscillate back and forth. Is the apparent motion simple harmonic motion?

Reasoning and Solution Since the bulbs are equally spaced and each bulb remains lit for the same amount of time, the apparent motion of a lighted bulb across the sign occurs at a constant speed. If the motion were simple harmonic motion, however, it would not have a constant speed (see Equation 10.7), because it would have zero speed at each end of the sign and increase to a maximum speed at the center of the sign. Therefore, although the apparent motion of a lighted bulb is oscillatory, *it is not simple harmonic motion.*

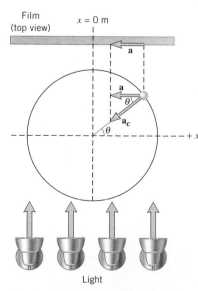

Figure 10.14 The acceleration **a** of the ball's shadow is the x component of the centripetal acceleration $\mathbf{a_c}$ of the ball on the reference circle.

ACCELERATION

In simple harmonic motion, the velocity is not constant; consequently, there must be an acceleration. This acceleration can also be determined with the aid of the reference-circle model. As Figure 10.14 shows, the ball on the reference circle moves in uniform circular motion, and, therefore, has a centripetal acceleration $\mathbf{a_c}$ that points toward the center of the circle. The acceleration **a** of the shadow is the x component of the centripetal acceleration; $a = -a_c \cos \theta$. The minus sign is needed because the acceleration of the shadow points to the left. Recalling that the centripetal acceleration is related to the angular speed ω by $a_c = r\omega^2$ (Equation 8.11) and using $r = A$, we find that $a_c = A\omega^2$. With this substitution, the acceleration in simple harmonic motion becomes

$$a = -A\omega^2 \cos \theta = -A\omega^2 \cos \omega t \qquad (\omega \text{ in rad/s}) \qquad (10.9)$$

The acceleration, like the velocity, does *not* have a constant value as time passes. The maximum magnitude of the acceleration is

$$a_{max} = A\omega^2 \qquad (\omega \text{ in rad/s}) \tag{10.10}$$

Although both the amplitude A and the angular frequency ω determine the maximum value, the frequency has a particularly strong effect, because it is squared. Example 5 shows that the acceleration can be remarkably large in a practical situation.

Example 5   The Loudspeaker Revisited—The Maximum Acceleration

The loudspeaker diaphragm in Figure 10.12 is vibrating at a frequency of $f = 1.0$ kHz, and the amplitude of the motion is $A = 0.20$ mm. (a) What is the maximum acceleration of the diaphragm, and (b) where does this maximum acceleration occur?

The physics of **a loudspeaker diaphragm.**

Reasoning The maximum acceleration a_{max} of an object vibrating in simple harmonic motion is $a_{max} = A\omega^2$ (ω in rad/s), according to Equation 10.10. Equation 10.6 shows that the angular frequency ω is related to the frequency f by $\omega = 2\pi f$.

Solution

(a) Using Equations 10.10 and 10.6, we find that the maximum acceleration of the vibrating diaphragm is

Problem solving insight
Do not confuse the vibrational frequency f with the angular frequency ω. The value for f is in hertz (cycles per second). The value for ω is in radians per second. The two are related by $\omega = 2\pi f$.

$$a_{max} = A\omega^2 = A(2\pi f)^2 = (0.20 \times 10^{-3} \text{ m})[2\pi(1.0 \times 10^3 \text{ Hz})]^2$$
$$= \boxed{7.9 \times 10^3 \text{ m/s}^2}$$

This is an incredible acceleration, being more than 800 times the acceleration due to gravity, and the diaphragm must be built to withstand it.

(b) The maximum acceleration occurs when the force acting on the diaphragm is a maximum. The maximum force arises when the diaphragm is at the ends of its path, where the displacement is greatest. Thus, the maximum acceleration occurs at $x = +A$ and $x = -A$ in Figure 10.12.

FREQUENCY OF VIBRATION

With the aid of Newton's second law ($\Sigma F = ma$), it is possible to determine the frequency at which an object of mass m vibrates on a spring. We assume that the mass of the spring itself is negligible and that the only force acting on the object in the horizontal direction is due to the spring—that is, the Hooke's law restoring force. Thus, the net force is $\Sigma F = -kx$, and Newton's second law becomes $-kx = ma$, where a is the acceleration of the object. The displacement and acceleration of an oscillating spring are, respectively, $x = A \cos \omega t$ (Equation 10.3) and $a = -A\omega^2 \cos \omega t$ (Equation 10.9). Substituting these expressions for x and a into the relation $-kx = ma$, we find that

$$-k(A \cos \omega t) = m(-A\omega^2 \cos \omega t)$$

which yields

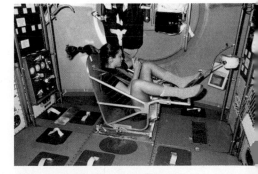

$$\omega = \sqrt{\frac{k}{m}} \qquad (\omega \text{ in rad/s}) \tag{10.11}$$

In this expression, the angular frequency ω must be in radians per second. Larger spring constants k and smaller masses m result in larger frequencies. Example 6 illustrates an application of Equation 10.11.

Example 6   A Body Mass Measurement Device

Figure 10.15 Astronaut Tamara Jernigan uses a body mass measurement device to measure her mass while in orbit. (Courtesy NASA.)

Astronauts who spend long periods of time in orbit periodically measure their body masses as part of their health-maintenance programs. On earth, it is simple to measure body weight W with a scale and convert it to mass m using the acceleration due to gravity, since $W = mg$. However, this procedure does not work in orbit, because both the scale and the astronaut are in free-fall and cannot press against each other (see Conceptual Example 12 in Chapter 5). Instead, astronauts use a body mass measurement device, as Figure 10.15 illustrates. This device consists of a spring-mounted chair in which the astronaut sits. The chair is then started oscillating in simple harmonic motion. The period of the motion is measured electronically and is automatically

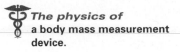
The physics of **a body mass measurement device.**

Need more practice?

Interactive LearningWare 10.1
Suppose that an object on a vertical spring oscillates up and down at a frequency of 5.00 Hz. By how much would this object, hanging from rest, stretch the spring?

Related Homework: Problem 21

Go to
www.wiley.com/college/cutnell
for an interactive solution.

converted into a value of the astronaut's mass, after the mass of the chair is taken into account. The spring used in one such device has a spring constant of 606 N/m, and the mass of the chair is 12.0 kg. The measured oscillation period is 2.41 s. Find the mass of the astronaut.

Reasoning The relation $\omega = \sqrt{k/m}$ (Equation 10.11) can be solved for the mass m in terms of the spring constant k and the angular frequency ω. The spring constant is known. Although the angular frequency is not known, it can be related to the given oscillation period of $T = 2.41$ s by using $\omega = 2\pi/T$ (Equation 10.4). The mass calculated using Equation 10.11 is the total mass of the astronaut and the chair, so that it will be necessary to subtract the mass of the chair to obtain the mass of the astronaut.

Solution Using Equations 10.11 and 10.4 gives

$$\omega = \frac{2\pi}{T} = \sqrt{\frac{k}{m}}$$

Solving for the mass m, we find that

$$m = \frac{kT^2}{4\pi^2} = \frac{(606 \text{ N/m})(2.41 \text{ s})^2}{4\pi^2} = 89.2 \text{ kg}$$

Accounting for the 12.0-kg mass of the chair reveals that the mass of the astronaut is

$$m_{\text{Astronaut}} = 89.2 \text{ kg} - 12.0 \text{ kg} = \boxed{77.2 \text{ kg}}$$

The physics of **detecting and measuring small amounts of chemicals.**

   Example 6 indicates that the mass of the vibrating object influences the frequency of simple harmonic motion. Electronic sensors are being developed that take advantage of this effect in detecting and measuring small amounts of chemicals. These sensors utilize tiny quartz crystals that vibrate when an electric current passes through them. If the crystal is coated with a substance that absorbs a particular chemical, then its mass increases as the chemical is absorbed and, according to the relation $f = \frac{1}{2\pi}\sqrt{k/m}$ (Equations 10.6 and 10.11), the frequency of the simple harmonic motion decreases. The change in frequency is detected electronically, and the sensor is calibrated to give the mass of the absorbed chemical.

✓**Check Your Understanding 2**

The drawing shows plots of the displacement x versus the time t for three objects undergoing simple harmonic motion. Which object, I, II, or III, has the greatest maximum velocity? *(The answer is given at the end of the book.)*

Background: The amplitude, period, frequency, and maximum speed of simple harmonic motion hold the key to answering this question.

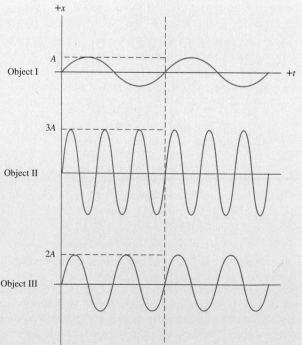

For similar questions (including calculational counterparts), consult Self-Assessment Test 10.1. The test is described next.

Self-Assessment Test 10.1

Test your understanding of the material in Sections 10.1 and 10.2:

• The Restoring Force of an Ideal Spring   • Simple Harmonic Motion

Go to **www.wiley.com/college/cutnell**

10.3 Energy and Simple Harmonic Motion

We saw in Chapter 6 that an object above the surface of the earth has gravitational potential energy. Therefore, when the object is allowed to fall, like the hammer of the pile driver in Figure 6.13, it can do work. A spring also has potential energy when the spring is stretched or compressed, which we refer to as ***elastic potential energy.*** Because of elastic potential energy, a stretched or compressed spring can do work on an object that is attached to the spring. For instance, Figure 10.16 shows a door-closing unit that is often found on screen doors. When the door is opened, a spring inside the unit is compressed and has elastic potential energy. When the door is released, the compressed spring expands and does the work of closing the door.

To find an expression for the elastic potential energy, we will determine the work done by the spring force on an object. Figure 10.17 shows an object attached to one end of a stretched spring. When the object is released, the spring contracts and pulls the object from its initial position x_0 to its final position x_f. The work W done by a constant force is given by Equation 6.1 as $W = (F \cos \theta)s$, where F is the magnitude of the force, s is the magnitude of the displacement ($s = x_0 - x_f$), and θ is the angle between the force and the displacement. The magnitude of the spring force is not constant, however. Equation 10.2 gives the spring force as $F = -kx$, and as the spring contracts, the magnitude of this force changes from kx_0 to kx_f. In using Equation 6.1 to determine the work, we can account for the changing magnitude by using an average magnitude $\overline{F}$ in place of the constant magnitude F. Because the dependence of the spring force on x is linear, the magnitude of the average force is just one-half the sum of the initial and final values, or $\overline{F} = \frac{1}{2}(kx_0 + kx_f)$. The work W_{elastic} done by the average spring force is, then,

$$W_{\text{elastic}} = (\overline{F} \cos \theta)s = \tfrac{1}{2}(kx_0 + kx_f) \cos 0° \,(x_0 - x_f)$$

$$W_{\text{elastic}} = \underbrace{\tfrac{1}{2}kx_0^2}_{\substack{\text{Initial elastic} \\ \text{potential energy}}} - \underbrace{\tfrac{1}{2}kx_f^2}_{\substack{\text{Final elastic} \\ \text{potential energy}}} \qquad (10.12)$$

In the calculation above, θ is $0°$, since the spring force points to the left in Figure 10.17, which is the same direction as the displacement. Equation 10.12 indicates that the work done by the spring force is equal to the difference between the initial and final values of the quantity $\frac{1}{2}kx^2$. The quantity $\frac{1}{2}kx^2$ is analogous to the quantity mgh, which we identified in Section 6.3 as the gravitational potential energy. Here, we identify the quantity $\frac{1}{2}kx^2$ as the elastic potential energy. Equation 10.13 indicates that the elastic potential energy is a maximum for a fully stretched or compressed spring and zero for a spring that is neither stretched nor compressed ($x = 0$ m).

The physics of a door-closing unit.

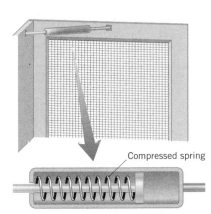

Compressed spring

Figure 10.16 A door-closing unit. The elastic potential energy stored in the compressed spring is used to close the door.

■ **DEFINITION OF ELASTIC POTENTIAL ENERGY**

The elastic potential energy PE_{elastic} is the energy that a spring has by virtue of being stretched or compressed. For an ideal spring that has a spring constant k and is stretched or compressed by an amount x relative to its unstrained length, the elastic potential energy is

$$PE_{\text{elastic}} = \tfrac{1}{2}kx^2 \qquad (10.13)$$

SI Unit of Elastic Potential Energy: joule (J)

Position when spring is unstrained

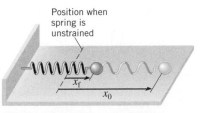

x_f

x_0

Figure 10.17 When the object is released, its displacement changes from an initial value of x_0 to a final value of x_f.

CONCEPTS AT A GLANCE

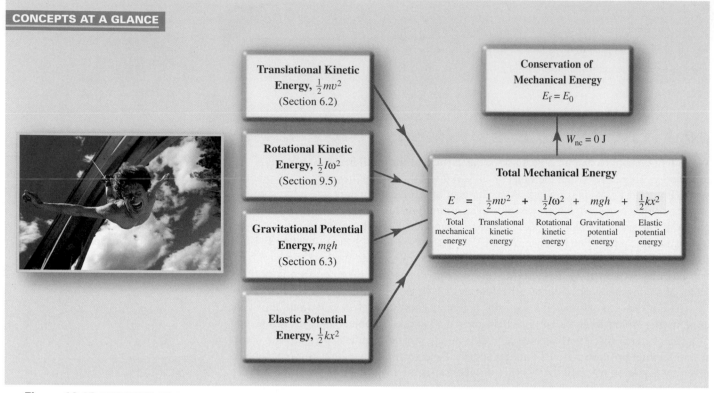

Figure 10.18 CONCEPTS AT A GLANCE The elastic potential energy is added to other energies to give the total mechanical energy, which is conserved if W_{nc}, the net work done by external nonconservative forces, is zero. As this bungee jumper oscillates up and down at the end of the elastic cord, the total mechanical energy remains constant, to the extent that the nonconservative forces of friction and air resistance are negligible. (© Terje Rakke/Stone/Getty Images)

▶ **CONCEPTS AT A GLANCE** The total mechanical energy E is a familiar idea that we originally defined to be the sum of the translational kinetic energy and the gravitational potential energy, as illustrated by the Concepts-at-a-Glance chart in Figure 6.15. Then, we included the rotational kinetic energy, as the chart in Figure 9.23 shows. We now expand the total mechanical energy to include the elastic potential energy, as the Concepts-at-a-Glance chart in Figure 10.18 indicates. ◀

In Equation 10.14 the elastic potential energy is included as part of the total mechanical energy:

$$\underbrace{E}_{\substack{\text{Total} \\ \text{mechanical} \\ \text{energy}}} = \underbrace{\tfrac{1}{2}mv^2}_{\substack{\text{Translational} \\ \text{kinetic} \\ \text{energy}}} + \underbrace{\tfrac{1}{2}I\omega^2}_{\substack{\text{Rotational} \\ \text{kinetic} \\ \text{energy}}} + \underbrace{mgh}_{\substack{\text{Gravitational} \\ \text{potential} \\ \text{energy}}} + \underbrace{\tfrac{1}{2}kx^2}_{\substack{\text{Elastic} \\ \text{potential} \\ \text{energy}}} \qquad (10.14)$$

As Section 6.5 discusses, the total mechanical energy is conserved when external nonconservative forces (such as friction) do no net work; that is, when $W_{nc} = 0$ J. Then, the final and initial values of E are the same: $E_f = E_0$. The principle of conservation of total mechanical energy is the subject of the next example.

Example 7   An Object on a Horizontal Spring

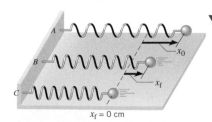

Figure 10.19 The total mechanical energy of this system is entirely elastic potential energy (A), partly elastic potential energy and partly kinetic energy (B), and entirely kinetic energy (C).

Figure 10.19 shows an object of mass $m = 0.200$ kg that is vibrating on a horizontal frictionless table. The spring has a spring constant $k = 545$ N/m. It is stretched initially to $x_0 = 4.50$ cm and then released from rest (see part A of the drawing). Determine the final translational speed v_f of the object when the final displacement of the spring is (a) $x_f = 2.25$ cm and (b) $x_f = 0$ cm.

Reasoning The conservation of mechanical energy indicates that, in the absence of friction (a nonconservative force), the final and initial total mechanical energies are the same (see Figure 10.18):

$$E_f = E_0$$

$$\tfrac{1}{2}mv_f^2 + \tfrac{1}{2}I\omega_f^2 + mgh_f + \tfrac{1}{2}kx_f^2 = \tfrac{1}{2}mv_0^2 + \tfrac{1}{2}I\omega_0^2 + mgh_0 + \tfrac{1}{2}kx_0^2$$

Since the object is moving on a horizontal table, the final and initial heights are the same:

$h_f = h_0$. The object is not rotating, so its angular speed is zero: $\omega_f = \omega_0 = 0$ rad/s. And, as the problem states, the initial translational speed of the object is zero, $v_0 = 0$ m/s. With these substitutions, the conservation-of-energy equation becomes

$$\tfrac{1}{2}mv_f^2 + \tfrac{1}{2}kx_f^2 = \tfrac{1}{2}kx_0^2$$

from which we can obtain v_f:

$$v_f = \sqrt{\frac{k}{m}(x_0^2 - x_f^2)}$$

Solution

(a) Since $x_0 = 0.0450$ m and $x_f = 0.0225$ m, the final translational speed is

$$v_f = \sqrt{\frac{545 \text{ N/m}}{0.200 \text{ kg}}[(0.0450 \text{ m})^2 \quad (0.0225 \text{ m})^2]} = \boxed{2.03 \text{ m/s}}$$

The total mechanical energy at this point is partly translational kinetic energy ($\tfrac{1}{2}mv_f^2 = 0.414$ J) and partly elastic potential energy ($\tfrac{1}{2}kx_f^2 = 0.138$ J). The total mechanical energy E is the sum of these two energies: $E = 0.414$ J $+ 0.138$ J $= 0.552$ J. Because the total mechanical energy remains constant during the motion, this value equals the initial total mechanical energy when the object is stationary and the energy is entirely elastic potential energy ($E_0 = \tfrac{1}{2}kx_0^2 = 0.552$ J).

(b) When $x_0 = 0.0450$ m and $x_f = 0$ m, we have

$$v_f = \sqrt{\frac{k}{m}(x_0^2 - x_f^2)} = \sqrt{\frac{545 \text{ N/m}}{0.200 \text{ kg}}[(0.0450 \text{ m})^2 - (0 \text{ m})^2]} = \boxed{2.35 \text{ m/s}}$$

Now the total mechanical energy is due entirely to the translational kinetic energy ($\tfrac{1}{2}mv_f^2 = 0.552$ J), since the elastic potential energy is zero. Note that the total mechanical energy is the same as it is in part (a). In the absence of friction, the simple harmonic motion of a spring converts the different types of energy between one form and another, the total always remaining the same.

Conceptual Example 8 takes advantage of energy conservation to illustrate what happens to the maximum speed, amplitude, and angular frequency of a simple harmonic oscillator when its mass is changed suddenly at a certain point in the motion.

Conceptual Example 8
Changing the Mass of a Simple Harmonic Oscillator

Figure 10.20a shows a box of mass m attached to a spring that has a force constant k. The box rests on a horizontal, frictionless surface. The spring is initially stretched to $x = A$ and then released from rest. The box then executes simple harmonic motion that is characterized by a maximum speed v_{max}, an amplitude A, and an angular frequency ω. When the box is passing through the point where the spring is unstrained ($x = 0$ m), a second box of the same mass m and speed v_{max} is attached to it, as in part b of the drawing. Discuss what happens to (a) the maximum speed, (b) the amplitude, and (c) the angular frequency of the subsequent simple harmonic motion.

Reasoning and Solution

(a) The maximum speed of an object in simple harmonic motion occurs when the object is passing through the point where the spring is unstrained ($x = 0$ m), as in Figure 10.20b. Since the second box is attached at this point with the same speed, *the maximum speed of the two-box system remains the same as that of the one-box system.*

(b) At the same speed, the maximum kinetic energy of the two boxes is twice that of a single box, since the mass is twice as much. Subsequently, when the two boxes move to the left and compress the spring, their kinetic energy is converted into elastic potential energy. Since the two boxes have twice as much kinetic energy as one box alone, the two will have twice as much elastic potential energy when they come to a halt at the extreme left. Here, we are using the principle of conservation of mechanical energy, which applies since friction is absent. But the elastic potential energy is proportional to the amplitude squared (A^2) of the motion, *so the amplitude of the two-box system is greater than that of the one-box system by a factor of $\sqrt{2}$.*

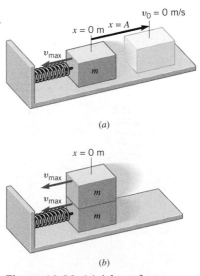

Figure 10.20 (a) A box of mass m, starting from rest at $x = A$, undergoes simple harmonic motion about $x = 0$ m. (b) When $x = 0$ m, a second box, with the same mass and speed, is attached.

Need more practice?

Interactive LearningWare 10.2

A rifle fires a 2.10×10^{-2}-kg pellet straight upward because the pellet rests on a compressed spring that is released when the trigger is pulled. The spring has a negligible mass and is compressed by 9.10×10^{-2} m from its unstrained length. The pellet rises to a height of 6.10 m above its position on the compressed spring. Ignoring air resistance, determine the spring constant.

Related Homework: *Problem 38*

Go to
www.wiley.com/college/cutnell
for an interactive solution.

(c) The angular frequency ω of a simple harmonic oscillator is given by Equation 10.11 as $\omega = \sqrt{k/m}$. Since the mass of the two-box system is twice that of the one-box system, *the angular frequency of the two-box system is smaller than that of the one-box system by a factor of $\sqrt{2}$.*

Related Homework: *Conceptual Question 6, Problem 32*

In the previous two examples, gravitational potential energy plays no role because the spring is horizontal. The next example illustrates that gravitational potential energy must be taken into account when a spring is oriented vertically.

Example 9   A Falling Ball on a Vertical Spring

A 0.20-kg ball is attached to a vertical spring, as in Figure 10.21. The spring constant of the spring is 28 N/m. The ball, supported initially so that the spring is neither stretched nor compressed, is released from rest. In the absence of air resistance, how far does the ball fall before being brought to a momentary stop by the spring?

Reasoning Since air resistance is absent, only the conservative forces of gravity and the spring act on the ball. Therefore, the principle of conservation of mechanical energy applies:

$$E_f = E_0$$

$$\tfrac{1}{2}mv_f^2 + \tfrac{1}{2}I\omega_f^2 + mgh_f + \tfrac{1}{2}kx_f^2 = \tfrac{1}{2}mv_0^2 + \tfrac{1}{2}I\omega_0^2 + mgh_0 + \tfrac{1}{2}kx_0^2$$

The problem states that the final and initial translational speeds of the ball are zero: $v_f = v_0 = 0$ m/s. The ball and spring do not rotate, so the final and initial angular speeds are also zero: $\omega_f = \omega_0 = 0$ rad/s. As Figure 10.21 indicates, the initial height of the ball is h_0, and the final height is $h_f = 0$ m. In addition, the spring is unstrained ($x_0 = 0$ m) to begin with, and so it has no elastic potential energy initially. With these substitutions, the conservation of mechanical energy equation reduces to

$$\tfrac{1}{2}kx_f^2 = mgh_0$$

This result shows that the initial gravitational potential energy (mgh_0) is converted into elastic potential energy ($\tfrac{1}{2}kx_f^2$). When the ball falls to its lowest point, its displacement is $x_f = -h_0$, where the minus sign indicates that the displacement is downward. Substituting this result into the equation above and solving for h_0 yields $h_0 = 2mg/k$.

Solution The distance that the ball falls before coming to a momentary halt is

$$h_0 = \frac{2mg}{k} = \frac{2(0.20 \text{ kg})(9.8 \text{ m/s}^2)}{28 \text{ N/m}} = \boxed{0.14 \text{ m}}$$

Problem solving insight
When evaluating the total mechanical energy E, always include a potential energy term for every conservative force acting on the system. In Example 9 there are two such terms, gravitational and elastic.

Concept Simulation 10.1

In this simulation you can control the initial velocity and mass of an object on a vertical spring. As the object oscillates, its vertical position and velocity are plotted as a function of time. You can also choose to display any or all of the following energies as a function of time: kinetic energy, gravitational potential energy, elastic potential energy, and total mechanical energy.

Related Homework: *Problem 25*

Go to **www.wiley.com/college/cutnell**

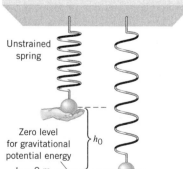

Figure 10.21 The ball is supported initially so that the spring is unstrained. After being released from rest, the ball falls through the distance h_0 before being momentarily stopped by the spring.

✔ **Check Your Understanding 3**

A block is attached to the end of a horizontal ideal spring and rests on a frictionless surface. The block is pulled so that the spring stretches relative to its unstrained length. In each of the following three cases, the spring is stretched initially by the same amount, but the block is given different initial speeds. Rank the amplitudes of the resulting simple harmonic motion in decreasing order (largest first). (a) The block is released from rest. (b) The block is given an initial speed v_0. (c) The block is given an initial speed $v_0/2$. *(The answer is given at the end of the book.)*

Background: Here, the pertinent themes are elastic potential energy, kinetic energy, energy conservation, and the amplitude of simple harmonic motion.

For similar questions (including calculational counterparts), consult Self-Assessment Test 10.2. This test is described at the end of Section 10.8.

10.4 The Pendulum

As Figure 10.22 shows, a *simple pendulum* consists of a particle of mass m, attached to a frictionless pivot P by a cable of length L and negligible mass. When the particle is pulled away from its equilibrium position by an angle θ and released, it swings back and forth. By attaching a pen to the bottom of the swinging particle and moving a strip of paper beneath it at a steady rate, we can record the position of the particle as time passes. The graphical record reveals a pattern that is similar (but not identical) to the sinusoidal pattern for simple harmonic motion.

The force of gravity is responsible for the back-and-forth rotation about the axis at P. The rotation speeds up as the particle approaches the lowest point on the arc and slows down on the upward part of the swing. Eventually the angular speed is reduced to zero, and the particle swings back. As Section 9.4 discusses, a net torque is required to change the angular speed. The gravitational force $m\mathbf{g}$ produces this torque. (The tension $\mathbf{T}$ in the cable creates no torque, because it points directly at the pivot P and, therefore, has a zero lever arm.) According to Equation 9.1, the magnitude of the torque τ is the product of the magnitude mg of the gravitational force and the lever arm ℓ, so that $\tau = -(mg)\ell$. The minus sign is included since the torque is a restoring torque: that is, it acts to reduce the angle θ [the angle θ is positive (counterclockwise), while the torque is negative (clockwise)]. The lever arm ℓ is the perpendicular distance between the line of action of $m\mathbf{g}$ and the pivot P. From Figure 10.22 it can be seen that ℓ is very nearly equal to the arc length s of the circular path when the angle θ is small (about 10° or less). Furthermore, if θ is expressed in radians, the arc length and the radius L of the circular path are related, according to $s = L\theta$ (Equation 8.1). Under these conditions, it follows that $\ell \approx s = L\theta$, and the torque created by gravity is

$$\tau \approx \underbrace{-mgL}_{k'}\,\theta$$

In the equation above, the term mgL has a constant value k', independent of θ. *For small angles,* then, the torque that restores the pendulum to its vertical equilibrium position is proportional to the angular displacement θ. The expression $\tau = -k'\theta$ has the same form as the Hooke's law restoring force for an ideal spring, $F = -kx$. Therefore, we expect the frequency of the back-and-forth movement of the pendulum to be given by an equation analogous to Equation 10.11 ($\omega = 2\pi f = \sqrt{k/m}$). In place of the spring constant k, the constant $k' = mgL$ will appear, and, as usual in rotational motion, in place of the mass m, the moment of inertia I will appear:

$$\omega = 2\pi f = \sqrt{\frac{mgL}{I}} \qquad \text{(small angles only)} \qquad (10.15)$$

The moment of inertia of a particle of mass m, rotating at a radius $r = L$ about an axis, is given by $I = mL^2$ (Equation 9.6). Substituting this expression for I into Equation 10.15 reveals for a simple pendulum that

Simple pendulum $\qquad \omega = 2\pi f = \sqrt{\dfrac{g}{L}} \qquad \text{(small angles only)} \qquad (10.16)$

The mass of the particle has been eliminated algebraically from this expression, so only the length L and the acceleration g due to gravity determine the frequency of a simple pendulum. If the angle of oscillation is large, the pendulum does not exhibit simple harmonic motion, and Equation 10.16 does not apply. Equation 10.16 provides the basis for using a pendulum to keep time, as Example 10 demonstrates.

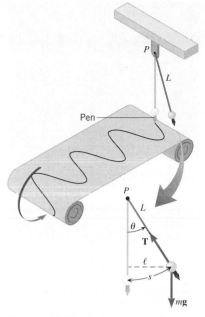

Figure 10.22 A simple pendulum swinging back and forth about the pivot P. If the angle θ is small, the swinging is approximately simple harmonic motion.

Concept Simulation 10.2

This simulation allows you to see the pendulum motion in Figure 10.22. A plot of the angular displacement versus time is synchronized with the motion, and, among other things, you can explore the effects of changing the length of the pendulum and the acceleration due to gravity.

Related Homework: *Conceptual Questions 11, 12, Problem 41*

Go to
www.wiley.com/college/cutnell

Example 10 Keeping Time

Determine the length of a simple pendulum that will swing back and forth in simple harmonic motion with a period of 1.00 s.

Reasoning When a simple pendulum is swinging back and forth in simple harmonic motion, its frequency f is given by Equation 10.16 as $f = \frac{1}{2\pi}\sqrt{g/L}$, where g is the acceleration due to gravity and L is the length of the pendulum. We also know from Equation 10.5 that the fre-

quency is the reciprocal of the period T, so $f = 1/T$. Thus, the equation above becomes $1/T = \frac{1}{2\pi}\sqrt{g/L}$. We can solve this equation for the length L of the pendulum.

Solution The length of the pendulum is

$$L = \frac{T^2 g}{4\pi^2} = \frac{(1.00 \text{ s})^2 (9.80 \text{ m/s}^2)}{4\pi^2} = \boxed{0.248 \text{ m}}$$

Figure 10.23 shows a clock that uses a pendulum to keep time.

Figure 10.23 This pendulum clock keeps time as the pendulum swings back and forth. (© Robert Mathena/ Fundamental Photographs)

It is not necessary that the object in Figure 10.22 be a particle. It may be an extended object, in which case the pendulum is called a ***physical pendulum.*** For small oscillations, Equation 10.15 still applies, but the moment of inertia I is no longer mL^2. The proper value for the rigid object must be used. (See Section 9.4 for a discussion of moment of inertia.) In addition, the length L for a physical pendulum is the distance between the axis at P and the center of gravity of the object. The next example deals with an important type of physical pendulum.

Example 11   Pendulum Motion and Walking

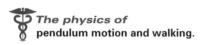

The physics of pendulum motion and walking.

When we walk, our legs alternately swing forward about the hip joint as a pivot. In this motion the leg is acting approximately as a physical pendulum. Treating the leg as a uniform rod of length $D = 0.80$ m, find the time it takes for the leg to swing forward.

Reasoning The time it takes for the leg to swing forward is one-half of the period T, which is related to the frequency f by $f = 1/T$ (Equation 10.5). For a physical pendulum the frequency is given by $f = \frac{1}{2\pi}\sqrt{mgL/I}$ (Equation 10.15), where the moment of inertia for a thin rod of length D rotating about an axis perpendicular to one end is given in Table 9.1 as $I = \frac{1}{3}mD^2$. In Equation 10.15 the length L is the distance between the pivot at the hip and the center of gravity of the leg. Since we are treating the leg as a thin uniform rod, the center of gravity is at the center and $L = 0.40$ m.

Solution Using Equations 10.5 and 10.15, we find that the period is

$$f = \frac{1}{T} = \frac{1}{2\pi}\sqrt{\frac{mgL}{I}} \quad \text{or} \quad T = 2\pi\sqrt{\frac{I}{mgL}}$$

Substituting the moment of inertia $I = \frac{1}{3}mD^2$ of the thin rod into this result gives

$$T = 2\pi\sqrt{\frac{\frac{1}{3}mD^2}{mgL}} = \frac{2\pi D}{\sqrt{3gL}} = \frac{2\pi (0.80 \text{ m})}{\sqrt{3(9.8 \text{ m/s}^2)(0.40 \text{ m})}} = 1.5 \text{ s}$$

The desired time is one-half of the period or $\boxed{0.75 \text{ s}}$.

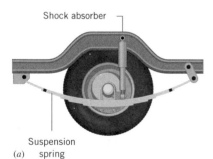

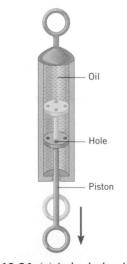

Figure 10.24 (a) A shock absorber mounted in the suspension system of an automobile and (b) a simplified, cutaway view of the shock absorber.

10.5 Damped Harmonic Motion

In simple harmonic motion, an object oscillates with a constant amplitude, because there is no mechanism for dissipating energy. In reality, however, friction or some other energy-dissipating mechanism is always present. In the presence of energy dissipation, the amplitude of oscillation decreases as time passes, and the motion is no longer simple harmonic motion. Instead, it is referred to as ***damped harmonic motion,*** the decrease in amplitude being called "damping."

One widely used application of damped harmonic motion is in the suspension system of an automobile. Figure 10.24a shows a shock absorber attached to a main suspension spring of a car. A shock absorber is designed to introduce damping forces, which reduce the vibrations associated with a bumpy ride. As part b of the drawing shows, a shock absorber consists of a piston in a reservoir of oil. When the piston moves in response to a bump in the road, holes in the piston head permit the piston to pass through the oil. Viscous forces that arise during this movement cause the damping.

Figure 10.25 illustrates the different degrees of damping that can exist. As applied to the example of a car's suspension system, these graphs show the vertical position of the

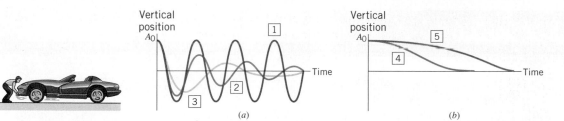

(a)   (b)

Figure 10.25 Damped harmonic motion. The degree of damping increases from curve 1 to curve 5. Curve 1 represents undamped or simple harmonic motion. Curves 2 and 3 show underdamped motion. Curve 4 represents critically damped harmonic motion. Curve 5 shows overdamped motion.

chassis after it has been pulled upward by an amount A_0 at time $t_0 = 0$ s and then released. Part *a* of the figure compares undamped or simple harmonic motion in curve 1 (red) to slightly damped motion in curve 2 (green). In damped harmonic motion, the chassis oscillates with decreasing amplitude and it eventually comes to rest. As the degree of damping is increased from curve 2 to curve 3 (gold), the car makes fewer oscillations before coming to a halt. Part *b* of the drawing shows that as the degree of damping is increased further, there comes a point when the car does not oscillate at all after it is released but, rather, settles directly back to its equilibrium position, as in curve 4 (blue). The smallest degree of damping that completely eliminates the oscillations is termed "critical damping," and the motion is said to be *critically damped.*

Figure 10.25*b* also shows that the car takes the longest time to return to its equilibrium position in curve 5 (purple), where the degree of damping is above the value for critical damping. When the damping exceeds the critical value, the motion is said to be *overdamped.* In contrast, when the damping is less than the critical level, the motion is said to be *underdamped* (curves 2 and 3). Typical automobile shock absorbers are designed to produce underdamped motion somewhat like that in curve 3.

The physics of a shock absorber.

Concept Simulation 10.3

Here, simple harmonic motion is compared to damped harmonic motion. For an object on a spring you can adjust the spring constant, the mass and initial position of the object, and the extent of the damping. Graphs of the object's position, velocity, and acceleration versus time are synchronized with the motion and allow you to see the effects of your adjustments.

Related Homework: Problems 17, 68

Go to **www.wiley.com/college/cutnell**

10.6 *Driven Harmonic Motion and Resonance*

In damped harmonic motion, a mechanism such as friction dissipates or reduces the energy of an oscillating system, with the result that the amplitude of the motion decreases in time. This section discusses the opposite effect—namely, the increase in amplitude that results when energy is continually added to an oscillating system.

To set an object on an ideal spring into simple harmonic motion, some agent must apply a force that stretches or compresses the spring initially. Suppose that this force is applied at all times, not just for a brief initial moment. The force could be provided, for example, by a person who simply pushes and pulls the object back and forth. The resulting motion is known as ***driven harmonic motion,*** because the additional force drives or controls the behavior of the object to a large extent. The additional force is identified as the ***driving force.***

Figure 10.26 illustrates one particularly important example of driven harmonic motion. Here, the driving force has the same frequency as the spring system and always points in the direction of the object's velocity. The frequency of the spring system is $f = (1/2\pi)\sqrt{k/m}$ and is called a natural frequency, because it is the frequency at which the spring system naturally oscillates. Since the driving force and the velocity always have the same direction, positive work is done on the object at all times, and the total mechanical energy of the system increases. As a result, the amplitude of the vibration becomes larger and will increase without limit, if there is no damping force to dissipate the energy being added by the driving force. The situation depicted in Figure 10.26 is known as ***resonance.***

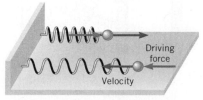

Figure 10.26 Resonance occurs when the frequency of the driving force (blue arrows) matches a frequency at which the object naturally vibrates. The red arrows represent the velocity of the object.

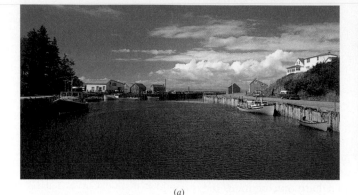

(a)   (b)

Figure 10.27 The Bay of Fundy at (a) high tide and (b) low tide. In some places the water level changes by almost 15 m. (© Everett Johnson/Leo de Wys, Inc.)

■ **RESONANCE**

Resonance is the condition in which a time-dependent force can transmit large amounts of energy to an oscillating object, leading to a large amplitude motion. In the absence of damping, resonance occurs when the frequency of the force matches a natural frequency at which the object will oscillate.

The role played by the frequency of a driving force is a critical one. The matching of this frequency with a natural frequency of vibration allows even a relatively weak force to produce a large amplitude vibration, because the effect of each push–pull cycle is cumulative.

The physics of high tides at the Bay of Fundy.

Resonance can occur with any object that can oscillate, and springs need not be involved. The greatest tides in the world occur in the Bay of Fundy, which lies between the Canadian provinces of New Brunswick and Nova Scotia. Figure 10.27 shows the enormous difference between the water level at high and low tides, a difference that in some locations averages about 15 m. This phenomenon is partly due to resonance. The time, or period, that it takes for the tide to flow into and ebb out of a bay depends on the size of the bay, the topology of the bottom, and the configuration of the shoreline. The ebb and flow of the water in the Bay of Fundy has a period of 12.5 hours, which is very close to the lunar tidal period of 12.42 hours. The tide then "drives" water into and out of the Bay of Fundy at a frequency (once per 12.42 hours) that nearly matches the natural frequency of the bay (once per 12.5 hours). The result is the extraordinary high tide in the bay. (You can create a similar effect in a bathtub full of water by moving back and forth in synchronism with the waves you're causing.)

10.7 *Elastic Deformation*
STRETCHING, COMPRESSION, AND YOUNG'S MODULUS

We have seen that a spring returns to its original shape when the force compressing or stretching it is removed. In fact, all materials become distorted in some way when they are squeezed or stretched, and many of them, such as rubber, return to their original shape when the squeezing or stretching is removed. Such materials are said to be "elastic." From an atomic viewpoint, elastic behavior has its origin in the forces that atoms exert on each other, and Figure 10.28 symbolizes these forces with the aid of springs. It is because of these atomic-level "springs" that a material tends to return to its initial shape once the forces that cause the deformation are removed.

The interatomic forces that hold the atoms of a solid together are particularly strong, so considerable force must be applied to stretch a solid object. Experiments have shown that the magnitude of the force can be expressed by the following relation, provided that the amount of stretching is small compared to the original length of the object:

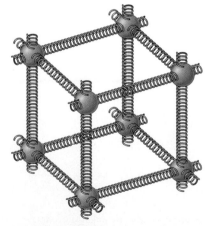

Figure 10.28 The forces between atoms act like springs. The atoms are represented by red spheres, and the springs between some atoms have been omitted for clarity.

$$F = Y\left(\frac{\Delta L}{L_0}\right)A \qquad (10.17)$$

As Figure 10.29 shows, F denotes the magnitude of the stretching force applied perpendicularly to the surface at the end, A is the cross-sectional area of the rod, ΔL is the increase in length, and L_0 is the original length. The term Y is a proportionality constant called *Young's modulus,* after Thomas Young (1773–1829). Solving Equation 10.17 for Y shows that Young's modulus has units of force per unit area (N/m^2). *It should be noted that the magnitude of the force in Equation 10.17 is proportional to the fractional increase in length $\Delta L/L_0$, rather than the absolute increase ΔL.* The magnitude of the force is also proportional to the cross-sectional area A, which need not be circular, but can have any shape (e.g., rectangular).

Table 10.1 reveals that the value of Young's modulus depends on the nature of the material. The values for metals are much larger than those for bone, for example. Equation 10.17 indicates that, for a given force, the material with the greater value of Y undergoes the smaller change in length. This difference between the changes in length is the reason why surgical implants (e.g., artificial hip joints), which are often made from stainless steel or titanium alloys, can lead to chronic deterioration of the bone that is in contact with the implanted prosthesis.

Forces that are applied as in Figure 10.29 and cause stretching are called "tensile" forces, because they create a tension in the material, much like the tension in a rope. Equation 10.17 also applies when the force compresses the material along its length. In this situation, the force is applied in a direction opposite to that shown in Figure 10.29, and ΔL stands for the amount by which the original length L_0 decreases. Table 10.1 indicates, for example, that bone has different values of Young's modulus for compression and tension, the value for tension being greater. Such differences are related to the structure of the material. The solid part of bone consists of collagen fibers (a protein material) distributed throughout hydroxyapatite (a mineral). The collagen acts like the steel rods in reinforced concrete and increases the value of Y for tension relative to that for compression.

Most solids have Young's moduli that are rather large, reflecting the fact that a large force is needed to change the length of a solid object by even a small amount, as Example 12 illustrates.

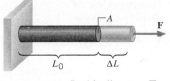

Figure 10.29 In this diagram, **F** denotes the stretching force, A the cross-sectional area, L_0 the original length of the rod, and ΔL the amount of stretch.

The physics of surgical implants.

The physics of bone structure.

Example 12 Bone Compression

In a circus act, a performer supports the combined weight (1640 N) of a number of colleagues (see Figure 10.30). Each thighbone (femur) of this performer has a length of 0.55 m and an effective cross-sectional area of 7.7×10^{-4} m^2. Determine the amount by which each thighbone compresses under the extra weight.

Table 10.1 *Values for the Young's Modulus of Solid Materials*

Material	Young's Modulus Y (N/m^2)
Aluminum	6.9×10^{10}
Bone	
Compression	9.4×10^{9}
Tension	1.6×10^{10}
Brass	9.0×10^{10}
Brick	1.4×10^{10}
Copper	1.1×10^{11}
Mohair	2.9×10^{9}
Nylon	3.7×10^{9}
Pyrex glass	6.2×10^{10}
Steel	2.0×10^{11}
Teflon	3.7×10^{8}
Titanium	1.2×10^{11}
Tungsten	3.6×10^{11}

Figure 10.30 The entire weight of the balanced group is supported by the legs of the performer who is lying on his back. (Image courtesy of Ringling Brothers and Barnum & Bailey®, THE GREATEST SHOW ON EARTH®.)

☤ **The physics of bone compression.**

Reasoning The additional weight supported by each thighbone is $F = \frac{1}{2}(1640 \text{ N}) = 820 \text{ N}$, and Table 10.1 indicates that Young's modulus for bone compression is $9.4 \times 10^9 \text{ N/m}^2$. Since the length and cross-sectional area of the thighbone are also known, we may use Equation 10.17 to find the amount by which the additional weight compresses the thighbone.

Solution The amount of compression ΔL of each thighbone is

$$\Delta L = \frac{FL_0}{YA} = \frac{(820 \text{ N})(0.55 \text{ m})}{(9.4 \times 10^9 \text{ N/m}^2)(7.7 \times 10^{-4} \text{ m}^2)} = \boxed{6.2 \times 10^{-5} \text{ m}}$$

This is a very small change, the fractional decrease being $\Delta L/L_0 = 0.000\ 11$.

SHEAR DEFORMATION AND THE SHEAR MODULUS

Table 10.2 *Values for the Shear Modulus of Solid Materials*

Material	Shear Modulus S (N/m^2)
Aluminum	2.4×10^{10}
Bone	1.2×10^{10}
Brass	3.5×10^{10}
Copper	4.2×10^{10}
Lead	5.4×10^9
Nickel	7.3×10^{10}
Steel	8.1×10^{10}
Tungsten	1.5×10^{11}

It is possible to deform a solid object in a way other than stretching or compressing it. For instance, place a book on a rough table and push on the top cover, as in Figure 10.31a. Notice that the top cover, and the pages below it, become shifted relative to the stationary bottom cover. The resulting deformation is called a **shear deformation** and occurs because of the combined effect of the force **F** applied (by the hand) to the top of the book and the force −**F** applied (by the table) to the bottom of the book. The directions of the forces are parallel to the covers of the book, each of which has an area A, as illustrated in part b of the drawing. These two forces have equal magnitudes, but opposite directions, so the book remains in equilibrium. Equation 10.18 gives the magnitude F of the force needed to produce an amount of shear ΔX for an object with thickness L_0:

$$F = S\left(\frac{\Delta X}{L_0}\right)A \qquad (10.18)$$

This equation is very similar to Equation 10.17. The constant of proportionality S is called the **shear modulus** and, like Young's modulus, has units of force per unit area (N/m^2). The value of S depends on the nature of the material, and Table 10.2 gives some representative values. Example 13 illustrates how to determine the shear modulus of a favorite dessert.

Example 13 J-E-L-L-O

A block of Jell-O is resting on a plate. Figure 10.32a gives the dimensions of the block. You are bored, impatiently waiting for dinner, and push tangentially across the top surface with a force of $F = 0.45$ N, as in part b of the drawing. The top surface moves a distance $\Delta X = 6.0 \times 10^{-3}$ m relative to the bottom surface. Use this idle gesture to measure the shear modulus of Jell-O.

Reasoning The finger applies a force that is parallel to the top surface of the Jell-O block. The shape of the block changes, because the top surface moves a distance ΔX relative to the bottom surface. The magnitude of the force required to produce this change in shape is given by Equation 10.18 as $F = S(\Delta X/L_0)A$. We know the values for all the variables in this relation except S, which can be determined.

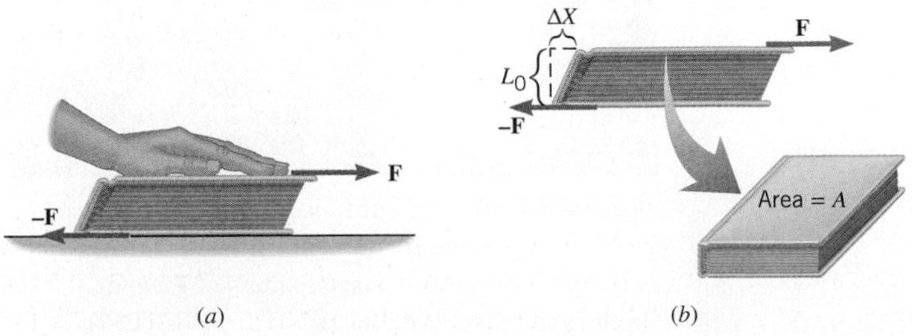

Figure 10.31 (a) An example of a shear deformation. The shearing forces **F** and −**F** are applied parallel to the top and bottom covers of the book. In general, shearing forces cause a solid object to change its shape. (b) The shear deformation is ΔX. The area of each cover is A, and the thickness of the book is L_0.

Solution Solving Equation 10.18 for the shear modulus S, we find that $S = FL_0/(A \, \Delta X)$, where $A = (0.070 \text{ m})(0.070 \text{ m})$ is the area of the top surface, and $L_0 = 0.030$ m is the thickness of the block:

$$S = \frac{FL_0}{A \, \Delta X} = \frac{(0.45 \text{ N})(0.030 \text{ m})}{(0.070 \text{ m})(0.070 \text{ m})(6.0 \times 10^{-3} \text{ m})} = \boxed{460 \text{ N/m}^2}$$

Jell-O can be deformed easily, so its shear modulus is significantly less than that of a more rigid material like steel (see Table 10.2).

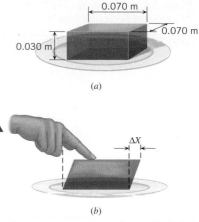

(a)

(b)

Figure 10.32 (a) A block of Jell-O and (b) a shearing force applied to it.

Although Equations 10.17 and 10.18 are algebraically similar, they refer to different kinds of deformations. The tensile force in Figure 10.29 is perpendicular to the surface whose area is A, whereas the shearing force in Figure 10.31 is parallel to that surface. Furthermore, the ratio $\Delta L/L_0$ in Equation 10.17 is different from the ratio $\Delta X/L_0$ in Equation 10.18. The distances ΔL and L_0 are parallel, whereas ΔX and L_0 are perpendicular. Young's modulus refers to a *change in length* of one dimension of a solid object as a result of tensile or compressive forces. The shear modulus refers to a *change in shape* of a solid object as a result of shearing forces.

VOLUME DEFORMATION AND THE BULK MODULUS

When a compressive force is applied along one dimension of a solid, the length of that dimension decreases. It is also possible to apply compressive forces so that the size of every dimension (length, width, and depth) decreases, leading to a decrease in volume, as Figure 10.33 illustrates. This kind of overall compression occurs, for example, when an object is submerged in a liquid, and the liquid presses inward everywhere on the object. The forces acting in such situations are applied perpendicular to every surface, and it is more convenient to speak of the perpendicular force per unit area, rather than the amount of any one force in particular. The magnitude of the perpendicular force per unit area is called the *pressure P*

■ **DEFINITION OF PRESSURE**

The pressure P is the magnitude F of the force acting perpendicular to a surface divided by the area A over which the force acts:

$$P = \frac{F}{A} \tag{10.19}$$

SI Unit of Pressure: N/m^2 = pascal (Pa)

Equation 10.19 indicates that the SI unit for pressure is the unit of force divided by the unit of area, or newton/meter^2 (N/m^2). This unit of pressure is often referred to as a *pascal* (Pa), named after the French scientist Blaise Pascal (1623–1662).

Suppose we change the pressure on an object by an amount ΔP, where, as usual, the "delta" notation ΔP represents the final pressure P minus the initial pressure P_0: $\Delta P = P - P_0$. Because of this change in pressure, the volume of the object changes by an amount $\Delta V = V - V_0$, where V and V_0 are the final and initial volumes, respectively. Such a pressure change occurs, for example, when a swimmer dives deeper into the water. Experiment reveals that the change ΔP in pressure needed to change the volume by an amount ΔV is directly proportional to the fractional change $\Delta V/V_0$ in the volume:

$$\Delta P = -B\left(\frac{\Delta V}{V_0}\right) \tag{10.20}$$

This relation is analogous to Equations 10.17 and 10.18, except that the area A in those equations does not appear here explicitly; the area is already taken into account by the concept of pressure (force per unit area). The proportionality constant B is known as the

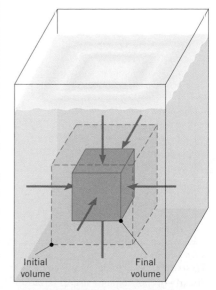

Figure 10.33 The arrows denote the forces that push perpendicularly on every surface of an object immersed in a liquid. The force per unit area is the pressure. When the pressure increases, the volume of the object decreases.

Table 10.3 *Values for the Bulk Modulus of Solid and Liquid Materials*

Material	Bulk Modulus B (N/m²)
Solids	
Aluminum	7.1×10^{10}
Brass	6.7×10^{10}
Copper	1.3×10^{11}
Lead	4.2×10^{10}
Nylon	6.1×10^{9}
Pyrex glass	2.6×10^{10}
Steel	1.4×10^{11}
Liquids	
Ethanol	8.9×10^{8}
Oil	1.7×10^{9}
Water	2.2×10^{9}

bulk modulus. The minus sign occurs because an increase in pressure (ΔP positive) always creates a decrease in volume (ΔV negative), and B is given as a positive quantity. Like Young's modulus and the shear modulus, the bulk modulus has units of force per unit area (N/m²), and its value depends on the nature of the material. Table 10.3 gives representative values of the bulk modulus.

 **Check Your Understanding 4**

Two rods are made from the same material. One has a circular cross section, and the other has a square cross section. The circle just fits within the square. When the same force is applied to stretch these rods, they each stretch by the same amount. Which rod is longer, or are they equal in length? *(The answer is given at the end of the book.)*

Background: Elastic deformation and Young's modulus hold the key to this question.

For similar questions (including calculational counterparts), consult Self-Assessment Test 10.2. This test is described at the end of Section 10.8.

10.8 Stress, Strain, and Hooke's Law

Table 10.4 *Stress and Strain Relations for Elastic Behavior*

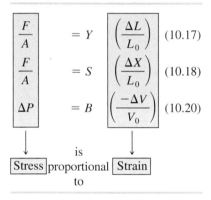

$$\frac{F}{A} = Y \left(\frac{\Delta L}{L_0} \right) \quad (10.17)$$

$$\frac{F}{A} = S \left(\frac{\Delta X}{L_0} \right) \quad (10.18)$$

$$\Delta P = B \left(\frac{-\Delta V}{V_0} \right) \quad (10.20)$$

$$\boxed{\text{Stress}} \underset{\text{to}}{\text{ is proportional }} \boxed{\text{Strain}}$$

Equations 10.17, 10.18, and 10.20 specify the amount of force needed for a given amount of elastic deformation, and they are repeated in Table 10.4 to emphasize their common features. The left side of each equation is the magnitude of the force per unit area required to cause an elastic deformation. In general, the ratio of the magnitude of the force to the area is called the *stress*. The right side of each equation involves the change in a quantity (ΔL, ΔX, or ΔV) divided by a quantity (L_0 or V_0) relative to which the change is compared. The terms $\Delta L/L_0$, $\Delta X/L_0$, and $\Delta V/V_0$ are unitless ratios, and each is referred to as the *strain* that results from the applied stress. In the case of stretch and compression, the strain is the fractional change in length, whereas in volume deformation it is the fractional change in volume. In shear deformation the strain refers to a change in shape of the object. Experiments show that these three equations, with constant values for Young's modulus, the shear modulus, and the bulk modulus, apply to a wide range of materials. Therefore, stress and strain are directly proportional to one another, a relationship first discovered by Robert Hooke (1635–1703) and now referred to as *Hooke's law.*

■ **HOOKE'S LAW FOR STRESS AND STRAIN**

Stress is directly proportional to strain.

SI Unit of Stress: newton per square meter = pascal (Pa)

SI Unit of Strain: Strain is a unitless quantity.

In reality, materials obey Hooke's law only up to a certain limit, as Figure 10.34 shows. As long as stress remains proportional to strain, a plot of stress versus strain is a straight line. The point on the graph where the material begins to deviate from straight-line behavior is called the "proportionality limit." Beyond the proportionality limit stress and strain are no longer directly proportional. However, if the stress does not exceed the "elastic limit" of the material, the object will return to its original size and shape once the stress is removed. The "elastic limit" is the point beyond which the object no longer returns to its original size and shape when the stress is removed; the object remains permanently deformed.

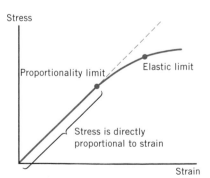

Figure 10.34 Hooke's law (stress is directly proportional to strain) is valid only up to the proportionality limit of a material. Beyond this limit, Hooke's law no longer applies. Beyond the elastic limit, the material remains deformed even when the stress is removed.

 Self-Assessment Test 10.2

Test your understanding of the material in Sections 10.3–10.8:

- Energy and Simple Harmonic Motion
- The Pendulum
- Damped Harmonic Motion and Resonance
- Elastic Deformation

Go to **www.wiley.com/college/cutnell**

10.9 *Concepts & Calculations*

This chapter has examined an important kind of vibratory motion known as simple harmonic motion. Specifically, it has discussed how the motion's displacement, velocity, and acceleration vary with time and explained what determines the frequency of the motion. In addition, it was seen that the elastic force is conservative, so that the total mechanical energy is conserved if nonconservative forces, such as friction and air resistance, are absent. We conclude now by presenting some examples that review important features of simple harmonic motion.

Concepts & Calculations Example 14
A Diver Vibrating in Simple Harmonic Motion

A 75-kg diver is standing at the end of a diving board while it is vibrating up and down in simple harmonic motion, as indicated in Figure 10.35. The diving board has an effective spring constant of $k = 4100$ N/m, and the vertical distance between the highest and lowest points in the motion is 0.30 m. (a) What is the amplitude of the motion? (b) Starting when the diver is at the highest point, what is his speed one-quarter of a period later? (c) If the vertical distance between his highest and lowest points were doubled to 0.60 m, what would be the time required for the diver to make one complete motional cycle?

Concept Questions and Answers How is the amplitude A related to the vertical distance between the highest and lowest points of the diver's motion?

> *Answer* The amplitude is the distance from the midpoint of the motion to either the highest or the lowest point. Thus, the amplitude is one-half the vertical distance between the highest and lowest points in the motion.

Starting from the top, where is the diver located one-quarter of a period later, and what can be said about his speed at this point?

> *Answer* The time for the diver to complete one motional cycle is defined as the period. In one cycle, the diver moves downward from the highest point to the lowest point and then moves upward and returns to the highest point. In a time equal to one-quarter of a period, the diver completes one-quarter of this cycle and, therefore, is halfway between the highest and lowest points. His speed is momentarily zero at the highest and lowest points and is a maximum at the halfway point.

If the amplitude of the motion were to double, would the period also double?

> *Answer* No. The period is the time to complete one cycle, and it is equal to the distance traveled during one cycle divided by the average speed. If the amplitude doubles, the distance also doubles. However, the average speed also doubles. We can verify this by examining Equation 10.7, which gives the diver's velocity as $v = -A\omega \sin \omega t$. The speed is the magnitude of this value, or $A\omega \sin \omega t$. Since the speed is proportional to the amplitude A, the speed at every point in the cycle also doubles when the amplitude doubles. Thus, the average speed doubles. However, the period, being the distance divided by the average speed, does not change.

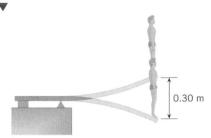

Figure 10.35 A diver at the end of a diving board is bouncing up and down in simple harmonic motion.

Solution

(a) Since the amplitude A is one-half the vertical distance between the highest and lowest points in the motion, $A = \frac{1}{2}(0.30 \text{ m}) = \boxed{0.15 \text{ m}}$.

(b) When the diver is halfway between the highest and lowest points, his speed is a maximum. The maximum speed of an object vibrating in simple harmonic motion is given by Equation 10.8 as $v_{max} = A\omega$, where A is the amplitude of the motion and ω is the angular frequency. The angular frequency can be determined from Equation 10.11 as $\omega = \sqrt{k/m}$, where k is the effective spring constant of the diving board and m is the mass of the diver. The maximum speed is

$$v_{max} = A\omega = A\sqrt{\frac{k}{m}} = (0.15 \text{ m})\sqrt{\frac{4100 \text{ N/m}}{75 \text{ kg}}} = \boxed{1.1 \text{ m/s}}$$

(c) The period is the same, regardless of the amplitude of the motion. From Equation 10.4 we know that the period T and the angular speed ω are related by $T = 2\pi/\omega$, where $\omega = \sqrt{k/m}$.

Thus, the period can be written as

$$T = \frac{2\pi}{\omega} = \frac{2\pi}{\sqrt{k/m}} = 2\pi\sqrt{\frac{m}{k}} = 2\pi\sqrt{\frac{75\ \text{kg}}{4100\ \text{N/m}}} = \boxed{0.85\ \text{s}}$$

As expected, the period does not depend on the amplitude of the motion.

Concepts & Calculations Example 15
Bungee Jumping and the Conservation of Mechanical Energy

A 68.0-kg bungee jumper is standing on a tall platform ($h_0 = 46.0$ m), as indicated in Figure 10.36. The bungee cord has an unstrained length of $L_0 = 9.00$ m, and when stretched, behaves like an ideal spring with a spring constant of $k = 66.0$ N/m. The jumper falls from rest, and the only forces acting on him are his weight and, for the latter part of the descent, the elastic force of the bungee cord. What is his speed (it is not zero) when he is at the following heights above the water (see the drawing): (a) $h_A = 37.0$ m and (b) $h_B = 15.0$ m?

Concept Questions and Answers Can we use the conservation of mechanical energy to find his speed at any point during the descent?

Answer Yes. His weight and the elastic force of the bungee cord are the only forces acting on him and are conservative forces. Therefore, the total mechanical energy remains constant (is conserved) during his descent.

What types of energy does he have when he is standing on the platform?

Answer Since he's at rest, he has neither translational nor rotational kinetic energy. The bungee cord is not stretched, so there is no elastic potential energy. Relative to the water, however, he does have gravitational potential energy, since he is 46.0 m above it.

What types of energy does he have at point A?

Answer Since he's moving downward, he possesses translational kinetic energy. He is not rotating though, so his rotational kinetic energy is zero. Because the bungee cord is still not stretched at this point, there is no elastic potential energy. But he still has gravitational potential energy relative to the water, because he is 37.0 m above it.

What types of energy does he have at point B?

Answer He has translational kinetic energy, because he's still moving downward. He's not rotating, so his rotational kinetic energy is zero. The bungee cord is stretched at this point, so there is elastic potential energy. He has gravitational potential energy, because he is still 15.0 m above the water.

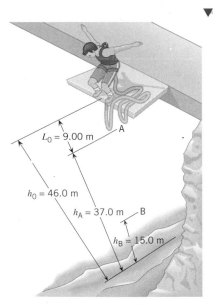

Figure 10.36 A bungee jumper jumps from a height of $h_0 = 46.0$ m. The length of the unstrained bungee cord is $L_0 = 9.00$ m.

Solution

(a) The total mechanical energy is the sum of the kinetic and potential energies, as expressed in Equation 10.14:

$$\underbrace{E}_{\substack{\text{Total}\\\text{mechanical}\\\text{energy}}} = \underbrace{\tfrac{1}{2}mv^2}_{\substack{\text{Translational}\\\text{kinetic}\\\text{energy}}} + \underbrace{\tfrac{1}{2}I\omega^2}_{\substack{\text{Rotational}\\\text{kinetic}\\\text{energy}}} + \underbrace{mgh}_{\substack{\text{Gravitational}\\\text{potential}\\\text{energy}}} + \underbrace{\tfrac{1}{2}kx^2}_{\substack{\text{Elastic}\\\text{potential}\\\text{energy}}}$$

The conservation of mechanical energy states that the total mechanical energy at point A is equal to that at the platform:

$$\underbrace{\tfrac{1}{2}mv_A^2 + \tfrac{1}{2}I\omega_A^2 + mgh_A + \tfrac{1}{2}kx_A^2}_{\text{Total mechanical energy at A}} = \underbrace{\tfrac{1}{2}mv_0^2 + \tfrac{1}{2}I\omega_0^2 + mgh_0 + \tfrac{1}{2}kx_0^2}_{\text{Total mechanical energy at the platform}}$$

While standing on the platform, the jumper is at rest, so $v_0 = 0$ m/s and $\omega_0 = 0$ rad/s. The bungee cord is not stretched, so $x_0 = 0$ m. At point A, the jumper is not rotating, $\omega_A = 0$ rad/s, and the bungee cord is still not stretched, $x_A = 0$ m. With these substitutions, the conservation of mechanical energy becomes

$$\tfrac{1}{2}mv_A^2 + mgh_A = mgh_0$$

Solving for the speed at point A yields

$$v_A = \sqrt{2g(h_0 - h_A)} = \sqrt{2(9.80\ \text{m/s}^2)(46.0\ \text{m} - 37.0\ \text{m})} = \boxed{13\ \text{m/s}}$$

(b) At point B the total mechanical energy is the same as it was on the platform, so

$$\underbrace{\tfrac{1}{2}mv_B^2 + \tfrac{1}{2}I\omega_B^2 + mgh_B + \tfrac{1}{2}kx_B^2}_{\text{Total mechanical energy at B}} = \underbrace{mgh_0}_{\substack{\text{Total mechanical} \\ \text{energy at} \\ \text{the platform}}}$$

We set $\omega_B = 0$ rad/s, since there is no rotational motion. Furthermore, the bungee cord stretches by an amount $x_B = h_0 - L_0 - h_B$ (see the drawing). Therefore, we have

$$v_B = \sqrt{2g(h_0 - h_B) - \left(\frac{k}{m}\right)(h_0 - L_0 - h_B)^2}$$

$$= \sqrt{2(9.80 \text{ m/s}^2)(46.0 \text{ m} - 15.0 \text{ m}) - \left(\frac{66.0 \text{ N/m}}{68.0 \text{ kg}}\right)(46.0 \text{ m} - 9.00 \text{ m} - 15.0 \text{ m})^2}$$

$$= \boxed{11.7 \text{ m/s}}$$

At the end of the problem set for this chapter, you will find homework problems that contain both conceptual and quantitative parts. These problems are grouped under the heading *Concepts & Calculations, Group Learning Problems.* They are designed for use by students working alone or in small learning groups. The conceptual part of each problem provides a convenient focus for group discussions.

Concept Summary

This summary presents an abridged version of the chapter, including the important equations and all available learning aids. For convenient reference, the learning aids (including the text's examples) are placed next to or immediately after the relevant equation or discussion. The following learning aids may be found on-line at **www.wiley.com/college/cutnell**:

Interactive LearningWare examples are solved according to a five-step interactive format that is designed to help you develop problem-solving skills.	**Concept Simulations** are animated versions of text figures or animations that illustrate important concepts. You can control parameters that affect the display, and we encourage you to experiment.
Interactive Solutions offer specific models for certain types of problems in the chapter homework. The calculations are carried out interactively.	**Self-Assessment Tests** include both qualitative and quantitative questions. Extensive feedback is provided for both incorrect and correct answers, to help you evaluate your understanding of the material.

Topic	*Discussion*	*Learning Aids*

10.1 The Ideal Spring and Simple Harmonic Motion

Force applied to an ideal spring

The force that must be applied to stretch or compress an ideal spring is

$$F_{\text{Applied}} = kx \qquad (10.1)$$

where k is the spring constant and x is the displacement of the spring from its unstrained length.

Examples 1, 2

Interactive Solution 10.9

Restoring force of an ideal spring

A spring exerts a restoring force on an object attached to the spring. The restoring force F produced by an ideal spring is

$$F = -kx \qquad (10.2)$$

where the minus sign indicates that the restoring force points opposite to the displacement of the spring.

Simple harmonic motion

Simple harmonic motion is the oscillatory motion that occurs when a restoring force of the form $F = -kx$ acts on an object. A graphical record of position versus time for an object in simple harmonic motion is sinusoidal. The ampli-

Amplitude

tude A of the motion is the maximum distance that the object moves away from its equilibrium position.

10.2 Simple Harmonic Motion and the Reference Circle

The period T of simple harmonic motion is the time required to complete one cycle of the motion, and the frequency f is the number of cycles per second that

Topic	Discussion	Learning Aids
	occurs. Frequency and period are related according to	
Period and frequency	$$f = \frac{1}{T} \qquad (10.5)$$	
Angular frequency	The frequency f (in Hz) is related to the angular frequency ω (in rad/s) according to $$\omega = 2\pi f \qquad (\omega \text{ in rad/s}) \qquad (10.6)$$	
Maximum speed	The maximum speed of an object in simple harmonic motion is $$v_{\max} = A\omega \qquad (\omega \text{ in rad/s}) \qquad (10.8)$$ where A is the amplitude of the motion.	**Examples 3, 4**
Maximum acceleration	The maximum acceleration of an object in simple harmonic motion is $$a_{\max} = A\omega^2 \qquad (\omega \text{ in rad/s}) \qquad (10.10)$$	**Example 5** **Interactive Solution 10.75**
Angular frequency of simple harmonic motion	The angular frequency of simple harmonic motion is $$\omega = \sqrt{\frac{k}{m}} \qquad (\omega \text{ in rad/s}) \qquad (10.11)$$	**Examples 6, 14** **Interactive LearningWare 10.1**

Use Self-Assessment Test 10.1 to evaluate your understanding of Sections 10.1 and 10.2.

	10.3 Energy and Simple Harmonic Motion	
Elastic potential energy	The elastic potential energy of an object attached to an ideal spring is $$PE_{\text{elastic}} = \tfrac{1}{2}kx^2 \qquad (10.13)$$	
Total mechanical energy	The total mechanical energy E of such a system is the sum of its translational and rotational kinetic energies, gravitational potential energy, and elastic potential energy: $$E = \tfrac{1}{2}mv^2 + \tfrac{1}{2}I\omega^2 + mgh + \tfrac{1}{2}kx^2 \qquad (10.14)$$	**Examples 7, 8, 9, 15** **Interactive LearningWare 10.2** **Concept Simulation 10.1**
Conservation of mechanical energy	If external nonconservative forces like friction do no net work, the total mechanical energy of the system is conserved: $$E_{\text{f}} = E_0$$	**Interactive Solutions 10.31, 10.35, 10.77**
	10.4 The Pendulum	
	A simple pendulum is a particle of mass m attached to a frictionless pivot by a cable whose length is L and whose mass is negligible. The small-angle ($\leq 10°$) back-and-forth swinging of a simple pendulum is simple harmonic motion, but large-angle movement is not. The frequency f of the motion is given by	**Concept Simulation 10.2**
Frequency of a simple pendulum	$$2\pi f = \sqrt{\frac{g}{L}} \qquad \text{(small angles only)} \qquad (10.16)$$	**Example 10**
	A physical pendulum consists of a rigid object, with moment of inertia I and mass m, suspended from a frictionless pivot. For small-angle displacements, the frequency f of simple harmonic motion for a physical pendulum is given by	
Frequency of a physical pendulum	$$2\pi f = \sqrt{\frac{mgL}{I}} \qquad \text{(small angles only)} \qquad (10.15)$$ where L is the distance between the axis of rotation and the center of gravity of the rigid object.	**Example 11**
	10.5 Damped Harmonic Motion	
Damped harmonic motion **Critical damping**	Damped harmonic motion is motion in which the amplitude of oscillation decreases as time passes. Critical damping is the minimum degree of damping that eliminates any oscillations in the motion as the object returns to its equilibrium position.	**Concept Simulation 10.3**
	10.6 Driven Harmonic Motion and Resonance	
Driven harmonic motion **Resonance**	Driven harmonic motion occurs when a driving force acts on an object along with the restoring force. Resonance is the condition under which the driving force can transmit large amounts of energy to an oscillating object, leading to large-	

Topic	Discussion	Learning Aids

amplitude motion. In the absence of damping, resonance occurs when the frequency of the driving force matches a natural frequency at which the object oscillates.

10.7 Elastic Deformation

One type of elastic deformation is stretch and compression. The magnitude F of the force required to stretch or compress an object of length L_0 and cross-sectional area A by an amount ΔL is (see Figure 10.29)

Young's modulus

$$F = Y\left(\frac{\Delta L}{L_0}\right)A \qquad (10.17) \quad \textbf{Example 12}$$

where Y is a constant called Young's modulus.

Another type of elastic deformation is shear. The magnitude F of the shearing force required to create an amount of shear ΔX for an object of thickness L_0 and cross-sectional area A is (see Figure 10.31)

Shear modulus

$$F = S\left(\frac{\Delta X}{L_0}\right)A \qquad (10.18) \quad \textbf{Example 13}$$

where S is a constant called the shear modulus.

A third type of elastic deformation is volume deformation, which has to do with pressure. The pressure P is the magnitude F of the force acting perpendicular to a surface divided by the area A over which the force acts:

Pressure

$$P = \frac{F}{A} \qquad (10.19)$$

The SI unit for pressure is N/m^2, a unit known as a pascal (Pa): $1\ Pa = 1\ N/m^2$.

The change ΔP in pressure needed to change the volume V_0 of an object by an amount ΔV is (see Figure 10.33)

Bulk modulus

$$\Delta P = -B\left(\frac{\Delta V}{V_0}\right) \qquad (10.20)$$

where B is a constant known as the bulk modulus.

10.8 Stress, Strain, and Hooke's Law

Stress and strain

Stress is the magnitude of the force per unit area applied to an object and causes strain. For stretch/compression, the strain is the fractional change $\Delta L/L_0$ in length. For shear, the strain reflects the change in shape of the object and is given by $\Delta X/L_0$ (see Figure 10.31). For volume deformation, the strain is the

Hooke's law

fractional change in volume $\Delta V/V_0$. Hooke's law states that stress is directly proportional to strain.

Use *Self-Assessment Test 10.2* **to evaluate your understanding of Sections 10.3–10.8.**

Conceptual Questions

1. Two people pull on a horizontal spring that is attached to an immovable wall. Then, they detach it from the wall and pull on opposite ends of the horizontal spring. They pull just as hard in each case. In which situation, if either, does the spring stretch more? Account for your answer.

2. The drawing shows identical springs that are attached to a box in two different ways. Initially, the springs are unstrained. The box is then pulled to the right and released. In each case the initial displacement of the box is the same. At the moment of release, which box, if either, experiences the greater net force due to the springs? Provide a reason for your answer.

3. In Figures 10.11 and 10.14, the shadow moves in simple harmonic motion. Where on the shadow's path is (a) the velocity equal to zero and (b) the acceleration equal to zero?

4. A steel ball is dropped onto a concrete floor. Over and over again, it rebounds to its original height. Is the motion simple harmonic motion? Justify your answer.

5. Ignoring the damping introduced by the shock absorbers, explain why the number of passengers in a car affects the vibration frequency of the car's suspension system.

6. Review Conceptual Example 8 before answering this question. A block is attached to a horizontal spring and slides back and forth in simple harmonic motion on a frictionless horizontal surface. A second identical block is suddenly attached to the first block. The attachment is accomplished by joining the blocks at one extreme end of the oscillation cycle. The velocities of the blocks are exactly matched at the instant of joining. Explain how (a) the amplitude, (b) the frequency, and (c) the maximum speed of the oscillation change.

7. A particle is oscillating in simple harmonic motion. The time required for the particle to travel through one complete cycle is equal to the period of the motion, no matter what the amplitude is. But how can this be, since larger amplitudes mean that the particle travels farther? Explain.

8. An electric saber saw consists of a blade that is driven back and forth by a pin mounted on the circumference of a rotating circular disk. As the disk rotates at a constant angular speed, the pin engages a slot and forces the blade back and forth, in the sequence (*a*), (*b*), (*c*), and (*d*) shown in the drawing. Is the motion of the blade simple harmonic motion? Explain.

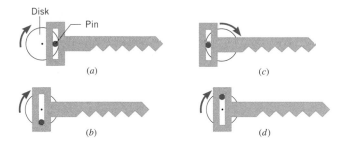

(*a*)   (*c*)

(*b*)   (*d*)

9. Is more elastic potential energy stored in a spring when the spring is compressed by one centimeter than when it is stretched by the same amount? Explain.

10. Suppose that a grandfather clock (a simple pendulum) is running slowly. That is, the time it takes to complete each cycle is longer than it should be. Should one shorten or lengthen the pendulum to make the clock keep the correct time? Why?

11. Consult **Concept Simulation 10.2** at **www.wiley.com/college/ cutnell** to review the concept that lies at the heart of the answer to this question. In principle, the motion of a simple pendulum and an object on an ideal spring can both be used to provide the basic time interval or period used in a clock. Which of the two kinds of clocks becomes more inaccurate when carried to the top of a high mountain? Justify your answer.

12. **Concept Simulation 10.2** at **www.wiley.com/college/cutnell** deals with the concept on which this question is based. Suppose you were kidnapped and held prisoner by space invaders in a completely isolated room, with nothing but a watch and a pair of shoes (including shoelaces of known length). Explain how you might determine whether this room is on earth or on the moon.

13. Two people are sitting on playground swings. One person is pulled back 4° from the vertical and released, while the other is pulled back 8° from the vertical and released. If the two swings are started together, will they both come back to the starting points at the same time? Justify your answer.

14. A car travels over a road that contains a series of equally spaced bumps. Explain why a particularly jarring ride can result if the horizontal velocity of the car, the bump spacing, and the oscillation frequency of the car's suspension system are properly "matched."

15. The drawing shows two cylinders. They are identical in all respects, except one is hollow. In

a setup like that in Figure 10.29, identical forces are applied to the right end of each cylinder. Which cylinder, if either, stretches the most? Why?

16. Young's modulus for steel is thousands of times greater than that for rubber. All other factors being equal, does this mean that steel stretches much more easily than rubber? Explain.

17. A trash compactor crushes empty aluminum cans, thereby reducing the total volume of the cans by 75%. Can the value given in Table 10.3 for the bulk modulus of aluminum be used to calculate the change in pressure generated in the trash compactor? Give a reason for your answer.

18. Both sides of the relation $F = S(\Delta X/L_0)A$ (Equation 10.18) can be divided by the area A to give F/A on the left side. Why can't this F/A term be called a pressure, such as the pressure that appears in $\Delta P = -B(\Delta V/V_0)$ (Equation 10.20)?

19. The block in the drawing rests on the ground. Which face, A, B, or C, experiences (a) the largest stress and (b) the smallest stress when the block is resting on it? Explain.

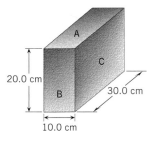

Problems

Note: Unless otherwise indicated, the values for Young's modulus Y, the shear modulus S, and the bulk modulus B are given, respectively, in Table 10.1, Table 10.2, and Table 10.3.

ssm Solution is in the Student Solutions Manual.   **www** Solution is available on the World Wide Web at www.wiley.com/college/cutnell

☤ This icon represents a biomedical application.

Section 10.1 The Ideal Spring and Simple Harmonic Motion

1. ssm A hand exerciser utilizes a coiled spring. A force of 89.0 N is required to compress the spring by 0.0191 m. Determine the force needed to compress the spring by 0.0508 m.

2. When the rubber band in a slingshot is stretched, it obeys Hooke's law. Suppose that the "spring constant" for the rubber band is $k = 44$ N/m. When the rubber band is pulled back with a force of 9.6 N, how far does it stretch?

3. A spring has a spring constant of 248 N/m. Find the magnitude of the force needed (a) to stretch the spring by 3.00×10^{-2} m from

its unstrained length and (b) to compress the spring by the same amount.

4. The graph shows the force F that an archer applies to the string of a long bow versus the string's displacement x. Drawing back this bow is analogous to stretching a spring. From the data in the graph determine the effective spring constant of the bow.

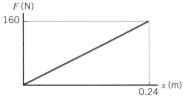

5. ssm A car is hauling a 92-kg trailer, to which it is connected by a spring. The spring constant is 2300 N/m. The car accelerates with an acceleration of 0.30 m/s². By how much does the spring stretch?

6. A 0.70-kg block is hung from and stretches a spring that is attached to the ceiling. A second block is attached to the first one, and the amount that the spring stretches from its unstrained length triples. What is the mass of the second block?

*7. ssm A small ball is attached to one end of a spring that has an unstrained length of 0.200 m. The spring is held by the other end, and the ball is whirled around in a horizontal circle at a speed of 3.00 m/s. The spring remains nearly parallel to the ground during the motion and is observed to stretch by 0.010 m. By how much would the spring stretch if it were attached to the ceiling and the ball allowed to hang straight down, motionless?

*8. Review Conceptual Example 2 as an aid in solving this problem. An object is attached to the lower end of a 100-coil spring that is hanging from the ceiling. The spring stretches by 0.160 m. The spring is then cut into two identical springs of 50 coils each. As the drawing shows, each spring is attached between the ceiling and the object. By how much does each spring stretch?

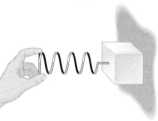

50-coil spring

*9. Interactive Solution 10.9 at **www.wiley.com/college/ cutnell** discusses a method used to solve this problem. To measure the static friction co-efficient between a 1.6-kg block and a vertical wall, the setup shown in the drawing is used. A spring (spring con-stant = 510 N/m) is attached to the block. Someone pushes on the end of the spring in a direction perpendicular to the wall until the block does not slip downward. If the spring in such a setup is com-pressed by 0.039 m, what is the coefficient of static friction?

*10. A 10.1-kg uniform board is wedged into a corner and held by a spring at a 50.0° angle, as the drawing shows. The spring has a spring constant of 176 N/m and is parallel to the floor. Find the amount by which the spring is stretched from its unstrained length.

*11. ssm In 0.750 s, a 7.00-kg block is pulled through a distance of 4.00 m on a frictionless horizontal surface, starting from rest. The block has a constant acceleration and is pulled by means of a horizontal spring that is attached to the block. The spring constant of the spring is 415 N/m. By how much does the spring stretch?

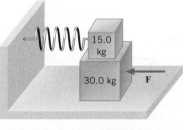

Problem 10

50.0°

12. A 30.0-kg block is resting on a flat horizontal table. On top of this block is resting a 15.0-kg block, to which a horizontal spring is attached, as the drawing illustrates. The spring constant of the spring is 325 N/m. The coefficient of kinetic friction between the lower block and the table is 0.600, and the coefficient of static friction between the two blocks is 0.900. A horizontal force **F is applied to the lower block as shown. This force is increasing in such a way as to keep the blocks

15.0 kg

30.0 kg   **F**

moving at a *constant speed*. At the point where the upper block begins to slip on the lower block, determine (a) the amount by which the spring is compressed and (b) the magnitude of the force **F**.

**13. A 15.0-kg block rests on a horizontal table and is attached to one end of a massless, horizontal spring. By pulling horizontally on the other end of the spring, someone causes the block to accelerate uniformly and reach a speed of 5.00 m/s in 0.500 s. In the process, the spring is stretched by 0.200 m. The block is then pulled at a *constant speed* of 5.00 m/s, during which time the spring is stretched by only 0.0500 m. Find (a) the spring constant of the spring and (b) the coefficient of kinetic friction between the block and the table.

Section 10.2 Simple Harmonic Motion and the Reference Circle

14. A loudspeaker diaphragm is producing a sound for 2.5 s by moving back and forth in simple harmonic motion. The angular fre-quency of the motion is 7.54×10^4 rad/s. How many times does the diaphragm move back and forth?

15. ssm The shock absorbers in the suspension system of a car are in such bad shape that they have no effect on the behavior of the springs attached to the axles. Each of the identical springs attached to the front axle supports 320 kg. A person pushes down on the mid-dle of the front end of the car and notices that it vibrates through five cycles in 3.0 s. Find the spring constant of either spring.

16. Refer to Conceptual Example 2 as an aid in solving this prob-lem. A 100-coil spring has a spring constant of 420 N/m. It is cut into four shorter springs, each of which has 25 coils. One end of a 25-coil spring is attached to a wall. An object of mass 46 kg is at-tached to the other end of the spring, and the system is set into hori-zontal oscillation. What is the angular frequency of the motion?

17. Concept Simulation 10.3 at **www.wiley.com/ college/cutnell** illustrates the concepts pertinent to this problem. An 0.80-kg object is attached to one end of a spring, as in

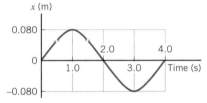

Figure 10.6, and the system is set into simple harmonic motion. The displacement x of the object as a function of time is shown in the drawing. With the aid of these data, determine (a) the amplitude A of the motion, (b) the angular frequency ω, (c) the spring constant k, (d) the speed of the object at $t = 1.0$ s, and (e) the magnitude of the ob-ject's acceleration at $t = 1.0$ s.

18. A computer to be used in a satellite must be able to withstand accelerations of up to 25 times the acceleration due to gravity. In a test to see whether it meets this specification, the computer is bolted to a frame that is vibrated back and forth in simple harmonic motion at a frequency of 9.5 Hz. What is the minimum amplitude of vibra-tion that must be used in this test?

*19. Objects of equal mass are oscillating up and down in simple harmonic motion on two different vertical springs. The spring con-stant of spring 1 is 174 N/m. The motion of the object on spring 1 has twice the amplitude as the motion of the object on spring 2. The magnitude of the maximum velocity is the same in each case. Find the spring constant of spring 2.

*20. When an object of mass m_1 is hung on a vertical spring and set into vertical simple harmonic motion, its frequency is 12.0 Hz. When another object of mass m_2 is hung on the spring along with m_1, the frequency of the motion is 4.00 Hz. Find the ratio m_2/m_1 of the masses.

*21. ssm Interactive LearningWare 10.1 at **www.wiley.com/col-lege/cutnell** reviews the concepts involved in this problem. A spring stretches by 0.018 m when a 2.8-kg object is suspended from its end. How much mass should be attached to this spring so that its frequency of vibration is $f = 3.0$ Hz?

* **22.** A 3.0-kg block is placed between two horizontal springs. Neither spring is strained when the block is located at the position labeled $x = 0$ m in the drawing. The block is then displaced a distance of 0.070 m from the position where $x = 0$ m and released from rest. (a) What is the speed of the block when it passes back through the $x = 0$ m position? (b) Determine the angular frequency ω of this system.

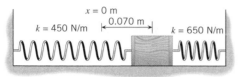

** **23.** A tray is moved horizontally back and forth in simple harmonic motion at a frequency of $f = 2.00$ Hz. On this tray is an empty cup. Obtain the coefficient of static friction between the tray and the cup, given that the cup begins slipping when the amplitude of the motion is 5.00×10^{-2} m.

Section 10.3 Energy and Simple Harmonic Motion

24. An archer pulls the bowstring back for a distance of 0.470 m before releasing the arrow. The bow and string act like a spring whose spring constant is 425 N/m. (a) What is the elastic potential energy of the drawn bow? (b) The arrow has a mass of 0.0300 kg. How fast is it traveling when it leaves the bow?

25. ssm Concept Simulation 10.1 at **www.wiley.com/college/cutnell** allows you to explore the concepts to which this problem relates. A 2.00-kg object is hanging from the end of a vertical spring. The spring constant is 50.0 N/m. The object is pulled 0.200 m downward and released from rest. Complete the table below by calculating the translational kinetic energy, the gravitational potential energy, the elastic potential energy, and the total mechanical energy E for each of the vertical positions indicated. The vertical positions h indicate distances above the point of release, where $h = 0$ m.

h (meters)	KE	PE (gravity)	PE (elastic)	E
0				
0.200				
0.400				

26. In preparation for shooting a ball in a pinball machine, a spring ($k = 675$ N/m) is compressed by 0.0650 m relative to its unstrained length. The ball ($m = 0.0585$ kg) is at rest against the spring at point A. When the spring is released, the ball slides (without rolling) to point B, which is 0.300 m higher than point A. How fast is the ball moving at B?

27. A spring is hung from the ceiling. A 0.450-kg block is then attached to the free end of the spring. When released from rest, the block drops 0.150 m before momentarily coming to rest. (a) What is the spring constant of the spring? (b) Find the angular frequency of the block's vibrations.

28. A vertical spring with a spring constant of 450 N/m is mounted on the floor. From directly above the spring, which is unstrained, a 0.30-kg block is dropped from rest. It collides with and sticks to the spring, which is compressed by 2.5 cm in bringing the block to a momentary halt. Assuming air resistance is negligible, from what height (in cm) above the compressed spring was the block dropped?

29. ssm A 1.00×10^{-2}-kg block is resting on a horizontal frictionless surface and is attached to a horizontal spring whose spring constant is 124 N/m. The block is shoved parallel to the spring axis and is given an initial speed of 8.00 m/s, while the spring is initially unstrained. What is the amplitude of the resulting simple harmonic motion?

30. A 3.2-kg block is hanging stationary from the end of a vertical spring that is attached to the ceiling. The elastic potential energy of this spring/mass system is 1.8 J. What is the elastic potential energy of the system when the 3.2-kg block is replaced by a 5.0-kg block?

31. Refer to Interactive Solution 10.31 at **www.wiley.com/college/cutnell** for help in solving this problem. A heavy-duty stapling gun uses a 0.140-kg metal rod that rams against the staple to eject it. The rod is pushed by a stiff spring called a "ram spring" ($k = 32\ 000$ N/m). The mass of this spring may be ignored. Squeezing the handle of the gun first compresses the ram spring by 3.0×10^{-2} m from its unstrained length and then releases it. Assuming that the ram spring is oriented vertically and is still compressed by 0.8×10^{-2} m when the downward-moving ram hits the staple, find the speed of the ram at the instant of contact.

* **32.** Review Conceptual Example 8 before starting this problem. A block is attached to a horizontal spring and oscillates back and forth on a frictionless horizontal surface at a frequency of 3.00 Hz. The amplitude of the motion is 5.08×10^{-2} m. At the point where the block has its maximum speed, it suddenly splits into two identical parts, only one part remaining attached to the spring. (a) What is the amplitude and the frequency of the simple harmonic motion that exists after the block splits? (b) Repeat part (a), assuming that the block splits when it is at one of its extreme positions.

* **33. ssm** A 1.1-kg object is suspended from a vertical spring whose spring constant is 120 N/m. (a) Find the amount by which the spring is stretched from its unstrained length. (b) The object is pulled straight down by an additional distance of 0.20 m and released from rest. Find the speed with which the object passes through its original position on the way up.

* **34.** An 86.0-kg climber is scaling the vertical wall of a mountain. His safety rope is made of nylon that, when stretched, behaves like a spring with a spring constant of 1.20×10^3 N/m. He accidentally slips and falls freely for 0.750 m before the rope runs out of slack. How much is the rope stretched when it breaks his fall and momentarily brings him to rest?

* **35.** Refer to Interactive Solution 10.35 at **www.wiley.com/college/cutnell** to review a method by which this problem can be solved. An 11.2-kg block and a 21.7-kg block are resting on a horizontal frictionless surface. Between the two is squeezed a spring (spring constant = 1330 N/m). The spring is compressed by 0.141 m from its unstrained length and is not attached permanently to either block. With what speed does each block move away after the mechanism keeping the spring squeezed is released and the spring falls away?

** **36.** A 1.00×10^{-2}-kg bullet is fired horizontally into a 2.50-kg wooden block attached to one end of a massless, horizontal spring ($k = 845$ N/m). The other end of the spring is fixed in place, and the spring is unstrained initially. The block rests on a horizontal, frictionless surface. The bullet strikes the block perpendicularly and quickly comes to a halt within it. As a result of this completely inelastic collision, the spring is compressed along its axis and causes the block/bullet to oscillate with an amplitude of 0.200 m. What is the speed of the bullet?

** **37. ssm** A 70.0-kg circus performer is fired from a cannon that is elevated at an angle of 40.0° above the horizontal. The cannon uses strong elastic bands to propel the performer, much in the same way that a slingshot fires a stone. Setting up for this stunt involves stretching the bands by 3.00 m from their unstrained length. At the point where the performer flies free of the bands, his height above the floor is the same as that of the net into which he is shot. He takes 2.14 s to travel the horizontal distance of 26.8 m between this point and the net. Ignore friction and air resistance and determine the effective spring constant of the firing mechanism.

** 38. **Interactive LearningWare 10.2** at **www.wiley.com/college/cutnell** explores the approach taken in problems such as this one. A spring is mounted vertically on the floor. The mass of the spring is negligible. A certain object is placed on the spring to compress it. When the object is pushed further down by just a bit and then released, one up/down oscillation cycle occurs in 0.250 s. However, when the object is pushed down by 5.00×10^{-2} m to point P and then released, the object flies entirely off the spring. To what height above point P does the object rise in the absence of air resistance?

Section 10.4 The Pendulum

39. If the period of a simple pendulum is to be 2.0 s, what should be its length?

40. A simple pendulum is made from a 0.65-m-long string and a small ball attached to its free end. The ball is pulled to one side through a small angle and then released from rest. After the ball is released, how much time elapses before it attains its greatest speed?

41. **ssm** **Concept Simulation 10.2** at **www.wiley.com/college/cutnell** allows you to explore the effect of the acceleration due to gravity on pendulum motion, which is the focus of this problem. Astronauts on a distant planet set up a simple pendulum of length 1.2 m. The pendulum executes simple harmonic motion and makes 100 complete vibrations in 280 s. What is the acceleration due to gravity?

42. A pendulum clock can be approximated as a simple pendulum of length 1.00 m and keeps accurate time at a location where $g = 9.83$ m/s². In a location where $g = 9.78$ m/s², what must be the new length of the pendulum, such that the clock continues to keep accurate time (that is, its period remains the same)?

* 43. **ssm** Pendulum A is a physical pendulum made from a thin, rigid, and uniform rod whose length is d. One end of this rod is attached to the ceiling by a frictionless hinge, so the rod is free to swing back and forth. Pendulum B is a simple pendulum whose length is also d. Obtain the ratio T_A/T_B of their periods for small-angle oscillations.

* 44. A pendulum is constructed from a thin, rigid, and uniform rod with a small sphere attached to the end opposite the pivot. This arrangement is a good approximation to a simple pendulum (period = 0.66 s), because the mass of the sphere (lead) is much greater than the mass of the rod (aluminum). When the sphere is removed, the pendulum no longer is a simple pendulum, but is then a physical pendulum. What is the period of the physical pendulum?

** 45. A point on the surface of a solid sphere (radius = R) is attached directly to a pivot on the ceiling. The sphere swings back and forth as a physical pendulum with a small amplitude. What is the length of a simple pendulum that has the same period as this physical pendulum? Give your answer in terms of R.

Section 10.7 Elastic Deformation, Section 10.8 Stress, Strain, and Hooke's Law

46. A tow truck is pulling a car out of a ditch by means of a steel cable that is 9.1 m long and has a radius of 0.50 cm. When the car just begins to move, the tension in the cable is 890 N. How much has the cable stretched?

47. **ssm** **www** A 3500-kg statue is placed on top of a cylindrical concrete ($Y = 2.3 \times 10^{10}$ N/m²) stand. The stand has a cross-

sectional area of 7.3×10^{-2} m² and a height of 1.8 m. By how much does the statue compress the stand?

48. A copper cube, 0.30 m on a side, is subjected to two shearing forces, each of magnitude $F = 6.0 \times 10^6$ N (see the drawing). Find the angle θ (in degrees), which is one measure of how the shape of the block has been altered by shear deformation.

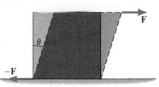

49. **ssm** Two metal beams are joined together by four rivets, as the drawing indicates. Each rivet has a radius of 5.0×10^{-3} m and is to be exposed to a shearing stress of no more than 5.0×10^8 Pa. What is the maximum tension **T** that can be applied to each beam, assuming that each rivet carries one-fourth of the total load?

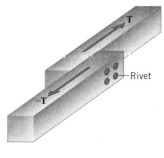

50. An 1800-kg car, being lifted at a steady speed by a crane, hangs at the end of a cable whose radius is 6.0×10^{-3} m. The cable is 15 m in length and stretches by 8.0×10^{-3} m because of the weight of the car. Determine (a) the stress, (b) the strain, and (c) Young's modulus for the cable.

51. The pressure increases by 1.0×10^4 N/m² for every meter of depth beneath the surface of the ocean. At what depth does the volume of a Pyrex glass cube, 1.0×10^{-2} m on an edge at the ocean's surface, decrease by 1.0×10^{-10} m³?

52. The femur is a bone in the leg whose minimum cross-sectional area is about 4.0×10^{-4} m². A compressional force in excess of 6.8×10^4 N will fracture this bone. (a) Find the maximum stress that this bone can withstand. (b) What is the strain that exists under a maximum-stress condition?

53. **ssm** A piece of aluminum is surrounded by air at a pressure of 1.01×10^5 Pa. The aluminum is placed in a vacuum chamber where the pressure is reduced to zero. Determine the fractional change $\Delta V/V_0$ in the volume of the aluminum.

54. When subjected to a force of compression, the length of a bone decreases by 2.7×10^{-5} m. When this same bone is subjected to a tensile force of the same magnitude, by how much does it stretch?

55. The shovel of a backhoe is controlled by hydraulic cylinders that are moved by oil under pressure. Determine the volume strain $\Delta V/V_0$ (including the algebraic sign) experienced by the oil, when the pressure increases from 1.8×10^5 Pa to 6.5×10^5 Pa while the shovel is digging a trench.

56. A copper cylinder and a brass cylinder are stacked end to end, as in the drawing. Each cylinder has a radius of 0.25 cm. A compressive force of $F = 6500$ N is applied to the right end of the brass cylinder. Find the amount by which the length of the stack decreases.

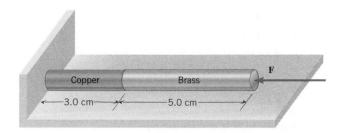

*57. A die is designed to punch holes of radii 1.00×10^{-2} m in a metal sheet that is 3.0×10^{-3} m thick, as the drawing illustrates. To punch through the sheet, the die must exert a shearing stress of 3.5×10^8 Pa. What force **F** must be applied to the die?

*58. ☤ A gymnast does a one-arm handstand. The humerus, which is the upper arm bone between the elbow and the shoulder joint, may be approximated as a 0.30-m-long cylinder with an outer radius of 1.00×10^{-2} m and a hollow inner core with a radius of 4.0×10^{-3} m. Excluding the arm, the mass of the gymnast is 63 kg. (a) What is the compressional strain of the humerus? (b) By how much is the humerus compressed?

*59. **ssm** A helicopter is lifting a 2100-kg jeep. The steel suspension cable is 48 m long and has a radius of 5.0×10^{-3} m. (a) Find the amount that the cable is stretched when the jeep is suspended motionless in the air. (b) What is the amount of cable stretch when the jeep is hoisted upward with an acceleration of 1.5 m/s²?

*60. A block of copper is securely fastened to the floor. A force of 1800 N is applied to the top surface of the block, as the drawing shows. Find (a) the amount by which the height of the block is changed and (b) the shear deformation of the block.

*61. **ssm www** An 8.0-kg stone at the end of a steel wire is being whirled in a circle at a constant tangential speed of 12 m/s. The stone is moving on the surface of a frictionless horizontal table. The wire is 4.0 m long and has a radius of 1.0×10^{-3} m. Find the strain in the wire.

*62. Two rods are identical in all respects except one: one rod is made from aluminum, and the other from tungsten. The rods are

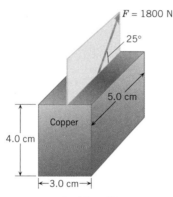

Problem 57

Problem 60

joined end to end, in order to make a single rod that is twice as long as either the aluminum or tungsten rod. What is the effective value of Young's modulus for this composite rod? That is, what value $Y_{\text{Composite}}$ of Young's modulus should be used in Equation 10.17 when applied to the composite rod? Note that the change $\Delta L_{\text{Composite}}$ in the length of the composite rod is the sum of the changes in length of the aluminum and tungsten rods.

*63. A 1.0×10^{-3}-kg spider is hanging vertically by a thread that has a Young's modulus of 4.5×10^9 N/m² and a radius of 13×10^{-6} m. Suppose that a 95-kg person is hanging vertically on an aluminum wire. What is the radius of the wire that would exhibit the same strain as the spider's thread, when the thread is stressed by the full weight of the spider?

*64. A square plate is 1.0×10^{-2} m thick, measures 3.0×10^{-2} m on a side, and has a mass of 7.2×10^{-2} kg. The shear modulus of the material is 2.0×10^{10} N/m². One of the square faces rests on a flat horizontal surface, and the coefficient of static friction between the plate and the surface is 0.90. A force is applied to the top of the plate, as in Figure 10.31a. Determine (a) the maximum possible amount of shear stress, (b) the maximum possible amount of shear strain, and (c) the maximum possible amount of shear deformation ΔX (see Figure 10.31b) that can be created by the applied force just before the plate begins to move.

65. ssm www A solid brass sphere is subjected to a pressure of 1.0×10^5 Pa due to the earth's atmosphere. On Venus the pressure due to the atmosphere is 9.0×10^6 Pa. By what fraction $\Delta r/r_0$ (including the algebraic sign) does the radius of the sphere change when it is exposed to the Venusian atmosphere? Assume that the change in radius is very small relative to the initial radius.

66. ☤ A cylindrically shaped piece of collagen (a substance found in the body in connective tissue) is being stretched by a force that increases from 0 to 3.0×10^{-2} N. The length and radius of the collagen are, respectively, 2.5 and 0.091 cm, and Young's modulus is 3.1×10^6 N/m². (a) If the stretching obeys Hooke's law, what is the spring constant k for collagen? (b) How much work is done by the variable force that stretches the collagen? (See Section 6.9 for a discussion of the work done by a variable force.)

Additional Problems

67. ssm www Atoms in a solid are not stationary, but vibrate about their equilibrium positions. Typically, the frequency of vibration is about $f = 2.0 \times 10^{12}$ Hz, and the amplitude is about 1.1×10^{-11} m. For a typical atom, what is its (a) maximum speed and (b) maximum acceleration?

68. In Concept Simulation 10.3 at **www.wiley.com/college/cutnell** you can explore the concepts that are important in this problem. A block of mass $m = 0.750$ kg is fastened to an unstrained horizontal spring whose spring constant is $k = 82.0$ N/m. The block is given a displacement of $+0.120$ m, where the + sign indicates that the displacement is along the $+x$ axis, and then released from rest. (a) What is the force (magnitude and direction) that the spring exerts on the block just before the block is released? (b) Find the angular frequency ω of the resulting oscillatory motion. (c) What is the maximum speed of the block? (d) Determine the magnitude of the maximum acceleration of the block.

69. A spiral staircase winds up to the top of a tower in an old castle. To measure the height of the tower, a rope is attached to the top

of the tower and hung down the center of the staircase. However, nothing is available with which to measure the length of the rope. Therefore, at the bottom of the rope a small object is attached so as to form a simple pendulum that just clears the floor. The period of the pendulum is measured to be 9.2 s. What is the height of the tower?

70. A person bounces up and down on a trampoline, while always staying in contact with it. The motion is simple harmonic motion, and it takes 1.90 s to complete one cycle. The height of each bounce above the equilibrium position is 45.0 cm. Determine (a) the amplitude and (b) the angular frequency of the motion. (c) What is the maximum speed attained by the person?

71. ssm A CD player is mounted on four cylindrical rubber blocks. Each cylinder has a height of 0.030 m and a cross-sectional area of 1.2×10^{-3} m², and the shear modulus for rubber is 2.6×10^6 N/m². If a horizontal force of magnitude 32 N is applied to the CD player, how far will the unit move sideways? Assume that each block is subjected to one-fourth of the force.

72. A person who weighs 670 N steps onto a spring scale in the bathroom, and the spring compresses by 0.79 cm. (a) What is the spring constant? (b) What is the weight of another person who compresses the spring by 0.34 cm?

73. The drawing shows a 160-kg crate hanging from the end of a steel bar. The length of the bar is 0.10 m, and its cross-sectional area is 3.2×10^{-4} m². Neglect the weight of the bar itself and determine (a) the shear stress on the bar and (b) the vertical deflection ΔY of the right end of the bar.

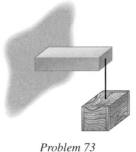

Problem 73

* **74.** In Figure 10.9, the radius of the reference circle is 0.500 m. Suppose the frequency of the simple harmonic motion of the shadow is 2.00 Hz. At time $t = 0.0500$ s, calculate (a) the displacement x, (b) the magnitude of the velocity, and (c) the magnitude of the acceleration of the shadow.

* **75. Interactive Solution 10.75** at **www.wiley.com/college/cutnell** presents a model for solving this problem. A spring (spring constant = 112 N/m) is mounted on the floor and is oriented vertically. A 0.400-kg block is placed on top of the spring and pushed down to start it oscillating in simple harmonic motion. The block is not attached to the spring. (a) Obtain the frequency (in Hz) of the motion. (b) Determine the amplitude at which the block will lose contact with the spring.

* **76.** A block rests on a frictionless horizontal surface and is attached to a spring. When set into simple harmonic motion, the block oscillates back and forth with an angular frequency of 7.0 rad/s. The drawing shows the position of the block when the spring is unstrained. This position is labeled "$x = 0$ m." The drawing also shows a small bottle located 0.080 m to the right of this position. The block is pulled to the right, stretching the spring by 0.050 m, and is then thrown to the left. In order for the block to knock over the bottle, it must be thrown with a speed exceeding v_0. Ignoring the width of the block, find v_0.

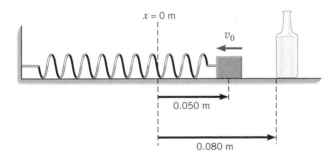

* **77.** Consult **Interactive Solution 10.77** at **www.wiley.com/college/ cutnell** to explore a model for solving this problem. A spring is compressed by 0.0620 m and is used to launch an object horizontally with a speed of 1.50 m/s. If the object were attached to the spring, at what angular frequency (in rad/s) would it oscillate?

* **78.** A piece of mohair from an Angora goat has a radius of 31×10^{-6} m. What is the least number of identical pieces of mohair that should be used to suspend a 75-kg person, so the strain $\Delta L/L_0$ experienced by each piece is less than 0.010? Assume that the tension is the same in all the pieces.

* **79.** The front spring of a car's suspension system has a spring constant of 1.50×10^6 N/m and supports a mass of 215 kg. The wheel has a radius of 0.400 m. The car is traveling on a bumpy road, on which the distance between the bumps is equal to the circumference of the wheel.

Due to resonance, the wheel starts to vibrate strongly when the car is traveling at a certain minimum linear speed. What is this speed?

** **80.** A 0.200-m uniform bar has a mass of 0.750 kg and is released from rest in the vertical position, as the drawing indicates. The spring is initially unstrained and has a spring constant of $k = 25.0$ N/m. Find the tangential speed with which end A strikes the horizontal surface.

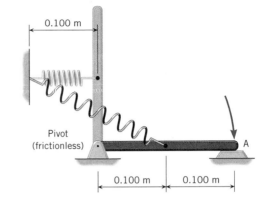

** **81. ssm** A steel wire is strung between two supports attached to a ceiling. Initially, there is no tension in the wire when it is horizontal. A 96-N picture is then hung from the center of the wire, as the drawing illustrates, so the ends of the wire make angles of 26° with respect to the horizontal. What is the radius of the wire?

** **82.** A copper rod (length = 2.0 m, radius = 3.0×10^{-3} m) hangs down from the ceiling. A 9.0-kg object is attached to the lower end of the rod. The rod acts as a "spring," and the object oscillates vertically with a small amplitude. Ignoring the rod's mass, find the frequency f of the simple harmonic motion.

** **83. ssm** The drawing shows a top view of a frictionless horizontal surface, where there are two springs with particles of mass m_1 and m_2 attached to them. Each spring has a spring constant of 120 N/m. The particles are pulled to the right and then released from the positions shown in the drawing. How much time passes before the particles are side by side for the first time at $x = 0$ m if (a) $m_1 = m_2 = 3.0$ kg and (b) $m_1 = 3.0$ kg and $m_2 = 27$ kg?

** **84.** The drawing shows two crates that are connected by a steel wire that passes over a pulley. The unstretched length of the wire is 1.5 m, and its cross-sectional area is 1.3×10^{-5} m². The pulley is frictionless and massless. When the crates are accelerating, determine the change in length of the wire. Ignore the mass of the wire.

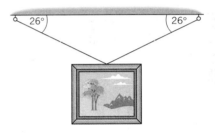

Position of unstrained spring ($x = 0$ m)

Problem 83

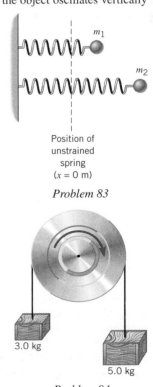

Problem 84

Concepts & Calculations   Group Learning Problems

Note: Each of these problems consists of Concept Questions followed by a related quantitative Problem. They are designed for use by students working alone or in small learning groups. The Concept Questions involve little or no mathematics and are intended to stimulate group discussions. They focus on the concepts with which the problems deal. Recognizing the concepts is the essential initial step in any problem-solving technique.

85. Concept Questions A spring lies on a horizontal table, and the left end is attached to a wall. The other end is connected to a box. The box is pulled to the right, stretching the spring. Static friction exists between the box and the table. When the box is released, it does not move. (a) Is the direction of the restoring force of the spring the same as or opposite to the direction of the static frictional force? Why? (b) How is the magnitude of the restoring force related to the magnitude of the static frictional force? (c) What factors determine the *maximum* static frictional force that can be applied to the box?

Problem A 0.80-kg box is attached to a horizontal spring whose spring constant is 59 N/m. The coefficient of static friction between the box and the table on which it rests is $\mu_S = 0.74$. How far can the spring be stretched from its unstrained position without the box moving?

86. Concept Question The drawing shows three situations in which a block is attached to a spring. The position labeled "0 m" represents the unstrained position of the spring. The block is moved from an initial position x_0 to a final position x_f. The displacement is *s*. For each case, determine the direction of the restoring force as the block is moved. (Note that the direction of this force may change during the movement.) Then, without doing any calculations, decide whether the total work done by the restoring force of the spring is positive, negative, or zero. Give your reason for each answer in terms of the directions of the displacement and the restoring force.

Problem Suppose the spring constant is $k = 46.0$ N/m. Using the data provided in the drawing, determine the total work done by the restoring force of the spring for each situation. Verify that your answers are consistent with your answers to the Concept Question.

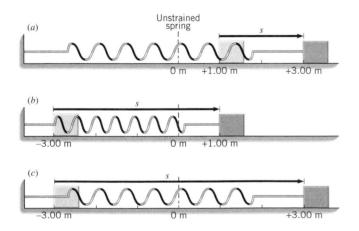

87. Concept Questions A spring is resting vertically on a table. A small box is dropped onto the top of the spring and compresses it. (a) Two forces are acting on the box: its weight and the restoring force of the spring. What law determines the relationship between these forces when the acceleration of the box is zero? (b) What is the speed of the box when the spring is fully compressed? (c) Many people say that the acceleration of the box is zero when the spring is fully compressed. Is this correct? Explain why or why not.

Problem Suppose the spring has a spring constant of 450 N/m and the box has a mass of 1.5 kg. The speed of the box just before it makes contact with the spring is 0.49 m/s. (a) Determine the magnitude of the spring's displacement at the instant when the acceleration of the box is zero. (b) What is the magnitude of the spring's displacement when the spring is fully compressed?

88. Concept Question The different types of energy possessed by a falling bungee jumper are discussed in Concepts & Calculations Example 15. When the jumper reaches his lowest point in the fall, which is above the water, what types of energy does he have? Provide a reason for each answer.

Problem Using the data given in Concepts & Calculations Example 15, determine how far he is from the water when he reaches the lowest point in the fall.

89. Concept Questions Two physical pendulums (not simple pendulums) are made from meter sticks that are suspended from the ceiling at one end. They are identical, except that one is made of wood and the other of metal. They are set into oscillation and execute simple harmonic motion. (a) How does the angular frequency of a physical pendulum depend on its mass? Take into account how the mass enters directly as well as indirectly through the moment of inertia. (b) Which pendulum, if either, has the greatest period? Why?

Problem The wood and metal pendulums have masses of 0.17 and 0.85 kg, respectively. Determine the periods of the two physical pendulums. Check to see that your answers are consistent with your answers to the Concept Questions.

90. Concept Questions One end of a piano wire is wrapped around a cylindrical tuning peg and the other end is fixed in place. The tuning peg is turned so as to stretch the wire. (a) For each turn of the tuning peg, by how much does the length of the wire change? Express your answer in terms of the radius of the tuning peg. (b) As the wire is stretched, do you expect the tension to increase, decrease, or remain the same? Why?

Problem The piano wire is made from steel ($Y = 2.0 \times 10^{11}$ N/m^2). It has a radius of 0.80 mm, and an unstrained length of 0.76 m. The radius of the tuning peg is 1.8 mm. Initially, there is no tension in the wire. Find the tension in the wire when the tuning peg is turned through two revolutions.

* **91. Concept Questions** A vertical ideal spring is mounted on the floor. A block is placed on the unstrained spring in two different ways. (a) The block is placed on the spring and not released until it rests stationary on the spring in its equilibrium position. As the block rests on the spring in equilibrium, what determines how much the spring is compressed? Explain. (b) Immediately after being placed on the unstrained spring, the block is released from rest and falls downward until it comes to a momentary halt. What determines how much the spring is compressed in this case, which is not an example of equilibrium? Account for your answer. (c) In which of the situations described in (a) and (b) is the compression of the spring greater? Give a reason for your answer.

Problem The spring constant of the spring is 170 N/m, and the mass of the block is 0.64 kg. Determine the amount by which the spring is compressed in Concept Questions (a) and (b). Verify that your answers are consistent with your answer to Concept Question (c).

*92. ☤ **Concept Questions** Depending on how you fall, you can break a bone easily. The severity of the break depends on how much energy the bone absorbs in the accident, and to evaluate this let us treat the bone as an ideal spring. (a) The applied force needed to change the length of a spring is given by Equation 10.1, where the change is denoted by the displacement x. To change the length of a bone, the necessary applied force is given by Equation 10.17, where the change is denoted by ΔL rather than x. Recognizing that these two equations are equivalent, explain which factors determine the effective spring constant k of a piece of bone, such as the thighbone (femur) in the leg. (b) The elastic potential energy for an ideal spring is given by Equation 10.13. Using Equation 10.1, discuss how the energy can be expressed in terms of the force applied to the spring. (c) When a person falls from rest from a height and strikes the ground stiff-legged and comes to rest without rotating, what determines the amount of energy that his legs must absorb as elastic potential energy? Justify your answer. You may ignore air resistance and friction.

Problem The maximum applied force of compression that one man's thighbone can endure without breaking is 7.0×10^4 N. The minimum effective cross-sectional area of the bone is 4.0×10^{-4} m^2, and its length is 0.55 m. The mass of the man is 65 kg. He falls straight down without rotating, strikes the ground stiff-legged on one foot, and comes to a halt without rotating. To see that it is easy to break a thighbone when falling in this fashion, find the maximum distance through which his center of gravity can fall without his breaking a bone.

Chapter 11

Fluids

These stunt planes, like all planes, remain aloft because the air exerts an upward force, called the lift, on the wings. Air is a fluid, and this chapter examines the forces and pressures that fluids exert when they are at rest and in motion. (© Baron Wolman/Stone/Getty Images)

11.1 *Mass Density*

Fluids are materials that can flow, and they include both gases and liquids. Air is the most common gas, and moves from place to place as wind. Water is the most familiar liquid and has many uses, from generating hydroelectric power to white-water rafting. The *mass density* of a liquid or gas is an important factor that determines its behavior as a fluid. As indicated below, the mass density is the mass per unit volume and is denoted by the Greek letter rho (ρ).

■ DEFINITION OF MASS DENSITY

The mass density ρ is the mass m of a substance divided by its volume V:

$$\rho = \frac{m}{V} \tag{11.1}$$

SI Unit of Mass Density: kg/m³

Equal volumes of different substances generally have different masses, so the density depends on the nature of the material, as Table 11.1 indicates. Gases have the smallest densities because gas molecules are relatively far apart and a gas contains a large fraction of empty space. In contrast, the molecules are much more tightly packed in liquids and solids, and the tighter packing leads to larger densities. The densities of gases are very sensitive to changes in temperature and pressure. However, for the range of temperatures and pressures encountered in this text, the densities of liquids and solids do not differ much from the values in Table 11.1.

It is the mass of a substance, not its weight, that enters into the definition of density. In situations where weight is needed, it can be calculated from the mass density, the volume, and the acceleration due to gravity, as Example 1 illustrates.

Example 1 Blood as a Fraction of Body Weight

The body of a man whose weight is 690 N contains about 5.2×10^{-3} m³ (5.5 qt) of blood. (a) Find the blood's weight and (b) express it as a percentage of the body weight.

Reasoning To find the weight W of the blood, we need the mass m, since $W = mg$, where g is the magnitude of the acceleration due to gravity. According to Table 11.1, the density of blood is 1060 kg/m³, so the mass of the blood can be found by using the given volume of 5.2×10^{-3} m³ in Equation 11.1.

Solution

(a) The mass and the weight of the blood are

$$m = \rho V = (1060 \text{ kg/m}^3)(5.2 \times 10^{-3} \text{ m}^3) = 5.5 \text{ kg} \tag{11.1}$$

$$W = mg = (5.5 \text{ kg})(9.80 \text{ m/s}^2) = \boxed{54 \text{ N}} \tag{4.5}$$

(b) The percentage of body weight contributed by the blood is

$$\text{Percentage} = \frac{54 \text{ N}}{690 \text{ N}} \times 100 = \boxed{7.8\%}$$

A convenient way to compare densities is to use the concept of *specific gravity.* The specific gravity of a substance is its density divided by the density of a standard reference material, usually chosen to be water at 4 °C.

$$\text{Specific gravity} = \frac{\text{Density of substance}}{\text{Density of water at 4 °C}} = \frac{\text{Density of substance}}{1.000 \times 10^3 \text{ kg/m}^3} \tag{11.2}$$

Being the ratio of two densities, specific gravity has no units. For example, Table 11.1 reveals that diamond has a specific gravity of 3.52, since the density of diamond is 3.52 times greater than the density of water at 4 °C.

Table 11.1 *Mass Densities[a] of Common Substances*

Substance	Mass Density ρ (kg/m³)
Solids	
Aluminum	2 700
Brass	8 470
Concrete	2 200
Copper	8 890
Diamond	3 520
Gold	19 300
Ice	917
Iron (steel)	7 860
Lead	11 300
Quartz	2 660
Silver	10 500
Wood (yellow pine)	550
Liquids	
Blood (whole, 37 °C)	1 060
Ethyl alcohol	806
Mercury	13 600
Oil (hydraulic)	800
Water (4 °C)	1.000×10^3
Gases	
Air	1.29
Carbon dioxide	1.98
Helium	0.179
Hydrogen	0.0899
Nitrogen	1.25
Oxygen	1.43

[a] Unless otherwise noted, densities are given at 0 °C and 1 atm pressure.

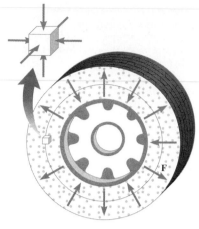

Figure 11.1 In colliding with the inner walls of the tire, the air molecules (blue dots) exert a force on every part of the wall surface. If a small cube were inserted inside the tire, the cube would experience forces (blue arrows) acting perpendicular to each of its six faces.

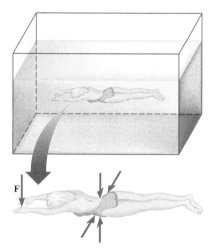

Figure 11.2 Water applies a force perpendicular to each surface within the water, including the walls and bottom of the swimming pool, and all parts of the swimmer's body.

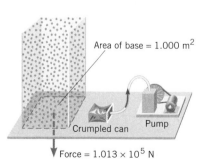

Figure 11.3 Atmospheric pressure at sea level is 1.013×10^5 Pa, which is sufficient to crumple a can if the inside air is pumped out.

The next two sections deal with the important concept of pressure. We will see that the density of a fluid is one factor determining the pressure that a fluid exerts.

11.2 Pressure

People who have fixed a flat tire know something about pressure. The final step in the job is to reinflate the tire to the proper pressure. The underinflated tire is soft because it contains an insufficient number of air molecules to push outward against the rubber and give the tire that solid feel. When air is added from a pump, the number of molecules and the collective force they exert are increased. The air molecules within a tire are free to wander throughout its entire volume, and in the course of their wandering they collide with one another and with the inner walls of the tire. The collisions with the walls allow the air to exert a force against every part of the wall surface, as Figure 11.1 shows. The pressure P exerted by a fluid is defined in Section 10.7 (Equation 10.19) as the magnitude F of the force acting perpendicular to a surface divided by the area A over which the force acts:

$$P = \frac{F}{A} \tag{11.3}$$

The SI unit for pressure is a newton/meter^2 (N/m^2), a combination that is referred to as a pascal (Pa). A pressure of 1 Pa is a very small amount. Many common situations involve pressures of approximately 10^5 Pa, an amount referred to as one *bar* of pressure. Alternatively, force can be measured in pounds and area in square inches, so another unit for pressure is pounds per square inch (lb/in.^{2}), often abbreviated as "psi."

Because of its pressure, the air in a tire applies a force to any surface with which the air is in contact. Suppose, for instance, that a small cube is inserted inside the tire. As Figure 11.1 shows, the air pressure causes a force to act perpendicularly on each face of the cube. In a similar fashion, a liquid such as water also exerts pressure. A swimmer, for example, feels the water pushing perpendicularly inward everywhere on her body, as Figure 11.2 illustrates. In general, a static fluid cannot produce a force parallel to a surface, for if it did, the surface would apply a reaction force to the fluid, consistent with Newton's action–reaction law. In response, the fluid would flow and would not then be static.

While fluid pressure can generate a force, pressure itself is not a vector quantity, as is the force. In the definition of pressure, $P = F/A$, the symbol F refers only to the magnitude of the force, so that pressure has no directional characteristic. The force generated by the pressure of a static fluid is always perpendicular to the surface that the fluid contacts, as Example 2 illustrates.

Example 2 The Force on a Swimmer

Suppose the pressure acting on the back of a swimmer's hand is 1.2×10^5 Pa, a realistic value near the bottom of the diving end of a pool. The surface area of the back of the hand is 8.4×10^{-3} m^2. (a) Determine the magnitude of the force that acts on it. (b) Discuss the direction of the force.

Reasoning From the definition of pressure in Equation 11.3, we can see that the magnitude of the force is the pressure times the area. The direction of the force is always perpendicular to the surface that the water contacts.

Solution

(a) A pressure of 1.2×10^5 Pa is 1.2×10^5 N/m^2. From Equation 11.3, we find

$$F = PA = (1.2 \times 10^5 \text{ N/m}^2)(8.4 \times 10^{-3} \text{ m}^2) = \boxed{1.0 \times 10^3 \text{ N}}$$

This is a rather large force, about 230 lb.

(b) In Figure 11.2, the hand (palm downward) is oriented parallel to the bottom of the pool. Since the water pushes perpendicularly against the back of the hand, the force **F** is directed downward in the drawing. This downward-acting force is balanced by an upward-acting force on the palm, so that the hand is in equilibrium. If the hand were rotated by 90°, the directions of these forces would also be rotated by 90°, always being perpendicular to the hand.

A person need not be under water to experience the effects of pressure. Walking about on land, we are at the bottom of the earth's atmosphere, which is a fluid and pushes inward on our bodies just like the water in a swimming pool. As Figure 11.3 indicates, there is enough air above the surface of the earth to create the following pressure at sea level:

Atmospheric pressure at sea level

$$1.013 \times 10^5 \ \text{Pa} = 1 \ \text{atmosphere}$$

This amount of pressure corresponds to 14.70 lb/in.^{2} and is referred to as one *atmosphere (atm)*. One atmosphere of pressure is a significant amount. Look, for instance, in Figure 11.3 at the results of pumping out the air from within a gasoline can. With no internal air to push outward, the inward push of the external air is unbalanced and is strong enough to crumple the can.

In contrast to the situation in Figure 11.3, reducing the pressure is sometimes beneficial. Lynx, for example, are especially well suited for hunting on snow because of their oversize paws (see Figure 11.4). The large paws function as snowshoes that distribute the weight over a large area. Thus, they reduce the weight per unit area, or the pressure that the cat applies to the surface, which helps to keep it from sinking into the snow.

Figure 11.4 Lynx have large paws that act as natural snowshoes. (© Daniel J. Cox/Stone/Getty Images)

The physics of lynx paws (see above).

11.3 *Pressure and Depth in a Static Fluid*

▶ CONCEPTS AT A GLANCE The deeper an underwater swimmer goes, the more strongly the water pushes on his body and the greater is the pressure that he experiences. To determine the relation between pressure and depth, we turn to Newton's second law. The Concepts-at-a-Glance chart in Figure 11.5 outlines our approach. This chart is an expanded version of the charts in Figures 4.27 and 10.4. In using the second law, we will focus on two external forces that act on the fluid. One is the gravitational force—that is, the weight of the fluid. The other is the collisional force that is responsible for fluid pressure, as the previous section discusses. Since the fluid is at rest, its acceleration is zero ($\mathbf{a} = 0 \ \text{m/s}^2$), and it is in equilibrium. By applying the second law in the form $\Sigma \mathbf{F} = 0$, we will derive a relation between pressure and depth. This relation is especially important because it leads to Pascal's principle (Section 11.5) and Archimedes' principle (Section 11.6), both of which are essential in describing the properties of static fluids. ◀

Figure 11.5 CONCEPTS AT A GLANCE The gravitational force and the collisional forces exerted by a fluid are used with Newton's second law to determine the relation between pressure and depth in a static ($\mathbf{a} = 0 \ \text{m/s}^2$) fluid. The deeper that these scuba divers go, the greater is the water pressure on them. (© Jeff Hunter/The Image Bank/Getty Images)

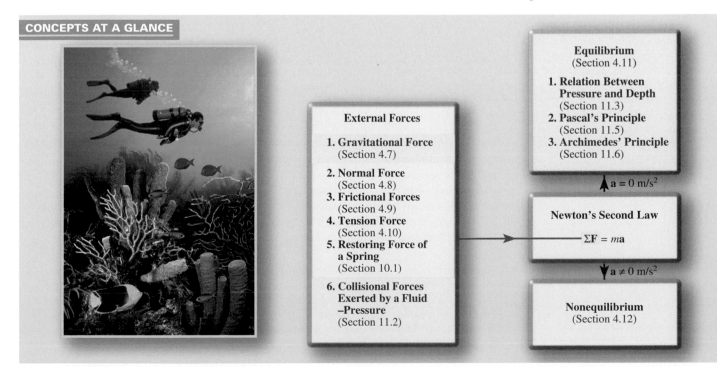

CONCEPTS AT A GLANCE

External Forces

1. **Gravitational Force** (Section 4.7)
2. **Normal Force** (Section 4.8)
3. **Frictional Forces** (Section 4.9)
4. **Tension Force** (Section 4.10)
5. **Restoring Force of a Spring** (Section 10.1)
6. **Collisional Forces Exerted by a Fluid –Pressure** (Section 11.2)

Newton's Second Law

$\Sigma \mathbf{F} = m\mathbf{a}$

Equilibrium (Section 4.11)

1. **Relation Between Pressure and Depth** (Section 11.3)
2. **Pascal's Principle** (Section 11.5)
3. **Archimedes' Principle** (Section 11.6)

$\mathbf{a} = 0 \ \text{m/s}^2$

$\mathbf{a} \neq 0 \ \text{m/s}^2$

Nonequilibrium (Section 4.12)

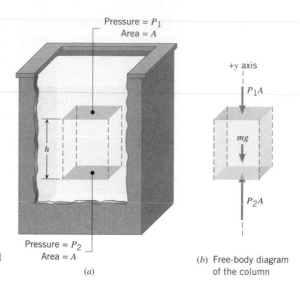

Pressure = P_1
Area = A

h

Pressure = P_2
Area = A

(a)

+y axis

P_1A

mg

P_2A

(b) Free-body diagram
of the column

Figure 11.6 (a) A container of fluid in which one column of the fluid is outlined. The fluid is at rest. (b) The free-body diagram, showing the vertical forces acting on the column.

Figure 11.6 shows a container of fluid and focuses attention on one column of the fluid. The free-body diagram in the figure shows all the vertical forces acting on the column. On the top face (area $= A$), the fluid pressure P_1 generates a downward force whose magnitude is P_1A. Similarly, on the bottom face, the pressure P_2 generates an upward force of magnitude P_2A. The pressure P_2 is greater than the pressure P_1 because the bottom face supports the weight of more fluid than the upper one does. In fact, the excess weight supported by the bottom face is exactly the weight of the fluid within the column. As the free-body diagram indicates, this weight is mg, where m is the mass of the fluid and g is the magnitude of the acceleration due to gravity. Since the column is in equilibrium, we can set the sum of the vertical forces equal to zero and find that

$$\Sigma F_y = P_2A - P_1A - mg = 0 \quad \text{or} \quad P_2A - P_1A + mg$$

The mass m is related to the density ρ and the volume V of the column by $m = \rho V$. Since the volume is the cross-sectional area A times the vertical dimension h, we have $m = \rho Ah$. With this substitution, the condition for equilibrium becomes $P_2A = P_1A + \rho Ahg$. The area A can be eliminated algebraically from this expression, with the result that

$$P_2 = P_1 + \rho gh \tag{11.4}$$

Equation 11.4 indicates that if the pressure P_1 is known at a higher level, the larger pressure P_2 at a deeper level can be calculated by adding the increment ρgh. In determining the pressure increment ρgh, we assumed that the density ρ is the same at any vertical distance h or, in other words, the fluid is incompressible. The assumption is reasonable for liquids, since the bottom layers can support the upper layers with little compression. In a gas, however, the lower layers are compressed markedly by the weight of the upper layers, with the result that the density varies with vertical distance. For example, the density of our atmosphere is larger near the earth's surface than it is at higher altitudes. When applied to gases, the relation $P_2 = P_1 + \rho gh$ can be used only when h is small enough that any variation in ρ is negligible.

A significant feature of Equation 11.4 is that the pressure increment ρgh is affected by the vertical distance h, but not by any horizontal distance within the fluid. Conceptual Example 3 helps to clarify this feature.

Conceptual Example 3   The Hoover Dam

Lake Mead is the largest wholly artificial reservoir in the United States and was formed after the completion of the Hoover Dam in 1936. As Figure 11.7a suggests, the water in the reservoir backs up behind the dam for a considerable distance (about 200 km or 120 miles). Suppose that all the water were removed, except for a relatively narrow vertical column in contact with the dam. Figure 11.7*b* shows a side view of this hypothetical situation, in which the water against the dam has the same depth as in Figure 11.7*a*. Would the Hoover Dam still be needed to contain the water in this hypothetical reservoir, or could a much less massive structure do the job?

Reasoning and Solution Since our hypothetical reservoir contains much less water than Lake Mead, it is tempting to say that a much less massive structure than the Hoover Dam would be needed. Not so, however. Hoover Dam would still be needed to contain our hypothetical reservoir. To see why, imagine a small square on the inner face of the dam, located beneath the water. The magnitude of the force on this square is the product of its area and the pressure of the water. But the pressure depends on the depth—that is, on the vertical distance *h* in Equation 11.4. The horizontal distance of the water behind the dam does not appear in this equation and, therefore, has no effect on the pressure. Consequently, at a given depth the small square experiences a force that is the same in both parts of Figure 11.7. Certainly the dam experiences greater forces generated by greater water pressures at greater depths. But no matter where on the inner face of the dam the square is, the force that the water applies to it depends only on the depth, not on the amount of water backed up behind the dam. ***Thus, the dam for our imaginary reservoir would sustain the same forces that the Hoover Dam sustains and would need to be equally large.***

(a)

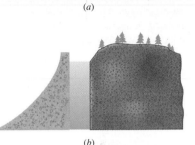
(b)

Figure 11.7 (*a*) The Hoover Dam in Nevada and Lake Mead behind it. (*b*) This drawing shows a hypothetical reservoir formed by removing most of the water from Lake Mead. Conceptual Example 3 compares the dam needed for this hypothetical reservoir with the Hoover Dam. (© Jim Richardson/Corbis Images)

The next example deals further with the relationship between pressure and depth given by Equation 11.4.

Example 4   The Swimming Hole

Figure 11.8 shows the cross section of a swimming hole. Points *A* and *B* are both located at a distance of *h* = 5.50 m below the surface of the water. Find the pressure at each of these two points.

Reasoning The pressure at point *B* is the same as that at point *A*, since both are located at the same vertical distance beneath the surface and only the vertical distance *h* affects the pressure increment *ρgh* in Equation 11.4. To understand this important feature more clearly, consider the path *AA'B'B* in Figure 11.8. The pressure decreases on the way up along the vertical segment *AA'* and increases by the same amount on the way back down along segment *B'B*. Since no change in pressure occurs along the horizontal segment *A'B'*, the pressure is the same at *A* and *B*.

Solution The pressure acting on the surface of the water is the atmospheric pressure of 1.01×10^5 Pa. Using this value as P_1 in Equation 11.4, we can determine a value for the pressure P_2 at either point *A* or *B*, both of which are located 5.50 m under the water. Table 11.1 gives the density of water as 1.000×10^3 kg/m^3.

$$P_2 = P_1 + \rho g h$$
$$P_2 = 1.01 \times 10^5 \text{ Pa} + (1.000 \times 10^3 \text{ kg/m}^3)(9.80 \text{ m/s}^2)(5.50 \text{ m}) \; = \; \boxed{1.55 \times 10^5 \text{ Pa}}$$

Problem solving insight
The pressure at any point in a fluid depends on the vertical distance *h* of the point beneath the surface. However, for a given vertical distance, the pressure is the same, no matter where the point is located horizontally in the fluid.

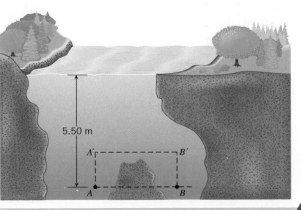

Figure 11.8 The pressures at points *A* and *B* are the same, since both points are located at the same vertical distance of 5.50 m beneath the surface of the water.

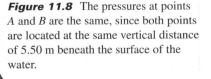

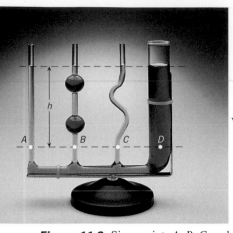

Figure 11.9 Since points A, B, C, and D are at the same distance h beneath the liquid surface, the pressure at each of them is the same. (© Richard Megna/Fundamental Photographs)

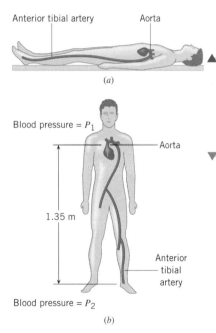

(a)

(b)

Figure 11.10 The blood pressure in the feet can exceed the blood pressure in the heart, depending on whether a person is (a) reclining horizontally or (b) standing.

The physics of
pumping water from a well.

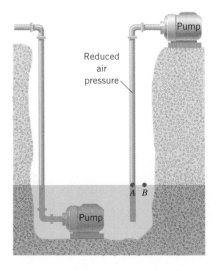

Figure 11.11 A water pump can be placed at the bottom of a well or at ground level. Conceptual Example 6 discusses how the depth of the well influences the choice of which method to use.

Figure 11.9 shows an irregularly shaped container of liquid. Reasoning similar to that used in Example 4 leads to the conclusion that the pressure is the same at points A, B, C, and D, since each is at the same vertical distance h beneath the surface. In effect, the arteries in our bodies constitute an irregularly shaped "container" for the blood. The next example examines the blood pressure at different places in this "container."

Example 5 Blood Pressure

Blood in the arteries is flowing, but as a first approximation, the effects of this flow can be ignored and the blood can be treated as a static fluid. Estimate the amount by which the blood pressure P_2 in the anterior tibial artery at the foot exceeds the blood pressure P_1 in the aorta at the heart when a person is (a) reclining horizontally as in Figure 11.10a and (b) standing as in Figure 11.10b.

Reasoning and Solution

(a) When the body is horizontal, there is little or no vertical separation between the feet and the heart. Since $h = 0$ m,

$$P_2 - P_1 = \rho g h = \boxed{0 \text{ Pa}} \tag{11.4}$$

(b) When an adult is standing up, the vertical separation between the feet and the heart is about 1.35 m, as Figure 11.10b indicates. Table 11.1 gives the density of blood as 1060 kg/m³, so that

$$P_2 - P_1 = \rho g h = (1060 \text{ kg/m}^3)(9.80 \text{ m/s}^2)(1.35 \text{ m}) = \boxed{1.40 \times 10^4 \text{ Pa}}$$

Sometimes fluid pressure places limits on how a job can be done. Conceptual Example 6 illustrates how fluid pressure restricts the height to which water can be pumped.

Conceptual Example 6 Pumping Water

Figure 11.11 shows two methods for pumping water from a well. In one method, the pump is submerged in the water at the bottom of the well, while in the other, it is located at ground level. If the well is shallow, either technique can be used. However, if the well is very deep, only one of the methods works. Which is it?

Reasoning and Solution To answer this question, we need to examine the nature of the job done by each pump. The pump at the bottom of the well literally pushes water up the pipe. For a very deep well, the column of water becomes very tall, and the pressure at the bottom of the pipe becomes large, due to the pressure increment $\rho g h$ in Equation 11.4. However, as long as the pump can push with sufficient strength to overcome the large pressure, it can shove the next increment of water into the pipe, so the method can be used for very deep wells. In contrast, the pump at ground level does not push water at all. It removes air from the pipe. As the pump reduces the air pressure within the pipe (see point *A*), the greater air pressure outside (see point *B*) pushes water up the pipe. But even the strongest pump can only remove *all* of the air. Once the air is completely removed, an increase in pump strength does not increase the height to which the water is pushed by the external air pressure. ***Thus, a ground-level pump can only cause water to rise to a certain maximum height, so it cannot be used for very deep wells.*** (Incidentally, the pump at ground level works just the way you do when you drink through a straw. You draw some of the air out of the straw, and the external air pressure pushes the liquid up into it.)

Related Homework: *Problems 25, 77*

✔ Check Your Understanding 1

A scuba diver is swimming under water, and the graph shows a plot of the water pressure acting on the diver as a function of time. In each of the three regions, A → B, B → C, and C → D, does the depth of the diver increase, decrease, or remain constant? (The answer is given at the end of the book.)

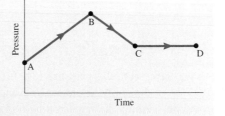

Background: The relationship between pressure and depth in a fluid is the focus of this question.

For similar questions (including calculational counterparts), consult Self-Assessment Test 11.1. This test is described at the end of Section 11.6.

11.4 *Pressure Gauges*

One of the simplest pressure gauges is the mercury barometer used for measuring atmospheric pressure. This device is a tube sealed at one end, filled completely with mercury, and then inverted, so that the open end is under the surface of a pool of mercury (see Figure 11.12). Except for a negligible amount of mercury vapor, the space above the mercury in the tube is empty, and the pressure P_1 is nearly zero there. The pressure P_2 at point A at the bottom of the mercury column is the same as that at point B—namely, atmospheric pressure—for these two points are at the same level. With $P_1 = 0$ Pa and $P_2 = P_{atm}$, it follows from Equation 11.4 that $P_{atm} = 0$ Pa $+ \rho g h$. Thus, the atmospheric pressure can be determined from the height h of the mercury in the tube, the density ρ of mercury, and the acceleration due to gravity. Usually weather forecasters report the pressure in terms of the height h, expressing it in millimeters or inches of mercury. For instance, using $P_{atm} = 1.013 \times 10^5$ Pa and $\rho = 13.6 \times 10^3$ kg/m³ for the density of mercury, we find that $h = P_{atm}/(\rho g) = 760$ mm (29.9 inches).* Slight variations from this value occur, depending on weather conditions and altitude.

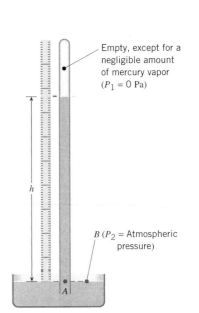

Figure 11.12 A mercury barometer.

Figure 11.13 shows another kind of pressure gauge, the open-tube manometer. The phrase "open-tube" refers to the fact that one side of the U-tube is open to atmospheric pressure. The tube contains a liquid, often mercury, and its other side is connected to the container whose pressure P_2 is to be measured. When the pressure in the container is equal to the atmospheric pressure, the liquid levels in both sides of the U-tube are the same. When the pressure in the container is greater than atmospheric pressure, as in Figure 11.13, the liquid in the tube is pushed downward on the left side and upward on the right side. The relation $P_2 = P_1 + \rho g h$ can be used to determine the container pressure. Atmospheric pressure exists at the top of the right column, so that $P_1 = P_{atm}$. The pressure P_2 is the same at points *A* and *B*, so we find that $P_2 = P_{atm} + \rho g h$, or

$$P_2 - P_{atm} = \rho g h$$

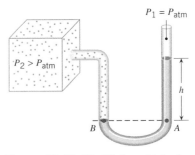

Figure 11.13 The U-shaped tube is called an open-tube manometer and can be used to measure the pressure P_2 in a container.

The height h is proportional to $P_2 - P_{atm}$, which is called the ***gauge pressure***. The gauge pressure is the amount by which the container pressure differs from atmospheric pressure. The actual value for P_2 is called the ***absolute pressure.***

Problem solving insight
When solving problems that deal with pressure, be sure to note the distinction between gauge pressure and absolute pressure.

---

* A pressure of one millimeter of mercury is sometimes referred to as one *torr*, to honor the inventor of the barometer, Evangelista Torricelli (1608–1647). Thus, one atmosphere of pressure is 760 torr.

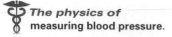

The physics of measuring blood pressure.

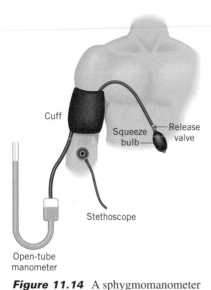

Cuff

Squeeze bulb

Release valve

Stethoscope

Open-tube manometer

Figure 11.14 A sphygmomanometer is used to measure blood pressure.

The sphygmomanometer is a familiar device for measuring blood pressure. As Figure 11.14 illustrates, a squeeze bulb can be used to inflate the cuff with air, which cuts off the flow of blood through the artery below the cuff. When the release valve is opened, the cuff pressure drops. Blood begins to flow again when the pressure created by the heart at the peak of its beating cycle exceeds the cuff pressure. Using a stethoscope to listen for the initial flow, the operator can measure the corresponding cuff gauge pressure with, for example, an open-tube manometer. This cuff gauge pressure is called the *systolic* pressure. Eventually, there comes a point when even the pressure created by the heart at the low point of its beating cycle is sufficient to cause blood to flow. Identifying this point with the stethoscope, the operator can measure the corresponding cuff gauge pressure, which is referred to as the *diastolic* pressure. The systolic and diastolic pressures are reported in millimeters of mercury, and values of 120 and 80, respectively, are typical of a young, healthy heart.

11.5 Pascal's Principle

As we have seen, the pressure in a fluid increases with depth, due to the weight of the fluid above the point of interest. A completely enclosed fluid may be subjected to an additional pressure by the application of an external force. For example, Figure 11.15a shows two interconnected cylindrical chambers. The chambers have different diameters and, together with the connecting tube, are completely filled with a liquid. The larger chamber is sealed at the top with a cap, while the smaller one is fitted with a movable piston. We now follow the approach outlined earlier in Figure 11.5 and begin by asking what determines the pressure P_1 at a point immediately beneath the piston. According to the definition of pressure, it is the magnitude F_1 of the external force divided by the area A_1 of the piston: $P_1 = F_1/A_1$. If it is necessary to know the pressure P_2 *at any deeper place in the liquid,* we just add to the value of P_1 the increment $\rho g h$, which takes into account the depth h below the piston: $P_2 = P_1 + \rho g h$. The important feature here is this: The pressure P_1 adds to the pressure $\rho g h$ due to the depth of the liquid at any point, whether that point is in the smaller chamber, the connecting tube, or the larger chamber. Therefore, if the applied pressure P_1 is increased or decreased, the pressure at any other point within the confined liquid changes correspondingly. This behavior is described by ***Pascal's principle.***

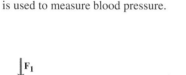

(a)

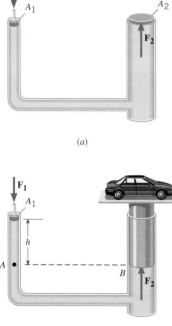

(b)

Figure 11.15 (a) An external force $\mathbf{F_1}$ is applied to the piston on the left. As a result, a force $\mathbf{F_2}$ is exerted on the cap on the chamber on the right. (b) The familiar hydraulic car lift.

■ **PASCAL'S PRINCIPLE**

Any change in the pressure applied to a completely enclosed fluid is transmitted undiminished to all parts of the fluid and the enclosing walls.

The usefulness of the arrangement in Figure 11.15a becomes apparent when we calculate the force F_2 applied by the liquid to the cap on the right side. The area of the cap is A_2 and the pressure there is P_2. As long as the tops of the left and right chambers are at the same level, the pressure increment $\rho g h$ is zero, so that the relation $P_2 = P_1 + \rho g h$ becomes $P_2 = P_1$. Consequently, $F_2/A_2 = F_1/A_1$, and

$$F_2 = F_1 \left(\frac{A_2}{A_1}\right) \tag{11.5}$$

If area A_2 is larger than area A_1, a large force $\mathbf{F_2}$ can be applied to the cap on the right chamber, starting with a smaller force $\mathbf{F_1}$ on the left. Depending on the ratio of the areas A_2/A_1, the force $\mathbf{F_2}$ can be large indeed, as in the familiar hydraulic car lift shown in part *b*. In this device the force $\mathbf{F_2}$ is not applied to a cap that seals the larger chamber, but, rather, to a movable plunger that lifts a car. Example 7 deals with a hydraulic car lift.

Example 7 A Car Lift

The physics of a hydraulic car lift.

In a hydraulic car lift, the input piston has a radius of $r_1 = 0.0120$ m and a negligible weight. The output plunger has a radius of $r_2 = 0.150$ m. The combined weight of the car and the plunger is $F_2 = 20\,500$ N. The lift uses hydraulic oil that has a density of 8.00×10^2 kg/m^3.

What input force F_1 is needed to support the car and the output plunger when the bottom surfaces of the piston and plunger are at (a) the same level and (b) the levels shown in Figure 11.15b with $h = 1.10$ m?

Reasoning When the bottom surfaces of the piston and plunger are at the same levels, as in part (a), Equation 11.5 applies. However, this equation does not apply in part (b), where the bottom surface of the output plunger is $h = 1.10$ m below the input piston. Therefore, our solution in part (b) will take into account the pressure increment $\rho g h$. In either case, we will see that the input force is less than the combined weight of the plunger and car.

Problem solving insight
Note that the relation $F_1 = F_2(A_1/A_2)$, which results from Pascal's principle, applies only when the points 1 and 2 lie at the same depth ($h = 0$ m) in the fluid.

Solution

(a) Using $A = \pi r^2$ for the circular areas of the piston and plunger, we rearrange Equation 11.5 and find that

$$F_1 = F_2\left(\frac{A_1}{A_2}\right) = F_2\left(\frac{\pi r_1^2}{\pi r_2^2}\right) = (20\ 500\ \text{N})\frac{(0.0120\ \text{m})^2}{(0.150\ \text{m})^2} = \boxed{131\ \text{N}}$$

(b) In Figure 11.15b, the bottom surface of the plunger at point B is at the same level as point A, which is at a depth h beneath the input piston. Therefore, we can apply Equation 11.4, $P_2 = P_1 + \rho g h$, with $P_2 = F_2/(\pi r_2^2)$ and $P_1 = F_1/(\pi r_1^2)$:

$$\frac{F_2}{\pi r_2^2} = \frac{F_1}{\pi r_1^2} + \rho g h$$

Solving for F_1 gives

$$F_1 = F_2\left(\frac{r_1^2}{r_2^2}\right) - \rho g h(\pi r_1^2)$$

$$= (20\ 500\ \text{N})\frac{(0.0120\ \text{m})^2}{(0.150\ \text{m})^2} - (8.00 \times 10^2\ \text{kg/m}^3)$$

$$\times (9.80\ \text{m/s}^2)(1.10\ \text{m})\,\pi\,(0.0120\ \text{m})^2 = \boxed{127\ \text{N}}$$

The answer here is less than in part (a) because the weight of the 1.10-m column of hydraulic oil provides some of the input force to support the car.

In a device such as a hydraulic car lift, the same amount of work is done by both the input and output forces in the absence of friction. The larger output force $\mathbf{F_2}$ moves through a smaller distance, while the smaller input force $\mathbf{F_1}$ moves through a larger distance. The work, being the product of the magnitude of the force and the distance, is the same in either case since mechanical energy is conserved.

An enormous variety of clever devices use hydraulic fluids, just as the car lift does. In a backhoe, for instance, the fluids multiply a small input force into the large output force required for digging (see Figure 11.16).

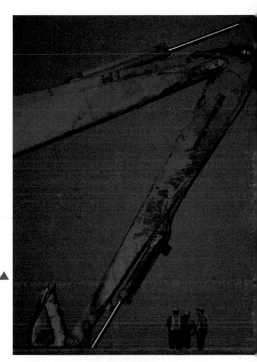

Figure 11.16 A backhoe uses a hydraulic fluid to generate a large output force, starting with a small input force. The output force is far greater than anything generated by humans. (© Richard Hamilton Smith/Corbis Images)

11.6 Archimedes' Principle

Anyone who has tried to push a beach ball under the water has felt how the water pushes back with a strong upward force. This upward force is called the ***buoyant force,*** and all fluids apply such a force to objects that are immersed in them. The buoyant force exists because fluid pressure is larger at greater depths.

In Figure 11.17 a cylinder of height h is being held under the surface of a liquid. The pressure P_1 on the top face generates the downward force P_1A, where A is the area of the face. Similarly, the pressure P_2 on the bottom face generates the upward force P_2A. Since the pressure is greater at greater depths, the upward force exceeds the downward force. Consequently, the liquid applies to the cylinder a net upward force, or buoyant force, whose magnitude F_B is

$$F_B = P_2A - P_1A = (P_2 - P_1)A = \rho g h A$$

We have substituted $P_2 - P_1 = \rho g h$ from Equation 11.4 into this result. In so doing, we find that the buoyant force equals $\rho g h A$. The quantity hA is the volume of liquid that the

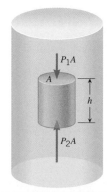

Figure 11.17 The fluid applies a downward force P_1A to the top face of the submerged cylinder and an upward force P_2A to the bottom face.

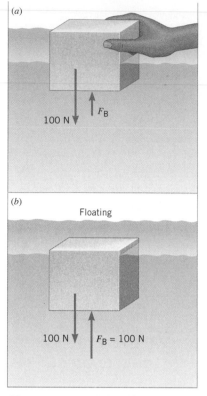

(a)

100 N ↓ F_B ↑

(b)

Floating

100 N ↓ $F_B = 100$ N ↑

Figure 11.18 (a) An object of weight 100 N is being immersed in a liquid. The deeper the object is, the more liquid it displaces, and the stronger the buoyant force is. (b) The buoyant force matches the 100-N weight, so the object floats.

cylinder moves aside or displaces in being submerged, and ρ denotes the density of the liquid, not the density of the material from which the cylinder is made. Therefore, ρhA gives the mass m of the displaced fluid, so that the buoyant force equals mg, the weight of the displaced fluid. The phrase "weight of the displaced fluid" refers to the weight of the fluid that would spill out, if the container were filled to the brim before the cylinder is inserted into the liquid. The "buoyant force" is not a new type of force. It is just the name given to the net upward force exerted by the fluid on the object.

The shape of the object in Figure 11.17 is not important. No matter what its shape, the buoyant force pushes it upward in accord with ***Archimedes' principle.*** It was an impressive accomplishment that the Greek scientist Archimedes (ca. 287–212 B.C.) discovered the essence of this principle so long ago.

■ **ARCHIMEDES' PRINCIPLE**

Any fluid applies a buoyant force to an object that is partially or completely immersed in it; the magnitude of the buoyant force equals the weight of the fluid that the object displaces:

$$\underbrace{F_B}_{\substack{\text{Magnitude of} \\ \text{buoyant force}}} = \underbrace{W_{\text{fluid}}}_{\substack{\text{Weight of} \\ \text{displaced fluid}}} \qquad (11.6)$$

The effect that the buoyant force has depends on its strength compared with the strengths of the other forces that are acting. For example, if the buoyant force is strong enough to balance the force of gravity, an object will float in a fluid. Figure 11.18 explores this possibility. In part a, a block that weighs 100 N displaces some liquid, and the liquid applies a buoyant force F_B to the block, according to Archimedes' principle. Nevertheless, if the block were released, it would fall further into the liquid because the buoyant force is not sufficiently strong to balance the weight of the block. In part b, however, enough of the block is submerged to provide a buoyant force that can balance the 100-N weight, so the block is in equilibrium and floats when released. If the buoyant force were not large enough to balance the weight, even with the block completely submerged, the block would sink. Even if an object sinks, there is still a buoyant force acting on it; it's just that the buoyant force is not large enough to balance the weight. Example 8 provides additional insight into what determines whether an object floats or sinks in a fluid.

Example 8 **A Swimming Raft**

A solid, square pinewood raft measures 4.0 m on a side and is 0.30 m thick. (a) Determine whether the raft floats in water, and (b) if so, how much of the raft is beneath the surface (see the distance h in Figure 11.19).

Reasoning To determine whether the raft floats, we will compare the weight of the raft to the maximum possible buoyant force and see whether there could be enough buoyant force to balance the weight. If so, then the value of the distance h can be obtained by utilizing the fact that the floating raft is in equilibrium, with the magnitude of the buoyant force equaling the raft's weight.

Solution

(a) The weight of the raft can be calculated from the density $\rho_{\text{pine}} = 550$ kg/m^3 (Table 11.1), the volume of the wood, and the acceleration due to gravity. The volume of the wood is $V_{\text{pine}} = 4.0$ m × 4.0 m × 0.30 m = 4.8 m^3, so that

$$\substack{\text{Weight} \\ \text{of raft}} = (\rho_{\text{pine}} V_{\text{pine}})g = (550 \text{ kg/m}^3)(4.8 \text{ m}^3)(9.80 \text{ m/s}^2) = 26\,000 \text{ N}$$

The maximum possible buoyant force occurs when the entire raft is under the surface, displacing 4.8 m^3 of water. According to Archimedes' principle, the weight of this volume of water is the maximum buoyant force F_B^{max}. It can be obtained using the density of water:

$$F_B^{\text{max}} = \rho_{\text{water}} V_{\text{pine}} g = (1.000 \times 10^3 \text{ kg/m}^3)(4.8 \text{ m}^3)(9.80 \text{ m/s}^2) = 47\,000 \text{ N}$$

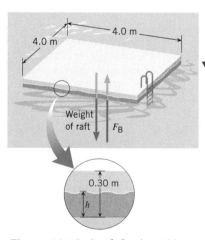

4.0 m   4.0 m

Weight of raft ↓ F_B ↑

0.30 m

h

Figure 11.19 A raft floating with a distance h beneath the surface of the water.

Problem solving insight
When using Archimedes' principle to find the buoyant force F_B that acts on an object, be sure to use the density of the displaced fluid, not the density of the object.

Since the maximum possible buoyant force exceeds the 26 000-N weight of the raft, the raft will float only partially submerged at a distance h beneath the water.

(b) We now find the value of h. The buoyant force balances the raft's weight, so $F_B = 26\,000$ N. But according to Equation 11.6, the magnitude of the buoyant force is also the weight of the displaced fluid, so $F_B = 26\,000$ N $= W_{fluid}$. Using the density of water, we can also express the weight of the displaced water as $W_{fluid} = \rho_{water}V_{water}g$, where the volume is $V_{water} = 4.0$ m $\times$ 4.0 m $\times h$. As a result,

$$26\,000 \text{ N} = W_{fluid} = \rho_{water}(4.0 \text{ m} \times 4.0 \text{ m} \times h)g$$

$$h = \frac{26\,000 \text{ N}}{\rho_{water}(4.0 \text{ m} \times 4.0 \text{ m})g}$$

$$= \frac{26\,000 \text{ N}}{(1.000 \times 10^3 \text{ kg/m}^3)(4.0 \text{ m} \times 4.0 \text{ m})(9.80 \text{ m/s}^2)} = \boxed{0.17 \text{ m}}$$

Need more practice?

🖱️**Interactive LearningWare 11.1**
A spring has a spring constant of 578 N/m. When used to suspend an object in air, the spring stretches by 0.0640 m. When used to suspend the same object in water (density = 1.00×10^3 kg/m³), the spring stretches by 0.0520 m. What is the volume of the part of the object that is under water?

Related Homework: *Problem 48*

Go to
www.wiley.com/college/cutnell
for an interactive solution.

In deciding whether the raft floats in part (a) of Example 8, we compared the raft's weight $(\rho_{pine}V_{pine})g$ to the maximum possible buoyant force $(\rho_{water}V_{pine})g$. The comparison depends only on the densities ρ_{pine} and ρ_{water}. The take-home message is this: Any object that is *solid throughout* will float in a liquid if the density of the object is less than or equal to the density of the liquid. For instance, at 0 °C ice has a density of 917 kg/m³, whereas water has a density of 1000 kg/m³. Therefore, ice floats in water.

Although a solid piece of a high-density material like steel will sink in water, such materials can, nonetheless, be used to make floating objects. A supertanker, for example, floats because it is *not* solid metal. It contains enormous amounts of empty space and, because of its shape, displaces enough water to balance its own large weight. Conceptual Example 9 focuses on the reason ships float.

Conceptual Example 9   How Much Water Is Needed to Float a Ship?

A ship floating in the ocean is a familiar sight. But is all that water really necessary? Can an ocean vessel float in the amount of water that a swimming pool contains, for instance?

Reasoning and Solution   In principle, a ship can float in much less than the amount of water that a swimming pool contains. To see why, look at Figure 11.20. Part *a* shows the ship floating in the ocean because it contains empty space within its hull and displaces enough water to balance its own weight. Part *b* shows the water that the ship displaces, which, according to Archimedes' principle, has a weight that equals the ship's weight. Although this wedge-shaped portion of water represents the water *displaced* by the ship, it is not the amount that must be present to float the ship, as part *c* illustrates. This part of the drawing shows a canal, the cross section of which matches the shape in part *b*. **All that is needed, in principle, is a thin section of water that separates the hull of the floating ship from the sides of the canal.** This thin section of water could have a very small volume indeed.

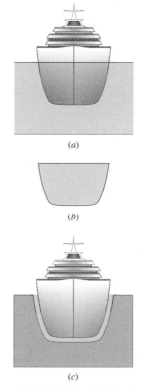

(a)

(b)

(c)

Figure 11.20 (*a*) A ship floating in the ocean. (*b*) This is the water that the ship displaces. (*c*) The ship floats here in a canal that has a cross section similar in shape to that in part *b*.

A useful application of Archimedes' principle can be found in car batteries. To alert the owner that recharging is necessary, some batteries include a state-of-charge indicator, such as the one illustrated in Figure 11.21. The battery includes a viewing port that looks down through a plastic rod, which extends into the battery acid. Attached to the end of this rod is a "cage," containing a green ball. The cage has holes in it that allow the acid to enter. When the battery is charged, the density of the acid is great enough that its buoyant force makes the ball rise to the top of the cage, to just beneath the plastic rod. The viewing port shows a green dot. As the battery discharges, the density of the acid decreases. Since the buoyant force is the weight of the acid displaced by the ball, the buoyant force also decreases. As a result, the ball sinks into one of the two chambers oriented at an angle beneath it (see Figure 11.21). With the ball no longer visible, the viewing port shows a dark or black dot, warning that the battery charge is low.

The physics of
a state-of-charge battery indicator.

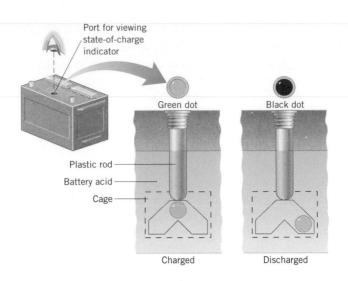

Port for viewing
state-of-charge
indicator

Green dot

Black dot

Plastic rod

Battery acid

Cage

Charged

Discharged

Figure 11.21 A state-of-charge indicator for a car battery.

Archimedes' principle has allowed us to determine how an object can float in a liquid. This principle also applies to gases, as the next example illustrates.

Example 10 A Goodyear Airship

The physics of a Goodyear airship.

Normally, a Goodyear airship, such as that in Figure 11.22, contains about 5.40×10^3 m³ of helium (He) whose density is 0.179 kg/m³. Find the weight of the load W_L that the airship can carry in equilibrium at an altitude where the density of air is 1.20 kg/m³.

Reasoning The airship and its load are in equilibrium. Thus, the buoyant force F_B applied to the airship by the surrounding air balances the weight W_{He} of the helium and the weight W_L of the load, including the solid parts of the airship. The free-body diagram in Figure 11.22b shows these forces.

Solution Because the forces in the free-body diagram balance, we have

$$W_{He} + W_L = F_B \quad \text{or} \quad W_L = F_B - W_{He}$$

According to Archimedes' principle, the buoyant force is the weight of the displaced air: $F_B = W_{air} = \rho_{air} V_{ship} g$. The weight of the helium is $W_{He} = \rho_{He} V_{ship} g$, where we have assumed that the volume of the helium and the volume of the airship have nearly the same value of $V_{ship} = 5.40 \times 10^3$ m³. Thus, we find that

$$W_L = \rho_{air} V_{ship} g - \rho_{He} V_{ship} g = (\rho_{air} - \rho_{He}) V_{ship} g$$
$$W_L = (1.20 \text{ kg/m}^3 - 0.179 \text{ kg/m}^3)(5.40 \times 10^3 \text{ m}^3) \times (9.80 \text{ m/s}^2) = \boxed{5.40 \times 10^4 \text{ N}}$$

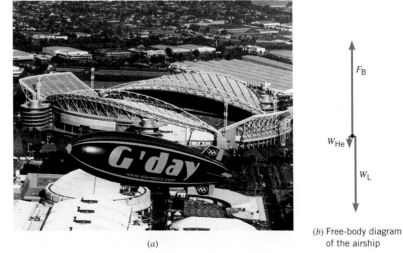

Figure 11.22 (*a*) A helium-filled Goodyear airship soars over the Sydney Olympic stadium in Australia at the summer 2000 Olympic games. The airship has been painted with the words 'G'day' on one side and 'Good Luck' on the other. (*b*) The free-body diagram of the airship, including the load weight W_L. (©AP/Wide World Photos)

F_B

W_{He}

W_L

(*a*)

(*b*) Free-body diagram of the airship

Figure 11.23 Two fluid particles in a stream. At different locations in the stream the particle velocities may be different, as indicated by v_1 and v_2.

11.7 Fluids in Motion

Fluids can move or flow in many ways. Water may flow smoothly and slowly in a quiet stream or violently over a waterfall. The air may form a gentle breeze or a raging tornado. To deal with such diversity, it helps to identify some of the basic types of fluid flow.

Fluid flow can be steady or unsteady. In *steady flow* the velocity of the fluid particles at any point is constant as time passes. For instance, in Figure 11.23 a fluid particle flows with a velocity of $v_1 = +2$ m/s past point 1. In steady flow every particle passing through this point has this same velocity. At another location the velocity may be different, as in a river, which usually flows fastest near its center and slowest near its banks. Thus, at point 2 in the figure, the fluid velocity is $v_2 = +0.5$ m/s, and if the flow is steady, all particles passing through this point have a velocity of $+0.5$ m/s. *Unsteady flow* exists whenever the velocity at a point in the fluid changes as time passes. *Turbulent flow* is an extreme kind of unsteady flow and occurs when there are sharp obstacles or bends in the path of a fast-moving fluid, as in the rapids in Figure 11.24. In turbulent flow, the velocity at a point changes erratically from moment to moment, both in magnitude and direction.

Fluid flow can be compressible or incompressible. Most liquids are nearly incompressible; that is, the density of a liquid remains almost constant as the pressure changes. To a good approximation, then, liquids flow in an incompressible manner. In contrast, gases are highly compressible. However, there are situations in which the density of a flowing gas remains constant enough that the flow can be considered incompressible.

Fluid flow can be viscous or nonviscous. A viscous fluid, such as honey, does not flow readily and is said to have a large viscosity. In contrast, water is less viscous and flows more readily; water has a smaller viscosity than honey. The flow of a viscous fluid is an energy-dissipating process. The viscosity hinders neighboring layers of fluid from sliding freely past one another. A fluid with zero viscosity flows in an unhindered manner with no dissipation of energy. Although no real fluid has zero viscosity at normal temperatures, some fluids have negligibly small viscosities. An incompressible, nonviscous fluid is called an *ideal fluid.*

When the flow is steady, *streamlines* are often used to represent the trajectories of the fluid particles. A streamline is a line drawn in the fluid such that a tangent to the streamline at any point is parallel to the fluid velocity at that point. Figure 11.25 shows the velocity vectors at three points along a streamline. The fluid velocity can vary (in both magnitude and direction) from point to point along a streamline, but at any given point, the velocity is constant in time, as required by the condition of steady flow. In fact, steady flow is often called *streamline flow.*

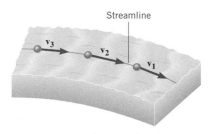

Figure 11.24 The flow of water in white-water rapids is an example of turbulent flow. (© Michael Kevin Daly/Corbis Images)

Figure 11.25 At any point along a streamline, the velocity vector of the fluid particle at that point is tangent to the streamline.

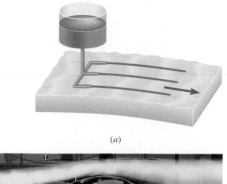

Figure 11.26 (a) In the steady flow of a liquid, a colored dye reveals the streamlines. (b) A smoke streamer reveals a streamline pattern for the air flowing around this pursuit cyclist, as he tests his bike for wind resistance in a wind tunnel. (© Nicholas Pinturas/Stone/Getty Images)

Figure 11.26a illustrates a method for making streamlines visible by using small tubes to release a colored dye into the moving liquid. The dye does not immediately mix with the liquid and is carried along a streamline. In the case of a flowing gas, such as that in a wind tunnel, streamlines are often revealed by smoke streamers, as part b of the figure shows.

In steady flow, the pattern of streamlines is steady in time, and, as Figure 11.26a indicates, no two streamlines cross one another. If they did cross, every particle arriving at the crossing point could go one way or the other. This would mean that the velocity at the crossing point would change from moment to moment, a condition that does not exist in steady flow.

11.8 The Equation of Continuity

Have you ever used your thumb to control the water flowing from the end of a hose, as in Figure 11.27? If so, you have seen that the water velocity increases when your thumb reduces the cross-sectional area of the hose opening. This kind of fluid behavior is described by the *equation of continuity.* This equation expresses the following simple idea: If a fluid enters one end of a pipe at a certain rate (e.g., 5 kilograms per second), then fluid must also leave at the same rate, assuming that there are no places between the entry and exit points to add or remove fluid. The mass of fluid per second (e.g., 5 kg/s) that flows through a tube is called the *mass flow rate.*

Figure 11.28 shows a small mass of fluid or fluid element (dark blue) moving along a tube. Upstream at position 2, where the tube has a cross-sectional area A_2, the fluid has a speed v_2 and a density ρ_2. Downstream at location 1, the corresponding quantities are A_1, v_1, and ρ_1. During a small time interval Δt, the fluid at point 2 moves a distance of $v_2 \Delta t$, as the drawing shows. The volume of fluid that has flowed past this point is the cross-sectional area times this distance, or $A_2 v_2 \Delta t$. The mass Δm_2 of this fluid element is the product of the density and volume: $\Delta m_2 = \rho_2 A_2 v_2 \Delta t$. Dividing Δm_2 by Δt gives the mass flow rate (the mass per second):

$$\text{Mass flow rate at position 2} = \frac{\Delta m_2}{\Delta t} = \rho_2 A_2 v_2 \qquad (11.7a)$$

Similar reasoning leads to the mass flow rate at position 1:

$$\text{Mass flow rate at position 1} = \frac{\Delta m_1}{\Delta t} = \rho_1 A_1 v_1 \qquad (11.7b)$$

Since no fluid can cross the sidewalls of the tube, the mass flow rates at positions 1 and 2 must be equal. But these positions were selected arbitrarily, so the mass flow rate has the same value everywhere in the tube, an important result known as the *equation of continuity.* The equation of continuity is an expression of the fact that mass is conserved (i.e., neither created nor destroyed) as the fluid flows.

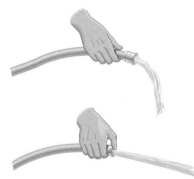

Figure 11.27 When the end of a hose is partially closed off, thus reducing its cross-sectional area, the fluid velocity increases.

■ **EQUATION OF CONTINUITY**

The mass flow rate ($\rho A v$) has the same value at every position along a tube that has a single entry and a single exit point for fluid flow. For two positions along such a tube

$$\rho_1 A_1 v_1 = \rho_2 A_2 v_2 \qquad (11.8)$$

where ρ = fluid density (kg/m^3)
$\quad A$ = cross-sectional area of tube (m^2)
$\quad v$ = fluid speed (m/s)

SI Unit of Mass Flow Rate: kg/s

The density of an incompressible fluid does not change during flow, so that $\rho_1 = \rho_2$, and the equation of continuity reduces to

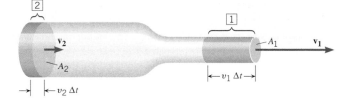

Figure 11.28 In general, a fluid flowing in a tube that has different cross-sectional areas A_1 and A_2 at positions 1 and 2 also has different velocities $\mathbf{v_1}$ and $\mathbf{v_2}$ at these positions.

Incompressible fluid $A_1 v_1 = A_2 v_2$ (11.9)

The quantity Av represents the volume of fluid per second that passes through the tube and is referred to as the ***volume flow rate Q:***

$$Q - \text{Volume flow rate} - Av \qquad (11.10)$$

Equation 11.9 shows that where the tube's cross-sectional area is large, the fluid speed is small, and, conversely, where the tube's cross-sectional area is small, the speed is large. Example 11 explores this behavior in more detail for the hose in Figure 11.27.

Example 11 A Garden Hose

A garden hose has an unobstructed opening with a cross-sectional area of 2.85×10^{-4} m^2, from which water fills a bucket in 30.0 s. The volume of the bucket is 8.00×10^{-3} m^3 (about two gallons). Find the speed of the water that leaves the hose through (a) the unobstructed opening and (b) an obstructed opening with half as much area.

Reasoning If we can determine the volume flow rate Q, the speed of the water can be obtained from Equation 11.10 as $v = Q/A$, since the area A is given. The volume flow rate can be found from the volume of the bucket and its fill time.

Solution

(a) The volume flow rate Q is equal to the volume of the bucket divided by the fill time. Therefore, the speed of the water is

$$v = \frac{Q}{A} = \frac{(8.00 \times 10^{-3}\text{ m}^3)/(30.0\text{ s})}{2.85 \times 10^{-4}\text{ m}^2} = \boxed{0.936\text{ m/s}} \qquad (11.10)$$

(b) Water can be considered incompressible, so the equation of continuity can be applied in the form $A_1 v_1 = A_2 v_2$. Since $A_2 = \frac{1}{2}A_1$, we find that

$$v_2 = \left(\frac{A_1}{A_2}\right)v_1 = \left(\frac{A_1}{\frac{1}{2}A_1}\right)(0.936\text{ m/s}) = \boxed{1.87\text{ m/s}} \qquad (11.9)$$

Concept Simulation 11.1

The speed of an incompressible fluid, such as water, changes as it flows through a hose whose cross-sectional area changes. In this simulation the user can select different cross-sectional areas and see how the speed of the water changes as it moves from one region to another. The equation of continuity states that the volume flow rate, $Q = Av$, remains constant at all points along the hose, where A is the cross-sectional area and v is the speed of the water. Thus, as A increases, v decreases, and vice versa.

Related Homework: *Conceptual Question 19, Problems 52, 87*

Go to **www.wiley.com/college/cutnell**

The next example applies the equation of continuity to the flow of blood.

Example 12 A Clogged Artery

In the condition known as atherosclerosis, a deposit or atheroma forms on the arterial wall and reduces the opening through which blood can flow. In the carotid artery in the neck, blood flows three times faster through a partially blocked region than it does through an unobstructed region. Determine the ratio of the effective radii of the artery at the two places.

Reasoning Blood, like most liquids, is incompressible, and the equation of continuity in the form of $A_1 v_1 = A_2 v_2$ (Equation 11.9) can be applied. In applying this equation, we use the fact that the area of a circle is πr^2.

The physics of a clogged artery.

Solution From Equation 11.9, it follows that

$$\underbrace{(\pi r_U^2)v_U}_{\substack{\text{Unobstructed} \\ \text{volume flow rate}}} = \underbrace{(\pi r_O^2)v_O}_{\substack{\text{Obstructed} \\ \text{volume flow rate}}}$$

The ratio of the radii is

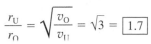
$$\frac{r_U}{r_O} = \sqrt{\frac{v_O}{v_U}} = \sqrt{3} = \boxed{1.7}$$

Problem solving insight
The equation of continuity in the form $A_1 v_1 = A_2 v_2$ applies only when the density of the fluid is constant. If the density is not constant, the equation of continuity is $\rho_1 A_1 v_1 = \rho_2 A_2 v_2$.

✔ **Check Your Understanding 3**

Water flows from left to right through the five sections (A, B, C, D, E) of the pipe shown in the drawing. In which section(s) does the water speed increase, decrease, and remain constant? Treat the water as an incompressible fluid. *(The answer is given at the end of the book.)*

	Speed Increases	Speed Decreases	Speed is Constant
a.	A, B	D, E	C
b.	D	B	A, C, E
c.	D, E	A, B	C
d.	B	D	A, C, E
e.	A, B	C, D	E

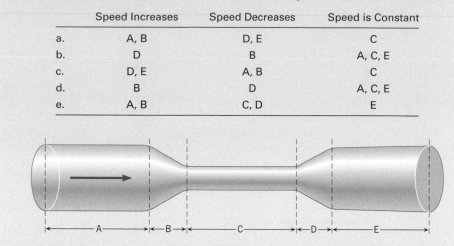

Background: The equation of continuity holds the key here. The fact that water can be treated as an incompressible fluid is important.

For similar questions (including calculational counterparts), consult Self-Assessment Test 11.2. This test is described at the end of Section 11.11.

Figure 11.29 CONCEPTS AT A GLANCE The work–energy theorem leads to Bernoulli's equation when the net work W_{nc} done by external nonconservative forces is not zero. Bernoulli's equation reveals that the pressure associated with moving air, such as this tornado in Miami (May 12, 1997), is lower than that of stationary air, such as that in the buildings. (© Reuters/RTV/Archive Photos/Getty Images)

11.9 *Bernoulli's Equation*

▶ CONCEPTS AT A GLANCE For *steady flow,* the speed, pressure, and elevation of an *incompressible and nonviscous* fluid are related by an equation discovered by Daniel Bernoulli (1700–1782). To derive ***Bernoulli's equation,*** we will use the work–energy theorem, as outlined in the Concepts-at-a-Glance chart in Figure 11.29. This theorem,

CONCEPTS AT A GLANCE

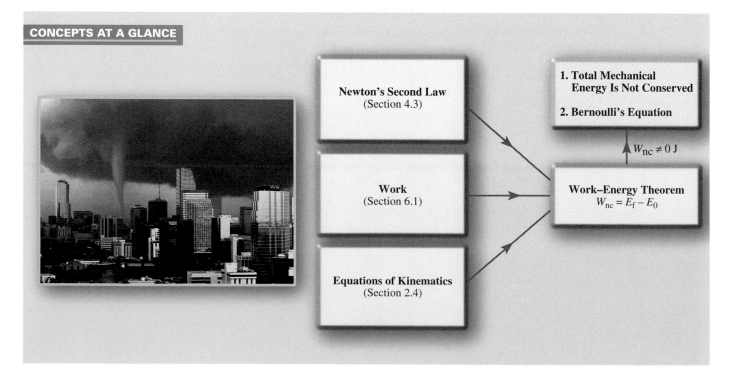

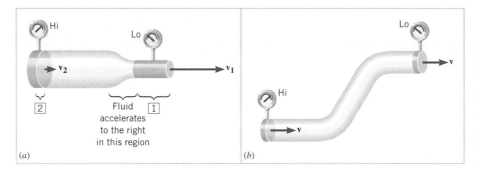

Figuro 11.30 (*a*) In this horizontal pipe, the pressure in region 2 is greater than that in region 1. The difference in pressures leads to the net force that accelerates the fluid to the right. (*b*) When the fluid changes elevation, the pressure at the bottom is greater than that at the top, assuming the cross-sectional area of the pipe is constant.

which is introduced in Chapter 6, states that the net work W_{nc} done on an object by external nonconservative forces is equal to the change in the total mechanical energy of the object (see Equation 6.8). As mentioned earlier, the pressure within a fluid is caused by collisional forces, which are nonconservative. Therefore, when a fluid is accelerated because of a difference in pressures, work is being done by nonconservative forces ($W_{nc} \neq 0$ J), and this work changes the total mechanical energy of the fluid from an initial value of E_0 to a final value of E_f. The total mechanical energy is not conserved. We will now see how the work–energy theorem leads directly to Bernoulli's equation. ◄

To begin with, let us make two observations about a moving fluid. First, whenever a fluid is flowing in a horizontal pipe and encounters a region of reduced cross-sectional area, the pressure of the fluid drops, as the pressure gauges in Figure 11.30*a* indicate. The reason for this follows from Newton's second law. When moving from the wider region 2 to the narrower region 1, the fluid speeds up or accelerates, consistent with the conservation of mass (as expressed by the equation of continuity). According to the second law, the accelerating fluid must be subjected to an unbalanced force. But there can be an unbalanced force only if the pressure in region 2 exceeds the pressure in region 1. We will see that the difference in pressures is given by Bernoulli's equation. The second observation is that if the fluid moves to a higher elevation, the pressure at the lower level is greater than the pressure at the higher level, as in Figure 11.30*b*. The basis for this observation is our previous study of static fluids, and Bernoulli's equation will confirm it, provided that the cross-sectional area of the pipe does not change.

To derive Bernoulli's equation, consider Figure 11.31*a*. This drawing shows a fluid element of mass m, upstream in region 2 of a pipe. Both the cross-sectional area and the elevation are different at different places along the pipe. The speed, pressure, and elevation in this region are v_2, P_2, and y_2, respectively. Downstream in region 1 these variables have the values v_1, P_1, and y_1. As Chapter 6 discusses, an object moving under the influence of gravity has a total mechanical energy E that is the sum of the kinetic energy KE and the gravitational potential energy PE: $E = \text{KE} + \text{PE} = \frac{1}{2}mv^2 + mgy$. When work

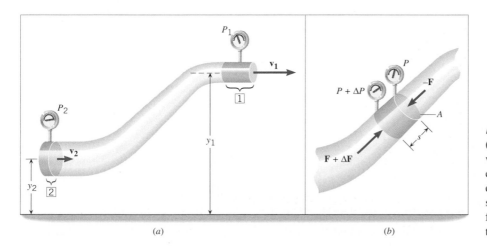

Figure 11.31 (*a*) A fluid element (dark blue) moving through a pipe whose cross-sectional area and elevation change. (*b*) The fluid element experiences a force $-\mathbf{F}$ on its top surface due to the fluid above it, and a force $\mathbf{F} + \Delta\mathbf{F}$ on its bottom surface due to the fluid below it.

W_{nc} is done on the fluid element by external nonconservative forces, the total mechanical energy changes. According to the work–energy theorem, the work equals the change in the total mechanical energy:

$$W_{nc} = E_1 - E_2 = \underbrace{(\tfrac{1}{2}mv_1^2 + mgy_1)}_{\substack{\text{Total mechanical} \\ \text{energy in region 1}}} - \underbrace{(\tfrac{1}{2}mv_2^2 + mgy_2)}_{\substack{\text{Total mechanical} \\ \text{energy in region 2}}} \qquad (6.8)$$

Figure 11.31*b* helps us understand how the work W_{nc} arises. On the top surface of the fluid element, the surrounding fluid exerts a pressure P. This pressure gives rise to a force of magnitude $F = PA$, where A is the cross-sectional area. On the bottom surface, the surrounding fluid exerts a slightly greater pressure, $P + \Delta P$, where ΔP is the pressure difference between the ends of the element. As a result, the force on the bottom surface has a magnitude of $F + \Delta F = (P + \Delta P)A$. The magnitude of the *net* force pushing the fluid element up the pipe is $\Delta F = (\Delta P)A$. When the fluid element moves through its own length s, the work done is the product of the magnitude of the net force and the distance: Work $= (\Delta F)s = (\Delta P)As$. The quantity As is the volume V of the element, so the work is $(\Delta P)V$. The total work done on the fluid element in moving it from region 2 to region 1 is the sum of the small increments of work $(\Delta P)V$ done as the element moves along the pipe. This sum amounts to $W_{nc} = (P_2 - P_1)V$, where $P_2 - P_1$ is the pressure difference between the two regions. With this expression for W_{nc}, the work–energy theorem becomes

$$W_{nc} = (P_2 - P_1)V = (\tfrac{1}{2}mv_1^2 + mgy_1) - (\tfrac{1}{2}mv_2^2 + mgy_2)$$

By dividing both sides of this result by the volume V, recognizing that m/V is the density ρ of the fluid, and rearranging terms, we obtain Bernoulli's equation.

■ **BERNOULLI'S EQUATION**

In the steady flow of a nonviscous, incompressible fluid of density ρ, the pressure P, the fluid speed v, and the elevation y at any two points (1 and 2) are related by

$$P_1 + \tfrac{1}{2}\rho v_1^2 + \rho g y_1 = P_2 + \tfrac{1}{2}\rho v_2^2 + \rho g y_2 \qquad (11.11)$$

Since the points 1 and 2 were selected arbitrarily, the term $P + \tfrac{1}{2}\rho v^2 + \rho g y$ has a constant value at all positions in the flow. For this reason, Bernoulli's equation is sometimes expressed as $P + \tfrac{1}{2}\rho v^2 + \rho g y = $ constant.

Equation 11.11 can be regarded as an extension of the earlier result that specifies how the pressure varies with depth in a static fluid ($P_2 = P_1 + \rho g h$), the terms $\tfrac{1}{2}\rho v_1^2$ and $\tfrac{1}{2}\rho v_2^2$ accounting for the effects of fluid speed. Bernoulli's equation reduces to the result for static fluids when the speed of the fluid is the same everywhere ($v_1 = v_2$), as it is when the cross-sectional area remains constant. Under such conditions, Bernoulli's equation is $P_1 + \rho g y_1 = P_2 + \rho g y_2$. After rearrangement, this result becomes

$$P_2 = P_1 + \rho g(y_1 - y_2) = P_1 + \rho g h$$

which is the result (Equation 11.4) for static fluids.

11.10 *Applications of Bernoulli's Equation*

When a moving fluid is contained in a horizontal pipe, all parts of it have the same elevation ($y_1 = y_2$), and Bernoulli's equation simplifies to

$$P_1 + \tfrac{1}{2}\rho v_1^2 = P_2 + \tfrac{1}{2}\rho v_2^2 \qquad (11.12)$$

Thus, the quantity $P + \tfrac{1}{2}\rho v^2$ remains constant throughout a horizontal pipe; if v increases, P decreases and vice versa. This is exactly the result that we deduced qualitatively from Newton's second law at the beginning of Section 11.9, and Conceptual Example 13 illustrates it.

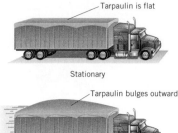

Tarpaulin is flat

Stationary

Conceptual Example 13 Tarpaulins and Bernoulli's Equation

A tarpaulin is a piece of canvas that is used to cover a cargo, like that pulled by the truck in Figure 11.32. When the truck is stationary the tarpaulin lies flat, but it bulges outward when the truck is speeding down the highway. Account for this behavior.

Tarpaulin bulges outward

Moving

Reasoning and Solution The behavior is a direct consequence of the pressure changes that Bernoulli's equation describes for flowing fluids. When the truck is stationary, the air outside and inside the cargo area is stationary, so the air pressure is the same in both places. This pressure applies the same force to the outer and inner surfaces of the canvas, with the result that the tarpaulin lies flat. When the truck is moving, the outside air rushes over the top surface of the canvas. In accord with Bernoulli's equation (Equation 11.12), the moving air has a lower pressure than the stationary air within the cargo area. *The greater inside pressure generates a greater force on the inner surface of the canvas, and the tarpaulin bulges outward.*

Figure 11.32 The tarpaulin that covers the cargo is flat when the truck is stationary but bulges outward when the truck is moving.

Related Homework: *Problem 57*

Example 14 applies Equation 11.12 to a dangerous physiological condition known as an aneurysm.

Example 14 An Enlarged Blood Vessel

An aneurysm is an abnormal enlargement of a blood vessel such as the aorta. Suppose that, because of an aneurysm, the cross-sectional area A_1 of the aorta increases to a value $A_2 = 1.7A_1$. The speed of the blood ($\rho = 1060$ kg/m^3) through a normal portion of the aorta is $v_1 = 0.40$ m/s. Assuming that the aorta is horizontal (the person is lying down), determine the amount by which the pressure P_2 in the enlarged region exceeds the pressure P_1 in the normal region.

⚕ The physics of an aneurysm.

Reasoning Bernoulli's equation (Equation 11.12), may be used to find the pressure difference between two points in a fluid moving horizontally. However, in order to use this relation we need to know the speed of the blood in the enlarged region of the artery, as well as the speed in the normal section. The speed in the enlarged region can be obtained from the equation of continuity (Equation 11.9), which relates the blood speeds in the enlarged and normal regions to the cross-sectional areas of these regions.

Solution For purposes of using Bernoulli's equation, we define region 2 to be the aneurysm. In the form for horizontal flow, this equation is $P_1 + \frac{1}{2}\rho v_1^2 = P_2 + \frac{1}{2}\rho v_2^2$, the difference in pressure being $P_2 - P_1 = \frac{1}{2}\rho(v_1^2 - v_2^2)$. From the equation of continuity, the speed v_2 of the blood in the aneurysm is given by $v_2 = (A_1/A_2)v_1$, where A_2 is the cross-sectional area of the enlarged artery, and A_1 and v_1 are the cross-sectional area and blood speed in the normal artery. Substituting this value for v_2 into Bernoulli's equation yields

$$P_2 - P_1 = \frac{1}{2}\rho(v_1^2 - v_2^2) = \frac{1}{2}\rho v_1^2\left[1 - \left(\frac{A_1}{A_2}\right)^2\right]$$

$$= \frac{1}{2}(1060 \text{ kg/m}^3)(0.40 \text{ m/s})^2\left[1 - \frac{A_1^2}{(1.7A_1)^2}\right] = \boxed{55 \text{ Pa}}$$

This result is positive, indicating that P_2 is greater than P_1. The excess pressure puts added stress on the already weakened tissue of the arterial wall at the aneurysm.

The impact of fluid flow on pressure is widespread. Figure 11.33, for instance, illustrates how household plumbing takes into account the implications of Bernoulli's equation. The U-shaped section of pipe beneath the sink is called a "trap," because it traps water, which serves as a barrier to prevent sewer gas from leaking into the house. Part *a* of the drawing shows poor plumbing. When water from the clothes washer rushes through the sewer pipe, the high-speed flow causes the pressure at point *A* to drop. The pressure at point *B* in the sink, however, remains at the higher

The physics of household plumbing.

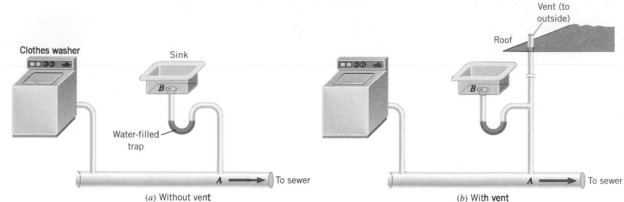

(a) Without vent   *(b) With **vent***

Figure 11.33 In a household plumbing system, a vent is necessary to equalize the pressures at points *A* and *B*, thus preventing the trap from being emptied. An empty trap allows sewer gas to enter the house.

The physics of airplane wings.

The physics of a curveball.

Need more practice?

🖱 **Interactive LearningWare 11.2**
A fountain sends a stream of water 5.00 m straight upward. The pipe from which the water exits has an effective cross-sectional area of $5.00 \times 10^{-4} \text{ m}^2$. How many gallons per minute does the fountain use? $(1 \text{ gal} = 3.79 \times 10^{-3} \text{ m}^3)$

Related Homework: Problem 60

Go to
www.wiley.com/college/cutnell
for an interactive solution.

atmospheric pressure. As a result of this pressure difference, the water is pushed out of the trap and into the sewer line, leaving no protection against sewer gas. A correctly designed system is vented to the outside of the house, as in Figure 11.33*b*. The vent ensures that the pressure at *A* remains the same as that at *B* (atmospheric pressure), even when water from the clothes washer is rushing through the pipe. Thus, the purpose of the vent is to prevent the trap from being emptied, not to provide an escape route for sewer gas.

One of the most spectacular examples of how fluid flow affects pressure is the dynamic lift on airplane wings. Figure 11.34*a* shows a wing (in cross section) moving to the right, with the air flowing leftward past the wing. Because of the shape of the wing, the air travels faster over the curved upper surface than it does over the flatter lower surface. According to Bernoulli's equation, the pressure above the wing is lower (faster moving air), while the pressure below the wing is higher (slower moving air). Thus, the wing is lifted upward. Part *b* of the figure shows the wing of an airplane.

The curveball, one of the main weapons in the arsenal of a baseball pitcher, is another illustration of the effects of fluid flow. Figure 11.35*a* shows a baseball moving to the right with no spin. The view is from above, looking down toward the ground. In this situation, air flows with the same speed around both sides of the ball, and the pressure is the same on both sides. No net force exists to make the ball curve to either side. However, when the ball is given a spin, the air close to its surface is dragged around with it; the air on one-half of the ball is speeded up (lower pressure), while that on the other half is slowed down (higher pressure). Part *b* of the picture illustrates the effects of a counterclockwise spin. The baseball experiences a net deflection force and curves on its way from the pitcher's mound to the plate, as part *c* shows.*

As a final application of Bernoulli's equation, Figure 11.36*a* shows a large tank from which water is emerging through a small pipe near the bottom. Bernoulli's equation can be used to determine the speed (called the efflux speed) at which the water leaves the pipe, as the next example shows.

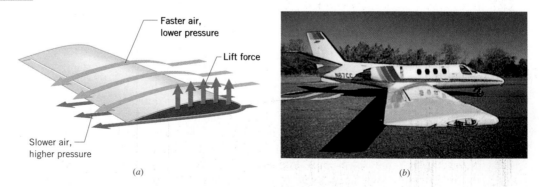

Figure 11.34 (*a*) Air flowing around an airplane wing. The wing is moving to the right. (*b*) The end of this wing has roughly the shape indicated in part *a*. (© Gary Gladstone/The Image Bank/ Getty Images)

(a)   *(b)*

* In the jargon used in baseball, the pitch shown in Figure 11.35 is called a "slider."

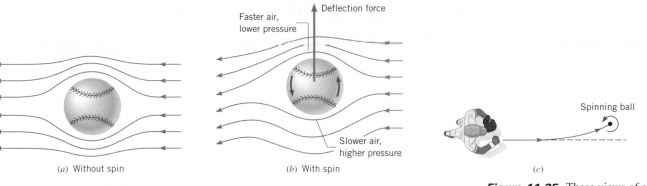

(a) Without spin   (b) With spin   (c)

Example 15 Efflux Speed

The tank in Figure 11.36a is open to the atmosphere at the top. Find an expression for the speed of the liquid leaving the pipe at the bottom.

Reasoning We assume that the liquid behaves as an ideal fluid. Therefore, we can apply Bernoulli's equation, and in preparation for doing so, we locate two points in the liquid in Figure 11.36a. Point 1 is just outside the efflux pipe, and point 2 is at the top surface of the liquid. The pressure at each of these points is equal to the atmospheric pressure, a fact that will be used to simplify Bernoulli's equation.

Solution Since the pressures at points 1 and 2 are the same, we have $P_1 = P_2$, and Bernoulli's equation becomes $\frac{1}{2}\rho v_1^2 + \rho g y_1 = \frac{1}{2}\rho v_2^2 + \rho g y_2$. The density ρ can be eliminated algebraically from this result, which can then be solved for the square of the efflux speed v_1:

$$v_1^2 = v_2^2 + 2g(y_2 - y_1) = v_2^2 + 2gh$$

We have substituted $h = y_2 - y_1$ for the height of the liquid above the efflux tube. If the tank is very large, the liquid level changes only slowly, and the speed at point 2 can be set equal to zero, so that $\boxed{v_1 = \sqrt{2gh}}$.

Figure 11.35 These views of a baseball are from above, looking down toward the ground, with the ball moving to the right. (a) Without spin, the ball does not curve to either side. (b) A spinning ball curves in the direction of the deflection force. (c) The spin in part b causes the ball to curve as shown here.

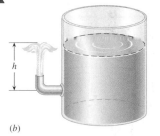

(a)

In Example 15 the liquid is assumed to be an ideal fluid, and the speed with which it leaves the pipe is the same as if the liquid had freely fallen through a height h (see Equation 2.9 with $x = h$ and $a = g$). This result is known as ***Torricelli's theorem.*** If the outlet pipe were pointed directly upward, as in Figure 11.36b, the liquid would rise to a height h equal to the fluid level above the pipe. However, if the liquid is not an ideal fluid, its viscosity cannot be neglected. Then, the efflux speed would be less than that given by Bernoulli's equation, and the liquid would rise to a height less than h.

✔ Check Your Understanding 4

Fluid is flowing from left to right through a pipe (see the drawing). Points A and B are at the same elevation, but the cross-sectional areas of the pipe are different. Points B and C are at different elevations, but the cross-sectional areas are the same. Rank the pressures at the three points, highest to lowest. *(The answer is given at the end of the book.)*

a. A and B (a tie), C
b. C, A and B (a tie)
c. B, C, A
d. C, B, A
e. A, B, C

Background: Bernoulli's equation plays the principal role. The equation of continuity is also pertinent.

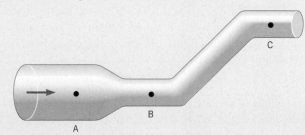

For similar questions (including calculational counterparts), consult Self-Assessment Test 11.2. This test is described at the end of Section 11.11.

(b)

Figure 11.36 (a) Bernoulli's equation can be used to determine the speed of the liquid leaving the small pipe. (b) An ideal fluid (no viscosity) will rise to the fluid level in the tank after leaving a vertical outlet nozzle.

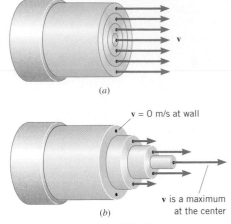

Figure 11.37 (*a*) In ideal (nonviscous) fluid flow, all fluid particles across the pipe have the same velocity. (*b*) In viscous flow, the speed of the fluid is zero at the surface of the pipe and increases to a maximum along the center axis.

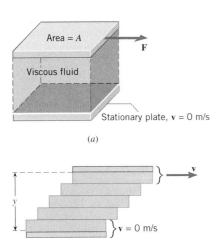

Figure 11.38 (*a*) A force **F** is applied to the top plate, which is in contact with a viscous fluid. (*b*) Because of the force **F**, the top plate and the adjacent layer of fluid move with a constant velocity **v**.

*11.11 Viscous Flow

In an ideal fluid there is no viscosity to hinder the fluid layers as they slide past one another. Within a pipe of uniform cross section, every layer of an ideal fluid moves with the same velocity, even the layer next to the wall, as Figure 11.37*a* shows. When viscosity is present, the fluid layers have different velocities, as part *b* of the drawing illustrates. The fluid at the center of the pipe has the greatest velocity. In contrast, the fluid layer next to the wall surface does not move at all because it is held tightly by intermolecular forces. So strong are these forces that if a solid surface moves, the adjacent fluid layer moves along with it and remains at rest *relative* to the moving surface.

To help introduce viscosity in a quantitative fashion, Figure 11.38*a* shows a viscous fluid between two parallel plates. The top plate is free to move while the bottom one is stationary. If the top plate is to move with a velocity **v** relative to the bottom plate, a force **F** is required. For a highly viscous fluid, like thick honey, a large force is needed; for a less viscous fluid, like water, a smaller one will do. As part *b* of the drawing suggests, we may imagine the fluid to be composed of many thin horizontal layers. When the top plate moves, the intermediate fluid layers slide over each other. The velocity of each layer is different, changing uniformly from **v** at the top plate to zero at the bottom plate. The resulting flow is called **laminar flow,** since a thin layer is often referred to as a lamina. As each layer moves, it is subjected to viscous forces from its neighbors. The purpose of the force **F** is to compensate for the effect of these forces, so that any layer can move with a constant velocity.

The amount of force required in Figure 11.38*a* depends on several factors. Larger areas *A*, being in contact with more fluid, require larger forces, so that the force is proportional to the contact area ($F \propto A$). For a given area, greater speeds require larger forces, with the result that the force is proportional to the speed ($F \propto v$). The force is also inversely proportional to the perpendicular distance *y* between the top and bottom plates ($F \propto 1/y$). The larger the distance *y*, the smaller is the force required to achieve a given speed with a given contact area. These three proportionalities can be expressed simultaneously in the following way: $F \propto Av/y$. Equation 11.13 expresses this relationship with the aid of a proportionality constant η (Greek letter *eta*), which is called the **coefficient of viscosity** or simply the **viscosity.**

■ **FORCE NEEDED TO MOVE A LAYER OF VISCOUS FLUID WITH A CONSTANT VELOCITY**

The magnitude of the tangential force **F** required to move a fluid layer at a constant speed *v*, when the layer has an area *A* and is located a perpendicular distance *y* from an immobile surface, is given by

$$F = \frac{\eta A v}{y} \qquad (11.13)$$

where η is the coefficient of viscosity.

SI Unit of Viscosity: Pa·s

Common Unit of Viscosity: poise (P)

By solving Equation 11.13 for the viscosity, $\eta = Fy/(vA)$, it can be seen that the SI unit for viscosity is N·m/[(m/s)·m²] = Pa·s. Another common unit for viscosity is the *poise* (P), which is used in the cgs system of units and is named after the French physician Jean Poiseuille (1799–1869; pronounced, approximately, as Pwah-zoy′). The following relation exists between the two units:

$$1 \text{ poise (P)} = 0.1 \text{ Pa·s}$$

Values of viscosity depend on the nature of the fluid. Under ordinary conditions, the viscosities of liquids are significantly *larger* than those of gases. Moreover, the viscosities of either liquids or gases depend markedly on temperature. Usually, the viscosities of liquids decrease as the temperature is increased. Anyone who has heated honey or oil, for example, knows that these fluids flow much more freely at an elevated temperature. In contrast, the viscosities of gases increase as the temperature is raised. An ideal fluid has $\eta = 0$ P.

Viscous flow occurs in a wide variety of situations, such as oil moving through a pipeline or a liquid being forced through the needle of a hypodermic syringe. Figure 11.39 identifies the factors that determine the volume flow rate Q (in m^3/s) of the fluid. First, a difference in pressures $P_2 - P_1$ must be maintained between any two locations along the pipe for the fluid to flow. In fact, Q is proportional to $P_2 - P_1$, a greater pressure difference leading to a larger flow rate. Second, a long pipe offers greater resistance to the flow than a short pipe does, and Q is inversely proportional to the length L. Because of this fact, long pipelines, such as the Alaskan pipeline, have pumping stations at various places along the line to compensate for a drop in pressure (see Figure 11.40). Third, high-viscosity fluids flow less readily than low-viscosity fluids, and Q is inversely proportional to the viscosity η. Finally, the volume flow rate is larger in a pipe of larger radius, other things being equal. The dependence on the radius R is a surprising one, Q being proportional to the fourth power of the radius, or R^4. If, for instance, the pipe radius is reduced to one-half of its original value, the volume flow rate is reduced to one-sixteenth of its original value, assuming the other variables remain constant. The mathematical relation for Q in terms of these parameters was discovered by Poiseuille and is known as *Poiseuille's law.*

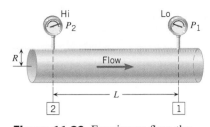

Figure 11.39 For viscous flow, the difference in pressures $P_2 - P_1$, the radius R and length L of the tube, and the viscosity η of the fluid influence the volume flow rate.

The physics of **pipeline pumping stations.**

■ **POISEUILLE'S LAW**

A fluid whose viscosity is η, flowing through a pipe of radius R and length L, has a volume flow rate Q given by

$$Q = \frac{\pi R^4 (P_2 - P_1)}{8 \eta L} \qquad (11.14)$$

where P_1 and P_2 are the pressures at the ends of the pipe.

Example 16 illustrates the use of Poiseuille's law.

Example 16 Giving an Injection

A hypodermic syringe is filled with a solution whose viscosity is 1.5×10^{-3} Pa·s. As Figure 11.41 shows, the plunger area of the syringe is 8.0×10^{-5} m^2, and the length of the needle is 0.025 m. The internal radius of the needle is 4.0×10^{-4} m. The gauge pressure in a vein is 1900 Pa (14 mm of mercury). What force must be applied to the plunger, so that 1.0×10^{-6} m^3 of solution can be injected in 3.0 s?

Reasoning The necessary force is the pressure applied to the plunger times the area of the plunger. Since viscous flow is occurring, the pressure is different at different points along the syringe. However, the barrel of the syringe is so wide that little pressure difference is required to sustain the flow up to point 2, where the fluid encounters the narrow needle. Consequently, the pressure applied to the plunger is nearly equal to the pressure P_2 at point 2. To find this pressure, we apply Poiseuille's law to the needle. This law indicates that $P_2 - P_1 = 8\eta LQ/(\pi R^4)$. We note that the pressure P_1 is given as a gauge pressure, which, in this case, is the amount of pressure in excess of atmospheric pressure. This causes no difficulty, because we need to find the amount of force in excess of that applied to the plunger by the atmosphere. The volume flow rate Q can be obtained from the time needed to inject the known volume of solution.

Solution The volume flow rate is $Q = (1.0 \times 10^{-6}\ m^3)/(3.0\ s) = 3.3 \times 10^{-7}\ m^3/s$. According to Poiseuille's law (Equation 11.14), the required pressure difference is

$$P_2 - P_1 = \frac{8\eta LQ}{\pi R^4} = \frac{8(1.5 \times 10^{-3}\ \text{Pa·s})(0.025\ m)(3.3 \times 10^{-7}\ m^3/s)}{\pi (4.0 \times 10^{-4}\ m)^4} = 1200\ \text{Pa}$$

Since $P_1 = 1900$ Pa, the pressure P_2 must be $P_2 = 1200$ Pa + 1900 Pa = 3100 Pa. The force that must be applied to the plunger is this pressure times the plunger area:

$$F = (3100\ \text{Pa})(8.0 \times 10^{-5}\ m^2) = \boxed{0.25\ \text{N}}$$

Figure 11.40 As oil flows along the Alaskan pipeline, the pressure drops because oil is a viscous fluid. Pumping stations are located along the pipeline to compensate for the drop in pressure. (© Ken Graham/Stone/Getty Images)

 *The physics of* **a hypodermic syringe.**

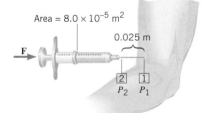

Figure 11.41 The difference in pressure $P_2 - P_1$ required to sustain the fluid flow through a hypodermic needle can be found with the aid of Poiseuille's law.

Self-Assessment Test 11.2

Test your knowledge of the important ideas in Sections 11.7–11.11:

- The Equation of Continuity
- Bernoulli's Equation
- Viscous Flow

Go to **www.wiley.com/college/cutnell**

11.12 Concepts & Calculations

In the behavior of fluids pressure plays an important role. As we have seen in this chapter, pressure is the magnitude of the force acting perpendicular to a surface divided by the area of the surface. Pressure should not be confused, however, with the force itself. The idea of a force is introduced in Chapter 4. The next example serves to emphasize that pressure and force are different concepts. In addition, it reviews the techniques that Chapters 4 and 9 discuss for analyzing objects that are in equilibrium.

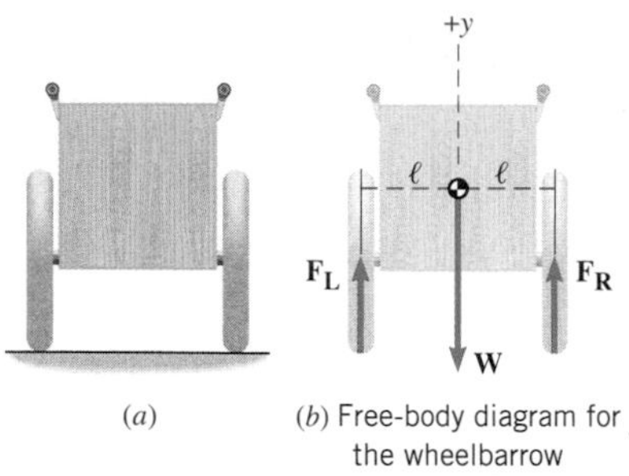

Figure 11.42 (*a*) Rear view of a wheelbarrow with balloon tires. (*b*) Free-body diagram of the wheelbarrow, showing its weight **W** and the forces $\mathbf{F_L}$ and $\mathbf{F_R}$ that the ground applies, respectively, to the left and right tires.

Concepts & Calculations Example 17 Pressure and Force

Figure 11.42*a* shows a rear view of a loaded two-wheeled wheelbarrow on a horizontal surface. It has balloon tires and a weight of $W = 684$ N, which is uniformly distributed. The left tire has a contact area with the ground of $A_L = 6.6 \times 10^{-4}\ \text{m}^2$, whereas the right tire is underinflated and has a contact area of $A_R = 9.9 \times 10^{-4}\ \text{m}^2$. Find the force and the pressure that each tire applies to the ground.

Concept Questions and Answers Force is a vector. Therefore, both a direction and a magnitude are needed to specify it. Are both a direction and a magnitude needed to specify a pressure?

> *Answer* No. Only a magnitude is needed to specify a pressure because pressure is a scalar quantity, not a vector quantity. The definition of pressure is the magnitude of the force acting perpendicular to a surface divided by the area of the surface. This definition does not include directional information.

The problem asks for the force that each tire applies to the ground. How is this force related to the force that the ground applies to each tire?

> *Answer* As Newton's law of action–reaction indicates, the force that a tire applies to the ground has the same magnitude and opposite direction as the force that the ground applies to the tire.

Do the left and right tires apply the same force to the ground?

> *Answer* Yes. Since the weight is uniformly distributed, each tire supports half the load. Therefore, each tire applies the same force to the ground.

Do the left and right tires apply the same pressure to the ground?

> *Answer* No. Pressure involves the contact area as well as the magnitude of the force. Since the force magnitudes are the same but the contact areas are different, the pressures are different, too.

Solution Since the wheelbarrow is at rest, it is in equilibrium and the net force acting on it must be zero, as Chapters 4 and 9 discuss. Referring to the free-body diagram in Figure 11.42*b* and taking upward as the positive direction, we have

$$\underbrace{\Sigma F_y}_{\text{Net force}} = F_L + F_R - W = 0$$

where F_L and F_R are, respectively, the forces that the ground applies to the left and right tires. At equilibrium the net torque acting on the wheelbarrow must also be zero, as Chapter 9 discusses. In the free-body diagram the weight acts at the center of gravity, which is at the center of the wheelbarrow. Using this point as the axis of rotation for torques, we can see that the forces F_L and F_R have the same lever arm ℓ, which is half the width of the wheelbarrow. For this axis, the weight W has a zero lever arm and, hence, a zero torque. Therefore, it does not contribute to the net torque. Remembering that counterclockwise torques are positive and setting the net torque equal to zero, we have

$$\underbrace{\Sigma\tau = F_R\ell - F_L\ell = 0}_{\text{Net torque}} \quad \text{or} \quad F_R = F_L$$

Substituting this result into the force equation gives

$$F_L + F_L - W = 0$$
$$F_L = F_R = \tfrac{1}{2}W = \tfrac{1}{2}(684 \text{ N}) = 342 \text{ N}$$

The value of 342 N is the magnitude of the upward force that the ground applies to each tire. Using Newton's action–reaction law, we conclude that each tire applies to the ground a $\boxed{\text{downward force of 342 N}}$. Using this value and the given contact areas, we find that the left and right tires apply the following pressures P_L and P_R to the ground:

$$P_L = \frac{F}{A_L} = \frac{342 \text{ N}}{6.6 \times 10^{-4} \text{ m}^2} = \boxed{5.2 \times 10^5 \text{ Pa}}$$

$$P_R = \frac{F}{A_R} = \frac{342 \text{ N}}{9.9 \times 10^{-4} \text{ m}^2} = \boxed{3.5 \times 10^5 \text{ Pa}}$$

▲

One manifestation of fluid pressure is the buoyant force that is applied to any object immersed in a fluid. Section 11.6 discusses this force, as well as Archimedes' principle, which specifies a convenient way to determine it. Example 18 focuses on the essence of this principle.

Concepts & Calculations Example 18   **The Buoyant Force**

A father (weight $W = 830$ N) and his daughter (weight $W = 340$ N) are spending the day at the lake. They are each sitting on a beach ball that is just submerged beneath the water (see Figure 11.43). Ignoring the weight of the air within the balls and the parts of their legs that are underwater, find the radius of each ball.

Concept Questions and Answers Each person and beach ball is in equilibrium, being stationary and having no acceleration. Thus, the net force acting on each is zero. What balances the downward-acting weight in each case?

Answer The downward-acting weight is balanced by the upward-acting buoyant force F_B that the water applies to the ball.

In which case is the buoyant force greater?

Answer The buoyant force acting on the father's beach ball is greater, since it must balance his greater weight.

In the situation described, what determines the magnitude of the buoyant force?

Answer According to Archimedes' principle, the magnitude of the buoyant force equals the weight of the fluid that the ball displaces. Since the ball is completely submerged, it displaces a volume of water that equals the ball's volume. The weight of this volume of water is the magnitude of the buoyant force.

Which beach ball has the larger radius?

Answer The father's ball has the larger volume and the larger radius. This follows because a larger buoyant force acts on that ball. For the buoyant force to be larger, that ball must displace a greater volume of water, according to Archimedes' principle. Therefore, the volume of that ball is larger, since the balls are completely submerged.

Solution For an object in equilibrium, the net force acting on the object must be zero. Therefore, taking upward to be the positive direction, we have

$$\underbrace{\Sigma F = F_B - W = 0}_{\text{Net force}}$$

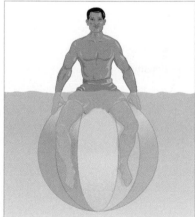

Figure 11.43 The two bathers are sitting on different-sized beach balls that are just submerged beneath the water.

Archimedes' principle specifies that the magnitude of the buoyant force is the weight of the water displaced by the ball. Using the definition of density given in Equation 11.1, the mass of the displaced water is $m = \rho V$, where $\rho = 1.00 \times 10^3$ kg/m^3 is the density of water (see Table 11.1) and V is the volume displaced. Since all of the ball is submerged, $V = \frac{4}{3}\pi r^3$, assuming that the ball remains spherical. The weight of the displaced water is $mg = \rho(\frac{4}{3}\pi r^3)g$. With this value for the buoyant force, the force equation becomes

$$F_B - W = \rho(\tfrac{4}{3}\pi r^3)g - W = 0$$

Solving for the radius, we find that

Father $r = \sqrt[3]{\dfrac{3W}{4\pi\rho g}} = \sqrt[3]{\dfrac{3(830\ \text{N})}{4\pi(1.00 \times 10^3\ \text{kg/m}^3)(9.80\ \text{m/s}^2)}} = \boxed{0.27\ \text{m}}$

Daughter $r = \sqrt[3]{\dfrac{3W}{4\pi\rho g}} = \sqrt[3]{\dfrac{3(340\ \text{N})}{4\pi(1.00 \times 10^3\ \text{kg/m}^3)(9.80\ \text{m/s}^2)}} = \boxed{0.20\ \text{m}}$

As expected, the radius of the father's beach ball is greater.

At the end of the problem set for this chapter, you will find homework problems that contain both conceptual and quantitative parts. These problems are grouped under the heading *Concepts & Calculations, Group Learning Problems*. They are designed for use by students working alone or in small learning groups. The conceptual part of each problem provides a convenient focus for group discussions.

Concept Summary

This summary presents an abridged version of the chapter, including the important equations and all available learning aids. For convenient reference, the learning aids (including the text's examples) are placed next to or immediately after the relevant equation or discussion. The following learning aids may be found on-line at **www.wiley.com/college/cutnell**:

Interactive LearningWare examples are solved according to a five-step interactive format that is designed to help you develop problem-solving skills.	**Concept Simulations** are animated versions of text figures or animations that illustrate important concepts. You can control parameters that affect the display, and we encourage you to experiment.
Interactive Solutions offer specific models for certain types of problems in the chapter homework. The calculations are carried out interactively.	**Self-Assessment Tests** include both qualitative and quantitative questions. Extensive feedback is provided for both incorrect and correct answers, to help you evaluate your understanding of the material.

Topic	*Discussion*	*Learning Aids*

11.1 Mass Density

Fluids are materials that can flow, and they include gases and liquids.

The mass density ρ of a substance is its mass m divided by its volume V:

Mass Density	$\rho = \dfrac{m}{V}$ (11.1)	**Example 1**

The specific gravity of a substance is its mass density divided by the density of water at 4 °C (1.000×10^3 kg/m^3):

Specific gravity	$\text{Specific gravity} = \dfrac{\text{Density of substance}}{1.000 \times 10^3\ \text{kg/m}^3}$ (11.2)	

11.2 Pressure

The pressure P exerted by a fluid is the magnitude F of the force acting perpendicular to a surface embedded in the fluid divided by the area A over which the force acts:

Pressure	$P = \dfrac{F}{A}$ (11.3)	**Examples 2, 17** **Interactive Solution 11.15**

The SI unit for measuring pressure is the pascal (Pa); 1 Pa = 1 N/m^2.

Atmospheric pressure	One atmosphere of pressure is 1.013×10^5 Pa or 14.7 lb/in.^{2}

Topic	Discussion	Learning Aids

11.3 Pressure and Depth in a Static Fluid

In the presence of gravity, the upper layers of a fluid push downward on the layers beneath, with the result that fluid pressure is related to depth. In an incompressible static fluid whose density is ρ, the relation is

$$P_2 = P_1 + \rho gh \qquad (11.4)$$ **Examples 3–6**

where P_1 is the pressure at one level, P_2 is the pressure at a level that is h meters deeper, and g is the magnitude of the acceleration due to gravity.

11.4 Pressure Gauges

Two basic types of pressure gauges are the mercury barometer and the open-tube manometer.

Gauge pressure and absolute pressure The gauge pressure is the amount by which a pressure P differs from atmospheric pressure. The absolute pressure is the actual value for P.

11.5 Pascal's Principle

Pascal's principle states that any change in the pressure applied to a completely   **Example 7** enclosed fluid is transmitted undiminished to all parts of the fluid and the en-   **Interactive Solution 11.33** closing walls.

11.6 Archimedes' Principle

The buoyant force is the upward force that a fluid applies to an object that is partially or completely immersed in it.

Archimedes' principle states that the magnitude of the buoyant force equals the weight of the fluid that the partially or completely immersed object displaces:

Archimedes' principle

$$\underbrace{F_B}_{\substack{\text{Magnitude of} \\ \text{buoyant force}}} = \underbrace{W_{\text{fluid}}}_{\substack{\text{Weight of} \\ \text{displaced fluid}}} \qquad (11.6)$$

Examples 8, 9, 10, 18
Interactive LearningWare 11.1

Use *Self-Assessment Test 11.1* to evaluate your understanding of Sections 11.1–11.6.

11.7 Fluids in Motion
11.8 The Equation of Continuity

Steady flow In steady flow, the velocity of the fluid particles at any point is constant as time passes.

Ideal fluid An incompressible, nonviscous fluid is known as an ideal fluid.

The mass flow rate of a fluid with a density ρ, flowing with a speed v in a pipe of cross-sectional area A, is the mass per second (kg/s) flowing past a point and is given by

Mass flow rate
$$\text{Mass flow rate} = \rho Av \qquad (11.7)$$

The equation of continuity expresses the fact that mass is conserved: what flows into one end of a pipe flows out the other end, assuming there are no additional entry or exit points in between. Expressed in terms of the mass flow rate, the equation of continuity is

Equation of continuity
$$\rho_1 A_1 v_1 = \rho_2 A_2 v_2 \qquad (11.8)$$

where the subscripts 1 and 2 denote two points along the pipe.

Incompressible fluid If a fluid is incompressible, the density at any two points is the same, $\rho_1 = \rho_2$. For an incompressible fluid, the equation of continuity becomes

$$A_1 v_1 = A_2 v_2 \qquad (11.9)$$

Examples 11, 12
Concept Simulation 11.1

The product Av is known as the volume flow rate Q:

Volume flow rate
$$Q = \text{Volume flow rate} = Av \qquad (11.10)$$

Topic	Discussion	Learning Aids

11.9 *Bernoulli's Equation*

11.10 *Applications of Bernoulli's Equation*

In the steady flow of an ideal fluid whose density is ρ, the pressure P, the fluid speed v, and the elevation y at any two points (1 and 2) in the fluid are related by Bernoulli's equation:

Examples 13, 14, 15

Interactive LearningWare 11.2

Bernoulli's equation

$$P_1 + \tfrac{1}{2}\rho v_1^2 + \rho g y_1 = P_2 + \tfrac{1}{2}\rho v_2^2 + \rho g y_2 \qquad (11.11)$$

Interactive Solution 11.63

When the flow is horizontal ($y_1 = y_2$), Bernoulli's equation indicates that higher fluid speeds are associated with lower fluid pressures.

11.11 *Viscous Flow*

The magnitude F of the tangential force required to move a fluid layer at a constant speed v, when the layer has an area A and is located a perpendicular distance y from an immobile surface, is given by

Force needed to move a layer of viscous fluid with a constant speed

$$F = \frac{\eta A v}{y} \qquad (11.13)$$

where η is the coefficient of viscosity.

A fluid whose viscosity is η, flowing through a pipe of radius R and length L, has a volume flow rate Q given by

Poiseuille's law

$$Q = \frac{\pi R^4 (P_2 - P_1)}{8 \eta L} \qquad (11.14)$$

Example 16

Interactive Solution 11.73

where P_1 and P_2 are the pressures at the ends of the pipe.

Use Self-Assessment Test 11.2 to evaluate your understanding of Sections 11.7–11.11.

Conceptual Questions

1. A pile of empty aluminum cans has a volume of 1.0 m³. The density of aluminum is 2700 kg/m³. Explain why the mass of the pile is *not* $\rho_{A1} V = (2700 \text{ kg/m}^3)(1.0 \text{ m}^3) = 2700$ kg.

2. A method for resealing a partially full bottle of wine under a vacuum uses a specially designed rubber stopper to close the bottle. A simple pump is attached to the stopper, and to remove air from the bottle, the plunger of the pump is pulled up and then released. After about 15 pull-and-release cycles the wine is under a partial vacuum. On the 15th pull-and-release cycle, it is harder to pull up the plunger than on the first cycle. Why?

3. A person could not balance her entire weight on the pointed end of a single nail, because it would penetrate her skin. However, she can lie safely on a "bed of nails." A "bed of nails" consists of many nails driven through a sheet of wood so that the pointed ends form a flat array. Why is the "bed of nails" trick safe?

4.  As you climb a mountain, your ears "pop" because of the changes in atmospheric pressure. In which direction does your eardrum move (a) as you climb up and (b) as you climb down? Give your reasoning.

5. A bottle of juice is sealed under partial vacuum, with a lid on which a red dot or "button" is painted. Around the button the following phrase is printed: "Button pops up when seal is broken." Explain why the button remains pushed in when the seal is intact.

6. A sealed container of water is full, except for a tube that is attached to it, as the drawing at right shows. Water is poured into this tube, and before the tube is full the container bursts. Explain why.

7. A closed tank is completely filled with water. A valve is then opened at the bottom of the tank and water begins to flow out. When the water stops flowing, will the tank be completely empty, or will there still be a noticeable amount of water in it? Explain your answer.

8. Could you use a straw to sip a drink on the moon? Explain.

9. Why does the cork fly out with a loud "pop" when a bottle of champagne is opened? To answer this question on an exam a student says, "Because the gas pressure in the bottle is about 0.3×10^5 Pa." Explain whether this refers to absolute or gauge pressure.

10. A scuba diver is below the surface of the water when a storm approaches, dropping the air pressure above the water. Would a sufficiently sensitive pressure gauge attached to his wrist register this drop in air pressure? Give your reasoning.

11. A steel beam is suspended completely under water by a cable that is attached to one end of the beam, so it hangs vertically. Another identical beam is also suspended completely under water, but by a cable that is attached to the center of the beam so it hangs horizontally. Which beam, if either, experiences the greater buoyant force? Provide a reason for your answer. Neglect any change in water density with depth.

12. A glass beaker, filled to the brim with water, is resting on a scale. A block is placed in the water, causing some of it to spill over. The water that spills is wiped away, and the beaker is still filled to the brim. How do the initial and final readings on the scale compare if the block is made from (a) wood and (b) iron? In both cases, explain your answer, using data from Table 11.1 and Archimedes' principle.

13. On a distant planet the acceleration due to gravity is less than it is on earth. Would you float more easily in water on this planet than on earth? Account for your answer.

14. Put an ice cube in a glass, and fill the glass to the brim with water. When the ice cube melts and the temperature of the water returns to its initial value, would the water level drop, remain the same, or rise (causing water to spill out)? Explain what you would observe in terms of Archimedes' principle.

15. As a person dives toward the bottom of a swimming pool, the pressure increases noticeably. Does the buoyant force also increase? Justify your answer. Neglect any change in water density with depth.

16. Suppose that an orbiting space station of the future had a swimming pool in it. If there is no artificial gravity, would a buoyant force be exerted on a swimmer? Explain.

17. The drawing shows a side view of part of a system of gates that is being installed to protect the city of Venice, Italy, against flooding due to very high tides. When not in use the hollow, triangular-shaped gate is filled with seawater and rests on the ocean floor. The gate is anchored to the ocean floor by a hinge mechanism, about which it can rotate. When a dangerously high tide is expected, water is pumped out of the gate, and the structure rotates into a position where it can protect against flooding. Explain why the gate rotates into position.

Not in use   In use

18. In steady flow, the velocity **v** of a fluid particle at any point is constant in time. On the other hand, a fluid accelerates when it moves into a region of smaller cross-sectional area. (a) Explain what causes the acceleration. (b) Explain why the condition of steady flow does not rule out such an acceleration.

19. Concept Simulation 11.1 at **www.wiley.com/college/cutnell** reviews the central idea in this question. The cross-sectional area of a stream of water becomes smaller as the water falls from a faucet. Account for this phenomenon in terms of the equation of continuity.

What would you expect to happen to the cross-sectional area when the water is shot upward, as it is in a fountain?

20. Have you ever had a large truck pass you from the opposite direction on a narrow two-lane road? You probably noticed that your car was pulled toward the truck as it passed. What can you conclude about the speed of the air between your car and the truck compared to that on the opposite side of the car? Provide a reason for your answer.

21. Can Bernoulli's equation describe the flow of water that is cascading down a rock-strewn spillway? Explain.

22. Hold two sheets of paper by adjacent corners, so that they hang downward. They should be parallel and slightly separated, so that you can see the floor through the gap between them. Blow air strongly down through the gap. Do the sheets move further apart, or do they come closer together? Discuss what you observe in terms of Bernoulli's equation.

23. Which way would you have to spin a baseball so that it curves upward on its way to the plate? In describing the spin, state how you are viewing the ball. Justify your answer.

24. You are traveling on a train with your window open. As the train approaches its rather high operating speed, your ears "pop." Your eardrums respond to a decrease or increase in the air pressure by "popping" outward or inward, respectively. Do your ears "pop" outward or inward on the train? Give your reasoning in terms of Bernoulli's equation.

25. A passenger is smoking in the backseat of a moving car. To remove the smoke, the driver opens his window just a bit. Explain why the smoke is drawn to and out of the driver's window.

26. The airport in Phoenix, Arizona, has occasionally been closed to large planes because of weather conditions that do not entail storms or low visibility. Instead, the conditions combine to create an unusually low air density. Using Bernoulli's equation, explain why such an air density would make it difficult for a large, heavy plane to take off, especially if the runway were not exceptionally long.

27. To change the oil in a car, you remove a plug beneath the engine and let the old oil run out. Your car has been sitting in the garage on a cold day. Before changing the oil, it is advisable to run the engine for a while. Why?

Problems

ssm Solution is in the Student Solutions Manual.   **www** Solution is available on the World Wide Web at www.wiley.com/college/cutnell
This icon represents a biomedical application.

Section 11.1 Mass Density

1. ssm Accomplished silver workers in India can pound silver into incredibly thin sheets, as thin as 3.00×10^{-7} m (about one-hundredth of the thickness of this sheet of paper). Find the area of such a sheet that can be formed from 1.00 kg of silver.

2. The *karat* is a dimensionless unit that is used to indicate the proportion of gold in a gold-containing alloy. An alloy that is one karat gold contains a weight of pure gold that is one part in twenty-four. What is the volume of gold in a 14.0-karat gold necklace whose weight is 1.27 N?

3. A pirate in a movie is carrying a chest (0.30 m × 0.30 m × 0.20 m) that is supposed to be filled with gold. To see how ridiculous this is, determine the weight (in newtons) of the gold. To judge how large this weight is, remember that 1 N = 0.225 lb.

4. A water bed has dimensions of 1.83 m × 2.13 m × 0.229 m. The floor of the bedroom will tolerate an additional weight of no more than 6660 N. Find the weight of the water in the bed and determine whether it should be purchased.

5. ssm The ice on a lake is 0.010 m thick. The lake is circular, with a radius of 480 m. Find the mass of the ice.

* **6.** An irregularly shaped chunk of concrete has a hollow spherical cavity inside. The mass of the chunk is 33 kg, and the volume enclosed by the outside surface of the chunk is 0.025 m³. What is the radius of the spherical cavity?

* **7.** A bar of gold measures 0.15 m × 0.050 m × 0.050 m. How many gallons of water have the same mass as this bar?

* **8.** A hypothetical spherical planet consists entirely of iron. What is the period of a satellite that orbits this planet just above its surface? Consult Table 11.1 as necessary.

** **9. ssm www** An antifreeze solution is made by mixing ethylene glycol ($\rho = 1116$ kg/m³) with water. Suppose the specific gravity of such a solution is 1.0730. Assuming that the total volume of the solution is the sum of its parts, determine the volume percentage of ethylene glycol in the solution.

Section 11.2 Pressure

10. A glass bottle of soda is sealed with a screw cap. The absolute pressure of the carbon dioxide inside the bottle is 1.80×10^5 Pa. Assuming that the top and bottom surfaces of the cap each have an area of 4.10×10^{-4} m², obtain the magnitude of the force that the screw thread exerts on the cap in order to keep it on the bottle. The air pressure outside the bottle is one atmosphere.

11. An airtight box has a removable lid of area 1.3×10^{-2} m² and negligible weight. The box is taken up a mountain where the air pressure outside the box is 0.85×10^5 Pa. The inside of the box is completely evacuated. What is the magnitude of the force required to pull the lid off the box?

12. United States currency is printed using intaglio presses that generate a printing pressure of 8.0×10^4 lb/in.² A $20 bill is 6.1 in. by 2.6 in. Calculate the magnitude of the force that the printing press applies to one side of the bill.

13. ssm High-heeled shoes can cause tremendous pressure to be applied to a floor. Suppose the radius of a heel is 6.00×10^{-3} m. At times during a normal walking motion, nearly the entire body weight acts perpendicular to the surface of such a heel. Find the pressure that is applied to the floor under the heel because of the weight of a 50.0-kg woman.

14. A person who weighs 625 N is riding a 98-N mountain bike. Suppose the entire weight of the rider and bike is supported equally by the two tires. If the gauge pressure in each tire is 7.60×10^5 Pa, what is the area of contact between each tire and the ground?

15. Interactive Solution 11.15 at **www.wiley.com/college/cutnell** presents a model for solving this problem. A solid concrete block weighs 169 N and is resting on the ground. Its dimensions are 0.400 m × 0.200 m × 0.100 m. A number of identical blocks are stacked on top of this one. What is the smallest number of whole blocks (including the one on the ground) that can be stacked so that their weight creates a pressure of at least two atmospheres on the ground beneath the first block?

* **16.** A cylinder is fitted with a piston, beneath which is a spring, as in the drawing. The cylinder is open at the top. Friction is absent. The spring constant of the spring is 3600 N/m. The piston has a negligible mass and a radius of 0.025 m. (a) When air beneath the piston is completely pumped out, how much does the atmospheric pressure cause the spring to compress? (b) How much work does the atmospheric pressure do in compressing the spring?

* **17. ssm** A cylinder (with circular ends) and a hemisphere are solid throughout and made from the same material. They are resting on the ground, the cylinder on one of its ends and the hemisphere on its flat side. The weight of each causes the same pressure to act on the ground. The cylinder is 0.500 m high. What is the radius of the hemisphere?

** **18.** A house has a roof (colored gray) with the dimensions shown in the drawing. Determine the magnitude and direction of the net force that the atmosphere applies to the roof when the outside pressure rises suddenly by 10.0 mm of mercury, before the pressure in the attic can adjust.

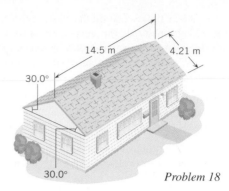

14.5 m 4.21 m 30.0° 30.0° *Problem 18*

Section 11.3 Pressure and Depth in a Static Fluid, Section 11.4 Pressure Gauges

19. ssm Some researchers believe that the dinosaur Barosaurus held its head erect on a long neck, much as a giraffe does. If so, fossil remains indicate that its heart would have been about 12 m below its brain. Assume that the blood has the density of water, and calculate the amount by which the blood pressure in the heart would have exceeded that in the brain. Size estimates for the single heart needed to withstand such a pressure range up to two tons. Alternatively, Barosaurus may have had a number of smaller hearts.

20. At a given instant, the blood pressure in the heart is 1.6×10^4 Pa. If an artery in the brain is 0.45 m above the heart, what is the pressure in the artery? Ignore any pressure changes due to blood flow.

21. The Mariana trench is located in the Pacific Ocean at a depth of about 11 000 m below the surface of the water. The density of seawater is 1025 kg/m³. (a) If an underwater vehicle were to explore such a depth, what force would the water exert on the vehicle's observation window (radius = 0.10 m)? (b) For comparison, determine the weight of a jetliner whose mass is 1.2×10^5 kg.

22. Measured along the surface of the water, a rectangular swimming pool has a length of 15 m. Along this length, the flat bottom of the pool slopes downward at an angle of 11° below the horizontal, from one end to the other. By how much does the pressure at the bottom of the deep end exceed the pressure at the bottom of the shallow end?

23. ssm A water tower is a familiar sight in many towns. The purpose of such a tower is to provide storage capacity and to provide sufficient pressure in the pipes that deliver the water to customers. The drawing shows a spherical reservoir that contains 5.25×10^5 kg of water when full. The reservoir is vented to the atmosphere at the top. For a full reservoir, find the gauge pressure that the water has at the faucet in (a) house A and (b) house B. Ignore the diameter of the delivery pipes.

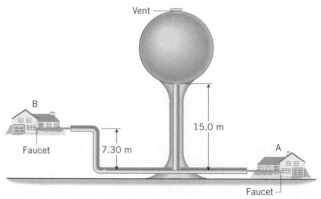

Vent

B Faucet 7.30 m 15.0 m A Faucet

24. ☤ The human lungs can function satisfactorily up to a limit where the pressure difference between the outside and inside of the lungs is one-twentieth of an atmosphere. If a diver uses a snorkel for breathing, how far below the water can she swim? Assume the diver is in salt water whose density is 1025 kg/m³.

25. As background for this problem, review Conceptual Example 6. A submersible pump is put under the water at the bottom of a well and is used to push water up through a pipe. What minimum output gauge pressure must the pump generate to make the water reach the nozzle at ground level, 71 m above the pump?

* **26.** A mercury barometer reads 747.0 mm on the roof of a building and 760.0 mm on the ground. Assuming a constant value of 1.29 kg/m³ for the density of air, determine the height of the building.

* **27.** ssm www Mercury is poured into a tall glass. Ethyl alcohol is then poured on top of the mercury until the height of the ethyl alcohol itself is 110 cm. The two fluids do not mix, and the air pressure at the top of the ethyl alcohol is one atmosphere. What is the absolute pressure at a point that is 7.10 cm below the ethyl alcohol–mercury interface?

* **28.** Figure 11.12 shows a mercury barometer. Consider two barometers, one using mercury and another using an unknown liquid. Suppose that the pressure above the liquid in each tube is maintained at the same value P, between zero and atmospheric pressure. The height of the unknown liquid is 16 times greater than the height of the mercury. Find the density of the unknown liquid.

* **29.** A 1.00-m-tall container is filled to the brim, partway with mercury and the rest of the way with water. The container is open to the atmosphere. What must be the depth of the mercury so that the absolute pressure on the bottom of the container is twice the atmospheric pressure?

** **30.** As the drawing illustrates, a pond has the shape of an inverted cone with the tip sliced off and has a depth of 5.00 m. The atmospheric pressure above the pond is 1.01×10^5 Pa. The circular top surface (radius = R_2) and circular bottom surface (radius = R_1) of the pond are both parallel to the ground. The magnitude of the force acting on the top surface is the same as the magnitude of the force acting on the bottom surface. Obtain (a) R_2 and (b) R_1.

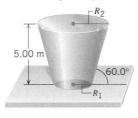

Section 11.5 Pascal's Principle

31. ssm The atmospheric pressure above a swimming pool changes from 755 to 765 mm of mercury. The bottom of the pool is a 12-m × 24-m rectangle. By how much does the force on the bottom of the pool increase?

32. In the hydraulic press used in a trash compactor, the radii of the input piston and the output plunger are 6.4×10^{-3} m and 5.1×10^{-2} m, respectively. The height difference between the input piston and the output plunger can be neglected. What force is applied to the trash when the input force is 330 N?

33. Interactive Solution 11.33 at **www.wiley.com/college/cutnell** presents a model for solving this problem. You can also prepare for this problem by reviewing Example 7. The hydraulic oil in a car lift has a density of 8.30×10^2 kg/m³. The weight of the input piston is negligible. The radii of the input piston and output plunger are 7.70 × 10⁻³ m and 0.125 m, respectively. What input force F is needed to support the 24 500-N combined weight of a car and the output plunger, when (a) the bottom surfaces of the piston and plunger are at the same level, and (b) the bottom surface of the output plunger is 1.30 m *above* that of the input piston?

34. A dentist's chair with a patient in it weighs 2100 N. The output plunger of a hydraulic system begins to lift the chair when the dentist's foot applies a force of 55 N to the input piston. Neglect any height difference between the plunger and the piston. What is the ratio of the radius of the plunger to the radius of the piston?

* **35.** ssm A dump truck uses a hydraulic cylinder, as the drawing illustrates. When activated by the operator, a pump injects hydraulic oil into the cylinder at an absolute pressure of 3.54×10^6 Pa and drives the output plunger, which has a radius of 0.150 m. Assuming the plunger remains perpendicular to the floor of the load bed, find the torque that the plunger creates about the axis identified in the drawing.

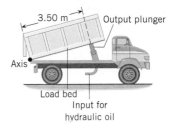

* **36.** The drawing shows a hydraulic system used with disc brakes. The force F is applied perpendicularly to the brake pedal. The pedal rotates about the axis shown in the drawing and causes a force to be applied perpendicularly to the input piston (radius = 9.50×10^{-3} m) in the master cylinder. The resulting pressure is transmitted by the brake fluid to the output plungers (radii = 1.90×10^{-2} m), which are covered with the brake linings. The linings are pressed against both sides of a disc attached to the rotating wheel. Suppose that the magnitude of F is 9.00 N. Assume that the input piston and the output plungers are at the same vertical level, and find the force applied to each side of the rotating disc.

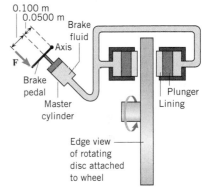

* **37.** The drawing shows a hydraulic chamber in which a spring (spring constant = 1600 N/m) is attached to the input piston, and a rock of mass 40.0 kg rests on the output plunger. The piston and plunger are nearly at the same height, and each has a negligible mass. By how much is the spring compressed from its unstrained position?

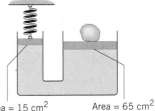

Section 11.6 Archimedes' Principle

38. A duck is floating on a lake with 25% of its volume beneath the water. What is the average density of the duck?

39. ssm What is the radius of a hydrogen-filled balloon that would carry a load of 5750 N (in addition to the weight of the hydrogen) when the density of air is 1.29 kg/m³?

40. Only a small part of an iceberg protrudes above the water, while the bulk lies below the surface. The density of ice is 917 kg/m³ and that of seawater is 1025 kg/m³. Find the percentage of the iceberg's volume that lies below the surface.

41. A paperweight, when weighed in air, has a weight of $W = 6.9$ N. When completely immersed in water, however, it has a weight of $W_{\text{in water}} = 4.3$ N. Find the volume of the paperweight.

42. What is the total mass of swimmers that the raft in Example 8 can carry and float with its top surface at water level?

43. ssm www A person can change the volume of his body by taking air into his lungs. The amount of change can be determined by weighing the person under water. Suppose that under water a person weighs 20.0 N with partially full lungs and 40.0 N with empty lungs. Find the change in body volume.

* **44.** What is the smallest number of whole logs (ρ = 725 kg/m³, radius = 0.0800 m, length = 3.00 m) that can be used to build a raft that will carry four people, each of whom has a mass of 80.0 kg?

* **45.** A hollow cubical box is 0.30 m on an edge. This box is floating in a lake with one-third of its height beneath the surface. The walls of the box have a negligible thickness. Water is poured into the box. What is the depth of the water in the box at the instant the box begins to sink?

* **46.** An object is solid throughout. When the object is completely submerged in ethyl alcohol, its apparent weight is 15.2 N. When completely submerged in water, its apparent weight is 13.7 N. What is the volume of the object?

** **47. ssm** A solid cylinder (radius = 0.150 m, height = 0.120 m) has a mass of 7.00 kg. This cylinder is floating in water. Then oil (ρ = 725 kg/m³) is poured on top of the water until the situation shown in the drawing results. How much of the height of the cylinder is in the oil?

** **48. Interactive LearningWare 11.1** at **www.wiley.com/college/cutnell** provides a review of the concepts that are important in this problem. A spring is attached to the bottom of an empty swimming pool, with the axis of the spring oriented vertically. An 8.00-kg block of wood (ρ = 840 kg/m³) is fixed to the top of the spring and compresses it. Then the pool is filled with water, completely covering the block. The spring is now observed to be stretched twice as much as it had been compressed. Determine the percentage of the block's total volume that is hollow. Ignore any air in the hollow space.

** **49.** One kilogram of glass (ρ = 2.60 × 10³ kg/m³) is shaped into a hollow spherical shell that just barely floats in water. What are the inner and outer radii of the shell? Do not assume the shell is thin.

Section 11.8 The Equation of Continuity

50. Oil is flowing with a speed of 1.22 m/s through a pipeline with a radius of 0.305 m. How many gallons of oil (1 gal = 3.79 × 10⁻³ m³) flow in one day?

51. ssm Water flows with a volume flow rate of 1.50 m³/s in a pipe. Find the water speed where the pipe radius is 0.500 m.

52. Concept Simulation 11.1 at **www.wiley.com/college/cutnell** reviews the concept that plays the central role in this problem. (a) The volume flow rate in an artery supplying the brain is 3.6 × 10⁻⁶ m³/s. If the radius of the artery is 5.2 mm, determine the average blood speed. (b) Find the average blood speed at a constriction in the artery if the constriction reduces the radius by a factor of 3. Assume that the volume flow rate is the same as that in part (a).

53. A room has a volume of 120 m³. An air-conditioning system is to replace the air in this room every twenty minutes, using ducts that have a square cross section. Assuming that air can be treated as an incompressible fluid, find the length of a side of the square if the air speed within the ducts is (a) 3.0 m/s and (b) 5.0 m/s.

* **54.** Three fire hoses are connected to a fire hydrant. Each hose has a radius of 0.020 m. Water enters the hydrant through an underground pipe of radius 0.080 m. In this pipe the water has a speed of 3.0 m/s. (a) How many kilograms of water are poured onto a fire in one hour? (b) Find the water speed in each hose.

* **55. ssm** A water line with an internal radius of 6.5 × 10⁻³ m is connected to a shower head that has 12 holes. The speed of the water in the line is 1.2 m/s. (a) What is the volume flow rate in the line? (b) At what speed does the water leave one of the holes (effective hole radius = 4.6 × 10⁻⁴ m) in the head?

Section 11.9 Bernoulli's Equation,
Section 11.10 Applications of Bernoulli's Equation

56. One way to administer an inoculation is with a "gun" that shoots the vaccine through a narrow opening. No needle is necessary, for the vaccine emerges with sufficient speed to pass directly into the tissue beneath the skin. The speed is high, because the vaccine (ρ = 1100 kg/m³) is held in a reservoir where a high pressure pushes it out. The pressure on the surface of the vaccine in one gun is 4.1 × 10⁶ Pa above the atmospheric pressure outside the narrow opening. The dosage is small enough that the vaccine's surface in the reservoir is nearly stationary during an inoculation. The vertical height between the vaccine's surface in the reservoir and the opening can be ignored. Find the speed at which the vaccine emerges.

57. ssm Review Conceptual Example 13 as an aid in understanding this problem. Suppose that a 15-m/s wind is blowing across the roof of your house. The density of air is 1.29 kg/m³. (a) Determine the reduction in pressure (below atmospheric pressure of stationary air) that accompanies this wind. (b) Explain why some roofs are "blown outward" during high winds.

58. The blood speed in a normal segment of a horizontal artery is 0.11 m/s. An abnormal segment of the artery is narrowed down by an arteriosclerotic plaque to one-fourth the normal cross-sectional area. What is the difference in blood pressures between the normal and constricted segments of the artery?

59. ssm An airplane wing is designed so that the speed of the air across the top of the wing is 251 m/s when the speed of the air below the wing is 225 m/s. The density of the air is 1.29 kg/m³. What is the lifting force on a wing of area 24.0 m²?

60. Interactive LearningWare 11.2 at **www.wiley.com/college/cutnell** reviews the approach taken in problems such as this one. A small crack occurs at the base of a 15.0-m-high dam. The effective crack area through which water leaves is 1.30 × 10⁻³ m². (a) Ignoring viscous losses, what is the speed of water flowing through the crack? (b) How many cubic meters of water per second leave the dam?

61. The water tower in the drawing is drained by a pipe that extends to the ground. The flow is nonviscous. (a) What is the absolute pressure at point 1 if the valve is *closed*, assuming that the top surface of the water at point 2 is at atmospheric pressure. (b) What is the absolute pressure at point 1 when the valve is opened and the water is flowing? Assume that the water speed at point 2 is negligible. (c) Assuming the effective cross-sectional area of the valve opening is 2.00 × 10⁻² m², find the volume flow rate at point 1.

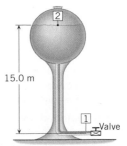

* **62.** In a very large closed tank, the absolute pressure of the air above the water is 6.01 × 10⁵ Pa. The water leaves the bottom of the tank through a nozzle that is directed straight upward. The opening of the nozzle is 4.00 m below the surface of the water. (a) Find the

speed at which the water leaves the nozzle. (b) Ignoring air resistance and viscous effects, determine the height to which the water rises.

* **63. Interactive Solution 11.63** at **www.wiley.com/college/cutnell** presents one method for modeling this problem. The construction of a flat rectangular roof (5.0 m × 6.3 m) allows it to withstand a maximum net outward force of 22 000 N. The density of the air is 1.29 kg/m^3. At what wind speed will this roof blow outward?

* **64.** A pump and its horizontal intake pipe are located 12 m beneath the surface of a reservoir. The speed of the water in the intake pipe causes the pressure there to decrease, in accord with Bernoulli's principle. Assuming nonviscous flow, what is the maximum speed with which water can flow through the intake pipe?

* **65. ssm** A Venturi meter is a device for measuring the speed of a fluid within a pipe. The drawing shows a gas flowing at speed v_2 through a horizontal section of pipe whose cross-sectional area is $A_2 = 0.0700$ m^2. The gas has a density of $\rho = 1.30$ kg/m^3. The Venturi meter has a cross-sectional area of $A_1 = 0.0500$ m^2 and has been substituted for a section of the larger pipe. The pressure difference between the two sections is $P_2 - P_1 = 120$ Pa. Find (a) the speed v_2 of the gas in the larger original pipe and (b) the volume flow rate Q of the gas.

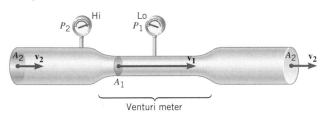

Venturi meter

* **66.** A liquid is flowing through a horizontal pipe whose radius is 0.0200 m. The pipe bends straight upward through a height of 10.0 m and joins another horizontal pipe whose radius is 0.0400 m. What volume flow rate will keep the pressures in the two horizontal pipes the same?

* **67.** An airplane has an effective wing surface area of 16 m^2 that is generating the lift force. In level flight the air speed over the top of the wings is 62.0 m/s, while the air speed beneath the wings is 54.0 m/s. What is the weight of the plane?

** **68.** A siphon tube is useful for removing liquid from a tank. The siphon tube is first filled with liquid, and then one end is inserted into the tank. Liquid then drains out the other end, as the drawing illustrates. (a) Using reasoning similar to that employed in obtaining Torricelli's theorem, derive an expression for the speed v of the fluid emerging from the tube. This expression should give v in terms of the vertical height y and the acceleration due to gravity g. (Note that this speed does

not depend on the depth d of the tube below the surface of the liquid.) (b) At what value of the vertical distance y will the siphon stop working? (c) Derive an expression for the absolute pressure at the highest point in the siphon (point A) in terms of the atmospheric pressure P_0, the fluid density ρ, g, and the heights h and y. (Note that the fluid speed at point A is the same as the speed of the fluid emerging from the tube, because the cross-sectional area of the tube is the same everywhere.)

** **69. ssm** A uniform rectangular plate is hanging vertically downward from a hinge that passes along its left edge. By blowing air at 11.0 m/s *over the top of the plate only*, it is possible to keep the plate in a horizontal position, as illustrated in part *a* of the drawing. To what value should the air speed be reduced so that the plate is kept at a 30.0° angle with respect to the vertical, as in part *b* of the drawing? (*Hint: Apply Bernoulli's equation in the form of Equation 11.12.*)

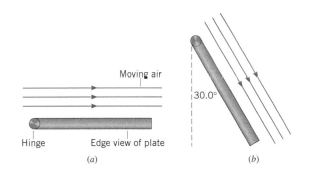

Moving air

Hinge   Edge view of plate

30.0°

(a)   (b)

Section 11.11 Viscous Flow

70. A blood vessel is 0.10 m in length and has a radius of 1.5 × 10^{-3} m. Blood ($\eta = 4 \times 10^{-3}$ Pa·s) flows at a rate of 1.0 × 10^{-7} m^3/s. Determine the difference in pressure that must be maintained between the two ends of the vessel.

71. ssm www A 1.3-m length of horizontal pipe has a radius of 6.4 × 10^{-3} m. Water flows with a volume flow rate of 9.0 × 10^{-3} m^3/s out of the right end of the pipe and into the air. What is the pressure in the flowing water at the left end of the pipe if the water behaves as (a) an ideal fluid and (b) a viscous fluid ($\eta = 1.00 \times 10^{-3}$ Pa·s)?

72. A cylindrical air duct in an air conditioning system has a length of 5.5 m and a radius of 7.2 × 10^{-2} m. A fan forces air ($\eta = 1.8 \times 10^{-5}$ Pa·s) through the duct, such that the air in a room (volume = 280 m^3) is replenished every ten minutes. Determine the difference in pressure between the ends of the air duct.

73. Interactive Solution 11.73 at **www.wiley.com/college/cutnell** illustrates a model for solving this problem. A pressure difference of 1.8 × 10^3 Pa is needed to drive water ($\eta = 1.0 \times 10^{-3}$ Pa·s) through a pipe whose radius is 5.1 × 10^{-3} m. The volume flow rate of the water is 2.8 × 10^{-4} m^3/s. What is the length of the pipe?

74. Poiseuille's law remains valid as long as the fluid flow is laminar. For sufficiently high speed, however, the flow becomes turbulent, even if the fluid is moving through a smooth pipe with no restrictions. It is found experimentally that the flow is laminar as long as the *Reynolds number* Re is less than about 2000: Re = $2\bar{v}\rho R/\eta$. Here $\bar{v}$, ρ, and η are, respectively, the average speed, density, and viscosity of the fluid, and R is the radius of the pipe. Calculate the highest average speed that blood ($\rho = 1060$ kg/m^3, $\eta = 4.0 \times 10^{-3}$ Pa·s) could have and still remain in laminar flow when it flows through the aorta ($R = 8.0 \times 10^{-3}$ m).

* **75. ssm** When an object moves through a fluid, the fluid exerts a viscous force **F** on the object that tends to slow it down. For a small sphere of radius R, moving slowly with a speed v, the magnitude of the viscous force is given by Stokes' law, $F = 6\pi\eta Rv$, where η is the viscosity of the fluid. (a) What is the viscous force on a sphere of radius $R = 5.0 \times 10^{-4}$ m falling through water ($\eta = 1.00 \times 10^{-3}$ Pa·s) when the sphere has a speed of 3.0 m/s? (b) The speed of the falling sphere increases until the viscous force

balances the weight of the sphere. Thereafter, no net force acts on the sphere, and it falls with a constant speed called the "terminal speed." If the sphere has a mass of 1.0×10^{-5} kg, what is its terminal speed?

* **76.** Water ($\eta = 1.00 \times 10^{-3}$ Pa·s) is flowing through a horizontal pipe with a volume flow rate of 0.014 m³/s. As the drawing shows, there are two vertical tubes that project from the pipe. From the data in the drawing, find the radius of the horizontal pipe.

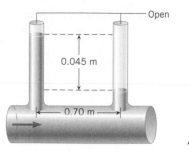

Problem 76

Additional Problems

77. Review Conceptual Example 6 as an aid in understanding this problem. Consider the pump on the right side of Figure 11.11, which acts to reduce the air pressure in the pipe. The air pressure outside the pipe is one atmosphere. Find the maximum depth from which this pump can extract water from the well.

78. Prairie dogs are burrowing rodents. They do not suffocate in their burrows, because the effect of air speed on pressure creates sufficient air circulation. The animals maintain a difference in the shapes of two entrances to the burrow, and because of this difference, the air ($\rho = 1.29$ kg/m³) blows past the openings at different speeds, as the drawing indicates. Assuming that the openings are at the same vertical level, find the difference in air pressure between the openings and indicate which way the air circulates.

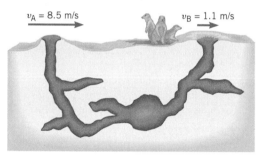

79. ssm A 0.10-m × 0.20-m × 0.30-m block is suspended from a wire and is completely under water. What buoyant force acts on the block?

80. The drawing shows an intravenous feeding. With the distance shown, nutrient solution ($\rho = 1030$ kg/m³) can just barely enter the blood in the vein. What is the gauge pressure of the venous blood? Express your answer in millimeters of mercury.

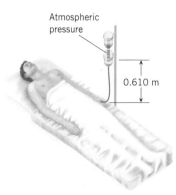

81. One end of a wire is attached to a ceiling, and a solid brass ball is tied to the lower end. The tension in the wire is 120 N. What is the radius of the brass ball?

82. Neutron stars consist only of neutrons and have unbelievably high densities. A typical mass and radius for a neutron star might be 2.7×10^{28} kg and 1.2×10^{3} m. (a) Find the density of such a star. (b) If a dime ($V = 2.0 \times 10^{-7}$ m³) were made from this material, how much would it weigh (in pounds)?

83. ssm A patient recovering from surgery is being given fluid intravenously. The fluid has a density of 1030 kg/m³, and 9.5×10^{-4} m³ of it flows into the patient every six hours. Find the mass flow rate in kg/s.

84. A blood transfusion is being set up in an emergency room for an accident victim. Blood has a density of 1060 kg/m³ and a viscosity of 4.0×10^{-3} Pa·s. The needle being used has a length of 3.0 cm and an inner radius of 0.25 mm. The doctor wishes to use a volume flow rate through the needle of 4.5×10^{-8} m³/s. What is the distance h above the victim's arm where the level of the blood in the transfusion bottle should be located? As an approximation, assume that the level of the blood in the transfusion bottle and the point where the needle enters the vein in the arm have the same pressure of one atmosphere. (In reality, the pressure in the vein is slightly above atmospheric pressure.)

85. ssm The main water line enters a house on the first floor. The line has a gauge pressure of 1.90×10^{5} Pa. (a) A faucet on the second floor, 6.50 m above the first floor, is turned off. What is the gauge pressure at this faucet? (b) How high could a faucet be before no water would flow from it, even if the faucet were open?

86. (a) The mass and radius of the sun are 1.99×10^{30} kg and 6.96×10^{8} m. What is its density? (b) If a solid object is made from a material that has the same density as the sun, would it sink or float in water? Why? (c) Would a solid object sink or float in water if it were made from a material whose density was the same as that of the planet Saturn (mass = 5.7×10^{26} kg, radius = 6.0×10^{7} m)? Provide a reason for your answer.

* **87. Concept Simulation 11.1** at **www.wiley.com/college/cutnell** reviews the central idea in this problem. In an adjustable nozzle for a garden hose, a cylindrical plug is aligned along the axis of the hose and can be inserted into the hose opening. The purpose of the plug is to change the speed of the water leaving the hose. The speed of the water passing around the plug is to be three times greater than the speed of the water before it encounters the plug. Find the ratio of the plug radius to the inside hose radius.

* **88.** A suitcase (mass $m = 16$ kg) is resting on the floor of an elevator. The part of the suitcase in contact with the floor measures 0.50 m by 0.15 m. The elevator is moving upward, the magnitude of its acceleration being 1.5 m/s². What pressure (in excess of atmospheric pressure) is applied to the floor beneath the suitcase?

* **89. ssm** A 1967 Kennedy half-dollar has a mass of 1.150×10^{-2} kg. The coin is a mixture of silver and copper, and in water weighs 0.1011 N. Determine the mass of silver in the coin.

*90. Two identical containers are open at the top and are connected at the bottom via a tube of negligible volume and a valve that is closed. Both containers are filled initially to the same height of 1.00 m, one with water, the other with mercury, as the drawing indicates. The valve is then opened. Water and mercury are immiscible. Determine the fluid level in the left container when equilibrium is reestablished.

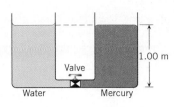

*91. ssm A tube is sealed at both ends and contains a 0.0100-m-long portion of liquid. The length of the tube is large compared to 0.0100 m. There is no air in the tube, and the vapor in the space above the liquid may be ignored. The tube is whirled around in a horizontal circle at a constant angular speed. The axis of the rotation passes through one end of the tube, and during the motion, the liquid collects at the other end. The pressure experienced by the liquid is the same as it would experience at the bottom of the tube, if the tube were completely filled with liquid and allowed to hang vertically. Find the angular speed (in rad/s) of the tube.

*92. A full can of soda has a mass of 0.416 kg. It contains 3.54×10^{-4} m³ of liquid. Assuming that the soda has the same density as water, find the volume of aluminum used to make the can.

*93. Water is running out of a faucet, falling straight down, with an initial speed of 0.50 m/s. At what distance below the faucet is the radius of the stream reduced to one-half its value at the faucet?

**94. A lighter-than-air balloon and its load of passengers and ballast are floating stationary above the earth. Ballast is weight (of negligi-ble volume) that can be dropped overboard to make the balloon rise. The radius of this balloon is 6.25 m. Assuming a constant value of 1.29 kg/m³ for the density of air, determine how much weight must be dropped overboard to make the balloon rise 105 m in 15.0 s.

**95. The drawing shows a cylinder fitted with a piston that has a mass m_1 of 0.500 kg and a radius of 2.50×10^{-2} m. The top of the piston is open to the atmosphere. The pressure beneath the piston is maintained at a reduced (but constant) value by means of the pump. As shown, a rope of negligible mass is at-

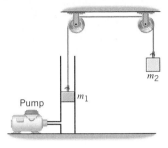

tached to the piston and passes over two massless pulleys. The other end of the rope is attached to a block that has a mass of $m_2 = 9.50$ kg. The block falls from rest down through a distance of 1.25 m in 3.30 s. Ignoring friction, find the absolute pressure beneath the piston.

**96. Two circular holes, one larger than the other, are cut in the side of a large water tank whose top is open to the atmosphere. The center of one of these holes is located twice as far beneath the surface of the water as the other. The volume flow rate of the water coming out of the holes is the same. (a) Decide which hole is located nearest the surface of the water. (b) Calculate the ratio of the radius of the larger hole to the radius of the smaller hole.

Concepts & Calculations   Group Learning Problems

Note: Each of these problems consists of Concept Questions followed by a related quantitative Problem. They are designed for use by students working alone or in small learning groups. The Concept Questions involve little or no mathematics and are intended to stimulate group discussions. They focus on the concepts with which the problems deal. Recognizing the concepts is the essential initial step in any problem-solving technique.

97. Concept Questions A gold prospector finds a solid rock that is composed solely of quartz and gold. (a) How is the total mass m_T of the rock related to the mass m_G of the gold and the mass m_Q of the quartz? (b) What is the relationship between the total volume V_T of the rock, the volume V_G of the gold, and the volume V_Q of the quartz? (c) How is the volume of a substance (gold or quartz) related to the mass of the substance and its density?

Problem The mass and volume of the rock are 12.0 kg and 4.00×10^{-3} m³. Find the mass of the gold in the rock.

98. Concept Questions A meat baster consists of a squeeze bulb attached to a plastic tube. When the bulb is squeezed and released, with the open end of the tube under the surface of the basting sauce, the sauce rises in the tube to a distance h, as the drawing shows. It can then be squirted over the meat. (a) Is the absolute pressure in the bulb in the drawing greater than or less than atmospheric pressure? (b) In a second trial, the distance h is somewhat less than it is in the drawing. Is the absolute pressure in the bulb in the second trial

greater or smaller than in the case shown in the drawing? Explain your answers.

Problem Using 1.013×10^5 Pa for the atmospheric pressure and 1200 kg/m³ for the density of the sauce, find the absolute pressure in the bulb when the distance h is (a) 0.15 m and (b) 0.10 m. Verify that your answers are consistent with your answers to the Concept Questions.

99. Concept Questions A hydrometer is a device used to measure the density of a liquid. It is a cylindrical tube weighted at one end, so that it floats with the heavier end downward. It is contained inside a large "medicine dropper," into which the liquid is drawn using a squeeze bulb (see the drawing). For use with your car, marks are put on the tube so that the level at which it floats indicates whether the liquid is battery acid (more dense) or antifreeze (less dense). (a) Compared to the weight W of the tube, how

much buoyant force is needed to make the tube float in either battery acid or antifreeze? (b) Is a greater volume of battery acid or a greater volume of antifreeze displaced by the hydrometer to provide the necessary buoyant force? (c) Which mark is farther up from the bottom of the tube? Justify your answers.

Problem The hydrometer has a weight of $W = 5.88 \times 10^{-2}$ N and a cross-sectional area of $A = 7.85 \times 10^{-5}$ m². How far from the bottom of the tube should the mark be put that denotes (a) battery acid ($\rho = 1280$ kg/m³) and (b) antifreeze ($\rho = 1073$ kg/m³)? Check to see that your answers are consistent with your answers to the Concept Questions.

100. Concept Questions Water flows straight down from an open faucet. The effects of air resistance and viscosity can be ignored. (a) After the water has fallen a bit below the faucet, is its speed less than, greater than, or the same as it was on leaving the faucet? (b) Is the volume flow rate in cubic meters per second less than, greater than, or the same as it was when the water left the faucet? (c) Is the cross-sectional area of the water stream less than, greater than, or the same as it was when the water left the faucet? Give your reasoning.

Problem The cross-sectional area of the faucet is $1.8 \times 10^{-4} \text{ m}^2$, and the speed of the water is 0.85 m/s as it leaves the faucet. Ignoring air resistance, find the cross-sectional area of the water stream at a point 0.10 m below the faucet. Make sure that your answer is consistent with your answers to the Concept Questions.

101. Concept Questions Water flowing out of a horizontal pipe emerges through a nozzle. The nozzle has a smaller radius than does the pipe. Treat the water as an ideal fluid. (a) What does the fact that the pipe is horizontal imply about the pipe's elevation y above the ground? (b) Is the speed at which the water flows in the pipe greater than, less than, or the same as the speed of the water emerging from the nozzle? (c) The emerging water is at atmospheric pressure. Is the absolute pressure of the water in the pipe greater than, less than, or the same as the atmospheric pressure? Explain your answers.

Problem The radius of the pipe is 1.9 cm, and the radius of the nozzle is 0.48 cm. The speed of the water in the pipe is 0.62 m/s. Determine the absolute pressure of the water in the pipe. Verify that your answer is consistent with your answers to the Concept Questions.

102. Concept Questions A ship is floating on a lake. Its hold is the interior space beneath its deck and is open to the atmosphere. The hull has a hole in it, which is below the water line, so water leaks into the hold. (a) How is the amount of water per second (in m³/s) entering the hold related to the speed of the entering water and the area of the hole? (b) Approximately how fast is the water at the *surface* of the lake moving? Justify your answer. (c) What causes the water to accelerate as it moves from the surface of the lake into the hole that is beneath the water line? Explain.

Problem The effective area of the hole is $8.0 \times 10^{-3} \text{ m}^2$ and is located 2.0 m beneath the surface of the lake. What volume of water per second leaks into the ship?

* **103. Concept Questions** A hot-air balloon is accelerating upward under the influence of two forces, its weight (including that of the hot air within the balloon) and the buoyant force. (a) How is the weight of the hot air determined from a knowledge of its density $\rho_{\text{hot air}}$ and the volume V of the balloon? (b) Does the buoyant force depend on the density of the hot air inside the balloon, the density of the cool air outside the balloon, or both? Provide a reason for your answer. (c) Draw a free-body diagram for the balloon, showing the forces that act on it. How is the upward acceleration of the balloon related to these forces and to its mass?

Problem The hot air inside the balloon has a density of $\rho_{\text{hot air}} = 0.93 \text{ kg/m}^3$, and that of the cool air outside is $\rho_{\text{cool air}} = 1.29 \text{ kg/m}^3$. What is the acceleration of the rising balloon? For simplicity, neglect the mass of the balloon fabric and the basket; consider only the mass of the hot air inside the balloon.

* **104. Concept Questions** Two hoses are connected to the same outlet using a Y-connector, as the drawing shows. The hoses A and B have the same length, but hose B has the larger radius. Each is open to the atmosphere at the end where the water exits. Water flows through both hoses as a viscous fluid, and Poiseuille's law $[Q = \pi R^4(P_2 - P_1)/(8\eta L)]$ applies to each. In this law, P_2 is the pressure upstream, P_1 is the pressure downstream, and Q is the volume flow rate. (a) For hoses A and B, is the value for the term $P_2 - P_1$ the same or different? (b) How is Q related to the radius of a hose and the speed of the water in the hose? Account for your answers.

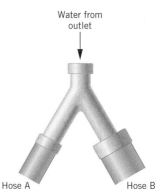

Water from outlet

Hose A   Hose B

Problem The ratio of the radius of hose B to the radius of hose A is $R_B/R_A = 1.50$. Find the ratio of the speed of the water in hose B to that in hose A.

Chapter 12

Temperature and Heat

After a long day in the hot sun, this road is hot enough to fry an egg. In other words, the road's temperature is high enough, and the road can supply the heat needed to cook the egg. Temperature and heat are both discussed in this chapter. (© Chip Simons/Taxi/Getty Images)

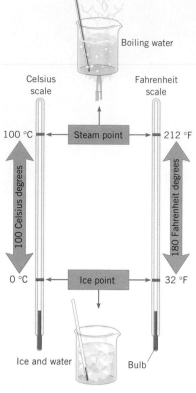

Figure 12.1 The Celsius and Fahrenheit temperature scales.

12.1 Common Temperature Scales

To measure temperature we use a thermometer. Many thermometers make use of the fact that materials usually expand with increasing temperature. For example, Figure 12.1 shows the common mercury-in-glass thermometer, which consists of a mercury-filled glass bulb connected to a capillary tube. When the mercury is heated, it expands into the capillary tube, the amount of expansion being proportional to the change in temperature. The outside of the glass is marked with an appropriate scale for reading the temperature.

A number of different temperature scales have been devised, two popular choices being the **Celsius** (formerly, centigrade) and **Fahrenheit scales.** Figure 12.1 illustrates these scales. Historically,* both scales were defined by assigning two temperature points on the scale and then dividing the distance between them into a number of equally spaced intervals. One point was chosen to be the temperature at which ice melts under one atmosphere of pressure (the "ice point"), and the other was the temperature at which water boils under one atmosphere of pressure (the "steam point"). On the Celsius scale, an ice point of 0 °C (0 degrees Celsius) and a steam point of 100 °C were selected. On the Fahrenheit scale, an ice point of 32 °F (32 degrees Fahrenheit) and a steam point of 212 °F were chosen. The Celsius scale is used worldwide, while the Fahrenheit scale is used mostly in the United States, often in home medical thermometers.

There is a subtle difference in the way the temperature of an object is reported, as compared to a *change* in its temperature. For example, the temperature of the human body is about 37 °C, where the symbol °C stands for "degrees Celsius." However, the *change* between two temperatures is specified in "Celsius degrees" (C°)—not in "degrees Celsius." Thus, if the body temperature rises to 39 °C, the change in temperature is 2 Celsius degrees or 2 C°, not 2 °C.

As Figure 12.1 indicates, the separation between the ice and steam points on the Celsius scale is divided into 100 Celsius degrees, while on the Fahrenheit scale the separation is divided into 180 Fahrenheit degrees. Therefore, the size of the Celsius degree is larger than that of the Fahrenheit degree by a factor of $\frac{180}{100}$, or $\frac{9}{5}$. Examples 1 and 2 illustrate how to convert between the Celsius and Fahrenheit scales using this factor.

Example 1 Converting from a Fahrenheit to a Celsius Temperature

A healthy person has an oral temperature of 98.6 °F. What would this reading be on the Celsius scale?

Reasoning and Solution A temperature of 98.6 °F is 66.6 Fahrenheit degrees above the ice point of 32.0 °F. Since 1 C° = $\frac{9}{5}$ F°, the difference of 66.6 F° is equivalent to

$$(66.6 \text{ F}°)\left(\frac{1 \text{ C}°}{\frac{9}{5} \text{ F}°}\right) = 37.0 \text{ C}°$$

Thus, the person's temperature is 37.0 Celsius degrees above the ice point. Adding 37.0 Celsius degrees to the ice point of 0 °C on the Celsius scale gives a Celsius temperature of $\boxed{37.0 \text{ °C}}$.

Example 2 Converting from a Celsius to a Fahrenheit Temperature

A time and temperature sign on a bank indicates that the outdoor temperature is −20.0 °C. Find the corresponding temperature on the Fahrenheit scale.

Reasoning and Solution The temperature of −20.0 °C is 20.0 Celsius degrees *below* the ice point of 0 °C. This number of Celsius degrees corresponds to

$$(20.0 \text{ C}°)\left(\frac{\frac{9}{5} \text{ F}°}{1 \text{ C}°}\right) = 36.0 \text{ F}°$$

* Today, the Celsius and Fahrenheit scales are defined in terms of the Kelvin temperature scale; Section 12.2 discusses the Kelvin scale.

The temperature, then, is 36.0 Fahrenheit degrees below the ice point. Subtracting 36.0 Fahrenheit degrees from the ice point of 32.0 °F on the Fahrenheit scale gives a Fahrenheit temperature of $\boxed{-4.0\ °F}$.

The reasoning strategy used in Examples 1 and 2 for converting between the Celsius and Fahrenheit scales is summarized below.

Reasoning Strategy

Converting Between Different Temperature Scales

1. Determine the magnitude of the difference between the stated temperature and the ice point on the initial scale.

2. Convert this number of degrees from one scale to the other scale by using the fact that $1\ C° = \frac{9}{5}F°$.

3. Add or subtract the number of degrees on the new scale to or from the ice point on the new scale.

✔ **Check Your Understanding 1**

On a new temperature scale the steam point is 348 °X, and the ice point is 112 °X. What is the temperature on this scale that corresponds to 28.0 °C? *(The answer is given at the end of the book.)*

Background: The reasoning strategy for converting between different temperature scales applies here.

For similar questions (including conceptual counterparts), consult Self-Assessment Test 12.1. This test is described at the end of Section 12.5.

12.2 The Kelvin Temperature Scale

Although the Celsius and Fahrenheit scales are widely used, the **Kelvin temperature scale** has greater scientific significance. It was introduced by the Scottish physicist William Thompson (Lord Kelvin, 1824–1907), and in his honor each degree on the scale is called a kelvin (K). By international agreement, the symbol K is not written with a degree sign (°), nor is the word "degrees" used when quoting temperatures. For example, a temperature of 300 K (not 300 °K) is read as "three hundred kelvins," not "three hundred degrees kelvin." The kelvin is the SI base unit for temperature.

Figure 12.2 compares the Kelvin and Celsius scales. The size of one kelvin is identical to that of one Celsius degree because there are one hundred divisions between the ice and steam points on both scales. As we will discuss shortly, experiments have shown that there exists a lowest possible temperature, below which no substance can be cooled. This lowest temperature is defined to be the zero point on the Kelvin scale and is referred to as absolute zero.

The ice point (0 °C) occurs at 273.15 K on the Kelvin scale. Thus, the Kelvin temperature T and the Celsius temperature T_c are related by

$$T = T_c + 273.15 \tag{12.1}$$

The number 273.15 in Equation 12.1 is an experimental result, obtained in studies that utilize a gas-based thermometer. When a gas confined to a fixed volume is heated, its pressure increases. Conversely, when the gas is cooled, its pressure decreases. For example, the air pressure in automobile tires can rise by as much as 20% after the car has been driven and the tires have become warm. The change in gas pressure with temperature is the basis for the **constant-volume gas thermometer.**

A constant-volume gas thermometer consists of a gas-filled bulb to which a pressure gauge is attached, as in Figure 12.3. The gas is often hydrogen or helium at a low density, and the pressure gauge can be a U-tube manometer filled with mercury. The bulb is placed in thermal contact with the substance whose temperature is being measured. The volume of the gas is held constant by raising or lowering the *right* column of the U-tube

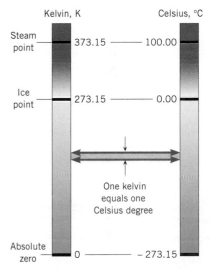

Figure 12.2 A comparison of the Kelvin and Celsius temperature scales.

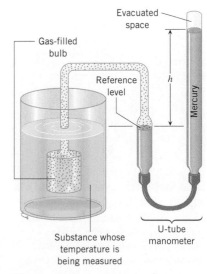

Figure 12.3 A constant-volume gas thermometer.

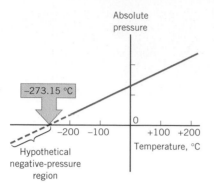

Figure 12.4 A plot of absolute pressure versus temperature for a low-density gas at constant volume. The graph is a straight line and, when extrapolated (dashed line), crosses the temperature axis at −273.15 °C.

Figure 12.5 (*a*) A thermocouple is made from two different types of wires, copper and constantan in this case. (*b*) A thermocouple junction between two different wires. (Courtesy Omega Engineering, Inc.)

manometer in order to keep the mercury level in the *left* column at the same reference level. The absolute pressure of the gas is proportional to the height *h* of the mercury on the right. As the temperature changes, the pressure changes and can be used to indicate the temperature, once the constant-volume gas thermometer has been calibrated.

Suppose the absolute pressure of the gas in Figure 12.3 is measured at different temperatures. If the results are plotted on a pressure-versus-temperature graph, a straight line is obtained, as in Figure 12.4. If the straight line is extended or extrapolated to lower and lower temperatures, the line crosses the temperature axis at −273.15 °C. In reality, no gas can be cooled to this temperature, because all gases liquify before reaching it. However, helium and hydrogen liquify at such low temperatures that they are often used in the thermometer. This kind of graph can be obtained for different amounts and types of low-density gases. In all cases, it is found that the straight line extrapolates back to −273.15° C on the temperature axis, which suggests that the value of −273.15° C has fundamental significance. The significance of this number is that it is the ***absolute zero point*** for temperature measurement. The phrase "absolute zero" means that temperatures lower than −273.15 °C cannot be reached by continually cooling a gas or any other substance. If lower temperatures could be reached, then further extrapolation of the straight line in Figure 12.4 would suggest that negative absolute gas pressures could exist. Such a situation would be impossible, because a negative absolute gas pressure has no meaning. Thus, the Kelvin scale is chosen so that its zero temperature point is the lowest temperature attainable.

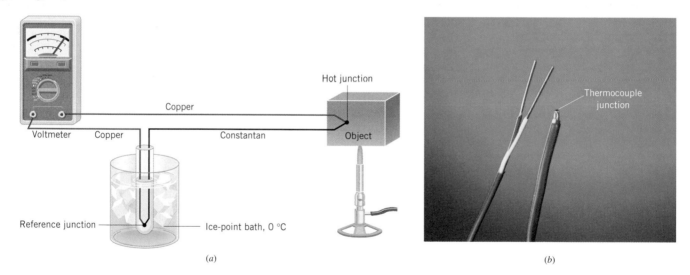

(*a*)   (*b*)

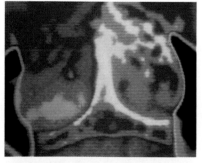

Figure 12.6 Invasive carcinoma (cancer) of the breast registers colors from red to yellow/white in this thermograph, indicating markedly elevated temperatures. (© Science Photo Library/Photo Researchers)

12.3 *Thermometers*

All thermometers make use of the change in some physical property with temperature. A property that changes with temperature is called a ***thermometric property.*** For example, the thermometric property of the mercury thermometer is the length of the mercury column, while in the constant-volume gas thermometer it is the pressure of the gas. Several other thermometers and their thermometric properties will now be discussed.

The *thermocouple* is a thermometer used extensively in scientific laboratories. It consists of thin wires of different metals, welded together at the ends to form two junctions, as Figure 12.5 illustrates. Often the metals are copper and constantan (a copper–nickel alloy). One of the junctions, called the "hot" junction, is placed in thermal contact with the object whose temperature is being measured. The other junction, termed the "reference" junction, is kept at a known constant temperature (usually an ice–water mixture at 0 °C). The thermocouple generates a voltage that depends on the *difference in tempera-*

ture between the two junctions. This voltage is the thermometric property and is measured by a voltmeter, as the drawing indicates. With the aid of calibration tables, the temperature of the hot junction can be obtained from the voltage. Thermocouples are used to measure temperatures as high as 2300 °C or as low as −270 °C.

Most substances offer resistance to the flow of electricity. Because this electrical resistance changes with temperature, electrical resistance is another thermometric property. *Electrical resistance thermometers* are often made from platinum wire, because platinum has excellent mechanical and electrical properties in the temperature range from −270 °C to +700 °C. The electrical resistance of platinum wire is known as a function of temperature. Thus, the temperature of a substance can be determined by placing the resistance thermometer in thermal contact with the substance and measuring the resistance of the platinum wire.

Radiation emitted by an object can also be used to indicate temperature. At low to moderate temperatures, the predominant radiation emitted is infrared. As the temperature is raised, the intensity of the radiation increases substantially. In one interesting application, an infrared camera registers the intensity of the infrared radiation produced at different locations on the human body. The camera is connected to a color monitor that displays the different infrared intensities as different colors. This "thermal painting" is called a *thermograph* or *thermogram.* Thermography is an important diagnostic tool in medicine. For example, breast cancer may be indicated in a thermograph by the elevated temperatures associated with malignant tissue. Figure 12.6 shows a thermograph used to diagnose breast cancer. Figure 12.7 shows thermographic images of a smoker's forearms before (left) and 5 minutes after (right) he has smoked a cigarette. After smoking, the forearms are cooler due to the effect of nicotine, which causes vasoconstriction (narrowing of the blood vessels) and reduces blood flow, a result that can lead to a higher risk from blood clotting. Temperatures in these images range from over 34 °C to about 28 °C and are indicated in decreasing order by the colors white, red, yellow, green, and blue.

Oceanographers and meteorologists use thermographs extensively to map the temperature distribution on the surface of the earth. For example, Figure 12.8 shows a satellite image of the sea-surface temperature of the Pacific Ocean. The region depicted in red is the 1997/98 El Niño, a large area of the ocean, approximately twice the width of the United States, where temperatures reached abnormally high values. This El Niño caused major weather changes in certain regions of the earth.

12.4 Linear Thermal Expansion

NORMAL SOLIDS

Have you ever found the metal lid on a glass jar too tight to open? One solution is to run hot water over the lid, which loosens because the metal expands more than the glass does. To varying extents, most materials expand when heated and contract when cooled. The increase in any one dimension of a solid is called **linear expansion,** linear in the sense that the expansion occurs along a line. Figure 12.9 illustrates the linear expansion of a rod whose length is L_0 when the temperature is T_0. When the temperature increases to $T_0 + \Delta T$, the length becomes $L_0 + \Delta L$, where ΔT and ΔL are the magnitudes of the changes in temperature and length, respectively. Conversely, when the temperature decreases to $T_0 - \Delta T$, the length decreases to $L_0 - \Delta L$.

For modest temperature changes, experiments show that the change in length is directly proportional to the change in temperature ($\Delta L \propto \Delta T$). In addition, the change in length is proportional to the initial length of the rod, a fact that can be understood with the aid of Figure 12.10. Part *a* of the drawing shows two identical rods. Each rod has a length L_0 and expands by ΔL when the temperature increases by ΔT. Part *b* shows the two heated rods combined into a single rod, for which the total expansion is the sum of the expansions of each part—namely, $\Delta L + \Delta L = 2\Delta L$. Clearly, the amount of expansion doubles if the rod is twice as long to begin with. In other words, the change in length is directly proportional to the original length ($\Delta L \propto L_0$). Equation 12.2 expresses the fact

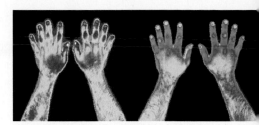

Figure 12.7 Thermogram showing a smoker's forearms before *(left)* and 5 minutes after *(right)* he has smoked a cigarette. Temperatures range from over 34 °C (white) to about 28 °C (blue). (© Dr. Arthur Tucker/Science Photo Library/Photo Researchers)

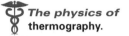

The physics of thermography.

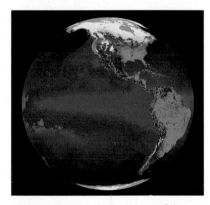

Figure 12.8 A thermogram of the 1997/98 El Niño (red), a large region of abnormally high temperatures in the Pacific Ocean. (Courtesy NOAA)

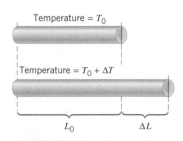

Figure 12.9 When the temperature of a rod is raised by ΔT, the length of the rod increases by ΔL.

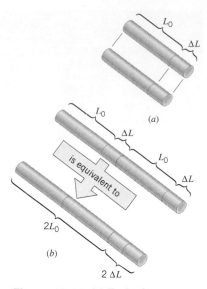

Figure 12.10 (*a*) Each of two identical rods expands by ΔL when heated. (*b*) When the rods are combined into a single rod of length $2L_0$, the "combined" rod expands by $2 \Delta L$.

that ΔL is proportional to both L_0 and ΔT ($\Delta L \propto L_0 \Delta T$) by using a proportionality constant α, which is called the ***coefficient of linear expansion.***

■ **LINEAR THERMAL EXPANSION OF A SOLID**

The length L_0 of an object changes by an amount ΔL when its temperature changes by an amount ΔT:

$$\Delta L = \alpha L_0 \Delta T \qquad (12.2)$$

where α is the coefficient of linear expansion.

Common Unit for the Coefficient of Linear Expansion: $\dfrac{1}{\text{C}°} = (\text{C}°)^{-1}$

Solving Equation 12.2 for α shows that $\alpha = \Delta L/(L_0 \Delta T)$. Since the length units of ΔL and L_0 algebraically cancel, the coefficient of linear expansion α has the unit of $(\text{C}°)^{-1}$ when the temperature difference ΔT is expressed in Celsius degrees (C°). Different materials with the same initial length expand and contract by different amounts as the temperature changes, so the value of α depends on the nature of the material. Table 12.1 shows some typical values. Coefficients of linear expansion also vary somewhat depending on the range of temperatures involved, but the values in Table 12.1 are adequate approximations. Example 3 deals with a situation where a dramatic effect due to thermal expansion can be observed, even though the change in temperature is small.

Table 12.1 *Coefficients of Thermal Expansion for Solids and Liquids[a]*

Substance	Coefficient of Thermal Expansion $(\text{C}°)^{-1}$	
	Linear (α)	Volumetric (β)
Solids		
Aluminum	23×10^{-6}	69×10^{-6}
Brass	19×10^{-6}	57×10^{-6}
Concrete	12×10^{-6}	36×10^{-6}
Copper	17×10^{-6}	51×10^{-6}
Glass (common)	8.5×10^{-6}	26×10^{-6}
Glass (Pyrex)	3.3×10^{-6}	9.9×10^{-6}
Gold	14×10^{-6}	42×10^{-6}
Iron or steel	12×10^{-6}	36×10^{-6}
Lead	29×10^{-6}	87×10^{-6}
Nickel	13×10^{-6}	39×10^{-6}
Quartz (fused)	0.50×10^{-6}	1.5×10^{-6}
Silver	19×10^{-6}	57×10^{-6}
Liquids[b]		
Benzene	—	1240×10^{-6}
Carbon tetrachloride	—	1240×10^{-6}
Ethyl alcohol	—	1120×10^{-6}
Gasoline	—	950×10^{-6}
Mercury	—	182×10^{-6}
Methyl alcohol	—	1200×10^{-6}
Water	—	207×10^{-6}

[a] The values for α and β pertain to a temperature near 20 °C.
[b] Since liquids do not have fixed shapes, the coefficient of linear expansion is not defined for them.

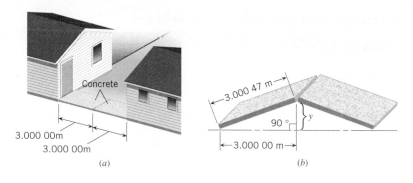

Figure 12.11 (*a*) Two concrete slabs completely fill the space between the buildings. (*b*) When the temperature increases, each slab expands, causing the sidewalk to buckle.

Example 3 Buckling of a Sidewalk

A concrete sidewalk is constructed between two buildings on a day when the temperature is 25 °C. The sidewalk consists of two slabs, each three meters in length and of negligible thickness (Figure 12.11*a*). As the temperature rises to 38 °C, the slabs expand, but no space is provided for thermal expansion. The buildings do not move, so the slabs buckle upward. Determine the vertical distance *y* in part *b* of the drawing.

Reasoning The expanded length of each slab is equal to its original length plus the change in length ΔL due to the rise in temperature. We know the original length, and Equation 12.2 can be used to find the change in length. Once the expanded length has been determined, the Pythagorean theorem can be employed to find the vertical distance *y* in Figure 12.11*b*.

Solution The change in temperature is $\Delta T = 38\ °C - 25\ °C = 13\ C°$, and the coefficient of linear expansion for concrete is given in Table 12.1. The change in length of each slab associated with this temperature change is

$$\Delta L = \alpha L_0 \Delta T = [12 \times 10^{-6}\ (C°)^{-1}](3.0\ m)(13\ C°) = 0.000\ 47\ m \quad (12.2)$$

The expanded length of each slab is, thus, 3.000 47 m. The vertical distance *y* can be obtained by applying the Pythagorean theorem to the right triangle in Figure 12.11*b*:

$$y = \sqrt{(3.000\ 47\ m)^2 - (3.000\ 00\ m)^2} = \boxed{0.053\ m}$$

Figure 12.12 An expansion joint in a bridge. (© Richard Choy/Peter Arnold, Inc.)

The buckling of a sidewalk is one consequence of not providing sufficient room for thermal expansion. To eliminate such problems, engineers incorporate expansion joints or spaces at intervals along bridge roadbeds, as Figure 12.12 shows.

Although Example 3 shows how thermal expansion can cause problems, there are also times when it can be useful. For instance, each year thousands of children are taken to emergency rooms suffering from burns caused by scalding tap water. Such accidents can be reduced with the aid of the antiscalding device shown in Figure 12.13. This device screws onto the end of a faucet and quickly shuts off the flow of water when it becomes too hot. As the water temperature rises, the actuator spring expands and pushes the plunger forward, shutting off the flow. When the water cools, the spring contracts and the water flow resumes.

The physics of an antiscalding device.

THERMAL STRESS

If the concrete slabs in Figure 12.11 had not buckled upward, they would have been subjected to immense forces from the buildings. The forces needed to prevent a solid object from expanding must be strong enough to counteract any change in length that would occur due to a change in temperature. Although the change in temperature may be small, the forces—and hence the stresses—can be enormous. They can, in fact, lead to serious structural damage. Example 4 illustrates just how large the stresses can be.

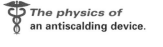

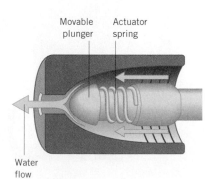

Figure 12.13 An antiscalding device.

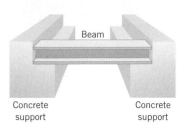

Figure 12.14 A steel beam is mounted between concrete supports with no room provided for thermal expansion of the beam.

Example 4   The Stress on a Steel Beam

A steel beam is used in the roadbed of a bridge. The beam is mounted between two concrete supports when the temperature is 23 °C, with no room provided for thermal expansion (Figure 12.14). What compressional stress must the concrete supports apply to each end of the beam, if they are to keep the beam from expanding when the temperature rises to 42 °C?

Reasoning Recall from Section 10.8 that the stress (force per unit cross-sectional area or F/A) required to change the length L_0 of an object by an amount ΔL is

$$\text{Stress} = \frac{F}{A} = Y \frac{\Delta L}{L_0} \qquad (10.17)$$

where Y is Young's modulus. If the steel beam were free to expand because of the change in temperature, the length would change by $\Delta L = \alpha L_0 \Delta T$. Because the concrete supports do not permit any expansion, they must supply a stress to compress the beam by an amount ΔL. Thus,

$$\text{Stress} = Y \frac{\Delta L}{L_0} = Y \frac{\alpha L_0 \Delta T}{L_0} = Y \alpha \Delta T$$

Solution For steel, the values of Young's modulus and the coefficient of linear expansion are $Y = 2.0 \times 10^{11} \text{ N/m}^2$ (Table 10.1) and $\alpha = 12 \times 10^{-6} \text{ (C°)}^{-1}$ (Table 12.1), respectively. The change in temperature from 23 to 42 °C is $\Delta T = 19 \text{ C°}$. The thermal stress is

$$\text{Stress} = Y\alpha\Delta T = (2.0 \times 10^{11} \text{ N/m}^2)[12 \times 10^{-6} \text{ (C°)}^{-1}](19 \text{ C°}) = \boxed{4.6 \times 10^7 \text{ N/m}^2}$$

This stress is huge. If the beam has a cross-sectional area of $A = 0.10 \text{ m}^2$, the force applied to each end by a concrete support is $F = (\text{Stress})A = 4.6 \times 10^6 \text{ N}$ (over 1 million lb).

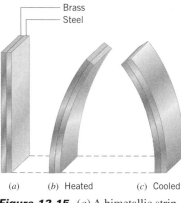

(a)   (b) Heated   (c) Cooled

Figure 12.15 (a) A bimetallic strip and how it behaves when (b) heated and (c) cooled.

THE BIMETALLIC STRIP

A **bimetallic strip** is made from two thin strips of metal that have *different* coefficients of linear expansion, as Figure 12.15a shows. Often brass [$\alpha = 19 \times 10^{-6} \text{ (C°)}^{-1}$] and steel [$\alpha = 12 \times 10^{-6} \text{ (C°)}^{-1}$] are selected. The two pieces are welded or riveted together. When the bimetallic strip is heated, the brass, having the larger value of α, expands more than the steel. Since the two metals are bonded together, the bimetallic strip bends into an arc as in part b, with the longer brass piece having a larger radius than the steel piece. When the strip is cooled, the bimetallic strip bends in the opposite direction, as in part c.

Bimetallic strips are frequently used as adjustable automatic switches in electrical appliances. Figure 12.16 shows an automatic coffee maker that turns off when the coffee is brewed to the selected strength. In part a, while the brewing cycle is on, electricity passes through the heating coil that heats the water. The electricity can flow because the contact mounted on the bimetallic strip touches the contact mounted on the "strength" adjustment

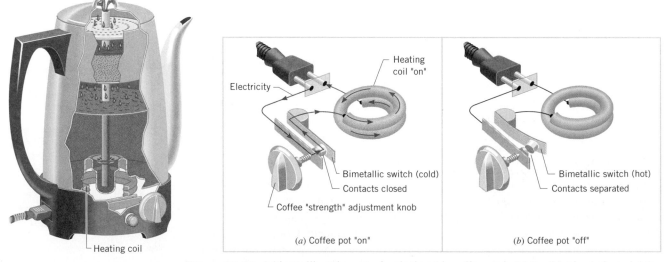

(a) Coffee pot "on"   (b) Coffee pot "off"

Figure 12.16 A bimetallic strip controls whether this coffee pot is (a) "on" (strip cool, straight) or (b) "off" (strip hot, bent).

knob, thus providing a continuous path for the electricity. When the bimetallic strip gets hot enough to bend away, as in part *b* of the drawing, the contacts separate. The electricity stops because it no longer has a continuous path along which to flow, and the brewing cycle is shut off. Turning the "strength" knob adjusts the brewing time by adjusting the distance through which the bimetallic strip must bend for the contact points to separate.

THE EXPANSION OF HOLES

An interesting example of linear expansion occurs when there is a hole in a piece of solid material. We know that the material itself expands when heated. But what about the hole? Does it expand, contract, or remain the same? Conceptual Example 5 provides some insight into the answer to this question.

Conceptual Example 5
Do Holes Expand or Contract When the Temperature Increases?

Figure 12.17*a* shows eight square tiles that are arranged to form a square pattern with a hole in the center. If the tiles are heated, what happens to the size of the hole?

Reasoning and Solution We can analyze this problem by disassembling the pattern into separate tiles, heating them, and then reassembling the pattern. Because each tile expands upon heating, it is evident from Figure 12.17*b* that the heated pattern expands and so does the hole in the center. In fact, if we had a ninth tile that was identical to and also heated like the others, it would fit exactly into the center hole, as Figure 12.17*c* indicates. Thus, not only does the hole in the pattern expand, but it also expands exactly as much as one of the tiles. Since the ninth tile is made of the same material as the others, we see that *the hole expands just as if it were made of the material of the surrounding tiles.* The thermal expansion of the hole and the surrounding material is analogous to a photographic enlargement; in both situations everything is enlarged, including holes.

Related Homework: *Problems 12, 18*

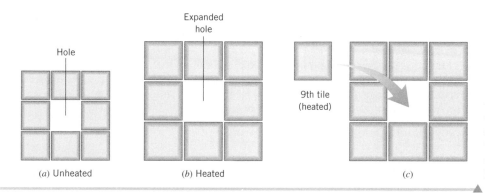

(*a*) Unheated   (*b*) Heated   (*c*)

Figure 12.17 (*a*) The tiles are arranged to form a square pattern with a hole in the center. (*b*) When the tiles are heated, the hole in the center, as well as the pattern, expands. (*c*) The expanded hole is the same size as a heated tile.

Instead of the separate tiles in Example 5, we could have used a square plate with a square hole in the center. The hole in the plate would have expanded just like the hole in the pattern of tiles. Furthermore, the same conclusion applies to a hole of any shape. Thus, it follows that *a hole in a piece of solid material expands when heated and contracts when cooled, just as if it were filled with the material that surrounds it.* If the hole is circular, the equation $\Delta L = \alpha L_0 \Delta T$ can be used to find the change in any linear dimension of the hole, such as its radius or diameter. Example 6 illustrates this type of linear expansion.

Problem solving insight

Example 6   A Heated Engagement Ring

A gold engagement ring has an inner diameter of 1.5×10^{-2} m and a temperature of 27 °C. The ring falls into a sink of hot water whose temperature is 49 °C. What is the change in the diameter of the hole in the ring?

Reasoning The hole expands as if it were filled with gold, so the change in the diameter is given by $\Delta L = \alpha L_0 \Delta T$, where $\alpha = 14 \times 10^{-6} (\text{C}°)^{-1}$ is the coefficient of linear expansion for gold (Table 12.1), L_0 is the original diameter, and ΔT is the change in temperature.

Solution The change in the ring's diameter is

$$\Delta L = \alpha L_0 \Delta T$$

$$= [14 \times 10^{-6} \ (\text{C}°)^{-1}](1.5 \times 10^{-2} \ \text{m})(49 \ °\text{C} - 27 \ °\text{C}) = \boxed{4.6 \times 10^{-6} \ \text{m}}$$

The previous two examples illustrate that holes expand like the surrounding material when heated. Therefore, holes in materials with larger coefficients of linear expansion expand more than those in materials with smaller coefficients of linear expansion. Conceptual Example 7 explores this aspect of thermal expansion.

Conceptual Example 7 **Expanding Cylinders**

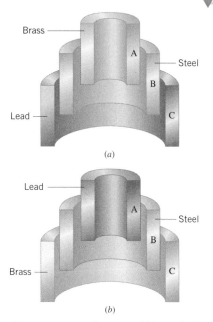

Brass — A — Steel

B

Lead — C

(a)

Lead — A — Steel

B

Brass — C

(b)

Figure 12.18 Conceptual Example 7 discusses the arrangements of the three cylinders shown in cutaway views in parts *a* and *b*.

Figure 12.18 shows a cross-sectional view of three cylinders, A, B, and C. Each is made from a different material: one is lead, one is brass, and one is steel. All three have the same temperature, and they barely fit inside each other. As the cylinders are heated to the same, but higher, temperature, cylinder C falls off, while cylinder A becomes tightly wedged to cylinder B. Which cylinder is made from which material?

Reasoning and Solution We need to consider how the outer and inner diameters of each cylinder change as the temperature is raised. With respect to the inner diameter, we will be guided by the fact that a hole expands as if it were filled with the surrounding material. According to Table 12.1, lead has the greatest coefficient of linear expansion, followed by brass, and then by steel. These data indicate that the outer and inner diameters of the lead cylinder change the most, while those of the steel cylinder change the least.

Since the steel cylinder expands the least, it cannot be the outer one, for if it were, the greater expansion of the middle cylinder would prevent the steel cylinder from falling off. The steel cylinder also cannot be the inner one, because then the greater expansion of the middle cylinder would allow the steel cylinder to fall out, contrary to what is observed. The only place left for the steel cylinder is in the middle, which leads to the two possibilities in Figure 12.18. In part *a*, lead is on the outside and will fall off as the temperature is raised, since lead expands more than steel. On the other hand, the inner brass cylinder expands more than the steel that surrounds it and becomes tightly wedged, as observed. Thus, one possibility is *A = brass, B = steel, and C = lead.*

In part *b* of the drawing, brass is on the outside. As the temperature is raised, brass expands more than steel, so the outer cylinder will again fall off. The inner lead cylinder has the greatest expansion and will be wedged against the middle steel cylinder. A second possible answer, then, is *A = lead, B = steel, and C = brass.*

Related Homework: *Conceptual Question 3*

✔ **Check Your Understanding 2**

A metal ball has a diameter that is slightly greater than the diameter of a hole that has been cut into a metal plate. The coefficient of linear thermal expansion for the metal from which the ball is made is greater than that for the metal of the plate. Which one or more of the following procedures can be used to make the ball pass through the hole? (a) Raise the temperatures of the ball and the plate by the same amount. (b) Lower the temperatures of the ball and the plate by the same amount. (c) Heat the ball and cool the plate. (d) Cool the ball and heat the plate. *(The answer is given at the end of the book.)*

Background: The coefficient of linear thermal expansion is the key concept here. The behavior of a hole in a piece of solid material that is heated or cooled is also important.

For similar questions (including calculational counterparts), consult Self-Assessment Test 12.1. This test is described at the end of Section 12.5.

12.5 *Volume Thermal Expansion*

The volume of a normal material increases as the temperature increases. Most solids and liquids behave in this fashion. By analogy with linear thermal expansion, the change in volume ΔV is proportional to the change in temperature ΔT and to the initial volume V_0, provided the change in temperature is not too large. These two proportionalities can be converted into Equation 12.3 with the aid of a proportionality constant β, known as the *coefficient of volume expansion.* The algebraic form of this equation is similar to that for linear expansion, $\Delta L = \alpha L_0 \Delta T$.

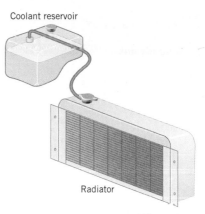

Figure 12.19 An automobile radiator and a coolant reservoir for catching the overflow from the radiator.

■ **VOLUME THERMAL EXPANSION**

The volume V_0 of an object changes by an amount ΔV when its temperature changes by an amount ΔT:

$$\Delta V = \beta V_0 \Delta T \qquad (12.3)$$

where β is the coefficient of volume expansion.

Common Unit for the Coefficient of Volume Expansion: $(C°)^{-1}$

The unit for β, like that for α, is $(C°)^{-1}$. Values for β depend on the nature of the material, and Table 12.1 lists some examples measured near 20 °C. The values of β for liquids are substantially larger than those for solids, because liquids typically expand more than solids, given the same initial volumes and temperature changes. Table 12.1 also shows that, for most solids, the coefficient of volume expansion is three times as much as the coefficient of linear expansion: $\beta = 3\alpha$.

If a cavity exists within a solid object, the volume of the cavity increases when the object expands, just as if the cavity were filled with the surrounding material. The expansion of the cavity is analogous to the expansion of a hole in a sheet of material. Accordingly, the change in volume of a cavity can be found using the relation $\Delta V = \beta V_0 \Delta T$, where β is the coefficient of volume expansion of the material that surrounds the cavity. Example 8 illustrates this point.

Example 8 An Automobile Radiator

A small plastic container, called the coolant reservoir, catches the radiator fluid that overflows when an automobile engine becomes hot (see Figure 12.19). The radiator is made of copper, and the coolant has a coefficient of volume expansion of $\beta = 4.10 \times 10^{-4}\ (C°)^{-1}$. If the radiator is filled to its 15-quart capacity when the engine is cold (6.0 °C), how much overflow from the radiator will spill into the reservoir when the coolant reaches its operating temperature of 92 °C?

Reasoning When the temperature increases, both the coolant and radiator expand. If they were to expand by the same amount, there would be no overflow. However, the liquid coolant expands more than the radiator, and the overflow volume is the amount of coolant expansion *minus* the amount of the radiator cavity expansion.

The physics of the overflow of an automobile radiator.

Problem solving insight
The way in which the level of a liquid in a container changes with temperature depends on the change in volume of both the liquid and the container.

Solution When the temperature increases by 86 C°, the coolant expands by an amount

$$\Delta V = \beta V_0 \Delta T = [4.10 \times 10^{-4}\ (C°)^{-1}](15\ \text{quarts})(86\ C°) = 0.53\ \text{quarts} \qquad (12.3)$$

The radiator cavity expands as if it were filled with copper [$\beta = 51 \times 10^{-6}\ (C°)^{-1}$; see Table 12.1]. The expansion of the radiator cavity is

$$\Delta V = \beta V_0 \Delta T = [51 \times 10^{-6}\ (C°)^{-1}](15\ \text{quarts})(86\ C°) = 0.066\ \text{quarts}$$

The overflow volume is 0.53 quarts $-$ 0.066 quarts = $\boxed{0.46\ \text{quarts}}$.

Although most substances expand when heated, a few do not. For instance, if water at 0 °C is heated, its volume *decreases* until the temperature reaches 4 °C. Above 4 °C water behaves normally, and its volume increases as the temperature increases. Because a given mass of water has a minimum volume at 4°C, the density (mass per unit volume) of water is greatest at 4 °C, as Figure 12.20 shows.

The fact that water has its greatest density at 4 °C, rather than at 0 °C, has important consequences for the way in which a lake freezes. When the air temperature drops, the

Need more practice?

Interactive LearningWare 12.1
A can is filled with liquid to 97.0% of its capacity. The temperature of the can and liquid is 0.0 °C. The material from which the can is made has a coefficient of volume expansion of $8.5 \times 10^{-5}\ (C°)^{-1}$. At a temperature of 100.0 °C, the can is filled to the brim. Determine the coefficient of volume expansion of the liquid.

Related Homework: Problems 30, 32

Go to
www.wiley.com/college/cutnell
for an interactive solution.

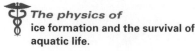

The physics of ice formation and the survival of aquatic life.

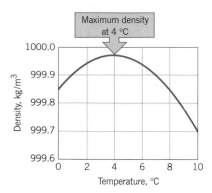

Figure 12.20 The density of water in the temperature range from 0 to 10 °C. At 4 °C water has a maximum density of 999.973 kg/m³. (This value is equivalent to the often-quoted density of 1.000 00 grams per milliliter.)

The physics of bursting water pipes.

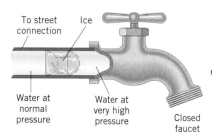

Figure 12.21 As water freezes and expands, enormous pressure is applied to the liquid water between the ice and the faucet.

surface layer of water is chilled. As the temperature of the surface layer drops toward 4 °C, this layer becomes more dense than the warmer water below. The denser water sinks and pushes up the deeper and warmer water, which in turn is chilled at the surface. This process continues until the temperature of the entire lake reaches 4 °C. Further cooling of the surface water below 4 °C makes it *less dense* than the deeper layers; consequently, the surface layer does not sink but stays on top. Continued cooling of the top layer to 0 °C leads to the formation of ice that floats on the water, because ice has a smaller density than water at any temperature. Below the ice, however, the water temperature remains above 0 °C. The sheet of ice acts as an insulator that reduces the loss of heat from the lake, especially if the ice is covered with a blanket of snow, which is also an insulator. As a result, lakes usually do not freeze solid, even during prolonged cold spells, so fish and other aquatic life can survive.

The fact that the density of ice is smaller than the density of water has an important consequence for home owners, who have to contend with the possibility of bursting water pipes during severe winters. Water often freezes in a section of pipe exposed to unusually cold temperatures. The ice can form an immovable plug that prevents the subsequent flow of water, as Figure 12.21 illustrates. When water (larger density) turns to ice (smaller density), its volume expands by 8.3%. Therefore, when more water freezes at the left side of the plug, the expanding ice pushes liquid back into the pipe leading to the street connection, and no damage is done. However, when ice forms on the right side of the plug, the expanding ice pushes liquid to the right. But it has nowhere to go if the faucet is closed. As ice continues to form and expand, the water pressure between the plug and faucet rises. Even a small increase in the amount of ice produces a large increase in the pressure. This situation is analogous to the thermal stress discussed in Example 4, where a small change in the length of the steel beam produces a large stress on the concrete supports. The entire section of pipe to the right of the blockage experiences the same elevated pressure, according to Pascal's principle (Section 11.5). Therefore, the pipe can burst at any point where it is structurally weak, even within the heated space of the building. If you should lose heat during the winter, there is a simple way to prevent pipes from bursting. Simply open the faucet, so it drips a little. The excessive pressure will be relieved.

Self-Assessment Test 12.1

Test your understanding of the material in Sections 12.1–12.5:

- Temperature Scales and Thermometers   • Linear Thermal Expansion
- Volume Thermal Expansion

Go to **www.wiley.com/college/cutnell**

12.6 *Heat and Internal Energy*

An object with a high temperature is said to be hot, and the word "hot" brings to mind the word "heat." *Heat* flows from a hotter object to a cooler object when the two are placed in contact. It is for this reason that a cup of hot coffee feels hot to the touch, while a glass of ice water feels cold. When the person in Figure 12.22a touches the coffee cup, heat flows from the hotter cup into the cooler hand. When the person touches the glass in part *b* of the drawing, heat again flows from hot to cold, in this case from the warmer hand into the colder glass. The response of the nerves in the hand to the arrival or departure of heat prompts the brain to identify the coffee cup as being hot and the glass as being cold.

But just what is heat? As the following definition indicates, heat is a form of energy, energy in transit from hot to cold.

■ **DEFINITION OF HEAT**

Heat is energy that flows from a higher-temperature object to a lower-temperature object because of the difference in temperatures.

SI Unit of Heat: joule (J)

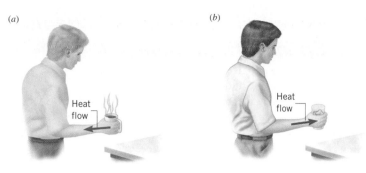

Figure 12.22 Heat is energy in transit from hot to cold. (*a*) Heat flows from the hotter coffee cup to the colder hand. (*b*) Heat flows from the warmer hand to the colder glass of ice water.

Being a kind of energy, heat is measured in the same units used for work, kinetic energy, and potential energy. Thus, the SI unit for heat is the joule.

The heat that flows from hot to cold in Figure 12.22 originates in the **internal energy** of the hot substance. The internal energy of a substance is the sum of the molecular kinetic energy (due to the random motion of the molecules), the molecular potential energy (due to forces that act between the atoms of a molecule and between molecules), and other kinds of molecular energy. When heat flows in circumstances where the work done is negligible, the internal energy of the hot substance decreases and the internal energy of the cold substance increases. Although heat may originate in the internal energy supply of a substance, *it is not correct to say that a substance contains heat.* The substance has internal energy, not heat. The word "heat" is used only when referring to the energy actually in transit from hot to cold.

▶ CONCEPTS AT A GLANCE   In the next two sections, we will consider some of the effects of heat. For instance, when preparing spaghetti for dinner, the first thing that the cook does is to place a pot of water on the stove. Heat from the stove causes the internal energy of the water to increase, as the Concepts-at-a-Glance chart in Figure 12.23 suggests. Associated with this increase in internal energy is a rise in temperature. After a while, however, the temperature reaches 100 °C and the water begins to boil. During boiling, any heat added to the water goes into producing steam, a process in which water changes from a liquid phase to a vapor phase. The next section investigates how the addition (or removal) of heat causes the temperature of a substance to change (see the upper-right portion of the chart in Figure 12.23). Then, Section 12.8 discusses the relationship between heat and phase changes (see the lower-right part of Figure 12.23), such as that which occurs when water boils. ◀

Figure 12.23 CONCEPTS AT A GLANCE When heat is added to or removed from a substance, its internal energy can change. This change can cause a change in temperature (Section 12.7) or a change in phase (Section 12.8). In this fire-breathing performance, the chemical energy released as heat during the burning process causes the temperature of the gas to increase. (© Ian Shaw/Stone/Getty Images)

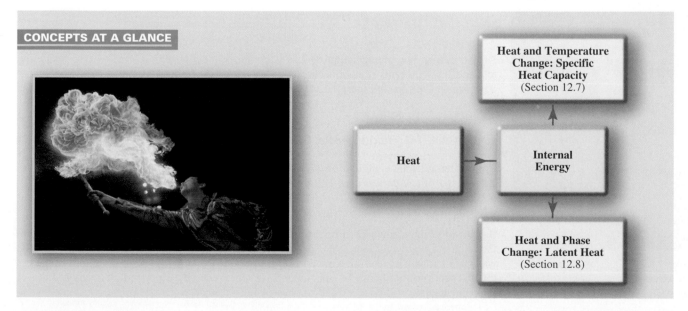

CONCEPTS AT A GLANCE

Heat → Internal Energy

Internal Energy → Heat and Temperature Change: Specific Heat Capacity (Section 12.7)

Internal Energy → Heat and Phase Change: Latent Heat (Section 12.8)

Table 12.2 *Specific Heat Capacities^a of Some Solids and Liquids*

Substance	Specific Heat Capacity, c $J/(kg \cdot C°)$
Solids	
Aluminum	9.00×10^2
Copper	387
Glass	840
Human body	3500
(37 °C, average)	
Ice (-15 °C)	2.00×10^3
Iron or steel	452
Lead	128
Silver	235
Liquids	
Benzene	1740
Ethyl alcohol	2450
Glycerin	2410
Mercury	139
Water (15 °C)	4186

^a Except as noted, the values are for 25 °C and 1 atm of pressure.

12.7 Heat and Temperature Change: Specific Heat Capacity

SOLIDS AND LIQUIDS

Greater amounts of heat are needed to raise the temperature of solids or liquids to higher values. A greater amount of heat is also required to raise the temperature of a greater mass of material. Similar comments apply when the temperature is lowered, except that heat must be removed. For limited temperature ranges, experiment shows that the amount of heat Q is directly proportional to the change in temperature ΔT and to the mass m. These two proportionalities are expressed below in Equation 12.4, with the help of a proportionality constant c that is referred to as the *specific heat capacity* of the material.

■ **HEAT SUPPLIED OR REMOVED IN CHANGING THE TEMPERATURE OF A SUBSTANCE**

The heat Q that must be supplied or removed to change the temperature of a substance of mass m by an amount ΔT is

$$Q = cm\Delta T \tag{12.4}$$

where c is the specific heat capacity of the substance.

Common Unit for Specific Heat Capacity: $J/(kg \cdot C°)$

Solving Equation 12.4 for the specific heat capacity shows that $c = Q/(m\,\Delta T)$, so the unit for specific heat capacity is $J/(kg \cdot C°)$. Table 12.2 reveals that the value of the specific heat capacity depends on the nature of the material. Examples 9 and 10 illustrate the use of Equation 12.4.

Example 9   A Hot Jogger

In a half hour, a 65-kg jogger can generate 8.0×10^5 J of heat. This heat is removed from the jogger's body by a variety of means, including the body's own temperature-regulating mechanisms. If the heat were not removed, how much would the body temperature increase?

Reasoning The increase in body temperature depends on the amount of heat Q generated by the jogger, her mass m, and the specific heat capacity c of the human body. Since numerical values are known for these three variables, we can determine the potential rise in temperature by using Equation 12.4.

Solution Table 12.2 gives the average specific heat capacity of the human body as 3500 $J/(kg \cdot C°)$. With this value, Equation 12.4 shows that

$$\Delta T = \frac{Q}{cm} = \frac{8.0 \times 10^5 \text{ J}}{[3500 \text{ J}/(kg \cdot C°)](65 \text{ kg})} = \boxed{3.5 \text{ C}°}$$

An increase in body temperature of 3.5 °C could be life-threatening. One way in which the jogger's body prevents it from occurring is to remove excess heat by perspiring. In contrast, dogs, such as the one in Figure 12.24, do not perspire but often pant to remove excess heat.

Example 10   Taking a Hot Shower

Cold water at a temperature of 15 °C enters a heater, and the resulting hot water has a temperature of 61 °C. A person uses 120 kg of hot water in taking a shower. (a) Find the energy needed to heat the water. (b) Assuming that the utility company charges $0.10 per kilowatt · hour for electrical energy, determine the cost of heating the water.

Reasoning The amount Q of heat needed to raise the water temperature can be found from the relation $Q = cm\,\Delta T$, since the specific heat capacity, mass, and temperature change of the water are known. To determine the cost of this energy, we multiply the cost per unit of energy ($0.10 per kilowatt · hour) by the amount of energy used, expressed in energy units of kilowatt · hours.

Figure 12.24 Dogs often pant to get rid of excess heat. (© James L. Stanfield/ National Geographic/Getty Images)

Solution

(a) The amount of heat needed to heat the water is

$$Q = cm \, \Delta T = [4186 \text{ J/(kg} \cdot \text{C}^\circ)](120 \text{ kg})(61 \text{ °C} - 15 \text{ °C}) = \boxed{2.3 \times 10^7 \text{ J}} \quad (12.4)$$

(b) The kilowatt · hour (kWh) is the unit of energy that utility companies use in your electric bill. To calculate the cost, we need to determine the number of joules in one kilowatt · hour. Recall that 1 kilowatt is 1000 watts (1 kW = 1000 W), 1 watt is 1 joule per second (1 W = 1 J/s; see Equation 6.10b), and 1 hour is equal to 3600 seconds (1 h = 3600 s). Thus,

$$1 \text{ kWh} = (1 \text{ kWh}) \left(\frac{1000 \text{ W}}{1 \text{ kW}} \right) \left(\frac{1 \text{ J/s}}{1 \text{ W}} \right) \left(\frac{3600 \text{ s}}{1 \text{ h}} \right) = 3.60 \times 10^6 \text{ J}$$

The number of kilowatt · hours of energy used to heat the water is

$$(2.3 \times 10^7 \text{ J}) \left(\frac{1 \text{ kWh}}{3.60 \times 10^6 \text{ J}} \right) = 6.4 \text{ kWh}$$

At a cost of $0.10 per kWh, the bill for the heat is $\boxed{\$0.64}$ or 64 cents.

GASES

As we will see in Section 15.6, the value of the specific heat capacity depends on whether the pressure or volume is held constant while energy in the form of heat is added to or removed from a substance. The distinction between constant pressure and constant volume is usually not important for solids and liquids but is significant for gases. As we will see in Section 15.6, a greater value for the specific heat capacity is obtained for a gas at constant pressure than for a gas at constant volume.

HEAT UNITS OTHER THAN THE JOULE

There are three heat units other than the joule in common use. One kilocalorie (1 kcal) was defined historically as the amount of heat needed to raise the temperature of one kilogram of water by one Celsius degree.* With $Q = 1.00$ kcal, $m = 1.00$ kg, and $\Delta T = 1.00$ C°, the equation $Q = cm \, \Delta T$ shows that such a definition is equivalent to a specific heat capacity for water of $c = 1.00$ kcal/(kg · C°). Similarly, one calorie (1 cal) was defined as the amount of heat needed to raise the temperature of one gram of water by one Celsius degree, which yields a value of $c = 1.00$ cal/(g · C°). (Nutritionists use the word "Calorie," with a capital C, to specify the energy content of foods; this use is unfortunate, since 1 Calorie = 1000 calories = 1 kcal.) The British thermal unit (Btu) is the other commonly used heat unit and was defined historically as the amount of heat needed to raise the temperature of one pound of water by one Fahrenheit degree.

It was not until the time of James Joule (1818–1889) that the relationship between energy in the form of work (in units of joules) and energy in the form of heat (in units of kilocalories) was firmly established. Joule's experiments revealed that the performance of mechanical work, like rubbing your hands together, can make the temperature of a substance rise, just as the absorption of heat can. His experiments and those of later workers have shown that

$$1 \text{ kcal} = 4186 \text{ joules} \quad \text{or} \quad 1 \text{ cal} = 4.186 \text{ joules}$$

Because of its historical significance, this conversion factor is known as the ***mechanical equivalent of heat.***

CALORIMETRY

In Section 6.8 we encountered the principle of conservation of energy, which states that energy can be neither created nor destroyed, but can only be converted from one form to another. There we dealt with kinetic and potential energies. In this chapter we have expanded our concept of energy to include heat, which is energy that flows from a higher-temperature

---

* From 14.5 to 15.5 °C.

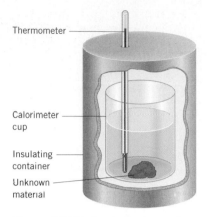

Thermometer

Calorimeter cup

Insulating container

Unknown material

Figure 12.25 A calorimeter can be used to measure the specific heat capacity of an unknown material.

Problem solving insight
In the equation "Heat gained = Heat lost," both sides must have the same algebraic sign. Therefore, when calculating heat contributions, always write any temperature changes as the higher minus the lower temperature.

Need more practice?

Interactive LearningWare 12.2
Water is moving with a speed of 5.00 m/s just before it passes over the top of a waterfall. At the bottom, 5.00 m below, the water flows away with a speed of 3.00 m/s. What is the largest amount by which the temperature of the water at the bottom could exceed the temperature of the water at the top?

Related Homework: Problem 46

Go to
www.wiley.com/college/cutnell
for an interactive solution.

object to a lower-temperature object because of the difference in temperature. No matter what its form, whether kinetic energy, potential energy, or heat, energy can be neither created nor destroyed. This fact governs the way objects at different temperatures come to an equilibrium temperature when they are placed in contact. If there is no heat loss to the external surroundings, the heat lost by the hotter objects equals the heat gained by the cooler ones, a process that is consistent with the conservation of energy. Just this kind of process occurs within a thermos. A perfect thermos would prevent any heat from leaking out or in. However, energy in the form of heat can flow *between* materials inside the thermos to the extent that they have different temperatures; for example, between ice cubes and warm tea. The transfer of energy continues until a common temperature is reached at thermal equilibrium.

The kind of heat transfer that occurs within a thermos of iced tea also occurs within a calorimeter, which is the experimental apparatus used in a technique known as ***calorimetry.*** Figure 12.25 shows that, like a thermos, a calorimeter is essentially an insulated container. It can be used to determine the specific heat capacity of a substance, as the next example illustrates.

Example 11   Measuring the Specific Heat Capacity

The calorimeter cup in Figure 12.25 is made from 0.15 kg of aluminum and contains 0.20 kg of water. Initially, the water and the cup have a common temperature of 18.0 °C. A 0.040-kg mass of unknown material is heated to a temperature of 97.0 °C and then added to the water. The temperature of the water, the cup, and the unknown material is 22.0 °C after thermal equilibrium is reestablished. Ignoring the small amount of heat gained by the thermometer, find the specific heat capacity of the unknown material.

Reasoning Since energy is conserved and there is negligible heat flow between the calorimeter and the outside surroundings, the heat gained by the cold water and the aluminum cup as they warm up is equal to the heat lost by the unknown material as it cools down. Each quantity of heat can be calculated using the relation $Q = cm\,\Delta T$, where we always write the change in temperature ΔT as the higher temperature minus the lower temperature. The equation "Heat gained = Heat lost" will then contain a single unknown quantity, the desired specific heat capacity.

Solution

$$\underbrace{(cm\,\Delta T)_{\text{Al}} + (cm\,\Delta T)_{\text{water}}}_{\substack{\text{Heat gained by} \\ \text{aluminum and water}}} = \underbrace{(cm\,\Delta T)_{\text{unknown}}}_{\substack{\text{Heat lost by} \\ \text{unknown material}}}$$

$$c_{\text{unknown}} = \frac{c_{\text{Al}}m_{\text{Al}}\,\Delta T_{\text{Al}} + c_{\text{water}}m_{\text{water}}\,\Delta T_{\text{water}}}{m_{\text{unknown}}\,\Delta T_{\text{unknown}}}$$

The changes in temperature for the three substances are $\Delta T_{\text{Al}} = \Delta T_{\text{water}} = 22.0\ °\text{C} - 18.0\ °\text{C} = 4.0\ \text{C}°$, and $\Delta T_{\text{unknown}} = 97.0\ °\text{C} - 22.0\ °\text{C} = 75.0\ \text{C}°$. Table 12.2 contains values for the specific heat capacities of aluminum and water. Substituting these data into the equation above, we find that

$$c_{\text{unknown}} = \frac{[9.00 \times 10^2\ \text{J/(kg} \cdot \text{C}°)](0.15\ \text{kg})(4.0\ \text{C}°) + [4186\ \text{J/(kg} \cdot \text{C}°)](0.20\ \text{kg})(4.0\ \text{C}°)}{(0.040\ \text{kg})(75.0\ \text{C}°)}$$

$$= \boxed{1300\ \text{J/(kg} \cdot \text{C}°)}$$

✔ **Check Your Understanding 3**

Consider a mass m of a material and a change ΔT in its temperature. Various possibilities for these variables are listed in the table below. Rank these possibilities in descending order (largest first), according to how much heat is needed to bring about the change in temperature. *(The answer is given at the end of the book.)*

	m (kg)	ΔT (C°)
(a)	2.0	15
(b)	1.5	40
(c)	3.0	25
(d)	2.5	20

(Continues)

12.8 Heat and Phase Change: Latent Heat

Surprisingly, there are situations in which the addition or removal of heat does not cause a temperature change. Consider a well-stirred glass of iced tea that has come to thermal equilibrium. Even though heat enters the glass from the warmer room, the temperature of the tea does not rise above 0 °C as long as ice cubes are present. Apparently the heat is being used for some purpose other than raising the temperature. In fact, the heat is being used to melt the ice, and only when all of it is melted will the temperature of the liquid begin to rise.

An important point illustrated by the iced tea example is that there is more than one type or phase of matter. For instance, some of the water in the glass is in the solid phase (ice) and some in the liquid phase. The gas or vapor phase is the third familiar phase of matter. In the gas phase, water is referred to as water vapor or steam. All three phases of water are present in the scene depicted in Figure 12.26.

Matter can change from one phase to another, and heat plays a role in the change. Figure 12.27 summarizes the various possibilities. A solid can *melt* or *fuse* into a liquid if heat is added, while the liquid can *freeze* into a solid if heat is removed. Similarly, a liquid can *evaporate* into a gas if heat is supplied, while the gas can *condense* into a liquid if heat is taken away. Rapid evaporation, with the formation of vapor bubbles within the liquid, is called *boiling*. Finally, a solid can sometimes change directly into a gas if heat is provided. We say that the solid *sublimes* into a gas. Examples of sublimation are (1) solid carbon dioxide, CO_2 (dry ice), turning into gaseous CO_2 and (2) solid naphthalene (moth balls) turning into naphthalene fumes. Conversely, if heat is removed under the right conditions, the gas will condense directly into a solid.

Figure 12.28 displays a graph that indicates what typically happens when heat is added to a material that changes phase. The graph records temperature versus heat added and refers to water at the normal atmospheric pressure of 1.01×10^5 Pa. The water starts off as ice at the subfreezing temperature of −30 °C. As heat is added, the temperature of the ice increases, in accord with the specific heat capacity of ice [2000 J/(kg·C°)]. Not until the temperature reaches the normal melting/freezing point of 0 °C does the water begin to change phase. Then, when heat is added, the solid changes into the liquid, the temperature staying at 0 °C until *all the ice has melted*. Once all the material is in the liquid phase, additional heat causes the temperature to increase again, now in accord with the specific heat capacity of liquid water [4186 J/(kg·C°)]. When the temperature reaches the normal boiling/condensing point of 100 °C, the water begins to change from the liquid to the gas phase and continues to do so as long as heat is added. The temperature remains at 100 °C *until all liquid is gone*. When all of the material is in the gas phase, additional heat once again causes the temperature to rise, this time according to the specific heat capacity of water vapor at constant atmospheric pressure [2020 J/(kg·C°)]. Conceptual Example 12 applies the information in Figure 12.28 to a familiar situation.

Figure 12.26 The three phases of water: ice is floating in liquid water, and water vapor (invisible) is present in the air. (© Paul Souders/Stone/Getty Images)

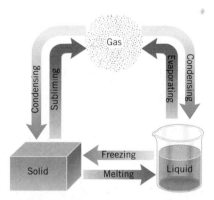

Figure 12.27 Three familiar phases of matter—solid, liquid, and gas—and the phase changes that can occur between any two of them.

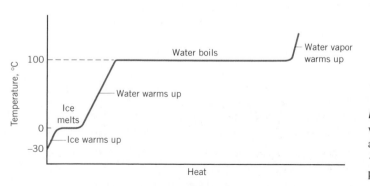

Figure 12.28 The graph shows the way the temperature of water changes as heat is added, starting with ice at −30 °C. The pressure is atmospheric pressure.

Conceptual Example 12　Saving Energy

Suppose you are cooking spaghetti for dinner, and the instructions say "boil the pasta in water for ten minutes." To cook spaghetti in an open pot with the least amount of energy, should you turn up the burner to its fullest so the water vigorously boils, or should you turn down the burner so the water barely boils?

Reasoning and Solution The spaghetti needs to cook at a temperature of 100 °C for ten minutes. It doesn't matter whether the water is vigorously boiling or barely boiling, because its temperature is still 100 °C. Remember, as long as there is boiling water in the pot, and the pot is open to one atmosphere of pressure, no amount of additional heat will cause the water temperature to rise above 100 °C. Additional heat only vaporizes the water and produces steam, which is of no use in cooking the spaghetti. So, save your money and turn down the heat, because *the least amount of energy is expended when the water barely boils.*

When a substance changes from one phase to another, the amount of heat that must be added or removed depends on the type of material and the nature of the phase change. The heat per kilogram associated with a phase change is referred to as *latent heat:*

■ HEAT SUPPLIED OR REMOVED IN CHANGING THE PHASE OF A SUBSTANCE

The heat Q that must be supplied or removed to change the phase of a mass m of a substance is

$$Q = mL \qquad (12.5)$$

where L is the latent heat of the substance.

SI Unit of Latent Heat: J/kg

The *latent heat of fusion* L_f refers to the change between solid and liquid phases, the *latent heat of vaporization* L_v applies to the change between liquid and gas phases, and the *latent heat of sublimation* L_s refers to the change between solid and gas phases.

Table 12.3 gives some typical values of latent heats of fusion and vaporization. For instance, the latent heat of fusion for water is $L_f = 3.35 \times 10^5$ J/kg. Thus, 3.35×10^5 J of heat must be supplied to melt one kilogram of ice at 0 °C into liquid water at 0 °C; conversely, this amount of heat must be removed from one kilogram of liquid water at 0 °C to freeze the liquid into ice at 0 °C. In comparison, the latent heat of vaporization for water has the much larger value of $L_v = 22.6 \times 10^5$ J/kg. When water boils at 100 °C, 22.6×10^5 J of heat must be supplied for each kilogram of liquid turned into steam. And when steam condenses at 100 °C, this amount of heat is released from each kilogram of

The physics of steam burns.

Table 12.3　*Latent Heats[a] of Fusion and Vaporization*

Substance	Melting Point (°C)	Latent Heat of Fusion, L_f (J/kg)	Boiling Point (°C)	Latent Heat of Vaporization, L_v (J/kg)
Ammonia	−77.8	33.2×10^4	−33.4	13.7×10^5
Benzene	5.5	12.6×10^4	80.1	3.94×10^5
Copper	1083	20.7×10^4	2566	47.3×10^5
Ethyl alcohol	−114.4	10.8×10^4	78.3	8.55×10^5
Gold	1063	6.28×10^4	2808	17.2×10^5
Lead	327.3	2.32×10^4	1750	8.59×10^5
Mercury	−38.9	1.14×10^4	356.6	2.96×10^5
Nitrogen	−210.0	2.57×10^4	−195.8	2.00×10^5
Oxygen	−218.8	1.39×10^4	−183.0	2.13×10^5
Water	0.0	33.5×10^4	100.0	22.6×10^5

[a] The values pertain to 1 atm pressure.

steam that changes back into liquid. Liquid water at 100 °C is hot enough by itself to cause a bad burn, and the additional effect of the large latent heat can cause severe tissue damage if condensation occurs on the skin.

By taking advantage of the latent heat of fusion, designers can now engineer clothing that can absorb or release heat to help maintain a comfortable and approximately constant temperature close to your body. As the photograph in Figure 12.29 shows, the fabric in this type of clothing is coated with microscopic balls of heat-resistant plastic that contain a substance known as a "phase-change material" (PCM). When you are enjoying your favorite winter sport, for example, it is easy to become overheated. The PCM prevents this by melting, absorbing excess body heat in the process. When you are taking a break and cooling down, however, the PCM freezes and releases heat to keep you warm. The temperature range over which the PCM can maintain a comfort zone is related to its melting/freezing temperature, which is determined by its chemical composition.

Examples 13 and 14 illustrate how to take into account the effect of latent heat when using the conservation-of-energy principle.

The physics of **high-tech clothing.**

Figure 12.29 This highly magnified image shows a fabric that has been coated with microscopic balls of heat-resistant plastic. The balls contain a substance known as a "phase-change material," the melting and freezing of which absorbs and releases heat. Clothing made from such fabrics can automatically adjust itself in reaction to your body heat and help maintain a constant temperature next to your skin. (Courtesy Outlast Technologies, Boulder, CO.)

Example 13 Ice-cold Lemonade

Ice at 0 °C is placed in a Styrofoam cup containing 0.32 kg of lemonade at 27 °C. The specific heat capacity of lemonade is virtually the same as that of water; that is, $c = 4186$ J/(kg·C°). After the ice and lemonade reach an equilibrium temperature, some ice still remains. The latent heat of fusion for water is $L_f = 3.35 \times 10^5$ J/kg. Assume that the mass of the cup is so small that it absorbs a negligible amount of heat, and ignore any heat lost to the surroundings. Determine the mass of ice that has melted.

Reasoning According to the principle of energy conservation, the heat gained by the melting ice equals the heat lost by the cooling lemonade. According to Equation 12.5, the heat gained by the melting ice is $Q = mL_f$, where m is the mass of the melted ice, and L_f is the latent heat of fusion for water. The heat lost by the lemonade is given by $Q = cm\,\Delta T$, where ΔT is the higher temperature of 27 °C minus the lower equilibrium temperature. The equilibrium temperature is 0 °C, because there is some ice remaining, and ice is in equilibrium with liquid water when the temperature is 0 °C.

Solution

$$\underbrace{(mL_f)_{ice}}_{\substack{\text{Heat gained} \\ \text{by ice}}} = \underbrace{(cm\,\Delta T)_{lemonade}}_{\substack{\text{Heat lost} \\ \text{by lemonade}}}$$

The mass m_{ice} of ice that has melted is

$$m_{ice} = \frac{(cm\,\Delta T)_{lemonade}}{L_f} = \frac{[4186\ \text{J/(kg·C°)}](0.32\ \text{kg})(27\ °C - 0\ °C)}{3.35 \times 10^5\ \text{J/kg}} = \boxed{0.11\ \text{kg}}$$

Example 14 Getting Ready for a Party

A 7.00-kg glass bowl [$c = 840$ J/(kg·C°)] contains 16.0 kg of punch at 25.0 °C. Two-and-a-half kilograms of ice [$c = 2.00 \times 10^3$ J/(kg·C°)] are added to the punch. The ice has an initial temperature of −20.0 °C, having been kept in a very cold freezer. The punch may be treated as if it were water [$c = 4186$ J/(kg·C°)], and it may be assumed that there is no heat flow between the punch bowl and the external environment. The latent heat of fusion for water is 3.35×10^5 J/kg. When thermal equilibrium is reached, all the ice has melted, and the final temperature of the mixture is above 0 °C. Determine this temperature.

Reasoning The final temperature can be found by using the conservation of energy: the heat gained is equal to the heat lost. Heat is gained (a) by the ice in warming up to the melting point, (b) by the ice in changing phase from a solid to a liquid, and (c) by the liquid that results from the ice warming up to the final temperature; heat is lost (d) by the punch and (e) by the bowl in cooling down. The heat gained or lost by each component in changing temperature can be determined from the relation $Q = cm\,\Delta T$. The heat gained when water changes phase from a solid to a liquid at 0 °C is $Q = mL_f$, where m is the mass of water and L_f is the latent heat of fusion.

Solution The heat gained or lost by each component is listed as follows:

(a) Heat gained when ice warms to 0.0 °C
$$= [2.00 \times 10^3 \text{ J/(kg·C°)}](2.50 \text{ kg})[0.0 \text{ °C} - (-20.0 \text{ °C})]$$

(b) Heat gained when ice melts at 0.0 °C
$$= (2.50 \text{ kg})(3.35 \times 10^5 \text{ J/kg})$$

(c) Heat gained when melted ice (liquid) warms to temperature T
$$= [4186 \text{ J/(kg·C°)}](2.50 \text{ kg})(T - 0.0 \text{ °C})$$

(d) Heat lost when punch cools to temperature T
$$= [4186 \text{ J/(kg·C°)}](16.0 \text{ kg})(25.0 \text{ °C} - T)$$

(e) Heat lost when bowl cools to temperature T
$$= [840 \text{ J/(kg·C°)}](7.00 \text{ kg})(25.0 \text{ °C} - T)$$

Setting the heat gained equal to the heat lost gives:

$$\underbrace{(a) + (b) + (c)}_{\text{Heat gained}} = \underbrace{(d) + (e)}_{\text{Heat lost}}$$

This equation can be solved to show that $\boxed{T = 11 \text{ °C}}$.

The physics of a dye-sublimation color printer.

An interesting application of the phase change between a solid and a gas is found in one kind of color printer used with computers. A dye-sublimation printer uses a thin plastic film coated with separate panels of cyan (blue), yellow, and magenta pigment or dye. A full spectrum of colors is produced by using combinations of tiny spots of these dyes. As Figure 12.30 shows, the coated film passes in front of a print head that extends across the width of the paper and contains 2400 heating elements. When a heating element is turned on, the dye in front of it absorbs heat and goes from a solid to a gas—it sublimes—with no liquid phase in between. A coating on the paper absorbs the gaseous dye on contact, producing a small spot of color. The intensity of the spot is controlled by the heating element, since each element can produce 256 different temperatures; the hotter the element, the greater the amount of dye transferred to the paper. The paper makes three separate passes across the print head, once for each of the dyes. The final result is an image of near-photographic quality.

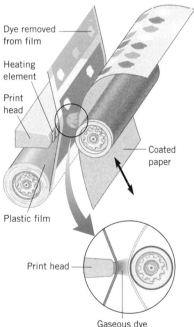

Figure 12.30 A dye-sublimation printer. As the plastic film passes in front of the print head, the heat from a given heating element causes one of three pigments or dyes on the film to sublime from a solid to a gas. The gaseous dye is absorbed onto the coated paper as a dot of color. The size of the dots on the paper has been exaggerated for clarity.

✓ **Check Your Understanding 4**

When ice cubes are used to cool a drink, both their mass and temperature are important in how effective they are. The table below lists several possibilities for the mass and temperature of the ice cubes used to cool a particular drink. Rank the possibilities in descending order (best first), according to their cooling effectiveness. *(The answer is given at the end of the book.)*

	Mass of ice cubes	Temperature of ice cubes
(a)	m	-6.0 °C
(b)	$\frac{1}{2}m$	-12 °C
(c)	$2m$	-3.0 °C

Background: Both the latent heat of phase change and the specific heat capacity must be considered in answering this question.

For similar questions (including calculational counterparts), consult Self-Assessment Test 12.2. This test is described at the end of Section 12.10.

*12.9 Equilibrium Between Phases of Matter

Under specific conditions of temperature and pressure, a substance can exist in equilibrium in more than one phase at the same time. Consider Figure 12.31, which shows a

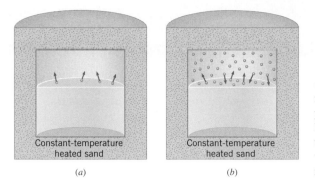

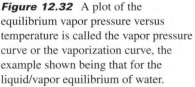

Figure 12.31 (*a*) Some of the molecules begin entering the vapor phase in the evacuated space above the liquid. (*b*) Equilibrium is reached when the number of molecules entering the vapor phase equals the number returning to the liquid.

container kept at a constant temperature by a large reservoir of heated sand. Initially the container is evacuated. Part *a* shows it just after it has been partially filled with a liquid and a few fast-moving molecules escape the liquid and form a vapor phase. These molecules pick up the required energy (the latent heat of vaporization) during collisions with neighboring molecules in the liquid. However, the reservoir of heated sand replenishes the energy carried away, thus maintaining the constant temperature. At first, the movement of molecules is predominantly from liquid to vapor, although some molecules in the vapor phase do reenter the liquid. As the molecules accumulate in the vapor, the number reentering the liquid eventually equals the number entering the vapor, at which point equilibrium is established, as in part *b*. From this point on, the concentration of molecules in the vapor phase does not change, and the vapor pressure remains constant. The pressure of the vapor that coexists in equilibrium with the liquid is called the ***equilibrium vapor pressure*** of the liquid.

The equilibrium vapor pressure does not depend on the volume of space above the liquid. If more space were provided, more liquid would vaporize, until equilibrium was reestablished at the same vapor pressure, assuming the same temperature is maintained. In fact, the equilibrium vapor pressure depends only on the temperature of the liquid; a higher temperature causes a higher pressure, as the graph in Figure 12.32 indicates for the specific case of water. Only when the temperature and vapor pressure correspond to a point on the curved line, which is called the ***vapor pressure curve*** or the ***vaporization curve,*** do liquid and vapor phases coexist in equilibrium.

To illustrate the use of a vaporization curve, let's see what happens when water boils in a pot that is *open to the air.* Assume that the air pressure acting on the water is 1.01×10^5 Pa (one atmosphere). When boiling occurs, bubbles of water vapor form throughout the liquid, rise to the surface, and break. For these bubbles to form and rise, the pressure of the vapor inside them must at least equal the air pressure acting on the surface of the water. According to Figure 12.32, a value of 1.01×10^5 Pa corresponds to a temperature of 100 °C. Consequently, water boils at 100 °C at one atmosphere of pressure. In general, a ***liquid boils at the temperature for which its vapor pressure equals the external pressure.*** Water will not boil, then, at sea level if the temperature is only 83 °C, because at

It takes less heat to boil water high on a mountain, because the boiling point becomes less than 100 °C as the air pressure decreases at higher elevations. (© Gregg Adams/Stone/Getty Images)

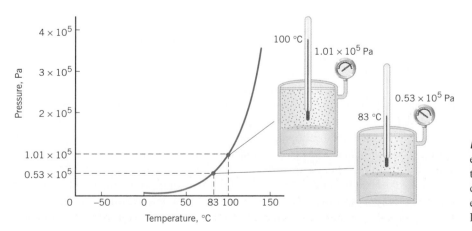

Figure 12.32 A plot of the equilibrium vapor pressure versus temperature is called the vapor pressure curve or the vaporization curve, the example shown being that for the liquid/vapor equilibrium of water.

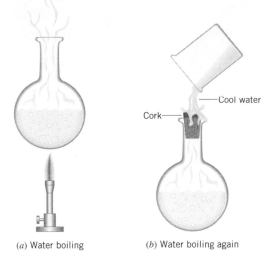

(a) Water boiling   (b) Water boiling again

Figure 12.33 (a) Water is boiling at a temperature of 100 °C and a pressure of one atmosphere. (b) The water boils at a temperature that is less than 100 °C, because the cool water reduces the pressure above the water in the flask.

this temperature the vapor pressure of water is only 0.53×10^5 Pa (see Figure 12.32), a value less than the external pressure of 1.01×10^5 Pa. However, water does boil at 83 °C on a mountain at an altitude of just under five kilometers, because the atmospheric pressure there is 0.53×10^5 Pa.

The fact that water can boil at a temperature less than 100 °C leads to an interesting phenomenon that Conceptual Example 15 discusses.

Conceptual Example 15   How to Boil Water That Is Cooling Down

Figure 12.33a shows water boiling in an open flask. Shortly after the flask is removed from the burner, the boiling stops. A cork is then placed in the neck of the flask to seal it. To restart the boiling, should you pour hot (but not boiling) water or cold water over the neck of the flask, as in part *b* of the drawing?

Reasoning and Solution When the open flask is removed from the burner, the water begins to cool. Boiling stops, because water cannot boil when its temperature is less than 100 °C and the external pressure is one atmosphere (1.01×10^5 Pa). Certainly, pouring hot water over the corked flask can rewarm the water within it to some extent. But since the water being poured is not boiling, its temperature must be less than 100 °C. Therefore, it cannot reheat the water within the flask to 100 °C and restart the boiling. However, when cold water is poured over the corked flask, it causes some of the water vapor inside to condense. Consequently, the pressure above the liquid in the flask drops. When the pressure drops to a certain level, boiling restarts. This occurs when the pressure in the flask becomes equal to the vapor pressure of the water at its current temperature (which is now less than 100 °C). ***Thus, it is possible to restart the boiling by pouring cold water over the neck of the flask.***

The physics of
spray cans.

The operation of spray cans is based on the equilibrium between a liquid and its vapor. Figure 12.34a shows that a spray can contains a liquid propellant that is mixed with the product (such as hair spray). Inside the can, propellant vapor forms over the liquid. A propellant is chosen that has an equilibrium vapor pressure that is greater than atmospheric pressure at room temperature. Consequently, when the nozzle of the can is pressed, as in part *b* of the drawing, the vapor pressure forces the liquid propellant and product up the tube in the can and out the nozzle as a spray. When the nozzle is released, the coiled spring reseals the can and the propellant vapor builds up once again to its equilibrium value.

As is the case for liquid/vapor equilibrium, a solid can be in equilibrium with its liquid phase only at specific conditions of temperature and pressure. For each temperature, there is a single pressure at which the two phases can coexist in equilibrium. A plot of the

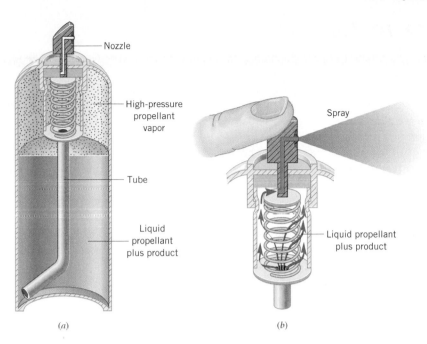

(a)   (b)

Figure 12.34 (*a*) A closed spray can containing liquid and vapor in equilibrium. (*b*) An open spray can.

equilibrium pressure versus equilibrium temperature is referred to as the ***fusion curve***, and Figure 12.35*a* shows a typical curve for a normal substance. A normal substance expands upon melting (e.g., carbon dioxide and sulfur). Since higher pressures make it more difficult for such materials to expand, a higher melting temperature is needed for a higher pressure, and the fusion curve slopes upward to the right. Part *b* of the picture illustrates the fusion curve for water, one of the few substances that contract when they melt. Higher pressures make it easier for such substances to melt. Consequently, a lower melting temperature is associated with a higher pressure, and the fusion curve slopes downward to the right.

It should be noted that just because two phases can coexist in equilibrium does not necessarily mean that they will. Other factors may prevent it. For example, water in an *open* bowl may never come into equilibrium with water vapor if air currents are present. Under such conditions the liquid, perhaps at a temperature of 25 °C, attempts to establish the corresponding equilibrium vapor pressure of 3.2×10^3 Pa. If air currents continually blow the water vapor away, however, equilibrium will never be established, and eventually the water will evaporate completely. Each kilogram of water that goes into the vapor phase takes along the latent heat of vaporization. Because of this heat loss, the remaining liquid would become cooler, except for the fact that the surroundings replenish the loss.

In the case of the human body, water is exuded by the sweat glands and evaporates from a much larger area than the surface of a typical bowl of water. The removal of heat along with the water vapor is called evaporative cooling and is one mechanism that the body uses to maintain its constant temperature.

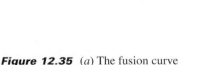 **The physics of evaporative cooling of the human body.**

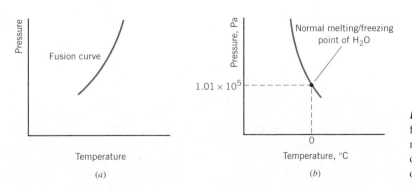

(a)

(b)

Figure 12.35 (*a*) The fusion curve for a normal substance that expands on melting. (*b*) The fusion curve for water, one of the few substances that contract on melting.

*12.10 Humidity

Air is a mixture of gases, including nitrogen, oxygen, and water vapor. The total pressure of the mixture is the sum of the partial pressures of the component gases. The **partial pressure** of a gas is the pressure it would exert if it alone occupied the entire volume at the same temperature as the mixture. The partial pressure of water vapor in air depends on weather conditions. It can be as low as zero or as high as the equilibrium vapor pressure of water at the given temperature.

The physics of relative humidity.

To provide an indication of how much water vapor is in the air, weather forecasters usually give the **relative humidity.** If the relative humidity is too low, the air contains such a small amount of water vapor that skin and mucous membranes tend to dry out. If the relative humidity is too high, especially on a hot day, we become very uncomfortable and our skin feels "sticky." Under such conditions, the air holds so much water vapor that the water exuded by sweat glands cannot evaporate efficiently. The relative humidity is defined as the ratio (expressed as a percentage) of the partial pressure of water vapor in the air to the equilibrium vapor pressure at a given temperature.

$$\begin{array}{c}\text{Percent}\\\text{relative}\\\text{humidity}\end{array} = \frac{\begin{array}{c}\text{Partial pressure}\\\text{of water vapor}\end{array}}{\begin{array}{c}\text{Equilibrium vapor pressure of}\\\text{water at the existing temperature}\end{array}} \times 100 \qquad (12.6)$$

The term in the denominator on the right of Equation 12.6 is given by the vaporization curve of water and is the pressure of the water vapor in equilibrium with the liquid. At a given temperature, the partial pressure of the water vapor in the air cannot exceed this value. If it did, the vapor would not be in equilibrium with the liquid and would condense as dew or rain to reestablish equilibrium.

When the partial pressure of the water vapor equals the equilibrium vapor pressure of water at a given temperature, the relative humidity is 100%. In such a situation, the vapor is said to be *saturated* because it is present in the maximum amount, as it would be above a pool of liquid at equilibrium in a closed container. If the relative humidity is less than 100%, the water vapor is said to be *unsaturated.* Example 16 demonstrates how to find the relative humidity.

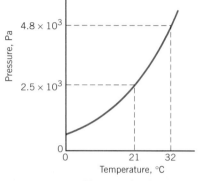

Figure 12.36 The vaporization curve of water.

Example 16 Relative Humidities

One day, the partial pressure of water vapor in the air is 2.0×10^3 Pa. Using the vaporization curve for water in Figure 12.36, determine the relative humidity if the temperature is (a) 32 °C and (b) 21 °C.

Reasoning and Solution

(a) According to Figure 12.36, the equilibrium vapor pressure of water at 32 °C is 4.8×10^3 Pa. Equation 12.6 reveals that the relative humidity is

$$\text{Relative humidity at 32 °C} = \frac{2.0 \times 10^3 \text{ Pa}}{4.8 \times 10^3 \text{ Pa}} \times 100 = \boxed{42\%}$$

(b) A similar calculation shows that

$$\text{Relative humidity at 21 °C} = \frac{2.0 \times 10^3 \text{ Pa}}{2.5 \times 10^3 \text{ Pa}} \times 100 = \boxed{80\%}$$

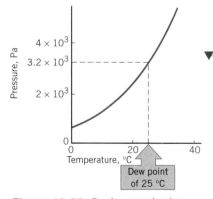

Figure 12.37 On the vaporization curve of water, the dew point is the temperature that corresponds to the actual partial pressure of water vapor in the air.

When air containing a given amount of water vapor is cooled, a temperature is reached in which the partial pressure of the vapor equals the equilibrium vapor pressure. This temperature is known as the **dew point.** For instance, Figure 12.37 shows that if the partial pressure of water vapor is 3.2×10^3 Pa, the dew point is 25 °C. This partial pressure would correspond to a relative humidity of 100%, if the ambient temperature were equal to the dew-point temperature. Hence, the dew point is the temperature below which water vapor in the air condenses in the form of liquid drops (dew or fog). The closer the actual temperature is to the dew point, the closer the relative humidity is to 100%. Thus,

The physics of fog formation.

for fog to form, the air temperature must drop below the dew point. Similarly, water condenses on the outside of a cold glass when the temperature of the air next to the glass falls below the dew point. And the cold coils in a home dehumidifier (see Figure 12.38) function very much in the same way that the cold glass does. The coils are kept cold by a circulating refrigerant. When the air blown across them by the fan cools below the dew point, water vapor condenses in the form of droplets, which collect in a receptacle.

The physics of a home dehumidifier.

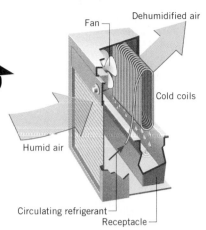

Self-Assessment Test 12.2

Test your understanding of the material in Sections 12.6–12.10:

- Heat and Specific Heat Capacity   • Latent Heat
- Phase Equilibrium and Humidity

Go to **www.wiley.com/college/cutnell**

12.11 Concepts & Calculations

This section contains examples that discuss one or more conceptual questions, followed by a related quantitative problem. Example 17 provides insight on the variables involved when the length and volume of an object change due to a temperature change. Example 18 discusses how different factors affect the temperature change of an object to which heat is being added.

Figure 12.38 The cold coils of a dehumidifier cool the air blowing across them below the dew point, and water vapor condenses out of the air.

Concepts & Calculations Example 17
Linear and Volume Thermal Expansion

Figure 12.39 shows three rectangular blocks made from the same material. The initial dimensions of each are expressed as multiples of D, where $D = 2.00$ cm. They are heated and their temperatures increase by 35.0 C°. The coefficients of linear and volume expansion are $\alpha = 1.50 \times 10^{-5}$ (C°)$^{-1}$ and $\beta = 4.50 \times 10^{-5}$ (C°)$^{-1}$, respectively. Determine the change in their (a) vertical heights and (b) volumes.

Concept Questions and Answers Does the change in the vertical height of a block depend only on its height, or does it also depend on its width and depth? Without doing any calculations, rank the blocks according to their change in height, largest first.

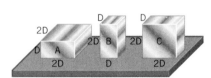

Answer According to Equation 12.2, $\Delta L = \alpha L_0 \Delta T$, the change ΔL in the height of an object depends on its original height L_0, the change in temperature ΔT, and the coefficient of linear expansion, α. It does not depend on the width or depth of the object. Since blocks B and C have twice the height of A, their heights will increase by twice as much as that of A. The heights of B and C, however, will increase by the same amount, even though C is twice as wide.

Does the change in the volume of a block depend only on its height, or does it also depend on its width and depth? Without doing any calculations, rank the blocks according to their greatest change in volume, largest first.

Figure 12.39 The temperatures of the three blocks are raised by the same amount. Which one(s) experience the greatest change in height and which the greatest change in volume?

Answer According to Equation 12.3, $\Delta V = \beta V_0 \Delta T$, the change ΔV in the volume of an object depends on its original volume V_0, the change in temperature ΔT, and the coefficient of volume expansion, β. Thus, the change in volume depends on height, width, and depth, because the original volume is the product of these three dimensions. The initial volumes of A, B, and C, are, respectively, $D \times 2D \times 2D = 4D^3$, $2D \times D \times D = 2D^3$, and $2D \times 2D \times D = 4D^3$. Blocks A and C have equal volumes, which are greater than that of B. Thus, we expect A and C to exhibit the greatest increase in volume, while B exhibits the smallest increase.

Solution

(a) The change in the height of each block is given by Equation 12.2 as

$$\Delta L_A = \alpha D\, \Delta T = [1.50 \times 10^{-5}\,(\text{C}°)^{-1}](2.00 \text{ cm})(35.0 \text{ C}°) = \boxed{1.05 \times 10^{-3} \text{ cm}}$$

$$\Delta L_B = \alpha(2D)\Delta T = [1.50 \times 10^{-5}\,(\text{C}°)^{-1}](2 \times 2.00 \text{ cm})(35.0 \text{ C}°) = \boxed{2.10 \times 10^{-3} \text{ cm}}$$

$$\Delta L_C = \alpha(2D)\Delta T = [1.50 \times 10^{-5}\,(\text{C}°)^{-1}](2 \times 2.00 \text{ cm})(35.0 \text{ C}°) = \boxed{2.10 \times 10^{-3} \text{ cm}}$$

As expected, the heights of B and C increase more than that of A.

(b) The change in the volume of each block is given by Equation 12.3 as

$$\Delta V_A = \beta(D \times 2D \times 2D)\Delta T$$
$$= [4.50 \times 10^{-5}\,(\text{C}°)^{-1}][4(2.00\text{ cm})^3](35.0\text{ C}°) = \boxed{5.04 \times 10^{-2}\text{ cm}^3}$$

$$\Delta V_B = \beta(2D \times D \times D)\Delta T$$
$$= [4.50 \times 10^{-5}\,(\text{C}°)^{-1}][2(2.00\text{ cm})^3](35.0\text{ C}°) = \boxed{2.52 \times 10^{-2}\text{ cm}^3}$$

$$\Delta V_C = \beta(2D \times 2D \times D)\Delta T$$
$$= [4.50 \times 10^{-5}\,(\text{C}°)^{-1}][4(2.00\text{ cm})^3](35.0\text{ C}°) = \boxed{5.04 \times 10^{-2}\text{ cm}^3}$$

As discussed earlier, the greatest change in volume occurs with A and C followed by B.

Concepts & Calculations Example 18
Heat and Temperature Changes

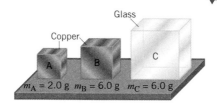

Figure 12.40 Which block has the greatest change in temperature when the same heat is supplied to each?

Objects A and B in Figure 12.40 are made from copper, but the mass of B is three times that of A. Object C is made from glass and has the same mass as B. The *same amount* of heat Q is supplied to each one: $Q = 14$ J. Determine the rise in temperature for each.

Concept Questions and Answers Which object, A or B, experiences the greater rise in temperature?

Answer Consider an extreme example. Suppose a cup and a swimming pool are filled with water. For the same heat input, would you intuitively expect the temperature of the cup to rise more than that of the pool? Yes, because the cup has less mass. Another way to arrive at this conclusion is to solve Equation 12.4 for the change in temperature: $\Delta T = Q/(cm)$. Since the heat Q and the specific heat capacity c are the same for A and B, ΔT is inversely proportional to the mass m. So smaller masses give rise to larger temperature changes. Therefore, A has a greater temperature change than B.

Which object, B or C, experiences the greater rise in temperature?

Answer Objects B and C are made from different materials. To see how the rise in temperature depends on the type of material, let's again use Equation 12.4: $\Delta T = Q/(cm)$. Since the heat Q and the mass m are the same for B and C, ΔT is inversely proportional to the specific heat capacity c. So smaller specific heat capacities give rise to larger temperature changes. Table 12.2 indicates that the specific heat capacities of copper and glass are 387 and 840 J/(kg·C°), respectively. Since B is made from copper, which has the smaller specific heat capacity, it has the greater temperature change.

Solution According to Equation 12.4, the temperature change for each object is

$$\Delta T_A = \frac{Q}{c_A m_A} = \frac{14\text{ J}}{[387\text{ J/(kg·C°)}](2.0 \times 10^{-3}\text{ kg})} = \boxed{18\text{ C}°}$$

$$\Delta T_B = \frac{Q}{c_B m_B} = \frac{14\text{ J}}{[387\text{ J/(kg·C°)}](6.0 \times 10^{-3}\text{ kg})} = \boxed{6.0\text{ C}°}$$

$$\Delta T_C = \frac{Q}{c_C m_C} = \frac{14\text{ J}}{[840\text{ J/(kg·C°)}](6.0 \times 10^{-3}\text{ kg})} = \boxed{2.8\text{ C}°}$$

As anticipated, ΔT_A is greater than ΔT_B, and ΔT_B is greater than ΔT_C.

At the end of the problem set for this chapter, you will find homework problems that contain both conceptual and quantitative parts. These problems are grouped under the heading *Concepts & Calculations, Group Learning Problems*. They are designed for use by students working alone or in small learning groups. The conceptual part of each problem provides a convenient focus for group discussions.

Concept Summary

This summary presents an abridged version of the chapter, including the important equations and all available learning aids. For convenient reference, the learning aids (including the text's examples) are placed next to or immediately after the relevant equation or discussion. The following learning aids may be found on-line at **www.wiley.com/college/cutnell**:

Interactive LearningWare examples are solved according to a five-step interactive format that is designed to help you develop problem-solving skills.	**Concept Simulations** are animated versions of text figures or animations that illustrate important concepts. You can control parameters that affect the display, and we encourage you to experiment.
Interactive Solutions offer specific models for certain types of problems in the chapter homework. The calculations are carried out interactively.	**Self-Assessment Tests** include both qualitative and quantitative questions. Extensive feedback is provided for both incorrect and correct answers, to help you evaluate your understanding of the material.

Topic	*Discussion*	*Learning Aids*
	12.1 Common Temperature Scales	
Celsius temperature scale **Fahrenheit temperature scale**	On the Celsius temperature scale, there are 100 equal divisions between the ice point (0 °C) and the steam point (100 °C). On the Fahrenheit temperature scale, there are 180 equal divisions between the ice point (32 °F) and the steam point (212 °F).	**Examples 1, 2**
	12.2 The Kelvin Temperature Scale	
Kelvin temperature scale	For scientific work, the Kelvin temperature scale is the scale of choice. One kelvin (K) is equal in size to one Celsius degree. However, the temperature T on the Kelvin scale differs from the temperature T_c on the Celsius scale by an additive constant of 273.15:	
	$$T = T_c + 273.15 \qquad (12.1)$$	
Absolute zero	The lower limit of temperature is called absolute zero and is designated as 0 K on the Kelvin scale.	
	12.3 Thermometers	
Thermometric property	The operation of any thermometer is based on the change in some physical property with temperature; this physical property is called a thermometric property. Examples of thermometric properties are the length of a column of mercury, electrical voltage, and electrical resistance.	
	12.4 Linear Thermal Expansion	
	Most substances expand when heated. For linear expansion, an object of length L_0 experiences a change ΔL in length when the temperature changes by ΔT:	
Linear thermal expansion	$$\Delta L = \alpha L_0 \Delta T \qquad (12.2)$$	**Example 3**
	where α is the coefficient of linear expansion.	
Thermal stress	For an object held rigidly in place, a thermal stress can occur when the object attempts to expand or contract. The stress can be large, even for small temperature changes.	**Example 4**
How a hole in a plate expands or contracts	When the temperature changes, a hole in a plate of solid material expands or contracts as if the hole were filled with the surrounding material.	**Examples 5, 6, 7**
	12.5 Volume Thermal Expansion	
	For volume expansion, the change ΔV in the volume of an object of volume V_0 is given by	
Volume thermal expansion	$$\Delta V = \beta V_0 \Delta T \qquad (12.3)$$	**Examples 8, 17**
	where β is the coefficient of volume expansion.	**Interactive LearningWare 12.1**
How a cavity expands or contracts	When the temperature changes, a cavity in a piece of solid material expands or contracts as if the cavity were filled with the surrounding material.	Interactive Solution 12.27

Use Self-Assessment Test 12.1 to evaluate your understanding of Sections 12.1–12.5.

Topic	Discussion	Learning Aids
	12.6 Heat and Internal Energy	
Internal energy Heat	The internal energy of a substance is the sum of the kinetic, potential, and other kinds of energy that the molecules of the substance have. Heat is energy that flows from a higher-temperature object to a lower-temperature object because of the difference in temperatures. The SI unit for heat is the joule (J).	
	12.7 Heat and Temperature Change: Specific Heat Capacity	
	The amount of heat Q that must be supplied or removed to change the temperature of a substance of mass m by an amount ΔT is	
Heat needed to change the temperature	$$Q = cm\Delta T \qquad (12.4)$$ where c is a constant known as the specific heat capacity.	**Examples 9, 10, 11, 18** **Interactive Solutions 12.43, 12.89**
Energy conservation and heat	When materials are placed in thermal contact within a perfectly insulated container, the principle of energy conservation requires that heat lost by warmer materials equals heat gained by cooler materials.	
	Heat is sometimes measured with a unit called the kilocalorie (kcal). The conversion factor between kilocalories and joules is known as the mechanical equivalent of heat:	
Mechanical equivalent of heat	$$1 \text{ kcal} = 4186 \text{ joules}$$	**Interactive LearningWare 12.2**
	12.8 Heat and Phase Change: Latent Heat	
	Heat must be supplied or removed to make a material change from one phase to another. The heat Q that must be supplied or removed to change the phase of a mass m of a substance is	
Heat needed to change the phase	$$Q = mL \qquad (12.5)$$ where L is the latent heat of the substance and has SI units of J/kg. The latent heats of fusion, vaporization, and sublimation refer, respectively, to the solid/liquid, the liquid/vapor, and the solid/vapor phase changes.	**Examples 12, 13, 14** **Interactive Solution 12.63**
	12.9 Equilibrium Between Phases of Matter	
	The equilibrium vapor pressure of a substance is the pressure of the vapor phase that is in equilibrium with the liquid phase. For a given substance, vapor pressure depends only on temperature. For a liquid, a plot of the equilibrium vapor	**Example 15**
Vapor pressure curve	pressure versus temperature is called the vapor pressure or vaporization curve.	
Fusion curve	The fusion curve gives the combinations of temperature and pressure for equilibrium between solid and liquid phases.	
	12.10 Humidity	
	The relative humidity is defined as follows:	
Relative humidity	$$\begin{matrix}\text{Percent} \\ \text{relative} \\ \text{humidity}\end{matrix} = \frac{\begin{matrix}\text{Partial pressure} \\ \text{of water vapor}\end{matrix}}{\begin{matrix}\text{Equilibrium vapor pressure of} \\ \text{water at the existing temperature}\end{matrix}} \times 100 \qquad (12.6)$$	**Example 16**
Dew point	The dew point is the temperature below which the water vapor in the air condenses. On the vaporization curve of water, the dew point is the temperature that corresponds to the actual pressure of water vapor in the air.	

Use Self-Assessment Test 12.2 to evaluate your understanding of Sections 12.6–12.10.

Conceptual Questions

1. For the highest accuracy, would you choose an aluminum or a steel tape rule for year-round outdoor use? Why?

2. The first international standard of length was a metal bar kept at the International Bureau of Weights and Measures. One meter of length was defined to be the distance between two fine lines engraved near the ends of the bar. Why was it important that the bar be kept at a constant temperature?

3. Conceptual Example 7 provides background for this question. A circular hole is cut through a flat aluminum plate. A spherical brass ball has a diameter that is slightly *smaller* than the diameter of the hole. The plate and the ball have the same temperature at all times. Should the plate and ball both be heated or both be cooled to *prevent* the ball from falling through the hole? Give your reasoning.

4. For added strength, many highways and buildings are constructed with reinforced concrete (concrete that is reinforced with embedded steel rods). Table 12.1 shows that the coefficient of linear expansion for concrete is the same as that for steel. Why is it important that these two coefficients be the same?

5. At a certain temperature, a rod is hung from an aluminum frame, as the drawing shows. A small gap exists between the rod and the floor. The frame and rod are heated uniformly. Explain whether the rod will ever touch the floor, assuming that the rod is made from (a) aluminum and (b) lead.

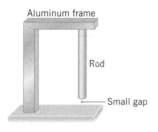

Aluminum frame

Rod

Small gap

6. A simple pendulum is made using a long thin metal wire. When the temperature drops, does the period of the pendulum increase, decrease, or remain the same? Account for your answer.

7. One type of cooking pot is made from stainless steel and has a copper coating over the outside of the bottom. At room temperature the bottom of this pot is flat, so that the pot sits flat on a kitchen counter top. When heated, however, this pot can be made to rock back and forth on the counter top with a gentle touch, because the bottom of the pot is bowed outward. With the aid of Table 12.1 account for the outward bowing.

8. A hot steel ring fits snugly over a cold brass cylinder. The temperatures of the ring and cylinder are, respectively, above and below room temperature. Account for the fact that it is nearly impossible to pull the ring off the cylinder once the assembly has reached room temperature.

9. For glass baking dishes, Pyrex glass is used instead of common glass. A cold Pyrex dish taken from the refrigerator can be put directly into a hot oven without cracking from thermal stress. A dish made from common glass would crack. With the aid of Table 12.1, explain why Pyrex is better in this respect than common glass.

10. Suppose liquid mercury and glass both had the same coefficient of volume expansion. Explain why such a mercury-in-glass thermometer would not work.

11. Is the buoyant force provided by warm water (above 4 °C) greater than, less than, or equal to that provided by cold water (also above 4 °C)? Explain your reasoning.

12. When the bulb of a mercury-in-glass thermometer is inserted into boiling water, the mercury column first drops slightly before it begins to rise. Account for this phenomenon. *(Hint: Consider what happens initially to the glass.)*

13. Two different objects are supplied with equal amounts of heat. Give the reason(s) why their temperature changes would not necessarily be the same.

14. Two objects are made from the same material. They have different masses and temperatures. If the two are placed in contact, which object will experience the greater temperature change? Explain.

15. Two identical mugs contain hot coffee from the same pot. One mug is full, while the other is only one-quarter full. Sitting on the kitchen table, which mug stays warmer longer? Explain.

16. Bicyclists often carry along a water bottle on hot days. When the bottle is wrapped in a wet sock, the water temperature stays near 75 °F for three or four hours, even when the air temperature is near 100 °F. Why does the water stay cool?

17. To help lower the high temperature of a sick patient, an alcohol rub is sometimes used. Isopropyl alcohol is rubbed over the patient's back, arms, legs, etc., and allowed to evaporate. Why does the procedure work?

18. Your head feels colder under an air-conditioning vent when your hair is wet than when it is dry. Why?

19. Suppose the latent heat of vaporization of H_2O were one-tenth its actual value. (a) Other things being equal, would it take the same time, a shorter time, or a longer time for a pot of water on a stove to boil away? (b) Would the evaporative cooling mechanism of the human body be as effective? Account for both answers.

20. Fruit blossoms are permanently damaged when the temperature drops below about −4 °C (a "hard freeze"). Orchard owners sometimes spray a film of water over the blossoms to protect them when a hard freeze is expected. From the point of view of phase changes, give a reason for the protection.

21. A camping stove is used to boil water on a mountain. Does it necessarily follow that the same stove can boil water at lower altitudes, such as at sea level? Provide a reason for your answer.

22. If a bowl of water is placed in a closed container and water vapor is pumped away rapidly enough, the remaining liquid will turn to ice. Explain why the ice appears.

23. Medical instruments are sterilized under the hottest possible temperatures. Explain why they are sterilized in an autoclave, which is a device that is essentially a pressure cooker and heats the instruments in water under a pressure greater than one atmosphere.

24. A bottle of carbonated soda is left outside in subfreezing temperatures, although it remains in the liquid form. When the soda is brought inside and opened, it immediately freezes. Explain why this could happen.

25. A bowl of water is covered tightly and allowed to sit at a constant temperature of 23 °C for a long time. What is the relative humidity in the space between the surface of the water and the cover? Justify your answer.

26. Is it possible for dew to form on Tuesday night and not on Monday night, even though Monday night is the cooler night? Incorporate the idea of the dew point into your answer.

27. Two rooms in a house have the same temperature. One of the rooms contains an indoor swimming pool. On a cold day the windows of one of the rooms are "steamed up." Which room is it? Explain.

28. A jar is half filled with boiling water. The lid is then screwed on the jar. After the jar has cooled to room temperature, the lid is difficult to remove. Why?

Problems

Note: For problems in this set, use the values of α and β given in Table 12.1, and the values of c, L_f, and L_v given in Tables 12.2 and 12.3, unless stated otherwise.

Section 12.1 Common Temperature Scales, Section 12.2 The Kelvin Temperature Scale, Section 12.3 Thermometers

1. ssm What's your normal body temperature? It may not be 98.6 °F, the oft-quoted average that was determined in the nineteenth century. A more recent study has reported an average temperature of 98.2 °F. What is the *difference* between these averages, expressed in Celsius degrees?

2. On the moon the surface temperature ranges from 375 K during the day to 1.00×10^2 K at night. What are these temperatures on the (a) Celsius and (b) Fahrenheit scales?

3. A personal computer is designed to operate over the temperature range from 50.0 to 104 °F. To what do these temperatures correspond (a) on the Celsius scale and (b) on the Kelvin scale?

4. Dermatologists often remove small precancerous skin lesions by freezing them quickly with liquid nitrogen, which has a temperature of 77 K. What is this temperature on the (a) Celsius and (b) Fahrenheit scales?

5. ssm A temperature of absolute zero occurs at −273.15 °C. What is this temperature on the Fahrenheit scale?

* 6. A copper–constantan thermocouple generates a voltage of 4.75×10^{-3} volts when the temperature of the hot junction is 110.0 °C and the reference junction is kept at 0.0 °C. If the voltage is proportional to the difference in temperature between the junctions, what is the temperature of the hot junction when the voltage is 1.90×10^{-3} volts?

* 7. ssm A constant-volume gas thermometer (see Figures 12.3 and 12.4) has a pressure of 5.00×10^3 Pa when the gas temperature is 0.00 °C. What is the temperature (in °C) when the pressure is 2.00×10^3 Pa?

* 8. Space invaders land on earth. On the invaders' temperature scale, the ice point is at 25 °I (I = invader), and the steam point is at 156 °I. The invaders' thermometer shows the temperature on earth to be 58 °I. Using logic similar to that in Example 1 in the text, what would this temperature be on the Celsius scale?

Section 12.4 Linear Thermal Expansion

9. A steel aircraft carrier is 370 m long when moving through the icy North Atlantic at a temperature of 2.0 °C. By how much does the carrier lengthen when it is traveling in the warm Mediterranean Sea at a temperature of 21 °C?

10. The Concorde is 62 m long when its temperature is 23 °C. In flight, the outer skin of this supersonic aircraft can reach 105 °C due to air friction. The coefficient of linear expansion of the skin is 2.0×10^{-5} (C °)$^{-1}$. Find the amount by which the Concorde expands.

11. ssm www Find the approximate length of the Golden Gate bridge if it is known that the steel in the roadbed expands by 0.53 m when the temperature changes from +2 to +32 °C.

12. Conceptual Example 5 provides background for this problem. A hole is drilled through a copper plate whose temperature is 11 °C. (a) When the temperature of the plate is increased, will the radius of the hole be larger or smaller than the radius at 11 °C? Why? (b) When the plate is heated to 110 °C, by what fraction $\Delta r/r_0$ will the radius of the hole change?

13. A commonly used method of fastening one part to another part is called "shrink fitting." A steel rod has a diameter of 2.0026 cm, and a flat plate contains a hole whose diameter is 2.0000 cm. The rod is cooled so that it just fits into the hole. When the rod warms up, the enormous thermal stress exerted by the plate holds the rod securely to the plate. By how many Celsius degrees should the rod be cooled?

14. A thin rod consists of two parts joined together. One-third of it is silver and two-thirds is gold. The temperature decreases by 26 C°. Determine the fractional decrease $\dfrac{\Delta L}{L_{0,\,\text{Silver}} + L_{0,\,\text{Gold}}}$ in the rod's length, where $L_{0,\,\text{Silver}}$ and $L_{0,\,\text{Gold}}$ are the initial lengths of the silver and gold rods.

15. ssm A rod made from a particular alloy is heated from 25.0 °C to the boiling point of water. Its length increases by 8.47×10^{-4} m. The rod is then cooled from 25.0 °C to the freezing point of water. By how much does the rod shrink?

* 16. As the drawing shows, two thin strips of metal are bolted together at one end and 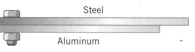 have the same temperature. One is steel, and the other is aluminum. The steel strip is 0.10% longer than the aluminum strip. By how much should the temperature of the strips be increased, so that the strips have the same length?

* 17. The brass bar and the aluminum bar in the drawing are each attached to an immovable wall. At 28 °C the air gap between the rods is 1.3×10^{-3} m. At what temperature will the gap be closed?

* 18. Consult Conceptual Example 5 for background pertinent to this problem. A lead sphere has a diameter that is 0.050% larger than the inner diameter of a steel ring when each has a temperature of 70.0 °C. Thus, the ring will not slip over the sphere. At what common temperature will the ring just slip over the sphere?

* 19. ssm www A simple pendulum consists of a ball connected to one end of a thin brass wire. The period of the pendulum is 2.0000 s. The temperature rises by 140 C°, and the length of the wire increases. Determine the period of the heated pendulum.

* 20. Concrete sidewalks are always laid in sections, with gaps between each section. For example, the drawing shows three identical 2.4-m sections, the outer two of which are against immovable walls.

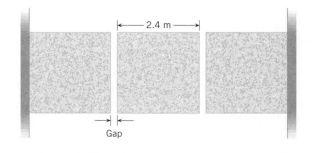

Gap

The two identical gaps between the sections are provided so that thermal expansion will not create the thermal stress that could lead to cracks. What is the minimum gap width necessary to account for an increase in temperature of 32 C°?

* **21.** A steel ruler is accurate when the temperature is 25 °C. When the temperature drops to −15 °C, the ruler no longer reads correctly, but it can be made to read correctly if a stress is applied to each end of the ruler. (a) Should the stress be a compression or a tension? Why? (b) What is the magnitude of the necessary stress?

** **22.** A steel ruler is calibrated to read true at 20.0 °C. A draftsman uses the ruler at 40.0 °C to draw a line on a 40.0 °C copper plate. As indicated on the warm ruler, the length of the line is 0.50 m. To what temperature should the plate be cooled, such that the length of the line truly becomes 0.50 m?

** **23. ssm** A wire is made by attaching two segments together, end to end. One segment is made of aluminum and the other is steel. The effective coefficient of linear expansion of the two-segment wire is $19 \times 10^{-6} \, (C°)^{-1}$. What fraction of the length is aluminum?

** **24.** An 85.0-N backpack is hung from the middle of an aluminum wire, as the drawing shows. The temperature of the wire then drops by 20.0 C°. Find the tension in the wire at the lower temperature. Assume that the distance between the supports does not change, and ignore any thermal stress.

Section 12.5 Volume Thermal Expansion

25. A copper kettle contains water at 24 °C. When the water is heated to its boiling point, the volume of the kettle expands by $1.2 \times 10^{-3} \, m^3$. Determine the volume of the kettle at 24 °C.

26. A swimming pool contains 110 m³ of water. The sun heats the water from 17 to 27 °C. What is the change in the volume of the water?

27. Interactive Solution 12.27 at **www.wiley.com/college/cutnell** presents a model for solving problems of this type. A thin spherical shell of silver has an inner radius of 2.0×10^{-2} m when the temperature is 18 °C. The shell is heated to 147 °C. Find the change in the interior volume of the shell.

28. A lead object and a quartz object each have the same initial volume. The volume of each increases by the same amount, because the temperature increases. If the temperature of the lead object increases by 4.0 C°, by how much does the temperature of the quartz object increase?

29. ssm When heated at constant pressure, a given volume of gas expands much more than an equal volume of liquid. The coefficient of volume expansion for air is about $3.7 \times 10^{-3} \, (C°)^{-1}$ at room temperature and one atmosphere of pressure. Find the ratio of the change in volume of air to the change in volume of water, assuming the same initial volumes and temperature changes.

30. Interactive LearningWare 12.1 at **www.wiley.com/college/cutnell** provides some useful background for this problem. Many hot-water heating systems have a reservoir tank connected directly to the pipeline, so as to allow for expansion when the water becomes hot. The heating system of a house has 76 m of copper pipe whose inside

radius is 9.5×10^{-3} m. When the water and pipe are heated from 24 to 78 °C, what must be the minimum volume of the reservoir tank to hold the overflow of water?

31. ssm Suppose that the steel gas tank in your car is completely filled when the temperature is 17 °C. How many gallons will spill out of the twenty-gallon tank when the temperature rises to 35 °C?

32. Consult **Interactive LearningWare 12.1** at **www.wiley.com/college/cutnell** for help in solving this problem. During an all-night cram session, a student heats up a one-half liter ($0.50 \times 10^{-3} \, m^3$) glass (Pyrex) beaker of cold coffee. Initially, the temperature is 18 °C, and the beaker is filled to the brim. A short time later when the student returns, the temperature has risen to 92 °C. The coefficient of volume expansion of coffee is the same as that of water. How much coffee (in cubic meters) has spilled out of the beaker?

* **33.** At the bottom of an old mercury-in-glass thermometer is a 45-mm³ reservoir filled with mercury. When the thermometer was placed under your tongue, the warmed mercury would expand into a very narrow cylindrical channel, called a capillary, whose radius was 1.7×10^{-2} mm. Marks were placed along the capillary that indicated the temperature. Ignore the thermal expansion of the glass and determine how far (in mm) the mercury would expand into the capillary when the temperature changed by 1.0 C°.

* **34.** A spherical brass shell has an interior volume of $1.60 \times 10^{-3} \, m^3$. Within this interior volume is a solid steel ball that has a volume of $0.70 \times 10^{-3} \, m^3$. The space between the steel ball and the inner surface of the brass shell is filled completely with mercury. A small hole is drilled through the brass, and the temperature of the arrangement is increased by 12 C°. What is the volume of the mercury that spills out of the hole?

* **35. ssm** The bulk modulus of water is $B = 2.2 \times 10^9 \, N/m^2$. What change in pressure ΔP (in atmospheres) is required to keep water from expanding when it is heated from 15 to 25 °C?

* **36.** A solid aluminum sphere has a radius of 0.50 m and a temperature of 75 °C. The sphere is then completely immersed in a pool of water whose temperature is 25 °C. The sphere cools, while the water temperature remains nearly at 25 °C, because the pool is very large. The sphere is weighed in the water immediately after being submerged (before it begins to cool) and then again after cooling to 25 °C. (a) Which weight is larger? Why? (b) Use Archimedes' principle to find the magnitude of the *difference* between the weights.

** **37. ssm www** Two identical thermometers made of Pyrex glass contain, respectively, identical volumes of mercury and methyl alcohol. If the expansion of the glass is taken into account, how many times greater is the distance between the degree marks on the methyl alcohol thermometer than that on the mercury thermometer?

** **38.** The column of mercury in a barometer (see Figure 11.12) has a height of 0.760 m when the pressure is one atmosphere and the temperature is 0.0 °C. Ignoring any change in the glass containing the mercury, what will be the height of the mercury column for the same one atmosphere of pressure when the temperature rises to 38.0 °C on a hot day? (*Hint: The pressure in the barometer is given by* Pressure = $\rho g h$, *and the density ρ of the mercury changes when the temperature changes.*)

Section 12.6 Heat and Internal Energy, Section 12.7 Heat and Temperature Change: Specific Heat Capacity

39. Blood can carry excess energy from the interior to the surface of the body, where the energy is dispersed in a number of ways. While a person is exercising, 0.6 kg of blood flows to the surface of the body and releases 2000 J of energy. The blood arriving at the surface has the temperature of the body interior, 37.0 °C.

Assuming that blood has the same specific heat capacity as water, determine the temperature of the blood that leaves the surface and returns to the interior.

40. A piece of glass has a temperature of 83.0 °C. Liquid that has a temperature of 43.0 °C is poured over the glass, completely covering it, and the temperature at equilibrium is 53.0 °C. The mass of the glass and the liquid is the same. Ignoring the container that holds the glass and liquid and assuming that the heat lost to or gained from the surroundings is negligible, determine the specific heat capacity of the liquid.

41. ssm If the price of electrical energy is $0.10 per kilowatt · hour, what is the cost of using electrical energy to heat the water in a swimming pool (12.0 m × 9.00 m × 1.5 m) from 15 to 27 °C?

42. Ideally, when a thermometer is used to measure the temperature of an object, the temperature of the object itself should not change. However, if a significant amount of heat flows from the object to the thermometer, the temperature will change. A thermometer has a mass of 31.0 g, a specific heat capacity of $c = 815$ J/(kg · C°), and a temperature of 12.0 °C. It is immersed in 119 g of water, and the final temperature of the water and thermometer is 41.5 °C. What was the temperature of the water before the insertion of the thermometer?

43. Review **Interactive Solution 12.43** at **www.wiley.com/ college/cutnell** for help in approaching this problem. When resting, a person has a metabolic rate of about 3.0×10^5 joules per hour. The person is submerged neck-deep into a tub containing 1.2×10^3 kg of water at 21.00 °C. If the heat from the person goes only into the water, find the water temperature after half an hour.

44. When you take a bath, how many kilograms of hot water (49.0 °C) must you mix with cold water (13.0 °C) so that the temperature of the bath is 36.0 °C? The total mass of water (hot plus cold) is 191 kg. Ignore any heat flow between the water and its external surroundings.

45. ssm At a fabrication plant, a hot metal forging has a mass of 75 kg and a specific heat capacity of 430 J/(kg · C°). To harden it, the forging is immersed in 710 kg of oil that has a temperature of 32 °C and a specific heat capacity of 2700 J/(kg · C°). The final temperature of the oil and forging at thermal equilibrium is 47 °C. Assuming that heat flows only between the forging and the oil, determine the initial temperature of the forging.

* **46.** Refer to **Interactive LearningWare 12.2** at **www.wiley.com/ college/cutnell** for a review of the concepts that play roles in this problem. The box of a well-known breakfast cereal states that one ounce of the cereal contains 110 Calories (1 food Calorie = 4186 J). If 2.0% of this energy could be converted by a weight lifter's body into work done in lifting a barbell, what is the heaviest barbell that could be lifted a distance of 2.1 m?

* **47.** In a passive solar house, the sun heats water stored in barrels to a temperature of 38 °C. The stored energy is then used to heat the house on cloudy days. Suppose that 2.4×10^8 J of heat is needed to maintain the inside of the house at 21 °C. How many barrels (1 barrel = 0.16 m³) of water are needed?

* **48.** The water in a swimming pool absorbs 2.00×10^9 J of heat from the sun. What is the change in volume of the water?

* **49. ssm** A rock of mass 0.20 kg falls from rest from a height of 15 m into a pail containing 0.35 kg of water. The rock and water have the same initial temperature. The specific heat capacity of the rock is 1840 J/(kg · C°). Ignore the heat absorbed by the pail itself, and determine the rise in the temperature of the rock and water.

** **50.** A steel rod ($\rho = 7860$ kg/m³) has a length of 2.0 m. It is bolted at both ends between immobile supports. Initially there is no tension in the rod, because the rod just fits between the supports. Find the tension that develops when the rod loses 3300 J of heat.

Section 12.8 Heat and Phase Change: Latent Heat

51. ssm How much heat must be added to 0.45 kg of aluminum to change it from a solid at 130 °C to a liquid at 660 °C (its melting point)? The latent heat of fusion for aluminum is 4.0×10^5 J/kg.

52. Assume that the pressure is one atmosphere and determine the heat required to produce 2.00 kg of water vapor at 100.0 °C, starting with (a) 2.00 kg of water at 100.0 °C and (b) 2.00 kg of liquid water at 0.0 °C.

53. Suppose that the amount of heat removed when 3.0 kg of water freezes at 0 °C were removed from ethyl alcohol at its freezing/melting point of −114 °C. How many kilograms of ethyl alcohol would freeze?

54. A person eats a container of yogurt. The Nutritional Facts label states that it contains 240 Calories (1 Calorie = 4186 J). What mass of perspiration would one have to lose to get rid of this energy? At body temperature, the latent heat of vaporization of water is 2.42×10^6 J/kg.

55. ssm Find the mass of water that vaporizes when 2.10 kg of mercury at 205 °C is added to 0.110 kg of water at 80.0 °C.

56. The latent heat of vaporization of H_2O at body temperature (37.0 °C) is 2.42×10^6 J/kg. To cool the body of a 75-kg jogger [average specific heat capacity = 3500 J/(kg · C°)] by 1.5 C°, how many kilograms of water in the form of sweat have to be evaporated?

57. A woman finds the front windshield of her car covered with ice at −12.0 °C. The ice has a thickness of 4.50×10^{-4} m, and the windshield has an area of 1.25 m². The density of ice is 917 kg/m³. How much heat is required to melt the ice?

58. A 0.200-kg piece of aluminum that has a temperature of −155 °C is added to 1.5 kg of water that has a temperature of 3.0 °C. At equilibrium the temperature is 0.0 °C. Ignoring the container and assuming that the heat exchanged with the surroundings is negligible, determine the mass of water that has been frozen into ice.

* **59.** Occasionally, huge icebergs are found floating on the ocean's currents. Suppose one such iceberg is 120 km long, 35 km wide, and 230 m thick. (a) How much heat would be required to melt this iceberg (assumed to be at 0 °C) into liquid water at 0 °C? The density of ice is 917 kg/m³. (b) The annual energy consumption by the United States in 1994 was 9.3×10^{19} J. If this energy were delivered to the iceberg every year, how many years would it take before the ice melted?

* **60.** To help keep his barn warm on cold days, a farmer stores 840 kg of solar-heated water ($L_f = 3.35 \times 10^5$ J/kg) in barrels. For how many hours would a 2.0-kW electric space heater have to operate to provide the same amount of heat as the water does when it cools from 10.0 to 0.0 °C and completely freezes?

* **61. ssm www** An unknown material has a normal melting/ freezing point of −25.0 °C, and the liquid phase has a specific heat capacity of 160 J/(kg · C°). One-tenth of a kilogram of the solid at −25.0 °C is put into a 0.150-kg aluminum calorimeter cup that contains 0.100 kg of glycerin. The temperature of the cup and the glycerin is initially 27.0 °C. All the unknown material

melts, and the final temperature at equilibrium is 20.0 °C. The calorimeter neither loses energy to nor gains energy from the external environment. What is the latent heat of fusion of the unknown material?

* **62.** Two grams of liquid water are at 0 °C, and another two grams are at 100 °C. Heat is removed from the water at 0 °C, completely freezing it at 0 °C. This heat is then used to vaporize some of the water at 100 °C. What is the mass (in grams) of the liquid water that remains?

* **63. Interactive Solution 12.63** at **www.wiley.com/college/cutnell** provides a model for solving problems such as this. A 42-kg block of ice at 0 °C is sliding on a horizontal surface. The initial speed of the ice is 7.3 m/s and the final speed is 3.5 m/s. Assume that the part of the block that melts has a very small mass and that all the heat generated by kinetic friction goes into the block of ice, and determine the mass of ice that melts into water at 0 °C.

* **64.** Equal masses of two different liquids have the same temperature of 25.0 °C. Liquid A has a freezing point of −68.0 °C and a specific heat capacity of 1850 J/(kg·C°). Liquid B has a freezing point of −96.0 °C and a specific heat capacity of 2670 J/(kg·C°). The same amount of heat must be removed from each liquid in order to freeze it into a solid at its respective freezing point. Determine the difference $L_{f, A} - L_{f, B}$ between the latent heats of fusion for these liquids.

* **65. ssm** It is claimed that if a lead bullet goes fast enough, it can melt completely when it comes to a halt suddenly, and all its kinetic energy is converted into heat via friction. Find the minimum speed of a lead bullet (initial temperature = 30.0 °C) for such an event to happen.

** **66.** A locomotive wheel is 1.00 m in diameter. A 25.0-kg steel band has a temperature of 20.0 °C and a diameter that is 6.00 × 10⁻⁴ m less than that of the wheel. What is the smallest mass of water vapor at 100 °C that can be condensed on the steel band to heat it, so that it will fit onto the wheel? Do not ignore the water that results from the condensation.

Section 12.9 Equilibrium Between Phases of Matter,
Section 12.10 Humidity

67. ssm Use the vapor pressure curve that accompanies this problem to determine the temperature at which liquid carbon dioxide exists in equilibrium with its vapor phase when the vapor pressure is 3.5×10^6 Pa.

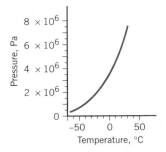

68. At a temperature of 10 °C the percent relative humidity is R_{10}, and at 40 °C it is R_{40}. At each of these temperatures the partial pressure of water vapor in the air is the same. Using the vapor pressure curve for water that accompanies this problem, determine the ratio R_{10}/R_{40} of the two humidity values.

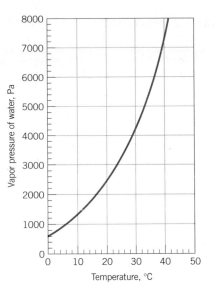

69. ssm What is the relative humidity on a day when the temperature is 30 °C and the dew point is 10 °C? Use the vapor pressure curve that accompanies problem 68.

70. The temperature of 2.0 kg of water is 100.0 °C, but the water is not boiling, because the external pressure acting on the water surface is 3.0 × 10⁵ Pa. Using the vapor pressure curve for water given in Figure 12.32, determine the amount of heat that must be added to the water to bring it to the point where it just begins to boil.

71. Using the vapor pressure curve for water that accompanies problem 68, find the partial pressure of water vapor on a day when the weather forecast gives the relative humidity as 56.0% and the temperature as 30.0 °C.

* **72.** A container is fitted with a movable piston of negligible mass and radius $r = 0.061$ m. Inside the container is liquid water in equilibrium with its vapor, as the drawing shows. The piston remains stationary with a 120-kg block on top of it. The air pressure acting on the top of the piston is one atmosphere. By using the vaporization curve for water in Figure 12.32, find the temperature of the water.

* **73. ssm** At a picnic, a glass contains 0.300 kg of tea at 30.0 °C, which is the air temperature. To make iced tea, someone adds 0.0670 kg of ice at 0.0 °C and stirs the mixture. When all the ice melts and the final temperature is reached, the glass begins to fog up, because water vapor condenses on the outer glass surface. Using the vapor pressure curve for water that accompanies problem 68, ignoring the specific heat capacity of the glass, and treating the tea as if it were water, estimate the relative humidity.

* **74.** The temperature of the air in a room is 36 °C. A person turns on a dehumidifier and notices that when the cooling coils reach 30 °C, water begins to condense on them. What is the relative humidity in the room? Use the vapor pressure curve that accompanies problem 68.

* **75.** A woman has been outdoors where the temperature is 10 °C. She walks into a 25 °C house, and her glasses "steam up." Using the vapor pressure curve for water that accompanies problem 68, find the smallest possible value for the relative humidity of the room.

** **76.** A tall column of water is open to the atmosphere. At a depth of 10.3 m below the surface, the water is boiling. What is the temperature at this depth? Use the vaporization curve for water in Figure 12.32, as needed.

Additional Problems

77. A steel section of the Alaskan pipeline had a length of 65 m and a temperature of 18 °C when it was installed. What is its change in length when the temperature drops to a frigid −45 °C?

78. A 10.0-kg block of ice has a temperature of −10.0 °C. The pressure is one atmosphere. The block absorbs 4.11×10^6 J of heat. What is the final temperature of the liquid water?

79. ssm Liquid nitrogen boils at a chilly −195.8 °C when the pressure is one atmosphere. A silver coin of mass 1.5×10^{-2} kg and temperature 25 °C is dropped into the boiling liquid. What mass of nitrogen boils off as the coin cools to −195.8 °C?

80. A thermos contains 150 cm³ of coffee at 85 °C. To cool the coffee, you drop two 11-g ice cubes into the thermos. The ice cubes are initially at 0 °C and melt completely. What is the final temperature of the coffee? Treat the coffee as if it were water.

81. A steel beam is used in the construction of a skyscraper. By what fraction $\Delta L/L_0$ does the length of the beam increase when the temperature changes from that on a cold winter day (−15 °F) to that on a summer day (+105 °F)?

82. When the temperature of a coin is raised by 75 C°, the coin's diameter increases by 2.3×10^{-5} m. If the original diameter is 1.8×10^{-2} m, find the coefficient of linear expansion.

83. ssm Suppose you are selling apple cider for two dollars a gallon when the temperature is 4.0 °C. The coefficient of volume expansion of the cider is 280×10^{-6} (C°)⁻¹. If the expansion of the container is ignored, how much more money (in pennies) would you make per gallon by refilling the container on a day when the temperature is 26 °C?

84. An ice chest at a beach party contains 12 cans of soda at 5.0 °C. Each can of soda has a mass of 0.35 kg and a specific heat capacity of 3800 J/(kg·C°). Someone adds a 6.5-kg watermelon at 27 °C to the chest. The specific heat capacity of watermelon is nearly the same as that of water. Ignore the specific heat capacity of the chest and determine the final temperature T of the soda and watermelon.

85. The relative humidity is 35% when the temperature is 27 °C. Using the vapor pressure curve for water that accompanies problem 68, determine the dew point.

* **86.** On the Rankine temperature scale, which is sometimes used in engineering applications, the ice point is at 491.67 °R and the steam point is at 671.67 °R. Determine a relationship (analogous to Equation 12.1) between the Rankine and Fahrenheit temperature scales.

* **87. ssm** Ice at −10.0 °C and steam at 130 °C are brought together at atmospheric pressure in a perfectly insulated container. After thermal equilibrium is reached, the liquid phase at 50.0 °C is present. Ig-

noring the container and the equilibrium vapor pressure of the liquid at 50.0 °C, find the ratio of the mass of steam to the mass of ice.

* **88.** When it rains, water vapor in the air condenses into liquid water, and energy is released. (a) How much energy is released when 0.0254 m (one inch) of rain falls over an area of 2.59×10^6 m² (one square mile)? (b) If the average energy needed to heat one home for a year is 1.50×10^{11} J, how many homes can be heated for a year with the energy determined in part (a)?

* **89. Interactive Solution 12.89** at **www.wiley.com/college/cutnell** deals with one approach to solving problems such as this. A 0.35-kg coffee mug is made from a material that has a specific heat capacity of 920 J/(kg·C°) and contains 0.25 kg of water. The cup and water are at 15 °C. To make a cup of coffeee, a small electric heater is immersed in the water and brings it to a boil in three minutes. Assume that the cup and water always have the same temperature and determine the minimum power rating of this heater.

* **90.** A brass spring has an unstrained length of 0.18 m and a spring constant of 1.3×10^4 N/m. The temperature increases from 21 to 135 °C. What is the magnitude of the compressional force that must be applied to the heated spring to bring it back to its original length?

* **91. ssm** An electric hot water heater takes in cold water at 13.0 °C and delivers hot water. The hot water has a constant temperature of 45.0 °C, when the "hot" faucet is left open all the time and the volume flow rate is 5.0×10^{-6} m³/s. What is the minimum power rating of the hot water heater?

** **92.** A steel bicycle wheel (without the rubber tire) is rotating freely with an angular speed of 18.00 rad/s. The temperature of the wheel changes from −100.0 to +300.0 °C. No net external torque acts on the wheel, and the mass of the spokes is negligible. (a) Does the angular speed increase or decrease as the wheel heats up? Why? (b) What is the angular speed at the higher temperature?

** **93.** A 1.5-kg steel sphere will not fit through a circular hole in a 0.85-kg aluminum plate, because the radius of the sphere is 0.10% larger than the radius of the hole. If both the sphere and the plate are always kept at the same temperature, how much heat must be put into the two so the ball just passes through the hole?

** **94.** An aluminum wire of radius 3.0×10^{-4} m is stretched between the ends of a concrete block, as the drawing illustrates. When the system (wire and concrete) is at 35 °C, the tension in the wire is 50.0 N. What is the tension in the wire when the system is heated to 185 °C?

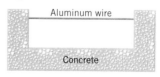

Concepts & Calculations   Group Learning Problems

Note: Each of these problems consists of Concept Questions followed by a related quantitative Problem. They are designed for use by students working alone or in small learning groups. The Concept Questions involve little or no mathematics and are intended to stimulate group discussions. They focus on the concepts with which the problems deal. Recognizing the concepts is the essential initial step in any problem-solving technique.

95. Concept Questions The drawing shows two thermometers, A and B, whose temperatures are measured in °A and °B. The ice and

boiling points of water are also indicated. (a) Is the size of one degree on the A scale larger or smaller than on the B scale? Why? (b) Is a given temperature on the A scale (e.g., +20 °A) hotter or colder than the same reading on the B scale (e.g., +20 °B)? Provide a reason for your answer.

Problem (a) Using the data in the drawing, determine the number of B° on the B scale that correspond to 1 A° on the A scale. (b) If the temperature of a substance reads +40.0 °A on the A scale, what would that temperature read on the B scale?

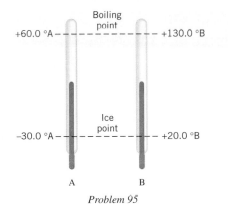

+60.0 °A — — — — — — — Boiling point — — — — — +130.0 °B

Ice point

−30.0 °A — — — — — — — — — — — — +20.0 °B

A   B

Problem 95

96. Concept Questions (a) When an object, such as a rod, is heated, what factors determine how much its length increases? (b) Suppose two rods are made from different materials, but their lengths change by the same amount when their temperatures change by the same amount. Are their initial lengths the same or different? Explain.

Problem One rod is made from lead and another from quartz. They are heated and experience the same change in temperature. If the initial length of the lead rod is 0.10 m, what is the initial length of the quartz rod? Be sure that your answer is consistent with your answer to the Concept Questions.

97. Concept Questions An aluminum can is filled to the brim with a liquid. The can and the liquid are heated so their temperatures change by the same amount. (a) In general, what factors determine how the volume of an object changes when it is heated? (b) Aluminum has a smaller coefficient of volume expansion than the liquid does. Which will expand more, the can or the liquid? Provide a reason for your answer. (c) How is the volume of liquid that spills over related to the changes in the volume of the can and the liquid?

Problem The can's initial volume at 5 °C is 3.5×10^{-4} m^3. The coefficient of volume expansion for aluminum is 69×10^{-6} (C°)$^{-1}$, according to Table 12.1. When the can and the liquid are heated to 78 °C, 3.6×10^{-6} m^3 of liquid spills over. What is the coefficient of volume expansion of the liquid?

98. Concept Questions (a) When heat is added to an object, what factors determine its change in temperature? (b) Suppose the same amount of heat is applied to two bars. They have the same mass, but experience different changes in temperature. Are the specific heat capacities the same for each bar? Why or why not?

Problem Two bars of identical mass are at 25 °C. One is made from glass and the other from another substance in Table 12.2. When identical amounts of heat are supplied to each, the glass bar reaches a temperature of 88 °C, while the other reaches 250.0 °C. What is the other substance?

99. Concept Questions Suppose you have two solid objects, A and B, made from different materials. They have the same mass, and each solid is at its melting temperature. You then add heat to melt them. (a) It takes less heat to melt A than B. Which has the larger latent heat of fusion? Why? (b) The mass of each is doubled. Does it require twice as much heat to melt them? Justify your answer.

Problem (a) Objects A and B have the same mass of 3.0 kg. They melt when 3.0×10^4 J of heat is added to A and 9.0×10^4 J is added to B. Determine the latent heats of fusion. (b) Find the heat

required to melt object A when its mass is 6.0 kg. Verify that your answers are consistent with your answers to the Concept Questions.

100. Concept Questions On a cool autumn morning, the relative humidity is 100%. (a) Does this mean that the partial pressure of water in the air equals atmospheric pressure? Why or why not? (b) Suppose the air warms up in the afternoon, but the partial pressure of water in the air does not change. Is the humidity still 100%? Provide a reason for your answer.

Problem The vapor pressure of water at 20 °C is 2500 Pa. (a) What percentage of atmospheric pressure is this? (b) What percentage of the total air pressure at 20 °C is due to water vapor if the relative humidity is 100%? (c) The vapor pressure of water at 35 °C is 5500 Pa. What is the relative humidity at this temperature if the partial pressure of water in the air has not changed from that at 20 °C?

* **101. Concept Questions** (a) A ball and a thin plate are made from different materials and have the same initial temperature. The ball does not fit through a hole in the plate, because the diameter of the ball is slightly larger than the diameter of the hole. However, the ball will pass through the hole when the ball and the plate are both heated to a common higher temperature. How do the coefficients of linear expansion of the ball and the plate compare? Explain. (b) The drawing shows three ball/plate arrangements like that in Concept Question (a). In each, the diameters of the balls are the same, and the diameters of the holes are the same. As the temperature increases, which ball falls through the hole first, which second, and which third? Account for your answers with the aid of data from Table 12.1.

Problem In each of the arrangements in the drawing the diameter of the ball is 1.0×10^{-5} m larger than the diameter of the hole, which has a diameter of 0.10 m. The initial temperature of each arrangement is 25.0 °C. At what temperature will the ball fall through the hole in each arrangement? Verify that your answer is consistent with your answer to Concept Question (b).

Gold   Steel   Quartz

Aluminum

Lead   Silver

Arrangement I   Arrangement II   Arrangement III

* **102. Concept Questions** (a) Two portions of the same liquid have the same mass, but different temperatures. They are mixed in a container that prevents the exchange of heat with the environment. Portion A has an initial temperature of T_{0A}, and portion B has an initial temperature of T_{0B}. Relative to these two temperatures, where on the temperature scale will the final equilibrium temperature of the mixture be located? Explain. (b) Three portions of the same liquid are mixed in a container that prevents the exchange of heat with the environment. Portion A has a mass m and a temperature of 94.0 °C, portion B has a mass m and a temperature of 78.0 °C, and portion C has a mass $2m$ and a temperature of 34.0 °C. Considering your answer to Concept Question (a) and without doing a heat-lost-equals-heat-gained calculation, deduce the final temperature of the mixture at equilibrium. Justify your answer.

Problem Using a heat-lost-equals-heat-gained calculation, determine the final equilibrium temperature of the mixture that results from the three portions described in Concept Question (b). Check to see that your answer agrees with your answer to Concept Question (b).

Chapter 13

The Transfer
of Heat

Polar bears lose relatively
small amounts of body heat, in
part because of their fur, which
consists of narrow, air-filled
tubes. This chapter considers
the processes by which heat is
lost or gained. (© Wayne R.
Bilenduke/The Image Bank/Getty
Images)

13.1 Convection

► CONCEPTS AT A GLANCE When heat is transferred to or from a substance, the internal energy of the substance can change, as we saw in Chapter 12. This change in internal energy is accompanied by a change in temperature or a change in phase, as suggested by the Concepts-at-a-Glance chart in Figure 12.23. The transfer of heat affects us in many ways. For instance, within our homes furnaces distribute heat on cold days, and air conditioners remove it on hot days. Our bodies constantly transfer heat in one direction or another, to prevent the adverse effects of hypo- and hyperthermia. And virtually all our energy originates in the sun and is transferred to us over a distance of 150 million kilometers through the void of space. Today's sunlight provides the energy to drive photosynthesis in the plants that provide our food and, hence, metabolic energy. Ancient sunlight nurtured the organic matter that became the fossil fuels of oil, natural gas, and coal. This chapter examines the three processes by which heat is transferred: convection, conduction, and radiation. The Concepts-at-a-Glance chart in Figure 13.1, which is a modification of that in Figure 12.23, illustrates how these three methods of heat transfer are related to our previous study of heat and internal energy. ◄

When part of a fluid is warmed, such as the air above a fire, the volume of the fluid expands, and the density decreases. According to Archimedes' principle (see Section 11.6), the surrounding cooler and denser fluid exerts a buoyant force on the warmer fluid and pushes it upward. As warmer fluid rises, the surrounding cooler fluid replaces it. This cooler fluid, in turn, is warmed and pushed upward. Thus, a continuous flow is established, which carries along heat. Whenever heat is transferred by the bulk movement of a gas or a liquid, the heat is said to be transferred by *convection*. The fluid flow itself is called a *convection current*.

■ **CONVECTION**

Convection is the process in which heat is carried from place to place by the bulk movement of a fluid.

The smoke rising from the volcanic eruption in Figure 13.2 is one visible result of convection. Figure 13.3 shows the less visible example of convection currents in a pan of

Figure 13.1 CONCEPTS AT A GLANCE Heat is transferred from one place to another by three methods: convection, conduction, and radiation. As discussed in Chapter 12, heat can change the internal energy of a substance, which results in a change in the temperature or phase of the substance. The amount of heat transferred from this Snowy Owl to its cold environment is a minimum, because losses due to convection, conduction, and radiation are kept small. (© David E. Myers/Stone/Getty Images)

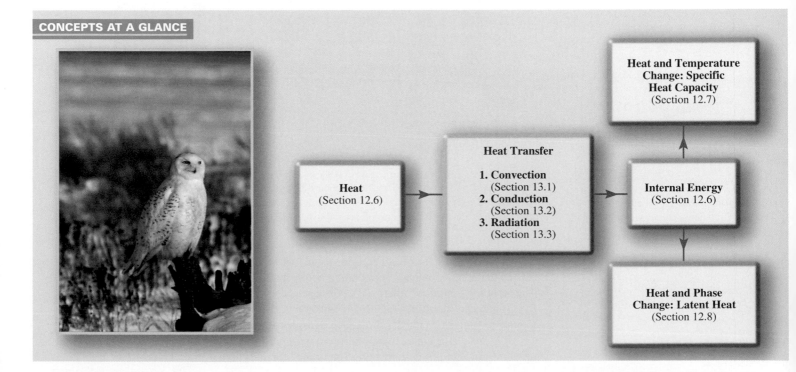

CONCEPTS AT A GLANCE

Heat
(Section 12.6)

Heat Transfer
1. **Convection**
(Section 13.1)
2. **Conduction**
(Section 13.2)
3. **Radiation**
(Section 13.3)

Internal Energy
(Section 12.6)

Heat and Temperature Change: Specific Heat Capacity
(Section 12.7)

Heat and Phase Change: Latent Heat
(Section 12.8)

water being heated on a gas burner. The currents distribute the heat from the burning gas to all parts of the water. Conceptual Example 1 deals with some of the important roles that convection plays in the home.

The physics of
heating and cooling by convection.

Figure 13.2 During a volcanic eruption, smoke at the top of the plume rises thousands of meters because of convection. (© Art Wolfe/The Image Bank/Getty Images)

Conceptual Example 1
Hot Water Baseboard Heating and Refrigerators

Hot water baseboard heating units are frequently used in homes, where they are mounted on the wall next to the floor, as in Figure 13.4*a*. In contrast, the cooling coil in a refrigerator is mounted near the top of the refrigerator, as in part *b* of the drawing. The locations for these heating and cooling devices are different, yet each location is designed to maximize the production of convection currents. Explain how.

Reasoning and Solution An important goal for a heating system is to distribute heat throughout a room. The analogous goal for the cooling coil is to remove heat from all of the space within a refrigerator. In each case, the heating or cooling device is positioned so that convection makes the goal achievable. The air above the baseboard unit is heated, like the air above a fire. Buoyant forces from the surrounding cooler air push the warm air upward. Cooler air near the ceiling is displaced downward and then warmed by the baseboard heating unit, leading to the convection current illustrated in Figure 13.4*a*. Had the heating unit been located near the ceiling, the warm air would have remained there, with very little convection to distribute the heat.

Within the refrigerator, the air in contact with the top-mounted coil is cooled, its volume decreases, and its density increases. The surrounding warmer and less dense air cannot provide sufficient buoyant force to support the colder air, which sinks downward. In the process, warmer air near the bottom is displaced upward and is then cooled by the coil, establishing the convection current shown in Figure 13.4*b*. Had the cooling coil been placed at the bottom of the refrigerator, stagnant, cool air would have collected there, with little convection to carry the heat from other parts of the refrigerator to the coil for removal.

Figure 13.3 Convection currents are set up when a pan of water is heated.

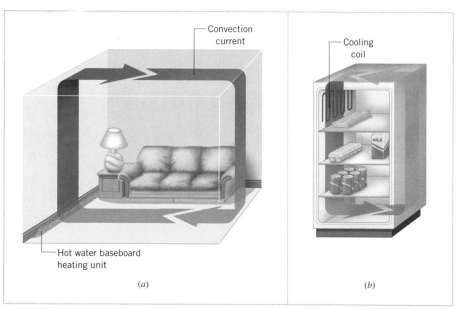

Convection current

Cooling coil

Hot water baseboard heating unit

(*a*)

(*b*)

Figure 13.4 (*a*) Air warmed by the baseboard heating unit is pushed to the top of the room by the cooler and denser air. (*b*) Air cooled by the cooling coil sinks to the bottom of the refrigerator. In both (*a*) and (*b*) a convection current is established.

Another example of convection occurs when the ground, heated by the sun's rays, warms the neighboring air. Surrounding cooler and denser air pushes the heated air upward. The resulting updraft or "thermal" can be quite strong, depending on the amount of heat that the ground can supply. As Figure 13.5 illustrates, these thermals can be used by glider pilots to gain considerable altitude. Birds such as eagles utilize thermals in a similar fashion.

It is usual for air temperature to decrease with increasing altitude, and the resulting upward convection currents are important for dispersing pollutants from industrial

The physics of
"thermals."

The physics of
an inversion layer.

sources and automobile exhaust systems. Sometimes, however, meteorological conditions cause a layer to form in the atmosphere where the temperature increases with increasing altitude. Such a layer is called an *inversion layer* because its temperature profile is inverted compared to the usual situation. An inversion layer arrests the normal upward convection currents, causing a stagnant-air condition in which the concentration of pollutants increases substantially. This leads to a smog layer that can often be seen hovering over large cities (Figure 13.6).

We have been discussing *natural convection,* in which a temperature difference causes the density at one place in a fluid to be different from that at another. Sometimes, natural convection is inadequate to transfer sufficient amounts of heat. In such cases *forced convection* is often used, and an external device such as a pump or a fan mixes the warmer and cooler portions of the fluid. Figure 13.7 shows an example of an automobile engine, where forced convection occurs in two ways. First, a pump circulates radiator fluid (water and antifreeze) through the engine to remove excess heat from the combustion process. Second, a radiator fan draws air through the radiator. Heat is transferred from the hotter radiator fluid to the cooler air, thereby cooling the fluid.

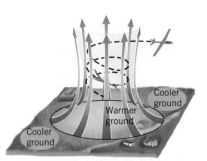

Figure 13.5 Updrafts, or thermals, are caused by the convective movement of air that the ground has warmed.

The physics of cooling by forced convection.

13.2 *Conduction*

Anyone who has fried a hamburger in an all-metal skillet knows that the metal handle becomes hot. Somehow, heat is transferred from the burner to the handle. Clearly, heat is not being transferred by the bulk movement of the metal or the surrounding air, so convection can be ruled out. Instead, heat is transferred directly through the metal by a process called *conduction.*

■ **CONDUCTION**

Conduction is the process whereby heat is transferred directly through a material, with any bulk motion of the material playing no role in the transfer.

One mechanism for conduction occurs when the atoms or molecules in a hotter part of the material vibrate or move with greater energy than those in a cooler part. By means of collisions, the more energetic molecules pass on some of their energy to their less energetic neighbors. For example, imagine a gas filling the space between two walls that face each other and are maintained at different temperatures. Molecules strike the hotter wall, absorb energy from it, and rebound with a greater kinetic energy than when they arrived. As these more energetic molecules collide with their less energetic neighbors, they transfer some of their energy to them. Eventually, this energy is passed on until it reaches the molecules next to the cooler wall. These molecules, in turn, collide with the wall, giving up some of their energy to it in the process. Through such molecular collisions, heat is conducted from the hotter to the cooler wall.

A similar mechanism for the conduction of heat occurs in metals. Metals are different from most substances in having a pool of electrons that are more or less free to wander throughout the metal. These free electrons can transport energy and allow metals to transfer heat very well. The free electrons are also responsible for the excellent electrical conductivity that metals have.

Those materials that conduct heat well are called *thermal conductors,* and those that conduct heat poorly are known as *thermal insulators.* Most metals are excellent thermal conductors; wood, glass, and most plastics are common thermal insulators. Thermal insulators have many important applications. Virtually all new housing construction incorporates thermal insulation in attics and walls to reduce heating and cooling costs. And the wooden or plastic handles on many pots and pans reduce the flow of heat to the cook's hand.

To illustrate the factors that influence the conduction of heat, Figure 13.8 displays a rectangular bar. The ends of the bar are in thermal contact with two bodies, one of which is kept at a constant higher temperature, while the other is kept at a constant lower temperature. Although not shown for the sake of clarity, the sides of the bar are insulated, so

Figure 13.6 In the absence of upward convection currents in the air, pollutants accumulate and form the smog layer that is easily visible near the top of this photograph of Los Angeles. (© Deborah Davis/Stone/Getty Images)

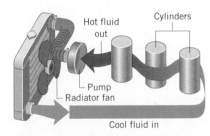

Figure 13.7 The forced convection generated by a pump circulates radiator fluid through an automobile engine to remove excess heat.

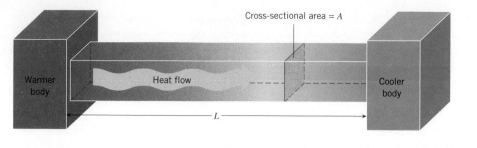

Figure 13.8 Heat is conducted through the bar when the ends of the bar are maintained at different temperatures. The heat flows from the warmer to the cooler end.

Table 13.1 Thermal Conductivities[a] of Selected Materials

Substance	Thermal Conductivity, k [J/(s·m·C°)]
Metals	
Aluminum	240
Brass	110
Copper	390
Iron	79
Lead	35
Silver	420
Steel (stainless)	14
Gases	
Air	0.0256
Hydrogen (H_2)	0.180
Nitrogen (N_2)	0.0258
Oxygen (O_2)	0.0265
Other Materials	
Asbestos	0.090
Body fat	0.20
Concrete	1.1
Diamond	2450
Glass	0.80
Goose down	0.025
Ice (0 °C)	2.2
Styrofoam	0.010
Water	0.60
Wood (oak)	0.15
Wool	0.040

[a] Except as noted, the values pertain to temperatures near 20 °C.

Concept Simulation 13.1

When heat is transferred through a bar, as illustrated in Figure 13.8, the amount of heat conducted per second depends on the length and cross-sectional area of the bar, as well as the temperature of each end. In this simulation the user can vary each of these parameters and see their effect on the energy flow.

Related Homework: Problems 1, 28

Go to
www.wiley.com/college/cutnell

Figure 13.9 Twice as much heat flows through two identical bars as through one.

the heat lost through them is negligible. The amount of heat Q conducted through the bar from the warmer end to the cooler end depends on a number of factors:

1. Q is proportional to the time t during which conduction takes place ($Q \propto t$). More heat flows in longer time periods.

2. Q is proportional to the temperature difference ΔT between the ends of the bar ($Q \propto \Delta T$). A larger difference causes more heat to flow. No heat flows when both ends have the same temperature, so that $\Delta T = 0$ C°.

3. Q is proportional to the cross-sectional area A of the bar ($Q \propto A$). Figure 13.9 helps to explain this fact by showing two identical bars (insulated sides not shown) placed between the warmer and cooler bodies. Clearly, twice as much heat flows through two bars as through one, since the cross-sectional area has been doubled.

4. Q is inversely proportional to the length L of the bar ($Q \propto 1/L$). Greater lengths of material conduct less heat. To experience this effect, put two insulated mittens (the pot holders that cooks keep around the stove) on the *same hand.* Then, touch a hot pot and notice that it feels cooler than when you wear only one mitten, signifying that less heat passes through the greater thickness ("length") of material.

These proportionalities can be stated together as $Q \propto (A\,\Delta T)t/L$. Equation 13.1 expresses this result with the aid of a proportionality constant k, which is called the ***thermal conductivity.***

■ **CONDUCTION OF HEAT THROUGH A MATERIAL**

The heat Q conducted during a time t through a bar of length L and cross-sectional area A is

$$Q = \frac{(kA\,\Delta T)t}{L} \qquad (13.1)$$

where ΔT is the temperature difference between the ends of the bar and k is the thermal conductivity of the material.

SI Unit of Thermal Conductivity: J/(s·m·C°)

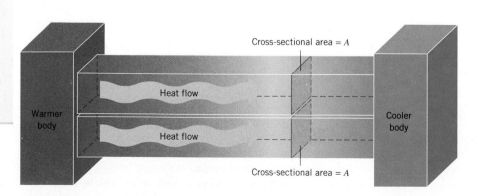

Since $k = QL/(tA\ \Delta T)$, the SI unit for thermal conductivity is $J \cdot m/(s \cdot m^2 \cdot C°)$ or $J/(s \cdot m \cdot C°)$. The SI unit of power is the joule per second (J/s) or watt (W), so the thermal conductivity is also given in units of $W/(m \cdot C°)$.

Different materials have different thermal conductivities, and Table 13.1 gives some representative values. Because metals are such good thermal conductors, they have large thermal conductivities. In comparison, liquids and gases generally have small thermal conductivities. In fact, in most fluids the heat transferred by conduction is negligible compared to that transferred by convection when there are strong convection currents. Air, for instance, with its small thermal conductivity, is an excellent thermal insulator when confined to small spaces where no appreciable convection currents can be established. Goose down, Styrofoam, and wool derive their fine insulating properties in part from the small dead-air spaces within them, as Figure 13.10 illustrates. We also take advantage of dead-air spaces when we dress "in layers" during very cold weather and put on several layers of relatively thin clothing rather than one thick layer. The air trapped between the layers acts as an excellent insulator.

Example 2 deals with the role that conduction through body fat plays in regulating body temperature.

The physics of dressing warm.

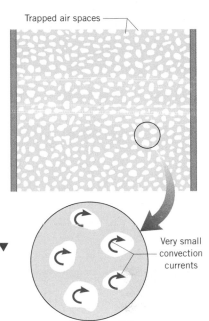

Example 2 Heat Transfer in the Human Body

When excessive heat is produced within the body, it must be transferred to the skin and dispersed if the temperature at the body interior is to be maintained at the normal value of 37.0 °C. One possible mechanism for transfer is conduction through body fat. Suppose that heat travels through 0.030 m of fat in reaching the skin, which has a total surface area of 1.7 m² and a temperature of 34.0 °C. Find the amount of heat that reaches the skin in half an hour (1800 s).

Reasoning and Solution In Table 13.1 the thermal conductivity of body fat is given as $k = 0.20\ J/(s \cdot m \cdot C°)$. According to Equation 13.1,

$$Q = \frac{(kA\ \Delta T)t}{L}$$

$$Q = \frac{[0.20\ J/(s \cdot m \cdot C°)](1.7\ m^2)(37.0\ °C - 34.0\ °C)(1800\ s)}{0.030\ m} = \boxed{6.1 \times 10^4\ J}.$$

For comparison, a jogger can generate over ten times this amount of heat in a half hour. Thus, conduction through body fat is not a particularly effective way of removing excess heat. Heat transfer via blood flow to the skin is more effective and has the added advantage that the body can vary the blood flow as needed (see Problem 7).

Figure 13.10 Styrofoam is an excellent thermal insulator because it contains many small, dead-air spaces. These small spaces inhibit heat transfer by convection currents, and air itself has a very low thermal conductivity.

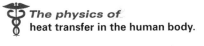

The physics of heat transfer in the human body.

Virtually all homes contain insulation in the walls to reduce heat loss. Example 3 illustrates how to determine this loss with and without insulation.

Example 3 Layered Insulation

One wall of a house consists of 0.019-m-thick plywood backed by 0.076-m-thick insulation, as Figure 13.11 shows. The temperature at the inside surface is 25.0 °C, while the temperature at the outside surface is 4.0 °C, both being constant. The thermal conductivities of the insulation and the plywood are, respectively, 0.030 and 0.080 $J/(s \cdot m \cdot C°)$, and the area of the wall is 35 m². Find the heat conducted through the wall in one hour (a) with the insulation and (b) without the insulation.

Reasoning The temperature T at the insulation–plywood interface (see Figure 13.11) must be determined before the heat conducted through the wall can be obtained. In calculating this temperature, we use the fact that no heat is accumulating in the wall because the inner and outer temperatures are constant. Therefore, the heat conducted through the insulation must equal the heat conducted through the plywood during the same time; that is, $Q_{insulation} = Q_{plywood}$. Each of the Q values can be expressed as $Q = (kA\ \Delta T)t/L$, according to Equation 13.1, leading to an

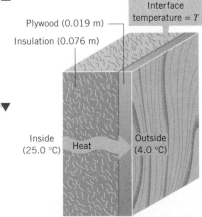

Figure 13.11 Heat flows through the insulation and plywood from the warmer inside to the cooler outside. The temperature of the insulation–plywood interface is T.

expression that can be solved for the interface temperature. Once a value for T is available, Equation 13.1 can be used to obtain the heat conducted through the wall.

Solution

(a) Using Equation 13.1 and the fact that $Q_{insulation} = Q_{plywood}$, we find that

$$\left[\frac{(kA\,\Delta T)t}{L}\right]_{insulation} = \left[\frac{(kA\,\Delta T)t}{L}\right]_{plywood}$$

$$\frac{[0.030\ \text{J/(s}\cdot\text{m}\cdot\text{C}°)]A(25.0\ °\text{C} - T)t}{0.076\ \text{m}} = \frac{[0.080\ \text{J/(s}\cdot\text{m}\cdot\text{C}°)]A(T - 4.0\ °\text{C})t}{0.019\ \text{m}}$$

Eliminating the area A and time t algebraically and solving this equation for T reveals that the temperature at the insulation–plywood interface is $T = 5.8\ °\text{C}$.

The heat conducted through the wall is either $Q_{insulation}$ or $Q_{plywood}$, since the two quantities are equal. Choosing $Q_{insulation}$ and using $T = 5.8\ °\text{C}$ in Equation 13.1, we find that

$$Q_{insulation} = \frac{[0.030\ \text{J/(s}\cdot\text{m}\cdot\text{C}°)](35\ \text{m}^2)(25.0\ °\text{C} - 5.8\ °\text{C})(3600\ \text{s})}{0.076\ \text{m}}$$

$$= \boxed{9.5 \times 10^5\ \text{J}}$$

(b) It is straightforward to use Equation 13.1 to calculate the amount of heat that would flow through the plywood in one hour if the insulation were absent:

$$Q_{plywood} = \frac{[0.080\ \text{J/(s}\cdot\text{m}\cdot\text{C}°)](35\ \text{m}^2)(25.0\ °\text{C} - 4.0\ °\text{C})(3600\ \text{s})}{0.019\ \text{m}}$$

$$= \boxed{110 \times 10^5\ \text{J}}$$

Without insulation, the heat loss is increased by a factor of about 12.

Fruit growers sometimes protect their crops by spraying them with water when overnight temperatures are expected to drop below freezing. Some fruit crops, like the blueberry plants in Figure 13.12, can withstand temperatures down to freezing (0 °C). However, as the temperature falls below freezing, the risk of damage rises significantly. When water is sprayed on the plants, it can freeze and form a covering of ice. When the water freezes it releases heat (see Section 12.8), some of which goes into warming the plant. In addition, both water and ice have relatively small thermal conductivities, as Table 13.1 indicates. Thus, they also protect the crop by acting as thermal insulators that reduce heat loss from the plants.

Although a layer of ice may be beneficial to blueberry plants, it is not so desirable inside a refrigerator, as Conceptual Example 4 discusses.

Figure 13.12 After a subfreezing night, this blueberry crop is being checked for freeze damage. The plants were sprayed the previous evening by sprinklers that put a coating of ice on them, thereby insulating the plants against the subfreezing temperatures. (© AP/Wide World Photos)

Conceptual Example 4   An Iced-up Refrigerator

In a refrigerator, heat is removed by a cold refrigerant fluid that circulates within a tubular space embedded within a metal plate, as Figure 13.13 illustrates. A good refrigerator cools food as quickly as possible. Decide whether the plate should be made from aluminum or stainless steel and whether the arrangement works better or worse when it becomes coated with a layer of ice.

Reasoning and Solution Figure 13.13 (see blowups) shows the metal cooling plate with and without a layer of ice. Without ice, heat passes by conduction through the metal to the refrigerant fluid within. For a given temperature difference across the thickness of metal, heat is transferred more quickly through the metal with the largest thermal conductivity. Table 13.1 indicates that the thermal conductivity of aluminum is more than 17 times larger than that of stainless steel. Therefore, *the plate should be made from aluminum.* When the plate becomes coated with ice, any heat that is removed by the refrigerant fluid must first be transferred by conduction through the ice before it encounters the aluminum plate. But the conduction of heat through ice occurs much less readily than through aluminum, because, as Table 13.1 indicates, ice has a much smaller thermal conductivity. Moreover, Equation 13.1 indicates that the heat conducted per unit time (Q/t) is inversely proportional to the thickness L of the ice. Thus, as ice builds up, the heat removed per unit time by the cooling plate decreases. *When covered with ice, the cooling plate works less well.*

Related Homework: *Problem 34*

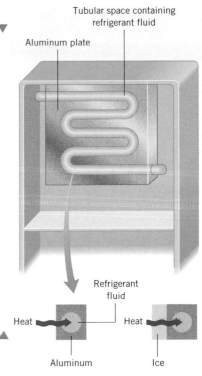

Figure 13.13 In a refrigerator, cooling is accomplished by a cold refrigerant fluid that circulates through a tubular space embedded within an aluminum plate. The arrangement works less well when the plate is coated with a layer of ice.

✔ **Check Your Understanding 1**

Two bars are placed between plates whose temperatures are T_{hot} and T_{cold} (see the drawing). The thermal conductivity of bar 1 is six times that of bar 2 ($k_1 = 6k_2$), but bar 1 has only one-third the cross-sectional area ($A_1 = \frac{1}{3}A_2$). Ignore any heat loss through the sides of the bars. Which statement below correctly describes the heat conducted by the bars in a given amount of time? *(The answer is given at the end of the book.)*

    a. Bar 1 conducts $\frac{1}{4}$ the heat as does bar 2; $Q_1 = \frac{1}{4}Q_2$.

    b. Bar 1 conducts $\frac{1}{8}$ the heat as does bar 2; $Q_1 = \frac{1}{8}Q_2$.

    c. Bar 1 conducts twice the heat as does bar 2; $Q_1 = 2Q_2$.

    d. Bar 1 conducts four times the heat as does bar 2; $Q_1 = 4Q_2$.

    e. Both bars conduct the same amount of heat; $Q_1 = Q_2$.

Background: The answer to this question depends on the factors that influence the transfer of heat by conduction.

For similar questions (including calculational counterparts), consult Self-Assessment Test 13.1. This test is described at the end of Section 13.3.

Need more practice?

🖱 **Interactive LearningWare 13.1**
A hollow, cylindrical glass tube is filled with hydrogen (H_2) gas. The tube has a length of 0.25 m and inner and outer radii of 0.020 and 0.023 m, respectively. One end of the tube is maintained at a temperature of 85 °C, while the other is kept at 15 °C. There is no heat loss through the curved surface. The thermal conductivity of glass is 0.80 J/(s·m·C°); the thermal conductivity of hydrogen is 0.18 J/(s·m·C°). Determine the total heat that flows through the tube (both the glass and the hydrogen) in a time of 55 s.

Related Homework: *Problem 8*

Go to
www.wiley.com/college/cutnell
for an interactive solution.

13.3   Radiation

Energy from the sun is brought to earth by large amounts of visible light waves, as well as by substantial amounts of infrared and ultraviolet waves. These waves belong to a class of waves known as electromagnetic waves, a class that also includes the microwaves used for cooking and the radio waves used for AM and FM broadcasts. The sunbather in Figure 13.14 feels hot because her body absorbs energy from the sun's electromagnetic waves. And anyone who has stood by a roaring fire or put a hand near an incandescent light bulb has experienced a similar effect. Thus, fires and light bulbs also emit electromagnetic waves, and when the energy of such waves is absorbed, it can have the same effect as heat.

    The process of transferring energy via electromagnetic waves is called *radiation,* and, unlike convection or conduction, it does not require a material medium. Electromagnetic waves from the sun, for example, travel through the void of space during their journey to earth.

Figure 13.14 Suntans are produced by ultraviolet rays. (© Angelo Cavalli/The Image Bank/Getty Images)

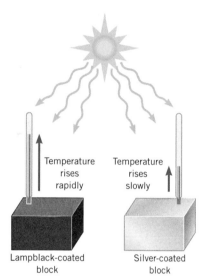

Figure 13.15 The temperature of the block coated with lampblack rises faster than the temperature of the block coated with silver because the black surface absorbs radiant energy from the sun at the greater rate.

The physics of summer clothing.

The physics of a white sifaka lemur warming up.

■ RADIATION

Radiation is the process in which energy is transferred by means of electromagnetic waves.

All bodies continuously radiate energy in the form of electromagnetic waves. Even an ice cube radiates energy, although so little of it is in the form of visible light that an ice cube cannot be seen in the dark. Likewise, the human body emits insufficient visible light to be seen in the dark. However, as Figures 12.6 and 12.7 illustrate, the infrared waves radiating from the body can be detected in the dark by electronic cameras. Generally, an object does not emit much visible light until the temperature of the object exceeds about 1000 K. Then a characteristic red glow appears, like that of a heating coil on an electric stove. When its temperature reaches about 1700 K, an object begins to glow white-hot, like the tungsten filament in an incandescent light bulb.

In the transfer of energy by radiation, the absorption of electromagnetic waves is just as important as their emission. The surface of an object plays a significant role in determining how much radiant energy the object will absorb or emit. The two blocks in sunlight in Figure 13.15, for example, are identical, except that one has a rough surface coated with lampblack (a fine black soot), while the other has a highly polished silver surface. As the thermometers indicate, the temperature of the black block rises at a much faster rate than that of the silvery block. This is because lampblack absorbs about 97% of the incident radiant energy, while the silvery surface absorbs only about 10%. The remaining part of the incident energy is reflected in each case. We see the lampblack as black in color because it reflects so little of the light falling on it, while the silvery surface looks like a mirror because it reflects so much light. Since the color black is associated with nearly complete absorption of visible light, the term *perfect blackbody* or, simply, *blackbody* is used when referring to an object that absorbs all the electromagnetic waves falling on it.

All objects emit and absorb electromagnetic waves simultaneously. When a body has the same constant temperature as its surroundings, the amount of radiant energy being absorbed must balance the amount being emitted in a given interval of time. The block coated with lampblack absorbs and emits the same amount of radiant energy, and the silvery block does too. In either case, if absorption were greater than emission, the block would experience a net gain in energy. As a result, the temperature of the block would rise and not be constant. Similarly, if emission were greater than absorption, the temperature would fall. Since absorption and emission are balanced, *a material that is a good absorber, like lampblack, is also a good emitter, and a material that is a poor absorber, like polished silver, is also a poor emitter.* A perfect blackbody, being a perfect absorber, is also a perfect emitter.

The fact that a black surface is both a good absorber and a good emitter is the reason people are uncomfortable wearing dark clothes during the summer. Dark clothes absorb a large fraction of the sun's radiation and then reemit it in all directions. About one-half of the emitted radiation is directed inward toward the body and creates the sensation of warmth. Light-colored clothes, in contrast, are cooler to wear, since they absorb and reemit relatively little of the incident radiation.

The use of light colors for comfort also occurs in nature. Most lemurs, for instance, are nocturnal and have dark fur like the one shown in Figure 13.16a. Since they are active at night, the dark fur poses no disadvantage in absorbing excessive sunlight. Figure 13.16b shows a species of lemur called the white sifaka, which lives in semiarid regions where there is little shade. The white color of the fur may help in thermoregulation, by reflecting sunlight during the hot part of the day. However, during the cool mornings, reflection of sunlight would be a hindrance in warming up. It is interesting to note that these lemurs have black skin and only sparse fur on their bellies, and that to warm up in the morning, they turn their dark bellies toward the sun. The dark color enhances the absorption of sunlight.

The amount of radiant energy Q emitted by a perfect blackbody is proportional to the radiation time interval t ($Q \propto t$). The longer the time, the greater is the amount of energy radiated. Experiment shows that Q is also proportional to the surface area A ($Q \propto A$). An object with a large surface area radiates more energy than one with a small surface area, other things being equal. Finally, experiment reveals that Q is proportional to the *fourth power of the Kelvin temperature T* ($Q \propto T^4$), so the emitted energy increases markedly with increasing temperature. If, for example, the Kelvin temperature of an object doubles, the object emits 2^4 or 16 times more energy. Combining these factors into a single proportionality, we see that $Q \propto T^4At$. This proportionality is converted into an equation by inserting a proportionality constant σ, known as the *Stefan–Boltzmann constant*. It has been found experimentally that $\sigma = 5.67 \times 10^{-8}$ J/(s·m²·K⁴):

$$Q = \sigma T^4At$$

The relationship above holds only for a perfect emitter. Most objects are not perfect emitters, however. Suppose that an object radiates only about 80% of the visible light energy that a perfect emitter would radiate, so Q (for the object) $= (0.80)\sigma T^4At$. The factor such as the 0.80 in this equation is called the ***emissivity e*** and is a dimensionless number between zero and one. The emissivity is the ratio of the energy an object actually radiates to the energy the object would radiate if it were a perfect emitter. For visible light, the value of e for the human body, for instance, varies between about 0.65 and 0.80, the smaller values pertaining to lighter skin colors. For infrared radiation, e is nearly one for all skin colors. For a perfect blackbody emitter, $e = 1$. Including the factor e on the right side of the expression $Q = \sigma T^4At$ leads to the ***Stefan–Boltzmann law of radiation.***

■ **THE STEFAN–BOLTZMANN LAW OF RADIATION**

The radiant energy Q, emitted in a time t by an object that has a Kelvin temperature T, a surface area A, and an emissivity e, is given by

$$Q = e\sigma T^4At \qquad (13.2)$$

where σ is the Stefan–Boltzmann constant and has a value of 5.67×10^{-8} J/(s·m²·K⁴).

In Equation 13.2, the Stefan–Boltzmann constant σ is a universal constant in the sense that its value is the same for all bodies, regardless of the nature of their surfaces. The emissivity e, however, depends on the condition of the surface.

Example 5 shows how the Stefan–Boltzmann law can be used to determine the size of a star.

Example 5   A Supergiant Star

The supergiant star Betelgeuse has a surface temperature of about 2900 K (about one-half that of our sun) and emits a radiant power (in joules per second, or watts) of approximately 4×10^{30} W (about 10 000 times as great as that of our sun). Assuming that Betelgeuse is a perfect emitter (emissivity $e = 1$) and spherical, find its radius.

Reasoning According to the Stefan–Boltzmann law, the power emitted is $Q/t = e\sigma T^4A$. A star with a relatively small temperature T can have a relatively large radiant power Q/t only if the area A is large. As we will see, Betelgeuse has a very large surface area, so its radius is enormous.

Solution Solving the Stefan–Boltzmann law for the area, we find

$$A = \frac{Q/t}{e\sigma T^4}$$

But the surface area of a sphere is $A = 4\pi r^2$, so $r = \sqrt{A/4\pi}$. Therefore, we have

$$r = \sqrt{\frac{Q/t}{4\pi e\sigma T^4}} = \sqrt{\frac{4 \times 10^{30}\ \text{W}}{4\pi(1)[5.67 \times 10^{-8}\ \text{J/(s·m²·K⁴)}](2900\ \text{K})^4}}$$

$$= \boxed{3 \times 10^{11}\ \text{m}}$$

For comparison, Mars orbits the sun at a distance of 2.28×10^{11} m. Betelgeuse is certainly a "supergiant."

(a)

(b)

Figure 13.16 (a) Most lemurs, like this one, are nocturnal and have dark fur. (© Wolfgang Kaehler/Corbis) (b) The species of lemur called the white sifaka, however, is active during the day and has white fur. (© Nigel Dennis/ Wildlife Pictures/Peter Arnold)

Problem solving insight
First solve an equation for the unknown in terms of the known variables. Then substitute numbers for the known variables, as this example shows.

The next example explains how to apply the Stefan–Boltzmann law when an object, such as a wood stove, simultaneously emits and absorbs radiant energy.

Example 6   A Wood-Burning Stove

The physics of a wood-burning stove.

A wood-burning stove stands unused in a room where the temperature is 18 °C (291 K). A fire is started inside the stove. Eventually, the temperature of the stove surface reaches a constant 198 °C (471 K), and the room warms to a constant 29 °C (302 K). The stove has an emissivity of 0.900 and a surface area of 3.50 m². Determine the *net* radiant power generated by the stove when the stove (a) is unheated and has a temperature equal to room temperature and (b) has a temperature of 198 °C.

Reasoning The stove emits more radiant power when heated than when unheated. In both cases, however, the Stefan–Boltzmann law can be used to determine the amount of power emitted. Power is the change in energy per unit time (Equation 6.10b), or *Q/t*. But in this problem we need to find the *net* power produced by the stove. The net power is the power the stove emits minus the power the stove absorbs. The power the stove absorbs comes from the walls, ceiling, and floor of the room, all of which emit radiation.

Solution

(a) Remembering that temperature must be expressed in kelvins when using the Stefan–Boltzmann law, we find that

$$\text{Power emitted by unheated stove at 18 °C} = \frac{Q}{t} = e\sigma T^4 A \tag{13.2}$$

Problem solving insight
In the Stefan–Boltzmann law of radiation, the temperature T must be expressed in kelvins, not in degrees Celsius or degrees Fahrenheit.

$$= (0.900)[5.67 \times 10^{-8} \text{ J/(s} \cdot \text{m}^2 \cdot \text{K}^4)](291 \text{ K})^4(3.50 \text{ m}^2) = 1280 \text{ W}$$

The fact that the unheated stove emits 1280 W of power and yet maintains a constant temperature means that the stove also absorbs 1280 W of radiant power from its surroundings. Thus, the *net* power generated by the unheated stove is zero:

$$\underset{\text{stove at 18 °C}}{\text{Net power generated by}} = \underset{\substack{\text{Power emitted} \\ \text{by stove at} \\ \text{18 °C}}}{1280 \text{ W}} - \underset{\substack{\text{Power emitted by} \\ \text{room at 18 °C and} \\ \text{absorbed by stove}}}{1280 \text{ W}} = \boxed{0 \text{ W}}$$

(b) The hot stove (198 °C or 471 K) emits more radiant power than it absorbs from the cooler room. The radiant power the stove emits is

$$\underset{\text{198 °C}}{\text{Power emitted by stove at}} = \frac{Q}{t} = e\sigma T^4 A$$

$$= (0.900)[5.67 \times 10^{-8} \text{ J/(s} \cdot \text{m}^2 \cdot \text{K}^4)](471 \text{ K})^4(3.50 \text{ m}^2) = 8790 \text{ W}$$

The radiant power the stove absorbs from the room is identical to the power that the stove would emit at the constant room temperature of 29 °C (302 K). The reasoning here is exactly like that in part (a):

$$\underset{\substack{\text{absorbed by stove}}}{\underset{\text{room at 29 °C and}}{\text{Power emitted by}}} = \frac{Q}{t} = e\sigma T^4 A$$

$$= (0.900)[5.67 \times 10^{-8} \text{ J/(s} \cdot \text{m}^2 \cdot \text{K}^4)](302 \text{ K})^4(3.50 \text{ m}^2) = 1490 \text{ W}$$

The *net* radiant power the stove produces from the fuel it burns is

$$\underset{\text{stove at 198 °C}}{\text{Net power generated by}} = \underset{\substack{\text{Power emitted} \\ \text{by stove at} \\ \text{198 °C}}}{8790 \text{ W}} - \underset{\substack{\text{Power emitted by} \\ \text{room at 29 °C and} \\ \text{absorbed by stove}}}{1490 \text{ W}} = \boxed{7300 \text{ W}}$$

Need more practice?

🖱 *Interactive LearningWare 13.2*
One day, the temperature of a radiator has to be 68 °C to keep the surrounding walls of a room at 24 °C. The next day, however, the temperature of the radiator needs to be only 49 °C to keep the walls at 24 °C because it is warmer outside. Assuming that the room is heated only by radiation, determine the ratio of the *net* power radiated by the unit on the colder day to that radiated on the warmer day.

Related Homework: *Problem 22*

Go to
www.wiley.com/college/cutnell
for an interactive solution.

Example 6 illustrates that when an object has a higher temperature than its surroundings, the object emits a net radiant power $P_{\text{net}} = (Q/t)_{\text{net}}$. The net power is the power the object emits minus the power it absorbs. Applying the Stefan–Boltzmann law as in Example 6 leads to the following expression for P_{net} when the temperature of the object is T and the temperature of the environment is T_0:

$$P_{net} = e\sigma A(T^4 - T_0^4) \tag{13.3}$$

✔ **Check Your Understanding 2**

Two identical cubes have the same temperature. One of them, however, is cut in two and the pieces are separated (see the drawing). What is true about the radiant energy emitted in a given time? *(The answer is given at the end of the book.)*

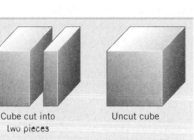

Cube cut into two pieces   Uncut cube

   a. The cube cut into two pieces emits twice as much radiant energy as does the uncut cube; $Q_{two\,pieces} = 2Q_{cube}$.
   b. The cube cut into two pieces emits more radiant energy than does the uncut cube according to $Q_{two\,pieces} = \frac{4}{3}Q_{cube}$.
   c. The cube cut into two pieces emits the same amount of radiant energy as does the uncut cube; $Q_{two\,pieces} = Q_{cube}$.
   d. The cube cut into two pieces emits one-half the radiant energy emitted by the uncut cube; $Q_{two\,pieces} = \frac{1}{2}Q_{cube}$.
   e. The cube cut into two pieces emits less radiant energy than does the uncut cube according to $Q_{two\,pieces} = \frac{1}{3}Q_{cube}$.

Background: The radiant energy emitted in a certain time by an object depends on its surface area, emissivity, and temperature. Only one of these variables changes when the object is cut into two pieces.

For similar questions (including calculational counterparts), consult Self-Assessment Test 13.1. This test is described next.

🖱 **Self-Assessment Test 13.1**

Test your understanding of the central ideas presented in Sections 13.1–13.3:

• Convection   • Conduction   • Radiation

Go to **www.wiley.com/college/cutnell**

13.4 Applications

To keep heating and air-conditioning bills to a minimum, it pays to use good thermal insulation in your home. Insulation inhibits convection between inner and outer walls and minimizes heat transfer by conduction. With respect to conduction, the logic behind home insulation ratings comes directly from Equation 13.1. According to this equation, the heat per unit time Q/t flowing through a thickness of material is $Q/t = kA\,\Delta T/L$. Keeping the value for Q/t to a minimum means using materials that have small thermal conductivities k and large thicknesses L. Construction engineers, however, prefer to use Equation 13.1 in the slightly different form shown below:

$$\frac{Q}{t} = \frac{A\,\Delta T}{L/k}$$

The term L/k in the denominator is called the R value of the insulation. For a building material, it is convenient to talk about an R value because it expresses in a single number the combined effects of thermal conductivity and thickness. Larger R values reduce the heat per unit time flowing through the material and, therefore, mean better insulation. It is also convenient to use R values to describe layered slabs formed by sandwiching together a number of materials with different thermal conductivities and different thicknesses. The R values for the individual layers can be added to give a single R value for the entire slab (see Problem 36). It should be noted, however, that R values are expressed using units of feet, hours, F°, and BTU for thickness, time, temperature, and heat, respectively.

The physics of
rating thermal insulation by *R* values.

When it is in the earth's shadow, an orbiting satellite is shielded from the intense electromagnetic waves emitted by the sun. But when a satellite moves out of the earth's

The physics of
regulating the temperature of an orbiting satellite.

Figure 13.17 Highly reflective metal foil covering this satellite minimizes temperature changes. (Courtesy NASA.)

The physics of
a thermos bottle *and*
a halogen cooktop stove.

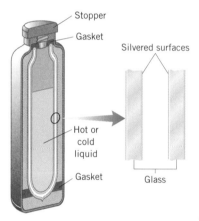

Figure 13.18 A thermos bottle minimizes energy transfer due to convection, conduction, and radiation.

Figure 13.19 In a halogen cooktop, quartz–iodine lamps emit a large amount of electromagnetic energy that is absorbed directly by a pot or pan.

shadow, the satellite experiences the full effect of these waves. As a result, the temperature within a satellite would decrease and increase sharply during an orbital period and sensitive electronic circuitry would suffer, unless precautions are taken. To minimize temperature fluctuations, satellites are often covered with a highly reflecting and, hence, poorly absorbing metal foil, as Figure 13.17 shows. By reflecting much of the sunlight, the foil minimizes temperature rises. Being a poor absorber, the foil is also a poor emitter and reduces radiant energy losses. Reducing these losses keeps the temperature from falling excessively when the satellite is in the earth's shadow.

A thermos bottle, sometimes referred to as a Dewar flask, reduces the rate at which hot liquids cool down or cold liquids warm up. A thermos usually consists of a double-walled glass vessel with silvered inner walls (see Figure 13.18) and accomplishes its job by minimizing heat transfer via convection, conduction, and radiation. The space between the walls is evacuated to minimize energy losses due to conduction and convection. The silvered surfaces reflect most of the radiant energy that would otherwise enter or leave the liquid in the thermos. Finally, little heat is lost through the glass or the rubberlike gaskets and stopper, since these materials have relatively small thermal conductivities.

Halogen cooktops use radiant energy to heat pots and pans. A halogen cooktop uses several quartz–iodine lamps, like the ones used for ultrabright automobile headlights. These lamps are electrically powered and are mounted below a ceramic top. (See Figure 13.19.) They radiate a great deal of electromagnetic energy, which passes through the ceramic top and is absorbed directly by the bottom of the pot. Consequently, the pot heats up very quickly, rivaling the time of a pot on an open gas burner.

13.5 *Concepts & Calculations*

Heat transfer by conduction is governed by Equation 13.1, as we have seen. Our first example illustrates a familiar application of this relation in the kitchen. It also gives us the opportunity to review the idea of latent heat of vaporization, which Section 12.8 discusses.

Concepts & Calculations Example 7 Boiling Water

Two pots are identical, except that in one case the flat bottom is aluminum and in the other it is copper. Each pot contains the same amount of boiling water and sits on a heating element that has a temperature of 155 °C. In the aluminum-bottom pot, the water boils away completely in 360 s. How long does it take the water in the copper-bottom pot to boil away completely?

Concept Questions and Answers Is the heat needed to boil away the water in the aluminum-bottom pot less than, greater than, or the same as the heat needed in the copper-bottom pot?

Answer The heat Q needed is the same in each case. When water boils, it changes from the liquid phase to the vapor phase. The heat needed to make the water boil away is $Q = mL_v$, according to Equation 12.5, where m is the mass of the water and L_v is the latent heat of vaporization for water. Since the amount of water in each pot is the same, the mass of water is the same in each case. Moreover, the latent heat is a characteristic of water and, therefore, is also the same in each case.

One of the factors in Equation 13.1 that influences the amount of heat conducted through the bottom of each pot is the temperature difference ΔT between the upper and lower surfaces of the pot's bottom. Is this temperature difference for the aluminum-bottom pot less than, greater than, or the same as that for the copper-bottom pot?

Answer For each pot the temperature difference is the same. At the upper surface of each pot bottom the temperature is 100.0 °C, because water boils at 100.0 °C under normal conditions of atmospheric pressure. The temperature remains at 100.0 °C until all the water is gone. For each pot the temperature at the lower surface of the pot bottom is 155 °C, the temperature of the heating element. Therefore, $\Delta T = 155\ °C - 100.0\ °C = 55\ C°$ for each pot.

Is the time required to boil away the water completely in the copper-bottom pot less than, greater than, or the same as that required for the aluminum-bottom pot?

Answer The time is less for the copper-bottom pot. The factors that influence the amount of heat conducted in a given time are the thermal conductivity, the area of the bottom, the temperature difference across the bottom, and the thickness of the bottom. All of these factors are the same for each pot except for the thermal conductivity, which is greater for copper (Cu) than for aluminum (Al) (see Table 13.1). The greater the thermal conductivity, the greater the heat that is conducted in a given time, other things being equal. Therefore, less time is required to boil the water away using the copper-bottom pot.

Solution Applying Equation 13.1 to the conduction of heat into both pots and using Equation 12.5 to express the heat needed to boil away the water, we have

$$Q_{Al} = \frac{(k_{Al}A\,\Delta T)t_{Al}}{L} = mL_v$$

$$Q_{Cu} = \frac{(k_{Cu}A\,\Delta T)t_{Cu}}{L} = mL_v$$

In these two equations the area A, the temperature difference ΔT, the thickness L of the pot bottom, the mass m of the water, and the latent heat of vaporization of water L_v have the same values. Therefore, we can set the two heats equal and obtain

$$\frac{(k_{Al}A\,\Delta T)t_{Al}}{L} = \frac{(k_{Cu}A\,\Delta T)t_{Cu}}{L} \quad \text{or} \quad k_{Al}t_{Al} = k_{Cu}t_{Cu}$$

Solving for t_{Cu} and taking values for the thermal conductivities from Table 13.1, we find

$$t_{Cu} = \frac{k_{Al}t_{Al}}{k_{Cu}} = \frac{[240 \text{ J/(s}\cdot\text{m}\cdot\text{C}°)](360 \text{ s})}{390 \text{ J/(s}\cdot\text{m}\cdot\text{C}°)} = \boxed{220 \text{ s}}$$

As expected, the boil-away time is less for the copper-bottom pot.

▲

Heat transfer by conduction is only one way in which heat gets from place to place. Heat transfer by radiation is another way, and it is governed by the Stefan–Boltzmann law of radiation, as Section 13.3 discusses. Example 8 deals with a case in which heat loss by radiation leads to freezing of water. It stresses the importance of the area from which the radiation occurs and also provides a review of the idea of latent heat of fusion, which Section 12.8 discusses.

Concepts & Calculations Example 8 Freezing Water

▼

One half of a kilogram of liquid water at 273 K (0 °C) is placed outside on a day when the temperature is 261 K (-12 °C). Assume that heat is lost from the water only by means of radiation and that the emissivity of the radiating surface is 0.60. How long does it take for the water to freeze into ice at 0 °C when the surface area from which the radiation occurs is (a) 0.035 m² (as it could be in a cup) and (b) 1.5 m² (as it could be if the water were spilled out to form a thin sheet)?

Concept Questions and Answers In case (a) is the heat that must be removed to freeze the water less than, greater than, or the same as in case (b)?

Answer The heat that must be removed is the same in both cases. When water freezes, it changes from the liquid phase to the solid phase. The heat that must be removed to make the water freeze is $Q = mL_f$, according to Equation 12.5, where m is the mass of the water and L_f is the latent heat of fusion for water. The mass is the same in both cases and so is L_f, since it is a characteristic of the water.

The loss of heat by radiation depends on the temperature of the radiating object. Does the temperature of the water change as the water freezes?

Answer No. The temperature of the water does not change as the freezing process takes place. The heat removed serves only to change the water from the liquid to the solid phase, as Section 12.8 discusses. Only after all the water has frozen does the temperature of the ice begin to fall below 0 °C.

The water both loses and gains heat by radiation. How, then, can heat transfer by radiation lead to freezing of the water?

Answer   The water freezes because it loses more heat by radiation than it gains. The gain occurs because the environment radiates heat and the water absorbs it. But the temperature of the environment is less than the temperature of the water. As a result, the environmental radiation is insufficient to offset completely the loss of heat due to radiation from the water.

Will it take longer for the water to freeze in case (a) when the area is smaller or in case (b) when the area is larger?

Answer   It will take longer when the area is smaller. This is because the amount of energy radiated in a given time is proportional to the area from which the radiation occurs. A smaller area means that less energy is radiated per second, so more time will be required to freeze the water by removing heat via radiation.

Solution   We use Equation 13.3 to take into account that the water both gains and loses heat via radiation. This expression gives the net power lost, the net power being the net heat divided by the time. Thus, we have

$$\frac{Q}{t} = e\sigma A(T^4 - T_0^{\,4}) \quad \text{or} \quad t = \frac{Q}{e\sigma A(T^4 - T_0^{\,4})}$$

Using Equation 12.5 to express the heat Q as $Q = mL_f$ and taking the latent heat of fusion for water from Table 12.3 ($L_f = 33.5 \times 10^4$ J/kg), we find

(a) *Smaller area*

$$t = \frac{mL_f}{e\sigma A(T^4 - T_0^{\,4})}$$

$$= \frac{(0.50 \text{ kg})(33.5 \times 10^4 \text{ J/kg})}{0.60[5.67 \times 10^{-8} \text{ J/(s} \cdot \text{m}^2 \cdot \text{K}^4)](0.035 \text{ m}^2)[(273 \text{ K})^4 - (261 \text{ K})^4]}$$

$$= \boxed{1.5 \times 10^5 \text{ s } (42 \text{ h})}$$

(b) *Larger area*

$$t = \frac{mL_f}{e\sigma A(T^4 - T_0^{\,4})}$$

$$= \frac{(0.50 \text{ kg})(33.5 \times 10^4 \text{ J/kg})}{0.60[5.67 \times 10^{-8} \text{ J/(s} \cdot \text{m}^2 \cdot \text{K}^4)](1.5 \text{ m}^2)[(273 \text{ K})^4 - (261 \text{ K})^4]}$$

$$= \boxed{3.6 \times 10^3 \text{ s } (1.0 \text{ h})}$$

As expected, the freezing time is longer when the area is smaller.

At the end of the problem set for this chapter, you will find homework problems that contain both conceptual and quantitative parts. These problems are grouped under the heading *Concepts & Calculations, Group Learning Problems.* They are designed for use by students working alone or in small learning groups. The conceptual part of each problem provides a convenient focus for group discussions.

Concept Summary

This summary presents an abridged version of the chapter, including the important equations and all available learning aids. For convenient reference, the learning aids (including the text's examples) are placed next to or immediately after the relevant equation or discussion. The following learning aids may be found on-line at **www.wiley.com/college/cutnell**:

Interactive LearningWare examples are solved according to a five-step interactive format that is designed to help you develop problem-solving skills.	**Concept Simulations** are animated versions of text figures or animations that illustrate important concepts. You can control parameters that affect the display, and we encourage you to experiment.
Interactive Solutions offer specific models for certain types of problems in the chapter homework. The calculations are carried out interactively.	**Self-Assessment Tests** include both qualitative and quantitative questions. Extensive feedback is provided for both incorrect and correct answers, to help you evaluate your understanding of the material.

Topic	Discussion	Learning Aids
	13.1 Convection	
	Convection is the process in which heat is carried from place to place by the bulk movement of a fluid.	
Natural convection	During natural convection, the warmer, less dense part of a fluid is pushed upward by the buoyant force provided by the surrounding cooler and denser part.	Example 1
Forced convection	Forced convection occurs when an external device, such as a fan or a pump, causes the fluid to move.	
	13.2 Conduction	
	Conduction is the process whereby heat is transferred directly through a material, with any bulk motion of the material playing no role in the transfer.	
Thermal conductors and thermal insulators	Materials that conduct heat well, such as most metals, are known as thermal conductors. Materials that conduct heat poorly, such as wood, glass, and most plastics, are referred to as thermal insulators.	
	The heat Q conducted during a time t through a bar of length L and cross-sectional area A is	Examples 2, 3, 4, 7
Conduction of heat through a material	$$Q = \frac{(kA\,\Delta T)t}{L} \qquad (13.1)$$ where ΔT is the temperature difference between the ends of the bar and k is the thermal conductivity of the material.	Concept Simulation 13.1 Interactive LearningWare 13.1 Interactive Solutions 13.13, 13.33
	13.3 Radiation	
	Radiation is the process in which energy is transferred by means of electromagnetic waves.	
Absorbers and emitters	All objects, regardless of their temperature, simultaneously absorb and emit electromagnetic waves. Objects that are good absorbers of radiant energy are also good emitters, and objects that are poor absorbers are also poor emitters.	
A perfect blackbody	An object that absorbs all the radiation incident upon it is called a perfect blackbody. A perfect blackbody, being a perfect absorber, is also a perfect emitter.	
	The radiant energy Q emitted during a time t by an object whose surface area is A and whose Kelvin temperature is T is given by the Stefan–Boltzmann law of radiation:	
Stefan–Boltzmann law of radiation	$$Q = e\sigma T^4 A t \qquad (13.2)$$	Example 5
Emissivity	where $\sigma = 5.67 \times 10^{-8}$ J/(s·m²·K⁴) is the Stefan–Boltzmann constant and e is the emissivity, a dimensionless number characterizing the surface of the object. The emissivity lies between 0 and 1, being zero for a nonemitting surface and one for a perfect blackbody.	
	The net radiant power is the power an object emits minus the power it absorbs. The net radiant power P_{net} emitted by an object of temperature T located in an environment of temperature T_0 is	Examples 6, 8
Net radiant power	$$P_{net} = e\sigma A(T^4 - T_0^4) \qquad (13.3)$$	Interactive LearningWare 13.2

Use Self-Assessment Test 13.1 to evaluate your understanding of Sections 13.1–13.3.

Conceptual Questions

1. One often hears about heat transfer by convection in gases and liquids, but not in solids. Why?

2. A heavy drape, hung close to a cold window, reduces heat loss considerably by interfering primarily with one of the three processes of heat transfer. Explain which one.

3. The *windchill factor* is a term used by weather forecasters. Roughly speaking, it refers to the fact that you feel colder when the wind is blowing than when it is not, even though the air temperature is the same in either case. Which of the three processes for heat transfer plays the principal role in the windchill factor? Explain your reasoning.

4. Often, motorists see the following warning sign on bridges: "Caution–Bridge surface freezes before road surface." Account for the warning in terms of heat transfer processes. Note that, unlike the road, a bridge has both surfaces exposed to the air.

5. A piece of Styrofoam and a piece of wood are joined together to form a layered slab. The two pieces have the same thickness and cross-sectional area. The exposed surfaces have constant temperatures. The temperature of the exposed Styrofoam surface is greater than the temperature of the exposed wood surface. Is the temperature of the Styrofoam–wood interface closer to the larger or smaller of the two temperatures? Give your reasoning, using Equation 13.1 and data from Table 13.1.

6. One way that heat is transferred from place to place inside the human body is by the flow of blood. Which one of the three heat transfer processes best describes this action of the blood? Justify your answer.

7. Some animals have hair strands that are hollow, air-filled tubes. Other animals have hair strands that are solid. Which kind of hair would be more likely to give an animal an advantage for surviving in very cold climates? Why?

8. A poker used in a fireplace is held at one end, while the other end is in the fire. Why are pokers made of iron rather than copper? Ignore the fact that iron may be cheaper and stronger.

9. In Alaska, a lack of snow allowed the ground to freeze down to a depth of about one meter, causing buried water pipes to freeze and burst. Why did a lack of snow lead to this situation?

10. Concrete walls often contain steel reinforcement bars. Does the steel enhance or degrade the insulating value of the concrete? Explain.

11. Grandma says that it is quicker to bake a potato if you put a nail into it. In fact, she is right. Justify her baking technique in terms of one of the three processes of heat transfer.

12. Several days after a snowstorm, the roof on a house is uniformly covered with snow. On a neighboring house, however, the snow on the roof has completely melted. Which house is probably better insulated? Give your reasoning.

13. One car has a metal body, while another has a plastic body. On a cold winter day these cars are parked side by side. If you put a bare hand on each car, the metal body feels colder. Why?

14. A high-quality pot is designed so heat can enter readily and be distributed evenly, while the rate of energy loss from the pot is kept to a minimum. Many high-quality pots have copper bases and polished stainless steel sides. Based on conduction and radiation principles, explain why this design is better than all-copper or all-steel units.

15. Two objects have the same size and shape. Object A has an emissivity of 0.3, and object B has an emissivity of 0.6. Each radiates the same power. Is the Kelvin temperature of A twice that of B? Give your reasoning.

16. A concave mirror can be used to start a fire by directing sunlight onto a small spot on a piece of paper. Explain why the mirror does not get as hot as the paper.

17. Two strips of material, A and B, are identical, except they have emissivities of 0.4 and 0.7, respectively. The strips are heated to the same temperature and have a red glow. A brighter glow signifies that more energy per second is being radiated. Which strip has the brighter glow? Explain.

18. A pot of water is being heated on an electric stove. The diameter of the pot is smaller than the diameter of the heating element on which the pot rests. The exposed outer edges of the heating element are glowing cherry red. When you lift the pot, you see that the part of the heating element beneath it is not glowing cherry red, indicating that it is cooler than the outer edges. Why are the outer edges hotter?

19. To keep your hands as warm as possible during skiing, should you wear mittens or gloves? (Mittens, except for the thumb, do not have individual finger compartments.) Give a reason for your answer.

20. Two identical hot cups of cocoa are sitting on a table. One has a metal spoon in it and one does not. After five minutes, which cup is cooler? Explain in terms of heat transfer processes.

21. (a) Would a hot solid cube cool more rapidly if it were left intact or cut in half? Explain your answer in terms of one or more of the three heat transfer processes. (b) Using reasoning similar to that used in answering part (a), decide which cools faster after cooking, one pound of wide and flat lasagna noodles or one pound of spaghetti noodles. Assume that both kinds of noodles are made from the same pasta and start out with the same temperature.

22. One day during the winter the sun has been shining all day. Toward sunset a light snow begins to fall. It collects without melting on a cement playground, but it melts immediately upon contact with a black asphalt road adjacent to the playground. Account for the fact that the snow collects in one place but not in the other.

23. If you were stranded in the mountains in cold weather, it would help to minimize energy losses from your body by curling up into the tightest ball possible. Which of the factors in Equation 13.2 are you using to the best advantage by curling into a ball? Why?

Problems

Note: For problems in this set, use the values for thermal conductivities given in Table 13.1 unless stated otherwise.

ssm Solution is in the Student Solutions Manual.   **www** Solution is available on the World Wide Web at www.wiley.com/college/cutnell
This icon represents a biomedical application.

Section 13.2 Conduction

1. ssm Concept Simulation 13.1 at **www.wiley.com/college/cutnell** illustrates the concepts pertinent to this problem. A person's body is covered with 1.6 m^2 of wool clothing. The thickness of the wool is 2.0×10^{-3} m. The temperature at the outside surface of the wool is 11 °C, and the skin temperature is 36 °C. How much heat per second does the person lose due to conduction?

2. The temperature in an electric oven is 160 °C. The temperature at the outer surface in the kitchen is 50 °C. The oven (surface area = 1.6 m^2) is insulated with material that has a thickness of 0.020 m and a thermal conductivity of 0.045 J/(s · m · C°). (a) How much energy is used to operate the oven for six hours? (b) At a price of $0.10 per kilowatt · hour for electrical energy, what is the cost of operating the oven?

3. In an electrically heated home, the temperature of the ground in contact with a concrete basement wall is 12.8 °C. The temperature at the inside surface of the wall is 20.0 °C. The wall is 0.10 m thick and has an area of 9.0 m^2. Assume that one kilowatt · hour of electrical energy costs $0.10. How many hours are required for one dollar's worth of energy to be conducted through the wall?

4. ⚕ The amount of heat per second conducted from the blood capillaries beneath the skin to the surface is 240 J/s. The energy is transferred a distance of 2.0×10^{-3} m through a body whose surface area is 1.6 m^2. Assuming that the thermal conductivity is that of body fat, determine the temperature difference between the capillaries and the surface of the skin.

5. ssm Due to a temperature difference ΔT, heat is conducted through an aluminum plate that is 0.035 m thick. The plate is then replaced by a stainless steel plate that has the same temperature difference and cross-sectional area. How thick should the steel plate be so that the same amount of heat per second is conducted through it?

6. A skier wears a jacket filled with goose down that is 15 mm thick. Another skier wears a wool sweater that is 5.0 mm thick. Both have the same surface area. Assuming that the temperature difference between the inner and outer surfaces of each garment is the same, calculate the ratio (wool/goose down) of the heat lost due to conduction during the same time interval.

7. ⚕ **ssm www** In the conduction equation $Q = (kA \, \Delta T)t/L$, the combination of factors kA/L is called the *conductance*. The human body has the ability to vary the conductance of the tissue beneath the skin by means of vasoconstriction and vasodilation, in which the flow of blood to the veins and capillaries underlying the skin is decreased and increased, respectively. The conductance can be adjusted over a range such that the tissue beneath the skin is equivalent to a thickness of 0.080 mm of Styrofoam or 3.5 mm of air. By what factor can the body adjust the conductance?

8. Interactive LearningWare 13.1 at **www.wiley.com/college/cutnell** explores the approach taken in problems such as this one. A composite rod is made from stainless steel and iron and has a length of 0.50 m. The cross section of this composite rod is shown in the drawing and consists of a square within a circle. The square cross section of the steel is 1.0 cm on a side. The temperature at one end of the rod is 78 °C, while it is 18 °C at the other end. Assuming that no heat exits through the cylindrical outer surface, find the total amount of heat conducted through the rod in two minutes.

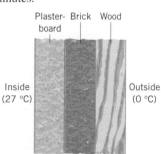
Iron
Stainless steel

* 9. Three building materials, plasterboard [$k = 0.30$ J/(s·m·C°)], brick [$k = 0.60$ J/(s·m·C°)], and wood [$k = 0.10$ J/(s·m·C°)], are sandwiched together as the drawing illustrates. The temperatures at the inside and outside surfaces are 27 °C and 0 °C, respectively. Each material has the same thickness and cross-sectional area. Find the temperature (a) at the plasterboard–brick interface and (b) at the brick–wood interface.

Plaster-board Brick Wood
Inside (27 °C)   Outside (0 °C)

* 10. In a house the temperature at the surface of a window is 25 °C. The temperature outside at the window surface is 5.0 °C. Heat is lost through the window via conduction, and the heat lost per second has a certain value. The temperature outside begins to fall, while the conditions inside the house remain the same. As a result, the heat lost per second increases. What is the temperature at the outside window surface when the heat lost per second doubles?

* 11. ssm Two rods, one of aluminum and the other of copper, are joined end to end. The cross-sectional area of each is 4.0×10^{-4} m^2,

and the length of each is 0.040 m. The free end of the aluminum rod is kept at 302 °C, while the free end of the copper rod is kept at 25 °C. The loss of heat through the sides of the rods may be ignored. (a) What is the temperature at the aluminum–copper interface? (b) How much heat is conducted through the unit in 2.0 s? (c) What is the temperature in the aluminum rod at a distance of 0.015 m from the hot end?

* 12. A copper rod has a length of 1.5 m and a cross-sectional area of 4.0×10^{-4} m^2. One end of the rod is in contact with boiling water and the other with a mixture of ice and water. What is the mass of ice per second that melts? Assume that no heat is lost through the side surface of the rod.

* 13. Consult **Interactive Solution 13.13** at **www.wiley.com/college/cutnell** to explore a model for solving this problem. One end of a brass bar is maintained at 306 °C, while the other end is kept at a constant, but lower temperature. The cross-sectional area of the bar is 2.6×10^{-4} m^2. Because of insulation, there is negligible heat loss through the sides of the bar. Heat flows through the bar, however, at the rate of 3.6 J/s. What is the temperature of the bar at a point 0.15 m from the hot end?

** 14. A 0.30-m-thick sheet of ice covers a lake. The air temperature at the ice surface is -15 °C. In five minutes, the ice thickens by a small amount. Assume that no heat flows from the ground below into the water and that the added ice is very thin compared to 0.30 m. Find the number of millimeters by which the ice thickens.

** 15. ssm www Two cylindrical rods have the same mass. One is made of silver (density = 10 500 kg/m^3), and one is made of iron (density = 7860 kg/m^3). Both rods conduct the same amount of heat per second when the same temperature difference is maintained across their ends. What is the ratio (silver-to-iron) of (a) the lengths and (b) the radii of these rods?

Section 13.3 Radiation

16. A person is standing outdoors in the shade where the temperature is 28 °C. (a) What is the radiant energy absorbed per second by his head when it is covered with hair? The surface area of the hair (assumed to be flat) is 160 cm^2 and its emissivity is 0.85. (b) What would be the radiant energy absorbed per second by the same person if he were bald and the emissivity of his head were 0.65?

17. ssm www How many days does it take for a perfect blackbody cube (0.0100 m on a side, 30.0 °C) to radiate the same amount of energy that a one-hundred-watt light bulb uses in one hour?

18. The filament of a light bulb has a temperature of 3.0×10^3 °C and radiates sixty watts of power. The emissivity of the filament is 0.36. Find the surface area of the filament.

19. An object emits 30 W of radiant power. If it were a perfect blackbody, other things being equal, it would emit 90 W of radiant power. What is the emissivity of the object?

20. The amount of radiant power produced by the sun is approximately 3.9×10^{26} W. Assuming the sun to be a perfect blackbody sphere with a radius of 6.96×10^8 m, find its surface temperature (in kelvins).

21. ssm A car parked in the sun absorbs energy at a rate of 560 watts per square meter of surface area. The car reaches a temperature at which it radiates energy at this same rate. Treating the car as a perfect radiator ($e = 1$), find the temperature.

22. ⚕ **Interactive LearningWare 13.2** at **www.wiley.com/college/cutnell** reviews the concepts that are involved in this problem. Suppose the skin temperature of a naked person is 34 °C when the person is standing inside a room whose temperature is 25 °C. The skin area of the individual is 1.5 m^2. (a) Assuming the emissivity is 0.80, find the net loss of radiant power from the body. (b) Determine the number of food Calories of energy (1 food Calorie = 4186 J) that is lost in one hour due to the net loss rate obtained in part (a). Metabolic conversion of food into energy replaces this loss.

23. The concrete wall of a building is 0.10 m thick. The temperature inside the building is 20.0 °C, while the temperature outside is 0.0 °C. Heat is conducted through the wall. When the building is unheated, the inside temperature falls to 0.0 °C, and heat conduction ceases. However, the wall does emit radiant energy when its temperature is 0.0 °C. The radiant energy emitted per second per square meter is the same as the heat lost per second per square meter due to conduction. What is the emissivity of the wall?

* **24.** A solid sphere has a temperature of 773 K. The sphere is melted down and recast into a cube that has the same emissivity and emits the same radiant power as the sphere. What is the cube's temperature?

** **25. ssm** A solid cylinder is radiating power. It has a length that is ten times its radius. It is cut into a number of smaller cylinders, each of which has the same length. Each small cylinder has the same temperature as the original cylinder. The total radiant power emitted by the pieces is twice that emitted by the original cylinder. How many smaller cylinders are there?

** **26.** A small sphere (emissivity = 0.90, radius − r_1) is located at the center of a spherical asbestos shell (thickness = 1.0 cm, outer radius = r_2). The thickness of the shell is small compared to the inner and outer radii of the shell. The temperature of the small sphere is 800.0 °C, while the temperature of the inner surface of the shell is 600.0 °C, both temperatures remaining constant. Assuming that $r_2/r_1 = 10.0$ and ignoring any air inside the shell, find the temperature of the outer surface of the shell.

Additional Problems

27. ssm One end of an iron poker is placed in a fire where the temperature is 502 °C, and the other end is kept at a temperature of 26 °C. The poker is 1.2 m long and has a radius of 5.0×10^{-3} m. Ignoring the heat lost along the length of the poker, find the amount of heat conducted from one end of the poker to the other in 5.0 s.

28. Concept Simulation 13.1 at **www.wiley.com/college/cutnell** illustrates the concepts pertinent to this problem. A refrigerator has a surface area of 5.3 m². It is lined with 0.075-m-thick insulation whose thermal conductivity is 0.030 J/(s·m·C°). The interior temperature is kept at 5 °C, while the temperature at the outside surface is 25 °C. How much heat per second is being removed from the unit?

29. A person eats a dessert that contains 260 Calories. (This "Calorie" unit, with a capital C, is the one used by nutritionists; 1 Calorie = 4186 J. See Section 12.7.) The skin temperature of a person is 36 °C and that of her environment is 21 °C. The emissivity of her skin is 0.75 and its surface area is 1.3 m². How much time would it take for her to emit a *net* radiant energy from her body that is equal to the energy contained in this dessert?

30. A person's body is producing energy internally due to metabolic processes. If the body loses more energy than metabolic processes are generating, its temperature will drop. If the drop is severe, it can be life-threatening. Suppose a person is unclothed and energy is being lost via radiation from a body surface area of 1.40 m², which has a temperature of 34 °C and an emissivity of 0.700. Suppose that metabolic processes are producing energy at a rate of 115 J/s. What is the temperature of the coldest room in which this person could stand and not experience a drop in body temperature?

31. ssm Review Example 6 before attempting this problem. Suppose the stove in that example had a surface area of only 2.00 m². What would its temperature (in kelvins) have to be so that it still generated a net power of 7300 W?

* **32.** Liquid helium is stored at its boiling-point temperature of 4.2 K in a spherical container (r = 0.30 m). The container is a perfect blackbody radiator. The container is surrounded by a spherical shield whose temperature is 77 K. A vacuum exists in the space between the container and the shield. The latent heat of vaporization for helium is 2.1×10^4 J/kg. What mass of liquid helium boils away through the venting valve in one hour?

* **33.** Refer to **Interactive Solution 13.33** at **www.wiley.com/college/cutnell** for help in solving this problem. In an aluminum pot, 0.15 kg of water at 100 °C boils away in four minutes. The bottom of the pot is 3.1×10^{-3} m thick and has a surface area of 0.015

m². To prevent the water from boiling too rapidly, a stainless steel plate has been placed between the pot and the heating element. The plate is 1.4×10^{-3} m thick, and its area matches that of the pot. Assuming that heat is conducted into the water only through the bottom of the pot, find the temperature at (a) the aluminum–steel interface and (b) the steel surface in contact with the heating element.

* **34.** Review Conceptual Example 4 before attempting this problem. To illustrate the effect of ice on the aluminum cooling plate, consider the drawing shown here and the data contained therein. Ignore any limitations due to significant figures. (a) Calculate the heat per second per square meter that is conducted through the ice–aluminum combination. (b) Calculate the heat per second per square meter that would be conducted through the aluminum if the ice were not present. Notice how much larger the answer is in (b) as compared to (a).

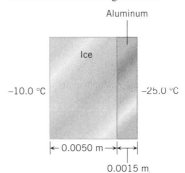

** **35. ssm www** The drawing shows a solid cylindrical rod made from a center cylinder of lead and an outer concentric jacket of copper. Except for its ends, the rod is insulated (not shown), so that the loss of heat from the curved surface is negligible. When a temperature difference is maintained between its ends, this rod conducts one-half the amount of heat that it would conduct if it were solid copper. Determine the ratio of the radii r_1/r_2.

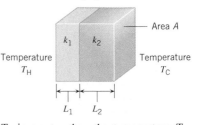

** **36.** The drawing shows a layered slab formed by joining together two pieces of different insulating materials that have thermal conductivities of k_1 and k_2. The thicknesses L_1 and L_2 are different, and the temperature T_H is greater than the temperature T_C. Show that the heat conducted through the slab in a time t is given by $Q = A(T_H - T_C)t/[(L_1/k_1) + (L_2/k_2)]$, so that the effective R value of the slab is $(L_1/k_1) + (L_2/k_2)$.

Concepts & Calculations   Group Learning Problems

Note: Each of these problems consists of Concept Questions followed by a related quantitative Problem. They are designed for use by students working alone or in small learning groups. The Concept Questions involve little or no mathematics and are intended to stimulate group discussions. They focus on the concepts with which the problems deal. Recognizing the concepts is the essential initial step in any problem-solving technique.

37. Concept Questions A pot of water is boiling on a stove under one atmosphere of pressure. Assume that heat enters the pot only through its bottom, which is copper and rests on a heating element. In a certain time, a mass m of water boils away. (a) What is the temperature of the boiling water and does it change during this time? (b) What determines the amount of heat needed to boil the water? (c) Is the temperature of the heating element in contact with the pot greater than, smaller than, or equal to 100 °C? Explain.

Problem In two minutes, the mass of water boiled away is $m =$ 0.45 kg. The radius of the pot bottom is $R = 6.5$ cm and the thickness is $L = 2.0$ mm. What is the temperature T_E of the heating element in contact with the pot? Verify that your answer is consistent with your answers to the Concept Questions.

38. Concept Questions A wall in a house contains a single window, the area and thickness of which are small relative to the area and thickness of the wall. The window consists of a single pane of glass. Treat the wall as a slab of the insulating material Styrofoam. The temperature difference between the inside and outside is the same for the wall and the window. Heat is lost via conduction through the window and the wall. Decide, for a given time period, whether each of the following factors, by itself, causes the window to lose more than, less than, or the same amount of heat as the wall: (a) the area, other things being equal, (b) the thickness, other things being equal, and (c) the thermal conductivity (see Table 13.1), other things being equal. Give your reasoning.

Problem The glass window pane has an area of 0.16 m² and a thickness of 2.0 mm. The Styrofoam area and thickness are 18 m² and 0.10 m, respectively. Of the total heat lost by the wall and the window, what is the percentage lost by the window?

39. Concept Questions Two objects are maintained at constant temperatures, one hot and one cold. Two identical bars can be attached end to end, as in part *a* of the drawing, or one on top of the other, as in part *b*. When either of these arrangements is placed between the hot and the cold objects for the same amount of time, heat Q flows from left to right. (a) Is the area through which the heat flows greater for arrangement *a* or arrangement *b*? (b) Is the thickness of the material through which the heat flows greater for arrangement *a* or arrangement *b*? (c) Is Q_a less than, greater than, or equal to Q_b? Account for your answers.

Problem Find the ratio Q_a/Q_b. Be sure that your answer is consistent with your answers to the Concept Questions.

(a)   (b)

40. Concept Questions Light bulb 1 operates with a higher filament temperature than light bulb 2, but both filaments have the same emissivity. (a) How is the power P expressed in terms of the energy Q radiated by a bulb and the time t during which the energy is radiated? (b) Does a higher filament temperature generate more radiated power or less radiated power? (c) Does a smaller area for radiation promote more radiated power or less radiated power? (d) Suppose that both bulbs radiate the same power. Is the filament area of bulb 1 greater than, less than, or the same as the filament area of bulb 2? Give your reasoning.

Problem Bulb 1 has a filament temperature of 2700 K, whereas bulb 2 has a filament temperature of 2100 K. Both bulbs radiate the same power. Find the ratio A_1/A_2 of the filament areas of the bulbs. Verify that your answer is consistent with your answers to the Concept Questions.

41. Concept Questions Via radiation, an object emits more power than it absorbs from the room in which it is located. (a) The object has a temperature T (in kelvins). According to the Stefan–Boltzmann law, the power radiated by the object is $Q/t = e\sigma T^4 A$, where A is the area from which the radiation is emitted. What is the expression for the power absorbed from the room, which has a temperature T_0 (in kelvins)? (b) Is the temperature of the object greater than, less than, or equal to the temperature of the room? Explain, assuming the temperatures are constant.

Problem The temperature of the room is 293 K. The object emits three times as much power as it absorbs from the room. What is the temperature of the object? Check to see that your answer is consistent with your answers to the Concept Questions.

42. Concept Questions Sirius B is a white star that has a much greater surface temperature than our sun does. Assume that both Sirius B and our sun are spherical and have the same emissivity. (a) Other things being equal, would the greater surface temperature imply that the power radiated by Sirius B is greater than, less than, or equal to the power radiated by our sun? (b) The fact is that Sirius B radiates much less power than our sun does. Considering this fact, is the surface area of Sirius B greater than, less than, or equal to the surface area of our sun? (c) Is the radius of Sirius B greater than, less than, or equal to the radius of our sun? Explain your answers.

Problem Our sun has a radius of 6.96×10^8 m. Sirius B radiates only 0.04 times the power radiated by our sun. Its surface temperature is four times that of our sun. Find the radius of Sirius B. Be sure that your answer is consistent with your answers to the Concept Questions.

* **43. Concept Questions** The block shown in the drawing has dimensions $L_0 \times 2L_0 \times 3L_0$. In drawings A, B, and C, heat is conducted through the block in three different directions. In each case, the same temperature difference exists between the opposite surfaces through which the heat passes, and the time during which the heat flows is the same. (a) The cross-sectional area of the opposite surfaces in C is greater than that in A. Does this fact alone mean that the heat conducted in C is greater than that conducted in A? Provide a reason for your answer. (b) The length of material through which heat is conducted is greater in A than in B. Does this fact alone imply that the heat conducted in A is smaller than that conducted in B? Why? (c) Rank the heat conducted in each of the three cases, largest first.

Problem The block in the drawing has a thermal conductivity of 250 J/(s·m·C°), and the length L_0 is 0.30 m. For each case—A, B, and C—the temperature of the warmer surface is 35 °C while that of the cooler surface is 19 °C. Determine the heat that flows in 5.0 s for each case. Be sure your answers are consistent with that in part (c) of the Concept Questions.

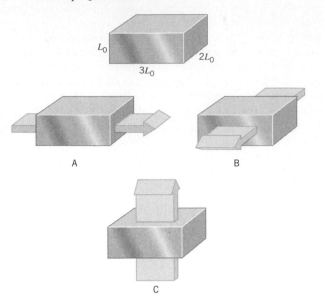

A

B

C

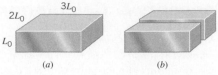

2L_0

3L_0

L_0

(a)

(b)

* **44. Concept Questions** Part (a) of the drawing shows a rectangular bar whose dimensions are $L_0 \times 2L_0 \times 3L_0$. The bar is at the same temperature as the room (not shown) in which it is located. (a) Is the *net* radiant power emitted by the bar greater than zero, equal to zero, or less than zero? Provide a reason for your answer. (b) The bar is then cut, lengthwise, into two equal pieces, as shown in part (b) of the drawing. The temperature of the bars does not change. Which situation, if either, *emits* more power into the room, the single bar in part (a) or the two bars in part (b) of the drawing? Why? (c) Which situation, if either, *absorbs* more power from the room, the single bar in part (a) or the two bars in part (b) of the drawing? Justify your reasoning.

Problem (a) What is the ratio of the power absorbed by the two bars in part (b) of the drawing to the single bar in part (a)? (b) Suppose the temperature of the single bar in part (a) is 450 K. What would the temperature (in kelvins) of the room and the two bars in part (b) have to be such that the two bars absorb the same power as the single bar in part (a)?

Chapter 14

The Ideal Gas Law and Kinetic Theory

If the air within these hot-air balloons behaves like an ideal gas, its pressure, volume, and temperature are related by the ideal gas law, the subject of this chapter. (© Harvey Lloyd/ Taxi/Getty Images)

14.1 Molecular Mass, the Mole, and Avogadro's Number

Often, we wish to compare the mass of one atom with another. To facilitate the comparison, a mass scale known as the **atomic mass scale** has been established. To set up this scale, a reference value (along with a unit) is chosen for one of the elements. The unit is called the **atomic mass unit** (symbol: u). By international agreement, the reference element is chosen to be the most abundant type or isotope* of carbon, which is called carbon-12. Its atomic mass† is defined to be exactly twelve atomic mass units, or 12 u. The relationship between the atomic mass unit and the kilogram is

$$1 \text{ u} = 1.6605 \times 10^{-27} \text{ kg}$$

The atomic masses of all the elements are listed in the periodic table, part of which is shown in Figure 14.1. The complete periodic table is given on the inside of the back cover. In general, the masses listed are average values and take into account the various isotopes of an element that exist naturally. For brevity, the unit "u" is often omitted from the table. For example, a magnesium atom (Mg) has an average atomic mass of 24.305 u, while the corresponding average value for the lithium atom (Li) is 6.941 u; thus, atomic magnesium is more massive than atomic lithium by a factor of (24.305 u)/(6.941 u) = 3.502. In the periodic table, the atomic mass of carbon (C) is given as 12.011 u, rather than exactly 12 u, because a small amount (about 1%) of the naturally occurring material is an isotope called carbon-13. The value of 12.011 u is an average that reflects the small contribution of carbon-13.

The molecular mass of a molecule is the sum of the atomic masses of its atoms. For instance, hydrogen and oxygen have atomic masses of 1.007 94 u and 15.9994 u, respectively, so the molecular mass of a water molecule (H_2O) is 2(1.007 94 u) + 15.9994 u = 18.0153 u.

Macroscopic amounts of materials contain large numbers of atoms or molecules. Even in a small volume of gas, 1 cm^3, for example, the number is enormous. It is convenient to express such large numbers in terms of a single unit, the **gram-mole,** or simply the **mole** (symbol: *mol*). **One gram-mole of a substance contains as many particles (atoms or molecules) as there are atoms in 12 grams of the isotope carbon-12.** Experiment shows that 12 grams of carbon-12 contain 6.022×10^{23} atoms. The number of atoms per mole is known as **Avogadro's number N_A,** after the Italian scientist Amedeo Avogadro (1776–1856):

$$N_A = 6.022 \times 10^{23} \text{ mol}^{-1}$$

Thus, the number of moles n contained in any sample is the number of particles N in the sample divided by the number of particles per mole N_A (Avogadro's number):

$$n = \frac{N}{N_A}$$

Although defined in terms of carbon atoms, the concept of a mole can be applied to any collection of objects by noting that one mole contains Avogadro's number of objects. Thus, one mole of atomic sulfur contains 6.022×10^{23} sulfur atoms, one mole of water contains 6.022×10^{23} H_2O molecules, and one mole of golf balls contains 6.022×10^{23} golf balls. Just as the meter is the SI base unit for length, the mole is the SI base unit for expressing "the amount of a substance."

The number n of moles contained in a sample can also be found from its mass. To see how, multiply and divide the right-hand side of the previous equation by the mass m_{particle} of a single particle, expressed in grams:

$$n = \frac{m_{\text{particle}} N}{m_{\text{particle}} N_A} = \frac{m}{\text{Mass per mole}}$$

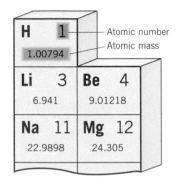

Figure 14.1 A portion of the periodic table showing the atomic number and atomic mass of each element. In the periodic table it is customary to omit the symbol "u" denoting the atomic mass unit.

---

* Isotopes are discussed in Section 31.1.
† In chemistry the expression "atomic weight" is frequently used in place of "atomic mass."

The numerator $m_{\text{particle}}N$ is the mass of a particle times the number of particles in the sample, which is the mass m of the sample. The denominator $m_{\text{particle}}N_A$ is the mass of a particle times the number of particles per mole, which is the mass per mole, expressed in grams per mole. The mass per mole of carbon-12 is 12 g/mol, since, by definition, 12 grams of carbon-12 contain one mole of atoms. On the other hand, the mass per mole of sodium (Na) is 22.9898 g/mol for the following reason: as indicated in Figure 14.1, a sodium atom is more massive than a carbon-12 atom by the ratio of their atomic masses, $(22.9898 \text{ u})/(12 \text{ u}) = 1.915\ 82$. Therefore, the mass per mole of sodium is $1.915\ 82$ times greater than that of carbon-12, or $(1.915\ 82)(12 \text{ g/mol}) = 22.9898$ g/mol. Note that the numerical value of the mass per mole of sodium (22.9898) is the same as the numerical value of its atomic mass. This is true in general, so *the mass per mole (in g/mol) of a substance has the same numerical value as the atomic or molecular mass of the substance (in atomic mass units).*

Problem solving insight

Since one gram-mole of a substance contains Avogadro's number of particles (atoms or molecules), the mass m_{particle} of a particle (in grams) can be obtained by dividing the mass per mole (in g/mol) by Avogadro's number:

$$m_{\text{particle}} = \frac{\text{Mass per mole}}{N_A}$$

Example 1 illustrates how to use the concepts of the mole, atomic mass, and Avogadro's number to determine the number of atoms and molecules present in two famous gemstones.

Example 1   The Hope Diamond and the Rosser Reeves Ruby

The physics of ▼ *gemstones.*

Figure 14.2a shows the Hope diamond (44.5 carats), which is almost pure carbon. Figure 14.2b shows the Rosser Reeves ruby (138 carats), which is primarily aluminum oxide (Al_2O_3). One carat is equivalent to a mass of 0.200 g. Determine (a) the number of carbon atoms in the diamond and (b) the number of Al_2O_3 molecules in the ruby.

Reasoning The number N of atoms (or molecules) in a sample is the number of moles n times the number of atoms per mole N_A (Avogadro's number); $N = nN_A$. We can determine the number of moles by dividing the mass of the sample m by the mass per mole of the substance.

Solution

(a) The mass of the Hope diamond is $m = (44.5 \text{ carats})[(0.200 \text{ g})/(1 \text{ carat})] = 8.90$ g. Since the average atomic mass of naturally occurring carbon is 12.011 u (see the periodic table on the inside of the back cover), the mass per mole of this substance is 12.011 g/mol. The number of moles of carbon in the Hope diamond is

$$n = \frac{m}{\text{Mass per mole}} = \frac{8.90 \text{ g}}{12.011 \text{ g/mol}} = 0.741 \text{ mol}$$

The number of carbon atoms in the Hope diamond is

$$N = nN_A = (0.741 \text{ mol})(6.022 \times 10^{23} \text{ atoms/mol}) = \boxed{4.46 \times 10^{23} \text{ atoms}}$$

(b) The mass of the Rosser Reeves ruby is $m = (138 \text{ carats})[(0.200 \text{ g})/(1 \text{ carat})] = 27.6$ g. The molecular mass of an aluminum oxide molecule (Al_2O_3) is the sum of the atomic masses of its atoms, which are 26.9815 u for aluminum and 15.9994 u for oxygen (see the periodic table on the inside of the back cover):

$$\text{Molecular mass} = \underbrace{2(26.9815 \text{ u})}_{\substack{\text{Mass of 2} \\ \text{aluminum} \\ \text{atoms}}} + \underbrace{3(15.9994 \text{ u})}_{\substack{\text{Mass of 3} \\ \text{oxygen} \\ \text{atoms}}} = 101.9612 \text{ u}$$

Thus, the mass per mole of Al_2O_3 is 101.9612 g/mol. Calculations like those in part (a) reveal that the Rosser Reeves ruby contains 0.271 mol or $\boxed{1.63 \times 10^{23} \text{ molecules of } Al_2O_3}$.

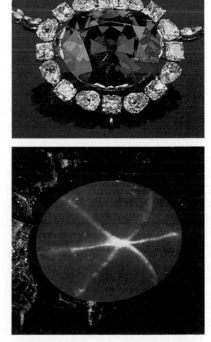

Figure 14.2 (a) The Hope diamond surrounded by 16 smaller diamonds. (b) The Rosser Reeves ruby. Both gems are on display at the Smithsonian Institution in Washington, D.C. (Courtesy Smithsonian Institution)

✔ **Check Your Understanding 1**

A gas mixture contains equal masses of the monatomic gases argon (atomic mass = 39.948 u) and neon (atomic mass = 20.179 u). They are the only gases in the mixture. Of the total number of atoms, what percentage is neon? *(The answer is given at the end of the book.)*

Background: The mole, atomic mass, and Avogadro's number are the concepts that must be used in answering this question.

For similar questions (including conceptual counterparts), consult Self-Assessment Test 14.1. This test is described at the end of Section 14.4.

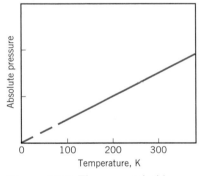

Figure 14.3 The pressure inside a constant-volume gas thermometer is directly proportional to the Kelvin temperature, which is characteristic of an ideal gas.

14.2 The Ideal Gas Law

An *ideal gas* is an idealized model for real gases that have sufficiently low densities. The condition of low density means that the molecules of the gas are so far apart that they do not interact (except during collisions that are effectively elastic). The ideal gas law expresses the relationship between the absolute pressure, the Kelvin temperature, the volume, and the number of moles of the gas.

In discussing the constant-volume gas thermometer, Section 12.2 has already explained the relationship between the absolute pressure and Kelvin temperature of a low-density gas. This thermometer utilizes a small amount of gas (e.g., hydrogen or helium) placed inside a bulb and kept at a constant volume. Since the density is low, the gas behaves as an ideal gas. Experiment reveals that a plot of gas pressure versus temperature is a straight line, as in Figure 12.4. This plot is redrawn in Figure 14.3, with the change that the temperature axis is now labeled in kelvins rather than in degrees Celsius. The graph indicates that the absolute pressure P is directly proportional to the Kelvin temperature T ($P \propto T$), for a fixed volume and a fixed number of molecules.

The relation between absolute pressure and the number of molecules of an ideal gas is simple. Experience indicates that it is possible to increase the pressure of a gas by adding more molecules; this is exactly what happens when a tire is pumped up. When the volume and temperature of a low-density gas are kept constant, doubling the number of molecules doubles the pressure. Thus, the absolute pressure of an ideal gas is proportional to the number of molecules or, equivalently, to the number of moles n of the gas ($P \propto n$).

To see how the absolute pressure of a gas depends on the volume of the gas, look at the partially filled balloon in Figure 14.4a. This balloon is "soft," because the pressure of the air is low. However, if all the air in the balloon is squeezed into a smaller "bubble," as

Figure 14.4 (*a*) The air pressure in the partially filled balloon can be increased by decreasing the volume of the balloon, as illustrated in (*b*).

(© Andy Washnik)

(*a*)   (*b*)

in part *b* of the figure, the "bubble" has a very tight feel. This tightness indicates that the pressure in the smaller volume is high enough to stretch the rubber substantially. Thus, it is possible to increase the pressure of a gas by reducing its volume, and if the number of molecules and the temperature are kept constant, the absolute pressure of an ideal gas is inversely proportional to its volume V ($P \propto 1/V$).

The three relations just discussed for the absolute pressure of an ideal gas can be expressed as a single proportionality, $P \propto nT/V$. This proportionality can be written as an equation by inserting a proportionality constant R, called the ***universal gas constant.*** The value of R has been determined experimentally to be 8.31 J/(mol · K) for any real gas with a density sufficiently low to ensure ideal gas behavior. The resulting equation is known as the ***ideal gas law.***

■ **IDEAL GAS LAW**

The absolute pressure P of an ideal gas is directly proportional to the Kelvin temperature T and the number of moles n of the gas and is inversely proportional to the volume V of the gas: $P = R(nT/V)$. In other words,

$$PV = nRT \tag{14.1}$$

where R is the universal gas constant and has the value of 8.31 J/(mol · K).

Sometimes, it is convenient to express the ideal gas law in terms of the total number of particles N, instead of the number of moles n. To obtain such an expression, we multiply and divide by Avogadro's number $N_A = 6.022 \times 10^{23}$ particles/mol* on the right in Equation 14.1 and recognize that the product nN_A is equal to the total number N of particles:

$$PV = nRT = nN_A \left(\frac{R}{N_A}\right) T = N \left(\frac{R}{N_A}\right) T$$

The constant term R/N_A is referred to as ***Boltzmann's constant,*** in honor of the Austrian physicist Ludwig Boltzmann (1844–1906), and is represented by the symbol k:

$$k = \frac{R}{N_A} = \frac{8.31 \text{ J/(mol·K)}}{6.022 \times 10^{23} \text{ mol}^{-1}} = 1.38 \times 10^{-23} \text{ J/K}$$

With this substitution, the ideal gas law becomes

$$PV = NkT \tag{14.2}$$

Example 2 presents an application of the ideal gas law.

Example 2 Oxygen in the Lungs

In the lungs, the respiratory membrane separates tiny sacs of air (absolute pressure = 1.00×10^5 Pa) from the blood in the capillaries. These sacs are called alveoli, and it is from them that oxygen enters the blood. The average radius of the alveoli is 0.125 mm, and the air inside contains 14% oxygen. Assuming that the air behaves as an ideal gas at body temperature (310 K), find the number of oxygen molecules in one of the sacs.

The physics of oxygen in the lungs.

Reasoning The pressure and temperature of the air inside an alveolus are known, and its volume can be determined since we know the radius. Thus, the ideal gas law in the form $PV = NkT$ can be used directly to find the number N of air particles inside one of the sacs. The number of oxygen molecules is 14% of the number of air particles.

Solution The volume of a spherical sac is $V = \frac{4}{3}\pi r^3$. Solving Equation 14.2 for the number of air particles, we have

$$N = \frac{PV}{kT} = \frac{(1.00 \times 10^5 \text{ Pa})[\frac{4}{3}\pi(0.125 \times 10^{-3} \text{ m})^3]}{(1.38 \times 10^{-23} \text{ J/K})(310 \text{ K})} = 1.9 \times 10^{14}$$

Problem solving insight
In the ideal gas law, the temperature T must be expressed on the Kelvin scale. The Celsius and Fahrenheit scales cannot be used.

The number of oxygen molecules is 14% of this value, or $0.14N = \boxed{2.7 \times 10^{13}}$.

* "Particles" is not an SI unit and is often omitted. Then, particles/mol = 1/mol = mol^{-1}.

With the aid of the ideal gas law, it can be shown that one mole of an ideal gas occupies a volume of 22.4 liters at a temperature of 273 K (0 °C) and a pressure of one atmosphere (1.013×10^5 Pa). These conditions of temperature and pressure are known as **standard temperature and pressure (STP).** Conceptual Example 3 discusses another interesting application of the ideal gas law.

The physics of **rising beer bubbles.**

Conceptual Example 3   Beer Bubbles on the Rise

The next time you get a chance, watch the bubbles rise in a glass of beer (see Figure 14.5). If you look carefully, you'll see them grow in size as they move upward, often doubling in volume by the time they reach the surface. Why does a bubble grow as it ascends?

Reasoning and Solution Beer bubbles contain mostly carbon dioxide (CO_2), a gas that is in the beer because of the fermentation process. The volume V of gas in a bubble is related to its temperature T, pressure P, and the number n of moles of CO_2 by the ideal gas law: $V = nRT/P$. Thus, one or more of these variables must be responsible for the growth of a bubble. Temperature can be eliminated immediately, since it is constant throughout the beer. What about the pressure? As a bubble rises, its depth decreases, and so does the fluid pressure. Since the volume is inversely proportional to pressure, part of the bubble growth is due to the decreasing pressure of the surrounding beer. However, some bubbles double in volume on the way up. To account for the doubling, there would have to be two atmospheres of pressure at the bottom of the glass, compared to the one atmosphere at the top. The pressure increment due to depth is $\rho g h$ according to Equation 11.4, so the extra pressure of one atmosphere at the bottom would mean 1.01×10^5 Pa $= \rho g h$. Solving for h with ρ equal to the density of water reveals that $h = 10.3$ m. Since most beer glasses are only about 0.2 m tall, we can rule out a change in pressure as the major cause of the change in volume. We are left with only one variable, the number of moles of CO_2 gas in the bubble. In fact, the number of moles does increase as the bubble rises. Each bubble acts as a nucleation site for CO_2 molecules, so as a bubble moves upward, it accumulates carbon dioxide from the surrounding beer and grows larger.

Related Homework: *Problem 22*

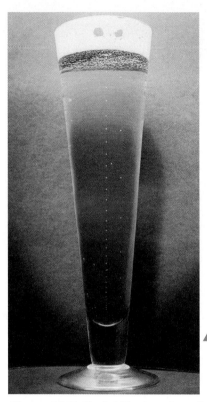

Figure 14.5 The bubbles in a glass of beer grow larger as they move upward. (Courtesy Richard Zare, Stanford University)

Historically, the work of several investigators led to the formulation of the ideal gas law. The Irish scientist Robert Boyle (1627–1691) discovered that at a constant temperature, the absolute pressure of a fixed mass (fixed number of moles) of a low-density gas is inversely proportional to its volume ($P \propto 1/V$). This fact is often called Boyle's law and can be derived from the ideal gas law by noting that $P = nRT/V = $ constant$/V$ when n and T are constants. Alternatively, if an ideal gas changes from an initial pressure and volume (P_i, V_i) to a final pressure and volume (P_f, V_f), it is possible to write $P_iV_i = nRT$ and $P_fV_f = nRT$. Since the right sides of these equations are equal, we may equate the left sides to give the following concise way of expressing **Boyle's law:**

Constant T, constant n $\qquad\qquad P_iV_i = P_fV_f \qquad\qquad$ (14.3)

Figure 14.6 illustrates how pressure and volume change according to Boyle's law for a fixed number of moles of an ideal gas at a constant temperature of 100 K. The gas begins with an initial pressure and volume of P_i and V_i and is compressed. The pressure increases as the volume decreases, according to $P = nRT/V$, until the final pressure and volume of P_f and V_f are reached. The curve that passes through the initial and final points is called an **isotherm,** meaning "same temperature." If the temperature had been 300 K, rather than 100 K, the compression would have occurred along the 300-K isotherm. Different isotherms do not intersect. Example 4 deals with an application of Boyle's law to scuba diving.

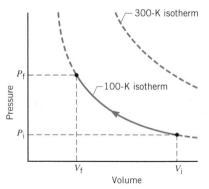

Figure 14.6 A pressure-versus-volume plot for a gas at a constant temperature is called an isotherm. For an ideal gas, each isotherm is a plot of the equation $P = nRT/V = $ constant$/V$.

The physics of **scuba diving.**

Example 4   Scuba* Diving

In scuba diving, a greater water pressure acts on a diver at greater depths. The air pressure inside the body cavities (e.g., lungs, sinuses) must be maintained at the same pressure as that of the surrounding water; otherwise they would collapse. A special valve automatically adjusts

* The word is an acronym for self-contained underwater breathing apparatus.

the pressure of the air breathed from a scuba tank to ensure that the air pressure equals the water pressure at all times. The scuba gear in Figure 14.7 consists of a 0.0150-m³ tank filled with compressed air at an absolute pressure of 2.02×10^7 Pa. Assuming that air is consumed at a rate of 0.0300 m³ per minute and that the temperature is the same at all depths, determine how long the diver can stay under seawater at a depth of (a) 10.0 m and (b) 30.0 m.

Reasoning The time (in minutes) that a scuba diver can remain under water is equal to the volume of air that is available divided by the volume per minute consumed by the diver. The available volume is the volume of air at the pressure P_2 breathed by the diver. This pressure is determined by the depth h beneath the surface, according to $P_2 = P_1 + \rho g h$ (Equation 11.4), where $P_1 = 1.01 \times 10^5$ Pa is the atmospheric pressure at the surface. Since we know the pressure and volume of air in the scuba tank, and since the temperature is constant, we can use Boyle's law to find the volume of air available at the pressure P_2.

Solution

(a) Using $\rho = 1025$ kg/m³ for the density of seawater, we find that the absolute pressure P_2 at the depth of $h = 10.0$ m is

$$P_2 = P_1 + \rho g h = 1.01 \times 10^5 \text{ Pa} + (1025 \text{ kg/m}^3)(9.80 \text{ m/s}^2)(10.0 \text{ m})$$
$$= 2.01 \times 10^5 \text{ Pa}$$

The pressure and volume of the air in the tank are $P_i = 2.02 \times 10^7$ Pa and $V_i = 0.0150$ m³, respectively. According to Boyle's law, the volume of air V_f available at a pressure of $P_f = 2.01 \times 10^5$ Pa is

$$V_f = \frac{P_i V_i}{P_f} = \frac{(2.02 \times 10^7 \text{ Pa})(0.0150 \text{ m}^3)}{2.01 \times 10^5 \text{ Pa}} = 1.51 \text{ m}^3 \qquad (14.3)$$

Of this volume, only 1.51 m³ − 0.0150 m³ = 1.50 m³ is available for breathing, because 0.0150 m³ of air always remains in the tank. At a consumption rate of 0.0300 m³/min, the compressed air will last for

$$t = \frac{1.50 \text{ m}^3}{0.0300 \text{ m}^3/\text{min}} = \boxed{50.0 \text{ min}}$$

(b) The calculation here is like that in part (a). Equation 11.4 indicates that at a depth of 30.0 m, the absolute water pressure is 4.02×10^5 Pa. Because this pressure is twice that at the 10.0-m depth, Boyle's law reveals that the volume of air provided by the tank is now only $V_f = 0.754$ m³. The air available for use is 0.754 m³ − 0.0150 m³ = 0.739 m³. At a consumption rate of 0.0300 m³/min, the air will last for $\boxed{t = 24.6 \text{ min}}$, so the deeper dive must have a shorter duration.

▲

Another investigator whose work contributed to the formulation of the ideal gas law was the Frenchman Jacques Charles (1746–1823). He discovered that at a constant pressure, the volume of a fixed mass (fixed number of moles) of a low-density gas is directly proportional to the Kelvin temperature ($V \propto T$). This relationship is known as Charles' law and can be obtained from the ideal gas law by noting that $V = nRT/P = (\text{constant})T$, if n and P are constant. Equivalently, when an ideal gas changes from an initial volume and temperature (V_i, T_i) to a final volume and temperature (V_f, T_f), it is possible to write $V_i/T_i = nR/P$ and $V_f/T_f = nR/P$. Thus, one way of stating *Charles' law* is

Constant P, constant n
$$\frac{V_i}{T_i} = \frac{V_f}{T_f} \qquad (14.4)$$

Figure 14.7 The air pressure inside the body cavities of a scuba diver must be maintained at the same pressure as that of the surrounding water. (© Chris McLaughlin/Corbis Images)

Problem solving insight
When using the ideal gas law, either directly or in the form of Boyle's law, remember that the pressure P must be the absolute pressure, not the gauge pressure.

✔ **Check Your Understanding 2**

Consider equal masses of the three monatomic gases argon (atomic mass = 39.948 u), krypton (atomic mass = 83.80 u), and xenon (atomic mass = 131.29 u). The pressure and volume of each is the same. Which gas has the greatest and which the smallest temperature? *(The answers are given at the end of the book.)*

Background: The mole, atomic mass, and the ideal gas law relate to this question.

For similar questions (including calculational counterparts), consult Self-Assessment Test 14.1. This test is described at the end of Section 14.4.

14.3 *Kinetic Theory of Gases*

▶ **CONCEPTS AT A GLANCE** As useful as it is, the ideal gas law provides no insight as to how pressure and temperature are related to properties of the molecules themselves, such as their masses and speeds. To show how such microscopic properties are related to the pressure and temperature of an ideal gas, this section examines the dynamics of molecular motion. The pressure that a gas exerts on the walls of a container is due to the force exerted by the gas molecules when they collide with the walls. Therefore, as the

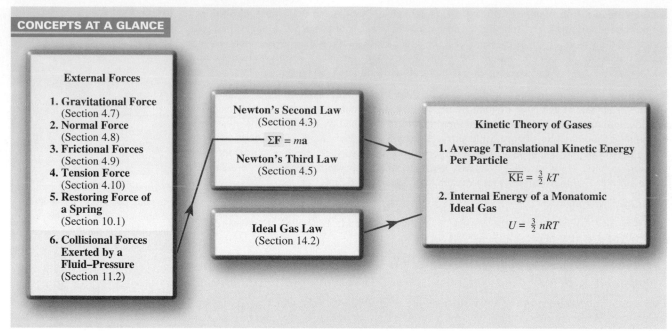

CONCEPTS AT A GLANCE

External Forces

1. **Gravitational Force**
   (Section 4.7)
2. **Normal Force**
   (Section 4.8)
3. **Frictional Forces**
   (Section 4.9)
4. **Tension Force**
   (Section 4.10)
5. **Restoring Force of a Spring**
   (Section 10.1)
6. **Collisional Forces Exerted by a Fluid–Pressure**
   (Section 11.2)

Newton's Second Law
(Section 4.3)

$$\Sigma F = ma$$

Newton's Third Law
(Section 4.5)

Ideal Gas Law
(Section 14.2)

Kinetic Theory of Gases

1. **Average Translational Kinetic Energy Per Particle**

   $$\overline{KE} = \tfrac{3}{2} kT$$

2. **Internal Energy of a Monatomic Ideal Gas**

   $$U = \tfrac{3}{2} nRT$$

Figure 14.8 **CONCEPTS AT A GLANCE** To formulate the kinetic theory of gases, Newton's second and third laws are brought together with the ideal gas law. One of the predictions of the theory is that the internal energy of an ideal gas is proportional to the Kelvin temperature of the gas.

Concepts-at-a-Glance chart in Figure 14.8 illustrates, we will begin by combining the notion of collisional forces exerted by a fluid (Section 11.2) with Newton's second and third laws of motion (Sections 4.3 and 4.5). These concepts will allow us to obtain an expression for the pressure in terms of microscopic properties. We will then combine this with the ideal gas law to show that the average translational kinetic energy $\overline{KE}$ of a particle in an ideal gas is $\overline{KE} = \tfrac{3}{2}kT$, where k is Boltzmann's constant and T is the Kelvin temperature. In the process, we will also see that the internal energy U of a monatomic ideal gas is $U = \tfrac{3}{2}nRT$, where n is the number of moles and R is the universal gas constant. ◄

THE DISTRIBUTION OF MOLECULAR SPEEDS

A macroscopic container filled with a gas at standard temperature and pressure contains a large number of particles (atoms or molecules). These particles are in constant, random motion, colliding with each other and with the walls of the container. In the course of one second, a particle undergoes many collisions, and each one changes the particle's speed and direction of motion. As a result, the atoms or molecules have different speeds. It is possible, however, to speak about an average particle speed. At any given instant, some particles have speeds less than, some near, and some greater than the average. For conditions of low gas density, the distribution of speeds within a large collection of molecules at a constant temperature was calculated by the Scottish physicist James Clerk Maxwell (1831–1879). Figure 14.9 displays the Maxwell speed distribution curves for O_2 gas at

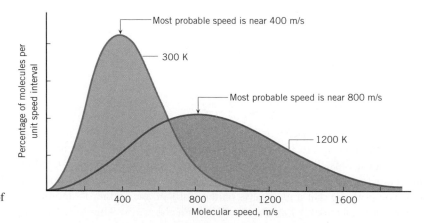

Figure 14.9 The Maxwell distribution curves for molecular speeds in oxygen gas at temperatures of 300 and 1200 K.

two different temperatures. When the temperature is 300 K, the maximum in the curve indicates that the most probable speed is about 400 m/s. At a temperature of 1200 K, the distribution curve shifts to the right, and the most probable speed increases to about 800 m/s.

KINETIC THEORY

If a ball is thrown against a wall, it exerts a force on the wall. As Figure 14.10 suggests, gas particles do the same thing, except that their masses are smaller and their speeds are greater. The number of particles is so great and they strike the wall so often that the effect of their individual impacts appears as a continuous force. Dividing the magnitude of this force by the area of the wall gives the pressure exerted by the gas.

To calculate the force, consider an ideal gas composed of N identical particles in a cubical container whose sides have length L. Except for elastic* collisions, these particles do not interact. Figure 14.11 focuses attention on one particle of mass m as it strikes the right wall perpendicularly and rebounds elastically. While approaching the wall, the particle has a velocity $+v$ and linear momentum $+mv$ (see Section 7.1 for a review of linear momentum). The particle rebounds with a velocity $-v$ and momentum $-mv$, travels to the left wall, rebounds again, and heads back toward the right. The time t between collisions with the right wall is the round-trip distance $2L$ divided by the speed of the particle; that is, $t = 2L/v$. According to Newton's second law of motion, in the form of the impulse–momentum theorem, the average force exerted on the particle by the wall is given by the change in the particle's momentum per unit time:

$$\text{Average force} = \frac{\text{Final momentum} - \text{Initial momentum}}{\text{Time between successive collisions}} \qquad (7.4)$$

$$= \frac{(-mv) - (+mv)}{2L/v} = \frac{-mv^2}{L}$$

According to Newton's law of action–reaction, the force applied to the wall by the particle is equal in magnitude to this value, but oppositely directed (i.e., $+mv^2/L$). The magnitude F of the *total* force exerted on the right wall is equal to the number of particles that collide with the wall during the time t multiplied by the average force exerted by each particle. Since the N particles move randomly in three dimensions, one-third of them on the average strike the right wall during the time t. Therefore, the total force is

$$F = \left(\frac{N}{3}\right)\left(\frac{m\overline{v^2}}{L}\right)$$

In this result v^2 has been replaced by $\overline{v^2}$, *the average value* of the squared speed. The collection of particles possesses a Maxwell distribution of speeds, so an average value for v^2 must be used, rather than a value for any individual particle. The square root of the quantity $\overline{v^2}$ is called the ***root-mean-square speed,*** or, for short, the *rms speed*; $v_{\text{rms}} = \sqrt{\overline{v^2}}$. With this substitution, the total force becomes

$$F = \left(\frac{N}{3}\right)\left(\frac{mv_{\text{rms}}^2}{L}\right)$$

Pressure is force per unit area, so the pressure P acting on a wall of area L^2 is

$$P = \frac{F}{L^2} = \left(\frac{N}{3}\right)\left(\frac{mv_{\text{rms}}^2}{L^3}\right)$$

Since the volume of the box is $V = L^3$, the equation above can be written as

$$PV = \tfrac{2}{3}N(\tfrac{1}{2}mv_{\text{rms}}^2) \qquad (14.5)$$

Equation 14.5 relates the macroscopic properties of the gas, its pressure and volume, to the microscopic properties of the constituent particles, their mass and speed. Since the

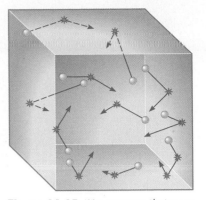

Figure 14.10 The pressure that a gas exerts is caused by the collisions of its molecules with the walls of the container.

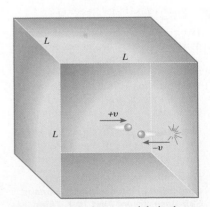

Figure 14.11 A gas particle is shown colliding elastically with the right wall of the container and rebounding from it.

---

* The term "elastic" is used here to mean that *on the average,* in a large number of particles, there is no gain or loss of translational kinetic energy because of collisions.

term $\frac{1}{2}mv_{rms}^2$ is the average translational kinetic energy $\overline{KE}$ of an individual particle, it follows that

$$PV = \tfrac{2}{3}N(\overline{KE})$$

This result is similar to the ideal gas law, $PV = NkT$. Both equations have identical terms on the left, so the terms on the right must be equal: $\frac{2}{3}N(\overline{KE}) = NkT$. Therefore,

$$\overline{KE} = \tfrac{1}{2}mv_{rms}^2 = \tfrac{3}{2}kT \qquad (14.6)$$

Equation 14.6 is significant, because it allows us to interpret temperature in terms of the motion of gas particles. This equation indicates that the Kelvin temperature is directly proportional to the average translational kinetic energy per particle in an ideal gas, no matter what the pressure and volume are. On the average, the particles have greater kinetic energies when the gas is hotter than when it is cooler. Conceptual Example 5 discusses a common misconception about the relation between kinetic energy and temperature.

Conceptual Example 5   Does a Single Particle Have a Temperature?

Each particle in a gas has kinetic energy. Furthermore, the equation $\frac{1}{2}mv_{rms}^2 = \frac{3}{2}kT$ establishes the relationship between the average kinetic energy per particle and the temperature of an ideal gas. Is it valid, then, to conclude that a single particle has a temperature?

Reasoning and Solution We know that a gas contains an enormous number of particles that are traveling with a distribution of speeds, such as those indicated by the graphs in Figure 14.9. Therefore, the particles do not all have the same kinetic energy, but possess a distribution of kinetic energies ranging from very nearly zero to extremely large values. If each particle had a temperature that was associated with its kinetic energy, there would be a whole range of different temperatures within the gas. This is not so, for a gas at thermal equilibrium has only one temperature (see Section 15.2), a temperature that would be registered by a thermometer placed in the gas. Thus, temperature is a property that characterizes the gas as a whole, a fact that is inherent in the relation $\frac{1}{2}mv_{rms}^2 = \frac{3}{2}kT$. The term v_{rms} is a kind of *average particle speed*. Therefore, $\frac{1}{2}mv_{rms}^2$ is the *average kinetic energy* per particle and is characteristic of the gas as a whole. Since the Kelvin temperature is proportional to $\frac{1}{2}mv_{rms}^2$, it is also a characteristic of the gas as a whole and cannot be ascribed to each gas particle individually. Thus, *a single gas particle does not have a temperature.*

If two ideal gases have the same temperature, the relation $\frac{1}{2}mv_{rms}^2 = \frac{3}{2}kT$ indicates that the average kinetic energy of each kind of gas particle is the same. In general, however, the rms speeds of the different particles are not the same, for the masses may be different. The next example illustrates these facts and shows how rapidly gas particles move at normal temperatures.

Example 6   The Speed of Molecules in Air

Air is primarily a mixture of nitrogen N_2 (molecular mass = 28.0 u) and oxygen O_2 (molecular mass = 32.0 u). Assume that each behaves as an ideal gas and determine the rms speeds of the nitrogen and oxygen molecules when the temperature of the air is 293 K.

Reasoning The rms speed can be obtained from Equation 14.6 as $v_{rms} = \sqrt{2\overline{KE}/m_{particle}}$, where $m_{particle}$ is the mass of a single particle (nitrogen or oxygen molecule). In this expression, the average kinetic energy is $\overline{KE} = \frac{3}{2}kT$ and is the same for both types of molecules at the same temperature. However, the masses of the nitrogen and oxygen molecules are different. The mass of each particle is equal to the mass per mole divided by the number N_A of particles per mole (Avogadro's number).

Solution The average kinetic energy per particle for both nitrogen and oxygen is

$$\overline{KE} = \tfrac{3}{2}kT = \tfrac{3}{2}(1.38 \times 10^{-23}\ \text{J/K})(293\ \text{K}) = 6.07 \times 10^{-21}\ \text{J} \qquad (14.6)$$

Since the molecular mass of nitrogen is 28.0 u, its mass per mole is 28.0 g/mol. Therefore, the mass of a nitrogen molecule is

$$m_{particle} = \frac{\text{Mass per mole}}{N_A} = \frac{28.0\ \text{g/mol}}{6.022 \times 10^{23}\ \text{mol}^{-1}} = 4.65 \times 10^{-23}\ \text{g} = 4.65 \times 10^{-26}\ \text{kg}$$

✔ **Check Your Understanding 3**

The pressure of a monatomic ideal gas is doubled, while its volume is reduced by a factor of four. What is the ratio of the new rms speed of the atoms to the initial rms speed? *(The answer is given at the end of the book.)*

Background: The ideal gas law and the kinetic theory of gases are required to answer this question.

For similar questions (including calculational counterparts), consult Self-Assessment Test 14.1. This test is described at the end of Section 14.4.

Similarly, we find the particle mass for oxygen to be 5.31×10^{-26} kg. The calculations of the rms speeds are shown below:

Nitrogen $\qquad v_{rms} = \sqrt{\dfrac{2(\overline{KE})}{m_{particle}}} = \sqrt{\dfrac{2(6.07 \times 10^{-21} \text{ J})}{4.65 \times 10^{-26} \text{ kg}}} = \boxed{511 \text{ m/s}}$

Oxygen $\qquad v_{rms} = \sqrt{\dfrac{2(\overline{KE})}{m_{particle}}} = \sqrt{\dfrac{2(6.07 \times 10^{-21} \text{ J})}{5.31 \times 10^{-26} \text{ kg}}} = \boxed{478 \text{ m/s}}$

For comparison, the speed of sound at a temperature of 293 K is 343 m/s (767 mi/h).

▲

> **Problem solving insight**
> The average translational kinetic energy is the same for all ideal-gas molecules at the same temperature, regardless of their masses. The rms translational speed of the molecules is not the same, however, but depends on the mass.

The equation $\overline{KE} = \frac{3}{2}kT$ has also been applied to particles much larger than atoms or molecules. The English botanist Robert Brown (1773–1858) observed through a microscope that pollen grains suspended in water move on very irregular, zigzag paths. This Brownian motion can also be observed with other particle suspensions, such as fine smoke particles in air. In 1905, Albert Einstein (1879–1955) showed that Brownian motion could be explained as a response of the large suspended particles to impacts from the moving molecules of the fluid medium (e.g., water or air). As a result of the impacts, the suspended particles have the same average translational kinetic energy as the fluid molecules—namely, $\overline{KE} = \frac{3}{2}kT$. But unlike the molecules, the particles are large enough to be seen through a microscope and, because of their relatively large mass, have a comparatively small average speed.

THE INTERNAL ENERGY OF A MONATOMIC IDEAL GAS

Chapter 15 deals with the science of thermodynamics, in which the concept of internal energy plays an important role. Using the results just developed for the average translational kinetic energy, we conclude this section by expressing the internal energy of a monatomic ideal gas in a form that is suitable for use later on.

The internal energy of a substance is the sum of the various kinds of energy that the atoms or molecules of the substance possess. A monatomic ideal gas is composed of single atoms. These atoms are assumed to be so small that the mass is concentrated at a point, with the result that the moment of inertia I about the center of mass is negligible. Thus, the rotational kinetic energy $\frac{1}{2}I\omega^2$ is also negligible. Vibrational kinetic and potential energies are absent, because the atoms are not connected by chemical bonds and, except for elastic collisions, do not interact. As a result, the internal energy U is the total translational kinetic energy of the N atoms that constitute the gas: $U = N(\frac{1}{2}mv_{rms}^2)$. Since $\frac{1}{2}mv_{rms}^2 = \frac{3}{2}kT$ according to Equation 14.6, the internal energy can be written in terms of the Kelvin temperature as

$$U = N(\tfrac{3}{2}kT)$$

Usually, U is expressed in terms of the number of moles n, rather than the number of atoms N. Using the fact that Boltzmann's constant is $k = R/N_A$, where R is the universal gas constant and N_A is Avogadro's number, and realizing that $N/N_A = n$, we find that

Monatomic ideal gas $\qquad\qquad U = \frac{3}{2}nRT \qquad\qquad (14.7)$

Thus, the internal energy depends on the number of moles and Kelvin temperature of the gas. In fact, it can be shown that the internal energy is proportional to the Kelvin temperature for *any type* of ideal gas (e.g., monatomic, diatomic, etc.). For example, when hot-air balloonists turn on the burner, they increase the temperature, and hence the internal energy per mole, of the air inside the balloon (see Figure 14.12).

14.4 Diffusion

You can smell the fragrance of a perfume at some distance from an open bottle because perfume molecules evaporate from the liquid, where they are relatively concentrated, and spread out into the air, where they are less concentrated. During their journey, they collide

Figure 14.12 Since air behaves approximately as an ideal gas, the internal energy per mole inside a hot-air balloon increases as the temperature rises. (© Donovan Reese/Stone/Getty Images)

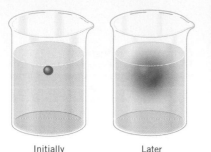

Figure 14.13 A drop of ink placed in water eventually becomes completely dispersed because of diffusion.

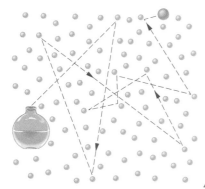

Figure 14.14 A perfume molecule collides with millions of air molecules during its journey, so the path has a zigzag shape. Although the air molecules are shown as stationary, they are also moving.

The physics of drug delivery systems.

with other molecules, so their paths resemble the zigzag paths characteristic of Brownian motion. The process in which molecules move from a region of higher concentration to one of lower concentration is called **diffusion.** Diffusion also occurs in liquids and solids, and Figure 14.13 illustrates ink diffusing through water. However, compared to the rate of diffusion in gases, the rate is generally smaller in liquids and even smaller in solids. The host medium, such as the air or water in the examples above, is referred to as the **solvent,** while the diffusing substance, like the perfume molecules or the ink in Figure 14.13, is known as the **solute.** Relatively speaking, diffusion is a slow process, even in a gas. Conceptual Example 7 illustrates why.

Conceptual Example 7   Why Diffusion Is Relatively Slow

In Example 6 we have seen that a gas molecule has a translational rms speed of hundreds of meters per second at room temperature. At such a speed, a molecule could travel across an ordinary room in just a fraction of a second. Yet, it often takes several seconds, and sometimes minutes, for the fragrance of a perfume to reach the other side of a room. Why does it take so long?

Reasoning and Solution When a perfume molecule diffuses through air, it makes millions of collisions each second with air molecules. As Figure 14.14 illustrates, the speed and direction of motion change abruptly as a result of each collision. Between collisions, the perfume molecule moves in a straight line at a constant speed. Although a perfume molecule does move very fast between collisions, it wanders only slowly away from the bottle because of the zigzag path resulting from the collisions. It would take a long time for a molecule to diffuse in this manner across a room. Usually, however, convection currents are present and carry the fragrance across the room in a matter of seconds or minutes.

Related Homework: *Problems 41, 42, 44*

Diffusion is the basis for drug delivery systems that bypass the need to administer medication orally or via injections. Figure 14.15 shows one such system, the transdermal patch. The word "transdermal" means "across the skin." Such patches, for example, are used to deliver nicotine in programs designed to help you stop smoking. The patch is attached to the skin using an adhesive, and the backing of the patch contains the drug within a reservoir. The concentration of the drug in the reservoir is relatively high, just like the concentration of perfume molecules above the liquid in the bottle in Figure 14.14. The drug diffuses slowly through a control membrane and directly into the skin, where its concentration is relatively low. Diffusion carries it into the blood vessels present in the skin. The purpose of the control membrane is to limit the rate of diffusion, which can also be adjusted by dissolving the drug in the reservoir in a neutral material to lower its initial concentration. Another diffusion-controlled drug delivery system utilizes capsules that are inserted surgically beneath the skin. Contraceptives are administered in this fashion, for instance. The drug in the capsule diffuses slowly into the bloodstream over extended periods that can be as long as a year.

Figure 14.15 Using diffusion, a transdermal patch delivers a drug directly into the skin, where it enters blood vessels. The backing contains the drug within the reservoir, and the control membrane limits the rate of diffusion into the skin. Another way to control the diffusion is to adjust the concentration of the drug in the reservoir by dissolving it in a neutral material.

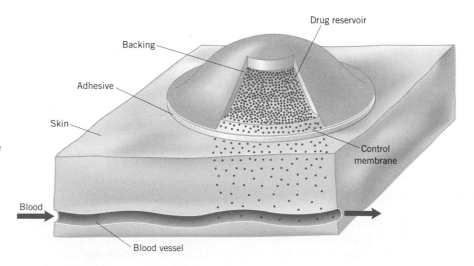

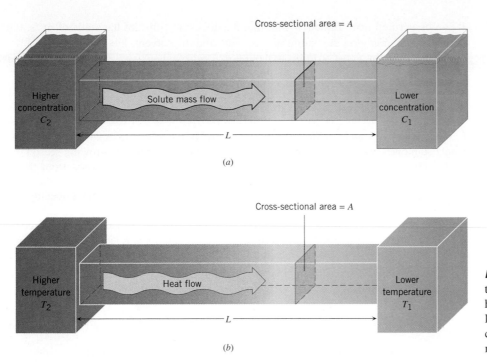

Figure 14.16 (*a*) Solute diffuses through the channel from the region of higher concentration to the region of lower concentration. (*b*) Heat is conducted along a bar whose ends are maintained at different temperatures.

The diffusion process can be described in terms of the arrangement in Figure 14.16*a*. A hollow channel of length L and cross-sectional area A is filled with a fluid. The left end of the channel is connected to a container in which the solute concentration C_2 is relatively high, while the right end is connected to a container in which the solute concentration C_1 is lower. These concentrations are defined as the total mass of the solute molecules divided by the volume of the solution (e.g., 0.1 kg/m^3). Because of the difference in concentration between the ends of the channel, $\Delta C = C_2 - C_1$, there is a net diffusion of the solute from the left end to the right end.

Figure 14.16*a* is similar to Figure 13.8 for the conduction of heat along a bar, which, for convenience, is reproduced in Figure 14.16*b*. When the ends of the bar are maintained at different temperatures, T_2 and T_1, the heat Q conducted along the bar in a time t is

$$Q = \frac{(kA\,\Delta T)t}{L} \tag{13.1}$$

where $\Delta T = T_2 - T_1$, and k is the thermal conductivity. Whereas conduction is the flow of heat from a region of higher temperature to a region of lower temperature, diffusion is the mass flow of solute from a region of higher concentration to a region of lower concentration. By analogy with Equation 13.1, it is possible to write an equation for diffusion: (1) replace Q by the mass m of solute that is diffusing through the channel, (2) replace $\Delta T = T_2 - T_1$ by the difference in concentrations $\Delta C = C_2 - C_1$, and (3) replace k by a constant known as the diffusion constant D. The resulting equation, first formulated by the German physiologist Adolf Fick (1829–1901), is referred to as **Fick's law of diffusion.**

■ **FICK'S LAW OF DIFFUSION**

The mass m of solute that diffuses in a time t through a solvent contained in a channel of length L and cross-sectional area A is

$$m = \frac{(DA\,\Delta C)t}{L} \tag{14.8}$$

where ΔC is the concentration difference between the ends of the channel and D is the diffusion constant.

SI Unit for the Diffusion Constant: m^2/s

It can be verified from Equation 14.8 that the diffusion constant has units of m^2/s, the exact value depending on the nature of the solute and the solvent. For example, the diffusion constant for ink in water is different from that for ink in benzene. Example 8 illustrates an important application of Fick's law.

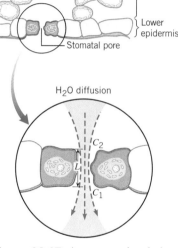

The physics of water loss from plant leaves.

Figure 14.17 A cross-sectional view of a leaf. Water vapor diffuses out of the leaf through a stomatal pore.

Example 8 Water Given Off by Plant Leaves

Large amounts of water can be given off by plants. It has been estimated, for instance, that a single sunflower plant can lose up to a pint of water a day during the growing season. Figure 14.17 shows a cross-sectional view of a leaf. Inside the leaf, water passes from the liquid phase to the vapor phase at the walls of the mesophyll cells. The water vapor then diffuses through the intercellular air spaces and eventually exits the leaf through small openings, called stomatal pores. The diffusion constant for water vapor in air is $D = 2.4 \times 10^{-5}$ m^2/s. A stomatal pore has a cross-sectional area of about $A = 8.0 \times 10^{-11}$ m^2 and a length of about $L = 2.5 \times 10^{-5}$ m. The concentration of water vapor on the interior side of a pore is roughly $C_2 = 0.022$ kg/m^3, while that on the outside is approximately $C_1 = 0.011$ kg/m^3. Determine the mass of water vapor that passes through a stomatal pore in one hour.

Reasoning and Solution Fick's law of diffusion shows that

$$m = \frac{(DA\ \Delta C)t}{L} \tag{14.8}$$

$$m = \frac{(2.4 \times 10^{-5}\ m^2/s)(8.0 \times 10^{-11}\ m^2)(0.022\ kg/m^3 - 0.011\ kg/m^3)(3600\ s)}{2.5 \times 10^{-5}\ m}$$

$$= \boxed{3.0 \times 10^{-9}\ kg}$$

This amount of water may not seem significant. However, a single leaf may have a million or so stomatal pores, so the water lost by an entire plant can be substantial.

✔ Check Your Understanding 4

The same solute is diffusing through the same solvent in each case referred to in the table below, which gives the length and cross-sectional area of the diffusion channel. In each case, the concentration difference between the ends of the diffusion channel is the same. Rank the diffusion rates (in kg/s) in descending order (largest first). *(The answer is given at the end of the book.)*

	Length	Cross-Sectional Area
(a)	$\frac{1}{2}L$	A
(b)	L	$\frac{1}{2}A$
(c)	$\frac{1}{3}L$	$2A$

Background: Fick's law of diffusion is the key.

For similar questions (including calculational counterparts), consult Self-Assessment Test 14.1. The test is described next.

Self-Assessment Test 14.1

Test your understanding of the material in this chapter:

• Molecular Mass, the Mole, and Avogadro's Number   • The Ideal Gas Law
• Kinetic Theory of Gases   • Diffusion

Go to **www.wiley.com/college/cutnell**

14.5 Concepts & Calculations

This chapter introduces the ideal gas law, which is a relation between the pressure, volume, temperature, and number of moles of an ideal gas. In Section 10.1, we examined how the compression of a spring depends on the force applied to it. Example 9 reviews how a gas produces a force and why an ideal gas at different temperatures causes a spring to compress by different amounts.

Concepts & Calculations Example 9
The Ideal Gas Law and Springs

Figure 14.18 shows three identical chambers containing a piston and a spring whose spring constant is $k = 5.8 \times 10^4$ N/m. The chamber in part a is completely evacuated, and the piston just touches its left end. In this position, the spring is unstrained. In part b of the drawing, 0.75 mol of ideal gas 1 is introduced into the chamber, and the spring compresses by $x_1 = 15$ cm. In part c, 0.75 mol of ideal gas 2 is introduced into the chamber, and the spring compresses by $x_2 = 24$ cm. Find the temperature of each gas.

Concept Questions and Answers Which gas exerts the greater force on the piston?

Answer We know that a greater force is required to compress a spring by a greater amount. Therefore, gas 2 exerts the greater force.

How is the force required to compress a spring related to the displacement of the spring from its unstrained position?

Answer According to Equation 10.1, the force F required to compress a spring is directly proportional to the displacement x of the spring from its unstrained position; $F = kx$, where k is the spring constant of the spring.

Which gas exerts the greater pressure on the piston?

Answer According to Equation 11.3, pressure is defined as the magnitude F of the force acting perpendicular to the surface of the piston divided by the area A of the piston; $P = F/A$. Since gas 2 exerts the greater force, and the area of the piston is the same for both gases, gas 2 exerts the greater pressure.

Which gas has the greater temperature?

Answer According to the ideal gas law, $T = PV/(nR)$, gas 2 has the greater temperature. Both gases contain the same number n of moles. However, gas 2 has both a greater pressure P and volume V. Thus, it has the greater temperature.

Solution We can use the ideal gas law in the form $T = PV/(nR)$ to determine the temperature T of each gas. First, however, we must find the pressure. According to Equation 11.3, the pressure is the magnitude F of the force that the gas exerts on the piston divided by the area A of the piston, so $P = F/A$. The force applied to the piston is related to the displacement x of the spring by Equation 10.1, $F = kx$. Thus, the pressure is $P = kx/A$. Using this expression for the pressure in the ideal gas law gives

$$T = \frac{PV}{nR} = \frac{\left(\dfrac{kx}{A}\right)V}{nR}$$

However, the cylindrical volume V of the gas is equal to the product of the displacement x and the area A, so $V = xA$. With this substitution, the temperature of the gas becomes

$$T = \frac{\left(\dfrac{kx}{A}\right)V}{nR} = \frac{\left(\dfrac{kx}{A}\right)(xA)}{nR} = \frac{kx^2}{nR}$$

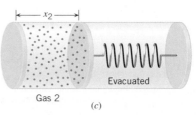

Figure 14.18 Two ideal gases at different temperatures compress the spring by different amounts.

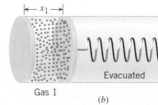

Unstrained spring

Evacuated

(a)

$\leftarrow x_1 \rightarrow$

Evacuated

Gas 1

(b)

$\leftarrow\!\!\!-\; x_2 \;-\!\!\!\rightarrow$

Evacuated

Gas 2

(c)

The temperatures of the gases are

Gas 1 $T_1 = \dfrac{kx_1^2}{nR} = \dfrac{(5.8 \times 10^4 \text{ N/m})(15 \times 10^{-2} \text{ m})^2}{(0.75 \text{ mol})[8.31 \text{ J/(mol}\cdot\text{K})]} = \boxed{210 \text{ K}}$

Gas 2 $T_2 = \dfrac{kx_2^2}{nR} = \dfrac{(5.8 \times 10^4 \text{ N/m})(24 \times 10^{-2} \text{ m})^2}{(0.75 \text{ mol})[8.31 \text{ J/(mol}\cdot\text{K})]} = \boxed{540 \text{ K}}$

As anticipated, gas 2, which compresses the spring more, has the higher temperature.

The kinetic theory of gases is important because it allows us to understand the relation between the macroscopic properties of a gas, such as pressure and temperature, and the microscopic properties of its particles, such as speed and mass. The following example reviews the essential features of this theory.

Concepts & Calculations Example 10
Hydrogen Atoms in Outer Space

In outer space the density of matter is extremely low, about one atom per cm^3. The matter is mainly hydrogen atoms ($m = 1.67 \times 10^{-27}$ kg) whose rms speed is 260 m/s. A cubical box, 2.0 m on a side, is placed in outer space, and the hydrogen atoms are allowed to enter. (a) What is the magnitude of the force that the atoms exert on one wall of the box? (b) Determine the pressure that the atoms exert. (c) Does outer space have a temperature, and, if so, what is it?

Concept Questions and Answers Why do hydrogen atoms exert a force on the walls of the box?

> *Answer*   Every time an atom collides with a wall and rebounds, the atom exerts a force on the wall. Imagine opening your hand so it is flat and having someone throw a ball straight at it. Your hand is like the wall, and as the ball rebounds, you can feel the force. Intuitively, you would expect the force to become greater as the speed and mass of the ball become greater. This is indeed the case.

Do the atoms generate a pressure on the walls of the box?

> *Answer*   Yes. Pressure is defined as the magnitude of the force exerted perpendicularly on a wall divided by the area of the wall. Since the atoms exert a force, they also produce a pressure.

Do hydrogen atoms in outer space have a temperature? If so, how is the temperature related to the microscopic properties of the atoms?

> *Answer*   Yes. The Kelvin temperature is proportional to the average kinetic energy of an atom. The average kinetic energy, in turn, is proportional to the mass of an atom and the square of the rms speed. Since we know both these quantities, we can determine the temperature of the gas.

Solution

(a) The magnitude F of the force exerted on a wall is given by (see Section 14.3)

$$F = \left(\frac{N}{3}\right)\left(\frac{mv_{\text{rms}}^2}{L}\right)$$

where N is the number of atoms in the box, m is the mass of a single atom, v_{rms} is the rms speed of the atoms, and L is the length of one side of the box. The volume of the cubical box is $(2.0 \times 10^2 \text{ cm})^3 = 8.0 \times 10^6 \text{ cm}^3$. The number N of atoms is equal to the number of atoms per cubic centimeter times the volume of the box in cubic centimeters:

$$N = \left(\frac{1 \text{ atom}}{\text{cm}^3}\right)(8.0 \times 10^6 \text{ cm}^3) = 8.0 \times 10^6$$

The magnitude of the force acting on one wall is

$$F = \left(\frac{N}{3}\right)\left(\frac{mv_{\text{rms}}^2}{L}\right) = \left(\frac{8.0 \times 10^6}{3}\right)\left[\frac{(1.67 \times 10^{-27} \text{ kg})(260 \text{ m/s})^2}{2.0 \text{ m}}\right]$$

$$= \boxed{1.5 \times 10^{-16} \text{ N}}$$

(b) The pressure is the magnitude of the force divided by the area A of a wall:

$$P = \frac{F}{A} = \frac{1.5 \times 10^{-16}\,\text{N}}{(2.0\,\text{m})^2} = \boxed{3.8 \times 10^{-17}\,\text{Pa}}$$

(c) According to Equation 14.6, the Kelvin temperature T of the hydrogen atoms is related to the average kinetic energy of an atom by $\frac{3}{2}kT = \frac{1}{2}mv_{\text{rms}}^2$, where k is Boltzmann's constant. Solving this equation for the temperature gives

$$T = \frac{mv_{\text{rms}}^2}{3k} = \frac{(1.67 \times 10^{-27}\,\text{kg})(260\,\text{m/s})^2}{3(1.38 \times 10^{-23}\,\text{J/K})} = \boxed{2.7\,\text{K}}$$

This is a frigid 2.7 kelvins above absolute zero.

▲

At the end of the problem set for this chapter, you will find homework problems that contain both conceptual and quantitative parts. These problems are grouped under the heading *Concepts & Calculations, Group Learning Problems*. They are designed for use by students working alone or in small learning groups. The conceptual part of each problem provides a convenient focus for group discussions.

Concept Summary

This summary presents an abridged version of the chapter, including the important equations and all available learning aids. For convenient reference, the learning aids (including the text's examples) are placed next to or immediately after the relevant equation or discussion. The following learning aids may be found on-line at **www.wiley.com/college/cutnell:**

Interactive LearningWare examples are solved according to a five-step interactive format that is designed to help you develop problem-solving skills.	**Concept Simulations** are animated versions of text figures or animations that illustrate important concepts. You can control parameters that affect the display, and we encourage you to experiment.
Interactive Solutions offer specific models for certain types of problems in the chapter homework. The calculations are carried out interactively.	**Self-Assessment Tests** include both qualitative and quantitative questions. Extensive feedback is provided for both incorrect and correct answers, to help you evaluate your understanding of the material.

Topic	*Discussion*	*Learning Aids*
	14.1 Molecular Mass, the Mole, and Avogadro's Number	
Atomic mass unit	Each element in the periodic table is assigned an atomic mass. One atomic mass unit (u) is exactly one-twelfth the mass of an atom of carbon-12. The molecular mass of a molecule is the sum of the atomic masses of its atoms.	
	The number of moles n contained in a sample is equal to the number of particles N (atoms or molecules) in the sample divided by the number of particles per mole N_A	
Number of moles	$$n = \frac{N}{N_A}$$	
Avogadro's number	where N_A is called Avogadro's number and has a value of $N_A = 6.022 \times 10^{23}$ particles per mole. The number of moles is also equal to the mass m of the sample (expressed in grams) divided by the mass per mole (expressed in grams per mole):	
Number of moles	$$n = \frac{m}{\text{Mass per mole}}$$	Example 1
Mass per mole	The mass per mole (in g/mol) of a substance has the same numerical value as the atomic or molecular mass of one of its particles (in atomic mass units).	
Mass of a particle	The mass m_{particle} of a particle (in grams) can be obtained by dividing the mass per mole (in g/mol) by Avogadro's number:	
	$$m_{\text{particle}} = \frac{\text{Mass per mole}}{N_A}$$	

Topic	Discussion	Learning Aids
	14.2 The Ideal Gas Law	
	The ideal gas law relates the absolute pressure P, the volume V, the number n of moles, and the Kelvin temperature T of an ideal gas according to	
Ideal gas law	$$PV = nRT \qquad (14.1)$$	
Universal gas constant	where $R = 8.31$ J/(mol $\cdot$ K) is the universal gas constant. An alternative form of the ideal gas law is	
Ideal gas law	$$PV = NkT \qquad (14.2)$$	Examples 2, 3, 9
Boltzmann's constant	where N is the number of particles and $k = \dfrac{R}{N_A}$ is Boltzmann's constant. A real gas behaves as an ideal gas when its density is low enough that its particles do not interact, except via elastic collisions.	Interactive Solutions 14.20, 14.21
	A form of the ideal gas law that applies when the number of moles and the temperature are constant is known as Boyle's law. Using the subscripts "i" and "f" to denote, respectively, initial and final conditions, we can write Boyle's law as	
Boyle's law	$$P_i V_i = P_f V_f \qquad (14.3)$$	Example 4
	A form of the ideal gas law that applies when the number of moles and the pressure are constant is called Charles' law:	
Charles' law	$$\frac{V_i}{T_i} = \frac{V_f}{T_f} \qquad (14.4)$$	
	14.3 Kinetic Theory of Gases	
Maxwell speed distribution	The distribution of particle speeds in an ideal gas at constant temperature is the Maxwell speed distribution (see Figure 14.9). The kinetic theory of gases indicates that the Kelvin temperature T of an ideal gas is related to the average translational kinetic energy $\overline{KE}$ of a particle according to	
Average translational kinetic energy	$$\overline{KE} = \tfrac{1}{2}mv_{rms}^2 = \tfrac{3}{2}kT \qquad (14.6)$$	Examples 5, 6, 10
Root-mean-square speed	where v_{rms} is the root-mean-square speed of the particles.	Interactive Solutions 14.35, 14.51
	The internal energy U of n moles of a monatomic ideal gas is	
Internal energy	$$U = \tfrac{3}{2}nRT \qquad (14.7)$$	
	The internal energy of any type of ideal gas (e.g., monatomic, diatomic) is proportional to its Kelvin temperature.	
	14.4 Diffusion	
	Diffusion is the process whereby solute molecules move through a solvent from a region of higher solute concentration to a region of lower solute concentration. Fick's law of diffusion states that the mass m of solute that diffuses in a time t through the solvent in a channel of length L and cross-sectional area A is	
Fick's law of diffusion	$$m = \frac{(DA\,\Delta C)t}{L} \qquad (14.8)$$	Examples 7, 8
	where ΔC is the solute concentration difference between the ends of the channel and D is the diffusion constant.	

Use Self-Assessment Test 14.1 to evaluate your understanding of Sections 14.1–14.4.

Conceptual Questions

1. (a) Which, if either, contains a greater number of molecules, a mole of hydrogen (H_2) or a mole of oxygen (O_2)? (b) Which one has more mass? Give reasons for your answers.

2. Suppose two different substances, A and B, have the same mass densities. (a) In general, does one mole of substance A have the same mass as one mole of substance B? (b) Does 1 m³ of substance A have the same mass as 1 m³ of substance B? Justify each answer.

3. A tightly sealed house has a large ceiling fan that blows air out of the house and into the attic. This fan is turned on, and the owners forget to open any windows or doors. What happens to the air pressure in the house after the fan has been on for a while, and does it become easier or harder for the fan to do its job? Explain.

4. Above the liquid in a can of hair spray is a gas at a relatively high pressure. The label on the can includes the warning "Do not store at high temperatures." Use the ideal gas law and explain why the warning is given.

5. Assuming that air behaves like an ideal gas, explain what happens to the pressure in a tightly sealed house when the electric furnace turns on for a while.

6. When you climb a mountain, your eardrums "pop" outward as the air pressure decreases. When you come down, they "pop" inward as the pressure increases. At the sea coast, there is a cave that can be entered only by swimming through a completely submerged passage and emerging into a pocket of air within the cave. The cave is not vented to the external atmosphere. As the tide comes in, the water level in the cave rises, and your eardrums "pop." Is this "popping" analogous to what happens as you climb up or down a mountain? Give your reasoning.

7. Atmospheric pressure decreases with increasing altitude. With this fact in mind, explain why helium-filled weather balloons are underinflated when they are launched from earth.

8. A slippery cork is being pressed into a very full (but not 100% full) bottle of wine. When released, the cork slowly slides back out. However, if some wine is removed from the bottle before the cork is inserted, the cork does not slide out. Account for these observations in terms of the ideal gas law.

9. A commonly used packing material consists of "bubbles" of air trapped between bonded layers of plastic, as the photograph shows. Using the ideal gas law, explain why this packing material offers less protection on cold days than on warm days.

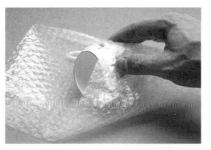

(© Andy Washnik)

10. The kinetic theory of gases assumes that a gas molecule rebounds with the same speed after colliding with the wall of a container. Suppose the speed after the collision is less than that before the collision. Would the pressure of the gas be greater than, equal to, or less than that predicted by kinetic theory? Account for your answer.

11. If the translational speed of each molecule in an ideal gas were tripled, would the Kelvin temperature also triple? If not, by what factor would the Kelvin temperature increase? Account for your answer.

12. If the temperature of an ideal gas is doubled from 50 to 100 °C, does the average translational kinetic energy of each particle double? If not, why not?

13. It is possible for both the pressure and volume of a monatomic ideal gas to change *without* causing the internal energy of the gas to change. Explain how this could occur.

14. Suppose that the atoms in a container of helium (He) have the same translational rms speed as the molecules in a container of argon (Ar). Treating each gas as an ideal gas, explain which, if either, has the greater temperature.

15. The ionosphere is the uppermost part of the earth's atmosphere. The temperature of the ionized gas in the ionosphere is about 1000 K. However, the density of the gas is extremely low, on the order of 10^{11} molecules/m^3. Although the temperature is very high, an astronaut in the ionosphere would not burn up. Why?

16. In the lungs, oxygen in very small sacs called alveoli diffuses into the blood. The diffusion occurs directly through the walls of the sacs. The walls are very thin, so the oxygen diffuses over a distance L that is quite small. Because there are so many alveoli, the effective area A across which diffusion occurs is very large. Use this information, together with Fick's law of diffusion, to explain why the mass of oxygen per second that diffuses into the blood is large.

Problems

Note: The pressures referred to in these problems are absolute pressures, unless indicated otherwise.

Section 14.1 Molecular Mass, the Mole, and Avogadro's Number

1. ssm A mass of 135 g of an element is known to contain 30.1×10^{23} atoms. What is the element?

2. Manufacturers of headache remedies routinely claim that their own brands are more potent pain relievers than the competing brands. The best way of making the comparison is to compare the number of molecules in the standard dosage. Tylenol uses 325 mg of acetaminophen ($C_8H_9NO_2$) as the standard dose, while Advil uses 2.00×10^2 mg of ibuprofen ($C_{13}H_{18}O_2$). Find the number of molecules of pain reliever in the standard doses of (a) Tylenol and (b) Advil.

3. The artificial sweetener NutraSweet is a chemical called aspartame ($C_{14}H_{18}N_2O_5$). What is (a) its molecular mass (in atomic mass units) and (b) the mass (in kg) of an aspartame molecule?

4. The active ingredient in the allergy medication Claritin contains carbon (C), hydrogen (H), chlorine (Cl), nitrogen (N), and oxygen (O). Its molecular formula is $C_{22}H_{23}ClN_2O_2$. The standard adult dosage utilizes 1.572×10^{19} molecules of this species. Determine the mass (in grams) of the active ingredient in the standard dosage.

5. ssm Hemoglobin has a molecular mass of 64 500 u. Find the mass (in kg) of one molecule of hemoglobin.

*6. A cylindrical glass of water (H_2O) has a radius of 4.50 cm and a height of 12.0 cm. The density of water is 1.00 g/cm^3. How many moles of water molecules are contained in the glass?

*7. ssm Estimate the spacing between the centers of neighboring atoms in a piece of solid aluminum, based on a knowledge of the density (2700 kg/m^3) and atomic mass (26.9815 u) of aluminum. *(Hint: Assume that the volume of the solid is filled with many small cubes, with one atom at the center of each.)*

*8. A sample of ethyl alcohol (C_2H_5OH) has a density of 806 kg/m^3 and a volume of 2.00×10^{-3} m^3. (a) Determine the mass (in kg) of a molecule of ethyl alcohol, and (b) find the number of molecules in the liquid.

Section 14.2 The Ideal Gas Law

9. A Goodyear blimp typically contains 5400 m³ of helium (He) at an absolute pressure of 1.1×10^5 Pa. The temperature of the helium is 280 K. What is the mass (in kg) of the helium in the blimp?

10. At the start of a trip, a driver adjusts the absolute pressure in her tires to be 2.81×10^5 Pa when the outdoor temperature is 284 K. At the end of the trip she measures the pressure to be 3.01×10^5 Pa. Ignoring the expansion of the tires, find the air temperature inside the tires at the end of the trip.

11. ssm A bicycle tire whose volume is 4.1×10^{-4} m³ has a temperature of 296 K and an absolute pressure of 4.8×10^5 Pa. A cyclist brings the pressure up to 6.2×10^5 Pa without changing the temperature or volume. How many moles of air must have been pumped into the tire?

12. A clown at a birthday party has brought along a helium cylinder, with which he intends to fill balloons. When full, each balloon contains 0.034 m³ of helium at an absolute pressure of 1.2×10^5 Pa. The cylinder contains helium at an absolute pressure of 1.6×10^7 Pa and has a volume of 0.0031 m³. The temperature of the helium in the tank and in the balloons is the same and remains constant. What is the maximum number of people who will get a balloon?

13. An ideal gas at 15.5 °C and a pressure of 1.72×10^5 Pa occupies a volume of 2.81 m³. (a) How many moles of gas are present? (b) If the volume is raised to 4.16 m³ and the temperature raised to 28.2 °C, what will be the pressure of the gas?

14. Oxygen for hospital patients is kept in special tanks, where the oxygen has a pressure of 65.0 atmospheres and a temperature of 288 K. The tanks are stored in a separate room, and the oxygen is pumped to the patient's room, where it is administered at a pressure of 1.00 atmosphere and a temperature of 297 K. What volume does 1.00 m³ of oxygen in the tanks occupy at the conditions in the patient's room?

15. ssm A young male adult takes in about 5.0×10^{-4} m³ of fresh air during a normal breath. Fresh air contains approximately 21% oxygen. Assuming that the pressure in the lungs is 1.0×10^5 Pa and air is an ideal gas at a temperature of 310 K, find the number of oxygen molecules in a normal breath.

16. A frictionless gas-filled cylinder is fitted with a movable piston, as the drawing shows. The block resting on the top of the piston determines the constant pressure that the gas has. The height h is 0.120 m when the temperature is 273 K and increases as the temperature increases. What is the value of h when the temperature reaches 318 K?

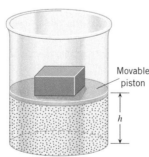

Movable piston

h

17. Two ideal gases have the same mass density and the same absolute pressure. One of the gases is helium (He), and its temperature is 175 K. The other gas is neon (Ne). What is the temperature of the neon?

18. On the sunlit surface of Venus, the atmospheric pressure is 9.0×10^6 Pa, and the temperature is 740 K. On the earth's surface the atmospheric pressure is 1.0×10^5 Pa, while the surface temperature can reach 320 K. These data imply that Venus has a "thicker" atmosphere at its surface than does the earth, which means that the number of molecules per unit volume (N/V) is greater on the surface of Venus than on the earth. Find the ratio $(N/V)_{\text{Venus}}/(N/V)_{\text{Earth}}$.

*19. ssm The relative humidity is 55% on a day when the temperature is 30.0 °C. Using the graph that accompanies Problem 68 in Chapter 12, determine the number of moles of water vapor per cubic meter of air.

*20. Interactive Solution 14.20 at **www.wiley.com/college/cutnell** offers one approach to this problem. One assumption of the ideal gas law is that the atoms or molecules themselves occupy a negligible volume. Verify that this assumption is reasonable by considering gaseous xenon (Xe). Xenon has an atomic radius of 2.0×10^{-10} m. For STP conditions, calculate the percentage of the total volume occupied by the atoms.

*21. Refer to **Interactive Solution 14.21** at **www.wiley.com/college/cutnell** for help with problems like this one. An apartment has a living room whose dimensions are 2.5 m × 4.0 m × 5.0 m. Assume that the air in the room is composed of 79% nitrogen (N_2) and 21% oxygen (O_2). At a temperature of 22 °C and a pressure of 1.01×10^5 Pa, what is the mass (in grams) of the air?

*22. Review Conceptual Example 3 before starting this problem. A bubble, located 0.200 m beneath the surface of the beer, rises to the top. The air pressure at the top is 1.01×10^5 Pa. Assume that the density of beer is the same as that of fresh water. If the temperature and number of moles of CO_2 remain constant as the bubble rises, find the ratio of its volume at the top to that at the bottom.

*23. ssm www A primitive diving bell consists of a cylindrical tank with one end open and one end closed. The tank is lowered into a freshwater lake, open end downward. Water rises into the tank, compressing the trapped air, whose temperature remains constant during the descent. The tank is brought to a halt when the distance between the surface of the water in the tank and the surface of the lake is 40.0 m. Atmospheric pressure at the surface of the lake is 1.01×10^5 Pa. Find the fraction of the tank's volume that is filled with air.

*24. The drawing shows two thermally insulated tanks. They are connected by a valve that is initially closed. Each tank contains neon gas at the pressure, temperature, and volume indicated in the drawing.

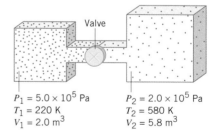

Valve

$P_1 = 5.0 \times 10^5$ Pa
$T_1 = 220$ K
$V_1 = 2.0$ m³

$P_2 = 2.0 \times 10^5$ Pa
$T_2 = 580$ K
$V_2 = 5.8$ m³

When the valve is opened, the contents of the two tanks mix, and the pressure becomes constant throughout. (a) What is the final temperature? Ignore any change in temperature of the tanks themselves. *(Hint: The heat gained by the gas in one tank is equal to that lost by the other.)* (b) What is the final pressure?

**25. A spherical balloon is made from a material whose mass is 3.00 kg. The thickness of the material is negligible compared to the 1.50-m radius of the balloon. The balloon is filled with helium (He) at a temperature of 305 K and just floats in air, neither rising nor falling. The density of the surrounding air is 1.19 kg/m³. Find the absolute pressure of the helium gas.

**26. A gas fills the right portion of a horizontal cylinder whose radius is 5.00 cm. The initial pressure of the gas is 1.01×10^5 Pa. A frictionless movable piston separates the gas from the left portion of the cylinder, which is evacuated and contains an ideal spring, as the draw-

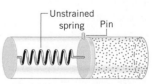

Unstrained spring   Pin

ing shows. The piston is initially held in place by a pin. The spring is initially unstrained, and the length of the gas-filled portion is 20.0 cm. When the pin is removed and the gas is allowed to expand, the length of the gas-filled chamber doubles. The initial and final temperatures are equal. Determine the spring constant of the spring.

** 27. ssm A cylindrical glass beaker of height 1.520 m rests on a table. The bottom half of the beaker is filled with a gas, and the top half is filled with liquid mercury that is exposed to the atmosphere. The gas and mercury do not mix because they are separated by a frictionless, movable piston of negligible mass and thickness. The initial temperature is 273 K. The temperature is increased until a value is reached when one-half of the mercury has spilled out. Ignore the thermal expansion of the glass and mercury, and find this temperature.

Section 14.3 Kinetic Theory of Gases

28. If the translational rms speed of the water vapor molecules (H_2O) in air is 648 m/s, what is the translational rms speed of the carbon dioxide molecules (CO_2) in the same air? Both gases are at the same temperature.

29. ssm The average value of the squared speed $\overline{v^2}$ does not equal the square of the average speed $(\overline{v})^2$. To verify this fact, consider three particles with the following speeds: $v_1 = 3.0$ m/s, $v_2 = 7.0$ m/s, and $v_3 = 9.0$ m/s. Calculate (a) $\overline{v^2} = \frac{1}{3}(v_1^2 + v_2^2 + v_3^2)$ and (b) $(\overline{v})^2 = [\frac{1}{3}(v_1 + v_2 + v_3)]^2$.

30. Near the surface of Venus, the rms speed of carbon dioxide molecules (CO_2) is 650 m/s. What is the temperature (in kelvins) of the atmosphere at that point?

31. The surface of the sun has a temperature of about 6.0×10^3 K. This hot gas contains hydrogen atoms ($m = 1.67 \times 10^{-27}$ kg). Find the rms speed of these atoms.

32. Two gas cylinders are identical. One contains the monatomic gas argon (Ar), and the other contains an equal mass of the monatomic gas krypton (Kr). The pressures in the cylinders are the same, but the temperatures are different. Determine the ratio $\dfrac{\overline{KE}_{Krypton}}{\overline{KE}_{Argon}}$ of the average kinetic energy of a krypton atom to the average kinetic energy of an argon atom.

33. ssm Suppose that a tank contains 680 m^3 of neon at an absolute pressure of 1.01×10^5 Pa. The temperature is changed from 293.2 to 294.3 K. What is the increase in the internal energy of the neon?

34. Initially, the translational rms speed of a molecule of an ideal gas is 463 m/s. The pressure and volume of this gas are kept constant, while the number of molecules is doubled. What is the final translational rms speed of the molecules?

* 35. Interactive Solution 14.35 at **www.wiley.com/college/cutnell** provides a model for problems of this type. The temperature near the surface of the earth is 291 K. A xenon atom (atomic mass = 131.29 u) has a kinetic energy equal to the average translational kinetic energy and is moving straight up. If the atom does not collide with any other atoms or molecules, how high up would it go before coming to rest? Assume that the acceleration due to gravity is constant throughout the ascent.

* 36. Helium (He), a monatomic gas, fills a 0.010-m^3 container. The pressure of the gas is 6.2×10^5 Pa. How long would a 0.25-hp engine have to run (1 hp = 746 W) to produce an amount of energy equal to the internal energy of this gas?

** 37. ssm www In a TV, electrons with a speed of 8.4×10^7 m/s strike the screen from behind, causing it to glow. The electrons come to a halt after striking the screen. Each electron has a mass of 9.11×10^{-31} kg, and there are 6.2×10^{16} electrons per second hit-

ting the screen over an area of 1.2×10^{-7} m^2. What is the pressure that the electrons exert on the screen?

** 38. When perspiration on the human body absorbs heat, some of the perspiration turns into water vapor. The latent heat of vaporization at body temperature (37 °C) is 2.42×10^6 J/kg. The heat absorbed is approximately equal to the average energy $\overline{E}$ given to a single water molecule (H_2O) times the number of water molecules that are vaporized. What is $\overline{E}$?

Section 14.4 Diffusion

39. A tube has a length of 0.015 m and a cross-sectional area of 7.0×10^{-4} m^2. The tube is filled with a solution of sucrose in water. The diffusion constant of sucrose in water is 5.0×10^{-10} m^2/s. A difference in concentration of 3.0×10^{-3} kg/m^3 is maintained between the ends of the tube. How much time is required for 8.0×10^{-13} kg of sucrose to be transported through the tube?

40. The diffusion constant for the amino acid glycine in water is 1.06×10^{-9} m^2/s. In a 2.0-cm-long tube with a cross-sectional area of 1.5×10^{-4} m^2 the mass rate of diffusion is $m/t = 4.2 \times 10^{-14}$ kg/s, because the glycine concentration is maintained at a value of 8.3×10^{-3} kg/m^3 at one end of the tube and at a lower value at the other end. What is the lower concentration?

41. Review Conceptual Example 7 before working this problem. For water vapor in air at 293 K, the diffusion constant is $D = 2.4 \times 10^{-5}$ m^2/s. As outlined in Problem 44(a), the time required for the first solute molecules to traverse a channel of length L is $t = L^2/(2D)$, according to Fick's law. Find the time t for $L = 0.010$ m. (b) For comparison, how long would a water molecule take to travel $L = 0.010$ m at the translational rms speed of water molecules (assumed to be an ideal gas) at a temperature of 293 K? (c) Explain why the answer to part (a) is so much longer than the answer to part (b).

42. Review Conceptual Example 7 before starting this problem. Ammonia, which has a strong smell, is diffusing through air contained in a tube of length L and cross-sectional area 4.0×10^{-4} m^2. The diffusion constant for ammonia in air 4.2×10^{-5} m^2/s. In a certain time, 8.4×10^{-8} kg of ammonia diffuses through the air when the difference in ammonia concentration between the ends of the tube is 3.5×10^{-2} kg/m^3. Find the speed at which ammonia diffuses through the air—that is, the length of the tube divided by the time to travel this length.

* 43. ssm www Carbon tetrachloride (CCl_4) is diffusing through benzene (C_6H_6), as the drawing illustrates. The concentration of CCl_4 at the left end of the tube is maintained at 1.00×10^{-2} kg/m^3, and the diffusion constant is 20.0×10^{-10} m^2/s. The CCl_4 enters the tube at a mass rate of 5.00×10^{-13} kg/s. Using these data and those shown in the drawing, find (a) the mass of CCl_4 per second that passes point A and (b) the concentration of CCl_4 at point A.

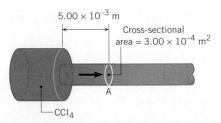

* 44. Review Conceptual Example 7 as background for this problem. It is possible to convert Fick's law into a form that is useful when the concentration is zero at one end of the diffusion channel ($C_1 = 0$ in Figure 14.16a). (a) Noting that AL is the volume V of the channel and that m/V is the average concentration of solute in the channel, show that Fick's law becomes $t = L^2/(2D)$. This form of Fick's law can be used to estimate the time required for the first solute molecules to traverse the channel. (b) A bottle of perfume is opened in a room where convection currents are absent. Assuming that the diffusion constant for perfume in air is 1.0×10^{-5} m^2/s, estimate the minimum time required for the perfume to be smelled 2.5 cm away.

** **45.** The drawing shows a container that is partially filled with 2.0 grams of water. The temperature is maintained at a constant 20 °C. The space above the liquid contains air that is completely saturated with water vapor. A tube of length 0.15 m and cross-sectional area 3.0×10^{-4} m^2 connects the water vapor at one end to air that remains completely dry at the other end. The diffusion constant for

water vapor in air is 2.4×10^{-5} m^2/s. How long does it take for the water in the container to evaporate completely? *(Hint: Refer to Problem 68 in Chapter 12 to find the pressure of the water vapor.)*

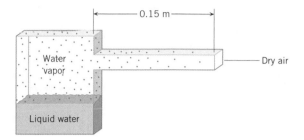

Additional Problems

46. It takes 0.16 g of helium (He) to fill a balloon. How many grams of nitrogen (N$_2$) would be required to fill the balloon to the same pressure, volume, and temperature?

47. ssm In a diesel engine, the piston compresses air at 305 K to a volume that is one-sixteenth of the original volume and a pressure that is 48.5 times the original pressure. What is the temperature of the air after the compression?

48. In a portable oxygen system, the oxygen (O$_2$) is contained in a cylinder whose volume is 0.0028 m^3. A full cylinder has an absolute pressure of 1.5×10^7 Pa when the temperature is 296 K. Find the mass (in kg) of oxygen in the cylinder.

49. ssm The diffusion constant of ethanol in water is 12.4×10^{-10} m^2/s. A cylinder has a cross-sectional area of 4.00 cm^2 and a length of 2.00 cm. A difference in ethanol concentration of 1.50 kg/m^3 is maintained between the ends of the cylinder. In one hour, what mass of ethanol diffuses through the cylinder?

50. Two moles of nitrogen (N$_2$) gas are placed in a container whose volume is 8.5×10^{-3} m^3. The pressure of the gas is 4.5×10^5 Pa. What is the average translational kinetic energy of a nitrogen molecule?

51. Consult **Interactive Solution 14.51** at **www.wiley.com/college/cutnell** to see how this problem can be solved. Very fine smoke particles are suspended in air. The translational rms speed of a smoke particle is 2.8×10^{-3} m/s, and the temperature is 301 K. Find the mass of a particle.

52. A closed cylindrical tank has a height of 0.80 m. The tank is initially filled with air at a pressure of 2.0 atm. Water is pumped into the tank until the air pressure reaches 6.0 atm. The temperature is constant. Determine the height of the compressed air.

* **53. ssm** A drop of water has a radius of 9.00×10^{-4} m. How many water molecules are in the drop?

* **54.** Compressed air can be pumped underground into huge caverns as a form of energy storage. The volume of a cavern is 5.6×10^5 m^3, and the pressure of the air in it is 7.7×10^6 Pa. Assume that air is a diatomic ideal gas whose internal energy U is given by $U = \frac{5}{2}nRT$. If one home uses 30.0 kW·h of energy per day, how many homes could this internal energy serve for one day?

* **55. ssm www** At the normal boiling point of a material, the liquid phase has a density of 958 kg/m^3, and the vapor phase has a density of 0.598 kg/m^3. Determine the ratio of the distance between neighboring molecules in the gas phase to that in the liquid phase. *(Hint: Assume that the volume of each phase is filled with many cubes, with one molecule at the center of each cube.)*

** **56.** In 10.0 s, 200 bullets strike and embed themselves in a wall. The bullets strike the wall perpendicularly. Each bullet has a mass of 5.0×10^{-3} kg and a speed of 1200 m/s. (a) What is the average change in momentum per second for the bullets? (b) Determine the average force exerted on the wall. (c) Assuming the bullets are spread out over an area of 3.0×10^{-4} m^2, obtain the average pressure they exert on this region of the wall.

** **57.** The mass of a hot-air balloon and its occupants is 320 kg (excluding the hot air inside the balloon). The air outside the balloon has a pressure of 1.01×10^5 Pa and a density of 1.29 kg/m^3. To lift off, the air inside the balloon is heated. The volume of the heated balloon is 650 m^3. The pressure of the heated air remains the same as that of the outside air. To what temperature (in kelvins) must the air be heated so that the balloon just lifts off? The molecular mass of air is 29 u.

Concepts & Calculations Group Learning Problems

Note: Each of these problems consists of Concept Questions followed by a related quantitative Problem. They are designed for use by students working alone or in small learning groups. The Concept Questions involve little or no mathematics and are intended to stimulate group discussions. They focus on the concepts with which the problems deal. Recognizing the concepts is the essential initial step in any problem-solving technique.

58. Concept Questions Suppose you know the mass per mole (in units of g/mol) of a substance. (a) Without doing any calculations, how can you determine the mass of one of its atoms (expressed in

atomic mass units)? (b) What information about the sample is obtained when you divide its mass by the mass per mole?

Problem Gold has a mass per mole of 196.967 g/mol. What is the mass of a single gold atom in (a) atomic mass units and (b) kilograms? (c) How many moles of gold atoms are in a 285-g sample?

59. Concept Questions The temperature of an ideal gas is doubled while the volume is kept constant. Does the absolute pressure of the gas double when the temperature that doubles is (a) the Kelvin temperature and (b) the Celsius temperature? Explain.

Problem Determine the ratio P_2/P_1 of the final pressure to the initial pressure when the temperature rises (a) from 35.0 to 70.0 K and (b) from 35.0 to 70.0 °C. Check to see that your answers are consistent with your answers to the Concept Questions.

60. Concept Question Four closed tanks, A, B, C, and D, each contain identical numbers of moles of an ideal gas. The table gives the absolute pressure and volume of the gas in each tank. Which tanks (if any) have the same temperature? Account for your answer.

	A	B	C	D
Absolute pressure (Pa)	25.0	30.0	20.0	2.0
Volume (m³)	4.0	5.0	5.0	75

Problem Each of the tanks contains 0.10 mol of gas. Using this number and the data in the table, compute the temperature of each gas. Verify that your answers are consistent with your answer to the Concept Question.

61. Concept Question Four containers are filled with monatomic ideal gases. For each container, the mass of an individual atom and the rms speed of the atoms are expressed in terms of m and v_{rms}, respectively (see the table). Rank the gases according to their temperatures, highest first. Justify your ranking.

	A	B	C	D
Mass	m	m	$2m$	$2m$
Rms speed	v_{rms}	$2v_{rms}$	v_{rms}	$2v_{rms}$

Problem Suppose that $m = 3.32 \times 10^{-26}$ kg, and $v_{rms} = 1223$ m/s. Find the temperature of the gas in each tank. Be sure that your answers are consistent with your answer to the Concept Question.

62. Concept Questions Imagine a gas, such as helium or neon, diffusing through air in an arrangement like that in Figure 14.16a. The diffusion rate is the number of gas atoms per second diffusing from the left end to the right end of the channel. The faster the atoms move, the greater is the diffusion rate. Therefore, the diffusion rate is proportional to the rms speed of the atoms. Recall, from the kinetic theory of gases, how the average kinetic energy of an atom is related to the temperature of the gas. (a) As the temperature of the gas increases, would you expect the diffusion rate to increase, decrease, or remain the same? Provide a reason for your answer. (b) For a given temperature, which atom, helium or neon, would you expect to have the greater diffusion rate? Why?

Problem The mass of a helium atom is 4.002 60 u and that of a neon atom is 20.179 u. (a) Suppose the temperature of the helium gas is raised to 580 from 290K. Find the ratio of the diffusion rate at the higher temperature to that at the lower temperature. (b) When

the temperature is 290K, determine the ratio of the diffusion rate of helium to that of neon. Check your answers for consistency with respect to your answers to the Concept Questions.

*** 63. Concept Questions** Consider a mixture of three different gases: 1.20 g of argon (molecular mass = 39.948 g/mol), 2.60 g of neon (molecular mass = 20.179 g/mol), and 3.20 g of helium (molecular mass = 4.0026 g/mol). (a) Explain how to calculate the number of moles of each species in terms of its mass and its molecular mass. (b) Of the total number of molecules in the mixture, each component is a certain percentage. Explain how to calculate that percentage. (c) Without doing any detailed calculations, determine which component has the greatest percentage. Give your reasoning. (d) Which component has the smallest percentage? Account for your answer.

Problem For the mixture described in the Concept Questions, determine the percentage of the total number of molecules that corresponds to each of the components. Verify that your answers are consistent with your answers to the Concept Questions.

*** 64. Concept Questions** The drawing shows a gas confined to a cylinder by a massless piston that is attached to an ideal spring. Outside the cylinder is a vacuum. The cross-sectional area of the piston is A. The initial pressure, volume, and temperature of the gas are, respectively, P_0, V_0, and T_0, and the spring is initially stretched by an amount x_0 with respect to its unstrained length. The gas is heated, so that its final pressure, volume, and temperature are P_f, V_f, and T_f and the spring is stretched by an amount x_f with respect to its unstrained length. (a) What is the relation between the magnitude of the force required to stretch an ideal spring and the amount of the stretch with respect to the unstrained length of the spring? (b) What are the magnitudes of the forces that the initial and final pressures apply to the piston and, hence, to the spring? Express your answers in terms of the pressures and the cross-sectional area of the piston. (c) According to the ideal gas law, how are the initial pressure, volume, and temperature related to the final pressure, volume, and temperature? (d) How is the final volume related to the initial volume, the cross-sectional area of the piston, and the initial and final amounts by which the spring is stretched? Account for your answer.

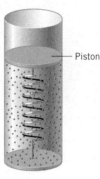
Piston

Problem The initial temperature and volume of the gas described in the Concept Questions are 273 K and 6.00×10^{-4} m³. The initial and final amounts by which the spring is stretched are, respectively, 0.0800 and 0.1000 m. The cross-sectional area of the piston is 2.50×10^{-3} m². What is the final temperature of the gas?

Chapter 15

Thermodynamics

This lava from the Kilauea Volcano in Hawaii is red hot because of heat absorbed deep within the earth. The laws that govern heat and work form the basis of thermodynamics, the subject of this chapter. (© G. Brad Lewis/Stone/Getty Images)

15.1 Thermodynamic Systems and Their Surroundings

We have studied heat (Chapter 12) and work (Chapter 6) as separate topics. Often, however, they occur simultaneously. In an automobile engine, for instance, fuel is burned at a relatively high temperature, some of its internal energy is used for doing the work of driving the pistons up and down, and the excess heat is removed by the cooling system to prevent overheating. ***Thermodynamics*** is the branch of physics that is built upon the fundamental laws that heat and work obey.

In thermodynamics the collection of objects on which attention is being focused is called the ***system,*** while everything else in the environment is called the ***surroundings.*** For example, the system in an automobile engine could be the burning gasoline, while the surroundings would then include the pistons, the exhaust system, the radiator, and the outside air. The system and its surroundings are separated by walls of some kind. Walls that permit heat to flow through them, such as those of the engine block, are called ***diathermal walls.*** Perfectly insulating walls that do not permit heat to flow between the system and its surroundings are known as ***adiabatic walls.***

To understand what the laws of thermodynamics have to say about the relationship between heat and work, it is necessary to describe the physical condition or ***state of a system.*** We might be interested, for instance, in the hot air within the balloon in Figure 15.1. The hot air itself would be the system, and the skin of the balloon provides the walls that separate this system from the surrounding cooler air. The state of the system would be specified by giving values for the pressure, volume, temperature, and mass of the hot air.

As this chapter discusses, there are four laws of thermodynamics. We begin with the one known as the zeroth law and then consider the remaining three.

Figure 15.1 The hot air in a balloon is one example of a thermodynamic system. (© Chase Swift/Corbis Images)

15.2 The Zeroth Law of Thermodynamics

The zeroth law of thermodynamics deals with the concept of ***thermal equilibrium.*** Two systems are said to be in thermal equilibrium if there is no net flow of heat between them when they are brought into thermal contact. For instance, you are *not* in thermal equilibrium with the water in Lake Michigan in January. Just dive into it, and you will find out how quickly your body loses heat to the frigid water. To help explain the central idea of the zeroth law of thermodynamics, Figure 15.2*a* shows two systems labeled A and B. Each is within a container whose adiabatic walls are made from insulation that prevents the flow of heat, and each has the same temperature, as indicated by the thermometers. In part *b*, one wall of each container is replaced by a thin silver sheet, and the two sheets are touched together. Silver has a large thermal conductivity, so heat flows through it readily and the silver sheets behave as diathermal walls. Even though the diathermal walls would permit it, no net flow of heat occurs in part *b*, indicating that the two systems are in thermal equilibrium. There is no net flow of heat because the two systems have the same temperature. We see, then, that ***temperature is the indicator of thermal equilibrium in the sense that there is no net flow of heat between two systems in thermal contact that have the same temperature.***

In Figure 15.2 the thermometer plays an important role. System A is in equilibrium with the thermometer, and so is system B. In each case, the thermometer registers the same temperature, thereby indicating that the two systems are equally hot. Consequently, systems A and B are found to be in thermal equilibrium with each other. In effect, the thermometer is a third system. The fact that system A and system B are each in thermal equilibrium with this third system at the same temperature means that they are in thermal equilibrium with each other. This finding is an example of the ***zeroth law of thermodynamics.***

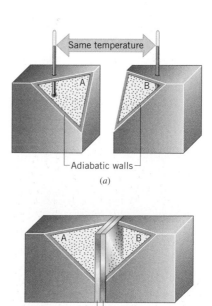

Figure 15.2 (*a*) Systems A and B are surrounded by adiabatic walls and register the same temperature on the thermometer. (*b*) When A is put into thermal contact with B through diathermal walls, no net flow of heat occurs between the systems.

CONCEPTS AT A GLANCE

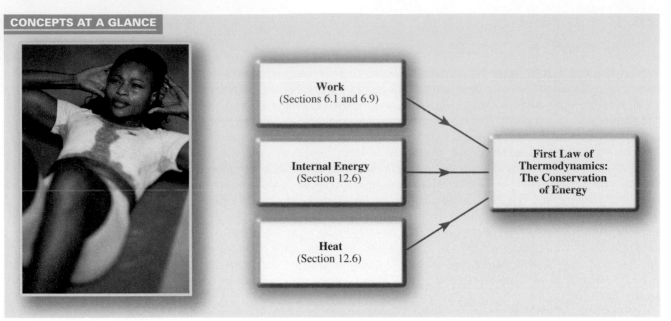

Figure 15.3 CONCEPTS AT A GLANCE
The concepts of work, heat, and the internal energy of a system are related by the first law of thermodynamics, which is an expression of the law of conservation of energy. The body of this person contains internal energy. Part of this energy is used for the workout, and part is lost in the form of heat carried off by evaporating sweat. (© Lori Adamski Peek/Stone/Getty Images)

■ **THE ZEROTH LAW OF THERMODYNAMICS**

Two systems individually in thermal equilibrium with a third system* are in thermal equilibrium with each other.

The zeroth law establishes temperature as the indicator of thermal equilibrium and implies that all parts of a system must be in thermal equilibrium if the system is to have a definable single temperature. In other words, there can be no flow of heat within a system that is in thermal equilibrium.

15.3 *The First Law of Thermodynamics*

▶ CONCEPTS AT A GLANCE In Chapter 6 we saw that forces can do work and that work can change the kinetic and potential energy of an object. For example, the atoms and molecules of a substance exert forces on one another. As a result, they have kinetic and potential energy. These and other kinds of molecular energy constitute the internal energy of a substance. When a substance participates in a process involving energy in the form of work and heat, the internal energy of the substance can change. The relationship between work, heat, and changes in the internal energy is known as the first law of thermodynamics and is illustrated in the Concepts-at-a-Glance chart in Figure 15.3. We will now see that the first law of thermodynamics is an expression of the conservation of energy. ◀

Suppose that a system gains heat Q and that this is the only effect occurring. Consistent with the law of conservation of energy, the internal energy of the system increases from an initial value of U_i to a final value of U_f, the change being $\Delta U = U_f - U_i = Q$. In writing this equation, we use the convention that *heat Q is positive when the system gains heat and negative when the system loses heat.* The internal energy of a system can also change because of work. If a system does work W on its surroundings and there is no heat flow, energy conservation indicates that the internal energy of the system decreases from U_i to U_f, the change now being $\Delta U = U_f - U_i = -W$. The minus sign is included because we follow the convention that *work is positive when it is done by the system and negative when it is done on the system.* A system can gain or lose energy simultaneously in the form of heat Q and work W. The change in internal

Problem solving insight

Problem solving insight

*The state of the third system is the same when it is in thermal equilibrium with either of the two systems. In Figure 15.2, for example, the mercury is at the same position in the thermometer in either system.

energy due to both factors is given by Equation 15.1. Thus, the *first law of thermodynamics* is just the conservation of energy principle applied to heat, work, and the change in the internal energy.

■ **THE FIRST LAW OF THERMODYNAMICS**

The internal energy of a system changes from an initial value U_i to a final value of U_f due to heat Q and work W:

$$\Delta U = U_f - U_i = Q - W \qquad (15.1)$$

Q is positive when the system gains heat and negative when it loses heat. W is positive when work is done by the system and negative when work is done on the system.

Example 1 illustrates the use of Equation 15.1 and the sign conventions for Q and W.

Example 1 Positive and Negative Work

Figure 15.4 illustrates a system and its surroundings. In part *a*, the system gains 1500 J of heat from its surroundings, and 2200 J of work is done *by* the system on the surroundings. In part *b*, the system also gains 1500 J of heat, but 2200 J of work is done *on* the system by the surroundings. In each case, determine the change in the internal energy of the system.

Reasoning In Figure 15.4*a* the system loses more energy in doing work than it gains in the form of heat, so the internal energy of the system decreases. Thus, we expect the change in the internal energy, $\Delta U = U_f - U_i$, to be negative. In part *b* of the drawing, the system gains energy in the form of both heat and work. The internal energy of the system increases, and we expect ΔU to be positive.

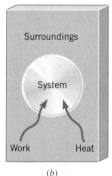

Figure 15.4 (*a*) The system gains energy in the form of heat but loses energy because work is done by the system. (*b*) The system gains energy in the form of heat and also gains energy because work is done on the system.

Solution

(a) The heat is positive, $Q = +1500$ J, since it is gained by the system. The work is positive, $W = +2200$ J, since it is done *by* the system. According to the first law of thermodynamics

$$\Delta U = Q - W = (+1500 \text{ J}) - (+2200 \text{ J}) = \boxed{-700 \text{ J}} \qquad (15.1)$$

The minus sign for ΔU indicates that the internal energy has decreased, as expected.

(b) The heat is positive, $Q = +1500$ J, since it is gained by the system. But the work is negative, $W = -2200$ J, since it is done *on* the system. Thus,

$$\Delta U = Q - W = (+1500 \text{ J}) - (-2200 \text{ J}) = \boxed{+3700 \text{ J}} \qquad (15.1)$$

The plus sign for ΔU indicates that the internal energy has increased, as expected.

Problem solving insight
When using the first law of thermodynamics, as expressed by Equation 15.1, be careful to follow the proper sign conventions for the heat Q and the work W.

In the first law of thermodynamics, the internal energy U, heat Q, and work W are energy quantities, and each is expressed in energy units such as joules. However, there is a fundamental difference between U, on the one hand, and Q and W on the other. The next example sets the stage for explaining this difference.

Example 2 An Ideal Gas

The temperature of three moles of a monatomic ideal gas is reduced from $T_i = 540$ K to $T_f = 350$ K by two different methods. In the first method 5500 J of heat flows into the gas, while in the second, 1500 J of heat flows into it. In each case find (a) the change in the internal energy and (b) the work done by the gas.

Reasoning Since the internal energy of a monatomic ideal gas is $U = \frac{3}{2}nRT$ (Equation 14.7) and since the number of moles n is fixed, only a change in temperature T can alter the internal energy. Because the change in T is the same in both methods, the change in U is also the same. From the given temperatures, the change ΔU in internal energy can be determined. Then, the first law of thermodynamics can be used with ΔU and the given heat values to calculate the work for each of the methods.

Solution

(a) Using Equation 14.7 for the internal energy of a monatomic ideal gas, we find for each method of adding heat that

$$\Delta U = \tfrac{3}{2}nR(T_f - T_i) = \tfrac{3}{2}(3.0 \text{ mol})[8.31 \text{ J/(mol·K)}](350 \text{ K} - 540 \text{ K}) = \boxed{-7100 \text{ J}}$$

(b) Since ΔU is now known and the heat is given in each method, Equation 15.1 can be used to determine the work:

1st method $W = Q - \Delta U = 5500 \text{ J} - (-7100 \text{ J}) = \boxed{12\ 600 \text{ J}}$

2nd method $W = Q - \Delta U = 1500 \text{ J} - (-7100 \text{ J}) = \boxed{8600 \text{ J}}$

In each method the gas does work, but it does more in the first method.

Problem solving insight

To understand the difference between U and either Q or W, consider the value for ΔU in Example 2. In both methods ΔU is the same. Its value is determined once the initial and final temperatures are specified because the internal energy of an ideal gas depends only on the temperature. Temperature is one of the variables (along with pressure and volume) that define the state of a system. ***The internal energy depends only on the state of a system, not on the method by which the system arrives at a given state.*** In recognition of this characteristic, internal energy is referred to as a ***function of state.**** In contrast, heat and work are not functions of state because they have different values for each different method used to make the system change from one state to another, as in Example 2.

✔ **Check Your Understanding 1**

A gas is enclosed within a chamber that is fitted with a frictionless piston. The piston is then pushed in, thereby compressing the gas. Which statement below regarding this process is consistent with the first law of thermodynamics? *(The answer is given at the end of the book.)*

    a. The internal energy of the gas will increase.
    b. The internal energy of the gas will decrease.
    c. The internal energy of the gas will not change.
    d. The internal energy of the gas may increase, decrease, or remain the same, depending on the amount of heat that the gas gains or loses.

Background: Answering this question depends on an understanding of the first law of thermodynamics. This law relates the concepts of internal energy, heat, and work.

For similar questions (including calculational counterparts), consult Self-Assessment Test 15.1. This test is described at the end of Section 15.6.

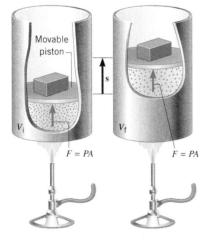

Figure 15.5 The substance in the chamber is expanding isobarically because the pressure is held constant by the external atmosphere and the weight of the piston and the block.

15.4 *Thermal Processes*

A system can interact with its surroundings in many ways, and the heat and work that come into play always obey the first law of thermodynamics. This section introduces four common thermal processes. In each case, the process is assumed to be ***quasi-static,*** which means that it occurs slowly enough that a uniform pressure and temperature exist throughout all regions of the system at all times.

    An isobaric process is one that occurs at constant pressure. For instance, Figure 15.5 shows a substance (solid, liquid, or gas) contained in a chamber fitted with a frictionless piston. The pressure P experienced by the substance is always the same and is determined by the external atmosphere and the weight of the piston and the block resting on it. Heating the substance makes it expand and do work W in lifting the piston and block through the displacement **s**. The work can be calculated from $W = Fs$ (Equation 6.1), where F is the magnitude of the force and s is the magnitude of the displacement. The force is generated by the pressure P acting on the bottom surface of the piston (area $= A$), according to

---

*The fact that an ideal gas is used in Example 2 does not restrict our conclusion. Had a real (nonideal) gas or other material been used, the only difference would have been that the expression for the internal energy would have been more complicated. It might have involved the volume V, as well as the temperature T, for instance.

$F = PA$ (Equation 10.19). With this substitution for F, the work becomes $W = (PA)s$. But the product $A \cdot s$ is the change in volume of the material, $\Delta V = V_f - V_i$, where V_f and V_i are the final and initial volumes, respectively. Thus, the expression for the work is

Isobaric process $W = P \, \Delta V = P(V_f - V_i)$ (15.2)

Consistent with our sign convention, this result predicts a positive value for the work done *by a system* when it expands isobarically (V_f exceeds V_i). Equation 15.2 also applies to an isobaric compression (V_f less than V_i). Then, the work is negative, since work must be done *on the system* to compress it. Example 3 emphasizes that $W = P \, \Delta V$ applies to any system, solid, liquid, or gas, as long as the pressure remains constant while the volume changes.

Example 3 Isobaric Expansion of Water

One gram of water is placed in the cylinder in Figure 15.5, and the pressure is maintained at 2.0×10^5 Pa. The temperature of the water is raised by 31 C°. In one case, the water is in the liquid phase and expands by the small amount of 1.0×10^{-8} m³. In another case, the water is in the gas phase and expands by the much greater amount of 7.1×10^{-5} m³. For the water in each case, find (a) the work done and (b) the change in the internal energy.

Reasoning In both cases the material expands, so work is done by the system (water) on the surroundings and is, therefore, positive. The work is done at a constant pressure, so it can be found from $W = P \, \Delta V$. Once the work is known, the first law of thermodynamics, $\Delta U = Q - W$, can be used to find the change in internal energy, provided a value for the heat Q can be found. The heat needed to raise the temperature can be obtained from $Q = cm \, \Delta T$ (Equation 12.4), where the specific heat capacity of liquid water is $c = 4186$ J/(kg·C°) (see Table 12.2) and the specific heat capacity of water vapor at constant pressure is $c_P = 2020$ J/(kg·C°).

Solution

(a) In both cases the process is isobaric, so the work is given by Equation 15.2:

$$W_{\text{liquid}} = P \, \Delta V = (2.0 \times 10^5 \text{ Pa})(1.0 \times 10^{-8} \text{ m}^3) = \boxed{0.0020 \text{ J}}$$

$$W_{\text{gas}} = P \, \Delta V = (2.0 \times 10^5 \text{ Pa})(7.1 \times 10^{-5} \text{ m}^3) = \boxed{14 \text{ J}}$$

Under comparable conditions, liquids (and solids) change volume much less than gases do, which leads to a much smaller work of expansion or compression for liquids than for gases.

(b) Using $\Delta U = Q - W$ (Equation 15.1) and $Q = cm \, \Delta T$ (Equation 12.4), we find that $\Delta U = cm \, \Delta T - W$. Therefore,

$$\Delta U_{\text{liquid}} = [4186 \text{ J/(kg·C°)}](0.0010 \text{ kg})(31 \text{ C°}) - 0.0020 \text{ J}$$

$$= 130 \text{ J} - 0.0020 \text{ J} = \boxed{130 \text{ J}}$$

$$\Delta U_{\text{gas}} = [2020 \text{ J/(kg·C°)}](0.0010 \text{ kg})(31 \text{ C°}) - 14 \text{ J}$$

$$= 63 \text{ J} - 14 \text{ J} = \boxed{49 \text{ J}}$$

For the liquid, virtually all the 130 J of heat serves to change the internal energy, since the volume change and the corresponding work of expansion are so small. In contrast, a significant fraction of the 63 J of heat added to the vapor causes work of expansion to be done, so that only 49 J is left to change the internal energy.

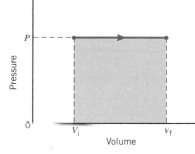

Figure 15.6 For an isobaric process, a pressure-versus-volume plot is a horizontal straight line, and the work done [$W = P(V_f - V_i)$] is the colored rectangular area under the graph.

It is often convenient to display thermal processes graphically. For instance, Figure 15.6 shows a plot of pressure versus volume for an isobaric expansion. Since the pressure is constant, the graph is a horizontal straight line, beginning at the initial volume V_i and ending at the final volume V_f. In terms of such a plot, the work $W = P(V_f - V_i)$ is the area under the graph, which is the shaded rectangle of height P and width $V_f - V_i$.

Another common thermal process is an *isochoric process, one that occurs at constant volume.* Figure 15.7a illustrates an isochoric process in which a substance (solid, liquid, or gas) is heated. The substance would expand if it could, but the rigid container keeps the volume constant, so the pressure–volume plot shown in Figure 15.7b is a vertical straight line. Because the volume is constant, the pressure inside rises, and the substance exerts more and more force on the walls. Although enormous forces can

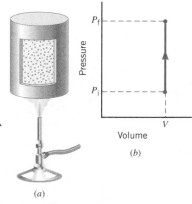

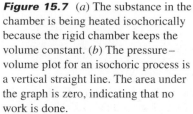

Figure 15.7 (a) The substance in the chamber is being heated isochorically because the rigid chamber keeps the volume constant. (b) The pressure–volume plot for an isochoric process is a vertical straight line. The area under the graph is zero, indicating that no work is done.

be generated in the closed container, no work is done, since the walls do not move. Consistent with zero work being done, the area under the vertical straight line in Figure 15.7*b* is zero. Since no work is done, the first law of thermodynamics indicates that the heat in an isochoric process serves only to change the internal energy: $\Delta U = Q - W = Q$.

A third important thermal process is an ***isothermal process, one that takes place at constant temperature.*** The next section illustrates the important features of an isothermal process when the system is an ideal gas.

Last, there is the ***adiabatic process, one that occurs without the transfer of heat.*** Since there is no heat transfer, Q equals zero, and the first law indicates that $\Delta U = Q - W = -W$. Thus, when work is done by a system adiabatically, W is positive and the internal energy of the system decreases by exactly the amount of the work done. When work is done on a system adiabatically, W is negative and the internal energy increases correspondingly. The next section discusses an adiabatic process for an ideal gas.

A process may be complex enough that it is not recognizable as one of the four just discussed. For instance, Figure 15.8 shows a process for a gas in which the pressure, volume, and temperature are changed along the straight line from X to Y. With the aid of integral calculus, it can be shown that ***the area under a pressure–volume graph is the work for any kind of process,*** so the area representing the work has been colored in the drawing. The volume increases, so that work is done by the gas. This work is positive by convention, as is the area. In contrast, if a process reduces the volume, work is done on the gas, and this work is negative by convention. Correspondingly, the area under the pressure–volume graph would be assigned a negative value. In Example 4, we determine the work for the case shown in Figure 15.8.

Example 4 Work and the Area Under a Pressure–Volume Graph

Determine the work for the process in which the pressure, volume, and temperature of a gas are changed along the straight line from X to Y in Figure 15.8.

Reasoning The work is given by the area (in color) under the straight line between X and Y. Since the volume increases, work is done by the gas on the surroundings, so the work is positive. The area can be found by counting squares in Figure 15.8 and multiplying by the area per square.

Solution We estimate that there are 8.9 colored squares in the drawing. The area of one square is $(2.0 \times 10^5 \text{ Pa})(1.0 \times 10^{-4} \text{ m}^3) = 2.0 \times 10^1 \text{ J}$, so the work is

$$W = +(8.9 \text{ squares})(2.0 \times 10^1 \text{ J/square}) = \boxed{+180 \text{ J}}$$

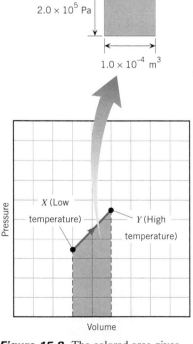

Figure 15.8 The colored area gives the work done by the gas for the process from X to Y.

✓ Check Your Understanding 2

The drawing shows a pressure-versus-volume plot for a three-step process: A → B, B → C, and C → A. For each step, the work can be positive, negative, or zero. Which answer below correctly describes the work for the three steps? *(The answer is given at the end of the book.)*

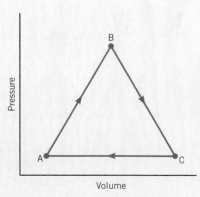

(Continues)

Work Done by the System			
	$A \rightarrow B$	$B \rightarrow C$	$C \rightarrow A$
a.	Positive	Negative	Negative
b.	Positive	Positive	Negative
c.	Negative	Negative	Positive
d.	Positive	Negative	Zero
e.	Negative	Positive	Zero

Background: For any kind of process, the work done can be evaluated by using a pressure-versus-volume plot. Recall that work is positive, negative, or zero, depending on how the volume changes.

For similar questions (including calculational counterparts), consult Self-Assessment Test 15.1. This test is described at the end of Section 15.6.

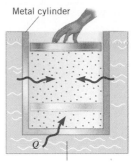

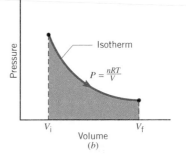

Figure 15.9 (*a*) The ideal gas in the cylinder is expanding isothermally at temperature *T*. The force holding the piston in place is reduced slowly, so the expansion occurs quasi-statically. (*b*) The work done by the gas is given by the colored area.

15.5 *Thermal Processes Using an Ideal Gas*

ISOTHERMAL EXPANSION OR COMPRESSION

When a system performs work isothermally, the temperature remains constant. In Figure 15.9*a*, for instance, a metal cylinder contains *n* moles of an ideal gas, and the large mass of hot water maintains the cylinder and gas at a constant Kelvin temperature *T*. The piston is held in place initially so the volume of the gas is V_i. As the external force applied to the piston is reduced quasi-statically, the gas expands to the final volume V_f. Figure 15.9*b* gives a plot of pressure ($P = nRT/V$) versus volume for the process. The solid red line in the graph is called an isotherm (meaning "constant temperature") and represents the relation between pressure and volume when the temperature is held constant. The work *W* done by the gas is *not* given by $W = P \, \Delta V = P(V_f - V_i)$ because the pressure is not constant. Nevertheless, the work is equal to the area under the graph. The techniques of integral calculus lead to the following result* for *W*:

Isothermal expansion or compression of an ideal gas

$$W = nRT \ln\left(\frac{V_f}{V_i}\right) \qquad (15.3)$$

Where does the energy for this work originate? Since the internal energy of any ideal gas is proportional to the Kelvin temperature ($U = \frac{3}{2}nRT$ for a monatomic ideal gas, for example), the internal energy remains constant throughout an isothermal process, and the change in internal energy is zero. The first law of thermodynamics becomes $\Delta U = 0 = Q - W$. In other words, $Q = W$, and the energy for the work originates in the hot water. Heat flows into the gas from the water, as Figure 15.9*a* illustrates. If the gas is compressed isothermally, Equation 15.3 still applies, and heat flows out of the gas into the water. The following example deals with the isothermal expansion of an ideal gas.

Example 5 Isothermal Expansion of an Ideal Gas

Two moles of the monatomic gas argon expand isothermally at 298 K, from an initial volume of $V_i = 0.025$ m³ to a final volume of $V_f = 0.050$ m³. Assuming that argon is an ideal gas, find (a) the work done by the gas, (b) the change in the internal energy of the gas, and (c) the heat supplied to the gas.

Reasoning and Solution

(a) The work done by the gas can be found from Equation 15.3:

$$W = nRT \ln\left(\frac{V_f}{V_i}\right) = (2.0 \text{ mol})[8.31 \text{ J/(mol} \cdot \text{K)}](298 \text{ K}) \ln\left(\frac{0.050 \text{ m}^3}{0.025 \text{ m}^3}\right) = \boxed{+3400 \text{ J}}$$

*In this result, "ln" denotes the natural logarithm to the base $e = 2.71828$. The natural logarithm is related to the common logarithm to the base ten by $\ln(V_f/V_i) = 2.303 \log(V_f/V_i)$.

Need more practice?

Interactive LearningWare 15.1
A bubble from the tank of a scuba diver contains 3.5×10^{-4} mol of gas. The bubble expands as it rises to the surface from a freshwater depth of 10.3 m. Assuming that the gas is an ideal gas and the temperature remains constant at 291 K, find the amount of heat that flows into the bubble.

Go to
www.wiloy.com/college/cutnell
for an interactive solution.

(b) The internal energy of a monatomic ideal gas is $U = \frac{3}{2}nRT$ (Equation 14.7) and does not change when the temperature is constant. Therefore, $\boxed{\Delta U = 0 \text{ J}}$.

(c) The heat Q supplied can be determined from the first law of thermodynamics:

$$Q = \Delta U + W = 0 \text{ J} + 3400 \text{ J} = \boxed{+3400 \text{ J}} \qquad (15.1)$$

ADIABATIC EXPANSION OR COMPRESSION

When a system performs work adiabatically, no heat flows into or out of the system. Figure 15.10a shows an arrangement in which n moles of an ideal gas do work under adiabatic conditions, expanding quasi-statically from an initial volume V_i to a final volume V_f. The arrangement is similar to that in Figure 15.9 for isothermal expansion. However, a different amount of work is done here, because the cylinder is now surrounded by insulating material that prevents the flow of heat, so $Q = 0$ J. According to the first law of thermodynamics, the change in internal energy is $\Delta U = Q - W = -W$. Since the internal energy of an ideal monatomic gas is $U = \frac{3}{2}nRT$ (Equation 14.7), it follows that $\Delta U = U_f - U_i = \frac{3}{2}nR(T_f - T_i)$, where T_i and T_f are the initial and final Kelvin temperatures. With this substitution, the relation $\Delta U = -W$ becomes

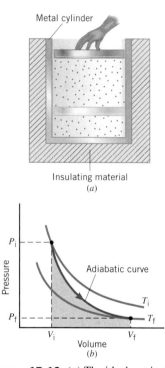

Metal cylinder

Insulating material

(a)

Pressure

P_i

Adiabatic curve

T_i

P_f

T_f

V_i

V_f

Volume

(b)

Figure 15.10 (a) The ideal gas in the cylinder is expanding adiabatically. The force holding the piston in place is reduced slowly, so the expansion occurs quasi-statically. (b) A plot of pressure versus volume yields the adiabatic curve shown in red, which intersects the isotherms (blue) at the initial temperature T_i and the final temperature T_f. The work done by the gas is given by the colored area.

Adiabatic expansion or compression of a monatomic ideal gas

$$W = \frac{3}{2}nR(T_i - T_f) \qquad (15.4)$$

When an ideal gas expands adiabatically, it does positive work, so W is positive in Equation 15.4. Therefore, the term $T_i - T_f$ is also positive, so the final temperature of the gas must be less than the initial temperature. The internal energy of the gas is reduced to provide the necessary energy to do the work, and because the internal energy is proportional to the Kelvin temperature, the temperature decreases. Figure 15.10b shows a plot of pressure versus volume for an adiabatic process. The adiabatic curve (red) intersects the isotherms (blue) at the higher initial temperature $[T_i = P_iV_i/(nR)]$ and the lower final temperature $[T_f = P_fV_f/(nR)]$. The colored area under the adiabatic curve represents the work done.

The reverse of an adiabatic expansion is an adiabatic compression (W is negative), and Equation 15.4 indicates that the final temperature exceeds the initial temperature. The energy provided by the agent doing the work increases the internal energy of the gas. As a result, the gas becomes hotter.

The equation that gives the adiabatic curve (red) between the initial pressure and volume (P_i, V_i) and the final pressure and volume (P_f, V_f) in Figure 15.10b can be derived using integral calculus. The result is

Adiabatic expansion or compression of an ideal gas

$$P_iV_i^{\gamma} = P_fV_f^{\gamma} \qquad (15.5)$$

where the exponent γ (Greek gamma) is the ratio of the specific heat capacities at constant pressure and constant volume, $\gamma = c_P/c_V$. Equation 15.5 applies in conjunction with the ideal gas law, for *each point* on the adiabatic curve satisfies the relation $PV = nRT$.

Table 15.1 summarizes the work done in the four types of thermal processes that we have been considering. For each process it also shows how the first law of thermodynamics depends on the work and other variables.

15.6 *Specific Heat Capacities*

In this section the first law of thermodynamics is used to gain an understanding of the factors that determine the specific heat capacity of a material. Remember, when the temperature of a substance changes as a result of heat flow, the change in temperature ΔT and

Table 15.1   Summary of Thermal Processes

Type of Thermal Process	Work Done	First Law of Thermodynamics ($\Delta U = Q - W$)
Isobaric (constant pressure)	$W = P(V_f - V_i)$	$\Delta U = Q - \underbrace{P(V_f - V_i)}_{W}$
Isochoric (constant volume)	$W = 0 \text{ J}$	$\Delta U = Q - \underbrace{0 \text{ J}}_{W}$
Isothermal (constant temperature)	$W = nRT \ln\left(\dfrac{V_f}{V_i}\right)$ (for an ideal gas)	$\underbrace{0 \text{ J}}_{\substack{\Delta U \text{ for an} \\ \text{ideal gas}}} = Q - \underbrace{nRT \ln\left(\dfrac{V_f}{V_i}\right)}_{W}$
Adiabatic (no heat flow)	$W = \frac{3}{2}nR(T_f - T_i)$ (for a monatomic ideal gas)	$\Delta U = \underbrace{0 \text{ J}}_{Q} - \underbrace{\frac{3}{2}nR(T_i - T_f)}_{W}$

the amount of heat Q are related according to $Q = cm\,\Delta T$ (Equation 12.4). In this expression c denotes the specific heat capacity in units of J/(kg·C°), and m is the mass in kilograms. It is more convenient now to express the amount of material as the number of moles n, rather than the number of kilograms. Therefore, we replace the expression $Q = cm\,\Delta T$ with the following analogous expression:

$$Q = Cn\,\Delta T \tag{15.6}$$

where the capital letter C (as opposed to the lowercase c) refers to the **molar specific heat capacity** in units of J/(mol·K). In addition, the unit for measuring the temperature change ΔT is the Kelvin (K) rather than the Celsius degree (C°), and $\Delta T = T_f - T_i$, where T_f and T_i are the final and initial temperatures. For gases it is necessary to distinguish between the molar specific heat capacities C_P and C_V, which apply, respectively, to conditions of constant pressure and constant volume. With the help of the first law of thermodynamics and an ideal gas as an example, it is possible to see why C_P and C_V differ.

To determine the molar specific heat capacities, we must first calculate the heat Q needed to raise the temperature of an ideal gas from T_i to T_f. According to the first law, $Q = \Delta U + W$. We also know that the internal energy of a monatomic ideal gas is $U = \frac{3}{2}nRT$ (Equation 14.7). As a result, $\Delta U = U_f - U_i = \frac{3}{2}nR(T_f - T_i)$. When the heating process occurs at constant pressure, the work done is given by Equation 15.2: $W = P\,\Delta V = P(V_f - V_i)$. For an ideal gas, $PV = nRT$, so the work becomes $W = nR(T_f - T_i)$. On the other hand, when the volume is constant, $\Delta V = 0 \text{ m}^3$, and the work done is zero. The calculation of the heat is summarized below:

$$Q = \Delta U + W$$

$$Q_{\text{constant pressure}} = \tfrac{3}{2}nR(T_f - T_i) + nR(T_f - T_i) = \tfrac{5}{2}nR(T_f - T_i)$$

$$Q_{\text{constant volume}} = \tfrac{3}{2}nR(T_f - T_i) + 0$$

The molar specific heat capacities can now be determined, since Equation 15.6 indicates that $C = Q/[n(T_f - T_i)]$:

Constant pressure for a monatomic ideal gas
$$C_P = \frac{Q_{\text{constant pressure}}}{n(T_f - T_i)} = \tfrac{5}{2}R \tag{15.7}$$

Constant volume for a monatomic ideal gas
$$C_V = \frac{Q_{\text{constant volume}}}{n(T_f - T_i)} = \tfrac{3}{2}R \tag{15.8}$$

The ratio γ of the specific heats is

Monatomic ideal gas
$$\gamma = \frac{C_P}{C_V} = \frac{\tfrac{5}{2}R}{\tfrac{3}{2}R} = \frac{5}{3} \tag{15.9}$$

For real monatomic gases near room temperature, experimental values of C_P and C_V give ratios very close to the theoretical value of $\frac{5}{3}$.

The difference between C_P and C_V arises because work is done when the gas expands in response to the addition of heat under conditions of constant pressure, whereas no work is done under conditions of constant volume. For a monatomic ideal gas, C_P exceeds C_V by an amount equal to R, the ideal gas constant:

$$C_P - C_V = R \qquad (15.10)$$

In fact, it can be shown that Equation 15.10 applies to any kind of ideal gas—monatomic, diatomic, etc.

Self-Assessment Test 15.1

Test your understanding of the central ideas presented in Sections 15.1–15.6:

- The Zeroth and First Laws of Thermodynamics   • Thermal Processes
- Specific Heat Capacities

Go to **www.wiley.com/college/cutnell**

15.7 *The Second Law of Thermodynamics*

Ice cream melts when left out on a warm day. A cold can of soda warms up on a hot day at a picnic. Ice cream and soda never become colder when left in a hot environment, for heat always flows spontaneously from hot to cold, and never from cold to hot. The spontaneous flow of heat is the focus of one of the most profound laws in all of science, the *second law of thermodynamics.*

■ **THE SECOND LAW OF THERMODYNAMICS: THE HEAT FLOW STATEMENT**

Heat flows spontaneously from a substance at a higher temperature to a substance at a lower temperature and does not flow spontaneously in the reverse direction.

▶ **CONCEPTS AT A GLANCE** It is important to realize that the second law of thermodynamics deals with a different aspect of nature than does the first law of thermodynamics. The second law is a statement about the natural tendency of heat to flow from hot to cold, whereas the first law deals with energy conservation and focuses on both heat and

Figure 15.11 CONCEPTS AT A GLANCE The first and second laws of thermodynamics are used to evaluate the performance of heat engines, as well as refrigerators, air conditioners, and heat pumps. The F-15 Eagle in this photograph uses two heat engines to propel itself forward. (© George Hall/Corbis Images)

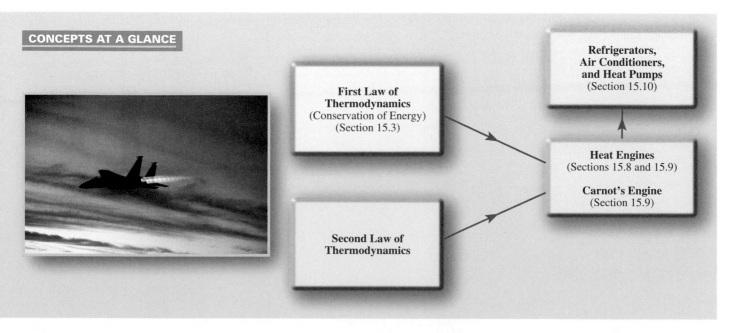

CONCEPTS AT A GLANCE

First Law of Thermodynamics (Conservation of Energy) (Section 15.3)

Second Law of Thermodynamics

Heat Engines (Sections 15.8 and 15.9)

Carnot's Engine (Section 15.9)

Refrigerators, Air Conditioners, and Heat Pumps (Section 15.10)

work. A number of important devices depend on heat and work in their operation, and to understand such devices both laws are needed. For instance, an automobile engine is a type of heat engine because it uses heat to produce work. In discussing heat engines, Sections 15.8 and 15.9 will bring together the first and second laws to analyze engine efficiency, as the Concepts-at-a-Glance chart in Figure 15.11 indicates. Then, in Section 15.10 we will see that refrigerators, air conditioners, and heat pumps also utilize heat and work and are closely related to heat engines. The way in which these three appliances operate also depends on both the first and second laws of thermodynamics. ◄

15.8 *Heat Engines*

A *heat engine* is any device that uses heat to perform work. It has three essential features:

The physics of a heat engine.

1. Heat is supplied to the engine at a relatively high input temperature from a place called the *hot reservoir.*

2. Part of the input heat is used to perform work by the *working substance* of the engine, which is the material within the engine that actually does the work (e.g., the gasoline–air mixture in an automobile engine).

3. The remainder of the input heat is rejected to a place called the *cold reservoir,* which has a temperature lower than the input temperature.

Figure 15.12 emphasizes these features in a schematic fashion. The symbol Q_H refers to the magnitude of the input heat, and the subscript H indicates the hot reservoir. The symbol Q_C refers to the magnitude of the rejected heat, and the subscript C denotes the cold reservoir. The symbol W stands for the magnitude of the work done. *These three symbols refer to magnitudes only, without reference to algebraic signs. Therefore, when these symbols appear in an equation, they do not have negative values assigned to them.*

Problem solving insight

To be highly efficient, a heat engine must produce a relatively large amount of work from as little input heat as possible. Thus, the *efficiency e* of a heat engine is defined as the ratio of the work W done by the engine to the input heat Q_H:

$$e = \frac{\text{Work done}}{\text{Input heat}} = \frac{W}{Q_H} \qquad (15.11)$$

If the input heat were converted entirely into work, the engine would have an efficiency of 1.00, since $W = Q_H$; such an engine would be 100% efficient. *Efficiencies are often quoted as percentages obtained by multiplying the ratio W/Q_H by a factor of 100.* Thus, an efficiency of 68% would mean that a value of 0.68 is used for the efficiency in Equation 15.11.

An engine, like any device, must obey the principle of conservation of energy. Some of the engine's input heat Q_H is converted into work W, and the remainder Q_C is rejected to the cold reservoir. If there are no other losses in the engine, the principle of energy conservation requires that

$$Q_H = W + Q_C \qquad (15.12)$$

Solving this equation for W and substituting the result into Equation 15.11 leads to the following alternative expression for the efficiency e of a heat engine:

$$e = \frac{Q_H - Q_C}{Q_H} = 1 - \frac{Q_C}{Q_H} \qquad (15.13)$$

Example 6 illustrates how the concepts of efficiency and energy conservation are applied to a heat engine.

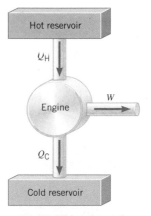

Figure 15.12 This schematic representation of a heat engine shows the input heat (magnitude = Q_H) that originates from the hot reservoir, the work (magnitude = W) that the engine does, and the heat (magnitude = Q_C) that the engine rejects to the cold reservoir.

Example 6 An Automobile Engine

An automobile engine has an efficiency of 22.0% and produces 2510 J of work. How much heat is rejected by the engine?

Reasoning Energy conservation indicates that the amount of heat rejected to the cold reservoir is the part of the input heat that is not converted into work. The amount Q_C rejected is

$Q_C = Q_H - W$, according to Equation 15.12. To use this equation, however, we need a value for the input heat Q_H, so we begin our solution by finding this value.

Solution From Equation 15.11 for the efficiency e, we find that $Q_H = W/e$. Substituting this result into Equation 15.12, we see that the rejected heat is

$$Q_C = Q_H - W = \frac{W}{e} - W = (2510 \text{ J})\left(\frac{1}{0.220} - 1\right) = \boxed{8900 \text{ J}}$$

Problem solving insight
When efficiency is stated as a percentage (e.g., 22.0%), it must be converted to a decimal fraction (e.g., 0.220) before being used in an equation.

In Example 6, less than one-quarter of the input heat is converted into work because the efficiency of the automobile engine is only 22.0%. If the engine were 100% efficient, all the input heat would be converted into work. Unfortunately, nature does not permit 100% efficient heat engines to exist, as the next section discusses.

15.9 *Carnot's Principle and the Carnot Engine*

What is it that allows a heat engine to operate with maximum efficiency? The French engineer Sadi Carnot (1796–1832) proposed that a heat engine has maximum efficiency when the processes within the engine are reversible. *A reversible process is one in which both the system and its environment can be returned to exactly the states they were in before the process occurred.*

In a reversible process, *both* the system and its environment can be returned to their initial states. Therefore, a process that involves an energy-dissipating mechanism, such as friction, cannot be reversible because the energy wasted due to friction would alter the system or the environment or both. There are also reasons other than friction why a process may not be reversible. For instance, the spontaneous flow of heat from a hot substance to a cold substance is irreversible, even though friction is not present. For heat to flow in the reverse direction, work must be done, as we will see in Section 15.10. The agent doing such work must be located in the environment of the hot and cold substances, and, therefore, the environment must change while the heat is moved back from cold to hot. Since the system and the environment cannot *both* be returned to their initial states, the process of spontaneous heat flow is irreversible. In fact, all spontaneous processes are irreversible, such as the explosion of an unstable chemical or the bursting of a bubble. When the word "reversible" is used in connection with engines, it does not just mean a gear that allows the engine to operate a device in reverse. All cars have a reverse gear, for instance, but no automobile engine is thermodynamically reversible, since friction exists no matter which way the car moves.

Today, the idea that the efficiency of a heat engine is a maximum when the engine operates reversibly is referred to as *Carnot's principle.*

■ **CARNOT'S PRINCIPLE: AN ALTERNATIVE STATEMENT OF THE SECOND LAW OF THERMODYNAMICS**

No irreversible engine operating between two reservoirs at constant temperatures can have a greater efficiency than a reversible engine operating between the same temperatures. Furthermore, all reversible engines operating between the same temperatures have the same efficiency.

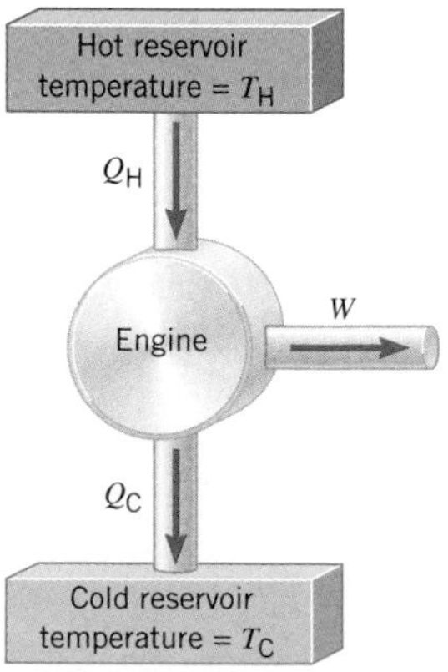

Figure 15.13 A Carnot engine is a reversible engine in which all input heat Q_H originates from a hot reservoir at a single temperature T_H, and all rejected heat Q_C goes into a cold reservoir at a single temperature T_C. The work done by the engine is W.

Carnot's principle is quite remarkable, for no mention is made of the working substance of the engine. It does not matter whether the working substance is a gas, a liquid, or a solid. As long as the process is reversible, the efficiency of the engine is a maximum. Furthermore, Carnot's principle does *not* state, or even imply, that a reversible engine has an efficiency of 100%.

It can be shown that if Carnot's principle were not valid, it would be possible for heat to flow spontaneously from a cold substance to a hot substance, in violation of the second law of thermodynamics. In effect, then, Carnot's principle is another way of expressing the second law.

No real engine operates reversibly. Nonetheless, the idea of a reversible engine provides a useful standard for judging the performance of real engines. Figure 15.13 shows a reversible engine, called a **Carnot engine,** that is particularly useful as an idealized model. An important feature of the Carnot engine is that all input heat Q_H originates from a hot reservoir at a *single temperature* T_H and all rejected heat Q_C goes into a cold reservoir *at a single temperature* T_C.

Carnot's principle implies that the efficiency of a reversible engine is independent of the working substance of the engine, and therefore can depend only on the temperatures of the hot and cold reservoirs. Since efficiency $e = 1 - Q_C/Q_H$ according to Equation 15.13, the ratio Q_C/Q_H can depend only on the reservoir temperatures. This observation led Lord Kelvin to propose a ***thermodynamic temperature scale.*** He proposed that the thermodynamic temperatures of the cold and hot reservoirs be defined such that their ratio is equal to Q_C/Q_H. Thus, the thermodynamic temperature scale is related to the heats absorbed and rejected by a Carnot engine, and is independent of the working substance. If a reference temperature is properly chosen, it can be shown that the thermodynamic temperature scale is identical to the Kelvin scale introduced in Section 12.2 and used in the ideal gas law. As a result, the ratio of the rejected heat Q_C to the input heat Q_H is

$$\frac{Q_C}{Q_H} = \frac{T_C}{T_H} \tag{15.14}$$

where the temperatures T_C and T_H *must be expressed in kelvins.*

The efficiency e_{Carnot} of a Carnot engine can be written in a particularly useful way by substituting Equation 15.14 into Equation 15.13 for the efficiency, $e = 1 - Q_C/Q_H$:

$$\begin{matrix}\text{Efficiency of a} \\ \text{Carnot engine}\end{matrix} = e_{Carnot} = 1 - \frac{T_C}{T_H} \tag{15.15}$$

This relation gives the *maximum possible efficiency* for a heat engine operating between two Kelvin temperatures T_C and T_H, and the next example illustrates its application.

Example 7 A Tropical Ocean as a Heat Engine

Water near the surface of a tropical ocean has a temperature of 298.2 K (25.0 °C), whereas water 700 m beneath the surface has a temperature of 280.2 K (7.0 °C). It has been proposed that the warm water be used as the hot reservoir and the cool water as the cold reservoir of a heat engine. Find the maximum possible efficiency for such an engine.

Reasoning The maximum possible efficiency is the efficiency that a Carnot engine would have (Equation 15.15) operating between temperatures of $T_H = 298.2$ K and $T_C = 280.2$ K.

Solution Using $T_H = 298.2$ K and $T_C = 280.2$ K in Equation 15.15, we find that

$$e_{Carnot} = 1 - \frac{T_C}{T_H} = 1 - \frac{280.2 \text{ K}}{298.2 \text{ K}} = \boxed{0.060 \ (6.0\%)}$$

In Example 7 the maximum possible efficiency is only 6.0%. The small efficiency arises because the Kelvin temperatures of the hot and cold reservoirs are so close. A greater efficiency is possible only when there is a greater difference between the reservoir temperatures. However, there are limits on how large the efficiency of a heat engine can be, as Conceptual Example 8 discusses.

Conceptual Example 8
Natural Limits on the Efficiency of a Heat Engine

Consider a hypothetical engine that receives 1000 J of heat as input from a hot reservoir and delivers 1000 J of work, rejecting no heat to a cold reservoir whose temperature is above 0 K. Decide whether this engine violates the first or the second law of thermodynamics, or both.

Reasoning and Solution The first law of thermodynamics is an expression of energy conservation. From the point of view of energy conservation, nothing is wrong with an engine that converts 1000 J of heat into 1000 J of work. Energy has been neither created nor destroyed; it

---

Concept Simulation 15.1

The amount of work W done by a Carnot engine, such as the one in Figure 15.13, is directly proportional to its efficiency e_{Carnot}, since $W = e_{Carnot}Q_H$. The efficiency, in turn, depends on the Kelvin temperatures of the hot and cold reservoirs. In this simulation, the input heat Q_H to the engine is fixed. The user can alter the Kelvin temperatures of the hot and cold reservoirs, thereby changing the efficiency of the engine and the amount of work it does.

Related Homework: *Conceptual Questions 14, 15, Problem 51*

Go to
www.wiley.com/college/cutnell

The physics of
extracting work from a warm ocean.

Problem solving insight
When determining the efficiency of a Carnot engine, be sure the temperatures T_C and T_H of the cold and hot reservoirs are expressed in kelvins; degrees Celsius or degrees Fahrenheit will not do.

has only been transformed from one form (heat) to another (work). This engine does, however, violate the second law of thermodynamics. Since all of the input heat is converted into work, the efficiency of the engine is 1, or 100%. But Equation 15.15, which is based on the second law, indicates that the maximum possible efficiency is $1 - T_C/T_H$, where T_C and T_H are the temperatures of the cold and hot reservoirs, respectively. Since we know that T_C is above 0 K, it is clear that the ratio T_C/T_H is greater than zero, so the maximum possible efficiency is less than 1, or less than 100%. It is important to understand that the first and second laws of thermodynamics address different aspects of nature. *It is the second law, not the first law, that limits the efficiencies of heat engines to values less than 100%.*

Example 8 has emphasized that *even a perfect heat engine has an efficiency that is less than 1.0 or 100%.* In this regard, we note that the maximum possible efficiency, as given by Equation 15.15, approaches 1.0 when T_C approaches absolute zero (0 K). However, experiments have shown that it is not possible to cool a substance to absolute zero (see Section 15.12), so nature does not permit the existence of a 100%-efficient heat engine. As a result, there will always be heat rejected to a cold reservoir whenever a heat engine is used to do work, even if friction and other irreversible processes are eliminated completely. This rejected heat is a form of thermal pollution. Thus, the second law of thermodynamics requires that at least some thermal pollution be generated whenever heat engines are used to perform work. This kind of thermal pollution can be reduced only if society reduces its dependence on heat engines to do work.

The physics of thermal pollution.

15.10 *Refrigerators, Air Conditioners, and Heat Pumps*

The natural tendency of heat is to flow from hot to cold, as indicated by the second law of thermodynamics. However, if work is used, heat can be *made* to flow from cold to hot, against its natural tendency. Refrigerators, air conditioners, and heat pumps are, in fact, devices that do just that. As Figure 15.14 *(left)* illustrates, these devices use work W to extract an amount of heat Q_C from the cold reservoir and deposit an amount of heat Q_H into the hot reservoir. Generally speaking, such a process is called a *refrigeration process.* A comparison of this drawing with Figure 15.13 shows that the directions of the arrows symbolizing heat and work in a refrigeration process are opposite to those in an engine process. Nonetheless, energy is conserved during a refrigeration process, just as it is in an engine process, so $Q_H = W + Q_C$. Moreover, if the process occurs reversibly, we have ideal devices that are called Carnot refrigerators, Carnot air conditioners, and Carnot heat pumps. For these ideal devices, the relation $Q_C/Q_H = T_C/T_H$ (Equation 15.14) applies, just as it does for the Carnot engine.

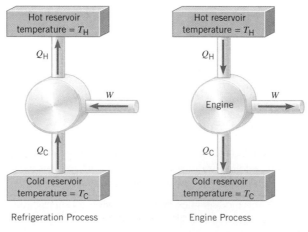

Figure 15.14 In the refrigeration process on the left, work W is used to remove heat Q_C from the cold reservoir and deposit heat Q_H into the hot reservoir. Compare this with the engine process on the right.

Refrigeration Process   Engine Process

In a *refrigerator,* the interior of the unit is the cold reservoir, while the warmer exterior is the hot reservoir. As Figure 15.15 illustrates, the refrigerator takes heat from the food inside and deposits it into the kitchen, along with the energy needed to do the work of making the heat flow from cold to hot. For this reason, the outside surfaces (usually the sides and back) of most refrigerators are warm to the touch while the units operate. Thus, a refrigerator warms the kitchen.

The physics of refrigerators.

An *air conditioner* is like a refrigerator, except that the room itself is the cold reservoir and the outdoors is the hot reservoir. Figure 15.16 shows a window unit, which cools a room by removing heat and depositing it outside, along with the work used to make the heat flow from cold to hot. Conceptual Example 9 considers a common misconception about refrigerators and air conditioners.

The physics of air conditioners.

Conceptual Example 9
You Can't Beat the Second Law of Thermodynamics

Is it possible to cool your kitchen by leaving the refrigerator door open or cool your bedroom by putting a window air conditioner on the floor by the bed?

Reasoning and Solution Whatever heat Q_C is removed from the air directly in front of the open refrigerator is deposited back into the kitchen at the rear of the unit. Moreover, according to the second law, work W is needed to move that heat from cold to hot, and the energy from this work is also deposited into the kitchen as additional heat. Thus, the open refrigerator puts into the kitchen an amount of heat $Q_H = Q_C + W$, which is more than it removes. *Rather than cooling the kitchen, the open refrigerator warms it up.* Putting a window air conditioner on the floor to cool your bedroom is similarly a no-win game. The heat pumped out the back of the air conditioner and into the bedroom is greater than the heat pulled into the front of the unit. Consequently, *the air conditioner actually warms the bedroom.*

Related Homework: *Problem 63*

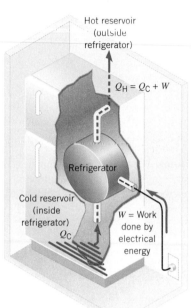

Figure 15.15 A refrigerator.

Need more practice?

Interactive LearningWare 15.2
The water in a glass on a table has a mass of 0.25 kg (about eight liquid ounces). It has the same temperature as the room, which is 27 °C. A Carnot refrigerator is operating in this room. The water is put in the refrigerator, the inside of which is maintained at 3 °C. How much work is done by the refrigerator in cooling the water from 27 to 3 °C?

Related Homework: *Problem 64*

Go to **www.wiley.com/college/cutnell** for an interactive solution.

The quality of a refrigerator or air conditioner is rated according to its coefficient of performance. Such appliances perform well when they remove a relatively large amount of heat Q_C from the cold reservoir with as little work W as possible. Therefore, the coefficient of performance is defined as the ratio of Q_C to W, and the greater this ratio is, the better the performance is:

Refrigerator or air conditioner

$$\text{Coefficient of performance} = \frac{Q_C}{W} \qquad (15.16)$$

Commercially available refrigerators and air conditioners have coefficients of performance in the range 2 to 6, depending on the temperatures involved. The coefficients of performance for these real devices are less than those for ideal, or Carnot, refrigerators and air conditioners.

In a sense, refrigerators and air conditioners operate like pumps. They pump heat "uphill" from a lower temperature to a higher temperature, just as a water pump forces water uphill from a lower elevation to a higher elevation. It would be appropriate to call them heat pumps. However, the name "heat pump" is reserved for the device illustrated in Figure 15.17, which is a home heating appliance. The *heat pump* uses work W to make

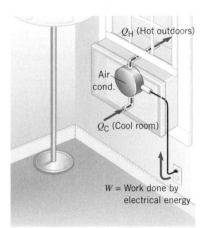

Figure 15.16 A window air conditioner removes heat from a room, which is the cold reservoir, and deposits heat outdoors, which is the hot reservoir.

The physics of heat pumps.

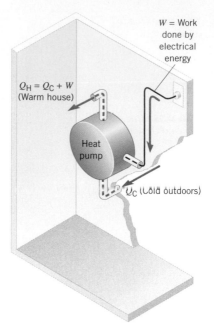

W = Work done by electrical energy

$Q_H = Q_C + W$
(Warm house)

Heat pump

Q_C (Cold outdoors)

Figure 15.17 In a heat pump the cold reservoir is the wintry outdoors, and the hot reservoir is the inside of the house.

heat Q_C from the wintry outdoors (the cold reservoir) flow up the temperature "hill" into a warm house (the hot reservoir). According to the conservation of energy, the heat pump deposits inside the house an amount of heat $Q_H = Q_C + W$. The air conditioner and the heat pump do closely related jobs. The air conditioner refrigerates the inside of the house and heats up the outdoors, while the heat pump refrigerates the outdoors and heats up the inside. These jobs are so closely related that most heat pump systems serve in a dual capacity, being equipped with a switch that converts them from heaters in the winter into air conditioners in the summer.

Heat pumps are popular for home heating in today's energy-conscious world, and it is easy to understand why. Suppose 1000 J of energy is available to use for home heating. Figure 15.18 shows that a conventional electric heating system uses this 1000 J to heat a coil of wire, just as in a toaster. A fan blows air across the hot coil, and forced convection carries the 1000 J of heat into the house. In contrast, the heat pump in Figure 15.17 does not use the 1000 J directly as heat. Instead, it uses the 1000 J to do the work W of pumping heat Q_C from the cooler outdoors into the warmer house. The heat pump delivers to the house an amount of energy $Q_H = Q_C + W$. With $W = 1000$ J, this becomes $Q_H = Q_C + 1000$ J, so that the heat pump delivers more than 1000 J of heat into the house, whereas the conventional electric heating system delivers only 1000 J. The next example shows how the basic relations $Q_H = W + Q_C$ and $Q_C/Q_H = T_C/T_H$ are used with heat pumps.

Example 10   A Heat Pump

An ideal or Carnot heat pump is used to heat a house to a temperature of $T_H = 294$ K (21 °C). How much work must be done by the pump to deliver $Q_H = 3350$ J of heat into the house when the outdoor temperature T_C is (a) 273 K (0 °C) and (b) 252 K (−21 °C)?

Reasoning The conservation of energy ($Q_H = W + Q_C$) applies to the heat pump. Thus, the work can be determined from $W = Q_H − Q_C$, provided we can obtain a value for Q_C, the heat taken by the pump from the outside. To determine Q_C, we use the fact that the pump is a Carnot heat pump and operates reversibly. Therefore, the relation $Q_C/Q_H = T_C/T_H$ (Equation 15.14) applies. Solving it for Q_C, we obtain $Q_C = Q_H(T_C/T_H)$. Using this result, we find that

$$W = Q_H - Q_C = Q_H - Q_H\left(\frac{T_C}{T_H}\right) = Q_H\left(1 - \frac{T_C}{T_H}\right)$$

Problem solving insight
When applying Equation 15.14 ($Q_C/Q_H = T_C/T_H$) to heat pumps, refrigerators, or air conditioners, be sure the temperatures T_C and T_H are expressed in kelvins; degrees Celsius or degrees Fahrenheit will not do.

Solution

(a) At an indoor temperature of $T_H = 294$ K and an outdoor temperature of $T_C = 273$ K, the work needed is

$$W = Q_H\left(1 - \frac{T_C}{T_H}\right) = (3350 \text{ J})\left(1 - \frac{273 \text{ K}}{294 \text{ K}}\right) = \boxed{240 \text{ J}}$$

(b) This solution is identical to that in part (a), except that it is now cooler outside, so $T_C = 252$ K. The necessary work is $W = \boxed{479 \text{ J}}$, which is more than in part (a). More work must be done because the heat is pumped up a greater temperature "hill" when the outside is colder than when it is warmer.

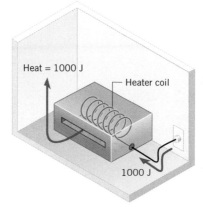

Heat = 1000 J

Heater coil

1000 J

Figure 15.18 This conventional electric heating system is delivering 1000 J of heat to the living room.

It is also possible to specify a coefficient of performance for heat pumps. However, unlike refrigerators and air conditioners, the job of a heat pump is to heat, not to cool. As a result, the coefficient of performance of a heat pump is the ratio of the heat Q_H delivered into the house to the work W required to deliver it:

Heat pump $\dfrac{\text{Coefficient of}}{\text{performance}} = \dfrac{Q_H}{W}$ (15.17)

The coefficient of performance depends on the indoor and outdoor temperatures. Commercial units have coefficients of about 3 to 4 under favorable conditions.

✔ **Check Your Understanding 3**

Each drawing represents a hypothetical heat engine or a hypothetical heat pump and shows the corresponding heats and work. Only one is allowed in nature. Which is it? *(The answer is given at the end of the book).*

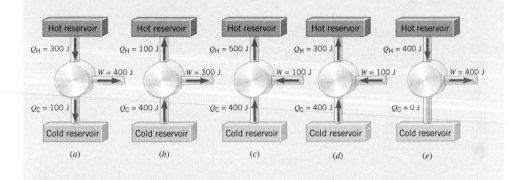

(a) (b) (c) (d) (e)

Background: The operation of any device must obey both the first and second laws of thermodynamics. One is a statement about the conservation of energy, and the other about the direction of heat flow.

For similar questions (including calculational counterparts), consult Self-Assessment Test 15.2. This test is described at the end of Section 15.11.

15.11 *Entropy*

A Carnot engine has the maximum possible efficiency for its operating conditions because the processes occurring within it are reversible. Irreversible processes, such as friction, cause real engines to operate at less than maximum efficiency, for they reduce our ability to use heat to perform work. As an extreme example, imagine that a hot object is placed in thermal contact with a cold object, so heat flows spontaneously, and hence irreversibly, from hot to cold. Eventually both objects reach the same temperature, and $T_C = T_H$. A Carnot engine using these two objects as heat reservoirs is unable to do work, because the efficiency of the engine is zero [$e_{Carnot} = 1 - (T_C/T_H) = 0$]. In general, irreversible processes cause us to lose some, but not necessarily all, of the ability to perform work. This partial loss can be expressed in terms of a concept called ***entropy.***

To introduce the idea of entropy we recall the relation $Q_C/Q_H = T_C/T_H$ that applies to a Carnot engine. This equation can be rearranged as $Q_C/T_C = Q_H/T_H$, which focuses attention on the heat Q divided by the Kelvin temperature T. The quantity Q/T is called the change in the entropy ΔS:

$$\Delta S = \left(\frac{Q}{T}\right)_R \qquad (15.18)$$

In this expression the temperature T must be in kelvins, and the subscript R refers to the word "reversible." It can be shown that Equation 15.18 applies to any process in which heat Q enters or leaves a system reversibly at a constant temperature. Such is the case for the heat that flows into and out of the reservoirs of a Carnot engine. Equation 15.18 indicates that the SI unit for entropy is a joule per kelvin (J/K).

Entropy, like internal energy, is a function of the state or condition of the system. Only the state of a system determines the entropy S that a system has. Therefore, the change in entropy ΔS is equal to the entropy of the final state of the system minus the entropy of the initial state.

We can now describe what happens to the entropy of a Carnot engine. As the engine operates, the entropy of the hot reservoir decreases, since heat Q_H departs at a Kelvin temperature T_H. The change in the entropy of the hot reservoir is $\Delta S_H = -Q_H/T_H$, where the minus sign is needed to indicate a decrease in entropy, since the symbol Q_H denotes only the magnitude of the heat. In contrast, the entropy of the cold reservoir increases by

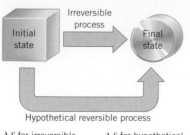

Figure 15.19 Although the relation $\Delta S = (Q/T)_R$ applies to reversible processes, it can be used as part of an indirect procedure to find the entropy change for an irreversible process. This drawing illustrates the procedure discussed in the text.

an amount $\Delta S_C = +Q_C/T_C$, for the rejected heat enters the cold reservoir at a Kelvin temperature T_C. The total change in entropy is

$$\Delta S_C + \Delta S_H = \frac{Q_C}{T_C} - \frac{Q_H}{T_H} = 0$$

because $Q_C/T_C = Q_H/T_H$ according to Equation 15.14.

The fact that the total change in entropy is zero for a Carnot engine is a specific illustration of a general result. It can be proved that when *any* reversible process occurs, the change in the entropy of the universe is zero; $\Delta S_{universe} = 0$ J/K for a reversible process. The word "universe" means that $\Delta S_{universe}$ takes into account the entropy changes of all parts of the system and all parts of the environment. ***Reversible processes do not alter the total entropy of the universe.*** To be sure, the entropy of one part of the universe may change because of a reversible process, but if so, the entropy of another part changes in the opposite way by the same amount.

What happens to the entropy of the universe when an *irreversible* process occurs is more complex, because the expression $\Delta S = (Q/T)_R$ does not apply directly. However, if a system changes irreversibly from an initial state to a final state, this expression can be used to calculate ΔS indirectly, as Figure 15.19 indicates. We imagine a hypothetical reversible process that causes the system to change between *the same initial and final states* and then find ΔS for this reversible process. The value obtained for ΔS also applies to the irreversible process that actually occurs, since only the nature of the initial and final states, and not the path between them, determines ΔS. Example 11 illustrates this indirect method and shows that spontaneous (irreversible) processes increase the entropy of the universe.

Example 11 The Entropy of the Universe Increases

Figure 15.20 shows 1200 J of heat flowing spontaneously through a copper rod from a hot reservoir at 650 K to a cold reservoir at 350 K. Determine the amount by which this irreversible process changes the entropy of the universe, assuming that no other changes occur.

Reasoning The hot-to-cold heat flow is irreversible, so the relation $\Delta S = (Q/T)_R$ is applied to a hypothetical process whereby the 1200 J of heat is taken reversibly from the hot reservoir and added reversibly to the cold reservoir.

Solution The total entropy change of the universe is the algebraic sum of the entropy changes for each reservoir:

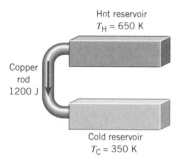

Figure 15.20 Heat flows spontaneously from a hot reservoir to a cold reservoir.

$$\Delta S_{universe} = -\underbrace{\frac{1200 \text{ J}}{650 \text{ K}}}_{\substack{\text{Entropy lost} \\ \text{by hot reservoir}}} + \underbrace{\frac{1200 \text{ J}}{350 \text{ K}}}_{\substack{\text{Entropy gained} \\ \text{by cold reservoir}}} = \boxed{+1.6 \text{ J/K}}$$

The irreversible process causes the entropy of the universe to increase by 1.6 J/K.

Example 11 is a specific illustration of a general result: ***Any irreversible process increases the entropy of the universe.*** In other words, $\Delta S_{universe} > 0$ J/K for an irreversible process. Reversible processes do not alter the entropy of the universe, whereas irreversible processes cause the entropy to increase. Therefore, the entropy of the universe continually increases, like time itself, and entropy is sometimes called "time's arrow." It can be shown that this behavior of the entropy of the universe provides a completely general statement of the second law of thermodynamics, which applies not only to heat flow but also to all kinds of other processes.

■ **THE SECOND LAW OF THERMODYNAMICS STATED IN TERMS OF ENTROPY**

The total entropy of the universe does not change when a reversible process occurs ($\Delta S_{universe} = 0$ J/K) and does increase when an irreversible process occurs ($\Delta S_{universe} > 0$ J/K).

When an irreversible process occurs and the entropy of the universe increases, the energy available for doing work decreases, as the next example illustrates.

Example 12 Energy Unavailable for Doing Work

Suppose that 1200 J of heat is used as input for an engine under two different conditions. In Figure 15.21a the heat is supplied by a hot reservoir whose temperature is 650 K. In part b of the drawing, the heat flows irreversibly through a copper rod into a second reservoir whose temperature is 350 K and then enters the engine. In either case, a 150-K reservoir is used as the cold reservoir. For each case, determine the maximum amount of work that can be obtained from the 1200 J of heat.

Reasoning According to Equation 15.11, the work W obtained from the engine is the product of its efficiency e and the input heat Q_H: $W = eQ_H = e(1200 \text{ J})$. The maximum amount of work is obtained when the efficiency is a maximum—that is, when the engine is a Carnot engine. The efficiency of a Carnot engine is given by Equation 15.15 as $e_{Carnot} = 1 - T_C/T_H$. Therefore, the efficiency may be determined from the Kelvin temperatures of the hot and cold reservoirs.

Solution

Before irreversible heat flow

$$e_{Carnot} = 1 - \frac{T_C}{T_H} = 1 - \frac{150 \text{ K}}{650 \text{ K}} = 0.77$$

$$W = (e_{Carnot})(1200 \text{ J}) = (0.77)(1200 \text{ J}) = \boxed{920 \text{ J}}$$

After irreversible heat flow

$$e_{Carnot} = 1 - \frac{T_C}{T_H} = 1 - \frac{150 \text{ K}}{350 \text{ K}} = 0.57$$

$$W = (e_{Carnot})(1200 \text{ J}) = (0.57)(1200 \text{ J}) = \boxed{680 \text{ J}}$$

When the 1200 J of input heat is taken from the 350-K reservoir instead of the 650-K reservoir, the efficiency of the Carnot engine is smaller. As a result, less work (680 J versus 920 J) can be extracted from the input heat.

Example 12 shows that 240 J less work (920 J − 680 J) can be performed when the input heat is obtained from the hot reservoir with the smaller temperature. In other words, the irreversible process of heat flow through the copper rod causes energy to become unavailable for doing work in the amount of $W_{unavailable} = 240 \text{ J}$. Example 11 shows that this irreversible process also causes the entropy of the universe to increase by an amount $\Delta S_{universe} = +1.6 \text{ J/K}$. These values for $W_{unavailable}$ and $\Delta S_{universe}$ are in fact related. If you multiply $\Delta S_{universe}$ by 150 K, which is the lowest Kelvin temperature in Example 12, you obtain $W_{unavailable} = (150 \text{ K}) \times (1.6 \text{ J/K}) = 240 \text{ J}$. This is one illustration of the following general result:

$$W_{unavailable} = T_0 \Delta S_{universe} \qquad (15.19)$$

where T_0 is the Kelvin temperature of the coldest heat reservoir. Since irreversible processes cause the entropy of the universe to increase, they cause energy to be degraded, in the sense that part of the energy becomes unavailable for the performance of work. In contrast, there is no penalty when reversible processes occur, because for them $\Delta S_{universe} = 0 \text{ J/K}$, and there is no loss of work.

Entropy can also be interpreted in terms of order and disorder. As an example, consider a block of ice (Figure 15.22) with each of its H_2O molecules fixed rigidly in place

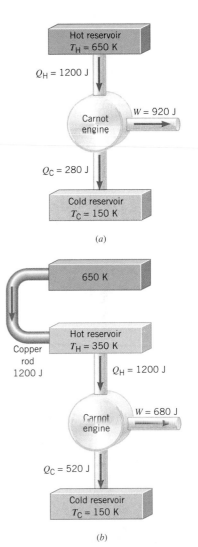

(a)

(b)

Figure 15.21 Heat in the amount of $Q_H = 1200 \text{ J}$ is used as input for an engine under two different conditions in parts a and b.

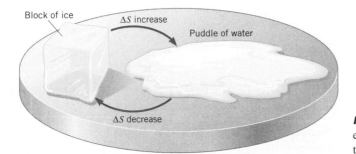

Figure 15.22 A block of ice is an example of an ordered system relative to a puddle of water.

in a highly structured and ordered arrangement. In comparison, the puddle of water into which the ice melts is disordered and unorganized, because the molecules in a liquid are free to move from place to place. Heat is required to melt the ice and produce the disorder. Moreover, heat flow into a system increases the entropy of the system, according to $\Delta S = (Q/T)_R$. We associate an increase in entropy, then, with an increase in disorder. Conversely, we associate a decrease in entropy with a decrease in disorder or a greater degree of order. Example 13 illustrates an order-to-disorder change and the increase of entropy that accompanies it.

Example 13 Order to Disorder

Find the change in entropy that results when a 2.3-kg block of ice melts slowly (reversibly) at 273 K (0 °C).

Reasoning Since the phase change occurs reversibly at a constant temperature, the change in entropy can be found by using Equation 15.18, $\Delta S = (Q/T)_R$, where Q is the heat absorbed by the melting ice. This heat can be determined by using the relation $Q = mL_f$ (Equation 12.5), where m is the mass and $L_f = 3.35 \times 10^5$ J/kg is the latent heat of fusion of water.

Solution Using Equation 15.18 and Equation 12.5, we find that the change in entropy is

$$\Delta S = \left(\frac{Q}{T}\right)_R = \frac{mL_f}{T} = \frac{(2.3 \text{ kg})(3.35 \times 10^5 \text{ J/kg})}{273 \text{ K}} = \boxed{+2.8 \times 10^3 \text{ J/K}}$$

a result that is positive, since the ice absorbs heat as it melts.

Figure 15.23 shows another order-to-disorder change that can be described in terms of entropy.

✓ Check Your Understanding 4

Two equal amounts of water are mixed together in an insulated container. The initial temperatures of the water are different, but the mixture reaches a uniform temperature. Do the energy and the entropy of the water increase, decrease, or remain constant as a result of the mixing process? *(The answer is given at the end of the book.)*

	Energy of the Water	Entropy of the Water
a.	Increases	Increases
b.	Decreases	Decreases
c.	Remains constant	Decreases
d.	Remains constant	Increases
e.	Remains constant	Remains constant

Background: This question focuses on energy, entropy, and the important idea of reversible versus irreversible processes.

For similar questions (including calculational counterparts), consult Self-Assessment 15.2. This test is described next.

Self-Assessment Test 15.2

Test your understanding of the central ideas presented in Sections 15.7–15.11:

• The Second Law of Thermodynamics   • Heat Engines and the Carnot Engine
• Refrigerators, Air Conditioners, and Heat Pumps   • Entropy

Go to **www.wiley.com/college/cutnell**

Figure 15.23 With the aid of 100 pounds of dynamite, demolition experts caused this hotel-casino in Las Vegas to go from an ordered state (lower entropy) to a disordered state (higher entropy). (© Reuters/Aaron Mayes/Archive Photos/Getty Images)

15.12 The Third Law of Thermodynamics

To the zeroth, first, and second laws of thermodynamics we add the third (and last) law. The *third law of thermodynamics* indicates that it is impossible to reach a temperature of absolute zero.

■ THE THIRD LAW OF THERMODYNAMICS

It is not possible to lower the temperature of any system to absolute zero ($T = 0$ K) in a finite number of steps.

This law, like the second law, can be expressed in a number of ways, but a discussion of them is beyond the scope of this text. The third law is needed to explain a number of experimental observations that cannot be explained by the other laws of thermodynamics.

15.13 Concepts & Calculations

The first law of thermodynamics is basically a restatement of the conservation-of-energy principle in terms of heat and work. Example 14 emphasizes this important fact by showing that the conservation principle and the first law provide the same approach to a problem. In addition, the example reviews the concept of latent heat of sublimation (see Section 12.8) and the ideal gas law (see Section 14.2).

Concepts & Calculations Example 14 The Sublimation of Zinc

The sublimation of zinc (mass per mole = 0.0654 kg/mol) occurs at a temperature of 6.00×10^2 K, and the latent heat of sublimation is 1.99×10^6 J/kg. The pressure remains constant during the sublimation. Assume that the zinc vapor can be treated as a monatomic ideal gas and that the volume of solid zinc is negligible compared to the corresponding vapor. What is the change in the internal energy of the zinc when 1.50 kg of zinc sublimates?

Concept Questions and Answers What is sublimation and what is the latent heat of sublimation?

Answer Sublimation is the process whereby a solid phase changes directly into a gas phase in response to the input of heat. The heat per kilogram needed to cause the phase change is called the latent heat of sublimation L_s. The heat Q needed to bring about the sublimation of a mass m of solid material is given by Equation 12.5 as $Q = mL_s$.

When a solid phase changes to a gas phase, does the volume of the material increase or decrease, and by how much?

Answer Gases, in general, have greater volumes than solids do, so the volume of the material increases. The increase in volume is $\Delta V = V_{gas} - V_{solid}$. Since the volume of the solid V_{solid} is negligibly small compared to the volume of the gas V_{gas}, we have $\Delta V = V_{gas}$. Using the ideal gas law as given in Equation 14.1, it follows that $V_{gas} = nRT/P$, so that $\Delta V = nRT/P$. In this result, n is the number of moles of material, R is the universal gas constant, and T is the Kelvin temperature.

As the material changes from a solid to a gas, does it do work on the environment or does the environment do work on it? How much work is involved?

Answer To make room for itself, the expanding material must push against the environment and, in so doing, does work on the environment. Since the pressure remains constant, the work done by the material is given by Equation 15.2 as $W = P\Delta V$. Since $\Delta V = nRT/P$, the work becomes $W = P(nRT/P) = nRT$.

In this problem we begin with heat Q, and it is used for two purposes: First, it makes the solid change into a gas, which entails a change ΔU in the internal energy of the material, $\Delta U = U_{gas} - U_{solid}$. Second, it allows the expanding material to do work W on the environment. According to the conservation-of-energy principle, how is Q related to ΔU and W?

Answer According to the conservation-of-energy principle, energy can neither be created nor destroyed, but can only be converted from one form to another (see Section 6.8). Therefore, part of the heat Q is used for ΔU and part for W, with the result that $Q = \Delta U + W$.

According to the first law of thermodynamics, how is Q related to ΔU and W?

Answer As indicated in Equation 15.1, the first law of thermodynamics is $\Delta U = Q - W$. Rearranging this equation gives $Q = \Delta U + W$, which is identical to the result obtained from the conservation-of-energy principle.

Solution Using the facts that $Q = \Delta U + W$, $Q = mL_s$, and $W = nRT$, we have that

$$Q = \Delta U + W \quad \text{or} \quad mL_s = \Delta U + nRT$$

Solving for ΔU gives

$$\Delta U = mL_s - nRT$$

In this result, n is the number of moles of the ideal gas. According to the discussion in Section 14.1, the number of moles of gaseous zinc is the mass m of the sample divided by the mass per mole of zinc or $n = m/(0.0654 \text{ kg/mol})$. Therefore, we find

$$\Delta U = mL_s - nRT$$

$$= (1.50 \text{ kg})\left(1.99 \times 10^6 \, \frac{\text{J}}{\text{kg}}\right) - \left(\frac{1.50 \text{ kg}}{0.0654 \text{ kg/mol}}\right)\left(8.31 \, \frac{\text{J}}{\text{mol} \cdot \text{K}}\right)(6.00 \times 10^2 \text{ K})$$

$$= \boxed{2.87 \times 10^6 \text{ J}}$$

Heat engines can be used to perform work, as we have seen in this chapter. The concept of work, however, was first introduced in Chapter 6, along with the idea of kinetic energy and the work–energy theorem. The next example reviews some of the main features of heat engines, as well as kinetic energy and the work–energy theorem.

Concepts & Calculations Example 15   The Work–Energy Theorem

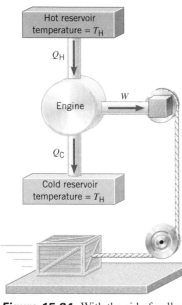

Hot reservoir temperature = T_H

Q_H

Engine W

Q_C

Cold reservoir temperature = T_H

Figure 15.24 With the aid of pulleys and a rope, a Carnot engine provides the work that is used to accelerate the crate from rest along a horizontal frictionless surface. See Example 15.

Each of two Carnot engines uses the same cold reservoir at a temperature of 275 K for its exhaust heat. Each engine receives 1450 J of input heat. The work from either of these engines is used to drive a pulley arrangement that uses a rope to accelerate a 125-kg crate from rest along a horizontal frictionless surface, as Figure 15.24 suggests. With engine 1 the crate attains a speed of 2.00 m/s, while with engine 2 it attains a speed of 3.00 m/s. Find the temperature of the hot reservoir for each engine.

Concept Questions and Answers With which engine is the change in the crate's kinetic energy greater?

Answer   The change is greater with engine 2. Kinetic energy is $KE = \frac{1}{2}mv^2$, according to Equation 6.2, where m is the mass of the crate and v is its speed. The change in the kinetic energy is the final minus the initial value, or $KE_f - KE_0$. Since the crate starts from rest, it has zero initial kinetic energy. Thus, the change is equal to the final kinetic energy. Since engine 2 gives the crate the greater final speed, it causes the greater change in kinetic energy.

Which engine does more work?

Answer   The work–energy theorem, as stated in Equation 6.3, indicates that the net work done on an object equals the change in the object's kinetic energy, or $W = KE_f - KE_0$. The net work is the work done by the net force. In Figure 15.24 the surface is horizontal, and the crate does not leave it. Therefore, the upward normal force that the surface applies to the crate must balance the downward weight of the crate. Furthermore, the surface is frictionless, so there is no friction force. The net force acting on the crate, then, consists of the single force due to the tension in the rope, which arises from the action of the engine. Thus, the work done by the engine is, in fact, the net work done on the crate. But we know that engine 2 causes the crate's kinetic energy to change by the greater amount, so that engine must do more work.

For which engine is the temperature of the hot reservoir greater?

Answer   The temperature of the hot reservoir for engine 2 is greater. We know that engine 2 does more work, but each engine receives the same 1450 J of input heat. Therefore, engine 2 derives more work from the input heat. In other words, it is more efficient. But the efficiency of a Carnot engine depends only on the Kelvin temperatures of its hot and cold reservoirs. Since both engines use the same cold reservoir whose temperature is 275 K, only the temperatures of the hot reservoirs are different. Higher temperatures for the hot reservoir are associated with greater efficiencies, so the temperature of the hot reservoir for engine 2 is greater.

Solution According to Equation 15.11, the efficiency e of a heat engine is the work W divided by the input heat Q_H, or $e = W/Q_H$. According to Equation 15.15, the efficiency of a

Carnot engine is $e_{Carnot} = 1 - T_C/T_H$, where T_C and T_H are, respectively, the Kelvin temperatures of the cold and hot reservoirs. Combining these two equations, we have

$$1 - \frac{T_C}{T_H} = \frac{W}{Q_H}$$

But W is the net work done on the crate, and it equals the change in the crate's kinetic energy, or $W = KE_f - KE_0 = \frac{1}{2}mv^2$, according to Equations 6.2 and 6.3. With this substitution the efficiency expression becomes

$$1 - \frac{T_C}{T_H} = \frac{\frac{1}{2}mv^2}{Q_H}$$

Solving for the temperature T_H, we find

$$T_H = \frac{T_C}{1 - \dfrac{mv^2}{2Q_H}}$$

As expected, the value for T_H for engine 2 is greater:

Engine 1
$$T_H = \frac{275 \text{ K}}{1 - \dfrac{(125 \text{ kg})(2.00 \text{ m/s})^2}{2(1450 \text{ J})}} = \boxed{332 \text{ K}}$$

Engine 2
$$T_H = \frac{275 \text{ K}}{1 - \dfrac{(125 \text{ kg})(3.00 \text{ m/s})^2}{2(1450 \text{ J})}} = \boxed{449 \text{ K}}$$

▲

At the end of the problem set for this chapter, you will find homework problems that contain both conceptual and quantitative parts. These problems are grouped under the heading *Concepts & Calculations, Group Learning Problems.* They are designed for use by students working alone or in small learning groups. The conceptual part of each problem provides a convenient focus for group discussions.

Concept Summary

This summary presents an abridged version of the chapter, including the important equations and all available learning aids. For convenient reference, the learning aids (including the text's examples) are placed next to or immediately after the relevant equation or discussion. The following learning aids may be found on-line at **www.wiley.com/college/cutnell**:

Interactive LearningWare examples are solved according to a five-step interactive format that is designed to help you develop problem-solving skills.	**Concept Simulations** are animated versions of text figures or animations that illustrate important concepts. You can control parameters that affect the display, and we encourage you to experiment.
Interactive Solutions offer specific models for certain types of problems in the chapter homework. The calculations are carried out interactively.	**Self-Assessment Tests** include both qualitative and quantitative questions. Extensive feedback is provided for both incorrect and correct answers, to help you evaluate your understanding of the material.

Topic	*Discussion*	*Learning Aids*

15.1 Thermodynamic Systems and Their Surroundings

A thermodynamic system is the collection of objects on which attention is being focused, and the surroundings are everything else in the environment. The state of a system is the physical condition of the system, as described by values for physical parameters, often pressure, volume, and temperature.

15.2 The Zeroth Law of Thermodynamics

Two systems are in thermal equilibrium if there is no net flow of heat between them when they are brought into thermal contact.

Topic	*Discussion*	*Learning Aids*

Thermal equilibrium and temperature

Temperature is the indicator of thermal equilibrium in the sense that there is no net flow of heat between two systems in thermal contact that have the same temperature.

Zeroth law of thermodynamics

The zeroth law of thermodynamics states that two systems individually in thermal equilibrium with a third system are in thermal equilibrium with each other.

15.3 The First Law of Thermodynamics

The first law of thermodynamics states that due to heat Q and work W, the internal energy of a system changes from its initial value of U_i to a final value of U_f according to

$$\Delta U = U_f - U_i = Q - W \qquad (15.1) \quad \textbf{Examples 1, 14}$$

Sign convention for Q and W

Q is positive when the system gains heat and negative when it loses heat. W is positive when work is done by the system and negative when work is done on the system.

The first law of thermodynamics is the conservation-of-energy principle applied to heat, work, and the change in the internal energy.

Function of state

The internal energy is called a function of state because it depends only on the state of the system and not on the method by which the system came to be in a given state.   **Example 2**

15.4 Thermal Processes

Quasi-static process

A thermal process is quasi-static when it occurs slowly enough that a uniform pressure and temperature exist throughout the system at all times.

Isobaric process

An isobaric process is one that occurs at constant pressure. The work W done when a system changes at a constant pressure P from an initial volume V_i to a final volume V_f is

$$W = P\,\Delta V = P(V_f - V_i) \qquad (15.2) \quad \textbf{Example 3}$$

Isochoric process

An isochoric process is one that takes place at constant volume, and no work is done in such a process.

Isothermal process

An isothermal process is one that takes place at constant temperature.

Adiabatic process

An adiabatic process is one that takes place without the transfer of heat.

Work done as the area under a pressure–volume graph

The work done in any kind of quasi-static process is given by the area under the corresponding pressure-versus-volume graph.   **Example 4**

15.5 Thermal Processes Using an Ideal Gas

When n moles of an ideal gas change quasi-statically from an initial volume V_i to a final volume V_f at a constant Kelvin temperature T, the work done is

Work done during an isothermal process

$$W = nRT \ln\left(\frac{V_f}{V_i}\right) \qquad (15.3)$$

Example 5

Interactive LearningWare 15.1

When n moles of a monatomic ideal gas change quasi-statically and adiabatically from an initial temperature T_i to a final temperature T_f, the work done is

Work done during an adiabatic process

$$W = \tfrac{3}{2}nR(T_i - T_f) \qquad (15.4)$$

During an adiabatic process, and in addition to the ideal gas law, an ideal gas obeys the relation

Adiabatic change in pressure and volume

$$P_i V_i^{\gamma} = P_f V_f^{\gamma} \qquad (15.5) \quad \textbf{Interactive Solution 15.27}$$

where $\gamma = c_P/c_V$ is the ratio of the specific heat capacities at constant pressure and constant volume.

15.6 Specific Heat Capacities

The molar specific heat capacity C of a substance determines how much heat Q is added or removed when the temperature of n moles of the substance changes by an amount ΔT:

$$Q = Cn\Delta T \qquad (15.6) \quad \textbf{Interactive Solution 15.85}$$

Topic	Discussion	Learning Aids
	For a monatomic ideal gas, the molar specific heat capacities at constant pressure and constant volume are, respectively,	

Specific heat capacities of a monatomic ideal gas

$$C_P = \tfrac{5}{2}R \qquad (15.7)$$

$$C_V = \tfrac{3}{2}R \qquad (15.8)$$

where R is the ideal gas constant. For any type of ideal gas, the difference between C_P and C_V is

$$C_P - C_V = R \qquad (15.10)$$

Use Self-Assessment Test 15.1 to evaluate your understanding of Sections 15.1–15.6.

15.7 The Second Law of Thermodynamics

The second law of thermodynamics can be stated in a number of equivalent forms. In terms of heat flow, the second law declares that heat flows spontaneously from a substance at a higher temperature to a substance at a lower temperature and does not flow spontaneously in the reverse direction.

15.8 Heat Engines

A heat engine produces work W from input heat Q_H that is extracted from a heat reservoir at a relatively high temperature. The engine rejects heat Q_C into a reservoir at a relatively low temperature.

The efficiency e of a heat engine is

Efficiency of a heat engine

$$e = \frac{\text{Work done}}{\text{Input heat}} = \frac{W}{Q_H} \qquad (15.11)$$

The conservation of energy requires that the input heat of magnitude Q_H must be equal to the work W done by the engine plus the heat of magnitude Q_C rejected to the cold reservoir:

Conservation of energy for a heat engine

$$Q_H = W + Q_C \qquad (15.12) \qquad \text{Example 6}$$

By combining Equation 15.12 with Equation 15.11, the efficiency of a heat engine can also be written as

$$e = 1 - \frac{Q_C}{Q_H} \qquad (15.13)$$

15.9 Carnot's Principle and the Carnot Engine

Reversible process

A reversible process is one in which both the system and its environment can be returned to exactly the states they were in before the process occurred.

Carnot's principle

Carnot's principle is an alternative statement of the second law of thermodynamics. It states that no irreversible engine operating between two reservoirs at constant temperatures can have a greater efficiency than a reversible engine operating between the same temperatures. Furthermore, all reversible engines operating between the same temperatures have the same efficiency.

A Carnot engine

A carnot engine is a reversible engine in which all input heat Q_H originates from a hot reservoir at a single Kelvin temperature T_H and all rejected heat Q_C goes into a cold reservoir at a single Kelvin temperature T_C. For a Carnot engine

$$\frac{Q_C}{Q_H} = \frac{T_C}{T_H} \qquad (15.14)$$

The efficiency e_{Carnot} of a Carnot engine is the maximum efficiency that an engine operating between two fixed temperatures can have:

Efficiency of a Carnot engine

$$e_{\text{Carnot}} = 1 - \frac{T_C}{T_H} \qquad (15.15)$$

Examples 7, 8, 15

Concept Simulation 15.1

15.10 Refrigerators, Air Conditioners, and Heat Pumps

Refrigerators, air conditioners, and heat pumps are devices that utilize work W to make heat of magnitude Q_C flow from a lower Kelvin temperature T_C to a **Example 9**

Topic	Discussion	Learning Aids
	higher Kelvin temperature T_H. In the process (the refrigeration process) they deposit heat of magnitude Q_H at the higher temperature. The principle of the conservation of energy requires that $Q_H = W + Q_C$.	Interactive LearningWare 15.2
	If the refrigeration process is ideal, in the sense that it occurs reversibly, the devices are called Carnot devices and the relation $Q_C/Q_H = T_C/T_H$ (Equation 15.14) holds.	Example 10
	The coefficient of performance of a refrigerator or an air conditioner is	
Coefficient of performance (refrigerator or air conditioner)	$$\text{Coefficient of performance} = \frac{Q_C}{W} \qquad (15.16)$$	
	The coefficient of performance of a heat pump is	
Coefficient of performance (heat pump)	$$\text{Coefficient of performance} = \frac{Q_H}{W} \qquad (15.17)$$	

15.11 Entropy

The change in entropy ΔS for a process in which heat Q enters or leaves a system reversibly at a constant Kelvin temperature T is

Topic	Discussion	Learning Aids
Change in entropy	$$\Delta S = \left(\frac{Q}{T}\right)_R \qquad (15.18)$$	Examples 11, 13

where the subscript R stands for "reversible."

Topic	Discussion	Learning Aids
The second law of thermodynamics (entropy statement)	The second law of thermodynamics can be stated in a number of equivalent forms. In terms of entropy, the second law states that the total entropy of the universe does not change when a reversible process occurs ($\Delta S_{universe} = 0$ J/K) and increases when an irreversible process occurs ($\Delta S_{universe} > 0$ J/K).	Interactive Solution 15.73
	Irreversible processes cause energy to be degraded in the sense that part of the energy becomes unavailable for the performance of work. The energy $W_{unavailable}$ that is unavailable for doing work because of an irreversible process is	Example 12
Unavailable work	$$W_{unavailable} = T_0\, \Delta S_{universe} \qquad (15.19)$$	
	where $\Delta S_{universe}$ is the total entropy change of the universe and T_0 is the Kelvin temperature of the coldest reservoir into which heat can be rejected.	
Entropy and disorder	Increased entropy is associated with a greater degree of disorder and decreased entropy with a lesser degree of disorder (more order).	

Use *Self-Assessment Test 15.2* to evaluate your understanding of *Sections 15.7–15.11.*

15.12 The Third Law of Thermodynamics

The third law of thermodynamics states that it is not possible to lower the temperature of any system to absolute zero ($T = 0$ K) in a finite number of steps.

Conceptual Questions

1. Ignore friction and assume that air behaves as an ideal gas. The plunger of a bicycle tire pump is pushed down rapidly with the end of the pump sealed so that no air escapes and there is little time for heat to flow through the cylinder wall. Explain why the cylinder of the pump becomes warm to the touch.

2. One hundred joules of heat is added to a gas, and the gas expands at constant pressure. Is it possible that the internal energy increases by 200 J? Account for your answer with the aid of the first law of thermodynamics.

3. A gas is compressed isothermally, and its internal energy increases. Is the gas an ideal gas? Justify your answer.

4. A material undergoes an isochoric process that is also adiabatic. Is the internal energy of the material at the end of the process greater than, less than, or the same as it was at the start? Justify your answer.

5. (a) Is it possible for the temperature of a substance to rise without heat flowing into it? (b) Does the temperature of a substance necessarily have to change because heat flows into or out of it? In each case, give your reasoning and use the example of an ideal gas.

6. The drawing shows a pressure–volume graph in which a gas expands at constant pressure from A to B, and then goes from B to C at constant volume. Complete the table below by deciding whether each of the remaining four entries should be positive (+), negative (−), or zero (0). Give a reason for each answer you choose.

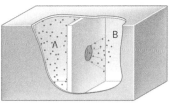

	ΔU	Q	W
$A \to B$	+		
$B \to C$		+	

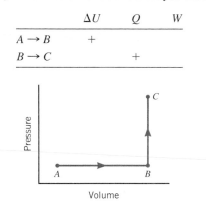

7. The drawing shows an arrangement for an adiabatic free expansion or "throttling" process. The process is adiabatic because the entire arrangement is contained within perfectly insulating walls. The gas in chamber A rushes suddenly into chamber B through a hole in the partition. Chamber B is initially evacuated, so the gas expands there under zero external pressure and the work $W = P\,\Delta V$ is zero. Assume that the gas is an ideal gas and explain how the final temperature of the gas after expansion compares to its initial temperature.

8. Suppose a material contracts when it is heated. Follow the same line of reasoning used in the text to reach Equations 15.7 and 15.8 and deduce which specific heat capacity for the material is larger, C_P or C_V.

9. When a solid melts at constant pressure, the volume of the resulting liquid does not differ much from the volume of the solid. Using what you know about the first law of thermodynamics and the latent heat of fusion, describe how the internal energy of the liquid compares to the internal energy of the solid.

10. Suppose you want to heat a gas so that its temperature will be as high as possible. Would you heat it under conditions of constant pressure or constant volume? Why?

11. Consider a hypothetical device that takes 10 000 J of heat from a hot reservoir and 5000 J of heat from a cold reservoir and produces 15 000 J of work. (a) Does this device violate the first law of thermodynamics? Explain. (b) Does the device violate the second law of thermodynamics? Explain.

12. If you saw an advertisement for an automobile that claimed the same gas mileage with and without the air conditioner operating, would you be suspicious? Explain, using what the second law of thermodynamics has to say about work and the direction of heat flow in an air conditioner.

13. The second law of thermodynamics, in the form of Carnot's principle, indicates that the most efficient heat engine operating between two temperatures is a reversible one. Does this mean that a reversible engine operating between the temperatures of 600 and 400 K must be more efficient than an *irreversible* engine operating between 700 and 300 K? Provide a reason for your answer.

14. **Concept Simulation 15.1** at **www.wiley.com/college/cutnell** allows you to explore the concepts that relate to this question. Three reversible engines, A, B, and C, use the same cold reservoir for their

exhaust heats. However, they use different hot reservoirs that have the following temperatures: (A) 1000 K, (B) 1100 K, and (C) 900 K. Rank these engines in order of increasing efficiency (smallest efficiency first). Account for your answer.

15. In **Concept Simulation 15.1** at **www.wiley.com/college/cutnell** you can explore the concepts that are important in this question. Suppose you wish to improve the efficiency of a Carnot engine. Compare the improvement to be realized via each of the following alternatives: (a) lower the Kelvin temperature of the cold reservoir by a factor of four, (b) raise the Kelvin temperature of the hot reservoir by a factor of four, (c) cut the Kelvin temperature of the cold reservoir in half and double the Kelvin temperature of the hot reservoir. Give your reasoning.

16. A refrigerator is kept in a garage that is not heated in the cold winter or air-conditioned in the hot summer. Does it cost more for this refrigerator to make a kilogram of ice cubes in the winter or in the summer? Give your reasoning.

17. Is it possible for a Carnot heat pump to have a coefficient of performance that is less than one? Justify your answer.

18. Air conditioners and refrigerators both remove heat from a cold reservoir and deposit it in a hot reservoir. Why, then, does an air conditioner cool the inside of a house while a refrigerator warms the house?

19. It has been said that heat pumps can't possibly deliver more energy into your house than they consume in operating. Can they or can't they? Explain.

20. A refrigerator is advertised as being easier to live with during the summer because it puts into your kitchen only the heat that it removes from the food. Does this advertising claim violate the second law of thermodynamics? Account for your answer.

21. On a summer day a window air conditioner cycles on and off, according to how the temperature within the room changes. Are you more likely to be able to fry an egg on the outside part of the unit when the unit is on or when it is off? Explain.

22. An event happens somewhere in the universe and, as a result, the entropy of an object changes by −5 J/K. Which one (or more) of the following is a possible value for the entropy change for the rest of the universe: −5 J/K, 0 J/K, +5 J/K, +10 J/K? Account for your choice(s) in terms of the second law of thermodynamics.

23. When water freezes from a less ordered liquid to a more ordered solid, its entropy decreases. Why doesn't this decrease in entropy violate the second law of thermodynamics?

24. In each of the following cases, which has the greater entropy: (a) a handful of popcorn kernels or the popcorn that results from them, (b) a salad before or after it has been tossed, (c) a messy apartment or a neat apartment? Why?

25. A glass of water contains a teaspoon of dissolved sugar. After a while, the water evaporates, leaving behind sugar crystals. The entropy of the sugar crystals is less than the entropy of the dissolved sugar because the sugar crystals are in a more ordered state. Why doesn't this process violate the second law of thermodynamics?

26. A builder uses lumber to construct a building, which is unfortunately destroyed in a fire. Thus, the lumber existed at one time or another in three different states: (1) as unused building material, (2) as a building, and (3) as a burned-out shell of a building. Rank these three states in order of decreasing entropy (largest first). Provide a reason for the ranking.

Problems

ssm Solution is in the Student Solutions Manual.   **www** Solution is available on the World Wide Web at www.wiley.com/college/cutnell
☤ This icon represents a biomedical application.

Section 15.3 The First Law of Thermodynamics

1. ssm The internal energy of a system changes because the system gains 165 J of heat and performs 312 J of work. In returning to its initial state, the system loses 114 J of heat. During this return process, (a) how much work is involved, and (b) is work done by the system or is work done on the system?

2. In moving out of a dormitory at the end of the semester, a student does 1.6×10^4 J of work. In the process, his internal energy decreases by 4.2×10^4 J. Determine each of the following quantities (including the algebraic sign): (a) W, (b) ΔU, and (c) Q.

3. A system does 164 J of work on its environment and gains 77 J of heat in the process. Find the change in the internal energy of (a) the system and (b) the environment.

4. Three moles of an ideal monatomic gas are at a temperature of 345 K. Then, 2438 J of heat are added to the gas, and 962 J of work are done on it. What is the final temperature of the gas?

5. ssm When one gallon of gasoline is burned in a car engine, 1.19×10^8 J of internal energy is released. Suppose that 1.00×10^8 J of this energy flows directly into the surroundings (engine block and exhaust system) in the form of heat. If 6.0×10^5 J of work is required to make the car go one mile, how many miles can the car travel on one gallon of gas?

* 6. ☤ In exercising, a weight lifter loses 0.150 kg of water through evaporation, the heat required to evaporate the water coming from the weight lifter's body. The work done in lifting weights is 1.40×10^5 J. (a) Assuming that the latent heat of vaporization of perspiration is 2.42×10^6 J/kg, find the change in the internal energy of the weight lifter. (b) Determine the minimum number of nutritional Calories of food (1 nutritional Calorie = 4186 J) that must be consumed to replace the loss of internal energy.

Section 15.4 Thermal Processes

7. The specific heat capacity of a material is 1100 J/(kg·C°). The temperature of 2.0 kg of this solid material is raised by 6.0 C°. Ignoring the work that corresponds to the small change in the volume of the material, determine the change in the internal energy of the material.

8. A gas, while expanding under isobaric conditions, does 480 J of work. The pressure of the gas is 1.6×10^5 Pa, and its initial volume is 1.5×10^{-3} m³. What is the final volume of the gas?

9. ssm The volume of a gas is changed along the curved line between A and B in the drawing. Do not assume that the curved line is an isotherm or that the gas is ideal. (a) Find the magnitude of the work for the process, and (b) determine whether the work is positive or negative.

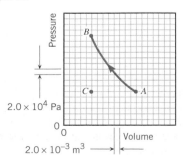

10. The pressure and volume of a gas are changed along the path $ABCA$. Determine the work done (including the algebraic sign) in each segment of the path: (a) A to B, (b) B to C, and (c) C to A.

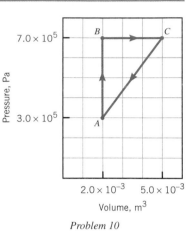

Problem 10

11. A gas is contained in a chamber such as that in Figure 15.5. Suppose the region outside the chamber is evacuated and the total mass of the block and the movable piston is 135 kg. When 2050 J of heat flows into the gas, the internal energy of the gas increases by 1730 J. What is the distance s through which the piston rises?

12. (a) Using the data presented in the accompanying pressure-versus-volume graph, estimate the magnitude of the work done when the system changes from A to B to C along the path shown. (b) Determine whether the work is done by the system or on the system and, hence, whether the work is positive or negative.

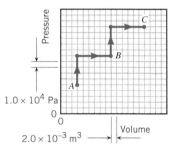

13. ssm A system gains 1500 J of heat, while the internal energy of the system increases by 4500 J and the volume decreases by 0.010 m³. Assume the pressure is constant and find its value.

* 14. Refer to the drawing in problem 9, where the curve between A and B is now an isotherm. An ideal gas begins at A and is changed along the horizontal line from A to C and then along the vertical line from C to B. (a) Find the heat for the process ACB and (b) determine whether it flows into or out of the gas.

* 15. ssm www A monatomic ideal gas expands isobarically. Using the first law of thermodynamics, prove that the heat Q is positive, so that it is impossible for heat to flow out of the gas.

* 16. Refer to the drawing that accompanies problem 12. When a system changes from A to B along the path shown on the pressure-versus-volume graph, it gains 2700 J of heat. What is the change in the internal energy of the system?

** 17. Water is heated in an open pan where the air pressure is one atmosphere. The water remains a liquid, which expands by a small amount as it is heated. Determine the ratio of the work done by the water to the heat absorbed by the water.

Section 15.5 Thermal Processes Using an Ideal Gas

18. The temperature of a monatomic ideal gas remains constant during a process in which 4700 J of heat flows out of the gas. How much work (including the proper + or − sign) is done?

19. ssm Three moles of an ideal gas are compressed from 5.5×10^{-2} to 2.5×10^{-2} m³. During the compression, 6.1×10^3 J of work is done on the gas, and heat is removed to keep the temperature of the gas constant at all times. Find (a) ΔU, (b) Q, and (c) the temperature of the gas.

20. Five moles of a monatomic ideal gas expand adiabatically, and its temperature decreases from 370 to 290 K. Determine (a) the work done (including the algebraic sign) by the gas, and (b) the change in its internal energy.

21. The pressure of a monatomic ideal gas ($\gamma = \frac{5}{3}$) doubles during an adiabatic compression. What is the ratio of the final volume to the initial volume?

22. One-half mole of a monatomic ideal gas expands adiabatically and does 610 J of work. By how many kelvins does its temperature change? Specify whether the change is an increase or a decrease.

23. ssm A monatomic ideal gas has an initial temperature of 405 K. This gas expands and does the same amount of work whether the expansion is adiabatic or isothermal. When the expansion is adiabatic, the final temperature of the gas is 245 K. What is the ratio of the final to the initial volume when the expansion is isothermal?

* **24.** The drawing refers to one mole of a monatomic ideal gas and shows a process that has four steps, two isobaric (A to B, C to D) and two isochoric (B to C, D to A). Complete the following table by calculating ΔU, W, and Q (including the algebraic signs) for each of the four steps.

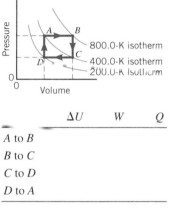

	ΔU	W	Q
A to B			
B to C			
C to D			
D to A			

* **25.** The pressure and volume of an ideal monatomic gas change from A to B to C, as the drawing shows. The curved line between A and C is an isotherm. (a) Determine the total heat for the process and (b) state whether the flow of heat is into or out of the gas.

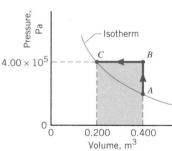

* **26.** A monatomic ideal gas expands from point A to point B along the path shown in the drawing. (a) Determine the work done by the

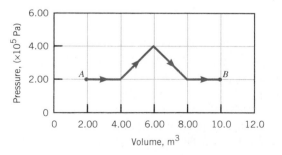

gas. (b) The temperature of the gas at point A is 185 K. What is its temperature at point B? (c) How much heat has been added to or removed from the gas during the process?

* **27.** Refer to **Interactive Solution 15.27** at **www.wiley.com/college/cutnell** for help in solving this problem. A diesel engine does not use spark plugs to ignite the fuel and air in the cylinders. Instead, the temperature required to ignite the fuel occurs because the pistons compress the air in the cylinders. Suppose air at an initial temperature of 21 °C is compressed adiabatically to a temperature of 688 °C. Assume the air to be an ideal gas for which $\gamma = \frac{7}{5}$. Find the compression ratio, which is the ratio of the initial volume to the final volume.

** **28.** One mole of a monatomic ideal gas has an initial pressure, volume, and temperature of P_0, V_0, and 438 K, respectively. It undergoes an isothermal expansion that triples the volume of the gas. Then, the gas undergoes an isobaric compression back to its original volume. Finally, the gas undergoes an isochoric increase in pressure, so that the final pressure, volume, and temperature are P_0, V_0, and 438 K, respectively. Find the total heat for this three-step process, and state whether it is absorbed by or given off by the gas.

** **29. ssm** The work done by one mole of a monatomic ideal gas ($\gamma = \frac{5}{3}$) in expanding adiabatically is 825 J. The initial temperature and volume of the gas are 393 K and 0.100 m^3. Obtain (a) the final temperature and (b) the final volume of the gas.

Section 15.6 Specific Heat Capacities

30. Heat is added to two identical samples of a monatomic ideal gas. In the first sample the heat is added while the volume of the gas is kept constant, and the heat causes the temperature to rise by 75 K. In the second sample, an identical amount of heat is added while the pressure (but not the volume) of the gas is kept constant. By how much does the temperature of this sample increase?

31. ssm How much heat is required to change the temperature of 1.5 mol of a monatomic ideal gas by 77 K if the pressure is held constant?

32. Argon is a monatomic gas whose atomic mass is 39.9 u. The temperature of eight grams of argon is raised by 75 K under conditions of constant pressure. Assuming that argon is an ideal gas, how much heat is required?

33. ssm The temperature of 2.5 mol of a monatomic ideal gas is 350 K. The internal energy of this gas is doubled by the addition of heat. How much heat is needed when it is added at (a) constant volume and (b) constant pressure?

34. The temperature of 2.5 mol of helium (a monatomic gas) is lowered by 35 K under conditions of constant volume. Assuming that helium behaves as an ideal gas, how much heat is removed from the gas?

35. Heat Q is added to a monatomic ideal gas at constant pressure. As a result, the gas does work W. Find the ratio Q/W.

* **36.** Even at rest, the human body generates heat. The heat arises because of the body's metabolism—that is, the chemical reactions that are always occurring in the body to generate energy. In rooms designed for use by large groups, adequate ventilation or air conditioning must be provided to remove this heat. Consider a classroom containing 200 students. Assume that the metabolic rate of generating heat is 130 W for each student and that the heat accumulates during a fifty-minute lecture. In addition, assume that the air has a molar specific heat of $C_V = \frac{5}{2}R$ and that the room (volume = 1200 m^3, initial pressure = 1.01 × 10^5 Pa, and initial temperature = 21 °C) is sealed shut. If all the heat generated by the students were absorbed by the air, by how much would the air temperature rise during a lecture?

* **37. ssm** A monatomic ideal gas expands at constant pressure. (a) What percentage of the heat being supplied to the gas is used to increase the internal energy of the gas? (b) What percentage is used for doing the work of expansion?

* **38.** Suppose a monatomic ideal gas is contained within a vertical cylinder that is fitted with a movable piston. The piston is frictionless and has a negligible mass. The area of the piston is 3.14×10^{-2} m², and the pressure outside the cylinder is 1.01×10^5 Pa. Heat (2093 J) is removed from the gas. Through what distance does the piston drop?

** **39.** One mole of neon, a monatomic gas, starts out at conditions of standard temperature and pressure. The gas is heated at constant volume until its pressure is tripled, then further heated at constant pressure until its volume is doubled. Assume that neon behaves as an ideal gas. For the entire process, find the heat added to the gas.

Section 15.8 Heat Engines

40. Due to a tune-up, the efficiency of an automobile engine increases by 5.0%. For an input heat of 1300 J, how much more work does the engine produce after the tune-up than before?

41. ssm In doing 16 600 J of work, an engine rejects 9700 J of heat. What is the efficiency of the engine?

42. Engine A discards 72% of its input heat into a cold reservoir. Engine B has twice the efficiency of engine A. What percentage of its input heat does engine B discard?

43. An engine has an efficiency of 64% and produces 5500 J of work. Determine (a) the input heat and (b) the rejected heat.

* **44.** Due to design changes, the efficiency of an engine increases from 0.23 to 0.42. For the same input heat Q_H, these changes increase the work done by the more efficient engine and reduce the amount of heat rejected to the cold reservoir. Find the ratio of the heat rejected to the cold reservoir for the improved engine to that for the original engine.

** **45. ssm** An engine has an efficiency e_1. The engine takes input heat Q_H from a hot reservoir and delivers work W_1. The heat rejected by this engine is used as input heat for a second engine, which has an efficiency e_2 and delivers work W_2. The overall efficiency of this two-engine device is the total work delivered ($W_1 + W_2$) divided by the input heat Q_H. Find an expression for the overall efficiency e in terms of e_1 and e_2.

Section 15.9 Carnot's Principle and the Carnot Engine

46. Five thousand joules of heat is put into a Carnot engine whose hot and cold reservoirs have temperatures of 500 and 200 K, respectively. How much heat is converted into work?

47. An engine has a hot reservoir temperature of 950 K and a cold reservoir temperature of 620 K. The engine operates at three-fifths maximum efficiency. What is the efficiency of the engine?

48. A Carnot engine operates with an efficiency of 27.0% when the temperature of its cold reservoir is 275 K. Assuming that the temperature of the hot reservoir remains the same, what must be the temperature of the cold reservoir in order to increase the efficiency to 32.0%?

49. ssm A Carnot engine has an efficiency of 0.700, and the temperature of its cold reservoir is 378 K. (a) Determine the temperature of its hot reservoir. (b) If 5230 J of heat is rejected to the cold reservoir, what amount of heat is put into the engine?

50. An engine does 18 500 J of work and rejects 6550 J of heat into a cold reservoir whose temperature is 285 K. What would be the smallest possible temperature of the hot reservoir?

51. Concept Simulation 15.1 at **www.wiley.com/college/cutnell** illustrates the concepts pertinent to this problem. A Carnot engine operates between temperatures of 650 and 350 K. To improve the efficiency of the engine, it is decided either to raise the temperature of the hot reservoir by 40 K or to lower the temperature of the cold reservoir by 40 K. Which change gives the greatest improvement? Justify your answer by calculating the efficiency in each case.

* **52.** From a hot reservoir at a temperature of T_1, Carnot engine A takes an input heat of 5550 J, delivers 1750 J of work, and rejects heat to a cold reservoir that has a temperature of 503 K. This cold reservoir at 503 K also serves as the hot reservoir for Carnot engine B, which uses the rejected heat of the first engine as input heat. Engine B also delivers 1750 J of work, while rejecting heat to an even colder reservoir that has a temperature of T_2. Find the temperatures (a) T_1 and (b) T_2.

* **53. ssm** A power plant taps steam superheated by geothermal energy to 505 K (the temperature of the hot reservoir) and uses the steam to do work in turning the turbine of an electric generator. The steam is then converted back into water in a condenser at 323 K (the temperature of the cold reservoir), after which the water is pumped back down into the earth where it is heated again. The output power (work per unit time) of the plant is 84 000 kilowatts. Determine (a) the maximum efficiency at which this plant can operate and (b) the minimum amount of rejected heat that must be removed from the condenser every twenty-four hours.

* **54.** Suppose the gasoline in a car engine burns at 631 °C, while the exhaust temperature (the temperature of the cold reservoir) is 139 °C and the outdoor temperature is 27 °C. Assume that the engine can be treated as a Carnot engine (a gross oversimplification). In an attempt to increase mileage performance, an inventor builds a second engine that functions between the exhaust and outdoor temperatures and uses the exhaust heat to produce additional work. Assume that the inventor's engine can also be treated as a Carnot engine. Determine the ratio of the total work produced by both engines to that produced by the first engine alone.

** **55. ssm** A nuclear-fueled electric power plant utilizes a so-called "boiling water reactor." In this type of reactor, nuclear energy causes water under pressure to boil at 285 °C (the temperature of the hot reservoir). After the steam does the work of turning the turbine of an electric generator, the steam is converted back into water in a condenser at 40 °C (the temperature of the cold reservoir). To keep the condenser at 40 °C, the rejected heat must be carried away by some means—for example, by water from a river. The plant operates at three-fourths of its Carnot efficiency, and the electrical output power of the plant is 1.2×10^9 watts. A river with a water flow rate of 1.0×10^5 kg/s is available to remove the rejected heat from the plant. Find the number of Celsius degrees by which the temperature of the river rises.

** **56.** The hot and cold reservoirs of a Carnot engine have temperatures of 845 and 395 K, respectively. The engine does the work of lifting a 15.0-kg block straight up from rest, so that at a height of 5.00 m the block has a speed of 8.50 m/s. How much heat must be put into the engine?

Section 15.10 Refrigerators, Air Conditioners, and Heat Pumps

57. A Carnot refrigerator maintains the food inside it at 276 K, while the temperature of the kitchen is 298 K. The refrigerator removes 3.00×10^4 J of heat from the food. How much heat is delivered to the kitchen?

58. The inside of a Carnot refrigerator is maintained at a temperature of 277 K, while the temperature in the kitchen is 299 K. Using

2500 J of work, how much heat can this refrigerator remove from its inside compartment?

59. ssm www The temperatures indoors and outdoors are 299 and 312 K, respectively. A Carnot air conditioner deposits 6.12×10^5 J of heat outdoors. How much heat is removed from the house?

60. A refrigerator operates between temperatures of 296 and 275 K. What would be its maximum coefficient of performance?

61. A Carnot heat pump operates between an outdoor temperature of 265 K and an indoor temperature of 298 K. Find its coefficient of performance.

62. A Carnot air conditioner maintains the temperature in a house at 297 K on a day when the temperature outside is 311 K. What is the coefficient of performance of the air conditioner?

* **63.** Review Conceptual Example 9 before attempting this problem. A window air conditioner has an average coefficient of performance of 2.0. This unit has been placed on the floor by the bed, in a futile attempt to cool the bedroom. During this attempt 7.6×10^4 J of heat is pulled in the front of the unit. The room is sealed and contains 3800 mol of air. Assuming that the molar specific heat capacity of the air is $C_V = \frac{5}{2}R$, determine the rise in temperature caused by operating the air conditioner in this manner.

* **64.** Two kilograms of liquid water at 0 °C is put into the freezer compartment of a Carnot refrigerator. The temperature of the compartment is -15 °C, and the temperature of the kitchen is 27 °C. If the cost of electrical energy is ten cents per kilowatt·hour, how much does it cost to make two kilograms of ice at 0 °C?

* **65. ssm www** A Carnot refrigerator transfers heat from its inside (6.0 °C) to the room air outside (20.0 °C). (a) Find the coefficient of performance of the refrigerator. (b) Determine the magnitude of the minimum work needed to cool 5.00 kg of water from 20.0 to 6.0 °C when it is placed in the refrigerator.

* **66.** An air conditioner keeps the inside of a house at a temperature of 19.0 °C when the outdoor temperature is 33.0 °C. Heat, leaking into the house at the rate of 10 500 joules per second, is removed by the air conditioner. Assuming that the air conditioner is a Carnot air conditioner, what is the work per second that must be done by the electrical energy in order to keep the inside temperature constant?

** **67. ssm** A Carnot engine uses hot and cold reservoirs that have temperatures of 1684 and 842 K, respectively. The input heat for this engine is Q_H. The work delivered by the engine is used to operate a Carnot heat pump. The pump removes heat from the 842-K reservoir and puts it into a hot reservoir at a temperature T'. The

amount of heat removed from the 842-K reservoir is also Q_H. Find the temperature T'.

Section 15.11 Entropy

68. Four kilograms of carbon dioxide sublimes from solid dry ice to a gas at a pressure of one atmosphere and a temperature of 194.7 K. The latent heat of sublimation is 5.77×10^5 J/kg. Find the change in entropy of the carbon dioxide.

69. On a cold day, 24 500 J of heat leaks out of a house. The inside temperature is 21 °C, and the outside temperature is -15 °C. What is the increase in the entropy of the universe that this heat loss produces?

70. Heat Q flows spontaneously from a reservoir at 394 K into a reservoir that has a lower temperature T. Because of the spontaneous flow, thirty percent of Q is rendered unavailable for work when a Carnot engine operates between the reservoir at temperature T and a reservoir at 248 K. Find the temperature T.

71. ssm Find the change in entropy of the H_2O molecules when (a) three kilograms of ice melts into water at 273 K and (b) three kilograms of water changes into steam at 373 K. (c) On the basis of the answers to parts (a) and (b), discuss which change creates more disorder in the collection of H_2O molecules.

* **72.** (a) Find the equilibrium temperature that results when one kilogram of liquid water at 373 K is added to two kilograms of liquid water at 283 K in a perfectly insulated container. (b) When heat is added to or removed from a solid or liquid of mass m and specific heat capacity c, the change in entropy can be shown to be $\Delta S = mc\ln(T_f/T_i)$, where T_i and T_f are the initial and final Kelvin temperatures. Use this equation to calculate the entropy change for each amount of water. Then combine the two entropy changes algebraically to obtain the total entropy change of the universe. Note that the process is irreversible, so the total entropy change of the universe is greater than zero. (c) Assuming that the coldest reservoir at hand has a temperature of 273 K, determine the amount of energy that becomes unavailable for doing work because of the irreversible process.

* **73.** Refer to **Interactive Solution 15.73** at **www.wiley.com/college/cutnell** to review a method by which this problem can be solved. (a) After 6.00 kg of water at 85.0 °C is mixed in a perfect thermos with 3.00 kg of ice at 0.0 °C, the mixture is allowed to reach equilibrium. Using the expression $\Delta S = mc\ln(T_f/T_i)$ [see problem 72] and the change in entropy for melting, find the change in entropy that occurs. (b) Should the entropy of the universe increase or decrease as a result of the mixing process? Give your reasoning and state whether your answer in part (a) is consistent with your answer here.

Additional Problems

74. Six grams of helium (molecular mass = 4.0 u) expands isothermally at 370 K and does 9600 J of work. Assuming that helium is an ideal gas, determine the ratio of the final volume of the gas to the initial volume.

75. ssm One-half mole of a monatomic ideal gas absorbs 1200 J of heat while 2500 J of work is done by the gas. (a) What is the temperature change of the gas? (b) Is the change an increase or a decrease?

76. The input heat for an engine is 2.41×10^4 J, and the rejected heat is 5.86×10^3 J. Find the work done by the engine.

77. A system undergoes a two-step process. In the first step, the internal energy of the system increases by 228 J when 166 J of work

is done on the system. In the second step, the internal energy of the system increases by 115 J when 177 J of work is done on the system. For the overall process, find the heat. What type of process is the overall process? Explain.

78. The water in a deep underground well is used as the cold reservoir of a Carnot heat pump that maintains the temperature of a house at 301 K. To deposit 14 200 J of heat in the house, the heat pump requires 800 J of work. Determine the temperature of the well water.

79. ssm A process occurs in which the entropy of a system increases by 125 J/K. During the process, the energy that becomes unavailable for doing work is zero. (a) Is this process reversible or irreversible? Give your reasoning. (b) Determine the change in the entropy of the surroundings.

80. Three moles of a monatomic ideal gas are heated at a constant volume of 1.50 m^3. The amount of heat added is 5.24 × 10^3 J. (a) What is the change in the temperature of the gas? (b) Find the change in its internal energy. (c) Determine the change in pressure.

81. A heat pump removes 2090 J of heat from the outdoors and delivers 3140 J of heat to the inside of a house. (a) How much work does the heat pump need? (b) What is the coefficient of performance of the heat pump?

82. A Carnot engine uses a hot reservoir consisting of a large amount of boiling water and a cold reservoir consisting of a large tub of ice and water. When 6800 J of heat is put into the engine and the engine produces work, how many kilograms of ice in the tub are melted due to the heat delivered to the cold reservoir?

83. ssm When a .22-caliber rifle is fired, the expanding gas from the burning gunpowder creates a pressure behind the bullet. This pressure causes the force that pushes the bullet through the barrel. The barrel has a length of 0.61 m and an opening whose radius is 2.8 × 10^{-3} m. A bullet (mass = 2.6 × 10^{-3} kg) has a speed of 370 m/s after passing through this barrel. Ignore friction and determine the average pressure of the expanding gas.

* 84. A piece of aluminum has a volume of 1.4 × 10^{-3} m^3. The coefficient of volume expansion for aluminum is $\beta = 69 \times 10^{-6}$ (C°)$^{-1}$. The temperature of this object is raised from 20 to 320 °C. How much work is done by the expanding aluminum if the air pressure is 1.01 × 10^5 Pa?

* 85. Interactive Solution 15.85 at **www.wiley.com/college/cutnell** offers one approach to this problem. A fifteen-watt heater is used to heat a monatomic ideal gas at a constant pressure of 7.60 × 10^5 Pa. During the process, the 1.40 × 10^{-3} m^3 volume of the gas increases by 25.0%. How long was the heater on?

* 86. How long would a 3.00-kW space heater have to run to put into a kitchen the same amount of heat as a refrigerator (coefficient of performance = 3.00) does when it freezes 1.50 kg of water at 20.0 °C into ice at 0.0 °C?

* 87. ssm A monatomic ideal gas ($\gamma = \frac{5}{3}$) is contained within a perfectly insulated cylinder that is fitted with a movable piston. The ini-

tial pressure of the gas is 1.50 × 10^5 Pa. The piston is pushed so as to compress the gas, with the result that the Kelvin temperature doubles. What is the final pressure of the gas?

* 88. Suppose that 31.4 J of heat is added to an ideal gas. The gas expands at a constant pressure of 1.40 × 10^4 Pa while changing its volume from 3.00 × 10^{-4} to 8.00 × 10^{-4} m^3. The gas is not monatomic, so the relation $C_P = \frac{5}{2}R$ does not apply. (a) Determine the change in the internal energy of the gas. (b) Calculate its molar specific heat capacity C_P.

* 89. A monatomic ideal gas is heated at a constant volume of 1.00 × 10^{-3} m^3, using a ten-watt heater. The pressure of the gas increases by 5.0 × 10^4 Pa. How long was the heater on?

* 90. The hot reservoir for a Carnot engine has a temperature of 890 K, while the cold reservoir has a temperature of 670 K. The heat input for this engine is 4800 J. The 670-K reservoir also serves as the hot reservoir for a second Carnot engine. This second engine uses the rejected heat of the first engine as input and extracts additional work from it. The rejected heat from the second engine goes into a reservoir that has a temperature of 420 K. Find the total work delivered by the two engines.

** 91. ssm www Engine A receives three times more input heat, produces five times more work, and rejects two times more heat than engine B. Find the efficiency of (a) engine A and (b) engine B.

** 92. The drawing shows an adiabatically isolated cylinder that is divided initially into two identical parts by an adiabatic partition. Both sides contain one mole of a monatomic ideal gas ($\gamma = \frac{5}{3}$), with the initial temperature being 525 K on the left and 275 K on the right. The partition is then allowed to move slowly (i.e., quasi-statically) to the right, until the pressures on each side of the partition are the same. Find the final temperatures on the (a) left and (b) right.

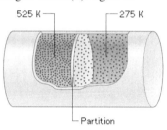

Concepts & Calculations Group Learning Problems

Note: Each of these problems consists of Concept Questions followed by a related quantitative Problem. They are designed for use by students working alone or in small learning groups. The Concept Questions involve little or no mathematics and are intended to stimulate group discussions. They focus on the concepts with which the problems deal. Recognizing the concepts is the essential initial step in any problem-solving technique.

93. Concept Questions A system does 4.8 × 10^4 J of work, and 7.6 × 10^4 J of heat flows into the system in the process. (a) Considered by itself, does the work increase or decrease the internal energy of the system? (b) Considered by itself, does the heat increase or decrease the internal energy? (c) Considering the work and heat together, does the internal energy of the system increase, decrease, or remain the same? Explain.

Problem Find the change in the internal energy of the system. Make sure that your answer is consistent with your answers to the Concept Questions.

94. Concept Questions A system gains a certain amount of energy in the form of heat at constant pressure, and the internal energy of the

system increases by an even greater amount. (a) Is any work done and, if so, is it done on or by the system? (b) If there is work, is it positive or negative, according to our convention? (c) Does the volume of the system increase, decrease, or remain the same? Give your reasoning.

Problem A system gains 2780 J of heat at a constant pressure of 1.26 × 10^5 Pa, and its internal energy increases by 3990 J. What is the change in volume of the system, and is it an increase or a decrease? Verify that your answer is consistent with your answers to the Concept Questions.

95. Concept Questions An ideal gas expands isothermally, doing work, while heat flows into the gas. (a) Does the internal energy of the gas increase, decrease, or remain the same? (b) Is the work done by the gas greater than, less than, or equal to the heat that flows into the gas? Account for your answers.

Problem Three moles of neon expand isothermally from 0.100 to 0.250 m^3. Into the gas flows 4.75 × 10^3 J of heat. Assuming that neon is an ideal gas, find its temperature.

96. Concept Questions A mountain climber, starting from rest, does work in climbing upward. At the top, she is again at rest. In the process, her body generates 4.1×10^6 J of energy via metabolic processes. In fact, her body acts like a heat engine, the efficiency of which is given by Equation 15.11 as $e = W/Q_H$, where W is the work and Q_H is the input heat. (a) Is the 4.1×10^6 J of energy equal to W or Q_H? (b) How is the work done in climbing upward related to the vertical height of the climb? Explain.

Problem The vertical height of the climb is 730 m. The climber has a mass of 52 kg. Find her efficiency as a heat engine.

97. Concept Questions Two Carnot engines, A and B, utilize the same hot reservoir, but engine A is less efficient than engine B. (a) Which engine produces more work for a given heat input? (b) Which engine has the lower cold-reservoir temperature? Give your reasoning.

Problem Carnot engine A has an efficiency of 0.60, and Carnot engine B has an efficiency of 0.80. Both engines utilize the same hot reservoir, which has a temperature of 650 K and delivers 1200 J of heat to each engine. Find the work produced by each engine and the temperatures of the cold reservoirs that they use. Check to see that your answers are consistent with your answers to the Concept Questions.

98. Concept Questions Two Carnot air conditioners, A and B, are removing heat from different rooms. The outside temperature is the same for both, but the room temperatures are different. The room serviced by unit A is kept colder than the room serviced by unit B. The heat removed from both rooms is the same. (a) Which unit requires the greater amount of work? (b) Which unit deposits the greater amount of heat outside? Explain.

Problem The outside temperature is 309.0 K. The room serviced by unit A is kept at a temperature of 294.0 K, while the room serviced by unit B is kept at 301.0 K. The heat removed from either room is 4330 J. For both units, find the work required and determine the heat deposited outside. Verify that your answers are consistent with your answers to the Concept Questions.

* **99. Concept Questions** An ideal gas is taken through the three processes ($A \rightarrow B$, $B \rightarrow C$, and $C \rightarrow A$) shown in the drawing. In general, for each process the internal energy U of the gas can change, because heat

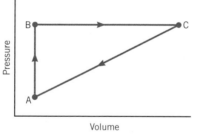

Q can be added to or removed from the gas and work W can be done by the gas or on the gas. (a) For the process $A \rightarrow B$, is work done by the gas or on the gas? Why or why not? (b) For the process $B \rightarrow C$, suppose that the change ΔU in the internal energy of the gas and the work W are known. Is it then possible to determine the heat Q that has been added to or removed from the gas? If so, explain how this could be done. (c) For the process $C \rightarrow A$, is it possible to find the change in the internal energy of the gas if the change in the internal energies for the processes $A \rightarrow B$ and $B \rightarrow C$ are known? If so, specify how this could be done.

Problem For the three processes shown in the drawing, fill in the five missing entries in the following table.

Process	ΔU	Q	W
$A \rightarrow B$	(b)	+561 J	(a)
$B \rightarrow C$	+4303 J	(c)	+2867 J
$C \rightarrow A$	(d)	(e)	−3740 J

* **100. Concept Questions** (a) A real (irreversible) engine operates between hot and cold reservoirs whose temperatures are T_H and T_C. The engine absorbs heat Q_H from the hot reservoir and performs work W. What is the expression that gives the change in entropy $\Delta S_{universe}$ of the universe associated with the operation of this engine? Express your answer in terms of T_H, T_C, Q_H, and Q_C. Note that Q_H and Q_C denote the magnitudes of the heats. (b) Would you expect the change in the entropy of the universe to be greater than, less than, or equal to, zero? Provide a reason for your answer. (c) Suppose that a reversible engine (a Carnot engine) operates between the same hot and cold temperatures as the irreversible engine described above. For the same input heat Q_H, would the reversible engine produce more, less, or the same work as the irreversible engine? Justify your answer. (d) In terms of $\Delta S_{universe}$, and possibly Q_H and Q_C, how could one determine the difference, if any, between the work done by the reversible engine and that done by the irreversible engine?

Problem (a) An irreversible engine operates between temperatures of 852 and 314 K. It absorbs 1285 J of heat from the hot reservoir and does 264 J of work. What is the change $\Delta S_{universe}$ in the entropy of the universe associated with the operation of this engine? (b) If the engine were reversible, how much work would it have done, assuming it operates between the same temperatures and absorbs the same heat as the irreversible engine? (c) Using the result of part (a), find the difference between the work produced by the reversible and irreversible engines.

Chapter 16

Waves and Sound

A water wave is one type of traveling wave. When waves traveling on the ocean encounter the shore, the surf can be spectacular as the water piles up on itself. (© Warren Bolster/Stone/Getty Images)

16.1 The Nature of Waves

Water waves have two features common to all waves:

1. A wave is a traveling disturbance.
2. A wave carries energy from place to place.

In Figure 16.1 the wave created by the motorboat travels across the lake and disturbs the fisherman. However, *there is no bulk flow of water* outward from the motorboat. The wave is not a bulk movement of water such as a river, but, rather, a disturbance traveling on the surface of the lake. Part of the wave's energy in Figure 16.1 is transferred to the fisherman and his boat.

We will consider two basic types of waves, transverse and longitudinal. Figure 16.2 illustrates how a transverse wave can be generated using a Slinky, a remarkable toy in the form of a long, loosely coiled spring. If one end of the Slinky is jerked up and down, as in part *a*, an upward pulse is sent traveling toward the right. If the end is then jerked down and up, as in part *b*, a downward pulse is generated and also moves to the right. If the end is continually moved up and down in simple harmonic motion, an entire wave is produced. As part *c* illustrates, the wave consists of a series of alternating upward and downward sections that propagate to the right, disturbing the vertical position of the Slinky in the process. To focus attention on the disturbance, a colored dot is attached to the Slinky in part *c* of the drawing. As the wave advances, the dot is displaced up and down in simple harmonic motion. The motion of the dot occurs perpendicular, or transverse, to the direction in which the wave travels. Thus, *a transverse wave is one in which the disturbance occurs perpendicular to the direction of travel of the wave.* Radio waves, light waves, and microwaves are transverse waves. Transverse waves also travel on the strings of instruments such as guitars and banjos.

A longitudinal wave can also be generated with a Slinky, and Figure 16.3 demonstrates how. When one end of the Slinky is pushed forward along its length (i.e., longitudinally) and then pulled back to its starting point, as in part *a*, a region where the coils are squeezed together or compressed is sent traveling to the right. If the end is pulled backward and then pushed forward to its starting point, as in part *b*, a region where the coils are pulled apart or stretched is formed and also moves to the right. If the end is continually moved back and forth in simple harmonic motion, an entire wave is created. As part *c* shows, the wave consists of a series of alternating compressed and stretched regions that travel to the right and disturb the separation between adjacent coils. A colored dot is once again attached to the Slinky to emphasize the vibratory nature of the disturbance. In response to the wave, the dot moves back and forth in simple harmonic motion along the line of travel of the wave. Thus, *a longitudinal wave is one in which the disturbance occurs parallel to the line of travel of the wave.* A sound wave is a longitudinal wave.

Figure 16.1 The wave created by the motorboat travels across the lake and disturbs the fisherman.

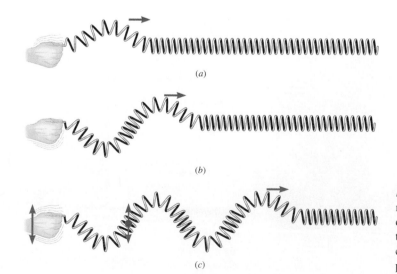

(a)

(b)

(c)

Figure 16.2 (*a*) An upward pulse moves to the right, followed by (*b*) a downward pulse. (*c*) When the end of the Slinky is moved up and down continuously, a transverse wave is produced.

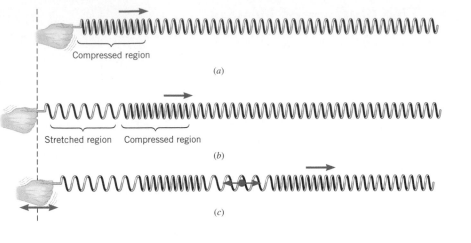

Figure 16.3 (*a*) A compressed region moves to the right, followed by (*b*) a stretched region. (*c*) When the end of the Slinky is moved back and forth continuously, a longitudinal wave is produced.

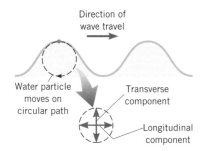

Figure 16.4 A water wave is neither transverse nor longitudinal, since water particles at the surface move clockwise on nearly circular paths as the wave moves from left to right.

Figure 16.5 CONCEPTS AT A GLANCE For periodic waves, the terms *cycle, amplitude, period,* and *frequency* have the same meaning as they do in simple harmonic motion. Sound, such as that produced by these musicians, is a periodic wave and is described using this terminology. (© Andy Sacks/Stone/ Getty Images)

Some waves are neither transverse nor longitudinal. For instance, in a water wave the motion of the water particles is not strictly perpendicular or strictly parallel to the line along which the wave travels. Instead, the motion includes both transverse and longitudinal components, since the water particles at the surface move on nearly circular paths, as Figure 16.4 indicates.

16.2 *Periodic Waves*

▶ CONCEPTS AT A GLANCE   The transverse and longitudinal waves that we have been discussing are called *periodic waves* because they consist of cycles or patterns that are produced over and over again by the source. In Figures 16.2 and 16.3 the repetitive patterns occur as a result of the simple harmonic motion of the left end of the Slinky. Every segment of the Slinky vibrates in simple harmonic motion. Sections 10.1 and 10.2 discuss the simple harmonic motion of an object on a spring and introduce the concepts of cycle, amplitude, period, and frequency. As the Concepts-at-a-Glance chart in Figure 16.5 illustrates, this same terminology is used to describe periodic waves, such as the sound waves we hear and the light waves we see (discussed in Chapter 24). ◀

Figure 16.6 uses a graphical representation of a transverse wave to review this terminology. One *cycle* of a wave is shaded in color in both parts of the drawing. A wave is a series of many cycles. In part *a* the vertical position of the Slinky is plotted on the vertical axis, and the corresponding distance along the length of the Slinky is plotted on the horizontal axis. Such a graph is equivalent to a photograph of the wave taken at one instant in

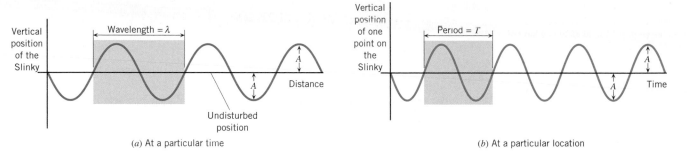

Vertical position of the Slinky | Wavelength = λ | A | Distance | Undisturbed position

(a) At a particular time

Vertical position of one point on the Slinky | Period = T | A | Time

(b) At a particular location

Figure 16.6 One cycle of the wave is shaded in color, and the amplitude of the wave is denoted as A.

time and shows the disturbance that exists at each point along the Slinky's length. As marked on this graph, the *amplitude A* is the maximum excursion of a particle of the medium from the particle's undisturbed position. The amplitude is the distance between a crest, or highest point on the wave pattern, and the undisturbed position; it is also the distance between a trough, or lowest point on the wave pattern, and the undisturbed position. The *wavelength* λ is the horizontal length of one cycle of the wave, as shown in Figure 16.6a. The wavelength is also the horizontal distance between two successive crests, two successive troughs, or any two successive equivalent points on the wave.

Part *b* of Figure 16.6 shows a graph in which time, rather than distance, is plotted on the horizontal axis. This graph is obtained by observing a single point on the Slinky. As the wave passes, the point under observation oscillates up and down in simple harmonic motion. As indicated on the graph, the *period T* is the time required for one complete up/down cycle, just as it is for an object vibrating on a spring. Equivalently, the period is the time required for the wave to travel a distance of one wavelength. The period T is related to the *frequency f,* just as it is for any example of simple harmonic motion:

$$f = \frac{1}{T} \qquad (10.5)$$

The period is commonly measured in seconds, and frequency is measured in cycles per second or hertz (Hz). If, for instance, one cycle of a wave takes one-tenth of a second to pass an observer, then ten cycles pass per second, as Equation 10.5 indicates [$f = 1/(0.1 \text{ s}) = 10$ cycles/s $= 10$ Hz].

A simple relation exists between the period, the wavelength, and the speed of a wave, a relation that Figure 16.7 helps to introduce. Imagine waiting at a railroad crossing, while a freight train moves by at a constant speed v. The train consists of a long line of identical boxcars, each of which has a length λ and requires a time T to pass, so the speed is $v = \lambda/T$. This same equation applies for a wave and relates the speed of the wave to the wavelength λ and the period T. Since the frequency of a wave is $f = 1/T$, the expression for the speed is

$$v = \frac{\lambda}{T} = f\lambda \qquad (16.1)$$

The terminology just discussed and the fundamental relations $f = 1/T$ and $v = f\lambda$ apply to longitudinal as well as to transverse waves. Example 1 illustrates how the wavelength of a wave is determined by the wave speed and the frequency established by the source.

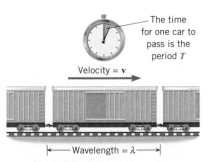

The time for one car to pass is the period T

Velocity = **v**

Wavelength = λ

Figure 16.7 A train moving at a constant speed serves as an analogy for a traveling wave.

Example 1 The Wavelengths of Radio Waves

AM and FM radio waves are transverse waves that consist of electric and magnetic disturbances. These waves travel at a speed of 3.00×10^8 m/s. A station broadcasts an AM radio wave whose frequency is 1230×10^3 Hz (1230 kHz on the dial) and an FM radio wave whose frequency is 91.9×10^6 Hz (91.9 MHz on the dial). Find the distance between adjacent crests in each wave.

Reasoning The distance between adjacent crests is the wavelength λ. Since the speed of each wave is $v = 3.00 \times 10^8$ m/s and the frequencies are known, the relation $v = f\lambda$ can be used to determine the wavelengths.

Problem solving insight
The equation $v = f\lambda$ applies to any kind of periodic wave.

Solution

AM
$$\lambda = \frac{v}{f} = \frac{3.00 \times 10^8 \text{ m/s}}{1230 \times 10^3 \text{ Hz}} = \boxed{244 \text{ m}}$$

FM
$$\lambda = \frac{v}{f} = \frac{3.00 \times 10^8 \text{ m/s}}{91.9 \times 10^6 \text{ Hz}} = \boxed{3.26 \text{ m}}$$

Notice that the wavelength of an AM radio wave is longer than two and one-half football fields!

16.3   The Speed of a Wave on a String

The properties of the material* or medium through which a wave travels determine the speed of the wave. For example, Figure 16.8 shows a transverse wave on a string and draws attention to four string particles that have been drawn as colored dots. As the wave moves to the right, each particle is displaced, one after the other, from its undisturbed position. In the drawing, particles 1 and 2 have already been displaced upward, while particles 3 and 4 are not yet affected by the wave. Particle 3 will be next to move because the section of string immediately to its left (i.e., particle 2) will pull it upward.

Figure 16.8 leads us to conclude that the speed with which the wave moves to the right depends on how quickly one particle of the string is accelerated upward in response to the net pulling force exerted by its adjacent neighbors. In accord with Newton's second law, a stronger net force results in a greater acceleration, and, thus, a faster-moving wave. The ability of one particle to pull on its neighbors depends on how tightly the string is stretched—that is, on the tension (see Section 4.10 for a review of tension). The greater the tension, the greater the pulling force the particles exert on each other, and the faster the wave travels, other things being equal. Along with the tension, a second factor influences the wave speed. According to Newton's second law, the inertia or mass of particle 3 in Figure 16.8 also affects how quickly it responds to the upward pull of particle 2. For a given net pulling force, a smaller mass has a greater acceleration than a larger mass. Therefore, other things being equal, a wave travels faster on a string whose particles have a small mass, or, as it turns out, on a string that has a small mass per unit length. The mass per unit length is called the *linear density* of the string. It is the mass m of the string divided by its length L, or m/L. The effects of the tension F and the mass per unit length are evident in the following expression for the speed v of a small-amplitude wave on a string:

$$v = \sqrt{\frac{F}{m/L}} \qquad (16.2)$$

The motion of transverse waves along a string is important in the operation of musical instruments, such as the guitar, the violin, and the piano. In these instruments, the strings are either plucked, bowed, or struck to produce transverse waves. Example 2 discusses the speed of the waves on the strings of a guitar.

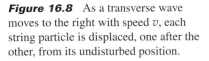

Concept Simulation 16.1

This simulation displays a wave traveling on a string and allows you to adjust the amplitude and frequency of the wave, as well as the thickness or mass of the string. The effect of your adjustments on the wavelength of the wave is displayed, as is the time for the wave to travel the length of the string. From the travel time you can assess how the speed of the wave depends on the mass of the string.

Go to
www.wiley.com/college/cutnell

Example 2   Waves Traveling on Guitar Strings

Transverse waves travel on the strings of an electric guitar after the strings are plucked (see Figure 16.9). The length of each string between its two fixed ends is 0.628 m, and the mass is

Figure 16.8 As a transverse wave moves to the right with speed v, each string particle is displaced, one after the other, from its undisturbed position.

---

* Electromagnetic waves (discussed in Chapter 24) can move through a vacuum, as well as through materials such as glass and water.

0.208 g for the highest pitched E string and 3.32 g for the lowest pitched E string. Each string is under a tension of 226 N. Find the speeds of the waves on the two strings.

Reasoning The speed of a wave on a guitar string, as expressed by Equation 16.2, depends on the tension F in the string and its linear density m/L. Since the tension is the same for both strings, and smaller linear densities give rise to greater speeds, we expect the wave speed to be greatest on the string with the smallest linear density.

The physics of **waves on guitar strings.**

Solution The speeds of the waves are given by Equation 16.2 as

High-pitched E $v = \sqrt{\dfrac{F}{m/L}} = \sqrt{\dfrac{226\ \text{N}}{(0.208 \times 10^{-3}\ \text{kg})/(0.628\ \text{m})}} = \boxed{826\ \text{m/s}}$

Low-pitched E $v = \sqrt{\dfrac{F}{m/L}} = \sqrt{\dfrac{226\ \text{N}}{(3.32 \times 10^{-3}\ \text{kg})/(0.628\ \text{m})}} = \boxed{207\ \text{m/s}}$

Notice how fast the waves move; the speeds correspond to 1850 and 463 mi/h.

Conceptual Example 3 offers additional insight into the nature of a wave as a traveling disturbance.

Conceptual Example 3   Wave Speed Versus Particle Speed

Is the speed of a transverse wave on a string the same as the speed at which a particle on the string moves (see Figure 16.10)?

Reasoning and Solution The particle speed v_{particle} specifies how fast the particle is moving as it oscillates up and down, and it is different from the wave speed. If the source of the wave (e.g., the hand in Figure 16.2c) vibrates in simple harmonic motion, each string particle vibrates in a like manner, with the same amplitude and frequency as the source. Moreover, the particle speed, unlike the wave speed, is not constant. As for any object in simple harmonic motion, the particle speed is greatest when the particle is passing through the undisturbed position of the string and zero when the particle is at its maximum displacement. According to Equation 10.7, the particle speed depends on the amplitude A, the angular frequency ω, and the time t through the relation $v_{\text{particle}} = A\omega \sin \omega t$, where the minus sign has been omitted because we are interested only in the speed, which is the magnitude of the particle's velocity. Thus, the speed of a string particle is determined by the *properties of the source* creating the wave and not by the properties of the string itself. In contrast, the speed of the wave is determined by the properties of the string—that is, the tension F and the mass per unit length m/L, according to Equation 16.2. We see, then, that *the two speeds, v_{wave} and v_{particle}, are not the same.*

Related Homework: *Conceptual Question 4, Problems 18, 99*

Figure 16.9 Plucking a guitar string generates transverse waves.

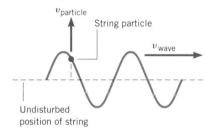

Figure 16.10 A transverse wave on a string is moving to the right with a constant speed v_{wave}. A string particle moves up and down in simple harmonic motion about the undisturbed position of the string. The speed of the particle v_{particle} changes from moment to moment as the wave passes.

✔ **Check Your Understanding 1**

String I and string II have the same length. However, the mass of string I is twice the mass of string II, and the tension in string I is eight times the tension in string II. A wave of the same amplitude and frequency travels on each of these strings. Which of the pictures in the drawing correctly shows the waves? *(The answer is given at the end of the book.)*

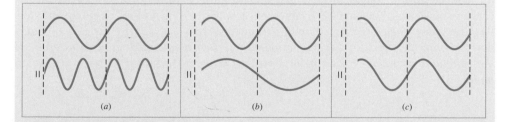

Background: This question deals with the relationship between the frequency, wavelength, and speed of a traveling wave. It also deals with the factors that determine the speed of a wave on a string.

For similar questions (including calculational counterparts), consult Self-Assessment Test 16.1. This test is described at the end of Section 16.4.

*16.4 The Mathematical Description of a Wave

When a wave travels through a medium, it displaces the particles of the medium from their undisturbed positions. Suppose a particle is located at a distance x from a coordinate origin. We would like to know the displacement y of this particle from its undisturbed position at any time t as the wave passes. For periodic waves that result from simple harmonic motion of the source, the expression for the displacement involves a sine or cosine, a fact that is not surprising. After all, in Chapter 10 simple harmonic motion is described using sinusoidal equations, and the graphs for a wave in Figure 16.6 look like a plot of displacement versus time for an object oscillating on a spring (see Figure 10.6).

Our tack will be to present the expression for the displacement and then show graphically that it gives a correct description. Equation 16.3 represents the displacement of a particle caused by a wave traveling in the $+x$ direction (to the right), with an amplitude A, frequency f, and wavelength λ. Equation 16.4 applies to a wave moving in the $-x$ direction (to the left).

Wave motion toward $+x$ $$y = A \sin\left(2\pi ft - \frac{2\pi x}{\lambda}\right)$$ (16.3)

Wave motion toward $-x$ $$y = A \sin\left(2\pi ft + \frac{2\pi x}{\lambda}\right)$$ (16.4)

These equations apply to transverse or longitudinal waves and assume that $y = 0$ m when $x = 0$ m and $t = 0$ s.

Consider a transverse wave moving in the $+x$ direction along a string. The term $(2\pi ft - 2\pi x/\lambda)$ in Equation 16.3 is called the *phase angle* of the wave. A string particle located at the origin ($x = 0$ m) exhibits simple harmonic motion with a phase angle of $2\pi ft$; that is, its displacement as a function of time is $y = A \sin(2\pi ft)$. A particle located at a distance x also exhibits simple harmonic motion, but its phase angle is

$$2\pi ft - \frac{2\pi x}{\lambda} = 2\pi f\left(t - \frac{x}{f\lambda}\right) = 2\pi f\left(t - \frac{x}{v}\right)$$

The quantity x/v is the time needed for the wave to travel the distance x. In other words, the simple harmonic motion that occurs at x is delayed by the time interval x/v compared to the motion at the origin.

Figure 16.11 shows the displacement y plotted as a function of position x along the string at a series of time intervals separated by one-fourth of the period $T(t = 0 \text{ s}, \frac{1}{4}T, \frac{2}{4}T, \frac{3}{4}T, T)$. These graphs are constructed by substituting the corresponding value for t into Equation 16.3, remembering that $f = 1/T$, and then calculating y at a series of values for x. The graphs are like photographs taken at various times as the wave moves to the right. For reference, the colored square on each graph marks the place on the wave that is located at $x = 0$ m when $t = 0$ s. As time passes, the colored square moves to the right, along with the wave. In a similar manner, it can be shown that Equation 16.4 represents a wave moving in the $-x$ direction. Note that the phase angles $(2\pi ft - 2\pi x/\lambda)$ in Equation 16.3 and $(2\pi ft + 2\pi x/\lambda)$ in Equation 16.4 are measured in *radians,* not degrees. **When a calculator is used to evaluate the functions sin $(2\pi ft - 2\pi x/\lambda)$ or sin $(2\pi ft + 2\pi x/\lambda)$, it must be set to its radian mode.**

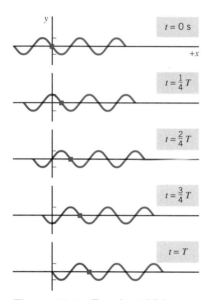

Figure 16.11 Equation 16.3 is plotted here at a series of times separated by one-fourth of the period T. The colored square in each graph marks the place on the wave that is located at $x = 0$ m when $t = 0$ s. As time passes, the wave moves to the right.

Problem solving insight

Self-Assessment Test 16.1

Check your understanding of the material in Sections 16.1–16.4:

• Periodic Waves   • The Speed of a Wave on a String
• The Mathematical Description of a Wave

Go to **www.wiley.com/college/cutnell**

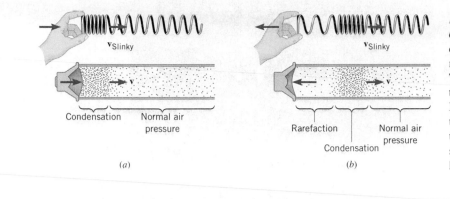

Figure 16.12 (*a*) When the speaker diaphragm moves outward, it creates a condensation. (*b*) When the diaphragm moves inward, it creates a rarefaction. The condensation and rarefaction on the Slinky are included for comparison. In reality, the velocity of the wave on the Slinky v_{Slinky} is much smaller than the velocity of sound in air **v**. For simplicity, the two waves are shown here to have the same velocity.

16.5 The Nature of Sound

LONGITUDINAL SOUND WAVES

Sound is a longitudinal wave that is created by a vibrating object, such as a guitar string, the human vocal cords, or the diaphragm of a loudspeaker. Moreover, sound can be created or transmitted only in a medium, such as a gas, liquid, or solid. As we will see, the particles of the medium must be present for the disturbance of the wave to move from place to place. Sound cannot exist in a vacuum.

To see how sound waves are produced and why they are longitudinal, consider the vibrating diaphragm of a loudspeaker. When the diaphragm moves outward, it compresses the air directly in front of it, as in Figure 16.12*a*. This compression causes the air pressure to rise slightly. The region of increased pressure is called a ***condensation,*** and it travels away from the speaker at the speed of sound. The condensation is analogous to the compressed region of coils in a longitudinal wave on a Slinky, which is included in Figure 16.12*a* for comparison. After producing a condensation, the diaphragm reverses its motion and moves inward, as in part *b* of the drawing. The inward motion produces a region known as a ***rarefaction,*** where the air pressure is slightly less than normal. The rarefaction is similar to the stretched region of coils in a longitudinal Slinky wave. Following immediately behind the condensation, the rarefaction also travels away from the speaker at the speed of sound. Figure 16.13 further emphasizes the similarity between a sound wave and a longitudinal Slinky wave. As the wave passes, the colored dots attached both to the Slinky and to an air molecule execute simple harmonic motion about their undisturbed positions. The colored arrows on either side of the dots indicate that the simple harmonic motion occurs parallel to the line of travel. The drawing also shows that the wavelength λ is the distance between the centers of two successive condensations; λ is also the distance between the centers of two successive rarefactions.

Figure 16.14 illustrates a sound wave spreading out in space after being produced by a loudspeaker. When the condensations and rarefactions arrive at the ear, they force the eardrum to vibrate at the same frequency as the speaker diaphragm. The vibratory motion of the eardrum is interpreted by the brain as sound. It should be emphasized that sound is not a mass movement of air, like the wind. As the condensations and rarefactions of the sound wave travel outward from the vibrating diaphragm in Figure 16.14, the individual air molecules are not carried along with the wave. Rather, each molecule executes simple harmonic motion about a fixed location. In doing so, one molecule collides with its neighbor and passes the condensations and rarefactions forward. The neighbor, in turn, repeats the process.

The physics of a loudspeaker diaphragm.

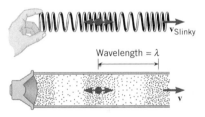

Figure 16.13 Both the wave on the Slinky and the sound wave are longitudinal. The colored dots attached to the Slinky and to an air molecule vibrate back and forth parallel to the line of travel of the wave.

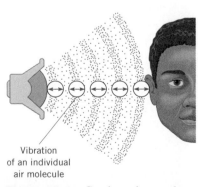

Figure 16.14 Condensations and rarefactions travel from the speaker to the listener, but the individual air molecules do not move with the wave. A given molecule vibrates back and forth about a fixed location.

THE FREQUENCY OF A SOUND WAVE

Each cycle of a sound wave includes one condensation and one rarefaction, and the ***frequency*** is the number of cycles per second that passes by a given location. For example, if the diaphragm of a speaker vibrates back and forth in simple harmonic motion at a frequency of 1000 Hz, then 1000 condensations, each followed by a rarefaction, are generated every second, thus forming a sound wave whose frequency is also 1000 Hz. A sound

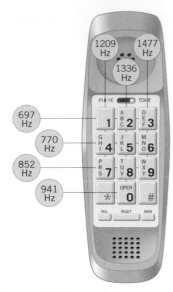

Figure 16.15 A push-button telephone and a schematic showing the two pure tones produced when each button is pressed.

The physics of push-button telephones.

Figure 16.16 Rhinoceroses call to one another using infrasonic sound waves. (Photo by Ron Garrison, courtesy Zoological Society of San Diego.)

with a single frequency is called a **pure tone.** Experiments have shown that a healthy young person hears all sound frequencies from approximately 20 to 20 000 Hz (20 kHz). The ability to hear the high frequencies decreases with age, however, and a normal middle-aged adult hears frequencies only up to 12 – 14 kHz.

Pure tones are used in push-button telephones, such as that shown in Figure 16.15. These phones simultaneously produce two pure tones when each button is pressed, a different pair of tones for each different button. The tones are transmitted electronically to the central telephone office, where they activate switching circuits that complete the call. For example, the drawing indicates that pressing the "5" button produces pure tones of 770 and 1336 Hz simultaneously, while the "9" button generates tones of 852 and 1477 Hz.

Sound can be generated whose frequency lies below 20 Hz or above 20 kHz, although humans normally do not hear it. Sound waves with frequencies below 20 Hz are said to be **infrasonic,** while those with frequencies above 20 kHz are referred to as **ultrasonic.** Rhinoceroses use infrasonic frequencies as low as 5 Hz to call one another (Figure 16.16), while bats use ultrasonic frequencies up to 100 kHz for locating their food sources and navigating (Figure 16.17).

Frequency is an objective property of a sound wave because frequency can be measured with an electronic frequency counter. A listener's perception of frequency, however, is subjective. The brain interprets the frequency detected by the ear primarily in terms of the subjective quality called **pitch.** A pure tone with a large (high) frequency is interpreted as a high-pitched sound, while a pure tone with a small (low) frequency is interpreted as a low-pitched sound. A piccolo produces high-pitched sounds, and a tuba produces low-pitched sounds.

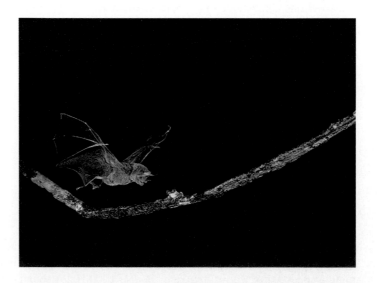

Figure 16.17 Bats use ultrasonic sound waves for navigating and locating food sources. (© Merlin D. Tuttle/Bat Conservation International/ Photo Researchers)

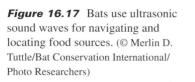

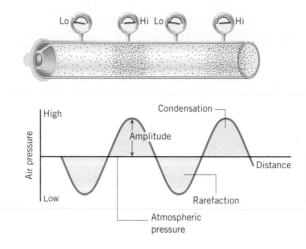

Figure 16.18 A sound wave is a series of alternating condensations and rarefactions. The graph shows that the condensations are regions of higher than normal air pressure, and the rarefactions are regions of lower than normal air pressure.

THE PRESSURE AMPLITUDE OF A SOUND WAVE

Figure 16.18 illustrates a pure-tone sound wave traveling in a tube. Attached to the tube is a series of gauges that indicate the pressure variations along the wave. The graph shows that the air pressure varies sinusoidally along the length of the tube. Although this graph has the appearance of a transverse wave, remember that the sound itself is a longitudinal wave. The graph also shows the *pressure amplitude* of the wave, which is the magnitude of the maximum change in pressure, measured relative to the undisturbed or atmospheric pressure. The pressure fluctuations in a sound wave are normally very small. For instance, in a typical conversation between two people the pressure amplitude is about 3×10^{-2} Pa, certainly a small amount compared with the atmospheric pressure of $1.01 \times 10^{+5}$ Pa. The ear is remarkable in being able to detect such small changes.

Loudness is an attribute of sound that depends primarily on the amplitude of the wave: the larger the amplitude, the louder the sound. The pressure amplitude is an objective property of a sound wave, since it can be measured. Loudness, on the other hand, is subjective. Each individual determines what is loud, depending on the acuteness of his or her hearing.

16.6 The Speed of Sound

GASES

Sound travels through gases, liquids, and solids at considerably different speeds, as Table 16.1 reveals. Near room temperature, the speed of sound in air is 343 m/s (767 mi/h) and is markedly greater in liquids and solids. For example, sound travels more than four times faster in water and more than seventeen times faster in steel than it does in air. In general, sound travels slowest in gases, faster in liquids, and fastest in solids.

Like the speed of a wave on a guitar string, the speed of sound depends on the properties of the medium. In a gas, it is only when molecules collide that the condensations and rarefactions of a sound wave can move from place to place. It is reasonable, then, to expect the speed of sound in a gas to have the same order of magnitude as the average molecular speed between collisions. For an ideal gas this average speed is the translational rms speed given by Equation 14.6: $v_{\text{rms}} = \sqrt{3kT/m}$, where T is the Kelvin temperature, m is the mass of a molecule, and k is Boltzmann's constant. Although the expression for v_{rms} overestimates the speed of sound, it does give the correct dependence on Kelvin temperature and particle mass. Careful analysis shows that the speed of sound in an ideal gas is given by

Ideal gas

$$v = \sqrt{\frac{\gamma k T}{m}}$$

(16.5)

Table 16.1 *Speed of Sound in Gases, Liquids, and Solids*

Substance	Speed (m/s)
Gases	
Air (0 °C)	331
Air (20 °C)	343
Carbon dioxide (0 °C)	259
Oxygen (0 °C)	316
Helium (0 °C)	965
Liquids	
Chloroform (20 °C)	1004
Ethyl alcohol (20 °C)	1162
Mercury (20 °C)	1450
Fresh water (20 °C)	1482
Seawater (20 °C)	1522
Solids	
Copper	5010
Glass (Pyrex)	5640
Lead	1960
Steel	5960

where $\gamma = c_P/c_V$ is the ratio of the specific heat capacity at constant pressure c_P to the specific heat capacity at constant volume c_V.

The factor γ is introduced in Section 15.5, where the adiabatic compression and expansion of an ideal gas is discussed. It appears in Equation 16.5 because the condensations and rarefactions of a sound wave are formed by adiabatic compressions and expansions of the gas. The regions that are compressed (the condensations) become slightly warmed, and the regions that are expanded (the rarefactions) become slightly cooled. However, no appreciable heat flows from a condensation to an adjacent rarefaction because the distance between the two (half a wavelength) is relatively large for most audible sound waves and a gas is a poor thermal conductor. Thus, the compression and expansion process is adiabatic. Example 4 illustrates the use of Equation 16.5.

Example 4 An Ultrasonic Ruler

The physics of an ultrasonic ruler.

Figure 16.19 shows an ultrasonic ruler that is used to measure the distance between itself and a target, such as a wall. To initiate the measurement, the ruler generates a pulse of ultrasonic sound that travels to the wall and, like an echo, reflects from it. The reflected pulse returns to the ruler, which measures the time it takes for the round-trip. Using a preset value for the speed of sound, the unit determines the distance to the wall and displays it on a digital readout. Suppose the round-trip travel time is 20.0 ms on a day when the air temperature is 23 °C. Assuming that air is an ideal gas for which $\gamma = 1.40$ and that the average molecular mass of air is 28.9 u, find the distance x to the wall.

Reasoning The distance between the ruler and the wall is $x = vt$, where v is the speed of sound and t is the time for the sound pulse to reach the wall. The time t is one-half the round-trip time, so $t = 10.0$ ms. The speed of sound in air can be obtained directly from Equation 16.5, provided the temperature and mass are expressed in the SI units of kelvins and kilograms, respectively.

Solution To convert the air temperature of 23 °C to the Kelvin temperature scale, we add 23 to 273.15 (see Equation 12.1): $T = 23 + 273.15 = 296$ K. The mass of a molecule (in kilograms) can be obtained from the conversion relation between atomic mass units and kilograms (see Section 14.1), 1 u $= 1.6605 \times 10^{-27}$ kg:

$$m = (28.9 \text{ u})\left(\frac{1.6605 \times 10^{-27} \text{ kg}}{1 \text{ u}}\right) = 4.80 \times 10^{-26} \text{ kg}$$

Problem solving insight
When using equation $v = \sqrt{\gamma kT/m}$ to calculate the speed of sound in an ideal gas, be sure to express the temperature T in kelvins and not in degrees Celsius or Fahrenheit.

For the speed of sound, we find

$$v = \sqrt{\frac{\gamma kT}{m}} = \sqrt{\frac{(1.40)(1.38 \times 10^{-23} \text{ J/K})(296 \text{ K})}{4.80 \times 10^{-26} \text{ kg}}} = 345 \text{ m/s} \qquad (16.5)$$

The distance to the wall is

$$x = vt = (345 \text{ m/s})(10.0 \times 10^{-3} \text{ s}) = \boxed{3.45 \text{ m}}$$

Figure 16.19 An ultrasonic ruler uses sound with a frequency greater than 20 kHz to measure the distance x to the wall. The blue arcs and blue arrow denote the outgoing sound wave, and the red arcs and red arrow denote the wave reflected from the wall.

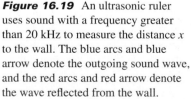

Sonar (**so**und **na**vigation **r**anging) is a technique for determining water depth and locating underwater objects, such as reefs, submarines, and schools of fish. The core of a sonar unit consists of an ultrasonic transmitter and receiver mounted on the bottom of a ship. The transmitter emits a short pulse of ultrasonic sound, and at a later time the reflected pulse returns and is detected by the receiver. The water depth is determined from the electronically measured round-trip time of the pulse and a knowledge of the speed of sound in water; the depth registers automatically on an appropriate meter. Such a depth measurement is similar to the distance measurement discussed for the ultrasonic ruler in Example 4.

The physics of sonar.

Conceptual Example 5 illustrates how the speed of sound in air can be used to estimate the distance to a thunderstorm, using a handy rule of thumb.

Conceptual Example 5 Lightning, Thunder, and a Rule of Thumb

There is a rule of thumb for estimating how far away a thunderstorm is. After you see a flash of lightning, count off the seconds until the thunder is heard. Divide the number of seconds by five. The result gives the approximate distance (in miles) to the thunderstorm. Why does this rule work?

Reasoning and Solution Figure 16.20 shows a lightning bolt and a person who is standing one mile (1.6×10^3 m) away. When the lightning occurs, light and sound (thunder) are produced very nearly at the same instant. Light travels so rapidly ($v_{light} = 3.0 \times 10^8$ m/s) that it reaches the observer almost instantaneously. Its travel time is only (1.6×10^3 m)/(3.0×10^8 m/s) = 5.3×10^{-6} s. In comparison, sound travels very slowly ($v_{sound} = 343$ m/s). The time for the thunder to reach the person is (1.6×10^3 m)/(343 m/s) = 5 s. Thus, the time interval between seeing the flash and hearing the thunder is about 5 seconds for every mile of travel. *This rule of thumb works because the speed of light is so much greater than the speed of sound* that the time needed for the light to reach the observer is negligible compared to the time needed for the sound.

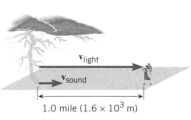

1.0 mile (1.6×10^3 m)

Figure 16.20 A lightning bolt from a thunderstorm generates a flash of light and sound (thunder). The speed of light is much greater than the speed of sound. Therefore, the light reaches the person first, followed about 5 seconds later by the sound.

LIQUIDS

In a liquid, the speed of sound depends on the density ρ and the *adiabatic* bulk modulus B_{ad} of the liquid:

Liquid
$$v = \sqrt{\frac{B_{ad}}{\rho}}$$
(16.6)

The bulk modulus is introduced in Section 10.7 in a discussion of the volume deformation of liquids and solids. There it is tacitly assumed that the temperature remains constant while the volume of the material changes; that is, the compression or expansion is isothermal. However, the condensations and rarefactions in a sound wave occur under *adiabatic* rather than isothermal conditions. Thus, the adiabatic bulk modulus B_{ad} must be used when calculating the speed of sound in liquids. Values of B_{ad} will be provided as needed in this text.

SOLID BARS

When sound travels through a long slender solid bar, the speed of the sound depends on the properties of the medium according to

Long slender solid bar
$$v = \sqrt{\frac{Y}{\rho}}$$
(16.7)

where Y is Young's modulus (defined in Section 10.7) and ρ is the density.

Need more practice?

🖱️**Interactive LearningWare 16.1**
A rhinoceros is calling to her mate using infrasonic sound whose frequency is 5.0 Hz. Her mate is 480 m away. The air temperature is 35 °C. Assume that air is an ideal gas for which $\gamma = 1.40$ and that the average mass of a molecule in the air is 4.80×10^{-26} kg. How many cycles of the sound wave are between the two animals?

Related Homework: Problem 36

Go to
www.wiley.com/college/cutnell
for an interactive solution.

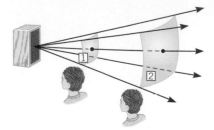

Figure 16.21 The power carried by a sound wave spreads out after leaving a source, such as a loudspeaker. Thus, the power passes perpendicularly through surface 1 and then through surface 2, which has the larger area.

Problem solving insight
Sound intensity *I* and sound power *P* are different concepts. They are related, however, since intensity equals power per unit area.

16.7 *Sound Intensity*

Sound waves carry energy that can be used to do work, like forcing the eardrum to vibrate. In an extreme case such as a sonic boom, the energy can be sufficient to cause damage to windows and buildings. The amount of energy transported per second by a sound wave is called the **power** of the wave and is measured in SI units of joules per second (J/s) or watts (W).

► **CONCEPTS AT A GLANCE** When a sound wave leaves a source, such as the loudspeaker in Figure 16.21, the power spreads out and passes through imaginary surfaces that have increasingly larger areas. For instance, the same sound power passes through the surfaces labeled 1 and 2 in the drawing. However, the power is spread out over a greater area in surface 2. The Concepts-at-a-Glance chart in Figure 16.22 indicates how we take this spreading-out effect into account. We will bring together the ideas of sound power and the area through which the power passes and, in the process, formulate the concept of sound intensity. The idea of wave intensity is not confined to sound waves. It will recur, for example, in Chapter 24 when we discuss another important type of waves, electromagnetic waves. ◄

The *sound intensity I* is defined as the sound power *P* that passes perpendicularly through a surface divided by the area *A* of that surface:

$$I = \frac{P}{A} \tag{16.8}$$

The unit of sound intensity is power per unit area, or W/m². The next example illustrates how the sound intensity changes as the distance from a loudspeaker changes.

Example 6   Sound Intensities

In Figure 16.21, 12×10^{-5} W of sound power passes perpendicularly through the surfaces labeled 1 and 2. These surfaces have areas of $A_1 = 4.0$ m² and $A_2 = 12$ m². Determine the sound intensity at each surface and discuss why listener 2 hears a quieter sound than listener 1.

Reasoning The sound intensity *I* is the sound power *P* passing perpendicularly through a surface divided by the area *A* of that surface. Since the same sound power passes through both surfaces and surface 2 has the greater area, the sound intensity is less at surface 2.

Figure 16.22 CONCEPTS AT A GLANCE The concept of sound intensity takes into account both the sound power and the area through which the power passes. These children are reacting to the high-intensity sound of rifle fire during a Memorial Day parade. (© AP/Wide World Photos)

Solution The sound intensity at each surface follows from Equation 16.8:

Surface 1 $I_1 = \dfrac{P}{A_1} = \dfrac{12 \times 10^{-5} \text{ W}}{4.0 \text{ m}^2} = \boxed{3.0 \times 10^{-5} \text{ W/m}^2}$

Surface 2 $I_2 = \dfrac{P}{A_2} = \dfrac{12 \times 10^{-5} \text{ W}}{12 \text{ m}^2} = \boxed{1.0 \times 10^{-5} \text{ W/m}^2}$

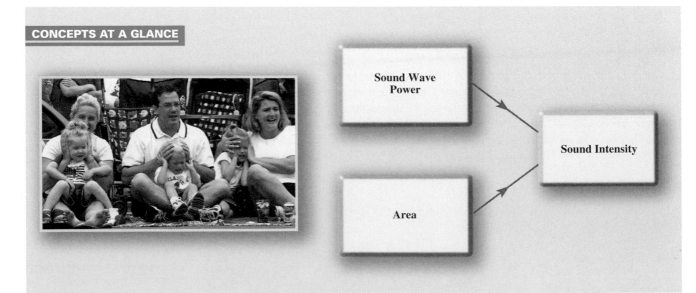

CONCEPTS AT A GLANCE

Sound Wave Power

Area

Sound Intensity

The sound intensity is less at the more distant surface, where the same power passes through a threefold greater area. The ear of a listener, with its fixed area, intercepts less power where the intensity, or power per unit area, is smaller. Thus, listener 2 intercepts less of the sound power than listener 1. With less power striking the ear, the sound is quieter.

For a 1000-Hz tone, the smallest sound intensity that the human ear can detect is about 1×10^{-12} W/m^2; this intensity is called the ***threshold of hearing.*** On the other extreme, continuous exposure to intensities greater than 1 W/m^2 can be painful and result in permanent hearing damage. The human ear is remarkable for the wide range of intensities to which it is sensitive.

If a source emits sound *uniformly in all directions,* the intensity depends on distance in a simple way. Figure 16.23 shows such a source at the center of an imaginary sphere (for clarity only a hemisphere is shown). The radius of the sphere is r. Since all the radiated sound power P passes through the spherical surface of area $A = 4\pi r^2$, the intensity at a distance r is

Spherically uniform radiation
$$I = \frac{P}{4\pi r^2}$$
(16.9)

From this we see that the intensity of a source that radiates sound uniformly in all directions varies as $1/r^2$. For example, if the distance increases by a factor of two, the sound intensity decreases by a factor of $2^2 = 4$. Example 7 illustrates the effect of the $1/r^2$ dependence of intensity on distance.

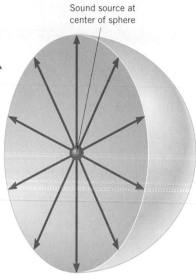

Sound source at center of sphere

Figure 16.23 The sound source at the center of the sphere emits sound uniformly in all directions. In this drawing, only a hemisphere is shown for clarity.

Example 7 Fireworks

During a fireworks display, a rocket explodes high in the air, as Figure 16.24 illustrates. Assume that the sound spreads out uniformly in all directions and that reflections from the ground can be ignored. When the sound reaches listener 2, who is $r_2 = 640$ m away from the explosion, the sound has an intensity of $I_2 = 0.10$ W/m^2. What is the sound intensity detected by listener 1, who is $r_1 = 160$ m away from the explosion?

Reasoning Listener 1 is four times closer to the explosion than listener 2. Therefore, the sound intensity detected by listener 1 is $4^2 = 16$ times greater than that detected by listener 2.

Solution The ratio of the sound intensities can be found using Equation 16.9:

$$\frac{I_1}{I_2} = \frac{\dfrac{P}{4\pi r_1^2}}{\dfrac{P}{4\pi r_2^2}} = \frac{r_2^2}{r_1^2} = \frac{(640 \text{ m})^2}{(160 \text{ m})^2} = 16$$

As a result, $I_1 = (16)I_2 = (16)(0.10 \text{ W/m}^2) = \boxed{1.6 \text{ W/m}^2}$.

Problem solving insight
Equation 16.9 can be used only when the sound spreads out uniformly in all directions and there are no reflections of the sound waves.

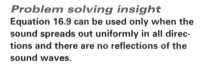

Figure 16.24 If an explosion in a fireworks display radiates sound uniformly in all directions, the intensity at any distance r is $I = P/(4\pi r^2)$, where P is the sound power of the explosion.

Equation 16.9 is valid only when no walls, ceilings, floors, etc. are present to reflect the sound and cause it to pass through the same surface more than once. Conceptual Example 8 demonstrates why this is so.

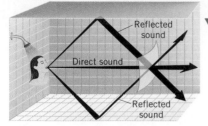

Figure 16.25 When someone sings in the shower, the sound power passing through part of an imaginary spherical surface (shown in blue) is the sum of the direct sound power and the reflected sound power.

Conceptual Example 8 Reflected Sound and Sound Intensity

Suppose the person singing in the shower in Figure 16.25 produces a sound power P. Sound reflects from the surrounding shower stall. At a distance r in front of the person, does Equation 16.9, $I = P/(4\pi r^2)$, underestimate, overestimate, or give the correct sound intensity?

Reasoning and Solution In arriving at Equation 16.9, it was assumed that the sound spreads out uniformly from the source and passes only once through the imaginary surface that surrounds it (see Figure 16.23). Only part of this imaginary surface is shown in Figure 16.25. The drawing illustrates three paths by which the sound passes through the surface. The "direct" sound travels directly along a path from its source to the surface. It is the intensity of the direct sound that is given by $I = P/(4\pi r^2)$. The remaining paths are two of the many that characterize the sound reflected from the shower stall. The total sound power that passes through the surface is the sum of the direct and reflected powers. Thus, the total sound intensity at a distance r from the source is greater than that of the direct sound alone, and ***the relation $I = P/(4\pi r^2)$ underestimates the sound intensity from the singing because it does not take into account the reflected sound.*** People like to sing in the shower because their voices sound so much louder due to the enhanced intensity caused by the reflected sound.

Related Homework: *Problems 52, 64*

16.8 Decibels

The ***decibel*** (dB) is a measurement unit used when comparing two sound intensities. The simplest method of comparison would be to compute the ratio of the intensities. For instance, we could compare $I = 8 \times 10^{-12}$ W/m^2 to $I_0 = 1 \times 10^{-12}$ W/m^2 by computing $I/I_0 = 8$ and stating that I is eight times greater than I_0. However, because of the way in which the human hearing mechanism responds to intensity, it is more appropriate to use a logarithmic scale for the comparison. For this purpose, the ***intensity level β*** (expressed in decibels) is defined as follows:

$$\beta = (10 \text{ dB}) \log\left(\frac{I}{I_0}\right) \tag{16.10}$$

where "log" denotes the logarithm to the base ten. I_0 is the intensity of the reference level to which I is being compared and is often the threshold of hearing, $I_0 = 1.00 \times 10^{-12}$ W/m^2. With the aid of a calculator, the intensity level can be evaluated for the values of I and I_0 given above:

$$\beta = (10 \text{ dB}) \log\left(\frac{8 \times 10^{-12} \text{ W/m}^2}{1 \times 10^{-12} \text{ W/m}^2}\right) = (10 \text{ dB}) \log 8 = (10 \text{ dB})(0.9) = 9 \text{ dB}$$

This result indicates that I is 9 decibels greater than I_0. Although β is called the "intensity level," it is *not* an intensity and does *not* have intensity units of W/m^2. In fact, the decibel, like the radian, is dimensionless.

Notice that if both I and I_0 are at the threshold of hearing, then $I = I_0$, and the intensity level is 0 dB according to Equation 16.10:

$$\beta = (10 \text{ dB}) \log\left(\frac{I_0}{I_0}\right) = (10 \text{ dB}) \log 1 = 0$$

since log 1 = 0. Thus, ***an intensity level of zero decibels does not mean that the sound intensity I is zero; it means that $I = I_0$.***

Intensity levels can be measured with a sound level meter, such as the one in Figure 16.26. The intensity level β is displayed on its scale, assuming that the threshold of hearing is 0 dB. Table 16.2 lists the intensities I and the associated intensity levels β for some common sounds, using the threshold of hearing as the reference level.

When a sound wave reaches a listener's ear, the sound is interpreted by the brain as loud or soft, depending on the intensity of the wave. Greater intensities give rise to louder sounds. However, the relation between intensity and loudness is not a simple proportionality, because doubling the intensity does *not* double the loudness, as we will now see.

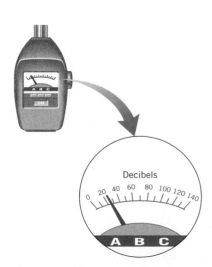

Figure 16.26 A sound level meter and a close-up view of its decibel scale.

Table 16.2 *Typical Sound Intensities and Intensity Levels Relative to the Threshold of Hearing*

	Intensity I (W/m^2)	Intensity Level β (dB)
Threshold of hearing	1.0×10^{-12}	0
Rustling leaves	1.0×10^{-11}	10
Whisper	1.0×10^{-10}	20
Normal conversation (1 meter)	3.2×10^{-6}	65
Inside car in city traffic	1.0×10^{-4}	80
Car without muffler	1.0×10^{-2}	100
Live rock concert	1.0	120
Threshold of pain	10	130

Suppose you are sitting in front of a stereo system that is producing an intensity level of 90 dB. If the volume control on the amplifier is turned up slightly to produce a 91-dB level, you would just barely notice the change in loudness. *Hearing tests have revealed that a one-decibel (1-dB) change in the intensity level corresponds to approximately the smallest change in loudness that an average listener with normal hearing can detect.* Since 1 dB is the smallest perceivable increment in loudness, a change of 3 dB—say, from 90 to 93 dB—is still a rather small change in loudness. Example 9 determines the factor by which the sound intensity must be increased to achieve such a change.

Example 9 Comparing Sound Intensities

Audio system 1 produces an intensity level of $\beta_1 = 90.0$ dB, and system 2 produces an intensity level of $\beta_2 = 93.0$ dB. The corresponding intensities (in W/m^2) are I_1 and I_2. Determine the ratio I_2/I_1.

Reasoning Intensity levels are related to intensities by logarithms (see Equation 16.10), and it is a property of logarithms (see Appendix D) that $\log A - \log B = \log (A/B)$. Subtracting the two intensity levels and using this property, we find that

$$\beta_2 - \beta_1 = (10 \text{ dB}) \log \left(\frac{I_2}{I_0} \right) - (10 \text{ dB}) \log \left(\frac{I_1}{I_0} \right) = (10 \text{ dB}) \log \left(\frac{I_2/I_0}{I_1/I_0} \right)$$

$$= (10 \text{ dB}) \log \left(\frac{I_2}{I_1} \right)$$

Solution Using the result just obtained, we find

$$93.0 \text{ dB} - 90.0 \text{ dB} = (10 \text{ dB}) \log \left(\frac{I_2}{I_1} \right)$$

$$0.30 = \log \left(\frac{I_2}{I_1} \right) \quad \text{or} \quad \frac{I_2}{I_1} = 10^{0.30} = \boxed{2.0}$$

Doubling the intensity changes the loudness by only a small amount (3 dB) and does not double it, so there is no simple proportionality between intensity and loudness.

To double the loudness of a sound, the intensity must be increased by more than a factor of two. *Experiment shows that if the intensity level increases by 10 dB, the new sound seems approximately twice as loud as the original sound.* For instance, a 70-dB intensity level sounds about twice as loud as a 60-dB level, and an 80-dB intensity level sounds about twice as loud as a 70-dB level. The factor by which the sound intensity must be increased to double the loudness can be determined by the method used in Example 9:

$$\beta_2 - \beta_1 = 10.0 \text{ dB} = (10 \text{ dB}) \left[\log \left(\frac{I_2}{I_0} \right) - \log \left(\frac{I_1}{I_0} \right) \right]$$

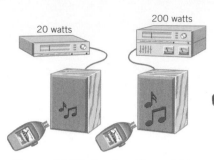

Figure 16.27 In spite of its tenfold greater power, the 200-watt audio system has only about double the loudness of the 20-watt system, when both are set for maximum volume.

Solving this equation reveals that $I_2/I_1 = 10.0$. Thus, increasing the sound intensity by a factor of ten will double the perceived loudness. Consequently, with both audio systems in Figure 16.27 set at maximum volume, the 200-watt system will sound only twice as loud as the much cheaper 20-watt system.

✓ **Check Your Understanding 3**

The drawing shows a source of sound and two observation points located at distances R_1 and R_2. The sound spreads uniformly from the source, and there are no reflecting surfaces in the environment. The sound heard at the distance R_2 is 6 dB quieter than that heard at the distance R_1. (a) What is the ratio I_2/I_1 of the sound intensities at the two distances? (b) What is the ratio R_2/R_1 of the distances? *(The answers are given at the end of the book.)*

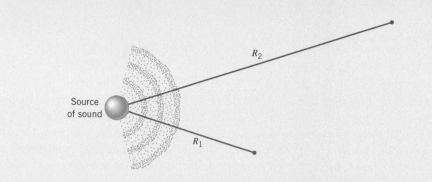

Background: The intensity level (expressed in decibels) plays the central role in part (a) of this question. In part (b) the facts that the source sends out the sound uniformly and that there are no reflecting surfaces are important. They tell you how the sound intensity depends on distance from the source.

For similar questions (including conceptual counterparts), consult Self-Assessment Test 16.2. This test is described at the end of Section 16.9.

Figure 16.28 CONCEPTS AT A GLANCE The Doppler effect arises when the source and the observer of the sound wave have different velocities with respect to the medium through which the sound travels. At a racing event, the Doppler effect creates the characteristic sound that you hear when the cars pass by. It is similar to that heard when a train blowing its horn passes you at a high speed. (© Simon Bruty/Stone/Getty Imates)

16.9 *The Doppler Effect*

Have you ever heard an approaching fire truck and noticed the distinct change in the sound of the siren as the truck passes? The effect is similar to what you get when you put together the two syllables "eee" and "yow" to produce "eee-yow." While the truck ap-

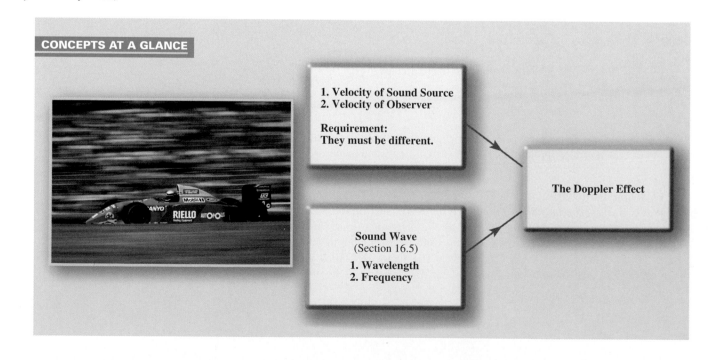

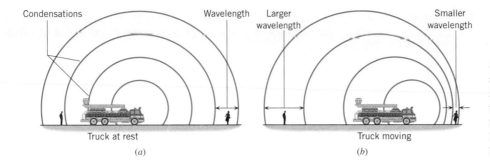

Figure 16.29 (*a*) When the truck is stationary, the wavelength of the sound is the same in front of and behind the truck. (*b*) When the truck is moving, the wavelength in front of the truck becomes smaller, while the wavelength behind the truck becomes larger.

proaches, the pitch of the siren is relatively high ("eee"), but as the truck passes and moves away, the pitch suddenly drops ("yow"). Something similar, but less familiar, occurs when an observer moves toward or away from a stationary source of sound. Such phenomena were first identified in 1842 by the Austrian physicist Christian Doppler (1803–1853) and are collectively referred to as the Doppler effect.

▶ CONCEPTS AT A GLANCE To explain why the Doppler effect occurs, we will bring together concepts that we have discussed previously — namely, the velocity of an object and the wavelength and frequency of a sound wave (Section 16.5). As the Concepts-at-a-Glance chart in Figure 16.28 indicates, we will combine the effects of the velocities of the source and observer of the sound with the definitions of wavelength and frequency. In so doing, we will learn that the ***Doppler effect*** is the change in frequency or pitch of the sound detected by an observer because the sound source and the observer have different velocities with respect to the medium of sound propagation. ◀

MOVING SOURCE

To see how the Doppler effect arises, consider the sound emitted by a siren on the stationary fire truck in Figure 16.29a. Like the truck, the air is assumed to be stationary with respect to the earth. Each solid blue arc in the drawing represents a condensation of the sound wave. Since the sound pattern is symmetrical, listeners standing in front of or behind the truck detect the same number of condensations per second and, consequently, hear the same frequency. Once the truck begins to move, the situation changes, as part *b* of the picture illustrates. Ahead of the truck, the condensations are now closer together, resulting in a decrease in the wavelength of the sound. This "bunching-up" occurs because the moving truck "gains ground" on a previously emitted condensation before emitting the next one. Since the condensations are closer together, the observer standing in front of the truck senses more of them arriving per second than she does when the truck is stationary. The increased rate of arrival corresponds to a greater sound frequency, which the observer hears as a higher pitch. Behind the moving truck, the condensations are farther apart than they are when the truck is stationary. This increase in the wavelength occurs because the truck pulls away from condensations emitted toward the rear. Consequently, fewer condensations per second arrive at the ear of an observer behind the truck, corresponding to a smaller sound frequency or lower pitch.

If the stationary siren in Figure 16.29a emits a condensation at the time $t = 0$ s, it will emit the next one at time T, where T is the period of the wave. The distance between these two condensations is the wavelength λ of the sound produced by the stationary source, as Figure 16.30a indicates. When the truck is moving with a speed v_s (the subscript "s" stands for the "source" of sound) toward a stationary observer, the siren also emits condensations at $t = 0$ s and at time T. However, prior to emitting the second condensation, the truck moves closer to the observer by a distance $v_s T$, as Figure 16.30b shows. As a result, the distance between successive condensations is no longer the wavelength λ created by the stationary siren, but, rather, a wavelength λ' that is shortened by the amount $v_s T$:

$$\lambda' = \lambda - v_s T$$

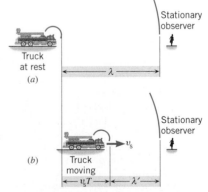

Figure 16.30 (*a*) When the fire truck is stationary, the distance between successive condensations is one wavelength λ. (*b*) When the truck moves with a speed v_s, the wavelength of the sound in front of the truck is shortened to λ'.

Let's denote the frequency perceived by the stationary observer as f_o, where the subscript "o" stands for "observer." According to Equation 16.1, f_o is equal to the speed of sound v divided by the shortened wavelength λ':

$$f_o = \frac{v}{\lambda'} = \frac{v}{\lambda - v_s T}$$

But for the stationary siren, we have $\lambda = v/f_s$ and $T = 1/f_s$, where f_s is the frequency of the sound emitted by the source (not the frequency f_o perceived by the observer). With the aid of these substitutions for λ and T, the expression for f_o can be arranged to give the following result:

Source moving toward stationary observer

$$f_o = f_s \left(\frac{1}{1 - \dfrac{v_s}{v}} \right)$$

(16.11)

Since the term $1 - v_s/v$ is in the denominator in Equation 16.11 and is less than one, the frequency f_o heard by the observer is *greater* than the frequency f_s emitted by the source. The difference between these two frequencies, $f_o - f_s$, is called the **Doppler shift,** and its magnitude depends on the ratio of the speed of the source v_s to the speed of sound v.

When the siren moves away from, rather than toward, the observer, the wavelength λ' becomes *greater* than λ according to

$$\lambda' = \lambda + v_s T$$

Notice the presence of the "+" sign in this equation, in contrast to the "−" sign that appeared earlier. The same reasoning that led to Equation 16.11 can be used to obtain an expression for the observed frequency f_o:

Source moving away from stationary observer

$$f_o = f_s \left(\frac{1}{1 + \dfrac{v_s}{v}} \right)$$

(16.12)

The denominator $1 + v_s/v$ in Equation 16.12 is greater than one, so the frequency f_o heard by the observer is *less* than the frequency f_s emitted by the source. The next example illustrates how large the Doppler shift is in a familiar situation.

Example 10 The Sound of a Passing Train

A high-speed train is traveling at a speed of 44.7 m/s (100 mi/h) when the engineer sounds the 415-Hz warning horn. The speed of sound is 343 m/s. What are the frequency and wavelength of the sound, as perceived by a person standing at a crossing, when the train is (a) approaching and (b) leaving the crossing?

Reasoning When the train approaches, the person at the crossing hears a sound whose frequency is greater than 415 Hz because of the Doppler effect. As the train moves away, the person hears a frequency that is less than 415 Hz. We may use Equations 16.11 and 16.12, respectively, to determine these frequencies. In either case, the observed wavelength can be obtained according to Equation 16.1 as the speed of sound divided by the observed frequency.

Solution

(a) When the train approaches, the observed frequency is

$$f_o = f_s \left(\frac{1}{1 - \dfrac{v_s}{v}} \right) = (415 \text{ Hz}) \left(\frac{1}{1 - \dfrac{44.7 \text{ m/s}}{343 \text{ m/s}}} \right) = \boxed{477 \text{ Hz}}$$

(16.11)

The observed wavelength is

$$\lambda' = \frac{v}{f_o} = \frac{343 \text{ m/s}}{477 \text{ Hz}} = \boxed{0.719 \text{ m}}$$

(16.1)

Need more practice?

🖱 **Interactive LearningWare 16.2**
A screeching hawk flies directly toward a bird-watcher at a speed of 11.0 m/s. The frequency heard by the stationary bird-watcher is 894 Hz when the speed of sound is 343 m/s. What is the frequency heard by the bird-watcher when the hawk flies away from the watcher at the same speed?

Related Homework: *Problem 84*

Go to
www.wiley.com/college/cutnell
for an interactive solution.

(b) When the train leaves the crossing, the observed frequency is

$$f_o = f_s \left(\frac{1}{1 + \dfrac{v_s}{v}} \right) = (415 \text{ Hz}) \left(\frac{1}{1 + \dfrac{44.7 \text{ m/s}}{343 \text{ m/s}}} \right) = \boxed{367 \text{ Hz}} \qquad (16.12)$$

In this case, the observed wavelength is

$$\lambda' = \frac{v}{f_o} = \frac{343 \text{ m/s}}{367 \text{ Hz}} = \boxed{0.935 \text{ m}}$$

MOVING OBSERVER

Figure 16.31 shows how the Doppler effect arises when the sound source is stationary and the observer moves, again assuming the air is stationary. The observer moves with a speed v_o ("o" stands for "observer") toward the stationary source and covers a distance $v_o t$ in a time t. During this time, the moving observer encounters all the condensations that he would if he were stationary, *plus an additional number.* The additional number of condensations encountered is the distance $v_o t$ divided by the distance λ between successive condensations, or $v_o t / \lambda$. Thus, the additional number of condensations encountered per second is v_o / λ. Since a stationary observer would hear a frequency f_s emitted by the source, the moving observer hears a higher frequency f_o given by

$$f_o = f_s + \frac{v_o}{\lambda} = f_s \left(1 + \frac{v_o}{f_s \lambda} \right)$$

Using the fact that $v = f_s \lambda$, we find that

Observer moving toward stationary source
$$f_o = f_s \left(1 + \frac{v_o}{v} \right) \qquad (16.13)$$

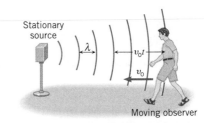

Figure 16.31 An observer moving with a speed v_o toward the stationary source intercepts more wave condensations per unit of time than does a stationary observer.

An observer moving *away from* a stationary source moves in the same direction as the sound wave and, as a result, intercepts *fewer* condensations per second than a stationary observer does. In this case, the moving observer hears a smaller frequency f_o that is given by

Observer moving away from stationary source
$$f_o = f_s \left(1 - \frac{v_o}{v} \right) \qquad (16.14)$$

The physical mechanism producing the Doppler effect in the case of the moving observer is different from that in the case of the moving source. When the source moves and the observer is stationary, the wavelength λ in Figure 16.30b changes, giving rise to the frequency f_o heard by the observer. On the other hand, when the observer moves and the source is stationary, the *wavelength λ in Figure 16.31 does not change.* Instead, a moving observer intercepts a different number of wave condensations per second than does a stationary observer and, therefore, detects a different frequency f_o.

GENERAL CASE

It is possible for *both* the sound source and the observer to move with respect to the medium of sound propagation. If the medium is stationary, Equations 16.11–16.14 may be combined to give the observed frequency f_o as

Source and observer both moving
$$f_o = f_s \left(\frac{1 \pm \dfrac{v_o}{v}}{1 \mp \dfrac{v_s}{v}} \right) \qquad (16.15)$$

In the numerator, the plus sign applies when the observer moves toward the source, and the minus sign applies when the observer moves away from the source. In the denominator, the minus sign is used when the source moves toward the observer, and the plus sign is used when the source moves away from the observer. The symbols v_o, v_s, and v denote numbers without an algebraic sign because the direction of travel has been taken into account by the plus and minus signs that appear directly in this equation.

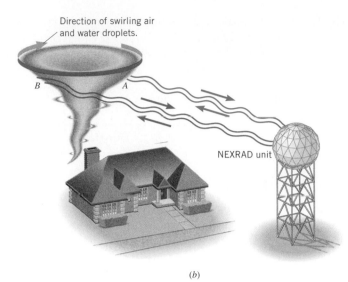

Direction of swirling air
and water droplets.

B   *A*

NEXRAD unit

(a)   (b)

Figure 16.32 (a) A tornado is one of
nature's most dangerous storms. (b)
The National Weather Service uses the
NEXRAD system, which is based on
Doppler-shifted radar, to identify the
storms that are likely to spawn
tornadoes. (© Paul and Lindamarie
Ambrose/Taxi/Getty Images)

✔ **Check Your Understanding 4**

When a truck is stationary, its horn produces a frequency of 500 Hz. You are driving your
car, and this truck is following behind. You hear its horn at a frequency of 520 Hz. (a) Who is
driving faster, you or the truck driver, or are you and the truck driver driving at the same
speed? (b) Refer to Equation 16.15 and decide which algebraic sign is to be used in the numerator and which in the denominator. *(The answers are given at the end of the book.)*

Background: The Doppler effect is the key here. However, the motions of both the car and
the truck influence the effect. Furthermore, the effect of one motion can partially or totally
offset the effect of the other motion, depending on the speeds of the car and the truck.

*For similar questions (including calculational counterparts), consult Self-Assessment Test
16.2. The test is described at the end of this section.*

The physics of
Next Generation Weather Radar.

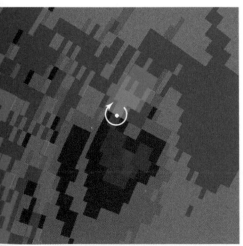

Figure 16.33 This color-enhanced
NEXRAD view of a tornado shows
winds moving toward (green) and away
from (red) a NEXRAD station, which
is below and to the right of the figure.
The white dot and arrow indicate the
the storm center and direction of wind
circulation. (Courtesy Kurt Hondl,
National Severe Storms Laboratory,
Norman, OK.)

NEXRAD

NEXRAD stands for **Nex**t Generation Weather **Rad**ar and is a nationwide system used by
the National Weather Service to provide dramatically improved early warning of severe
storms, such as the tornado in Figure 16.32. The system is based on radar waves, which
are a type of electromagnetic wave (see Chapter 24) and, like sound waves, can exhibit
the Doppler effect. The Doppler effect is at the heart of NEXRAD. As the drawing illustrates, a tornado is a swirling mass of air and water droplets. Radar pulses are sent out by
a NEXRAD unit, whose protective covering is shaped like a soccer ball. The waves reflect from the water droplets and return to the unit, where the frequency is observed and
compared to the outgoing frequency. For instance, droplets at point *A* in the drawing are
moving toward the unit, and the radar waves reflected from them have their frequency
Doppler-shifted to higher values. Droplets at point *B*, however, are moving away from the
unit. The frequency of the waves reflected from these droplets is Doppler-shifted to lower
values. Computer processing of the Doppler frequency shifts leads to color-enhanced
views on display screens (see Figure 16.33). These views reveal the direction and magnitude of the wind velocity and can identify, from distances up to 140 mi, the swirling air
masses that are likely to spawn tornadoes. The equations that specify the Doppler frequency shifts are different from those given for sound waves by Equations 16.11–16.15.
The reason for the difference is that radar waves propagate from one place to another by a
different mechanism than that of sound waves (see Section 24.5).

🖱 **Self-Assessment Test 16.2**

Check your understanding of the material in Sections 16.5–16.9:

- The Nature of Sound, the Speed of Sound, and Sound Intensity   • Decibels
- The Doppler Effect

Go to **www.wiley.com/college/cutnell**

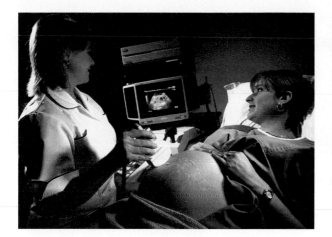
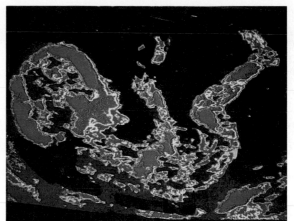

Figure 16.34 An ultrasonic scanner can be used to produce an image of the fetus as it develops in the uterus. (*Left,* © Deep Light Productions/Science Photo Library/Photo Researchers; *right,* © Howard Sochurek)

16.10 *Applications of Sound in Medicine*

When ultrasonic waves are used in medicine for diagnostic purposes, high-frequency sound pulses are produced by a transmitter and directed into the body. As in sonar, reflections occur. They occur each time a pulse encounters a boundary between two tissues that have different densities or a boundary between a tissue and the adjacent fluid. By scanning ultrasonic waves across the body and detecting the echoes generated from various internal locations, it is possible to obtain an image or sonogram of the inner anatomy. Ultrasonic imaging is employed extensively in obstetrics to examine the developing fetus (Figure 16.34). The fetus, surrounded by the amniotic sac, can be distinguished from other anatomical features so that fetal size, position, and possible abnormalities can be detected.

The physics of ultrasonic imaging.

Ultrasound is also used in other medically related areas. For instance, malignancies in the liver, kidney, brain, and pancreas can be detected with ultrasound. Yet another application involves monitoring the real-time movement of pulsating structures, such as heart valves ("echocardiography") and large blood vessels.

When ultrasound is used to form images of internal anatomical features or foreign objects in the body, the wavelength of the sound wave must be about the same size as, or smaller than, the object to be located. Therefore, high frequencies in the range from 1 to 15 MHz (1 MHz = 1 megahertz = 1×10^6 Hz) are the norm. For instance, the wavelength of 5-MHz ultrasound is $\lambda = v/f = 0.3$ mm, if a value of 1540 m/s is used for the speed of sound through tissue. A sound wave with a frequency higher than 5 MHz and a correspondingly shorter wavelength is required for locating objects smaller than 0.3 mm.

Ultrasound also has applications other than imaging. Neurosurgeons use a device called a **c**avitron **u**ltrasonic **s**urgical **a**spirator (CUSA) to remove brain tumors once thought to be inoperable. Ultrasonic sound waves cause the slender tip of the CUSA probe (see Figure 16.35) to vibrate at approximately 23 kHz. The probe shatters any section of the tumor that it touches, and the fragments are flushed out of the brain with a saline solution. Because the tip of the probe is small, the surgeon can selectively remove small bits of malignant tissue without damaging the surrounding healthy tissue.

The physics of the cavitron ultrasonic surgical aspirator.

Another application of ultrasound is in a new type of bloodless surgery, which can eliminate abnormal cells, such as those in benign hyperplasia of the prostate gland. Still in the experimental phase, this technique is known as HIFU (**h**igh-**i**ntensity **f**ocused **u**ltrasound). It is analogous to focusing the sun's electromagnetic waves by using a magnifying glass and producing a small region where the energy carried by the waves can cause localized heating. Ultrasonic waves can be used in a similar fashion. The waves enter directly through the skin and come into focus inside the body over a region that is sufficiently well defined to be surgically useful. Within this region the energy of the waves causes localized heating, leading to a temperature of about 56 °C (normal body temperature is 37 °C), which is sufficient to kill abnormal cells. The killed cells are eventually removed by the body's natural processes.

The physics of bloodless surgery with HIFU.

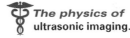

Figure 16.35 Neurosurgeons use a cavitron ultrasonic surgical aspirator (CUSA) to "cut out" brain tumors without adversely affecting the surrounding healthy tissue.

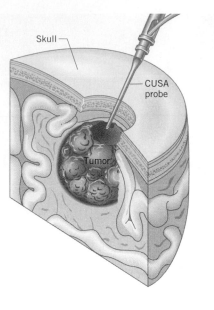

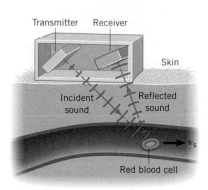

Figure 16.36 A Doppler flow meter measures the speed of red blood cells.

The physics of the Doppler flow meter.

The Doppler flow meter is a particularly interesting medical application of the Doppler effect. This device measures the speed of blood flow, using transmitting and receiving elements that are placed directly on the skin, as in Figure 16.36. The transmitter emits a continuous sound whose frequency is typically about 5 MHz. When the sound is reflected from the red blood cells, its frequency is changed in a kind of Doppler effect because the cells are moving. The receiving element detects the reflected sound, and an electronic counter measures its frequency, which is Doppler-shifted relative to the transmitter frequency. From the change in frequency the speed of the blood flow can be determined. Typically, the change in frequency is around 600 Hz for flow speeds of about 0.1 m/s. The Doppler flow meter can be used to locate regions where blood vessels have narrowed, since greater flow speeds occur in the narrowed regions, according to the equation of continuity (see Section 11.8). In addition, the Doppler flow meter can be used to detect the motion of a fetal heart as early as 8–10 weeks after conception.

*16.11 The Sensitivity of the Human Ear

The physics of hearing.

Although the ear is capable of detecting sound intensities as small as 1×10^{-12} W/m^2, it is *not* equally sensitive to all frequencies, as Figure 16.37 shows. This figure displays a series of graphs known as the **Fletcher–Munson curves,** after H. Fletcher and M. Munson, who first determined them in 1933. In these graphs the audible sound frequencies are plotted on the horizontal axis, and the sound intensity levels (in decibels) are plotted on the vertical axis. Each curve is a *constant loudness* curve because it shows the sound intensity level needed at each frequency to make the sound appear to have the same loudness. For example, the lowest (red) curve represents the threshold of hearing. It shows the intensity levels at which sounds of different frequencies just become audible. The graph indicates that the intensity level of a 100-Hz sound must be about 37 dB greater than the intensity level of a 1000-Hz sound to be at the threshold of hearing. Therefore, the ear is *less sensitive* to a 100-Hz sound than it is to a 1000-Hz sound. In general, Figure 16.37 reveals that the ear is most sensitive in the range of about 1–5 kHz, and becomes progressively less sensitive at higher and lower frequencies.

Each curve in Figure 16.37 represents a different loudness, and each is labeled according to its intensity level at 1000 Hz. For instance, the curve labeled "60" represents all sounds that have the same loudness as that of a 1000-Hz sound whose intensity level is 60 dB. These constant-loudness curves become flatter as the loudness increases, the relative flatness indicating that the ear is nearly equally sensitive to all frequencies when the sound is loud. Thus, when you listen to loud sounds, you hear the low frequencies, the middle frequencies, and the high frequencies about equally well. However, when you listen to quiet sounds, the high and low frequencies seem to be absent, because the ear is relatively insensitive to these frequencies under such conditions.

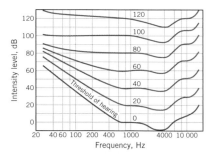

Figure 16.37 Each curve represents the intensity levels at which sounds of various frequencies have the same loudness. The curves are labeled by their intensity levels at 1000 Hz and are known as the Fletcher–Munson curves.

16.12 Concepts & Calculations

One of the important concepts that we encountered in this chapter is a transverse wave. For instance, transverse waves travel along a guitar string when it is plucked or along a violin string when it is bowed. The next example reviews how the travel speed depends on the properties of the string and on the tension in it.

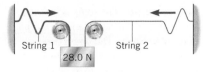

Figure 16.38 A wave travels on each of the two strings. The strings have different masses and lengths and together support the 28.0-N box. Which is the faster wave?

Concepts & Calculations Example 11
What Determines the Speed of a Wave on a String?

Figure 16.38 shows waves traveling on two strings. Each string is attached to a wall at one end and to a box that has a weight of 28.0 N at the other end. String 1 has a mass of 8.00 g and a length of 4.00 cm, and string 2 has a mass of 12.0 g and a length of 8.00 cm. Determine the speed of the wave on each string.

Concept Questions and Answers Is the tension the same in each string?

Answer Yes, the tension is the same. The two strings support the box, so the tension in each string is one-half the weight of the box. The fact that the strings have different masses and lengths does not affect the tension, which is determined only by the weight of the hanging box.

Is the speed of each wave the same?

Answer Not necessarily. The speed of a wave on a string depends on both the tension and the linear density, as Equation 16.2 indicates. The tension is the same in both strings, but if the linear densities of the strings are different, the speeds are different.

String 1 has a smaller mass and, hence, less inertia than string 2. Does this mean that the speed of the wave on string 1 is greater than that on string 2?

Answer Maybe yes, maybe no. The speed of a wave depends on the linear density of the string, which is its mass divided by its length. Depending on the lengths of the strings, string 1 could have a larger linear density and, hence, smaller speed, than string 2. The solution below illustrates this point.

Solution The speed of a wave on a string is given by Equation 16.2 as $v = \sqrt{F/(m/L)}$, where F is the tension and m/L is the mass per unit length, or linear density. Since both strings support the box, the tension in each is one-half the weight of the box, or $F = \frac{1}{2}(28.0 \text{ N}) = 14.0 \text{ N}$. The linear densities of the strings are

$$\frac{m_1}{L_1} = \frac{8.00 \text{ g}}{4.00 \text{ cm}} = 2.00 \text{ g/cm} = 0.200 \text{ kg/m}$$

$$\frac{m_2}{L_2} = \frac{12.0 \text{ g}}{8.00 \text{ cm}} = 1.50 \text{ g/cm} = 0.150 \text{ kg/m}$$

The speed of each wave is

$$v_1 = \sqrt{\frac{F}{m_1/L_1}} = \sqrt{\frac{14.0 \text{ N}}{0.200 \text{ kg/m}}} = \boxed{8.37 \text{ m/s}}$$

$$v_2 = \sqrt{\frac{F}{m_2/L_2}} = \sqrt{\frac{14.0 \text{ N}}{0.150 \text{ kg/m}}} = \boxed{9.66 \text{ m/s}}$$

The next example illustrates how the Doppler effect arises when an observer is moving away from or toward a stationary source of sound. In fact, we will see that it's possible for both situations to occur at the same time.

Concepts & Calculations Example 12
The Doppler Effect for a Moving Observer

A siren, mounted on a tower, emits a sound whose frequency is 2140 Hz. A person is driving a car away from the tower at a speed of 27.0 m/s. As Figure 16.39 illustrates, the sound reaches the person by two paths: the sound reflected from a building in front of the car, and the sound coming directly from the siren. The speed of sound is 343 m/s. What frequency does the person hear for the (a) reflected and (b) direct sounds?

Figure 16.39 The sound from the siren reaches the car by a reflected path and a direct path. The direct and reflected sound waves, as well as the motion of the car, are assumed to lie along the same line. Because of the Doppler effect, the driver hears a different frequency for each sound.

Concept Questions and Answers One way that the Doppler effect can arise is that the wavelength of the sound changes. For either the direct or the reflected sound, does the wavelength change?

Answer   No. The wavelength changes only when the source of the sound is moving, as illustrated in Figure 16.30b. The siren is stationary, so the wavelength does not change.

Why does the driver hear a frequency for the reflected sound that is different than 2140 Hz, and is it greater than or smaller than 2140 Hz?

Answer   The car and the reflected sound are traveling in opposite directions, the car to the right and the reflected sound to the left. The driver intercepts more wave cycles per second than if the car were stationary. Consequently, the driver hears a frequency greater than 2140 Hz.

Why does the driver hear a frequency for the direct sound that is different than 2140 Hz, and is it greater than or smaller than 2140 Hz?

Answer   The car and the direct sound are traveling in the same direction. As the sound passes the car, the number of wave cycles per second intercepted by the driver is now less than if the car were stationary. Thus, the driver hears a frequency that is less than 2140 Hz.

Solution

(a) For the reflected sound, the frequency f_o that the driver (the "observer") hears is equal to the frequency f_s of the waves emitted by the siren *plus* an additional number of cycles per second because the car and the reflected sound are moving in opposite directions. The additional number of cycles per second is v_o/λ, where v_o is the speed of the car and λ is the wavelength of the sound (see the subsection "Moving Observer" in Section 16.9). According to Equation 16.1, the wavelength is equal to the speed of sound v divided by the frequency of the siren, $\lambda = v/f_s$. Thus, the frequency heard by the driver can be written as

$$f_o = f_s + \underbrace{\frac{v_o}{\lambda}}_{\substack{\text{Additional number of} \\ \text{cycles intercepted} \\ \text{per second}}} = f_s + \frac{v_o}{v/f_s}$$

$$= f_s \left(1 + \frac{v_o}{v} \right) = (2140 \text{ Hz}) \left(1 + \frac{27.0 \text{ m/s}}{343 \text{ m/s}} \right) = \boxed{2310 \text{ Hz}}$$

(b) For the direct sound, the frequency f_o that the driver hears is equal to the frequency f_s of the waves emitted by the siren *minus* v_o/λ, because the car and direct sound are moving in the same direction:

$$f_o = f_s - \frac{v_o}{\lambda} = f_s - \frac{v_o}{v/f_s}$$

$$= f_s \left(1 - \frac{v_o}{v} \right) = (2140 \text{ Hz}) \left(1 - \frac{27.0 \text{ m/s}}{343 \text{ m/s}} \right) = \boxed{1970 \text{ Hz}}$$

As expected, for the reflected wave, the driver hears a frequency greater than 2140 Hz, while for the direct sound he hears a frequency less than 2140 Hz.

At the end of the problem set for this chapter, you will find homework problems that contain both conceptual and quantitative parts. These problems are grouped under the heading *Concepts & Calculations, Group Learning Problems*. They are designed for use by students working alone or in small learning groups. The conceptual part of each problem provides a convenient focus for group discussions.

Concept Summary

This summary presents an abridged version of the chapter, including the important equations and all available learning aids. For convenient reference, the learning aids (including the text's examples) are placed next to or immediately after the relevant equation or discussion. The following learning aids may be found on-line at **www.wiley.com/college/cutnell**:

Interactive LearningWare examples are solved according to a five-step interactive format that is designed to help you develop problem-solving skills.	**Concept Simulations** are animated versions of text figures or animations that illustrate important concepts. You can control parameters that affect the display, and we encourage you to experiment.
Interactive Solutions offer specific models for certain types of problems in the chapter homework. The calculations are carried out interactively.	**Self-Assessment Tests** include both qualitative and quantitative questions. Extensive feedback is provided for both incorrect and correct answers, to help you evaluate your understanding of the material.

Topic	*Discussion*	*Learning Aids*

16.1 The Nature of Waves

Transverse wave
Longitudinal wave

A wave is a traveling disturbance and carries energy from place to place. In a transverse wave, the disturbance occurs perpendicular to the direction of travel of the wave. In a longitudinal wave, the disturbance occurs parallel to the line along which the wave travels.

16.2 Periodic Waves

Cycle
Amplitude

Wavelength
Period
Frequency

A periodic wave consists of cycles or patterns that are produced over and over again by the source of the wave. The amplitude of the wave is the maximum excursion of a particle of the medium from the particle's undisturbed position. The wavelength λ is the distance along the length of the wave between two successive equivalent points, such as two crests or two troughs. The period T is the time required for the wave to travel a distance of one wavelength. The frequency f (in hertz) is the number of wave cycles per second that passes an observer and is the reciprocal of the period (in seconds):

Relation between frequency and period

$$f = \frac{1}{T} \qquad (10.5)$$

Relation between speed, frequency, and wavelength

The speed v of a wave is related to its wavelength and frequency according to

$$v = f\lambda \qquad (16.1) \quad \textbf{Example 1}$$

16.3 The Speed of a Wave on a String

The speed of a wave depends on the properties of the medium in which the wave travels. For a transverse wave on a string that has a tension F and a mass per unit length m/L, the wave speed is

Speed of a wave on a string

$$v = \sqrt{\frac{F}{m/L}} \qquad (16.2)$$

Examples 2, 3, 11

Concept Simulation 16.1

Linear density

The mass per unit length is also called the linear density.

Interactive Solution 16.17

16.4 The Mathematical Description of a Wave

When a wave of amplitude A, frequency f, and wavelength λ moves in the $+x$ direction through a medium, the wave causes a displacement y of a particle at position x according to

$$y = A \sin\left(2\pi ft - \frac{2\pi x}{\lambda}\right) \qquad (16.3)$$

For a wave moving in the $-x$ direction, the expression is

$$y = A \sin\left(2\pi ft + \frac{2\pi x}{\lambda}\right) \qquad (16.4)$$

Use Self-Assessment Test 16.1 to evaluate your understanding of Sections 16.1–16.4.

16.5 The Nature of Sound

Condensation

Sound is a longitudinal wave that can be created only in a medium; it cannot exist in a vacuum. Each cycle of a sound wave includes one condensation (a re-

Topic	Discussion	Learning Aids
Rarefaction	gion of greater than normal pressure) and one rarefaction (a region of less than normal pressure).	
Infrasonic frequency Ultrasonic frequency Pitch	A sound wave with a single frequency is called a pure tone. Frequencies less than 20 Hz are called infrasonic. Frequencies greater than 20 kHz are called ultrasonic. The brain interprets the frequency detected by the ear primarily in terms of the subjective quality known as pitch. A high-pitched sound is one with a large frequency (e.g., piccolo). A low-pitched sound is one with a small frequency (e.g., tuba).	
Pressure amplitude Loudness	The pressure amplitude of a sound wave is the magnitude of the maximum change in pressure, measured relative to the undisturbed pressure. The pressure amplitude is associated with the subjective quality of loudness. The larger the pressure amplitude, the louder the sound.	

16.6 *The Speed of Sound*

The speed of sound v depends on the properties of the medium. In an ideal gas, the speed of sound is

$$v = \sqrt{\frac{\gamma k T}{m}} \qquad (16.5)$$

Examples 4, 5

Interactive LearningWare 16.1

where $\gamma = c_P/c_V$ is the ratio of the specific heat capacities at constant pressure and constant volume, k is Boltzmann's constant, T is the Kelvin temperature, and m is the mass of a molecule of the gas. In a liquid, the speed of sound is

Speed of sound in a liquid

$$v = \sqrt{\frac{B_{\text{ad}}}{\rho}} \qquad (16.6)$$

where B_{ad} is the adiabatic bulk modulus and ρ is the mass density. In a solid that has a Young's modulus of Y and the shape of a long slender bar, the speed of sound is

Speed of sound in solid bar

$$v = \sqrt{\frac{Y}{\rho}} \qquad (16.7)$$

16.7 *Sound Intensity*

Intensity

The intensity I of a sound wave is the power P that passes perpendicularly through a surface divided by the area A of the surface

$$I = \frac{P}{A} \qquad (16.8)$$

Example 6

Interactive Solution 16.55

Threshold of hearing

The SI unit for intensity is watts per square meter (W/m^2). The smallest sound intensity that the human ear can detect is known as the threshold of hearing and is about 1×10^{-12} W/m^2 for a 1-kHz sound. When a source radiates sound uniformly in all directions and no reflections are present, the intensity of the sound is inversely proportional to the square of the distance from the source, according to

Spherically uniform radiation

$$I = \frac{P}{4\pi r^2} \qquad (16.9)$$ **Examples 7, 8**

16.8 *Decibels*

The intensity level β (in decibels) is used to compare a sound intensity I to the sound intensity I_0 of a reference level:

Intensity level in decibels

$$\beta = (10 \text{ dB}) \log\left(\frac{I}{I_0}\right) \qquad (16.10)$$ **Example 9**

The decibel, like the radian, is dimensionless. An intensity level of zero decibels means that $I = I_0$. One decibel is approximately the smallest change in loudness that an average listener with healthy hearing can detect. An increase of ten decibels in the intensity level corresponds approximately to a doubling of the loudness of the sound.

16.9 *The Doppler Effect*

The Doppler effect is the change in frequency detected by an observer because the sound source and the observer have different velocities with respect to the

Topic	Discussion	Learning Aids

medium of sound propagation. If the observer and source move with speeds v_o and v_s, respectively, and if the medium is stationary, the frequency f_o detected by the observer is

The Doppler effect

$$f_o = f_s \left(\frac{1 \pm \dfrac{v_o}{v}}{1 \mp \dfrac{v_s}{v}} \right)$$ (16.15)

Examples 10, 12

Interactive LearningWare 16.2

Interactive Solution 16.77

where f_s is the frequency of the sound emitted by the source and v is the speed of sound. In the numerator, the plus sign applies when the observer moves toward the source, and the minus sign applies when the observer moves away from the source. In the denominator, the minus sign is used when the source moves toward the observer, and the plus sign is used when the source moves away from the observer.

Use Self-Assessment Test 16.2 to evaluate your understanding of Sections 16.5–16.9.

Conceptual Questions

1. Considering the nature of a water wave (see Figure 16.4), describe the motion of a fishing float on the surface of a lake when a wave passes beneath the float. Is it really correct to say that the float bobs straight up and down? Explain.

2. "Domino Toppling" is one entry in the *Guinness Book of World Records*. The event consists of lining up an incredible number of dominoes and then letting them topple, one after another. Is the disturbance that propagates along the line of dominoes transverse, longitudinal, or partly both? Explain.

3. Suppose that a longitudinal wave moves along a Slinky at a speed of 5 m/s. Does one coil of the Slinky move through a distance of 5 m in one second? Justify your answer.

4. Examine Conceptual Example 3 before addressing this question. A wave moves on a string with a constant velocity. Does this mean that the particles of the string always have zero acceleration? Justify your answer.

5. A wire is strung tightly between two immovable posts. Discuss how an increase in temperature affects the speed of a transverse wave on this wire. Give your reasoning, ignoring any change in the mass per unit length of the wire.

6. A rope of mass *m* is hanging down from the ceiling. Nothing is attached to the loose end of the rope. A transverse wave is traveling on the rope. As the wave travels up the rope, does the speed of the wave increase, decrease, or remain the same? Give a reason for your choice.

7. One end of each of two identical strings is attached to a wall. Each string is being pulled tight by someone at the other end. A transverse pulse is sent traveling along one of the strings. A bit later an identical pulse is sent traveling along the other string. What, if anything, can be done to make the second pulse catch up with and pass the first pulse? Account for your answer.

8. In Section 4.10 the concept of a "massless" rope is discussed. Would it take any time for a transverse wave to travel the length of a massless rope? Justify your answer.

9. In a traveling sound wave, are there any particles that are *always* at rest as the wave passes by? Justify your answer.

10. Do you expect an echo to return to you more quickly or less quickly on a hot day than on a cold day, other things being equal? Account for your answer.

11. A loudspeaker produces a sound wave. Does the wavelength of the sound increase, decrease, or remain the same, when the wave travels from air into water? Justify your answer. *(Hint: The frequency does not change as the sound enters the water.)*

12. JELL-O starts out as a liquid and then sets to a gel. What would you expect to happen to the speed of sound in this material as the JELL-O sets? Does it increase, decrease, or remain the same? Give your reasoning.

13. Some animals rely on an acute sense of hearing for survival, and the visible part of the ear on such animals is often relatively large. Explain how this anatomical feature helps to increase the sensitivity of the animal's hearing for low-intensity sounds.

14. A source is emitting sound uniformly in all directions. There are no reflections anywhere. A *flat* surface faces the source. Is the sound intensity the same at all points on the surface? Give your reasoning.

15. If two people talk simultaneously and each creates an intensity level of 65 dB at a certain point, does the total intensity level at this point equal 130 dB? Account for your answer.

16. Two cars, one behind the other, are traveling in the same direction at the same speed. Does either driver hear the other's horn at a frequency that is different from that heard when both cars are at rest? Justify your answer.

17. A source of sound produces the same frequency underwater as it does in air. This source has the same velocity in air as it does underwater. The observer of the sound is stationary, both in air and underwater. Is the Doppler effect greater in air or underwater when the source (a) approaches and (b) moves away from the observer? Explain.

18. A music fan at a swimming pool is listening to a radio on a diving platform. The radio is playing a constant frequency tone when this fellow, clutching his radio, jumps off. Describe the Doppler effect heard by (a) a person left behind on the platform and (b) a person down below floating on a rubber raft. In each case, specify (1) whether the observed frequency is greater or smaller than the frequency produced by the radio, (2) whether the observed frequency is constant, and (3) how the observed frequency changes during the fall, if it does change. Give your reasoning.

19. When a car is at rest, its horn emits a frequency of 600 Hz. A person standing in the middle of the street hears the horn with a frequency of 580 Hz. Should the person jump out of the way? Account for your answer.

20. The text discusses how the Doppler effect arises when (1) the observer is stationary and the source moves and (2) the observer moves and the source is stationary. A car is speeding toward a large wall and sounds its horn. Is the Doppler effect present in the echo that the driver hears? If it is present, from which of the above situations does it arise, (1) or (2) or both? Explain.

Problems

Section 16.1 The Nature of Waves, Section 16.2 Periodic Waves

1. ssm A person standing in the ocean notices that after a wave crest passes by, ten more crests pass in a time of 120 s. What is the frequency of the wave?

2. Light is an electromagnetic wave and travels at a speed of 3.00×10^8 m/s. The human eye is most sensitive to yellow-green light, which has a wavelength of 5.45×10^{-7} m. What is the frequency of this light?

3. A longitudinal wave with a frequency of 3.0 Hz takes 1.7 s to travel the length of a 2.5-m Slinky (see Figure 16.3). Determine the wavelength of the wave.

4. Consider the freight train in Figure 16.7. Suppose 15 boxcars pass by in a time of 12.0 s and each has a length of 14.0 m. (a) What is the frequency at which each boxcar passes? (b) What is the speed of the train?

5. ssm In Figure 16.2c the hand moves the end of the Slinky up and down through two complete cycles in one second. The wave moves along the Slinky at a speed of 0.50 m/s. Find the distance between two adjacent crests on the wave.

6. A person lying on an air mattress in the ocean rises and falls through one complete cycle every five seconds. The crests of the wave causing the motion are 20.0 m apart. Determine (a) the frequency and (b) the speed of the wave.

7. ssm Suppose the amplitude and frequency of the transverse wave in Figure 16.2c are, respectively, 1.3 cm and 5.0 Hz. Find the *total vertical distance* (in cm) through which the colored dot moves in 3.0 s.

*8. A jetskier is moving at 8.4 m/s in the direction in which the waves on a lake are moving. Each time he passes over a crest, he feels a bump. The bumping frequency is 1.2 Hz, and the crests are separated by 5.8 m. What is the wave speed?

*9. The speed of a transverse wave on a string is 450 m/s, and the wavelength is 0.18 m. The amplitude of the wave is 2.0 mm. How much time is required for a particle of the string to move through a total distance of 1.0 km?

*10. A 3.49-rad/s ($33\frac{1}{3}$ rpm) record has a 5.00-kHz tone cut in the groove. If the groove is located 0.100 m from the center of the record (see drawing), what is the wavelength in the groove?

**11. A water-skier is moving at a speed of 12.0 m/s. When she skis in the same direction as a traveling wave, she springs upward every 0.600 s because of the wave crests. When she skis in the direction oppo-

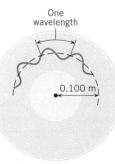

One wavelength

0.100 m

Problem 10

site to that in which the wave moves, she springs upward every 0.500 s in response to the crests. The speed of the skier is greater than the speed of the wave. Determine (a) the speed and (b) the wavelength of the wave.

Section 16.3 The Speed of a Wave on a String

12. The mass of a string is 5.0×10^{-3} kg, and it is stretched so that the tension in it is 180 N. A transverse wave traveling on this string has a frequency of 260 Hz and a wavelength of 0.60 m. What is the length of the string?

13. ssm The linear density of the A string on a violin is 7.8×10^{-4} kg/m. A wave on the string has a frequency of 440 Hz and a wavelength of 65 cm. What is the tension in the string?

14. A vibrator moves one end of a rope up and down to generate a wave. The tension in the rope is 58 N. The frequency is then doubled. To what value must the tension be adjusted, so the new wave has the same wavelength as the old one?

15. A transverse wave is traveling with a speed of 300 m/s on a horizontal string. If the tension in the string is increased by a factor of four, what is the speed of the wave?

16. Two wires are parallel, and one is directly above the other. Each has a length of 50.0 m and a mass per unit length of 0.020 kg/m. However, the tension in wire A is 6.00×10^2 N, and the tension in wire B is 3.00×10^2 N. Transverse wave pulses are generated simultaneously, one at the left end of wire A and one at the right end of wire B. The pulses travel toward each other. How much time does it take until the pulses pass each other?

17. Consult **Interactive Solution 16.17** at **www.wiley.com/college/cutnell** in order to review a model for solving this problem. To measure the acceleration due to gravity on a distant planet, an astronaut hangs a 0.055-kg ball from the end of a wire. The wire has a length of 0.95 m and a linear density of 1.2×10^{-4} kg/m. Using electronic equipment, the astronaut measures the time for a transverse pulse to travel the length of the wire and obtains a value of 0.016 s. The mass of the wire is negligible compared to the mass of the ball. Determine the acceleration due to gravity.

*18. Review Conceptual Example 3 before starting this problem. The amplitude of a transverse wave on a string is 4.5 cm. The ratio of the maximum particle speed to the speed of the wave is 3.1. What is the wavelength (in cm) of the wave?

*19. ssm www The drawing at right shows a frictionless incline and pulley. The two blocks are connected by a wire (mass per unit length = 0.0250 kg/m) and remain stationary. A transverse wave on the wire has a speed of

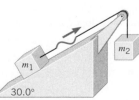

75.0 m/s. Neglecting the weight of the wire relative to the tension in the wire, find the masses m_1 and m_2 of the blocks.

** **20.** A copper wire, whose cross-sectional area is 1.1×10^{-6} m², has a linear density of 7.0×10^{-3} kg/m and is strung between two walls. At the ambient temperature, a transverse wave travels with a speed of 46 m/s on this wire. The coefficient of linear expansion for copper is 17×10^{-6} (C°)$^{-1}$, and Young's modulus for copper is 1.1×10^{11} N/m². What will be the speed of the wave when the temperature is lowered by 14 C°? Ignore any change in the linear density caused by the change in temperature.

** **21.** The drawing shows a 15.0-kg ball being whirled in a circular path on the end of a string. The motion occurs on a frictionless, horizontal table. The angular speed of the ball is $\omega = 12.0$ rad/s. The string has a mass of 0.0230 kg. How much time does it take for a wave on the string to travel from the center of the circle to the ball?

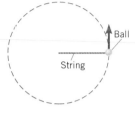

Section 16.4 The Mathematical Description of a Wave

(Note: The phase angles $(2\pi ft - 2\pi x/\lambda)$ and $(2\pi ft + 2\pi x/\lambda)$ are measured in radians, not degrees.)

22. A wave traveling in the $+x$ direction has an amplitude of 0.35 m, a speed of 5.2 m/s, and a frequency of 14 Hz. Write the equation of the wave in the form given by either Equation 16.3 or 16.4.

23. ssm A wave has the following properties: amplitude = 0.37 m, period = 0.77 s, wave speed = 12 m/s. The wave is traveling in the $-x$ direction. What is the mathematical expression (similar to Equation 16.3 or 16.4) for the wave?

24. The drawing shows two graphs that represent a transverse wave on a string. The wave is moving in the $+x$ direction. Using the information contained in these graphs, write the mathematical expression (similar to Equation 16.3 or 16.4) for the wave.

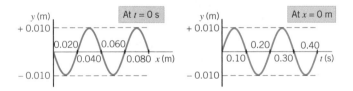

25. A wave causes a displacement y that is given in meters according to $y = (0.45) \sin (8.0\pi t + \pi x)$, where t and x are expressed in seconds and meters, respectively. (a) Find the amplitude, the frequency, the wavelength, and the speed of the wave. (b) Is this wave traveling in the $+x$ or $-x$ direction?

* **26.** The tension in a string is 15 N, and its linear density is 0.85 kg/m. A wave on the string travels toward the $-x$ direction; it has an amplitude of 3.6 cm and a frequency of 12 Hz. What are the (a) speed and (b) wavelength of the wave? (c) Write down a mathematical expression (like Equation 16.3 or 16.4) for the wave, substituting numbers for the variables A, f, and λ.

* **27. ssm** A transverse wave is traveling on a string. The displacement y of a particle from its equilibrium position is given by $y = (0.021$ m$) \sin (25t - 2.0x)$. Note that the phase angle $25t - 2.0x$ is in radians, t is in seconds, and x is in meters. The linear density of the string is 1.6×10^{-2} kg/m. What is the tension in the string?

** **28.** A transverse wave on a string has an amplitude of 0.20 m and a frequency of 175 Hz. Consider the particle of the string at $x = 0$ m. It begins with a displacement of $y = 0$ m when $t = 0$ s, according to

Equation 16.3 or 16.4. How much time passes between the first two instants when this particle has a displacement of $y = 0.10$ m?

Section 16.5 The Nature of Sound,
Section 16.6 The Speed of Sound

29. ssm The speed of a sound in a container of hydrogen at 201 K is 1220 m/s. What would be the speed of sound if the temperature were raised to 405 K? Assume that hydrogen behaves like an ideal gas.

30. For research purposes a sonic buoy is tethered to the ocean floor and emits an infrasonic pulse of sound. The period of this sound is 71 ms. Determine the wavelength of the sound.

31. The distance between a loudspeaker and the left ear of a listener is 2.70 m. (a) Calculate the time required for sound to travel this distance if the air temperature is 20 °C. (b) Assuming that the sound frequency is 523 Hz, how many wavelengths of sound are contained in this distance?

32. Have you ever listened for an approaching train by kneeling next to a railroad track and putting your ear to the rail? Young's modulus for steel is $Y = 2.0 \times 10^{11}$ N/m², and the density of steel is $\rho = 7860$ kg/m³. On a day when the temperature is 20 °C, how many times greater is the speed of sound in the rail than in the air?

33. ssm At 20 °C the densities of fresh water and ethyl alcohol are, respectively, 998 and 789 kg/m³. Find the ratio of the adiabatic bulk modulus of fresh water to the adiabatic bulk modulus of ethyl alcohol at 20 °C.

34. The wavelength of a sound wave in air is 2.74 m at 20 °C. What is the wavelength of this sound wave in fresh water at 20 °C? *(Hint: The frequency of the sound is the same in both media.)*

35. An explosion occurs at the end of a pier. The sound reaches the other end of the pier by traveling through three media: air, fresh water, and a slender metal handrail. The speeds of sound in air, water, and the handrail are 343, 1482, and 5040 m/s, respectively. The sound travels a distance of 125 m in each medium. (a) Through which medium does the sound arrive first, second, and third? (b) After the first sound arrives, how much later do the second and third sounds arrive?

36. Consult **Interactive LearningWare 16.1** at **www.wiley.com/college/cutnell** for insight into this problem. At what temperature is the speed of sound in helium (ideal gas, $\gamma = 1.67$, atomic mass = 4.003 u) the same as its speed in oxygen at 0 °C?

37. As the drawing illustrates, a siren can be made by blowing a jet of air through 20 equally spaced holes in a rotating disk. The time it takes for successive holes to move past the air jet is the period of the sound. The siren is to produce a 2200-Hz tone. What must be the angular speed ω (in rad/s) of the disk?

* **38.** As the drawing shows, one microphone is located at the origin, and a second microphone is located on the $+y$ axis. The microphones are separated by a distance of $D = 1.50$ m. A source of

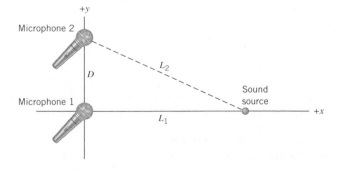

sound is located on the $+x$ axis, its distances from microphones 1 and 2 being L_1 and L_2, respectively. The speed of sound is 343 m/s. The sound reaches microphone 1 first, and then, 1.46 ms later, it reaches microphone 2. Find the distances L_1 and L_2.

* **39. ssm** A long slender bar is made from an unknown material. The length of the bar is 0.83 m, its cross-sectional area is 1.3×10^{-4} m², and its mass is 2.1 kg. A sound wave travels from one end of the bar to the other end in 1.9×10^{-4} s. From which one of the materials listed in Table 10.1 is the bar most likely to be made?

* **40.** When an earthquake occurs, two types of sound waves are generated and travel through the earth. The primary, or P, wave has a speed of about 8.0 km/s and the secondary, or S, wave has a speed of about 4.5 km/s. A seismograph, located some distance away, records the arrival of the P wave and then, 78 s later, records the arrival of the S wave. Assuming that the waves travel in a straight line, how far is the seismograph from the earthquake?

* **41. ssm** A hunter is standing on flat ground between two vertical cliffs that are directly opposite one another. He is closer to one cliff than to the other. He fires a gun and, after a while, hears three echoes. The second echo arrives 1.6 s after the first, and the third echo arrives 1.1 s after the second. Assuming that the speed of sound is 343 m/s and that there are no reflections of sound from the ground, find the distance between the cliffs.

* **42.** A sound wave travels twice as far in neon (Ne) as it does in krypton (Kr) in the same time interval. Both neon and krypton can be treated as monatomic ideal gases. The atomic mass of neon is 20.2 u, and that of krypton is 83.8 u. The temperature of the krypton is 293 K. What is the temperature of the neon?

* **43.** A monatomic ideal gas ($\gamma = 1.67$) is contained within a box whose volume is 2.5 m³. The pressure of the gas is 3.5×10^5 Pa. The total mass of the gas is 2.3 kg. Find the speed of sound in the gas.

* **44.** At a height of ten meters above the surface of a freshwater lake, a sound pulse is generated. The echo from the bottom of the lake returns to the point of origin 0.140 s later. The air and water temperatures are 20 °C. How deep is the lake?

** **45.** A jet is flying horizontally, as the drawing shows. When the plane is directly overhead at B, a person on the ground hears the sound coming from A in the drawing. The average temperature of the air is 20 °C. If the speed of the plane at A is 164 m/s, what is its speed at B, assuming that it has a constant acceleration?

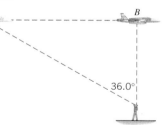

** **46.** The sonar unit on a boat is designed to measure the depth of fresh water ($\rho = 1.00 \times 10^3$ kg/m³, $B_{ad} = 2.20 \times 10^9$ Pa). When the boat moves into salt water ($\rho = 1025$ kg/m³, $B_{ad} = 2.37 \times 10^9$ Pa), the sonar unit is no longer calibrated properly. In salt water, the sonar unit indicates the water depth to be 10.0 m. What is the actual depth of the water?

** **47. ssm www** As a prank, someone drops a water-filled balloon out of a window. The balloon is released from rest at a height of 10.0 m above the ears of a man who is the target. Then, because of a guilty conscience, the prankster shouts a warning after the balloon is released. The warning will do no good, however, if shouted after the balloon reaches a certain point, even if the man could react infinitely quickly. Assuming that the air temperature is 20 °C and ignoring the effect of air resistance on the balloon, determine how far above the man's ears this point is.

Section 16.7 Sound Intensity

48. A typical adult ear has a surface area of 2.1×10^{-3} m². The sound intensity during a normal conversation is about 3.2×10^{-6} W/m² at the listener's ear. Assume that the sound strikes the surface of the ear perpendicularly. How much power is intercepted by the ear?

49. ssm A loudspeaker has a circular opening with a radius of 0.0950 m. The electrical power needed to operate the speaker is 25.0 W. The average sound intensity at the opening is 17.5 W/m². What percentage of the electrical power is converted by the speaker into sound power?

50. The average sound intensity inside a busy restaurant is 3.2×10^{-5} W/m². How much energy goes into each ear (area = 2.1×10^{-3} m²) during a one-hour meal?

51. Suppose that sound is emitted uniformly in all directions by a public address system. The intensity at a location 22 m away from the sound source is 3.0×10^{-4} W/m². What is the intensity at a spot that is 78 m away?

52. Suppose in Conceptual Example 8 (see Figure 16.25) that the person is producing 1.1 mW of sound power. Some of the sound is reflected from the floor and ceiling. The intensity of this reflected sound at a distance of 3.0 m from the source is 4.4×10^{-6} W/m². What is the total sound intensity due to both the direct and reflected sounds, at this point?

53. ssm www At a distance of 3.8 m from a siren, the sound intensity is 3.6×10^{-2} W/m². Assuming that the siren radiates sound uniformly in all directions, find the total power radiated.

* **54.** Deep ultrasonic heating is used to promote healing of torn tendons. It is produced by applying ultrasonic sound to the body. The sound transducer (generator) is circular with a radius of 1.8 cm, and it produces a sound intensity of 5.9×10^3 W/m². How much time is required for the transducer to emit 4800 J of sound energy?

* **55.** Review **Interactive Solution 16.55** at **www.wiley.com/college/ cutnell** for one approach to this problem. A dish of lasagna is being heated in a microwave oven. The effective area of the lasagna that is exposed to the microwaves is 2.2×10^{-2} m². The mass of the lasagna is 0.35 kg, and its specific heat capacity is 3200 J/(kg·C°). The temperature rises by 72 C° in 8.0 minutes. What is the intensity of the microwaves in the oven?

* **56.** When a helicopter is hovering 1450 m directly overhead, an observer on the ground measures a sound intensity I. Assume that sound is radiated uniformly from the helicopter and that ground reflections are negligible. How far must the helicopter fly in a straight line parallel to the ground before the observer measures a sound intensity of $\frac{1}{4}I$?

** **57. ssm** A rocket, starting from rest, travels straight up with an acceleration of 58.0 m/s². When the rocket is at a height of 562 m, it produces sound that eventually reaches a ground-based monitoring station directly below. The sound is emitted uniformly in all directions. The monitoring station measures a sound intensity I. Later, the station measures an intensity $\frac{1}{3}I$. Assuming that the speed of sound is 343 m/s, find the time that has elapsed between the two measurements.

Section 16.8 Decibels

58. A middle-aged man typically has poorer hearing than a middle-aged woman. In one case a woman can just begin to hear a musical tone, while a man can just begin to hear the tone only when its intensity level is increased by 7.8 dB relative to that for the woman. What is the ratio of the sound intensity just detected by the man to that just detected by the woman?

59. The bellow of a territorial bull hippopotamus has been measured at 115 dB above the threshold of hearing. What is the sound intensity?

60. A recording engineer works in a soundproofed room that is 44.0 dB quieter than the outside. If the sound intensity in the room is 1.20×10^{-10} W/m², what is the intensity outside?

61. ssm Humans can detect a difference in sound intensity levels as small as 1.0 dB. What is the ratio of the sound intensities?

62. The equation $\beta = (10 \text{ dB}) \log (I/I_0)$, which defines the decibel, is sometimes written in terms of power P (in watts) rather than intensity I (in watts/meter²). The form $\beta = (10 \text{ dB}) \log (P/P_0)$ can be used to compare two power levels in terms of decibels. Suppose that stereo amplifier A is rated at $P = 250$ watts per channel, and amplifier B has a rating of $P_0 = 45$ watts per channel. (a) Expressed in decibels, how much more powerful is A compared to B? (b) Will A sound more than twice as loud as B? Justify your answer.

63. For information, read Problem 62 before working this problem. Stereo manufacturers express the power output of a stereo amplifier using the decibel, abbreviated as dBW, where the "W" indicates that a reference power level of $P_0 = 1.00$ W has been used. If an amplifier has a power rated at 17.5 dBW, how many watts of power can this amplifier deliver?

* **64. ssm** Review Conceptual Example 8 as background for this problem. A loudspeaker is generating sound in a room. At a certain point, the sound waves coming directly from the speaker (without reflecting from the walls) create an intensity level of 75.0 dB. The waves reflected from the walls create, by themselves, an intensity level of 72.0 dB at the same point. What is the total intensity level? *(Hint: The answer is not 147.0 dB.)*

* **65.** In a discussion person A is talking 1.5 dB louder than person B, and person C is talking 2.7 dB louder than person A. What is the ratio of the sound intensity of person C to the sound intensity of person B?

* **66.** The sound intensity level of a person speaking normally is about 65 dB above the threshold of hearing. What is the minimum number of people speaking simultaneously, each with this intensity level, that is necessary to produce a sound intensity level at least 78 dB above the threshold of hearing?

* **67.** A portable radio is sitting at the edge of a balcony 5.1 m above the ground. The unit is emitting sound uniformly in all directions. By accident, it falls from rest off the balcony and continues to play on the way down. A gardener is working in a flower bed directly below the falling unit. From the instant the unit begins to fall, how much time is required for the sound intensity level heard by the gardener to increase by 10.0 dB?

** **68.** A source emits sound uniformly in all directions. A radial line is drawn from this source. On this line, determine the positions of two points, 1.00 m apart, such that the intensity level at one point is 2.00 dB greater than that at the other.

** **69. ssm** Suppose that when a certain sound intensity level (in dB) triples, the sound intensity (in W/m²) also triples. Determine this sound intensity level.

Section 16.9 The Doppler Effect

70. You are riding your bicycle directly away from a stationary source of sound and hear a frequency that is 1.0% lower than the emitted frequency. The speed of sound is 343 m/s. What is your speed?

71. The security alarm on a parked car goes off and produces a frequency of 960 Hz. The speed of sound is 343 m/s. As you drive toward this parked car, pass it, and drive away, you observe the frequency to change by 95 Hz. At what speed are you driving?

72. Suppose you are stopped for a traffic light, and an ambulance approaches you from behind with a speed of 18 m/s. The siren on the ambulance produces sound with a frequency of 955 Hz. The speed of sound in air is 343 m/s. What is the wavelength of the sound reaching your ears?

73. ssm A speeder looks in his rearview mirror. He notices that a police car has pulled behind him and is matching his speed of 38 m/s. The siren on the police car has a frequency of 860 Hz when the police car and the listener are stationary. The speed of sound is 343 m/s. What frequency does the speeder hear when the siren is turned on in the moving police car?

74. A bird is flying directly toward a stationary bird-watcher and emits a frequency of 1250 Hz. The bird-watcher, however, hears a frequency of 1290 Hz. What is the speed of the bird, expressed as a percentage of the speed of sound?

* **75. ssm** An aircraft carrier has a speed of 13.0 m/s relative to the water. A jet is catapulted from the deck and has a speed of 67.0 m/s relative to the water. The engines produce a 1550-Hz whine, and the speed of sound is 343 m/s. What is the frequency of the sound heard by the crew on the ship?

* **76.** A bungee jumper jumps from rest and screams with a frequency of 589 Hz. The air temperature is 20 °C. What is the frequency heard by the people on the ground below when she has fallen a distance of 11.0 m? Assume that the bungee cord has not yet taken effect, so she is in free-fall.

* **77.** Refer to **Interactive Solution 16.77** at **www.wiley.com/college/cutnell** for one approach to this type of problem. Two trucks travel at the same speed. They are far apart on adjacent lanes and approach each other essentially head-on. One driver hears the horn of the other truck at a frequency that is 1.14 times the frequency he hears when the trucks are stationary. The speed of sound is 343 m/s. At what speed is each truck moving?

* **78.** Two submarines are underwater and approaching each other head-on. Sub A has a speed of 12 m/s and sub B has a speed of 8 m/s. Sub A sends out a 1550-Hz sonar wave that travels at a speed of 1522 m/s. (a) What is the frequency detected by sub B? (b) Part of the sonar wave is reflected from B and returns to A. What frequency does A detect for this reflected wave?

* **79. ssm** A motorcycle starts from rest and accelerates along a straight line at 2.81 m/s². The speed of sound is 343 m/s. A siren at the starting point remains stationary. How far has the motorcycle gone when the driver hears the frequency of the siren at 90.0% of the value it has when the motorcycle is stationary?

** **80.** A microphone is attached to a spring that is suspended from the ceiling, as the drawing indicates. Directly below on the floor is a stationary 440-Hz source of sound. The microphone vibrates up and down in simple harmonic motion with a period of 2.0 s. The difference between the maximum and minimum sound frequencies detected by the microphone is 2.1 Hz. Ignoring any reflections of sound in the room and using 343 m/s for the speed of sound, determine the amplitude of the simple harmonic motion.

Sound source

Additional Problems

81. An amplified guitar has a sound intensity level that is 14 dB greater than the same unamplified sound. What is the ratio of the amplified intensity to the unamplified intensity?

82. A sonar unit on a submarine sends out a pulse of sound into seawater. The pulse returns 1.30 s later. What is the distance to the object that reflects the pulse back to the submarine?

83. ssm Argon (molecular mass = 39.9 u) is a monatomic gas. Assuming that it behaves like an ideal gas at 298 K ($\gamma = 1.67$), find (a) the rms speed of argon atoms and (b) the speed of sound in argon.

84. Interactive LearningWare 16.2 at **www.wiley.com/college/cutnell** provides some pertinent background for this problem. A convertible moves toward you and then passes you; all the while, its loudspeakers are producing a sound. The speed of the car is a constant 9.00 m/s, and the speed of sound is 343 m/s. What is the ratio of the frequency you hear while the car is approaching to the frequency you hear while the car is moving away?

85. ssm www The middle C string on a piano is under a tension of 944 N. The period and wavelength of a wave on this string are 3.82 ms and 1.26 m, respectively. Find the linear density of the string.

86. Tsunamis are fast-moving waves often generated by underwater earthquakes. In the deep ocean their amplitude is barely noticeable, but upon reaching shore, they can rise up to the astonishing height of a six-story building. One tsunami, generated off the Aleutian islands in Alaska, had a wavelength of 750 km and traveled a distance of 3700 km in 5.3 h. (a) What was the speed (in m/s) of the wave? For reference, the speed of a 747 jetliner is about 250 m/s. Find the wave's (b) frequency and (c) period.

87. ssm When a person wears a hearing aid, the sound intensity level increases by 30.0 dB. By what factor does the sound intensity increase?

88. A rocket in a fireworks display explodes high in the air. The sound spreads out uniformly in all directions. The intensity of the sound is 2.0×10^{-6} W/m^2 at a distance of 120 m from the explosion. Find the distance from the source at which the intensity is 0.80×10^{-6} W/m^2.

89. The right-most key on a piano produces a sound wave that has a frequency of 4185.6 Hz. Assuming that the speed of sound in air is 343 m/s, find the corresponding wavelength.

90. The displacement (in meters) of a wave is given according to $y = 0.26 \sin(\pi t - 3.7\pi x)$, where t is in seconds and x is in meters. (a) Is the wave traveling in the $+x$ or $-x$ direction? (b) What is the displacement y when $t = 38$ s and $x = 13$ m?

91. ssm From a vantage point very close to the track at a stock car race, you hear the sound emitted by a moving car. You detect a frequency that is 0.86 times smaller than that emitted by the car when it is stationary. The speed of sound is 343 m/s. What is the speed of the car?

92. A sound wave is incident on a pool of fresh water. The sound enters the water perpendicularly and travels a distance of 0.45 m before striking a 0.15-m-thick copper block lying on the bottom. The sound passes through the block, reflects from the bottom surface of the block, and returns to the top of the water along the same path. How much time elapses between when the sound enters and when it leaves the water?

93. ssm A listener doubles his distance from a source that emits sound uniformly in all directions. By how many decibels does the sound intensity level change?

94. A rocket engine emits 2.0×10^5 J of sound energy every second. The sound is emitted uniformly in all directions. What is the sound intensity level, measured relative to the threshold of hearing, at a distance of 85 m away from the engine?

* **95.** Two identical rifles are shot at the same time, and the sound intensity level is 80.0 dB. What would be the sound intensity level if only one rifle were shot? *(Hint: The answer is not 40.0 dB.)*

* **96.** In Figure 16.3*c* the colored dot exhibits simple harmonic motion as the longitudinal wave passes. The wave has an amplitude of 5.4×10^{-3} m and a frequency of 4.0 Hz. Find the maximum acceleration of the dot.

* **97. ssm** Two blocks are connected by a wire that has a mass per unit length of 8.50×10^{-4} kg/m. One block has a mass of 19.0 kg, and the other has a mass of 42.0 kg. These blocks are being pulled across a horizontal frictionless floor by a horizontal force **P** that is applied to the less massive block. A transverse wave travels on the wire between the blocks with a speed of 352 m/s (relative to the wire). The mass of the wire is negligible compared to the mass of the blocks. Find the magnitude of **P**.

* **98.** A steel cable of cross-sectional area 2.83×10^{-3} m^2 is kept under a tension of 1.00×10^4 N. The density of steel is 7860 kg/m^3 (this is *not* the linear density). At what speed does a transverse wave move along the cable?

* **99.** Review Conceptual Example 3 before starting this problem. A horizontal wire is under a tension of 315 N and has a mass per unit length of 6.50×10^{-3} kg/m. A transverse wave with an amplitude of 2.50 mm and a frequency of 585 Hz is traveling on this wire. As the wave passes, a particle of the wire moves up and down in simple harmonic motion. Obtain (a) the speed of the wave and (b) the maximum speed with which the particle moves up and down.

* **100.** When one person shouts at a football game, the sound intensity level at the center of the field is 60.0 dB. When all the people shout together, the intensity level increases to 109 dB. Assuming that each person generates the same sound intensity at the center of the field, how many people are at the game?

** **101.** In a mixture of argon (atomic mass = 39.9 u) and neon (atomic mass = 20.2 u), the speed of sound is 363 m/s at 3.00×10^2 K. Assume that both monatomic gases behave as ideal gases. Find the percentage of the atoms that are argon and the percentage that are neon.

** **102.** Civil engineers use a transit theodolite when surveying. A modern version of this device determines distance by measuring the time required for an ultrasonic pulse to reach a target, reflect from it, and return. Effectively, such a theodolite is calibrated properly when it is programmed with the speed of sound appropriate for the ambient air temperature. (a) Suppose the round-trip time for the pulse is 0.580 s on a day when the air temperature is 293 K, the temperature for which the instrument is calibrated. How far is the target from the theodolite? (b) Assume that air behaves as an ideal gas. If the air temperature were 298 K, rather than the calibration temperature of 293 K, what percentage error would there be in the distance measured by the theodolite?

Concepts & Calculations *Group Learning Problems*

Note: Each of these problems consists of Concept Questions followed by a related quantitative Problem. They are designed for use by students working alone or in small learning groups. The Concept Questions involve little or no mathematics and are intended to stimulate group discussions. They focus on the concepts with which the problems deal. Recognizing the concepts is the essential initial step in any problem-solving technique.

103. Concept Questions The drawing shows a snapshot of two waves traveling to the right at the same speed. (a) Rank the waves according to their wavelengths, largest first. (b) Which wave, if either, has the higher frequency? (c) If a particle were attached to each wave, like that in Figure 16.10, which particle would have the greater maximum speed as it moves up and down? Justify your answers.

Problem (a) From the data in the drawing, determine the wavelength of each wave. (b) If the speed of the waves is 12 m/s, calculate the frequency of each one. (c) What is the maximum speed for a particle attached to each wave? Check that your answers are consistent with those for the Concept Questions.

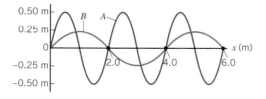

104. Concept Questions Example 4 in the text discusses an ultrasonic ruler that displays the distance between the ruler and an object, such as a wall. The ruler sends out a pulse of ultrasonic sound and measures the time it takes for the pulse to reflect from the object and return. The ruler uses this time, along with a preset value for the speed of sound in air, to determine the distance. Suppose you use this ruler underwater, rather than in air. (a) Is the speed of sound in water greater than, less than, or equal to the speed of sound in air? (b) Is the reading on the ruler greater than, less than, or equal to the actual distance? Provide reasons for your answers.

Problem The actual distance from the ultrasonic ruler to an object is 25.0 m. The adiabatic bulk modulus and density of seawater are $B_{ad} = 2.31 \times 10^9$ Pa and $\rho = 1025$ kg/m^3, respectively. Assume that the ruler uses a preset value of 343 m/s for the speed of sound in air, and determine the distance reading on its display. Verify that your answer is consistent with your answers to the Concept Questions.

105. Concept Question Suppose you are part of a team that is trying to break the sound barrier with a jet-powered car, which means that it must travel faster than the speed of sound in air. Would you attempt this feat early in the morning when the temperature is cool, later in the afternoon when the temperature is warmer, or does it even matter what the temperature is?

Problem In the morning, the air temperature is 0 °C and the speed of sound is 331 m/s. What must be the speed of your car if it is to break the sound barrier when the temperature has risen to 43 °C in the afternoon? Assume that air behaves like an ideal gas.

106. Concept Questions A source of sound is located at the center of two concentric spheres, parts of which are shown in the drawing. The source emits sound uniformly in all directions. On the spheres are drawn three small patches that may, or may not,

have equal areas. However, the same sound power passes through each patch. (a) Rank the sound intensity at each patch, greatest first. (b) Rank the area of each patch, largest first. Provide reasons for your answers.

Problem The source produces 2.3 W of sound power, and the radii of the concentric spheres are $r_A = 0.60$ m and $r_B = 0.80$ m. (a) Determine the sound intensity at each of the three patches. (b) The sound power that passes through each of the patches is 1.8×10^{-3} W. Find the area of each patch. Verify that your answers are consistent with those to the Concept Questions.

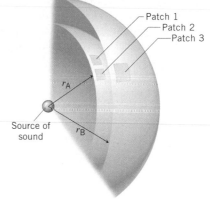

107. Concept Questions The threshold of hearing for an average young person is 0 dB. Two individuals who are not average have thresholds of hearing that are $\beta_1 = -8.00$ dB and $\beta_2 = +12.0$ dB. (a) Is the sound intensity at a level of -8.00 dB greater than, or less than, that at 0 dB? (b) Which individual has the better hearing? Why?

Problem How many times greater (or less) is the sound intensity when person 1 ($\beta_1 = -8.00$ dB) hears it at the threshold of hearing compared to when person 2 ($\beta_2 = +12.0$ dB) hears it at the threshold of hearing? Check to see that your answer is consistent with your answers to the Concept Questions.

108. Concept Questions The table shows three situations in which the Doppler effect may arise. The first two columns indicate the velocities of the sound source and the observer, where the length of each arrow is proportional to the speed. For each situation, fill in the empty columns by deciding whether the wavelength of the sound and the frequency heard by the observer increase, decrease, or remain the same compared to the case when there is no Doppler effect. Provide a reason for each answer.

	Velocity of Sound Source (Toward the Observer)	Velocity of Observer (Toward the Source)	Wavelength	Frequency Heard by Observer
(a)	0 m/s	0 m/s		
(b)	$\longrightarrow$	0 m/s		
(c)	$\longrightarrow$	$\longleftarrow$		

Problem The siren on an ambulance is emitting a sound whose frequency is 2450 Hz. The speed of sound is 343 m/s. (a) If the ambulance is stationary and you (the "observer") are sitting in a parked car, what is the wavelength of the sound and the frequency heard by you? (b) Suppose the ambulance is moving toward you at a speed of 26.8 m/s. Determine the wavelength of the sound and the frequency heard by you. (c) If the ambulance is moving toward you at a speed of 26.8 m/s and you are moving toward it at a speed of 14.0 m/s, find the wavelength of the sound and the frequency that you hear. Be sure that your answers are consistent with your answers to the Concept Questions.

* **109. Concept Questions** A uniform rope of mass m and length L is hanging straight down from the ceiling. (a) Is the tension in the rope greater near the top or near the bottom of the rope? Why? (b) A small-amplitude transverse wave is sent up the rope from the bottom end. Is the speed of the wave greater near the bottom or near the top of the rope? Explain. (c) Consider a section of the rope between the bottom end and a point that is a distance y meters above the bottom. What is the weight of this section? Express your answer in terms of m, L, y, and g (the acceleration due to gravity).

Problem (a) For the rope described in the Concept Questions, derive an expression that gives the speed of the wave on the rope in terms of the distance y above the bottom end and the acceleration g due to gravity. (b) Use the expression that you have derived to calculate the speeds at distances of 0.50 m and 2.0 m above the bottom end of the rope. Be sure that your answers are consistent with your answer to Concept Question (b).

* **110. Concept Questions** A wireless transmitting microphone is mounted on a small platform, which can roll down an incline, away from a speaker that is mounted at the top of the incline. The speaker broadcasts a fixed-frequency tone. (a) The platform is positioned in front of the speaker and released from rest. Describe how the velocity of the platform changes and why. (b) How is the changing velocity related to the acceleration of the platform? (c) Describe how the frequency detected by the microphone changes. Explain why the frequency changes as you have described. (d) Which equation given in the chapter applies to this situation? Justify your answer.

Problem The speaker broadcasts a tone that has a frequency of 1.000×10^4 Hz, and the speed of sound is 343 m/s. At a time of 1.5 s following the release of the platform, the microphone detects a frequency of 9939 Hz. At a time of 3.5 s following the release of the platform, the microphone detects a frequency of 9857 Hz. What is the acceleration (assumed constant) of the platform?

Chapter 17

The Principle of Linear Superposition and Interference Phenomena

Virtually all musical instruments, such as the guitar in this photograph, produce their sound in a way that involves the principle of linear superposition. This chapter discusses the principle and a number of other topics that relate to it.
(© Neal Preston/Corbis Images)

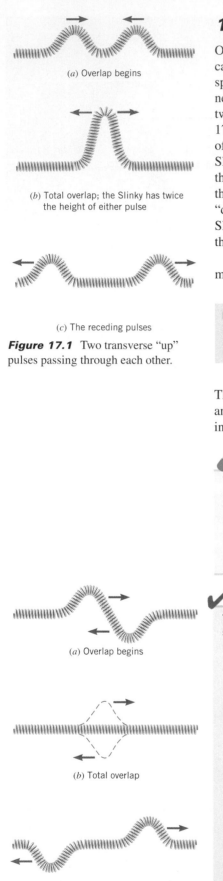

(a) Overlap begins

(b) Total overlap; the Slinky has twice
the height of either pulse

(c) The receding pulses

Figure 17.1 Two transverse "up"
pulses passing through each other.

17.1 The Principle of Linear Superposition

Often, two or more sound waves are present at the same place at the same time, as is the case with sound waves when everyone is talking at a party or when music plays from the speakers of a stereo system. To illustrate what happens when several waves pass simultaneously through the same region, let's first consider Figures 17.1 and 17.2, which show two transverse pulses of equal heights moving toward each other along a Slinky. In Figure 17.1 both pulses are "up," while in Figure 17.2 one is "up" and the other is "down." Part *a* of each drawing shows the two pulses beginning to overlap. The pulses merge, and the Slinky assumes a shape that is *the sum of the shapes of the individual pulses.* Thus, when the two "up" pulses overlap completely, as in Figure 17.1*b*, the Slinky has a pulse height that is twice the height of an individual pulse. Likewise, when the "up" pulse and the "down" pulse overlap exactly, as in Figure 17.2*b*, they momentarily cancel, and the Slinky becomes straight. In either case, the two pulses move apart after overlapping, and the Slinky once again conforms to the shapes of the individual pulses.

The adding together of individual pulses to form a resultant pulse is an example of a more general concept called the **principle of linear superposition.**

■ **THE PRINCIPLE OF LINEAR SUPERPOSITION**

When two or more waves are present simultaneously at the same place, the resultant disturbance is the sum of the disturbances from the individual waves.

This principle can be applied to all types of waves, including sound waves, water waves, and electromagnetic waves such as light. It embodies one of the most important concepts in physics, and the remainder of this chapter deals with examples related to it.

Concept Simulation 17.1

This simulation shows how two pulses, traveling in opposite directions on a string, are added together to form a resultant pulse. The user can choose whether the pulses are the same or different sizes, and whether they both point "up," or one points "up" and the other "down."

Related Homework: Problems 1, 50

Go to **www.wiley.com/college/cutnell**

✔ **Check Your Understanding 1**

The drawing shows two pulses traveling toward each other at $t = 0$ s. Each pulse has a constant speed of 1 cm/s. When $t = 2$ s, what is the height of the resultant pulse at (a) $x = 2$ cm, (b) $x = 4$ cm, and (c) $x = 6$ cm? *(The answers are given at the end of the book.)*

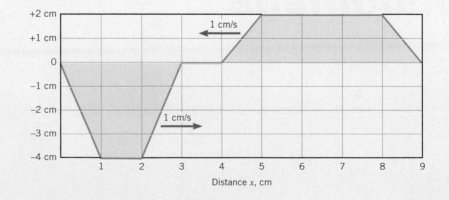

Background: Understanding the principle of linear superposition is the key to answering this question.

For similar questions (including conceptual counterparts), consult Self-Assessment Test 17.1, which is described at the end of Section 17.4.

(a) Overlap begins

(b) Total overlap

(c) The receding pulses

Figure 17.2 Two transverse pulses, one "up" and one "down," passing through each other.

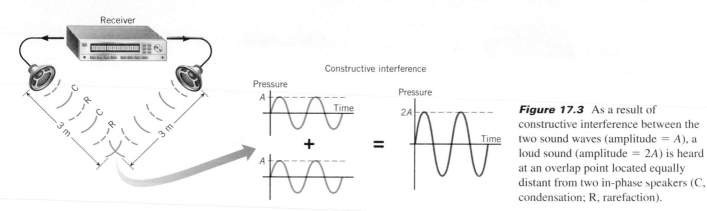

Figure 17.3 As a result of constructive interference between the two sound waves (amplitude = *A*), a loud sound (amplitude = 2*A*) is heard at an overlap point located equally distant from two in-phase speakers (C, condensation; R, rarefaction).

17.2 *Constructive and Destructive Interference of Sound Waves*

Suppose that the sounds from two speakers overlap in the middle of a listening area, as in Figure 17.3, and that each speaker produces a sound wave of the same amplitude and frequency. For convenience, the wavelength of the sound is chosen to be $\lambda = 1$ m. In addition, assume the diaphragms of the speakers vibrate in phase; that is, they move outward together and inward together. If the distance of each speaker from the overlap point is the same (3 m in the drawing), the condensations (C) of one wave always meet the condensations of the other when the waves come together; similarly, rarefactions (R) always meet rarefactions. According to the principle of linear superposition, the combined pattern is the sum of the individual patterns. As a result, the pressure fluctuations at the overlap point have twice the amplitude *A* that the individual waves have, and a listener at this spot hears a louder sound than that coming from either speaker alone. When two waves always meet condensation-to-condensation and rarefaction-to-rarefaction (or crest-to-crest and trough-to-trough), they are said to be *exactly in phase* and to exhibit *constructive interference.*

Now consider what happens if one of the speakers is moved. The result is surprising. In Figure 17.4, the left speaker is moved away* from the overlap point by a distance equal to one-half of the wavelength, or 0.5 m. Therefore, at the overlap point, a condensation arriving from the left meets a rarefaction arriving from the right. Likewise, a rarefaction arriving from the left meets a condensation arriving from the right. According to the principle of linear superposition, the net effect is a mutual cancellation of the two waves. The condensations from one wave offset the rarefactions from the other, leaving only a *constant air pressure.* A constant air pressure, devoid of condensations and rarefactions, means that a listener detects no sound. When two waves always meet condensation-to-rarefaction (or crest-to-trough), they are said to be *exactly out of phase* and to exhibit *destructive interference.*

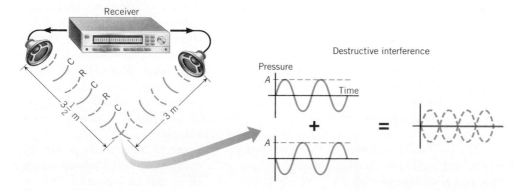

Figure 17.4 The speakers in this drawing vibrate in phase. However, the left speaker is one-half of a wavelength ($\frac{1}{2}$ m) farther from the overlap point than the right speaker. Because of destructive interference, no sound is heard at the overlap point (C, condensation; R, rarefaction).

* When the left speaker is moved back, its sound intensity and, hence, its pressure amplitude decrease at the overlap point. In this chapter assume that the power delivered to the left speaker by the receiver is increased slightly to keep the amplitudes equal at the overlap point.

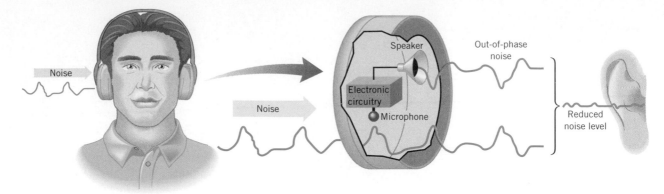

Figure 17.5 Noise-canceling headphones utilize destructive interference.

The physics of noise-canceling headphones.

Problem solving insight

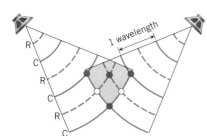

Figure 17.6 Two sound waves overlap in the shaded region. The solid lines denote the middle of the condensations (C), and the dashed lines denote the middle of the rarefactions (R). Constructive interference occurs at each solid dot (●) and destructive interference at each open dot (○).

When two waves meet, they interfere constructively if they always meet exactly in phase and destructively if they always meet exactly out of phase. In either case, this means that the wave patterns do not shift relative to one another as time passes. Sources that produce waves in this fashion are called *coherent sources.*

Destructive interference is the basis of a useful technique for reducing the loudness of undesirable sounds. For instance, Figure 17.5 shows a pair of noise-canceling headphones. Small microphones are mounted inside the headphones and detect noise such as the engine noise that an airplane pilot would hear. The headphones also contain circuitry to process the electronic signals from the microphones and reproduce the noise in a form that is exactly out of phase compared to the original. This out-of-phase version is played back through the headphone speakers and, because of destructive interference, combines with the original noise to produce a quieter background.

It should be apparent that if the left speaker in Figure 17.4 were moved away from the overlap point by *another* one-half wavelength ($3\frac{1}{2}$ m $+ \frac{1}{2}$ m $= 4$ m), the two waves would again be in phase, and constructive interference would occur. The listener would hear a loud sound because the left wave travels one whole wavelength ($\lambda = 1$ m) farther than the right wave and, at the overlap point, condensation meets condensation and rarefaction meets rarefaction. In general, the important issue is the *difference* in the distances traveled by each wave in reaching the overlap point:

> *For two wave sources vibrating in phase, a difference in path lengths that is zero or an integer number (1, 2, 3, . . .) of wavelengths leads to constructive interference; a difference in path lengths that is a half-integer number ($\frac{1}{2}$, $1\frac{1}{2}$, $2\frac{1}{2}$, . . .) of wavelengths leads to destructive interference.*

Interference effects can also be detected if the two speakers are fixed in position and the listener moves about the room. Consider Figure 17.6, where the sound waves spread outward from each speaker, as indicated by the concentric circular arcs. Each solid arc represents the middle of a condensation, and each dashed arc represents the middle of a rarefaction. Where the two waves overlap, there are places of constructive interference and places of destructive interference. Constructive interference occurs at any spot where two condensations or two rarefactions intersect, and the drawing shows four such places as solid dots. A listener stationed at any one of these locations hears a loud sound. On the other hand, destructive interference occurs at any place where a condensation and a rarefaction intersect, such as the two open dots in the picture. A listener situated at a point of destructive interference hears no sound. At locations where neither constructive nor destructive interference occurs, the two waves partially reinforce or partially cancel, depending on the position relative to the speakers. Thus, it is possible for a listener to walk about the overlap region and hear marked variations in loudness.

The individual sound waves from the speakers in Figure 17.6 carry energy, and the energy delivered to the overlap region is the sum of the energies of the individual waves. This fact is consistent with the principle of conservation of energy, which we first encountered in Section 6.8. This principle states that energy can neither be created nor destroyed, but can only be converted from one form to another. One of the interesting consequences of interference is that the energy is redistributed, so there are places within the overlap region where the sound is loud and other places where there is no sound at all. Interference, so to speak, "robs Peter to pay Paul," but energy is always conserved in the process. Example 1 illustrates how to decide what a listener hears.

Example 1 What Does a Listener Hear?

In Figure 17.7 two in-phase loudspeakers, A and B, are separated by 3.20 m. A listener is stationed at point C, which is 2.40 m in front of speaker B. The triangle ABC is a right triangle. Both speakers are playing identical 214-Hz tones, and the speed of sound is 343 m/s. Does the listener hear a loud sound or no sound?

Reasoning The listener will hear either a loud sound or no sound, depending on whether the interference occurring at point C is constructive or destructive. To determine which it is, we need to find the difference in the distances traveled by the two sound waves that reach point C and see whether the difference is an integer or half-integer number of wavelengths. In either event, the wavelength can be found from the relation $\lambda = v/f$ (Equation 16.1).

Solution Since the triangle ABC is a right triangle, the distance AC is given by the Pythagorean theorem as $\sqrt{(3.20 \text{ m})^2 + (2.40 \text{ m})^2} = 4.00$ m. The distance BC is given as 2.40 m. Thus, the difference in the travel distances for the waves is 4.00 m − 2.40 m = 1.60 m. The wavelength of the sound is

$$\lambda = \frac{v}{f} = \frac{343 \text{ m/s}}{214 \text{ Hz}} = 1.60 \text{ m} \qquad (16.1)$$

Since the difference in the distances is one wavelength, constructive interference occurs at point C, and the *listener hears a loud sound.*

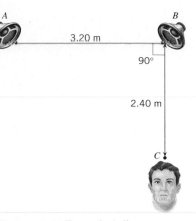

Figure 17.7 Example 1 discusses whether this setup leads to constructive or destructive interference at point C for 214-Hz sound waves.

Problem solving insight
To decide whether two sources of sound produce constructive or destructive interference at a point, determine the difference in path lengths between each source and that point and compare it to the wavelength of the sound.

Up to this point, we have been assuming that the speaker diaphragms vibrate synchronously, or in phase; that is, they move outward together and inward together. This may not be the case, however, and Conceptual Example 2 considers what happens then.

Conceptual Example 2 Out-of-Phase Speakers

To make a speaker operate, two wires must be connected between the speaker and the receiver (amplifier), as in Figure 17.8. To ensure that the diaphragms of two speakers vibrate in phase, it is necessary to make these connections in exactly the same way. If the wires for one speaker are not connected just as they are for the other speaker, the two diaphragms will vibrate out of phase. Whenever one diaphragm moves outward, the other will move inward, and vice versa. Suppose that in Figures 17.3 and 17.4 the connections are made so that the speaker diaphragms vibrate out of phase, everything else remaining the same. In each case, what kind of interference would result at the overlap point?

Reasoning and Solution Since the diaphragms are vibrating out of phase, one of them is now moving exactly opposite to the way it was moving originally. Let us assume that it is the one on the left in the drawings. The effect of this change is that every condensation originating from the left speaker becomes a rarefaction and every rarefaction becomes a condensation. As a result, in Figure 17.3 a rarefaction from the left now meets a condensation from the right at the overlap point, and *destructive interference results.* In Figure 17.4, a condensation from the left now meets a condensation from the right at the overlap point, *leading to constructive interference.* These results are opposite to those shown in the figures.

Instructions for connecting stereo systems specifically warn users to avoid out-of-phase vibration of the speaker diaphragms. One way to check your system is to play some music that is rich in low-frequency bass tones. Use the monaural mode on your receiver so the same sound comes from each speaker, and slide the speakers toward each other. If the diaphragms are moving in phase, the bass sound will either remain the same or slightly improve as the speakers come together. If the diaphragms are moving out of phase, the bass sound will decrease noticeably because of destructive interference when the speakers are right next to each other. In this event, simply interchange the wires to the terminals on one (but not both) of the speakers.

Figure 17.8 A loudspeaker is connected to a receiver (amplifier) by two wires. (© Andy Washnik)

The physics of
wiring stereo speakers.

The phenomena of constructive and destructive interference are exhibited by all types of waves, not just sound waves. We will encounter interference effects again in Chapter 27, in connection with light waves.

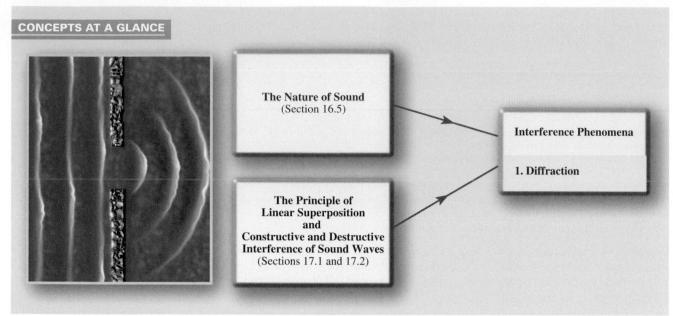

CONCEPTS AT A GLANCE

The Nature of Sound
(Section 16.5)

The Principle of
Linear Superposition
and
Constructive and Destructive
Interference of Sound Waves
(Sections 17.1 and 17.2)

Interference Phenomena

1. Diffraction

Figure 17.9 CONCEPTS AT A GLANCE By combining our knowledge of sound waves with the principle of linear superposition, we can understand the interference phenomenon of diffraction. In the illustration a water wave moves to the right and passes through an opening in a breakwater. To the right of the breakwater, the bending of the wave is an example of diffraction.

17.3 Diffraction

▶ CONCEPTS AT A GLANCE Section 16.5 discusses the fact that sound is a pressure wave created by a vibrating object, such as a loudspeaker. The previous two sections of this chapter have examined what happens when two sound waves are present simultaneously at the same place; according to the principle of linear superposition, a resultant disturbance is formed from the sum of the individual waves. This principle reveals that overlapping sound waves exhibit interference effects, whereby the sound energy is redistributed within the overlap region. As the Concepts-at-a-Glance chart in Figure 17.9 shows, we will now use the principle of linear superposition to explore another interference effect, that of diffraction. ◀

When a wave encounters an obstacle or the edges of an opening, it bends around them. For instance, a sound wave produced by a stereo system bends around the edges of an open doorway, as Figure 17.10*a* illustrates. If such bending did not occur, sound could be heard outside the room only at locations directly in front of the doorway, as part *b* of the drawing suggests. (It is assumed that no sound is transmitted directly through the walls.) The bending of a wave around an obstacle or the edges of an opening is called *diffraction.* All kinds of waves exhibit diffraction.

To demonstrate how the bending of waves arises, Figure 17.11 shows an expanded view of Figure 17.10*a*. When the sound wave reaches the doorway, the air in the doorway is set into longitudinal vibration. In effect, each molecule of the air in the doorway becomes a source of a sound wave in its own right, and, for purposes of illustra-

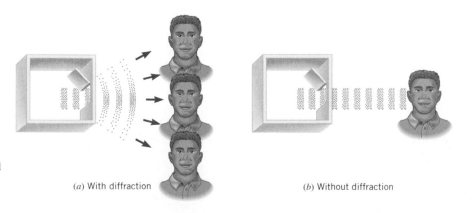

Figure 17.10 (*a*) The bending of a sound wave around the edges of the doorway is an example of diffraction. The source of the sound within the room is not shown. (*b*) If diffraction did not occur, the sound wave would not bend as it passes through the doorway.

(*a*) With diffraction

(*b*) Without diffraction

tion, the drawing shows two of the molecules. Each produces a sound wave that expands outward in three dimensions, much like a water wave does in two dimensions when a stone is dropped into a pond. The sound waves generated by all the molecules in the doorway must be added together to obtain the total sound wave at any location outside the room, in accord with the principle of linear superposition. However, even considering only the waves from the two molecules in the picture, it is clear that the expanding wave patterns reach locations off to either side of the doorway. The net effect is a "bending," or diffraction, of the sound around the edges of the opening. Further insight into the origin of diffraction can be obtained with the aid of Huygens' principle (see Section 27.5).

When the sound waves generated by every molecule in the doorway are added together, it is found that there are places where the intensity is a maximum and places where it is zero, in a fashion similar to that discussed in the previous section. Analysis shows that at a great distance from the doorway the intensity is a maximum directly opposite the center of the opening. As the distance to either side of the center increases, the intensity decreases and reaches zero, then rises again to a maximum, falls again to zero, rises back to a maximum, and so on. Only the maximum at the center is a strong one. The other maxima are weak and become progressively weaker at greater distances from the center. In Figure 17.11 the angle θ defines the location of the first minimum intensity point on either side of the center. Equation 17.1 gives θ in terms of the wavelength λ and the width D of the doorway and assumes that the doorway can be treated like a slit whose height is very large compared to its width:

Single slit—first minimum
$$\sin \theta = \frac{\lambda}{D} \qquad (17.1)$$

Waves also bend around the edges of openings other than single slits. Particularly important is the diffraction of sound by a circular opening, such as that in a loudspeaker. In this case, the angle θ is related to the wavelength λ and the diameter D of the opening by

Circular opening —first minimum
$$\sin \theta = 1.22 \frac{\lambda}{D} \qquad (17.2)$$

An important point to remember about Equations 17.1 and 17.2 is that the extent of the diffraction depends on the ratio of the wavelength to the size of the opening. If the ratio λ/D is small, then θ is small and little diffraction occurs. The waves are beamed in the forward direction as they leave an opening, much like the light from a flashlight. Such sound waves are said to have "narrow dispersion." Since high-frequency sound has a relatively small wavelength, it tends to have a narrow dispersion. On the other hand, for larger values of the ratio λ/D, the angle θ is larger. The waves spread out over a larger region and are said to have a "wide dispersion." Low-frequency sound, with its relatively large wavelength, typically has a wide dispersion.

In a stereo loudspeaker, a wide dispersion of the sound is desirable. Example 3 illustrates, however, that there are limitations to the dispersion that can be achieved, depending on the loudspeaker design.

Example 3 Designing a Loudspeaker for Wide Dispersion

A 1500-Hz sound and a 8500-Hz sound each emerges from a loudspeaker through a circular opening whose diameter is 0.30 m (see Figure 17.12). Assuming that the speed of sound in air is 343 m/s, find the diffraction angle θ for each sound.

Reasoning The diffraction angle θ for each sound wave is given by $\sin \theta = 1.22(\lambda/D)$. However, it will first be necessary to calculate the wavelengths of the sounds from $\lambda = v/f$ (Equation 16.1).

Solution The wavelengths of the two sounds are

$$\lambda_{1500} = \frac{343 \text{ m/s}}{1500 \text{ Hz}} = 0.23 \text{ m} \quad \text{and} \quad \lambda_{8500} = \frac{343 \text{ m/s}}{8500 \text{ Hz}} = 0.040 \text{ m}$$

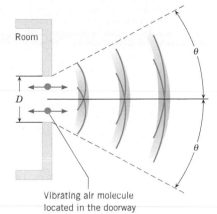

Figure 17.11 Each vibrating molecule of the air in the doorway generates a sound wave that expands outward and bends, or diffracts, around the edges of the doorway. Because of interference effects among the sound waves produced by all the molecules, the sound intensity is mostly confined to the region defined by the angle θ on either side of the doorway.

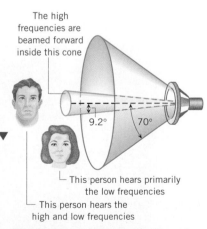

Figure 17.12 Because the dispersion of high frequencies is less than that of low frequencies, you should be directly in front of the speaker to hear both the high and low frequencies equally well.

Figure 17.13 Small-diameter speakers, called tweeters, are used to produce high-frequency sound. The small diameter helps to promote a wider dispersion of the sound. (© Christopher Gould/The Image Bank/Getty Images)

The physics of tweeter loudspeakers. (See above.)

Figure 17.14 CONCEPTS AT A GLANCE This figure, which is an extension of the concept chart in Figure 17.9, shows that the principle of linear superposition can also be used to explain the phenomenon of beats. Here, concert pianist Mordecai Shehori tunes his piano with the aid of software and beats. (© Andrea Mohin/New York Times Pictures)

The diffraction angles can now be determined:

1500-Hz sound
$$\sin \theta = 1.22 \frac{\lambda_{1500}}{D} = 1.22 \left(\frac{0.23 \text{ m}}{0.30 \text{ m}} \right) = 0.94 \tag{17.2}$$
$$\theta = \sin^{-1} 0.94 = \boxed{70°}$$

8500-Hz sound
$$\sin \theta = 1.22 \frac{\lambda_{8500}}{D} = 1.22 \left(\frac{0.040 \text{ m}}{0.30 \text{ m}} \right) = 0.16 \tag{17.2}$$
$$\theta = \sin^{-1} 0.16 = \boxed{9.2°}$$

Figure 17.12 illustrates these results. With a 0.30-m opening, the dispersion of the higher-frequency sound is limited to only 9.2°. To increase the dispersion, a smaller opening is needed. It is for this reason that loudspeaker designers use a small-diameter speaker called a *tweeter* to generate the high-frequency sound, as Figure 17.13 indicates.

As we have seen, diffraction is an interference effect, one in which some of the wave's energy is directed into regions that would otherwise not be accessible. Energy, of course, is conserved during this process, because energy is only redistributed during diffraction; no energy is created or destroyed.

17.4 Beats

▶ CONCEPTS AT A GLANCE In situations where waves with the *same frequency* overlap, we have seen how the principle of linear superposition leads to constructive and destructive interference and how it explains diffraction. We will see in this section that two overlapping waves with *slightly different frequencies* give rise to the phenomenon of beats. However, as the Concepts-at-a-Glance chart in Figure 17.14 illustrates, the principle of linear superposition again provides an explanation of what happens when the waves overlap. This chart is an expansion of that in Figure 17.9. ◀

A tuning fork has the property of producing a single-frequency sound wave when struck with a sharp blow. Figure 17.15 shows sound waves coming from two tuning forks placed side by side. The tuning forks in the drawing are identical, and each is designed to produce a 440-Hz tone. However, a small piece of putty has been attached to one fork, whose frequency is lowered to 438 Hz because of the added mass. When the forks are sounded simultaneously, the loudness of the resulting sound rises and falls periodically—

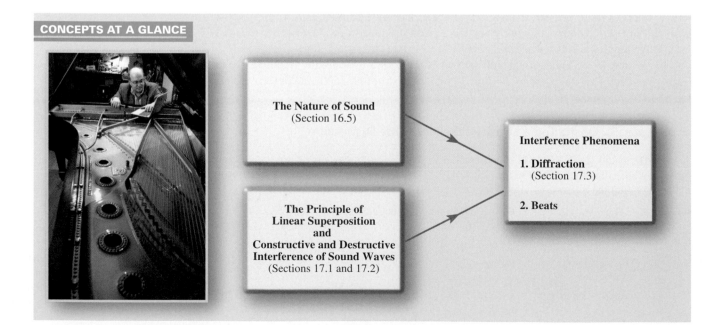

CONCEPTS AT A GLANCE

The Nature of Sound
(Section 16.5)

The Principle of
Linear Superposition
and
Constructive and Destructive
Interference of Sound Waves
(Sections 17.1 and 17.2)

Interference Phenomena

1. Diffraction
(Section 17.3)

2. Beats

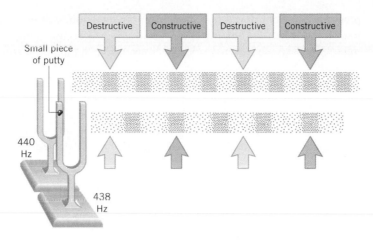

Figure 17.15 Two tuning forks have slightly different frequencies of 440 and 438 Hz. The phenomenon of beats occurs when the forks are sounded simultaneously. The sound waves are not drawn to scale.

faint, then loud, then faint, then loud, and so on. The periodic variations in loudness are called *beats* and result from the interference between two sound waves with slightly different frequencies.

For clarity, Figure 17.15 shows the condensations and rarefactions of the sound waves separately. In reality, however, the waves spread out and overlap. In accord with the principle of linear superposition, the ear detects the combined total of the two. Notice that there are places where the waves interfere constructively and places where they interfere destructively. When a region of constructive interference reaches the ear, a loud sound is heard. When a region of destructive interference arrives, the sound intensity drops to zero (assuming each of the waves has the same amplitude). The number of times per second that the loudness rises and falls is the ***beat frequency*** and is the ***difference*** between the two sound frequencies. Thus, in the situation illustrated in Figure 17.15, an observer hears the sound loudness rise and fall at the rate of 2 times per second (440 Hz − 438 Hz).

Figure 17.16 helps to explain why the beat frequency is the difference between the two frequencies. The drawing displays graphical representations of the pressure patterns of a 10-Hz wave and a 12-Hz wave, along with the pressure pattern that results when the two overlap. These frequencies have been chosen for convenience, even though they lie below the audio range and are inaudible. Audible sound waves behave in exactly the same way. The top two drawings, in blue, show the pressure variations in a one-second interval of each wave. The third drawing, in red, shows the result of adding together the blue patterns according to the principle of linear superposition. Notice that the amplitude in the red drawing is not constant, as it is in the individual waves. Instead, the amplitude changes from a minimum to a maximum, back to a minimum, and so on. When such

Concept Simulation 17.2

In this simulation you can see and hear how two waves of different frequencies interfere to produce beats. The user can vary the frequency of each wave and hear the intensity of the resulting sound wave rising and falling at the beat frequency. The simulation also displays graphs of the individual sound waves, as well as that of the resulting sound wave. In the graph of the resulting sound wave, the change in the amplitude with time is clearly evident, and it is this change that is associated with the beat frequency.

Related Homework: *Problem 18*

Go to
www.wiley.com/college/cutnell

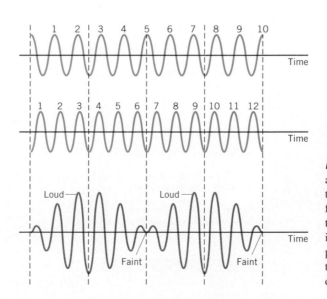

Figure 17.16 A 10-Hz sound wave and a 12-Hz sound wave, when added together, produce a wave with a beat frequency of 2 Hz. The drawings show the pressure patterns (in blue) of the individual waves and the pressure pattern (in red) that results when the two overlap. The time interval shown is one second.

pressure variations reach the ear and occur in the audible frequency range, they produce a loud sound when the amplitude is a maximum and a faint sound when the amplitude is a minimum. Two loud–faint cycles, or beats, occur in the one-second interval shown in the drawing, corresponding to a beat frequency of 2 Hz. Thus, the beat frequency is the difference between the frequencies of the individual waves, or 12 Hz − 10 Hz = 2 Hz.

The physics of **tuning a musical instrument**.

Musicians often tune their instruments by listening to a beat frequency. For instance, a guitar player plucks an out-of-tune string along with a tone from a source known to have the correct frequency. The guitarist adjusts the tension in the string until the beats vanish, ensuring that the string is vibrating at the correct frequency.

✔ Check Your Understanding 2

A tuning fork of unknown frequency and a tuning fork of frequency of 384 Hz produce 6 beats in 2 seconds. When a small piece of putty is attached to the tuning fork of unknown frequency, as in Figure 17.15, the beat frequency decreases. What is the frequency of that tuning fork? *(The answer is given at the end of the book.)*

Background: This question deals with the concept of beat frequency, its relationship to the frequencies of the sounds producing the beats, and how it changes when one of them changes.

For similar questions (including conceptual counterparts), consult Self-Assessment Test 17.1, which is described next.

🖱 Self-Assessment Test 17.1

Test your understanding of the material in Sections 17.1–17.4:

- The Principle of Linear Superposition   • Interference of Sound Waves
- Diffraction   • Beats

Go to **www.wiley.com/college/cutnell**

Figure 17.17 CONCEPTS AT A GLANCE This chart, which is a continuation of those in Figures 17.9 and 17.14, indicates that transverse and longitudinal standing waves can be explained by the principle of linear superposition. Violins produce sound by utilizing transverse standing waves that form when a string is bowed. (© EyeWire)

17.5 *Transverse Standing Waves*

▶ CONCEPTS AT A GLANCE A standing wave is another interference effect that can occur when two waves overlap. Standing waves can arise with transverse waves, such as those on a guitar string, and also with longitudinal sound waves, such as those in a flute. In any case, the principle of linear superposition provides an explanation of the effect, just as it does for diffraction and beats. The Concepts-at-a-Glance chart in Figure 17.17

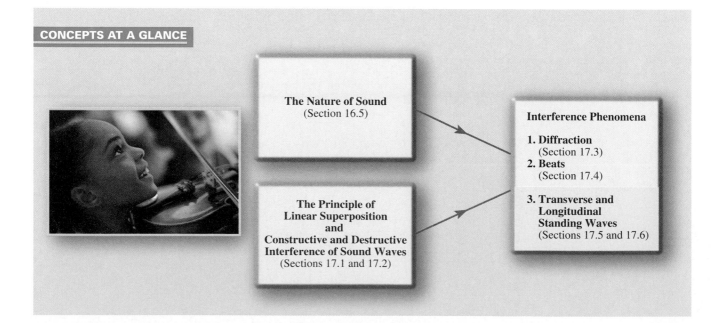

CONCEPTS AT A GLANCE

The Nature of Sound
(Section 16.5)

The Principle of Linear Superposition and Constructive and Destructive Interference of Sound Waves
(Sections 17.1 and 17.2)

Interference Phenomena

1. **Diffraction**
   (Section 17.3)
2. **Beats**
   (Section 17.4)

3. **Transverse and Longitudinal Standing Waves**
   (Sections 17.5 and 17.6)

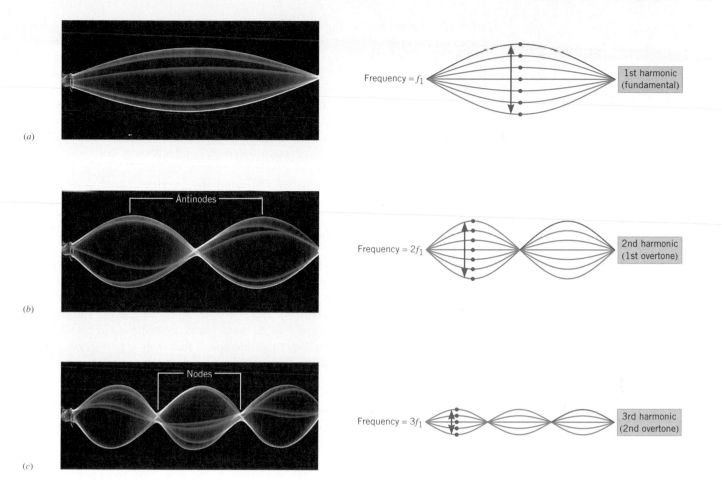

(a)

(b)

(c)

Frequency = f_1 — 1st harmonic (fundamental)

Frequency = $2f_1$ — 2nd harmonic (1st overtone)

Frequency = $3f_1$ — 3rd harmonic (2nd overtone)

Figure 17.18 Vibrating a string at certain unique frequencies sets up transverse standing wave patterns, such as the three shown in the photographs on the left. Each drawing on the right shows the various shapes that the string assumes at various times as it vibrates. The red dots attached to the strings focus attention on the maximum vibration that occurs at an antinode. In each of the drawings, one-half of a wave cycle is outlined in red. (© Richard Megna/Fundamental Photographs)

summarizes the role that this simple but powerful principle has played in this chapter. It is an expansion of the charts in Figures 17.9 and 17.14. ◄

Figure 17.18 shows some of the essential features of transverse standing waves. In this figure the left end of each string is vibrated back and forth, while the right end is attached to a wall. Regions of the string move so fast that they appear only as a blur in the photographs. Each of the patterns shown is called a ***transverse standing wave pattern.*** Notice that the patterns include special places called nodes and antinodes. The ***nodes*** are places that do not vibrate at all, and the ***antinodes*** are places where maximum vibration occurs. To the right of each photograph is a series of superimposed drawings that help us to visualize the motion of the string as it vibrates in a standing wave pattern. These drawings freeze the shape of the string at various times and emphasize the maximum vibration that occurs at an antinode with the aid of a red dot attached to the string.

Each standing wave pattern is produced at a unique frequency of vibration. These frequencies form a series, the smallest frequency f_1 corresponding to the one-loop pattern and the larger frequencies being integer multiples of f_1, as Figure 17.18 indicates. Thus, if f_1 is 10 Hz, the frequency needed to establish the 2-loop pattern is $2f_1$ or 20 Hz, while that needed to create the 3-loop pattern is $3f_1$ or 30 Hz, and so on. The frequencies in this series (f_1, $2f_1$, $3f_1$, etc.) are called ***harmonics.*** The lowest frequency f_1 is called the first harmonic, and the higher frequencies are designated as: the second harmonic ($2f_1$), the third harmonic ($3f_1$), and so forth. The harmonic number (1st, 2nd, 3rd, etc.) corresponds to the number of loops in the standing wave pattern. The frequencies in this series are also referred to as the fundamental frequency, the first overtone, the second overtone, and so on. Thus, frequencies above the fundamental are ***overtones*** (see Figure 17.18).

Standing waves arise because identical waves travel on the string in *opposite directions* and combine in accord with the principle of linear superposition. A standing wave is said to be *standing* because it does not travel in one direction or the other, as do the indi-

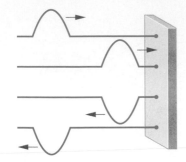

Figure 17.19 In reflecting from the wall, a forward-traveling half-cycle becomes a backward-traveling half-cycle that is inverted.

vidual waves that produce it. Figure 17.19 shows why there are waves traveling in both directions on the string. At the top of the picture, one-half of a wave cycle (the remainder of the wave is omitted for clarity) is moving toward the wall on the right. When the half-cycle reaches the wall, it causes the string to pull upward on the wall. Consistent with Newton's action–reaction law, the wall pulls downward on the string, and a downward-pointing half-cycle is sent back toward the left. Thus, the wave reflects from the wall. Upon arriving back at the point of origin, the wave reflects again, this time from the hand vibrating the string. For small vibration amplitudes, the hand is essentially fixed and behaves as the wall does in causing reflections. Repeated reflections at both ends of the string create a multitude of wave cycles traveling in both directions.

As each new cycle is formed by the vibrating hand, previous cycles that have reflected from the wall arrive and reflect again from the hand. Unless the timing is right, however, the new cycles and the reflected cycles tend to offset one another, and the formation of a standing wave is inhibited. Think about pushing someone on a swing and timing your pushes so that the effect of one push reinforces that of another. Such reinforcement in the case of the wave cycles leads to a large-amplitude standing wave. Suppose the string has a length L and its left end is being vibrated at a frequency f_1. The time required to create a new wave cycle is the period T of the wave, where $T = 1/f_1$ (Equation 10.5). On the other hand, the time needed for a cycle to travel from the hand to the wall and back, a distance of $2L$, is $2L/v$, where v is the wave speed. Reinforcement between new and reflected cycles occurs if these two times are equal; that is, if $1/f_1 = 2L/v$. Thus, a standing wave is established when the string is vibrated with a frequency of $f_1 = v/(2L)$.

Repeated reinforcement between newly created and reflected cycles causes a large-amplitude standing wave to develop on the string, *even when the hand itself vibrates with only a small amplitude*. Thus, the motion of the string is a resonance effect, analogous to that discussed in Section 10.6 for an object attached to a spring. The frequency f_1 at which resonance occurs is sometimes called a ***natural frequency*** of the string, similar to the frequency at which an object oscillates on a spring.

There is a difference between the resonance of the string and the resonance of a spring system, however. An object on a spring has only a single natural frequency, whereas the string has a *series* of natural frequencies. The series arises because a reflected wave cycle need not return to its point of origin in time to reinforce *every* newly created cycle. Reinforcement can occur, for instance, on *every other* new cycle, as it does if the string is vibrated at twice the frequency f_1, or $f_2 = 2f_1$. Likewise, if the vibration frequency is $f_3 = 3f_1$, reinforcement occurs on *every third* new cycle. Similar arguments apply for any frequency $f_n = nf_1$, where n is an integer. As a result, the series of natural frequencies that lead to standing waves on a string fixed at both ends is given by

String fixed at both ends
$$f_n = n\left(\frac{v}{2L}\right) \qquad n = 1, 2, 3, 4, \ldots \qquad (17.3)$$

It is also possible to obtain Equation 17.3 in another way. In Figure 17.18, one-half of a wave cycle is outlined in red for each of the harmonics, to show that each loop in a standing wave pattern corresponds to one-half a wavelength. Since the two fixed ends of the string are nodes, the length L of the string must contain an integer number n of half-wavelengths: $L = n(\frac{1}{2}\lambda_n)$ or $\lambda_n = 2L/n$. Using this result for the wavelength in the relation $f_n\lambda_n = v$ shows that $f_n(2L/n) = v$, which can be rearranged to give Equation 17.3.

Problem solving insight
The distance between two successive nodes (or between two successive anti-nodes) of a standing wave is equal to one-half of a wavelength.

Concept Simulation 17.3

In this simulation you can explore how standing waves are set up on a string. With the string fixed at both ends, the fundamental frequency of this standing wave is $f_1 = 2$ Hz. The user can increase the frequency of vibration up to 15 Hz. At the fundamental frequency or any multiple of it (2, 4, 6 Hz, . . .) a standing wave is set up on the string. At other, in-between, frequencies, the string vibrates erratically, and a standing wave is not established.

Go to **www.wiley.com/college/cutnell**

Standing waves on a string play an important role in the way many musical instruments produce sound. For instance, a guitar string is stretched between two supports and, when plucked, vibrates according to the series of natural frequencies given by Equation 17.3. The next two examples deal with how this series of frequencies governs the playing and the design of a guitar.

Example 4 Playing a Guitar

The heaviest string on an electric guitar has a linear density of $m/L = 5.28 \times 10^{-3}$ kg/m and is stretched with a tension of $F = 226$ N. This string produces the musical note E when vibrating along its entire length in a standing wave at the fundamental frequency of 164.8 Hz. (a) Find the length L of the string between its two fixed ends (see Figure 17.20a). (b) A guitar player wants the string to vibrate at a fundamental frequency of 2×164.8 Hz $= 329.6$ Hz, as it must if the musical note E is to be sounded one octave higher in pitch. To accomplish this, he presses the string against the proper fret and then plucks the string (see part b of the drawing). Find the distance L between the fret and the bridge of the guitar.

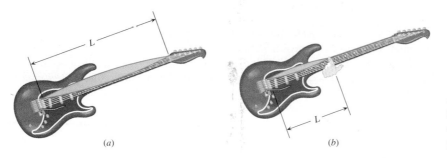

(a) (b)

Figure 17.20 These drawings show the standing waves (in blue) that exist on a guitar string under different playing conditions.

Reasoning The fundamental frequency f_1 is given by Equation 17.3 with $n = 1$: $f_1 = v/(2L)$. Since f_1 is known in both parts (a) and (b), the length L in each case can be calculated directly from this expression, once the speed v is known. The speed, in turn, is related to the tension F and the linear density m/L according to Equation 16.2.

Solution

(a) The speed is

$$v = \sqrt{\frac{F}{m/L}} = \sqrt{\frac{226 \text{ N}}{5.28 \times 10^{-3} \text{ kg/m}}} = 207 \text{ m/s} \qquad (16.2)$$

According to $f_1 = v/(2L)$, the length of the string is

$$L = \frac{v}{2f_1} = \frac{207 \text{ m/s}}{2(164.8 \text{ Hz})} = \boxed{0.628 \text{ m}}$$

(b) The distance L that locates the fret can be determined exactly as in part (a) by using the wave speed $v = 207$ m/s and noting that the frequency is now $f_1 = 329.6$ Hz: $\boxed{L = 0.314 \text{ m}}$. This length is exactly half that determined in part (a) because the frequencies have a ratio of $2:1$.

Need more practice?

Interactive LearningWare 17.2
Electric guitars often come equipped with a "whammy bar," which is a lever that allows the performer to adjust the tension in the strings. In this way, the performer can "dive bomb." To dive bomb means to produce a tone that begins with a high frequency and ends with a low frequency and to go through all the frequencies in between. Suppose that a performer dive bombs the frequency in half. By what factor has the whammy bar reduced the tension in the strings?

Go to
www.wiley.com/college/cutnell for an interactive solution.

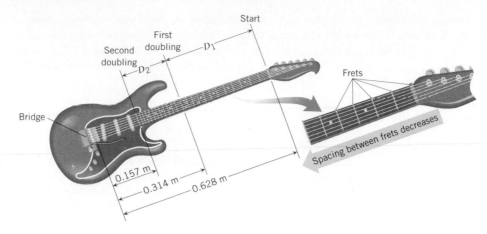

Figure 17.21 The spacing between the frets on the neck of a guitar decreases going down the neck toward the bridge.

Conceptual Example 5   The Frets on a Guitar

The physics of the frets on a guitar.

Figure 17.21 shows the frets on the neck of a guitar. They allow the player to produce a complete sequence of musical notes using a single string. Starting with the fret at the top of the neck, each successive fret indicates where the player should press to get the next note in the sequence. Musicians call the sequence the chromatic scale, and every thirteenth note in it corresponds to one octave, or a doubling of the sound frequency. The spacing between the frets is greatest at the top of the neck and decreases with each additional fret further on down. The spacing eventually becomes smaller than the width of a finger, limiting the number of frets that can be used. Why does the spacing between the frets decrease going down the neck?

Reasoning and Solution Our reasoning is based on Equation 17.3, with $n = 1$ [$f_1 = v/(2L)$]. The value of n is 1 because a string vibrates mainly at its fundamental frequency when plucked, as mentioned in Example 4. This equation indicates that the fundamental frequency f_1 is inversely proportional to the length L between a given fret and the bridge of the guitar. Thus, in Example 4 we found that the E-string had a length of 0.628 m, corresponding to a frequency of $f_1 = 164.8$ Hz. We also found that the length between the bridge and the fret that must be pressed to double this frequency to 329.6 Hz is one-half of 0.628 m, or 0.314 m. To understand why the spacing between frets decreases going down the neck, consider the fret that must be pressed to double the frequency again, from 329.6 Hz to 659.2 Hz. The length between the bridge and this fret would be one-half of 0.314 m, or 0.157 m. The three lengths of 0.628 m, 0.314 m, and 0.157 m are shown in Figure 17.21. The distances D_1 and D_2 are also shown and give the spacings between the starting fret at the top of the neck, the fret for the first doubling, and the fret for the second doubling. Clearly, D_1 is greater than D_2, and the frets near the top of the neck have more space between them than those further down.

Related Homework: *Problem 32*

✔ **Check Your Understanding 3**

A standing wave that corresponds to the fourth harmonic is set up on a string that is fixed at both ends. (a) How many loops are in this standing wave? (b) How many nodes (excluding the nodes at the ends of the string) does this standing wave have? (c) Is there a node or an antinode at the midpoint of the string? (d) If the frequency of this standing wave is 440 Hz, what is the frequency of the lowest-frequency standing wave that could be set up on this string? *(The answers are given at the end of the book.)*

Background: A transverse standing wave is the featured player here. A vibrating string can have a series of natural frequencies, and the fourth harmonic is one of them.

For similar questions (including calculational counterparts), consult Self-Assessment Test 17.2, which is described at the end of Section 17.6.

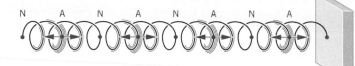

17.6 Longitudinal Standing Waves

Standing wave patterns can also be formed from longitudinal waves. For example, when sound reflects from a wall, the forward- and backward-going waves can produce a standing wave. Figure 17.22 illustrates the vibrational motion in a longitudinal standing wave on a Slinky. As in a transverse standing wave, there are nodes and antinodes. At the nodes the coils of the Slinky do not vibrate at all; that is, they have no displacement. At the antinodes the coils vibrate with maximum amplitude. The red dots in Figure 17.22 indicate the lack of vibration at a node and the maximum vibration at an antinode. At an antinode the coils have a maximum displacement. The vibration occurs along the line of travel of the individual waves, as is to be expected for longitudinal waves. In a standing wave of sound, the molecules or atoms of the medium behave as the red dots do.

Musical instruments in the wind family depend on longitudinal standing waves in producing sound. Since wind instruments (trumpet, flute, clarinet, pipe organ, etc.) are modified tubes or columns of air, it is useful to examine the standing waves that can be set up in such tubes. Figure 17.23 shows two cylindrical columns of air that are open at both ends. Sound waves, originating from a tuning fork, travel up and down within each tube, since they reflect from the ends of the tubes, even though the ends are open. If the frequency f of the tuning fork matches one of the natural frequencies of the air column, the downward- and upward-traveling waves combine to form a standing wave, and the sound of the tuning fork becomes markedly louder. To emphasize the longitudinal nature of the standing wave patterns, the left side of each pair of drawings in Figure 17.23 replaces the air in the tubes with Slinkies, on which the nodes and antinodes are indicated with red dots. As an additional aid in visualizing the standing waves, the right side of each pair of drawings shows blurred blue patterns within each tube. These patterns symbolize the amplitude of the vibrating air molecules at various locations. Wherever the pattern is widest, the amplitude of vibration is greatest (a displacement antinode), and wherever the pattern is narrowest there is no vibration (a displacement node).

To determine the natural frequencies of the air columns in Figure 17.23, notice that there is a displacement antinode at each end of the open tube because the air molecules there are free to move.* As in a transverse standing wave, the distance between two successive antinodes is one-half of a wavelength, so the length L of the tube must be an integer number n of half-wavelengths: $L = n(\frac{1}{2}\lambda_n)$ or $\lambda_n = 2L/n$. Using this wavelength in the relation $f_n = v/\lambda_n$ shows that the natural frequencies f_n of the tube are

Tube open at both ends
$$f_n = n\left(\frac{v}{2L}\right) \qquad n = 1, 2, 3, 4, \ldots \qquad (17.4)$$

At these frequencies, large-amplitude standing waves develop within the tube due to resonance. Example 6 illustrates how Equation 17.4 is involved when a flute is played.

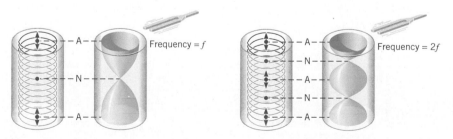

Figure 17.23 A pictorial representation of longitudinal standing waves on a Slinky (left side of each pair) and in a tube of air (right side of each pair) that is open at both ends (A, antinode; N, node).

* In reality, the antinode does not occur exactly at the open end. However, if the tube's diameter is small compared to its length, little error is made in assuming that the antinode is located right at the end.

*The physics of
a flute.*

Figure 17.24 The length L of a flute between the mouthpiece and the end of the instrument determines the fundamental frequency of the lowest playable note.

Example 6 Playing a Flute

When all the holes are closed on one type of flute, the lowest note it can sound is a middle C, whose fundamental frequency is 261.6 Hz. (a) The air temperature is 293 K, and the speed of sound is 343 m/s. Assuming the flute is a cylindrical tube open at both ends, determine the distance L in Figure 17.24—that is, the distance from the mouthpiece to the end of the tube. (This distance is only approximate, since the antinode does not occur exactly at the mouthpiece.) (b) A flautist can alter the length of the flute by adjusting the extent to which the head joint is inserted into the main stem of the instrument. If the air temperature rises to 305 K, to what length must the flute be adjusted to play a middle C?

Reasoning The fundamental frequency f_1 is given by Equation 17.4 with $n = 1$: $f_1 = v/(2L)$. This expression can be used to calculate the length as $L = v/(2 f_1)$. When the speed of sound v changes, as it does when the temperature changes, the length of the flute must be changed. The effect of temperature on the speed of sound in air is given by $v = \sqrt{\gamma kT/m}$ (Equation 16.5), assuming air behaves as an ideal gas. Thus, the speed is proportional to the square root of the Kelvin temperature ($v \propto \sqrt{T}$), a fact that we can use to find the speed at the higher temperature.

Solution

(a) At a temperature of 293 K, when the speed of sound is $v = 343$ m/s, the length of the flute is

$$L = \frac{v}{2f_1} = \frac{343 \text{ m/s}}{2(261.6 \text{ Hz})} = \boxed{0.656 \text{ m}}$$

(b) Since $v \propto \sqrt{T}$, it follows that

$$\frac{v_{305 \text{ K}}}{v_{293 \text{ K}}} = \frac{\sqrt{305 \text{ K}}}{\sqrt{293 \text{ K}}} = 1.02$$

As a result, $v_{305 \text{ K}} = 1.02(v_{293 \text{ K}}) = 1.02(343 \text{ m/s}) = 3.50 \times 10^2$ m/s. The adjusted flute length is

$$L = \frac{v}{2f_1} = \frac{3.50 \times 10^2 \text{ m/s}}{2(261.6 \text{ Hz})} = \boxed{0.669 \text{ m}}$$

Thus, to play in tune at the higher temperature, a flautist must lengthen the flute by 0.013 m.

Standing waves can also exist in a tube with only one end open, as the patterns in Figure 17.25 indicate. Note the difference between these patterns and those in Figure 17.23. Here the standing waves have a displacement antinode at the open end and a displacement node at the closed end, where the air molecules are not free to move. Since the distance between a node and an adjacent antinode is one-fourth of a wavelength, the length L of the tube must be an odd number of quarter-wavelengths: $L = 1(\frac{1}{4}\lambda)$ and $L = 3(\frac{1}{4}\lambda)$ for the two standing wave patterns in Figure 17.25. In general, then, $L = n(\frac{1}{4}\lambda_n)$, where n is any odd integer ($n = 1, 3, 5, \ldots$). From this result it follows that $\lambda_n = 4L/n$, and the natural frequencies f_n can be obtained from the relation $f_n = v/\lambda_n$:

Tube open at only one end $f_n = n\left(\dfrac{v}{4L}\right)$ $n = 1, 3, 5, \ldots$ (17.5)

A tube open at only one end can develop standing waves only at the odd harmonic frequencies f_1, f_3, f_5, etc. In contrast, a tube open at both ends can develop standing waves at all harmonic frequencies f_1, f_2, f_3, etc. Moreover, the fundamental frequency f_1 of a tube

Figure 17.25 A pictorial representation of the longitudinal standing waves on a Slinky (left side of each pair) and in a tube of air (right side of each pair) that is open only at one end (A, antinode; N, node).

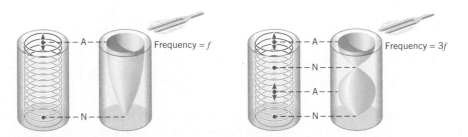

open at only one end (Equation 17.5) is one-half that of a tube open at both ends (Equation 17.4). In other words, a tube open only at one end needs to be only one-half as long as a tube open at both ends in order to produce the *same* fundamental frequency.

Energy is also conserved when a standing wave is produced, either on a string or in a tube of air. The energy of the standing wave is the sum of the energies of the individual waves that comprise the standing wave. Once again, interference redistributes the energy of the individual waves to create locations of greatest energy (displacement antinodes) and locations of no energy (displacement nodes).

✔ **Check Your Understanding 4**

A cylindrical bottle, partially filled with water, is open at the top. When you blow across the top of the bottle a standing wave is set up inside it. Is there a node or an antinode (a) at the top of the bottle and (b) at the surface of the water? (c) If the standing wave is vibrating at its fundamental frequency, what is the distance between the top of the bottle and the surface of the water? Express your answer in terms of the wavelength λ of the standing wave. (d) If you take a sip, is the fundamental frequency of the standing wave raised, lowered, or does it remain the same? *(The answers are given at the end of the book.)*

Background: The properties of a longitudinal standing wave are featured here. They depend on the length of the "tube" and whether the tube is open at both ends or open at only one end.

For similar questions (including calculational counterparts), consult Self-Assessment Test 17.2, which is described next.

🖱 **Self-Assessment Test 17.2**

Test your understanding of the material in Sections 17.5 and 17.6:

• Transverse Standing Waves • Longitudinal Standing Waves

Go to www.wiley.com/college/cutnell

---

*17.7 Complex Sound Waves

Musical instruments produce sound in a way that depends on standing waves. Examples 4 and 5 illustrate the role of transverse standing waves on the string of an electric guitar, while Example 6 stresses the role of longitudinal standing waves in the air column within a flute. In each example sound is produced at the fundamental frequency of the instrument.

In general, however, a musical instrument does not produce just the fundamental frequency when it plays a note, but simultaneously generates a number of harmonics as well. Different instruments, such as a violin and a trumpet, generate harmonics to different extents, and the harmonics give the instruments their characteristic sound qualities or timbres. Suppose, for instance, that a violinist and a trumpet player both sound concert A, a note whose fundamental frequency is 440 Hz. Even though both instruments are playing the same note, most people can distinguish the sound of the violin from that of the trumpet. The instruments sound different because the relative amplitudes of the harmonics (880 Hz, 1320 Hz, etc.) that the instruments create are different.

The sound wave corresponding to a note produced by a musical instrument or a singer is called a ***complex sound wave*** because it consists of a mixture of the fundamental and harmonic frequencies. The pattern of pressure fluctuations in a complex wave can be obtained by using the principle of linear superposition, as Figure 17.26 indicates. This drawing shows a bar graph in which the heights of the bars give the relative amplitudes of the harmonics contained in a note such as a singer might produce. When the individual pressure patterns for each of the three harmonics are added together, they yield the complex pressure pattern shown at the top of the picture.*

---

* In carrying out the addition, we assume that each individual pattern begins at zero at the origin when the time equals zero.

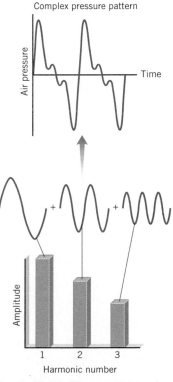

Figure 17.26 The topmost graph shows the pattern of pressure fluctuations such as a singer might produce. The pattern is the sum of the first three harmonics. The relative amplitudes of the harmonics correspond to the heights of the vertical bars in the bar graph.

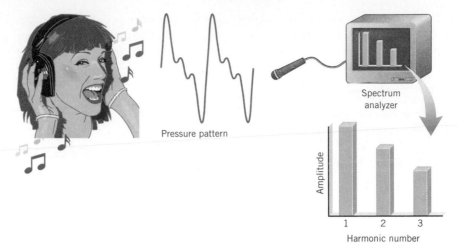

Figure 17.27 A microphone detects a complex sound wave produced by a singer's voice, and a spectrum analyzer determines the amplitude and frequency of each harmonic present in the wave.

The physics of
a spectrum analyzer.

In practice, a bar graph such as that in Figure 17.26 is determined with the aid of an electronic instrument known as a spectrum analyzer. When the note is produced, the complex sound wave is detected by a microphone that converts the wave into an electrical signal. The electrical signal, in turn, is fed into the spectrum analyzer, as Figure 17.27 illustrates. The spectrum analyzer then determines the amplitude and frequency of each harmonic present in the complex wave and displays the results on its screen.

17.8 *Concepts & Calculations*

Diffraction is the bending of a traveling wave around an obstacle or around the edges of an opening and is one of the consequences of the principle of linear superposition. As Equations 17.1 and 17.2 indicate, the extent of diffraction when a sound wave passes through an opening depends on the ratio of the wavelength of the sound to the width or diameter of the opening. Example 7 compares diffraction in two different media and reviews some of the fundamental properties of sound waves.

Concepts & Calculations Example 7
Diffraction in Two Different Media

A sound wave with a frequency of 15 kHz emerges through a circular opening that has a diameter of 0.20 m. Find the diffraction angle θ when the sound travels in air at a speed of 343 m/s and in water at a speed of 1482 m/s.

Concept Questions and Answers The diffraction angle for a circular opening is given by Equation 17.2 as $\sin \theta = 1.22\lambda/D$, where λ is the wavelength of the sound and D is the diameter of the opening. How is the wavelength related to the frequency of the sound?

Answer According to Equation 16.1, the wavelength is given by $\lambda = v/f$, where v is the speed of sound and f is the frequency.

Is the wavelength of the sound in air greater than, smaller than, or equal to the wavelength in water?

Answer According to Equation 16.1, the wavelength is proportional to the speed v for a given value of the frequency f. Since sound travels at a slower speed in air than in water, the wavelength in air is smaller than the wavelength in water.

Is the diffraction angle in air greater than, smaller than, or equal to the diffraction angle in water?

Answer The extent of diffraction is determined by λ/D, the ratio of the wavelength to the diameter of the opening. Smaller ratios lead to less diffraction or smaller diffraction angles. The wavelength in air is smaller than in water, and the diameter of the opening is the same in both cases. Therefore, the ratio λ/D is smaller in air than in water, and the diffraction angle in air is smaller than the diffraction angle in water.

Solution Using $\sin \theta = 1.22\lambda/D$ (Equation 17.2) and $\lambda = v/f$ (Equation 16.1), we have

$$\sin \theta = 1.22 \frac{\lambda}{D} = 1.22 \frac{v}{fD}$$

Applying this result for air and for water, we find

Air $\quad \theta = \sin^{-1}\left(1.22 \dfrac{v}{fD}\right) = \sin^{-1}\left[1.22 \dfrac{(343 \text{ m/s})}{(15\ 000 \text{ Hz})(0.20 \text{ m})}\right] = \boxed{8.0°}$

Water $\quad \theta = \sin^{-1}\left(1.22 \dfrac{v}{fD}\right) = \sin^{-1}\left[1.22 \dfrac{(1482 \text{ m/s})}{(15\ 000 \text{ Hz})(0.20 \text{ m})}\right] = \boxed{37°}$

As expected, the diffraction angle in air is smaller.

▲

The next example deals with standing waves of sound in a gas. One of the factors that affect the formation of standing waves is the speed at which the individual waves travel. This example reviews how the speed of sound depends on the properties of the gas.

Concepts & Calculations Example 8 Standing Waves of Sound

▼

Two tubes of gas are identical and are open at only one end. One tube contains neon (Ne) and the other krypton (Kr). Both are monatomic gases, have the same temperature, and may be assumed to be ideal gases. The fundamental frequency of the tube containing neon is 481Hz. What is the fundamental frequency of the tube containing krypton?

Concept Questions and Answers For a gas-filled tube open at only one end, the fundamental frequency ($n = 1$) is given by Equation 17.5 as $f_1 = v/(4L)$, where v is the speed of sound and L is the length of the tube. How is the speed related to the properties of the gas?

Answer According to Equation 16.5, the speed is given by $v = \sqrt{\gamma kT/m}$, where γ is the ratio of the specific heat capacities at constant pressure and constant volume, k is Boltzmann's constant, T is the Kelvin temperature, and m is the mass of an atom of the gas.

All of the factors that affect the speed of sound are the same except the atomic mass. The periodic table located on the inside of the back cover gives the atomic masses of neon and krypton as 20.179 u and 83.80 u, respectively. Is the speed of sound in krypton greater than, smaller than, or equal to the speed of sound in neon?

Answer According to Equation 16.5 the speed of sound is $v = \sqrt{\gamma kT/m}$, so the speed is inversely proportional to the square root of the mass m of an atom. Thus, the speed is smaller when the mass is greater. Since krypton has the greater mass, the speed of sound in krypton is smaller than in neon.

Is the fundamental frequency of the tube containing krypton greater than, smaller than, or equal to that of the tube containing neon?

Answer The fundamental frequency is given by Equation 17.5 as $f_1 = v/(4L)$. Since the speed of sound in krypton is smaller than in neon, the fundamental frequency of the krypton-filled tube is smaller than that of the neon-filled tube.

Solution Using $f_1 = v/(4L)$ (Equation 17.5) and $v = \sqrt{\gamma kT/m}$ (Equation 16.5), we have

$$f_1 = \frac{v}{4L} = \frac{1}{4L}\sqrt{\frac{\gamma kT}{m}}$$

Applying this result to both tubes and taking the ratio of the frequencies, we obtain

$$\frac{f_{1,\text{ Kr}}}{f_{1,\text{ Ne}}} = \frac{\dfrac{1}{4L}\sqrt{\gamma kT/m_\text{Kr}}}{\dfrac{1}{4L}\sqrt{\gamma kT/m_\text{Ne}}} = \sqrt{\frac{m_\text{Ne}}{m_\text{Kr}}}$$

Solving for $f_{1, \text{Kr}}$ gives

$$f_{1, \text{Kr}} = f_{1, \text{Ne}} \sqrt{\frac{m_{\text{Ne}}}{m_{\text{Kr}}}} \doteq (481 \text{ Hz}) \sqrt{\frac{20.179 \text{ u}}{83.80 \text{ u}}} = \boxed{236 \text{ Hz}}$$

As expected, the fundamental frequency for the krypton-filled tube is smaller than that for the neon-filled tube.

▲

At the end of the problem set for this chapter, you will find homework problems that contain both conceptual and quantitative parts. These problems are grouped under the heading *Concepts & Calculations, Group Learning Problems.* They are designed for use by students working alone or in small learning groups. The conceptual part of each problem provides a convenient focus for group discussions.

Concept Summary

This summary presents an abridged version of the chapter, including the important equations and all available learning aids. For convenient reference, the learning aids (including the text's examples) are placed next to or immediately after the relevant equation or discussion. The following learning aids may be found on-line at **www.wiley.com/college/cutnell**:

Interactive LearningWare examples are solved according to a five-step interactive format that is designed to help you develop problem-solving skills.	Concept Simulations are animated versions of text figures or animations that illustrate important concepts. You can control parameters that affect the display, and we encourage you to experiment.
Interactive Solutions offer specific models for certain types of problems in the chapter homework. The calculations are carried out interactively.	Self-Assessment Tests include both qualitative and quantitative questions. Extensive feedback is provided for both incorrect and correct answers, to help you evaluate your understanding of the material.

Topic	*Discussion*	*Learning Aids*
	17.1 The Principle of Linear Superposition	
Principle of linear superposition	The principle of linear superposition states that when two or more waves are present simultaneously at the same place, the resultant disturbance is the sum of the disturbances from the individual waves.	Concept Simulation 17.1
	17.2 Constructive and Destructive Interference of Sound Waves	
Constructive and destructive interference	Constructive interference occurs at a point when two waves meet there crest-to-crest and trough-to-trough, thus reinforcing each other. Destructive interference occurs when the waves meet crest-to-trough and cancel each other.	
In-phase and out-of-phase waves	When waves meet crest-to-crest and trough-to-trough, they are exactly in phase. When they meet crest-to-trough, they are exactly out of phase.	
Conditions for constructive and destructive interference	For two wave sources vibrating in phase, a difference in path lengths that is zero or an integer number (1, 2, 3, . . .) of wavelengths leads to constructive interference; a difference in path lengths that is a half-integer number ($\frac{1}{2}$, $1\frac{1}{2}$, $2\frac{1}{2}$, . . .) of wavelengths leads to destructive interference.	**Examples 1, 2** **Interactive Solution 17.53**
	17.3 Diffraction	
Diffraction	Diffraction is the bending of a wave around an obstacle or the edges of an opening. The angle through which the wave bends depends on the ratio of the wavelength λ of the wave to the width D of the opening; the greater the ratio λ/D, the greater the angle.	

Topic	Discussion	Learning Aids
	When a sound wave of wavelength λ passes through an opening, the place where the intensity of the sound is a minimum relative to the center of the opening is specified by the angle θ. If the opening is a rectangular slit of width D, such as a doorway, the angle is	
Single slit—first minimum	$$\sin \theta = \frac{\lambda}{D} \qquad (17.1)$$	
	If the opening is a circular opening of diameter D, such as that in a loudspeaker, the angle is	
Circular opening—first minimum	$$\sin \theta = 1.22 \frac{\lambda}{D} \qquad (17.2)$$	Examples 3, 7

17.4 Beats

Topic	Discussion	Learning Aids
Beats	Beats are the periodic variations in amplitude that arise from the linear superposition of two waves that have slightly different frequencies. When the waves are sound waves, the variations in amplitude cause the loudness to vary at the beat frequency, which is the difference between the frequencies of the waves.	Concept Simulation 17.2
Beat frequency		Interactive LearningWare 17.1

Use *Self-Assessment Test 17.1* to evaluate your understanding of Sections 17.1–17.4.

17.5 Transverse Standing Waves

Topic	Discussion	Learning Aids
Standing waves	A standing wave is the pattern of disturbance that results when oppositely traveling waves of the same frequency and amplitude pass through each other. A standing wave has places of minimum and maximum vibration called, respectively, nodes and antinodes.	
Nodes and antinodes		
Natural frequencies Harmonics	Under resonance conditions, standing waves can be established only at certain natural frequencies. The frequencies in this series (f_1, $2f_1$, $3f_1$, etc.) are called harmonics. The lowest frequency f_1 is called the first harmonic, the next frequency $2f_1$ is the second harmonic, and so on.	
	For a string that is fixed at both ends and has a length L, the natural frequencies are	Examples 4, 5
String fixed at both ends	$$f_n = n\left(\frac{v}{2L}\right) \qquad n = 1, 2, 3, 4, \ldots \qquad (17.3)$$	Interactive LearningWare 17.2
	where v is the speed of the wave on the string and n is a positive integer.	

17.6 Longitudinal Standing Waves

Topic	Discussion	Learning Aids
	For a gas in a cylindrical tube open at both ends, the natural frequencies of vibration are	
Tube open at both ends	$$f_n = n\left(\frac{v}{2L}\right) \qquad n = 1, 2, 3, 4, \ldots \qquad (17.4)$$	Example 6
	where v is the speed of sound in the gas and L is the length of the tube.	
	For a gas in a cylindrical tube open at only one end, the natural frequencies of vibration are	
Tube open at only one end	$$f_n = n\left(\frac{v}{4L}\right) \qquad n = 1, 3, 5, 7, \ldots \qquad (17.5)$$	Example 8 Interactive Solution 17.49

Use *Self-Assessment Test 17.2* to evaluate your understanding of Sections 17.5 and 17.6.

17.7 Complex Sound Waves

A complex sound wave consists of a mixture of a fundamental frequency and overtone frequencies.

Conceptual Questions

1. Does the principle of linear superposition imply that two sound waves, passing through the same place at the same time, always create a louder sound than either wave alone? Explain.

2. Suppose you are sitting at the overlap point between the two speakers in Figure 17.4. Because of destructive interference, you hear no sound, even though both speakers are emitting identical sound waves. One of the two speakers is suddenly shut off. Describe what you would hear.

3. Review Figure 17.3. As you walk along a line that is perpendicular to the line between the speakers and passes through the overlap point, you do not observe the loudness to change from loud to faint to loud. However, as you walk along a line through the overlap point and parallel to the line between the speakers, you do observe the loudness to alternate between faint and loud. Explain why your observations are different in the two cases.

4. At an open-air rock concert the music is played through speakers that are on the stage. The speakers point straight forward and not toward the sides. You are wandering about and notice that the sounds of the female vocalists can be heard in front of the stage but not off to the sides. The rhythmic bass, however, can be heard both in front of the stage and off to the sides. How does the phenomenon of diffraction explain your observations?

5. Refer to Example 1 in Section 16.2. Which type of radio wave, AM or FM, diffracts more readily around a given obstacle? Give your reasoning.

6. A tuning fork has a frequency of 440 Hz. The string of a violin and this tuning fork, when sounded together, produce a beat frequency of 1 Hz. From these two pieces of information alone, is it possible to determine the exact frequency of the violin string? Explain.

7. When the regions of constructive and destructive interference in Figure 17.15 move past a listener's ear, a beat frequency of 2 Hz is heard. Suppose that the tuning forks in the drawing were sounded underwater and that the listener is also underwater. The forks vibrate at 438 and 440 Hz, just as they do in air. However, sound travels four times faster in water than in air. Explain why the listener hears a beat frequency of 2 Hz, just as he does in air, even though the regions of constructive and destructive interference are moving past his ear four times faster.

8. The tension in a guitar string is doubled. Does the frequency of oscillation also double? If not, by what factor does the frequency change? Specify whether the change is an increase or a decrease.

9. A string is attached to a wall and vibrates back and forth, as in Figure 17.18. The vibration frequency and length of the string are fixed. The tension in the string is changed, and it is observed that at certain values of the tension a standing wave pattern develops. Account for the fact that no standing waves are observed once the tension is increased beyond a certain value.

10. A string is vibrating back and forth as in Figure 17.18a. The tension in the string is decreased by a factor of four, with the frequency and the length of the string remaining the same. Draw the new standing wave pattern that develops on the string. Give your reasoning.

11. A rope is hanging vertically straight down. The top end is being vibrated back and forth. Standing waves can develop on the rope analogous to those on a horizontal rope. (a) There is a node at the top end. Is there a node or an antinode at the bottom end? Explain. (b) The separation between successive nodes is not the same everywhere on the rope, as it would be if the rope were horizontal. Is the separation greater near the top or near the bottom? Taking into account the mass of the rope, give your reasoning.

12. In Figure 17.23 the tubes are filled with air. Suppose, instead, that the tube at the bottom of the drawing is filled with a gas in which the speed of sound is twice what it is in air. The frequency of the tuning fork remains unchanged. Draw the standing wave pattern that would exist in the tube. Justify your drawing.

13. Standing waves can ruin the acoustics of a concert hall if there is excessive reflection of the sound waves that the performers generate. For example, suppose a performer generates a 2093-Hz tone. If a large-amplitude standing wave is present, it is possible for a listener to move a distance of only 4.1 cm and hear the loudness of the tone change from loud to faint. Account for this observation in terms of standing waves, pointing out why the distance is 4.1 cm.

14. The natural frequencies of a wind instrument change when it is brought inside from the cold outdoors. Why do the frequencies change, and is the change an increase or a decrease? Justify your answers.

15. The tones produced by a typical orchestra are complex sound waves, and most have fundamental frequencies less than 5000 Hz. However, a high-quality stereo system must be able to reproduce frequencies up to 20 000 Hz accurately. Explain why.

Problems

Section 17.1 The Principle of Linear Superposition, Section 17.2 Constructive and Destructive Interference of Sound Waves

1. In **Concept Simulation 17.1** at **www.wiley.com/college/cutnell** you can explore how two pulses, traveling in opposite directions, combine to form a resultant pulse. Two pulses are traveling toward each other, each having a speed of 1 cm/s. At $t = 0$ s, their positions are shown in the drawing. When $t = 1$ s, what is the height of the resultant pulse at (a) $x = 3$ cm and at (b) $x = 4$ cm?

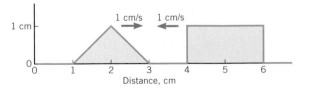

2. Two speakers, one directly behind the other, are each generating a 245-Hz sound wave. What is the smallest separation distance between the speakers that will produce destructive interference at a listener standing in front of them? The speed of sound is 343 m/s.

3. ssm The drawing graphs a string on which two rectangular pulses are traveling at a constant speed of 1 cm/s at time $t = 0$ s. Using the principle of linear superposition, draw the shape of the string's pulses at $t = 1$ s, 2 s, 3 s, and 4 s.

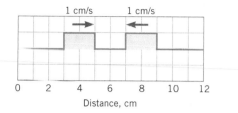

4. In Figure 17.7, suppose that the separation between speakers A and B is 5.00 m and the speakers are vibrating in phase. They are playing identical 125-Hz tones, and the speed of sound is 343 m/s. What is the largest possible distance between speaker B and the observer at C, such that he observes destructive interference?

5. ssm Two loudspeakers are vibrating in phase. They are set up as in Figure 17.7, and point C is located as shown there. The speed of sound is 343 m/s. The speakers play the same tone. What is the smallest frequency that will produce destructive interference at point C?

6. The sound produced by the loudspeaker in the drawing has a frequency of 12 000 Hz and arrives at the microphone via two different paths. The sound travels through the left tube LXM, which has a fixed length. Simultaneously, the sound trav-

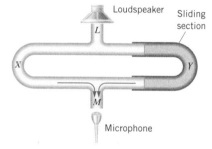

els through the right tube LYM, the length of which can be changed by moving the sliding section. At M, the sound waves coming from the two paths interfere. As the length of the path LYM is changed, the sound loudness detected by the microphone changes. When the sliding section is pulled out by 0.020 m, the loudness changes from a maximum to a minimum. Find the speed at which sound travels through the gas in the tube.

7. ssm www The drawing shows a loudspeaker A and point C, where a listener is positioned. A second loudspeaker B is located somewhere to the right of A. Both speakers vibrate in phase and

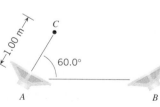

are playing a 68.6-Hz tone. The speed of sound is 343 m/s. What is the closest to speaker A that speaker B can be located, so that the listener hears no sound?

* **8.** The two speakers in the drawing are vibrating in phase, and a listener is standing at point P. Does constructive or destructive interference occur at P when the speakers produce sound waves whose frequency is (a) 1466 Hz and (b) 977 Hz? Justify your answers with appropriate calculations. Take the speed of sound to be 343 m/s.

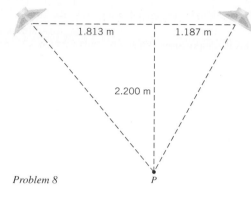

Problem 8

** **9.** Speakers A and B are vibrating in phase. They are directly facing each other, are 7.80 m apart, and are each playing a 73.0-Hz tone. The speed of sound is 343 m/s. On the line between the speakers there are three points where constructive interference occurs. What are the distances of these three points from speaker A?

Section 17.3 Diffraction

10. The entrance to a large lecture room consists of two side-by-side doors, one hinged on the left and the other hinged on the right. Each door is 0.700 m wide. Sound of frequency 607 Hz is coming through the entrance from within the room. The speed of sound is 343 m/s. What is the diffraction angle θ of the sound after it passes through the doorway when (a) one door is open and (b) both doors are open?

11. ssm A speaker has a diameter of 0.30 m. (a) Assuming that the speed of sound is 343 m/s, find the diffraction angle θ for a 2.0-kHz tone. (b) What speaker diameter D should be used to generate a 6.0-kHz tone whose diffraction angle is as wide as that for the 2.0-kHz tone in part (a)?

12. Sound emerges through a doorway, as in Figure 17.11. The width of the doorway is 77 cm, and the speed of sound is 343 m/s. Find the diffraction angle θ when the frequency of the sound is (a) 5.0 kHz and (b) 5.0×10^2 Hz.

13. Sound exits a diffraction horn loudspeaker through a rectangular opening like a small doorway. A person is sitting at an angle θ off to the side of a diffraction horn that has a width D of 0.060 m. The speed of sound is 343 m/s. This individual does not hear a sound wave that has a frequency of 8100 Hz. When she is sitting at an angle $\theta/2$, there is a different frequency that she does not hear. What is it?

* **14.** A row of seats is parallel to a stage at a distance of 8.7 m from it. At the center and front of the stage is a diffraction horn loudspeaker. This speaker sends out its sound through an opening that is like a small doorway with a width D of 7.5 cm. The speaker is playing a tone that has a frequency of 1.0×10^4 Hz. The speed of sound is 343 m/s. What is the distance between two seats, located near the center of the row, at which the tone cannot be heard?

* **15.** A 3.00-kHz tone is being produced by a speaker with a diameter of 0.175 m. The air temperature changes from 0 to 29 °C. Assuming air to be an ideal gas, find the *change* in the diffraction angle θ.

Section 17.4 Beats

16. Two out-of-tune flutes play the same note. One produces a tone that has a frequency of 262 Hz, while the other produces 266 Hz. When a tuning fork is sounded together with the 262-Hz tone, a beat frequency of 1 Hz is produced. When the same tuning fork is sounded together with the 266-Hz tone, a beat frequency of 3 Hz is produced. What is the frequency of the tuning fork?

17. ssm Two pure tones are sounded together. The drawing shows the pressure variations of the two sound waves, measured with respect to atmospheric pressure. What is the beat frequency?

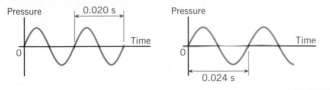

18. In **Concept Simulation 17.2** at **www.wiley.com/college/cutnell** you can explore the concepts that are important in this problem. A 440.0-Hz tuning fork is sounded together with an out-of-tune guitar string, and a beat frequency of 3 Hz is heard. When the string is tightened, the frequency at which it vibrates increases, and the beat frequency is heard to decrease. What was the original frequency of the guitar string?

19. ssm Two ultrasonic sound waves combine and form a beat frequency that is in the range of human hearing. The frequency of one of the ultrasonic waves is 70 kHz. What is (a) the smallest possible and (b) the largest possible value for the frequency of the other ultrasonic wave?

20. A tuning fork vibrates at a frequency of 524 Hz. An out-of-tune piano string vibrates at 529 Hz. How much time separates successive beats?

* **21.** A sound wave is traveling in seawater, where the adiabatic bulk modulus and density are 2.31×10^9 Pa and 1025 kg/m^3, respectively. The wavelength of the sound is 3.35 m. A tuning fork is struck underwater and vibrates at 440.0 Hz. What would be the beat frequency heard by an underwater swimmer?

** **22.** Two loudspeakers are mounted on a merry-go-round whose radius is 9.01 m. When stationary, the speakers both play a tone whose frequency is 100.0 Hz. As the drawing illustrates, they are situated at opposite ends of a diameter. The speed of sound is 343.00 m/s, and the merry-go-round revolves once every 20.0 s. What is the beat frequency that is detected by the listener when the merry-go-round is near the position shown?

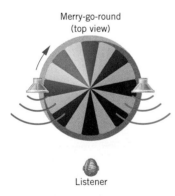

Merry-go-round
(top view)

Listener

Section 17.5 Transverse Standing Waves

23. ssm The A string on a string bass is tuned to vibrate at a fundamental frequency of 55.0 Hz. If the tension in the string were increased by a factor of four, what would be the new fundamental frequency?

24. A string of length 0.28 m is fixed at both ends. The string is plucked and a standing wave is set up that is vibrating at its second harmonic. The traveling waves that make up the standing wave have a speed of 140 m/s. What is the frequency of vibration?

25. The G string on a guitar has a fundamental frequency of 196 Hz and a length of 0.62 m. This string is pressed against the proper fret to produce the note C, whose fundamental frequency is 262 Hz. What is the distance L between the fret and the end of the string at the bridge of the guitar (see Figure 17.20*b*)?

26. A 41-cm length of wire has a mass of 6.0 g. It is stretched between two fixed supports and is under a tension of 160 N. What is the fundamental frequency of this wire?

27. ssm On a cello, the string with the largest linear density (1.56×10^{-2} kg/m) is the C string. This string produces a fundamental frequency of 65.4 Hz and has a length of 0.800 m between the two fixed ends. Find the tension in the string.

28. The fundamental frequency of a string fixed at both ends is 256 Hz. How long does it take for a wave to travel the length of this string?

29. A string has a linear density of 8.5×10^{-3} kg/m and is under a tension of 280 N. The string is 1.8 m long, is fixed at both ends, and is vibrating in the standing wave pattern shown in the drawing. Determine the (a) speed, (b) wavelength, and (c) frequency of the traveling waves that make up the standing wave.

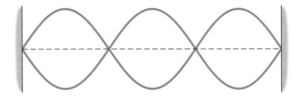

* **30.** Two strings have different lengths and linear densities, as the drawing shows. They are joined together and stretched so that the tension in each string is 190.0 N. The free ends of the joined string are fixed in place. Find the lowest frequency that permits standing waves in both strings with a node at the junction. The standing wave pattern in each string may have a different number of loops.

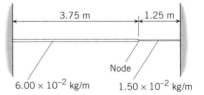

* **31. ssm** The E string on an electric bass guitar has a length of 0.628 m and, when producing the note E, vibrates at a fundamental frequency of 41.2 Hz. Players sometimes add to their instruments a device called a "D-tuner." This device allows the E string to be used to produce the note D, which has a fundamental frequency of 36.7 Hz. The D-tuner works by extending the length of the string, keeping all other factors the same. By how much does a D-tuner extend the length of the E string?

** **32.** Review Conceptual Example 5 before attempting this problem. As the drawing shows, the length of a guitar string is 0.628 m. The

frets are numbered for convenience. A performer can play a musical scale on a single string because the spacing *between the frets* is designed according to the following rule: When the string is pushed against any fret j, the fundamental frequency of the shortened string is larger by a factor of the twelfth root of two ($\sqrt[12]{2}$) than it is when the string is pushed against the fret $j - 1$. Assuming that the tension in the string is the same for any note, find the spacing (a) between fret 1 and fret 0 and (b) between fret 7 and fret 6.

** **33. ssm www** The drawing shows an arrangement in which a block (mass = 15.0 kg) is held in position on a frictionless incline by a cord (length = 0.600 m). The mass per unit length of the cord is 1.20×10^{-2} kg/m, so the mass of the cord is negligible compared to the mass of the block. The cord is being vibrated at a frequency of 165 Hz (vibration source not shown in the drawing). What are the values of the angle θ between 15.0° and 90.0° at which a standing wave exists on the cord?

Section 17.6 Longitudinal Standing Waves, Section 17.7 Complex Sound Waves

34. An organ pipe is open at both ends. It is producing sound at its third harmonic, the frequency of which is 262 Hz. The speed of sound is 343 m/s. What is the length of the pipe?

35. The fundamental frequency of a vibrating system is 400 Hz. For each of the following systems, give the three lowest frequencies (excluding the fundamental) at which standing waves can occur: (a) a string fixed at both ends, (b) a cylindrical pipe with both ends open, and (c) a cylindrical pipe with only one end open.

36. ⚕ Sound enters the ear, travels through the auditory canal, and reaches the eardrum. The auditory canal is approximately a tube open at only one end. The other end is closed by the eardrum. A typical length for the auditory canal in an adult is about 2.9 cm. The speed of sound is 343 m/s. What is the fundamental fre-

quency of the canal? (Interestingly, the fundamental frequency is in the frequency range where human hearing is most sensitive.)

37. ssm A tube of air is open at only one end and has a length of 1.5 m. This tube sustains a standing wave at its third harmonic. What is the distance between one node and the adjacent antinode?

38. A tube is open only at one end. A certain harmonic produced by the tube has a frequency of 450 Hz. The next higher harmonic has a frequency of 750 Hz. The speed of sound in air is 343 m/s. (a) What is the integer n that describes the harmonic whose frequency is 450 Hz? (b) What is the length of the tube?

39. ssm Both neon (Ne) and helium (He) are monatomic gases and can be assumed to be ideal gases. The fundamental frequency of a tube of neon is 268 Hz. What is the fundamental frequency of the tube if the tube is filled with helium, all other factors remaining the same?

* **40.** Two loudspeakers face each other, vibrate in phase, and produce identical 440-Hz tones. A listener walks from one speaker toward the other at a constant speed and hears the loudness change (loud–soft–loud) at a frequency of 3.0 Hz. The speed of sound is 343 m/s. What is the walking speed?

* **41.** A person hums into the top of a well and finds that standing waves are established at frequencies of 42, 70.0, and 98 Hz. The frequency of 42 Hz is not necessarily the fundamental frequency. The speed of sound is 343 m/s. How deep is the well?

* **42.** A tube, open at both ends, contains an unknown ideal gas for which $\gamma = 1.40$. At 293 K, the shortest tube in which a standing wave can be set up with a 294-Hz tuning fork has a length of 0.248 m. Find the mass of a gas molecule.

* **43. ssm www** A vertical tube is closed at one end and open to air at the other end. The air pressure is 1.01×10^5 Pa. The tube has a length of 0.75 m. Mercury (mass density = 13 600 kg/m³) is poured into it to shorten the effective length for standing waves. What is the absolute pressure at the bottom of the mercury column, when the fundamental frequency of the shortened, air-filled tube is equal to the third harmonic of the original tube?

** **44.** A tube, open at only one end, is cut into two shorter (nonequal) lengths. The piece that is open at both ends has a fundamental frequency of 425 Hz, while the piece open only at one end has a fundamental frequency of 675 Hz. What is the fundamental frequency of the original tube?

Additional Problems

45. If the string in Figure 17.18 is vibrating at a frequency of 4.0 Hz and the distance between two successive nodes is 0.30 m, what is the speed of the waves on the string?

46. The range of human hearing is roughly from twenty hertz to twenty kilohertz. Based on these limits and a value of 343 m/s for the speed of sound, what are the lengths of the longest and shortest pipes (open at both ends and producing sound at their fundamental frequencies) that you expect to find in a pipe organ?

47. ssm Review Example 1 in the text. Speaker A is moved further to the left, while ABC remains a right triangle. What is the separation between the speakers when constructive interference occurs again at point C?

48. When a guitar string is sounded along with a 440-Hz tuning fork, a beat frequency of 5 Hz is heard. When the same string is sounded along with a 436-Hz tuning fork, the beat frequency is 9 Hz. What is the frequency of the string?

49. Refer to **Interactive Solution 17.49** at **www.wiley.com/college/cutnell** to review a method by which this problem can be solved. The fundamental frequencies of two air columns are the same. Column A is open at both ends, while column B is open at only one end. The length of column A is 0.70 m. What is the length of column B?

50. Concept Simulation 17.1 at **www.wiley.com/college/cutnell** illustrates the concept that is pertinent to this problem. The drawing graphs a string on which two pulses (half up and half down) are traveling at a constant speed of 1 cm/s at $t = 0$ s. Using the principle of linear superposition, draw the shape of the string's pulses at $t = 1$ s, 2 s, 3 s, and 4 s.

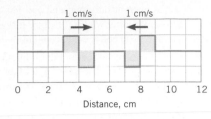

Distance, cm

51. ssm Suppose the strings on a violin are stretched with the same tension and each has the same length between its two fixed ends. The musical notes and corresponding fundamental frequencies of two of these strings are G (196.0 Hz) and E (659.3 Hz). The linear density of the E string is 3.47×10^{-4} kg/m. What is the linear density of the G string?

* **52.** The drawing shows two strings that have the same length and linear density. The left end of each string is attached to a wall, while

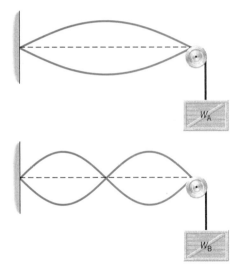

the right end passes over a pulley and is connected to objects of different weights (W_A and W_B). Different standing waves are set up on each string, but their frequencies are the same. If $W_A = 44$ N, what is W_B?

* **53.** Refer to **Interactive Solution 17.53** at **www.wiley.com/ college/cutnell** to review a method by which this problem can be solved. Two loudspeakers on a concert stage are vibrating in phase. A listener is 50.5 m from the left speaker and 26.0 m from the right one. The listener can respond to all frequencies from 20 to 20 000 Hz, and the speed of sound is 343 m/s. What are the two lowest frequencies that can be heard loudly due to constructive interference?

** **54.** Two tuning forks X and Y have different frequencies and produce an 8-Hz beat frequency when sounded together. When X is sounded along with a 392-Hz tone, a 3-Hz beat frequency is detected. When Y is sounded along with the 392-Hz tone, a 5-Hz beat frequency is heard. What are the fre-quencies f_X and f_Y when (a) f_X is greater than f_Y and (b) f_X is less than f_Y?

** **55. ssm www** The note that is three octaves above middle C is supposed to have a fundamental frequency of 2093 Hz. On a certain piano the steel wire that produces this note has a cross-sectional area of 7.85×10^{-7} m². The wire is stretched between two pegs. When the piano is tuned properly to produce the correct frequency at 25.0 °C, the wire is under a tension of 818.0 N. Suppose the temperature drops to 20.0 °C. In addition, as an approximation, assume that the wire is kept from contracting as the temperature drops. Con-sequently, the tension in the wire changes. What beat frequency is produced when this piano and another instrument (properly tuned) sound the note simultaneously?

Concepts & Calculations *Group Learning Problems*

Note: Each of these problems consists of Concept Questions followed by a related quantitative Problem. They are designed for use by students working alone or in small learning groups. The Concept Questions involve little or no mathematics and are intended to stimulate group discussions. They focus on the concepts with which the problems deal. Recognizing the concepts is the essential initial step in any problem-solving technique.

56. Concept Questions Both drawings show the same square, at one corner of which an observer O is stationed. Two loudspeakers are located at corners of the square, either as in drawing 1 or as in drawing 2. The speakers produce the same single-frequency tone in either drawing and are in phase. Constructive interference occurs in drawing 1, but destructive interference occurs in drawing 2. (a) Will only certain frequencies lead to the constructive interference in drawing 1, or will it occur for any frequency at all? (b) Will only certain frequencies lead to the destructive interference in drawing 2, or will it occur for any frequency at all? Justify your answers.

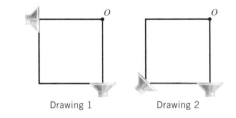

Drawing 1   Drawing 2

Problem One side of the square has a length of $L = 0.75$ m. The speed of sound is 343 m/s. Find the single smallest frequency that will produce both constructive interference in drawing 1 and destructive interference in drawing 2.

57. Concept Questions (a) When sound emerges from a loudspeaker, is the diffraction angle determined by the wavelength, the diameter of the speaker, or a combination of these two factors? (b) How is the wavelength of a sound related to its frequency? Explain your answers.

Problem The following two lists give diameters and sound frequencies for three loudspeakers. Pair each diameter with a frequency, so that the diffraction angle is the same for each of the speakers. The speed of sound is 343 m/s. Find the common diffraction angle.

Diameter, D	Frequency, f
0.050 m	6.0 kHz
0.10 m	4.0 kHz
0.15 m	12.0 kHz

58. Concept Questions Two cars have identical horns, each emitting a frequency f_s. One of the cars is moving toward a bystander waiting at a corner, and the other is parked. The two horns sound simultaneously. (a) From the moving horn, does the bystander hear a frequency that is greater than, less than, or equal to f_s? (b) From the stationary horn, does the bystander hear a frequency that is greater than, less than, or equal to f_s? (c) Does the bystander hear a beat frequency from the combined sound of the two horns? Account for your answers.

Problem The frequency that the horns emit is $f_s = 395$ Hz. The speed of the moving car is 12.0 m/s and the speed of sound is 343 m/s. What is the beat frequency heard by the bystander?

59. Concept Questions Two wires are stretched between two fixed supports and have the same length. On wire A there is a second-harmonic standing wave whose frequency is 660 Hz. However, the same frequency of 660 Hz is the third harmonic on wire B. (a) Is the fundamental frequency of wire A greater than, less than, or equal to the fundamental frequency of wire B? Explain. (b) How is the fundamental frequency related to the length L of the wire and the speed v at which individual waves travel back and forth on the wire? (c) Do the individual waves travel on wire A with a greater, smaller, or the same speed as on wire B? Give your reasoning.

Problem The common length of the wires is 1.2 m. Find the speed at which individual waves travel on each wire. Verify that your answer is consistent with your answers to the Concept Questions.

60. Concept Questions A copper block is suspended in air from a wire in Part 1 of the drawing. A container of mercury is then raised up around the block as in Part 2. (a) The fundamental frequency of the wire is given by Equation 17.3 with $n = 1$: $f_1 = v/(2L)$. How is the speed v at which individual waves travel on the wire related to the tension in the wire? (b) Is the tension in the wire in Part 2 less than, greater than, or equal to the tension in Part 1? (c) Is the funda-

mental frequency of the wire in Part 2 less than, greater than, or equal to the fundamental frequency in Part 1? Justify each of your answers.

Problem In Part 2 of the drawing one-half of the block's volume is submerged in the mercury. The density of copper is 8890 kg/m³, and the density of mercury is 13 600 kg/m³. Find the ratio of the fundamental frequency of the wire in Part 2 to the fundamental frequency in Part 1. Check to see that your answer is consistent with your answers to the Concept Questions.

61. Concept Questions One method for measuring the speed of sound uses standing waves. A cylindrical tube is open at both ends, and one end admits sound from a tuning fork. A movable plunger is inserted into the other end. The distance between the end of the tube where the tuning fork is and the plunger is L. For a fixed frequency, the plunger is moved until the smallest value of L is measured that allows a standing wave to be formed. (a) When a standing wave is formed in the tube, is there a node or an antinode at the end of the tube where the tuning fork is? (b) When a standing wave is formed, is there a node or an antinode at the plunger? (c) How is the smallest value of L related to the wavelength of the sound? Explain your answers.

Problem The tuning fork produces a 485-Hz tone, and the smallest value observed for L is 0.264 m. What is the speed of the sound in the gas in the tube?

* **62. Concept Questions** A listener is standing in front of two speakers that are producing sound of the same frequency and amplitude, except that they are vibrating out of phase. Initially, the distance between the listener and each speaker is the same (see the drawing). (a) What is meant by the phrase "the speakers are vibrating out of phase?" (b) When the sound waves reach the listener, would they exhibit constructive or destructive interference? Provide a reason for your answer. (c) When the listener begins to move sideways (to the right in the drawing), would the sound intensity increase or decrease? Why?

Problem As the listener moves sideways, the sound intensity gradually changes. When the distance x in the drawing is 0.92 m, the change reaches the maximum amount (either loud to soft, or soft to loud). Using the data shown in the drawing, determine the frequency of the sound emitted by the speakers.

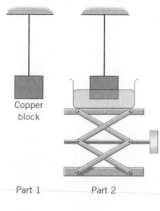

Part 1   Part 2

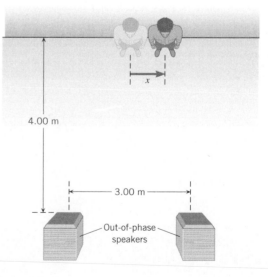

4.00 m

3.00 m

Out-of-phase speakers

* **63. Concept Questions** Standing waves are set up on two strings fixed at each end, as shown in the drawing. The tensions in the strings are the same, and each string has the same mass per unit length. However, one string is longer than the other. (a) Do the waves on the longer string have a larger speed, a smaller speed, or the same speed as those on the shorter string? Justify your answer. (b) Will the longer string vibrate at a higher frequency, a lower frequency, or the same frequency as the shorter string? Provide a reason for your answer. (c) Will the beat frequency produced by the two standing waves increase or decrease if the longer string is increased in length? Why?

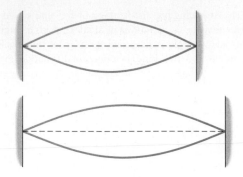

Problem The two strings in the drawing have the same tension and mass per unit length, but they differ in length by 0.57 cm. The waves on the shorter string propagate with a speed of 41.8 m/s, and the fundamental frequency of the shorter string is 225 Hz. Determine the beat frequency produced by the two standing waves.

Appendix A
Powers of Ten and Scientific Notation

In science, very large and very small decimal numbers are conveniently expressed in terms of powers of ten, some of which are listed below:

$$10^3 = 10 \times 10 \times 10 = 1000 \qquad 10^{-3} = \frac{1}{10 \times 10 \times 10}$$
$$= 0.001$$

$$10^2 = 10 \times 10 = 100 \qquad 10^{-2} = \frac{1}{10 \times 10} = 0.01$$

$$10^1 = 10 \qquad 10^{-1} = \frac{1}{10} = 0.1$$

$$10^0 = 1$$

Using powers of ten, we can write the radius of the earth in the following way, for example:

$$\text{Earth radius} = 6\,380\,000 \text{ m} = 6.38 \times 10^6 \text{ m}$$

The factor of ten raised to the sixth power is ten multiplied by itself six times, or one million, so the earth's radius is 6.38 million meters. Alternatively, the factor of ten raised to the sixth power indicates that the decimal point in the term 6.38 is to be moved six places *to the right* to obtain the radius as a number without powers of ten.

For numbers less than one, negative powers of ten are used. For instance, the Bohr radius of the hydrogen atom is

$$\text{Bohr radius} = 0.000\,000\,000\,0529 \text{ m} = 5.29 \times 10^{-11} \text{ m}$$

The factor of ten raised to the minus eleventh power indicates that the decimal point in the term 5.29 is to be moved eleven places *to the left* to obtain the radius as a number without powers of ten. Numbers expressed with the aid of powers of ten are said to be in *scientific notation.*

Calculations that involve the multiplication and division of powers of ten are carried out as in the following examples:

$$(2.0 \times 10^6)(3.5 \times 10^3) = (2.0 \times 3.5) \times 10^{6+3} = 7.0 \times 10^9$$

$$\frac{9.0 \times 10^7}{2.0 \times 10^4} = \left(\frac{9.0}{2.0}\right) \times 10^7 \times 10^{-4}$$

$$= \left(\frac{9.0}{2.0}\right) \times 10^{7-4} = 4.5 \times 10^3$$

The general rules for such calculations are

$$\frac{1}{10^n} = 10^{-n} \tag{A-1}$$

$$10^n \times 10^m = 10^{n+m} \qquad \text{(Exponents added)} \tag{A-2}$$

$$\frac{10^n}{10^m} = 10^{n-m} \qquad \text{(Exponents subtracted)} \tag{A-3}$$

where n and m are any positive or negative number.

Scientific notation is convenient because of the ease with which it can be used in calculations. Moreover, scientific notation provides a convenient way to express the significant figures in a number, as Appendix B discusses.

Appendix B
Significant Figures

The number of *significant figures* in a number is the number of digits whose values are known with certainty. For instance, a person's height is measured to be 1.78 m, with the measurement error being in the third decimal place. All three digits are known with certainty, so that the number contains three significant figures. If a zero is given as the last digit to the right of the decimal point, the zero is presumed to be significant. Thus, the number 1.780 m contains four significant figures. As another example, consider a distance of 1500 m. This number contains only two significant figures, the one and the five. The zeros immediately to the left of the unexpressed decimal point are not counted as significant figures. However, zeros located between significant figures are significant, so a distance of 1502 m contains four significant figures.

Scientific notation is particularly convenient from the point of view of significant figures. Suppose it is known that a certain distance is fifteen hundred meters, to four significant figures. Writing the number as 1500 m presents a problem because it implies that only two significant figures are known. In contrast, the scientific notation of 1.500×10^3 m has the advantage of indicating that the distance is known to four significant figures.

When two or more numbers are used in a calculation, the number of significant figures in the answer is limited by the number of significant figures in the original data. For instance, a rectangular garden with sides of 9.8 m and 17.1 m has an area of (9.8 m)(17.1 m). A calculator gives 167.58 m^2 for this product. However, one of the original lengths is known only to two significant figures, so the final answer is limited to only two significant figures and should be rounded off to 170 m^2. In general, *when numbers are multiplied or divided, the number of significant figures in the final answer equals the smallest number of significant figures in any of the original factors.*

The number of significant figures in the answer to an addition or a subtraction is also limited by the original data. Consider the total distance along a biker's trail that consists of three segments with the distances shown as follows:

$$
\begin{array}{ll}
 & 2.5 \ \ \text{km} \\
 & 11 \ \ \ \ \text{km} \\
 & \underline{5.26 \ \text{km}} \\
\text{Total} & 18.76 \ \text{km}
\end{array}
$$

The distance of 11 km contains no significant figures to the right of the decimal point. Therefore, neither does the sum of the three distances,

and the total distance should not be reported as 18.76 km. Instead, the answer is rounded off to 19 km. In general, ***when numbers are added or subtracted, the last significant figure in the answer occurs in the last column (counting from left to right) containing a number that results from a combination of digits that are all significant.*** In the answer of 18.76 km, the eight is the sum of $2 + 1 + 5$, each digit being significant. However, the seven is the sum of $5 + 0 + 2$, and the zero is not significant, since it comes from the 11-km distance, which contains no significant figures to the right of the decimal point.

Appendix C
Algebra

C.1 **Proportions and Equations**

Physics deals with physical variables and the relations between them. Typically, variables are represented by the letters of the English and Greek alphabets. Sometimes, the relation between variables is expressed as a proportion or inverse proportion. Other times, however, it is more convenient or necessary to express the relation by means of an equation, which is governed by the rules of algebra.

If two variables are ***directly proportional*** and one of them doubles, then the other variable also doubles. Similarly, if one variable is reduced to one-half its original value, then the other is also reduced to one-half its original value. In general, if x is directly proportional to y, then increasing or decreasing one variable by a given factor causes the other variable to change in the same way by the same factor. This kind of relation is expressed as $x \propto y$, where the symbol $\propto$ means "is proportional to."

Since the proportional variables x and y always increase and decrease by the same factor, the ratio of x to y must have a constant value, or $x/y = k$, where k is a constant, independent of the values for x and y. Consequently, a proportionality such as $x \propto y$ can also be expressed in the form of an equation: $x = ky$. The constant k is referred to as a ***proportionality constant.***

If two variables are ***inversely proportional*** and one of them increases by a given factor, then the other decreases by the same factor. An inverse proportion is written as $x \propto 1/y$. This kind of proportionality is equivalent to the following equation: $xy = k$, where k is a proportionality constant, independent of x and y.

C.2 **Solving Equations**

Some of the variables in an equation typically have known values, and some do not. It is often necessary to solve the equation so that a variable whose value is unknown is expressed in terms of the known quantities. ***In the process of solving an equation, it is permissible to manipulate the equation in any way, as long as a change made on one side of the equals sign is also made on the other side.*** For example, consider the equation $v = v_0 + at$. Suppose values for v, v_0, and a are available, and the value of t is required. To solve the equation for t, we begin by subtracting v_0 from *both* sides:

$$
\begin{array}{rcl}
v & = & v_0 + at \\
\underline{-v_0} & = & \underline{-v_0} \\
v - v_0 & = & at
\end{array}
$$

Next, we divide both sides of $v - v_0 = at$ by the quantity a:

$$
\frac{v - v_0}{a} = \frac{at}{a} = (1)t
$$

On the right side, the a in the numerator divided by the a in the denominator equals one, so that

$$
t = \frac{v - v_0}{a}
$$

It is always possible to check the correctness of the algebraic manipulations performed in solving an equation by substituting the answer back into the original equation. In the previous example, we substitute the answer for t into $v = v_0 + at$:

$$
v = v_0 + a\left(\frac{v - v_0}{a}\right) = v_0 + (v - v_0) = v
$$

The result $v = v$ implies that our algebraic manipulations were done correctly.

Algebraic manipulations other than addition, subtraction, multiplication, and division may play a role in solving an equation. The same basic rule applies, however: Whatever is done to the left side of an equation must also be done to the right side. As another example, suppose it is necessary to express v_0 in terms of v, a, and x, where $v^2 = v_0^2 + 2ax$. By subtracting $2ax$ from both sides, we isolate v_0^2 on the right:

$$
\begin{array}{rcl}
v^2 & = & v_0^2 + 2ax \\
\underline{-2ax} & = & \underline{-2ax} \\
v^2 - 2ax & = & v_0^2
\end{array}
$$

To solve for v_0, we take the positive and negative square root of *both* sides of $v^2 - 2ax = v_0^2$:

$$
v_0 = \pm\sqrt{v^2 - 2ax}
$$

C.3 **Simultaneous Equations**

When more than one variable in a single equation is unknown, additional equations are needed if solutions are to be found for all of the unknown quantities. Thus, the equation $3x + 2y = 7$ cannot be solved by itself to give unique values for both x and y. However, if x and y also (i.e., simultaneously) obey the equation $x - 3y = 6$, then both unknowns can be found.

There are a number of methods by which such simultaneous equations can be solved. One method is to solve one equation for x in terms of y and substitute the result into the other equation to obtain an expression containing only the single unknown variable y. The equation $x - 3y = 6$, for instance, can be solved for x by adding $3y$ to each side, with the result that $x = 6 + 3y$. The substitution of this expression for x into the equation $3x + 2y = 7$ is shown below:

$$3x + 2y = 7$$
$$3(6 + 3y) + 2y = 7$$
$$18 + 9y + 2y = 7$$

We find, then, that $18 + 11y = 7$, a result that can be solved for y:

$$
\begin{array}{rcr}
18 + 11y = & & 7 \\
-18 & & -18 \\
\hline
11y = & & -11
\end{array}
$$

Dividing both sides of this result by 11 shows that $y = -1$. The value of $y = -1$ can be substituted in either of the original equations to obtain a value for x:

$$
\begin{array}{rcl}
x - 3y & = & 6 \\
x - 3(-1) & = & 6 \\
x + 3 & = & 6 \\
-3 & & -3 \\
\hline
x & = & 3
\end{array}
$$

C.4 The Quadratic Formula

Equations occur in physics that include the square of a variable. Such equations are said to be *quadratic* in that variable, and often can be put into the following form:

$$ax^2 + bx + c = 0 \qquad \text{(C-1)}$$

where a, b, and c are constants independent of x. This equation can be solved to give the **quadratic formula**, which is

$$x = \frac{-b \pm \sqrt{b^2 - 4ac}}{2a} \qquad \text{(C-2)}$$

The $\pm$ in the quadratic formula indicates that there are two solutions. For instance, if $2x^2 - 5x + 3 = 0$, then $a = 2$, $b = -5$, and $c = 3$. The quadratic formula gives the two solutions as follows:

Solution 1:
Plus sign

$$
\begin{aligned}
x &= \frac{-b + \sqrt{b^2 - 4ac}}{2a} \\
&= \frac{-(-5) + \sqrt{(-5)^2 - 4(2)(3)}}{2(2)} \\
&= \frac{+5 + \sqrt{1}}{4} = \frac{3}{2}
\end{aligned}
$$

Solution 2:
Minus sign

$$
\begin{aligned}
x &= \frac{-b - \sqrt{b^2 - 4ac}}{2a} \\
&= \frac{-(-5) - \sqrt{(-5)^2 - 4(2)(3)}}{2(2)} \\
&= \frac{+5 - \sqrt{1}}{4} = 1
\end{aligned}
$$

Appendix D
Exponents and Logarithms

Appendix A discusses powers of ten, such as 10^3, which means ten multiplied by itself three times, or $10 \times 10 \times 10$. The three is referred to as an **exponent**. The use of exponents extends beyond powers of ten. In general, the term y^n means the factor y is multiplied by itself n times. For example, y^2, or y squared, is familiar and means $y \times y$. Similarly, y^5 means $y \times y \times y \times y \times y$.

The rules that govern algebraic manipulations of exponents are the same as those given in Appendix A (see Equations A-1, A-2, and A-3) for powers of ten:

$$\frac{1}{y^n} = y^{-n} \qquad \text{(D-1)}$$

$$y^n y^m = y^{n+m} \qquad \text{(Exponents added)} \qquad \text{(D-2)}$$

$$\frac{y^n}{y^m} = y^{n-m} \qquad \text{(Exponents subtracted)} \qquad \text{(D-3)}$$

To the three rules above we add two more that are useful. One of these is

$$y^n z^n = (yz)^n \qquad \text{(D-4)}$$

The following example helps to clarify the reasoning behind this rule:

$$3^2 5^2 = (3 \times 3)(5 \times 5) = (3 \times 5)(3 \times 5) = (3 \times 5)^2$$

The other additional rule is

$$(y^n)^m = y^{nm} \qquad \text{(Exponents multiplied)} \qquad \text{(D-5)}$$

To see why this rule applies, consider the following example:

$$(5^2)^3 = (5^2)(5^2)(5^2) = 5^{2+2+2} = 5^{2\times3}$$

Roots, such as a square root or a cube root, can be represented with fractional exponents. For instance,

$$\sqrt{y} = y^{1/2} \quad \text{and} \quad \sqrt[3]{y} = y^{1/3}$$

In general, the nth root of y is given by

$$\sqrt[n]{y} = y^{1/n} \qquad \text{(D-6)}$$

The rationale for Equation D-6 can be explained using the fact that $(y^n)^m = y^{nm}$. For instance, the fifth root of y is the number that, when multiplied by itself five times, gives back y. As shown below, the term $y^{1/5}$ satisfies this definition:

$$(y^{1/5})(y^{1/5})(y^{1/5})(y^{1/5})(y^{1/5}) = (y^{1/5})^5 = y^{(1/5)\times5} = y$$

Logarithms are closely related to exponents. To see the connection between the two, note that it is possible to express any number y as another number B raised to the exponent x. In other words,

$$y = B^x \qquad \text{(D-7)}$$

The exponent x is called the **logarithm** of the number y. The number B is called the **base number.** One of two choices for the base number is usually used. If $B = 10$, the logarithm is known as the *common logarithm,* for which the notation "log" applies:

Common logarithm $\qquad y = 10^x \quad$ or $\quad x = \log y \qquad$ (D-8)

If $B = e = 2.718 \ldots$, the logarithm is referred to as the *natural logarithm,* and the notation "ln" is used:

Natural logarithm $\qquad y = e^z \quad$ or $\quad z = \ln y \qquad$ (D-9)

The two kinds of logarithms are related by

$$\ln y = 2.3026 \log y \qquad \text{(D-10)}$$

Both kinds of logarithms are often given on calculators.

The logarithm of the product or quotient of two numbers A and C can be obtained from the logarithms of the individual numbers according to the rules below. These rules are illustrated here for natural logarithms, but they are the same for any kind of logarithm.

$$\ln (AC) = \ln A + \ln C \qquad \text{(D-11)}$$

$$\ln \left(\frac{A}{C} \right) = \ln A - \ln C \qquad \text{(D-12)}$$

Thus, the logarithm of the product of two numbers is the sum of the individual logarithms, and the logarithm of the quotient of two numbers is the difference between the individual logarithms. Another useful rule concerns the logarithm of a number A raised to an exponent n:

$$\ln A^n = n \ln A \qquad \text{(D-13)}$$

Rules D-11, D-12, and D-13 can be derived from the definition of the logarithm and the rules governing exponents.

Appendix E
Geometry and Trigonometry

E.1 *Geometry*

ANGLES

Two angles are equal if
1. They are vertical angles (see Figure E1).
2. Their sides are parallel (see Figure E2).

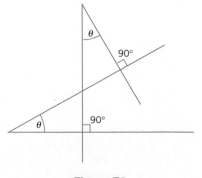

Figure E1   **Figure E2**

3. Their sides are mutually perpendicular (see Figure E3).

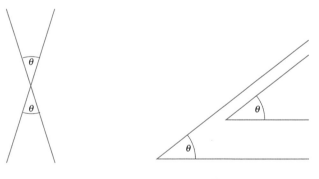

Figure E3

TRIANGLES

1. The **sum of the angles** of any triangle is 180° (see Figure E4).

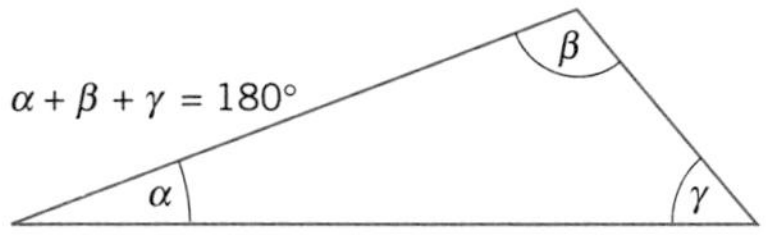

$\alpha + \beta + \gamma = 180°$

Figure E4

2. A **right triangle** has one angle that is 90°.
3. An **isosceles triangle** has two sides that are equal.
4. An **equilateral triangle** has three sides that are equal. Each angle of an equilateral triangle is 60°.
5. Two triangles are **similar** if two of their angles are equal (see Figure E5). The corresponding sides of similar triangles are proportional to each other:

$$\frac{a_1}{a_2} = \frac{b_1}{b_2} = \frac{c_1}{c_2}$$

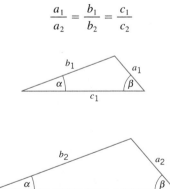

Figure E5

6. Two similar triangles are **congruent** if they can be placed on top of one another to make an exact fit.

CIRCUMFERENCES, AREAS, AND VOLUMES OF SOME COMMON SHAPES

1. Triangle of base b and altitude h (see Figure E6):

$$\text{Area} = \tfrac{1}{2}bh$$

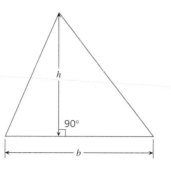

Figure E6

2. Circle of radius r:

$$\text{Circumference} = 2\pi r$$
$$\text{Area} = \pi r^2$$

3. Sphere of radius r:

$$\text{Surface area} = 4\pi r^2$$
$$\text{Volume} = \tfrac{4}{3}\pi r^3$$

4. Right circular cylinder of radius r and height h (see Figure E7):

$$\text{Surface area} = 2\pi r^2 + 2\pi rh$$
$$\text{Volume} = \pi r^2 h$$

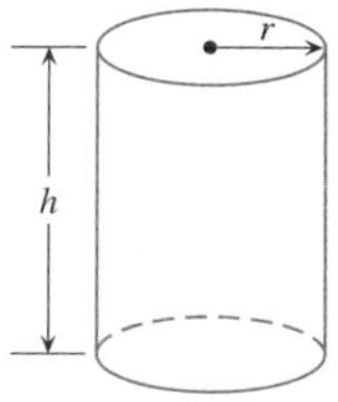

Figure E7

E.2 *Trigonometry*

BASIC TRIGONOMETRIC FUNCTIONS

1. For a right triangle, the sine, cosine, and tangent of an angle θ are defined as follows (see Figure E8):

$$\sin \theta = \frac{\text{Side opposite } \theta}{\text{Hypotenuse}} = \frac{h_\text{o}}{h}$$

$$\cos \theta = \frac{\text{Side adjacent to } \theta}{\text{Hypotenuse}} = \frac{h_\text{a}}{h}$$

$$\tan \theta = \frac{\text{Side opposite } \theta}{\text{Side adjacent to } \theta} = \frac{h_\text{o}}{h_\text{a}}$$

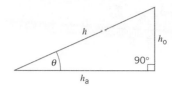

Figure E8

2. The secant ($\sec \theta$), cosecant ($\csc \theta$), and cotangent ($\cot \theta$) of an angle θ are defined as follows:

$$\sec \theta = \frac{1}{\cos \theta} \qquad \csc \theta = \frac{1}{\sin \theta} \qquad \cot \theta = \frac{1}{\tan \theta}$$

TRIANGLES AND TRIGONOMETRY

1. The **Pythagorean theorem** states that the square of the hypotenuse of a right triangle is equal to the sum of the squares of the other two sides (see Figure E8):

$$h^2 = h_\text{o}^2 + h_\text{a}^2$$

2. The *law of cosines* and the *law of sines* apply to any triangle, not just a right triangle, and they relate the angles and the lengths of the sides (see Figure E9):

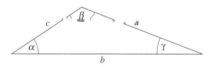

Figure E9

Law of cosines	$c^2 = a^2 + b^2 - 2ab \cos \gamma$
Law of sines	$\dfrac{a}{\sin \alpha} = \dfrac{b}{\sin \beta} = \dfrac{c}{\sin \gamma}$

OTHER TRIGONOMETRIC IDENTITIES

1. $\sin (-\theta) = -\sin \theta$
2. $\cos (-\theta) = \cos \theta$
3. $\tan (-\theta) = -\tan \theta$
4. $(\sin \theta) / (\cos \theta) = \tan \theta$
5. $\sin^2 \theta + \cos^2 \theta = 1$
6. $\sin (\alpha \pm \beta) = \sin \alpha \cos \beta \pm \cos \alpha \sin \beta$

$$\text{If } \alpha = 90°, \sin (90° \pm \beta) = \cos \beta$$
$$\text{If } \alpha = \beta, \sin 2\beta = 2 \sin \beta \cos \beta$$

7. $\cos (\alpha \pm \beta) = \cos \alpha \cos \beta \mp \sin \alpha \sin \beta$

$$\text{If } \alpha = 90°, \cos (90° \pm \beta) = \mp \sin \beta$$
$$\text{If } \alpha = \beta, \cos 2\beta = \cos^2 \beta - \sin^2 \beta = 1 - 2 \sin^2 \beta$$

Appendix F
Selected Isotopes[a]

Atomic No. Z	Element	Symbol	Atomic Mass No. A	Atomic Mass u	% Abundance, or Decay Mode If Radioactive	Half-Life (if Radioactive)
0	(Neutron)	n	1	1.008 665	β^-	10.37 min
1	Hydrogen	H	1	1.007 825	99.985	
	Deuterium	D	2	2.014 102	0.015	
	Tritium	T	3	3.016 050	β^-	12.33 yr
2	Helium	He	3	3.016 030	0.000 138	
			4	4.002 603	≈ 100	
3	Lithium	Li	6	6.015 121	7.5	
			7	7.016 003	92.5	
4	Beryllium	Be	7	7.016 928	EC, γ	53.29 days
			9	9.012 182	100	
5	Boron	B	10	10.012 937	19.9	
			11	11.009 305	80.1	
6	Carbon	C	11	11.011 432	β^+, EC	20.39 min
			12	12.000 000	98.90	
			13	13.003 355	1.10	
			14	14.003 241	β^-	5730 yr
7	Nitrogen	N	13	13.005 738	β^+, EC	9.965 min
			14	14.003 074	99.634	
			15	15.000 108	0.366	
8	Oxygen	O	15	15.003 065	β^+, EC	122.2 s
			16	15.994 915	99.762	
			18	17.999 160	0.200	
9	Fluorine	F	18	18.000 937	EC, β^+	1.8295 h
			19	18.998 403	100	
10	Neon	Ne	20	19.992 435	90.51	
			22	21.991 383	9.22	
11	Sodium	Na	22	21.994 434	β^+, EC, γ	2.602 yr
			23	22.989 767	100	
			24	23.990 961	β^-, γ	14.659 h
12	Magnesium	Mg	24	23.985 042	78.99	
13	Aluminum	Al	27	26.981 539	100	
14	Silicon	Si	28	27.976 927	92.23	
			31	30.975 362	β^-, γ	2.622 h
15	Phosphorus	P	31	30.973 762	100	
			32	31.973 907	β^-	14.282 days
16	Sulfur	S	32	31.972 070	95.02	
			35	34.969 031	β^-	87.51 days
17	Chlorine	Cl	35	34.968 852	75.77	
			37	36.965 903	24.23	
18	Argon	Ar	40	39.962 384	99.600	
19	Potassium	K	39	38.963 707	93.2581	
			40	39.963 999	β^-, EC, γ	1.277×10^9 yr
20	Calcium	Ca	40	39.962 591	96.941	
21	Scandium	Sc	45	44.955 910	100	
22	Titanium	Ti	48	47.947 947	73.8	

[a] Data for atomic masses are taken from *Handbook of Chemistry and Physics,* 66th ed., CRC Press, Boca Raton, FL. The masses are those for the neutral atom, including the Z electrons. Data for percent abundance, decay mode, and half-life are taken from E. Browne and R. Firestone, *Table of Radioactive Isotopes,* V. Shirley, Ed., Wiley, New York, 1986. α = alpha particle emission, β^- = negative beta emission, β^+ = positron emission, γ = γ-ray emission, EC = electron capture.

APPENDIX F Selected Isotopes *(continued)*

Atomic No. Z	Element	Symbol	Atomic Mass No. A	Atomic Mass u	% Abundance, or Decay Mode If Radioactive	Half-Life (if Radioactive)
23	Vanadium	V	51	50.943 962	99.750	
24	Chromium	Cr	52	51.940 509	83.789	
25	Manganese	Mn	55	54.938 047	100	
26	Iron	Fe	56	55.934 939	91.72	
27	Cobalt	Co	59	58.933 198	100	
			60	59.933 819	β^-, γ	5.271 yr
28	Nickel	Ni	58	57.935 346	68.27	
			60	59.930 788	26.10	
29	Copper	Cu	63	62.939 598	69.17	
			65	64.927 793	30.83	
30	Zinc	Zn	64	63.929 145	48.6	
			66	65.926 034	27.9	
31	Gallium	Ga	69	68.925 580	60.1	
32	Germanium	Ge	72	71.922 079	27.4	
			74	73.921 177	36.5	
33	Arsenic	As	75	74.921 594	100	
34	Selenium	Se	80	79.916 520	49.7	
35	Bromine	Br	79	78.918 336	50.69	
36	Krypton	Kr	84	83.911 507	57.0	
			89	88.917 640	β^-, γ	3.16 min
			92	91.926 270	β^-, γ	1.840 s
37	Rubidium	Rb	85	84.911 794	72.165	
38	Strontium	Sr	86	85.909 267	9.86	
			88	87.905 619	82.58	
			90	89.907 738	β^-	28.5 yr
			94	93.915 367	β^-, γ	1.235 s
39	Yttrium	Y	89	88.905 849	100	
40	Zirconium	Zr	90	89.904 703	51.45	
41	Niobium	Nb	93	92.906 377	100	
42	Molybdenum	Mo	98	97.905 406	24.13	
43	Technecium	Tc	98	97.907 215	β^-, γ	4.2×10^6 yr
44	Ruthenium	Ru	102	101.904 348	31.6	
45	Rhodium	Rh	103	102.905 500	100	
46	Palladium	Pd	106	105.903 478	27.33	
47	Silver	Ag	107	106.905 092	51.839	
			109	108.904 757	48.161	
48	Cadmium	Cd	114	113.903 357	28.73	
49	Indium	In	115	114.903 880	95.7; β^-	4.41×10^{14} yr
50	Tin	Sn	120	119.902 200	32.59	
51	Antimony	Sb	121	120.903 821	57.3	
52	Tellurium	Te	130	129.906 229	38.8; β^-	2.5×10^{21} yr
53	Iodine	I	127	126.904 473	100	
			131	130.906 114	β^-, γ	8.040 days
54	Xenon	Xe	132	131.904 144	26.9	
			136	135.907 214	8.9	
			140	139.921 620	β^-, γ	13.6 s
55	Cesium	Cs	133	132.905 429	100	
			134	133.906 696	β^-, EC, γ	2.062 yr
56	Barium	Ba	137	136.905 812	11.23	
			138	137.905 232	71.70	
			141	140.914 363	β^-, γ	18.27 min
57	Lanthanum	La	139	138.906 346	99.91	

APPENDIX F Selected Isotopes *(continued)*

Atomic No. Z	Element	Symbol	Atomic Mass No. A	Atomic Mass u	% Abundance, or Decay Mode If Radioactive	Half-Life (if Radioactive)
58	Cerium	Ce	140	139.905 433	88.48	
59	Prascodymium	Pr	141	140.907 647	100	
60	Neodymium	Nd	142	141.907 719	27.13	
61	Promethium	Pm	145	144.912 743	EC, α, γ	17.7 yr
62	Samarium	Sm	152	151.919 729	26.7	
63	Europium	Eu	153	152.921 225	52.2	
64	Gadolinium	Gd	158	157.924 099	24.84	
65	Terbium	Tb	159	158.925 342	100	
66	Dysprosium	Dy	164	163.929 171	28.2	
67	Holmium	Ho	165	164.930 319	100	
68	Erbium	Er	166	165.930 290	33.6	
69	Thulium	Tm	169	168.934 212	100	
70	Ytterbium	Yb	174	173.938 859	31.8	
71	Lutetium	Lu	175	174.940 770	97.41	
72	Hafnium	Hf	180	179.946 545	35.100	
73	Tantalum	Ta	181	180.947 992	99.988	
74	Tungsten (wolfram)	W	184	183.950 928	30.67	
75	Rhenium	Re	187	186.955 744	62.60; β^-	4.6×10^{10} yr
76	Osmium	Os	191	190.960 920	β^-, γ	15.4 days
			192	191.961 467	41.0	
77	Iridium	Ir	191	190.960 584	37.3	
			193	192.962 917	62.7	
78	Platinum	Pt	195	194.964 766	33.8	
79	Gold	Au	197	196.966 543	100	
			198	197.968 217	β^-, γ	2.6935 days
80	Mercury	Hg	202	201.970 617	29.80	
81	Thallium	Tl	205	204.974 401	70.476	
			208	207.981 988	β^-, γ	3.053 min
82	Lead	Pb	206	205.974 440	24.1	
			207	206.975 872	22.1	
			208	207.976 627	52.4	
			210	209.984 163	α, β^-, γ	22.3 yr
			211	210.988 735	β^-, γ	36.1 min
			212	211.991 871	β^-, γ	10.64 h
			214	213.999 798	β^-, γ	26.8 min
83	Bismuth	Bi	209	208.980 374	100	
			211	210.987 255	α, β^-, γ	2.14 min
			212	211.991 255	β^-, α, γ	1.0092 h
84	Polonium	Po	210	209.982 848	α, γ	138.376 days
			212	211.988 842	α, γ	45.1 s
			214	213.995 176	α, γ	163.69 μs
			216	216.001 889	α, γ	150 ms
85	Astatine	At	218	218.008 684	α, β^-	1.6 s
86	Radon	Rn	220	220.011 368	α, γ	55.6 s
			222	222.017 570	α, γ	3.825 days
87	Francium	Fr	223	223.019 733	α, β^-, γ	21.8 min
88	Radium	Ra	224	224.020 186	α, γ	3.66 days
			226	226.025 402	α, γ	1.6×10^3 yr
			228	228.031 064	β^-, γ	5.75 yr
89	Actinium	Ac	227	227.027 750	α, β^-, γ	21.77 yr
			228	228.031 015	β^-, γ	6.13 h

APPENDIX F Selected Isotopes *(continued)*

Atomic No. Z	Element	Symbol	Atomic Mass No. A	Atomic Mass u	% Abundance, or Decay Mode If Radioactive	Half-Life (if Radioactive)
90	Thorium	Th	228	228.028 715	α, γ	1.913 yr
			231	231.036 298	β^-, γ	1.0633 days
			232	232.038 054	100; α, γ	1.405×10^{10} yr
			234	234.043 593	β^-, γ	24.10 days
91	Protactinium	Pa	231	231.035 880	α, γ	3.276×10^4 yr
			234	234.043 303	β^-, γ	6.70 h
			237	237.051 140	β^-, γ	8.7 min
92	Uranium	U	232	232.037 130	α, γ	68.9 yr
			233	233.039 628	α, γ	1.592×10^5 yr
			235	235.043 924	0.7200; α, γ	7.037×10^8 yr
			236	236.045 562	α, γ	2.342×10^7 yr
			238	238.050 784	99.2745; α, γ	4.468×10^9 yr
			239	239.054 289	β^-, γ	23.47 min
93	Neptunium	Np	239	239.052 933	β^-, γ	2.355 days
94	Plutonium	Pu	239	239.052 157	α, γ	2.411×10^4 yr
			242	242.058 737	α, γ	3.763×10^5 yr
95	Americium	Am	243	243.061 375	α, γ	7.380×10^3 yr
96	Curium	Cm	245	245.065 483	α, γ	8.5×10^3 yr
97	Berkelium	Bk	247	247.070 300	α, γ	1.38×10^3 yr
98	Californium	Cf	249	249.074 844	α, γ	350.6 yr
99	Einsteinium	Es	254	254.088 019	α, γ, β^-	275.7 days
100	Fermium	Fm	253	253.085 173	EC, α, γ	3.00 days
101	Mendelevium	Md	255	255.091 081	EC, α	27 min
102	Nobelium	No	255	255.093 260	EC, α	3.1 min
103	Lawrencium	Lr	257	257.099 480	α, EC	646 ms
104	Rutherfordium	Rf	261	261.108 690	α	1.08 min
105	Dubnium	Db	262	262.113 760	α	34 s

Answers to Check Your Understanding

Chapter 1
CYU 1: b and d
CYU 2: (a) 11 m
(b) 5 m
CYU 3: (a) A_x is −, A_y is +
(b) B_x is +, B_y is −
(c) R_x is +, R_y is +

CYU 4: a

Chapter 2
CYU 1: His average velocity is 2.7 m/s, due east. His average speed is 8.0 m/s.
CYU 2: c
CYU 3: 1.73
CYU 4: b

Chapter 3
CYU 1: b
CYU 2: c
CYU 3: a and c
CYU 4: 1. +70 m/s
2. +30 m/s
3. +40 m/s
4. −60 m/s

Chapter 4
CYU 1: c
CYU 2: d
CYU 3: 43°
CYU 4: a
CYU 5: a

Chapter 5
CYU 1: (a) The velocity is due south, and the acceleration is due west.
(b) The velocity is due west, and the acceleration is due north.
CYU 2: (a) 4r
(b) 4r
CYU 3: less than
CYU 4: (a) less than
(b) equal to

Chapter 6
CYU 1: b
CYU 2: false
CYU 3: b and d
CYU 4: c

Chapter 7
CYU 1: b
CYU 2: a
CYU 3: d
CYU 4: above the halfway point

Chapter 8
CYU 1: B, C, A
CYU 2: 1.0 rev/s
CYU 3: 0.30 m
CYU 4: 8.0 m/s^2

Chapter 9
CYU 1: 0°, 45°, 90°
CYU 2: (a) C
(b) A
(c) B
CYU 3: A, B, C
CYU 4: (a) Both have the same translational speed.
(b) Both have the same translational speed.

Chapter 10
CYU 1: 180 N/m
CYU 2: II
CYU 3: b, c, a
CYU 4: The rod with the square cross section is longer.

Chapter 11
CYU 1: increase, decrease, remain constant
CYU 2: c
CYU 3: d
CYU 4: e

Chapter 12
CYU 1: 178 °X
CYU 2: b and d
CYU 3: c, b, d, a
CYU 4: c, a, b

Chapter 13
CYU 1: c
CYU 2: b

Chapter 14
CYU 1: 66.4%
CYU 2: Xenon has the greatest and argon the smallest temperature.
CYU 3: $\dfrac{v_{\text{rms, new}}}{v_{\text{rms, initial}}} = 0.707$
CYU 4: c, a, b

Chapter 15
CYU 1: d
CYU 2: b
CYU 3: c
CYU 4: d

Chapter 16
CYU 1: a
CYU 2: CO and N_2
CYU 3: (a) $\frac{1}{4}$
(b) 2
CYU 4: (a) The truck driver is driving faster.
(b) minus sign in both places

Chapter 17
CYU 1: (a) 0 cm
(b) −2 cm
(c) +2 cm
CYU 2: 387 Hz
CYU 3: (a) 4
(b) 3
(c) node
(d) 110 Hz
CYU 4: (a) antinode
(b) node
(c) $\frac{1}{4}\lambda$
(d) lowered

Answers to Odd-Numbered Problems

Chapter 1

1. (a) 5×10^{-3} g
  (b) 5 mg
  (c) 5×10^3 μg
3. (a) 5700 s
  (b) 86 400 s
5. 1.37 lb
7. (a) correct
  (b) not correct
  (c) not correct
  (d) correct
  (e) correct
9. 3.13×10^8 m^3
11. 5.5 km
13. 80.1 km, 25.9° south of west
15. 3.73 m
17. 35.3°
19. 99° opposite the 190-cm side, 3.0×10^1 degrees opposite the 95-cm side, 51° opposite the 150-cm side
21. 200 N due east or 600 N due west
23. 0.90 km, 56° north of west
25. (a) 5.31 km, south
  (b) 5.31 km, north
27. (a) 5600 newtons
  (b) along the dashed line
29. (a) 8.6 units
  (b) 34.9° north of west
  (c) 8.6 units
  (d) 34.9° south of west
31. (a) 15.8 m/s
  (b) 6.37 m/s
33. (a) 147 km
  (b) 47.9 km
35. (a) 5.70×10^2 newtons
  (b) 33.6° south of west
37. (a) 25.0°
  (b) 34.8 newtons
39. (a) 322 newtons
  (b) 209 newtons
  (c) 279 newtons
41. 7.1 m, 9.9° north of east
43. 3.00 m, 42.8° above the $-x$ axis
45. (a) 2.7 km
  (b) 6.0×10^1 degrees north of east
47. (a) 10.4 units
  (b) 12.0 units
49. (a) 178 units
  (b) 164 units
51. (a) 64 m
  (b) 37° south of east
53. (a) C
  (b) B
55. 30.2 m, 10.2°
57. 222 m, 55.8° below the $-x$ axis
59. 6.88 km, 26.9°
61. 1.2×10^2 m
63. 288 units due west and 156 units due north

Chapter 2

1. 0.80 s
3. (a) 12.4 km
  (b) 8.8 km, due east
5. (a) 464 m/s
  (b) 1040 mi/h
7. 52 m
9. 7.2×10^3 m
11. 2.1 s
13. (a) 4.0 s
  (b) 4.0 s
15. 3.44 m/s, due west
17. +30.0 m/s
19. 4.5 m
21. (a) 11.4 s
  (b) 44.7 m/s^2
23. (a) 1.7×10^2 cm/s^2
  (b) 0.15 s
25. -3.1 m/s^2
27. 0.74 m/s
29. 39.2 m
31. 0.87 m/s^2, in the same direction as the velocity
33. 1.2
35. (a) 13 m/s
  (b) 0.93 m/s^2
37. 91.5 m/s
39. -1.5 m/s^2
41. 1.1 s
43. 6.12 s
45. 1.7 s
47. 8.96 m
49. 0.767 m/s
51. 2.0×10^1 m
53. 10.6 m
55. 6.0 m below the top of the cliff
57. The answer is in graphical form.
59. 1.9 m/s^2 (segment A), 0 m/s^2 (segment B), 3.3 m/s^2 (segment C)
61. -8.3 km/h^2
63. (a) 6.6 s
  (b) 5.3 m/s
65. 44.1 m/s
67. (a) 1.5 m/s^2
  (b) 1.5 m/s^2
  (c) The car travels 76 m farther than the jogger.
69. 5×10^4 y
71. -22 m/s^2
73. (a) 2.67×10^4 m
  (b) 6.74 m/s, due north
75. 0.40 s
77. (a) 11 s
  (b) 3.0×10^1 m
79. 14 s

Chapter 3

1. 8.8×10^2 m
3. 8600 m

5. 242 m/s
7. 5.4 m/s
9. 27.0°
11. (a) 2.99×10^4 m/s
  (b) 2.69×10^4 m/s
13. 2.40 m
15. (a) 1.78 s
  (b) 20.8 m/s
17. 14.1 m/s
19. (a) 6.0×10^1 m
  (b) 290 m
21. 30.0 m
23. (a) 85 m
  (b) 610 m
25. (a) 1.1 s
  (b) 1.3 s
27. 24 buses
29. 1.7 s
31. 48 m
33. 11 m/s
35. 14.7 m/s
37. 14.9 m
39. 21.9 m/s, 40.0°
41. 0.141° and 89.860°
43. $D = 850$ m, $H = 31$ m
45. $v_{0B} = 8.79$ m/s, $\theta_B = 81.5°$
47. (a) 41 m/s, due east
  (b) 41 m/s, due west
49. 31 s
51. (a) 2.0×10^3 s
  (b) 1.8×10^3 m
53. 6.3 m/s, 18° north of east
55. 7.63 m/s, 26.4° north of east
57. 63.9 m/s, 85° west of south
59. 3.05 m/s, 14.8° north of west
61. 4.42 s
63. 14.6 s
65. (a) 239 m/s, 57.1° with respect to the horizontal
  (b) 239 m/s, 57.1° with respect to the horizontal
67. 5.17 s
69. 5.2 m/s, 52° west of south
71. 42°
73. 27.2°

Chapter 4

1. 93 N
3. 3.5×10^4 N
5. 130 N
7. 37 N
9. (a) 3.6 N
  (b) 0.40 N
11. 1.83 m/s^2, left
13. 30.9 m/s^2, 27.2° above the $+x$ axis
15. 0.78 m, 21° south of east
17. 18.4 N, 68° north of east
19. 1.8×10^{-7} N

21. (a) $W = 1.13 \times 10^3$ N, $m = 115$ kg
(b) $W = 0$ N, $m = 115$ kg
23. 0.223 m/s²
25. 1.76×10^{24} kg
27. (a) 3.75 m/s²
(b) 2.4×10^2 N
29. 4.7 kg
31. 0.0050
33. 0.414 L
35. 7.3×10^2 N
37. The block will move. $a = 3.72$ m/s²
39. 0.444
41. (a) 390 N
(b) 7.7 m/s, direction is toward second base
43. (a) 0.980 m/s², direction is opposite to the direction of motion
(b) 29.5 m
45. 68°
47. (a) 7.40×10^5 N
(b) 1.67×10^9 N
49. 9.70 N for each force
51. 929 N
53. 1.00×10^2 N, 53.1° south of east
55. 62 N
57. 16.3 N
59. 286 N
61. 406 N
63. 18.0 m/s², 56.3° above the $+x$ axis
65. 1730 N, due west
67. (a) 2.99 m/s²
(b) 129 N
69. 6.6 m/s
71. (a) 1.3 N
(b) 6.5 N
73. 29 400 N
75. 0.14 m/s²
77. (a) 4.25 m/s²
(b) 1080 N
79. 0.265 m
81. (a) $\Delta T_A = 0$ N, $\Delta T_B = -4.7$ N, $\Delta T_C = 0$ N
(b) $\Delta T_A = 0$ N, $\Delta T_B = 0$ N, $\Delta T_C = +4.7$ N
83. 1.2 s
85. (a) 13.7 N
(b) 1.37 m/s²
87. (a) 10.5 m/s²
(b) 1.07
89. 39 N
91. (a) 447 N
(b) 241 N
93. (a) 1610 N
(b) 2640 N
95. 4290 N
97. (a) 1.04×10^3 N
(b) 1.04×10^3 N
(c) 2.45 m/s²
(d) 1.74×10^{-22} m/s²
99. (a) 3.56 m/s²
(b) 281 N
101. 0.141
103. 8.7 s

105. (a) 5.2 m/s²
(b) 11 m/s²
107. 1.9×10^2 N
109. (a) 0.60 m/s²
(b) 104 N (left string), 230 N (right string)
111. 0.665

Chapter 5

1. 160 s
3. 6.9 m/s²
5. 332 m
7. (a) 5.0×10^1 m/s²
(b) zero
(c) 2.0×10^1 m/s²
9. 2.2
11. 0.68 m/s
13. (a) 0.189 N
(b) 4.00
15. 12 m/s
17. 3500 N
19. (a) 3510 N
(b) 14.9 m/s
21. 2.0×10^1 m/s
23. 184 m
25. 2.12×10^6 N
27. 1.33×10^4 m/s
29. 4.20×10^4 m/s
31. 1/27
33. 2.45×10^4 N
35. (a) 912 m
(b) 228 m
(c) 2.50 m/s²
37. (a) 1.70×10^3 N
(b) 1.66×10^3 N
39. 4.72×10^3 m
41. 8.48 m/s
43. (a) 1.2×10^4 N
(b) 1.7×10^4 N
45. 61°
47. 0.71
49. 2.9×10^4 N
51. 2.0×10^2 m/s²
53. 0.250
55. 23 N at 19.0 m/s and 77 N at 38.0 m/s

Chapter 6

1. -2.6×10^6 J
3. (a) 2980 J
(b) 3290 J
5. 42.8°
7. (a) 54.9 N
(b) 1060 J
(c) -1060 J
(d) 0 J
9. 45 N
11. 2.07×10^3 N
13. (a) 3.1×10^3 J
(b) 2.2×10^2 J
15. (a) 38 J
(b) 3.8×10^3 N
17. 6.4×10^5 J

19. 18%
21. 1.4×10^{11} J
23. 10.9 m/s
25. (a) -3.0×10^4 J
(b) The resistive force is not a conservative force.
27. 2.39×10^5 J
29. 5.24×10^5 J
31. 2.3×10^4 J
33. (a) 28.3 m/s
(b) 28.3 m/s
(c) 28.3 m/s
35. 4.13 m
37. 4.8 m/s
39. 3.29 m/s
41. 6.33 m
43. 40.8 kg
45. -4.51×10^4 J
47. 16.5 m
49. -1.21×10^6 J
51. (a) 2.8 J
(b) 35 N
53. 4.17 m/s
55. (a) 3.3×10^4 W
(b) 5.1×10^4 W
57. 3.6×10^6 J
59. 3.0×10^3 W
61. 6.7×10^2 N
63. (a) 93 J
(b) 0 J
(c) 2.3 m/s
65. (a) Bow 1 requires more work.
(b) 25 J
67. (a) 1.50×10^2 J
(b) 7.07 m/s
69. (a) At $h = 20.0$ m, KE = 0 J, PE = 392 J, and E = 392 J. At $h = 10.0$ m, KE = 196 J, PE = 196 J, and E = 392 J. At $h = 0$ m, KE = 392 J, PE = 0 J, and E = 392 J.
71. 2.2×10^3 J
73. (a) 1.80×10^3 J
(b) -1.20×10^3 J
75. 2450 N
77. 1.7 m/s
79. 13.5 m

Chapter 7

1. -8.7 kg·m/s
3. (a) $+1.7$ kg·m/s
(b) $+570$ N
5. 11 N·s, in the same direction as the average force
7. $+69$ N
9. 3.7 N·s
11. 4.28 N·s, upward
13. 960 N
15. -1.5×10^{-4} m/s
17. (a) Bonzo, since he has the recoil velocity with the smaller magnitude
(b) 1.7

19. (a) 77.9 m/s
  (b) 45.0 m/s
21. $m_1 = 1.00$ kg, $m_2 = 1.00$ kg
23. +547 m/s
25. 84 kg
27. 3.00 m
29. +9.09 m/s
31. (a) −0.400 m/s (5.00-kg ball),
     +1.60 m/s (7.50-kg ball)
  (b) +0.800 m/s
33. (a) +8.9 m/s
  (b) -3.6×10^4 N·s
  (c) 5.9 m
35. (a) 73.0°
  (b) 4.28 m/s
37. (a) +3.17 m/s
  (b) 0.0171
39. (a) 5.56 m/s
  (b) −2.83 m/s (1.50 kg ball), +2.73 m/s
     (4.60-kg ball)
  (c) 0.409 m (1.50-kg ball), 0.380 m
     (4.60-kg ball)
41. 4.67×10^6 m
43. 6.46×10^{-11} m
45. (a) −1.5 m/s
  (b) +1.1 m/s
47. +9.3 m/s
49. 3.0 m/s
51. +4500 m/s, in the same direction as the
   rocket before the explosion
53. 1.5×10^{10} m/s
55. 0.34 s
57. 0.097 m

Chapter 8

1. 13 rad/s
3. 63.7 grad
5. 6.4×10^{-3} rad/s^2
7. 492 rad/s
9. 825 m
11. 336 m/s
13. 6.05 m
15. 25 rev
17. 157.3 rad/s
19. (a) 1.2×10^4 rad
  (b) 1.1×10^2 s
21. 28 rad/s
23. 12.5 s
25. 1.95×10^4 rad
27. 7.37 s
29. 157 m/s
31. 22 rev/s
33. (a) 4.66×10^2 m/s
  (b) 70.6°
35. (a) 1.25 m/s
  (b) 7.98 rev/s
37. 14.8 rad/s
39. (a) 9.00 m/s^2
  (b) radially inward
41. (a) 2.5 m/s^2
  (b) 3.1 m/s^2
43. $1/\sqrt{3}$
45. 1.00 rad

47. 8.71 rad/s^2
49. 693 rad
51. 28.0 rad/s
53. 11.8 rad
55. 20.6 rad
57. 380 m/s^2
59. (a) 4.00×10^1 rad
  (b) 15.0 rad/s
61. 0.62 m
63. 0.300 m/s
65. 7.45 s
67. 2 revolutions

Chapter 9

1. 4.2 N·m
3. 843 N
5. 1.3
7. (a) $\tau = FL$
  (b) $\tau = FL$
  (c) $\tau = FL$
9. 0.667 m
11. 196 N (force on each hand), 96 N (force
   on each foot)
13. 0.591
15. (a) 27 N, to the left
  (b) 27 N, to the right
  (c) 27 N, to the right
  (d) 143 N, downward and to the left
     (79° below the horizontal)
17. 1200 N, to the left
19. (a) 1.60×10^5 N
  (b) 4.20×10^5 N
21. 37.6°
23. (a) 1.21×10^3 N
  (b) 1.01×10^3 N, downward
25. 51.4 N
27. 1.7 m
29. 1.25 kg·m^2
31. (a) −11 N·m
  (b) −9.2 rad/s^2
33. (a) 0.131 kg·m^2
  (b) 3.6×10^{-4} kg·m^2
  (c) 0.149 kg·m^2
35. (a) 5.94 rad/s^2
  (b) 44.0 N
37. 0.78 N
39. (a) 2.67 kg·m^2
  (b) 1.16 m
41. 2.12 s
43. (a) $v_{T1} = 12.0$ m/s, $v_{T2} = 9.00$ m/s,
     $v_{T3} = 18.0$ m/s
  (b) 1.08×10^3 J
  (c) 60.0 kg·m^2
  (d) 1.08×10^3 J
45. 6.1×10^5 rev/min
47. 2/7
49. 3/4
51. 2.0 m
53. 4.4 kg·m^2
55. 0.26 rad/s
57. 8% increase
59. 0.17 m
61. 0.25 m

63. (a) 0.14 rad/s
  (b) A net external torque must be applied
     in a direction opposite to the angular
     deceleration.
65. 0.060 kg·m^2
67. (a) 7.40×10^2 N
  (b) 0.851 N
69. 34 m/s
71. 69 N

Chapter 10

1. 237 N
3. (a) 7.44 N
  (b) 7.44 N
5. 0.012 m
7. 2.29×10^{-3} m
9. 0.79
11. 0.240 m
13. (a) 1.00×10^3 N/m
  (b) 0.340
15. 3.5×10^4 N/m
17. (a) 0.080 m
  (b) 1.6 rad/s
  (c) 2.0 N/m
  (d) 0 m/s
  (e) 0.20 m/s^2
19. 696 N/m
21. 4.3 kg
23. 0.806
25.

h (meters)	KE	PE (gravity)	PE (elastic)	E
0	0 J	0 J	8.76 J	8.76 J
0.200	1.00 J	3.92 J	3.84 J	8.76 J
0.400	0 J	7.84 J	0.92 J	8.76 J

27. (a) 58.8 N/m
  (b) 11.4 rad/s
29. 7.18×10^{-2} m
31. 14 m/s
33. (a) 9.0×10^{-2} m
  (b) 2.1 m/s
35. 1.25 m/s (11.2-kg block),
   0.645 m/s (21.7-kg block)
37. 2.37×10^3 N/m
39. 0.99 m
41. 6.0 m/s^2
43. 0.816
45. $7R/5$
47. 3.7×10^{-5} m
49. 1.6×10^5 N
51. 260 m
53. 1.4×10^{-6}
55. -2.8×10^{-4}
57. 6.6×10^4 N
59. (a) 6.3×10^{-2} m
  (b) 7.3×10^{-2} m
61. 4.6×10^{-4}
63. 1.0×10^{-3} m
65. -4.4×10^{-5}
67. (a) 140 m/s
  (b) 1.7×10^{15} m/s^2
69. 21 m
71. 7.7×10^{-5} m

73. (a) 4.9×10^6 N/m^2
   (b) 6.0×10^{-6} m
75. (a) 2.66 Hz
   (b) 0.0350 m
77. 24.2 rad/s
79. 33.4 m/s
81. 4.0×10^{-5} m
83. (a) 0.25 s
   (b) 0.75 s

Chapter 11

1. 317 m^2
3. 3400 N
5. 6.6×10^6 kg
7. 1.9 gal
9. 63%
11. 1.1×10^3 N
13. 4.33×10^6 Pa
15. 24 blocks
17. 0.750 m
19. 1.2×10^5 Pa
21. (a) 3.5×10^6 N
   (b) 1.2×10^6 N
23. (a) 2.45×10^5 Pa
   (b) 1.73×10^5 Pa
25. 7.0×10^5 Pa
27. 1.19×10^5 Pa
29. 0.74 m
31. 3.8×10^5 N
33. (a) 93.0 N
   (b) 94.9 N
35. 8.50×10^5 N·m
37. 5.7×10^{-2} m
39. 4.89 m
41. 2.7×10^{-4} m^3
43. 2.04×10^{-3} m^3
45. 0.20 m
47. 7.6×10^{-2} m
49. 5.28×10^{-2} m and 6.20×10^{-2} m
51. 1.91 m/s
53. (a) 0.18 m
   (b) 0.14 m
55. (a) 1.6×10^{-4} m^3/s
   (b) 2.0×10^1 m/s
57. (a) 150 Pa
   (b) The pressure inside the roof is greater
      than that outside the roof, so there is a
      net outward force.
59. 1.92×10^5 N
61. (a) 2.48×10^5 Pa
   (b) 1.01×10^5 Pa
   (c) 0.342 m^3/s
63. 33 m/s
65. (a) 14 m/s
   (b) 0.98 m^3/s
67. 9600 N
69. 7.78 m/s
71. (a) 1.01×10^5 Pa
   (b) 1.19×10^5 Pa
73. 1.7 m
75. (a) 2.8×10^{-5} N
   (b) 1.0×10^1 m/s

77. 10.3 m
79. 59 N
81. 7.0×10^{-2} m
83. 4.5×10^{-5} kg/s
85. (a) 1.26×10^5 Pa
   (b) 19.4 m
87. 0.816
89. 6.3×10^{-3} kg
91. 31.3 rad/s
93. 0.19 m
95. 5.75×10^4 Pa

Chapter 12

1. 0.2 C°
3. (a) 10.0 °C and 40.0 °C
   (b) 283.2 K and 313.2 K
5. -459.67 °F
7. -164 °C
9. 0.084 m
11. 1500 m
13. 110 C°
15. -2.82×10^{-4} m
17. 49 °C
19. 2.0027 s
21. (a) tension
   (b) 9.6×10^7 N/m^2
23. 0.6
25. 3.1×10^{-3} m^3
27. 2.5×10^{-7} m^3
29. 18
31. 0.33 gal
33. 9.0 mm
35. 45 atm
37. 6.9
39. 36.2 °C
41. $230
43. 21.03 °C
45. 940 °C
47. 21 barrels
49. 0.016 C°
51. 3.9×10^5 J
53. 9.3 kg
55. 9.49×10^{-3} kg
57. 1.85×10^5 J
59. (a) 3.0×10^{20} J
   (b) 3.2 years
61. 1.9×10^4 J/kg
63. 2.6×10^{-3} kg
65. 3.50×10^2 m/s
67. 0 °C
69. 33%
71. 2400 Pa
73. 28%
75. 39%
77. 4.9×10^{-2} m
79. 3.9×10^{-3} kg
81. 8.0×10^{-4}
83. one penny
85. 10 °C
87. 0.223
89. 650 W

91. 6.7×10^2 W
93. 1.3×10^5 J

Chapter 13

1. 8.0×10^2 J/s
3. 14 h
5. 2.0×10^{-3} m
7. 17
9. (a) 21 °C
   (b) 18 °C
11. (a) 130 °C
   (b) 830 J
   (c) 237 °C
13. 287 °C
15. (a) 2.0
   (b) 0.61
17. 14.5 da
19. 0.3
21. 320 K
23. 0.70
25. 12
27. 12 J
29. 1.2×10^4 s
31. 532 K
33. (a) 101.2 °C
   (b) 110.6 °C
35. 0.74

Chapter 14

1. Aluminum
3. (a) 294.307 u
   (b) 4.887×10^{-25} kg
5. 1.07×10^{-22} kg
7. 2.6×10^{-10} m
9. 1.0×10^3 kg
11. 2.3×10^{-2} mol
13. (a) 201 mol
   (b) 1.21×10^5 Pa
15. 2.5×10^{21}
17. 882 K
19. 0.93 mol/m^3
21. 5.9×10^4 g
23. 0.205
25. 6.19×10^5 Pa
27. 308 K
29. (a) 46.3 m^2/s^2
   (b) 40.1 m^2/s^2
31. 1.2×10^4 m/s
33. 3.9×10^5 J
35. 2820 m
37. 4.0×10^1 Pa
39. 11 s
41. (a) 2.1 s
   (b) 1.6×10^{-5} s
43. (a) 5.00×10^{-13} kg/s
   (b) 5.8×10^{-3} kg/m^3
45. 2.3×10^6 s
47. 925 K
49. 1.34×10^{-7} kg
51. 1.6×10^{-15} kg
53. 1.02×10^{20}

55. 11.7
57. 440 K

Chapter 15

1. (a) -261 J
   (b) Work is done on the system.
3. (a) -87 J
   (b) $+87$ J
5. 32 miles
7. 13 000 J
9. (a) 3100 J
   (b) negative
11. 0.24 m
13. 3.0×10^5 Pa
15. The answer is a proof.
17. 4.99×10^{-6}
19. (a) 0 J
   (b) -6.1×10^3 J
   (c) 310 K
21. 0.66
23. 1.81
25. (a) -8.00×10^4 J
   (b) Heat flows out of the gas.
27. 19.3
29. (a) 327 K
   (b) 0.132 m^3
31. 2400 J
33. (a) 1.1×10^4 J
   (b) 1.8×10^4 J
35. 5/2
37. (a) 60.0%
   (b) 40.0%
39. 2.38×10^4 J
41. 0.631
43. (a) 8600 J
   (b) 3100 J
45. $e = e_1 + e_2 - e_1e_2$
47. 0.21
49. (a) 1260 K
   (b) 1.74×10^4 J
51. lowering the temperature of the cold
   reservoir
53. (a) 0.360
   (b) 1.3×10^{13} J
55. 5.7 C°
57. 3.24×10^4 J
59. 5.86×10^5 J
61. 9.03
63. 1.4 K
65. (a) 2.0×10^1
   (b) 1.5×10^4 J
67. 1.26×10^3 K
69. 11.6 J/K
71. (a) $+3.68 \times 10^3$ J/K
   (b) $+1.82 \times 10^4$ J/K
   (c) The vaporization process creates
      more disorder.
73. (a) $+8.0 \times 10^2$ J/K
   (b) The entropy of the universe should
      increase.

75. (a) -2.1×10^2 K
   (b) decrease
77. 0 J, The process is an adiabatic
   process.
79. (a) reversible
   (b) -125 J/K
81. (a) 1050 J
   (b) 2.99
83. 1.2×10^7 Pa
85. 44.3 s
87. 8.49×10^5 Pa
89. 7.5 s
91. (a) 5/9
   (b) 1/3

Chapter 16

1. 0.083 Hz
3. 0.49 m
5. 0.25 m
7. 78 cm
9. 5.0×10^1 s
11. (a) 1.1 m/s
   (b) 6.55 m
13. 64 N
15. 600 m/s
17. 7.7 m/s^2
19. $m_1 = 28.7$ kg, $m_2 = 14.3$ kg
21. 3.26×10^{-3} s
23. $y = (0.37 \text{ m}) \sin (2.6 \, \pi t + 0.22 \, \pi x)$,
   where x and y are in meters and t is in
   seconds.
25. (a) $A = 0.45$ m, $f = 4.0$ Hz, $\lambda = 2.0$ m,
   $v = 8.0$ m/s
   (b) $-x$ direction
27. 2.5 N
29. 1730 m/s
31. (a) 7.87×10^{-3} s
   (b) 4.12
33. 2.06
35. (a) metal wave first, water wave second,
   air wave third
   (b) Second sound arrives 0.059 s
      later, and third sound arrives
      0.339 s later.
37. 690 rad/s
39. tungsten
41. 650 m
43. 8.0×10^2 m/s
45. 239 m/s
47. 0.404 m
49. 1.98%
51. 2.4×10^{-5} W/m^2
53. 6.5 W
55. 7.6×10^3 W/m^2
57. 2.6 s
59. 0.316 W/m^2
61. 1.3
63. 56.2 W
65. 2.6
67. 0.84 s

69. 2.39 dB
71. 17 m/s
73. 860 Hz
75. 1350 Hz
77. 22 m/s
79. 209 m
81. 25
83. (a) 431 m/s
   (b) 322 m/s
85. 8.68×10^{-3} kg/m
87. 1000
89. 8.19×10^{-2} m
91. 56 m/s
93. -6.0 dB
95. 77.0 dB
97. 153 N
99. (a) 2.20×10^2 m/s
   (b) 9.19 m/s
101. 57% argon, 43% neon

Chapter 17

1. (a) 2 cm
   (b) 1 cm
3. The answer is a series of
   drawings.
5. 107 Hz
7. 3.89 m
9. 3.90 m, 1.55 m, 6.25 m
11. (a) 44°
   (b) 0.10 m
13. 1.5×10^4 Hz
15. 3.7°
17. 8 Hz
19. (a) 50 kHz
   (b) 90 kHz
21. 8 Hz
23. 1.10×10^2 Hz
25. 0.46 m
27. 171 N
29. (a) 180 m/s
   (b) 1.2 m
   (c) 150 Hz
31. 0.077 m
33. 20.8° and 53.1°
35. (a) $f_2 = 800$ Hz, $f_3 = 1200$ Hz,
   $f_4 = 1600$ Hz
   (b) $f_2 = 800$ Hz, $f_3 = 1200$ Hz,
   $f_4 = 1600$ Hz
   (c) $f_3 = 1200$ Hz, $f_5 = 2000$ Hz,
   $f_7 = 2800$ Hz
37. 0.50 m
39. 602 Hz
41. 6.1 m
43. 1.68×10^5 Pa
45. 2.4 m/s
47. 5.06 m
49. 0.35 m
51. 3.93×10^{-3} kg/m
53. 28 Hz and 42 Hz
55. 12 Hz

Index